Military Laser Technology and Systems

For a listing of recent titles in the
Artech House Applied Photonics Library,
turn to the back of this book.

Military Laser Technology and Systems

David H. Titterton

ARTECH
HOUSE
BOSTON | LONDON
artechhouse.com

Library of Congress Cataloging-in-Publication Data
A catalog record for this book is available from the U.S. Library of Congress.

British Library Cataloguing in Publication Data
A catalog record for this book is available from the British Library.

ISBN-13: 978-1-60807-778-6

Cover design by John Gomes

685 Canton Street
Norwood, MA 02062

10 9 8 7 6 5 4 3 2 1

Contents

Preface *xix*

Part I Fundamentals 1

CHAPTER 1

Military Laser Technology and Systems 3

1.1 Introduction 3
1.2 Laser Development 6
1.3 Laser Systems Development 8
1.4 Aim of the Book 10
1.5 Organisation of the Book 10
1.5.1 Part I: Lasers and Fundamentals 11
1.5.2 Part II: Military Laser Systems 11
1.5.3 Part III: Safe Use, Education, and Training 12
References 12
Selected Bibliography 14

CHAPTER 2

Laser Action, Components and Definition of Terms 15

2.1 Introduction 15
2.2 Laser Classification 16
2.3 Laser Action and Stimulated Emission 17
2.3.1 Gain Medium, Pumping and Population Inversion 19
2.3.2 Laser Transitions 22
2.3.3 Laser Modes 23
2.3.4 Near- and Far-Field Beam Profiles 24
2.4 Oscillators and Amplifiers 25
2.5 Beam Generation and Its Control 29
2.5.1 Q-Switching 30
2.5.2 Cavity Dumping 33
2.5.3 Mode Locking 33
2.5.4 Gain Switching 34
2.5.5 Other Techniques 34
2.6 Other Laser Components 35
2.6.1 Mirrors 35

2.6.2 Coatings 35
2.6.3 Windows 36
2.7 Direct and Indirect Wavelength Generation 36
2.7.1 Optical Parametric Processes 40
2.7.2 Optical Parametric Oscillators 40
2.7.3 Phase Matching 42
2.7.4 Quasi Phase Matching 44
2.7.5 Harmonic Generation 45
2.7.6 Raman Conversion 46
2.7.7 Sum and Difference Frequency Mixing 47
2.7.8 Super-Continuum Generation 47
2.8 Performance Characteristic Terms and Parameters 47
2.8.1 Beam Quality and Divergence 48
2.8.2 Power 50
2.8.3 Radiant Intensity 50
2.8.4 Efficiency 50
2.8.5 Time to Full Brightness 51
2.8.6 Pulse Length 52
2.8.7 Pulse-Repetition Frequency 52
2.8.8 Duty Cycle and Duty Factor 52
2.8.9 Quantum Defect 53
2.8.10 Coherency and Speckle 54
2.8.11 Beam Stability 55
2.8.12 Line Width 55
2.8.13 Tuneability and Diversity 55
2.8.14 Polarisation 56
2.9 Summary 56
Selected Bibliography 57

CHAPTER 3

Laser Devices for Military Applications 59

3.1 Introduction 59
3.2 Military Laser-System Considerations 60
3.3 Laser-System Specifications and Requirements Definition 62
3.4 Solid-State Lasers 64
3.4.1 Future Challenges for Solid-State Devices 67
3.4.2 Rod Lasers 68
3.4.3 Slab Lasers 69
3.4.4 Slod Lasers 72
3.4.5 Plastic Lasers 73
3.4.6 Ceramic Lasers 75
3.5 Fibre Lasers 76
3.6 Disk Lasers 84
3.7 Vibronic Lasers 86
3.8 Liquid-Phase Lasers 88
3.9 Gas-Phase Lasers 92

3.9.1 Carbon-Dioxide Lasers 95
3.9.2 Chemical Lasers 96
3.9.3 Chemical Oxygen Iodine Laser 99
3.9.4 Explosively Pumped Iodine 101
3.9.5 Alkali Lasers 101
3.10 Semiconductor Lasers 102
3.10.1 Optically Pumped Semiconductor Lasers 107
3.10.2 Quantum-Cascade Laser 110
3.10.3 Interband-Cascade Lasers 113
3.10.4 W Structures 114
3.10.5 Strain-Engineered Diodes 114
3.10.6 Optically Pumped Semiconductor-Disk Lasers 116
3.11 Free-Electron Laser 118
3.12 Cryogenic Lasers 120
3.13 Cannon Lasers 121
3.14 Super-Continuum Lasers 123
3.14.1 Super-Continuum Generation Using Photonic-Crystal Fibre and Planar Waveguides 124
3.14.2 Filament Ultraviolet Generation 124
3.15 Waveguide Lasers 126
3.16 Laser Performance Summary 128
3.17 Summary 128
Selected Bibliography 131

CHAPTER 4

Beam-Director Technology 135

4.1 Introduction 135
4.2 Design Considerations for Beam Directors 137
4.3 Generic Beam-Director/Pointer System for Directed-Energy Systems 139
4.3.1 Gimbal 141
4.3.2 Gimbal Functions 141
4.3.3 Blind Arcs (Blockages) 142
4.3.4 Optical-Train Design 142
4.4 Threat-Alerting System 143
4.5 Target-Tracking Subsystem 144
4.5.1 Target-Tracking Algorithms 146
4.5.2 Aim-Point Refinement Techniques 147
4.6 Pointing Element 147
4.7 Target Sightline Stabilisation 148
4.8 System Processor 149
4.8.1 Built-In Test and Boresighting Functions 151
4.9 Other Functions 152
4.9.1 Beam Control 152
4.9.2 Battle-Damage Assessment 153
4.10 Video Processor 154
4.11 Power Supply 155

4.12 Environmental Cover 155
4.13 Laser-Integration Requirements 157
4.14 Generic Beam-Director System for Laser-Based Sensing Systems 158
4.15 Advanced Beam-Steering Techniques 159
4.16 Summary 161
Selected Bibliography 163

CHAPTER 5

Laser-Beam Propagation 165

5.1 Introduction 165
5.2 The Earth's Atmosphere 166
5.3 Beam-Attenuation Mechanisms 166
5.3.1 Absorption 167
5.3.2 Scattering 167
5.3.3 Extinction 169
5.4 Atmospheric Transmission 169
5.4.1 Weather 171
5.4.2 Humidity 171
5.4.3 Haze, Fog, Cloud and Rain 172
5.4.4 Summary of Transmission Effects on Beam Propagation 173
5.5 Models for Prediction of Transmission 174
5.5.1 Lowtran Model 175
5.5.2 Modtran Model 175
5.5.3 Fascode (Laser) Model 176
5.6 Turbulence 176
5.6.1 Turbulence Mitigation (Aperture Averaging) 180
5.6.2 Turbulence Mitigation (Adaptive Optics) 181
5.6.3 Turbulence Mitigation (Lucky Imaging) 182
5.6.4 Wind Effects 183
5.6.5 Summary of Turbulence and Wind Effects 184
5.7 Thermal Blooming 184
5.8 Ionisation 188
5.9 Nonlinear Propagation (Filaments) 188
5.10 Beam Propagation Through Wakes and Plumes 190
5.10.1 Propagation Characteristics 191
5.10.2 Impact on Installed Performance 191
5.11 Aero-Optical Effects 192
5.11.1 Aero-Optical Effects Around a Laser-DIRCM Transmitter 193
5.12 Summary 193
Selected Bibliography 195

CHAPTER 6

Radiometry, Testing and Evaluation of Laser Systems 197

6.1 Introduction 197
6.2 Radiometry 198
6.2.1 Photometry 200

6.2.2 Candela and Luminous Intensity 202
6.2.3 Radiant Flux and Luminous Flux 202
6.2.4 Radiant Intensity and Luminous Intensity 203
6.2.5 Irradiance and Illuminance 203
6.2.6 Radiance and Luminance 203
6.2.7 Radiant Exitance and Luminous Exitance 204
6.2.8 Radiant Exposure and Luminous Exposure 204
6.2.9 Radiant Energy and Luminous Energy 205
6.2.10 Total Radiant Flux and Total Luminous Flux 205
6.2.11 Radiance Temperature and Colour Temperature 205
6.3 System Considerations for Testing and Evaluation 206
6.3.1 Development Test and Evaluation (DT&E) 208
6.3.2 Operational Test and Evaluation (OT&E) 209
6.3.3 Technology Readiness Levels 209
6.4 Approaches to Testing and Evaluation 210
6.4.1 Introduction 210
6.4.2 Test and Evaluation Strategy 211
6.4.3 Test Design 212
6.4.4 Test Plan 212
6.4.5 Flight Trials 215
6.4.6 Documentation and Compliance 215
6.4.7 Simulation and Mathematical Modelling 215
6.4.8 Scale of Testing 216
6.4.9 Technology Insertion 217
6.4.10 Test Equipment 218
6.4.11 Laboratory Testing 222
6.4.12 Range and Field Trials 224
6.4.13 Platform Trials 226
6.5 Summary 228
Selected Bibliography 229

Part II Military Systems 231

CHAPTER 7

Laser Directed-Energy-Weapon Systems 233

7.1 Introduction 233
7.2 Historical Development 234
7.3 Military Utility 237
7.4 Summary of Potential Applications 239
7.5 Benefits of Laser Technology for Novel Systems 240
7.6 Laser-Beam Target-Interaction Effects 242
7.7 In-Band Damage Route 242
7.8 Out-of-Band Damage Route 244
7.9 Target-Defeat Techniques and Effects 244
7.9.1 In-Band Defeat 244
7.9.2 Out-of-Band Defeat 245

7.10 Basic Laser-Requirement Characteristics of In-Band Novel Weapon Systems 245
7.11 Basic Laser Characteristics for Out-of-Band Laser-Weapon Systems 247
7.12 Summary of Laser Requirements for Laser-Weapon Systems 248
7.13 Target Engagement 248
7.14 Laser-Weapon Subsystem Definition and Operation 250
7.15 Beam Control 253
7.16 Potential Laser Technology for an In-Band Novel Weapon System 254
7.16.1 Visible Band 254
7.16.2 Mid-Wave Infrared 254
7.16.3 Far-Infrared Waveband 254
7.17 Potential Laser Technology for an Out-of-Band Novel Weapon System 254
7.18 Tactical High-Power Laser-Weapon Mass and Volume Estimates 255
7.19 Tactical Laser-Weapon-System Considerations 256
7.20 Beam-Director Technology 257
7.21 Laser-Integration Requirements 257
7.22 System Processor 258
7.23 Battle-Damage Indication and Assessment 258
7.24 Concept of Operation 259
7.25 Future Requirements 259
7.25.1 Out-of-Band Systems 259
7.25.2 In-Band Systems 261
7.26 Myths 261
7.27 Summary 262
Selected Bibliography 263

CHAPTER 8

Laser-Based Electro-Optical Countermeasures 265

8.1 Introduction 265
8.2 Military Utility 268
8.3 System Operation 269
8.4 Defeat Mechanism 270
8.5 Pros and Cons of Laser-Based Dazzle Systems 271
8.6 System-Configuration Options 272
8.7 Subsystem Description and Functions 273
8.8 Target Cue 274
8.9 Laser-Source Characteristics for Electro-Optical Countermeasure (Dazzle) Systems 274
8.10 Summary of Laser-Source Requirements 275
8.10.1 Laser Source Options: Visible Band 275
8.10.2 Laser Source Options: Near-Infrared and Short-Wave Infrared Band 276
8.10.3 Laser Source Options: Mid-Wave Infrared Band 277
8.10.4 Laser Source Options: Far-Infrared Band 277

8.11 Beam Director 278
8.12 Laser-Integration Requirements 278
8.13 System Processor 278
8.14 Target Tracking and Stabilisation 279
8.15 Future Requirements 280
8.16 Summary 281
Selected Bibliography 282

CHAPTER 9
Laser Directed-Infrared Countermeasures (Laser Jammers) 283
9.1 Introduction 283
9.2 Military Utility 286
9.3 Historical development 286
9.3.1 Jammer Operation and Development 286
9.3.2 Laser-Source Technology 287
9.3.3 Defeat Mechanism 288
9.3.4 Jam-to-Signal Ratio 290
9.3.5 Concept of Operation 290
9.3.6 Principal Directed-Infrared Countermeasure Subsystems 293
9.3.7 Threat-Warning Subsystem 293
9.3.8 Beam Director 294
9.3.9 Directed-Infrared Countermeasure Processor 295
9.3.10 Infrared-Laser Requirements 295
9.3.11 Candidate Laser Technologies 296
9.3.12 Laser Integration 296
9.4 Open- and Closed-Loop Infrared-Countermeasure Systems 297
9.4.1 Open-Loop Systems 297
9.4.2 Closed-Loop IRCM (CLIRCM) 299
9.5 Future Requirements 301
9.6 Summary 303
Selected Bibliography 304

CHAPTER 10
Laser Range Finders 305
10.1 Introduction 306
10.2 Historical Development 307
10.3 Military Utility 308
10.4 System Operation 308
10.5 Laser-Performance Requirements 310
10.6 Subsystem Description and Function 312
10.7 Single-Photon Range Finding and Surveillance 312
10.7.1 Introduction 312
10.7.2 System Operation 313
10.7.3 Subsystem Description and Function 314
10.7.4 Laser Requirements 315

10.8 Future Requirements 315
10.9 Summary 316
Selected Bibliography 317

CHAPTER 11
Laser-Based Targeting 319
11.1 Introduction 319
11.2 Laser Targeting: Burst Illumination/Active Imaging 320
11.2.1 Introduction 320
11.2.2 Historical Development 321
11.2.3 Military Utility 322
11.2.4 System Operation 322
11.2.5 Laser Performance Requirements 324
11.2.6 Subsystem Description and Function 324
11.2.7 Future Requirements 328
11.2.8 Summary 328
11.3 Active Spectral Imaging 329
11.3.1 Introduction 330
11.3.2 Military Utility 331
11.3.3 System Configuration 332
11.3.4 Laser Performance Requirements 332
11.3.5 Subsystem Description and Function 333
11.3.6 Future Requirements 334
11.3.7 Summary 334
Selected Bibliography 335

CHAPTER 12
Laser Designation, Guidance and Fuzing 337
12.1 Introduction 337
12.2 Laser Designation 339
12.2.1 Introduction 339
12.2.2 Historical Development 340
12.2.3 Military Utility 341
12.2.4 System Operation 342
12.2.5 Laser-Performance Requirements 344
12.2.6 Subsystem Description and Function- 344
12.2.7 Future Requirements 346
12.2.8 Summary 347
12.3 Laser Beam-Rider Guidance 347
12.3.1 Introduction 347
12.3.2 Historical Development 348
12.3.3 Military Utility 349
12.3.4 Mode of Operation 349
12.3.5 The Weapon System: Firing Post 350
12.3.6 Guidance-Beam Generation 351
12.3.7 Laser Characteristics 352

12.3.8 Future Requirements 353
12.4 Laser Fuzing 353
12.4.1 Introduction 353
12.4.2 Mode of Operation 354
12.4.3 Principal Subsystems 356
12.4.4 Laser Characteristics 357
12.4.5 Future Requirements 357
12.4.6 Summary 358
Selected Bibliography 358

CHAPTER 13

Laser-Based Remote Sensing 359

13.1 Introduction 359
13.2 Introduction to Laser-Radar Systems 361
13.3 Historical Development 363
13.4 Military Utility 364
13.5 Basic Operation of a Laser-Radar System 366
13.6 Laser-Radar Subsystem Components 371
13.7 A Simple (Flash) Laser-Radar System 372
13.8 Dimensions of Laser-Radar Systems 373
13.9 Automatic Target Recognition 373
13.10 Laser-Radar Seeker 377
13.11 Bathymetry and Topographical Mapping 379
13.12 Obstacle Avoidance 383
13.13 Vibrometry 384
13.14 Acousto-Optic Mine Detection 387
13.15 Remote Detection of Hazardous Substances 388
13.15.1 Detection Using Infrared Absorption Spectroscopy 389
13.15.2 Remote-Sensing System Operation 390
13.15.3 Detection Using Raman Spectroscopy 392
13.15.4 Super-Continuum Techniques 398
13.15.5 Other Detection Techniques 398
13.16 Evolving Remote-Detection Techniques 399
13.16.1 Photo-Acoustic Spectroscopy 399
13.16.2 Laser-Induced Breakdown Spectroscopy 399
13.16.3 Stand-Off Terahertz Spectroscopy 401
13.16.4 Terahertz-Induced Fluorescence 402
13.17 Noncooperative Identification or Identification Friend or Foe 404
13.18 Future System Requirements 404
13.19 Summary 405
Selected Bibliography 405

CHAPTER 14

Free-Space Optical Communication Networks 407

14.1 Introduction 407
14.2 Basic Principles of Communications 408

14.2.1 Fundamentals 409
14.2.2 Military Utility 411
14.2.3 System Operation 414
14.3 Free-Space Laser-Based Communications 424
14.3.1 Summary of Point-to-Point Communication Operation 426
14.4 Short-Range Optical Communication System 427
14.4.1 Introduction 427
14.4.2 Short-Range Ultraviolet Nonline-of-Sight Communications 427
14.4.3 System Performance and Limitations 429
14.4.4 Laser Requirements 429
14.5 Retro-Reflective Optical Communications 429
14.5.1 Introduction 429
14.5.2 Military Utility 430
14.5.3 System Operation 430
14.5.4 Laser Requirements 431
14.6 Quantum Cryptography 431
14.6.1 Introduction 431
14.6.2 Military Utility 433
14.6.3 System Operation 433
14.6.4 Laser Requirements 433
14.6.5 Future Requirements 435
14.7 Summary 435
14.7.1 Future Requirements for Optical Communication Systems 436
Selected Bibliography 436

CHAPTER 15

Navigation Sensors 439

15.1 Introduction 439
15.2 Historical Development 440
15.3 Military Utility 441
15.4 Fundamental Principles 443
15.5 Ring-Laser Gyroscope 443
15.5.1 Description of the Ring-Laser Gyroscope 445
15.5.2 Lock-In Phenomenon 446
15.5.3 Alleviation of Lock-In 446
15.5.4 Sources of Error 447
15.5.5 Laser-Source Requirements 448
15.5.6 Ring-Laser Performance 448
15.6 Fibre-Optic Gyroscope 448
15.6.1 Principle of Operation 448
15.6.2 Description of the Fibre Gyroscope 449
15.6.3 Open-Loop Operation 450
15.6.4 Closed-Loop Operation 451
15.6.5 Summary of Laser Requirements 454
15.6.6 Summary of Performance 454
15.7 Integrated Fibre-Optic Gyroscope 454

15.7.1 Sources of Error 455
15.8 Other Optical Sensors 455
15.8.1 Laser Requirements 456
15.9 Applications 456
15.9.1 Artillery Pointing 456
15.10 Summary 457
15.10.1 Future Requirements 458
Selected Bibliography 458

CHAPTER 16

Other Emerging Military Applications 459

16.1 Laser Decoys 459
16.1.1 Introduction 459
16.1.2 System Operation 460
16.1.3 Laser Parameters 462
16.2 Direct-Fire Simulation 462
16.2.1 Introduction 462
16.2.2 Laser Requirements 462
16.2.3 MILES 463
16.2.4 System Operation 463
16.3 Search and Rescue 463
16.3.1 Introduction 463
16.3.2 Laser Requirements 464
16.4 Illumination Aid 464
16.4.1 Introduction 464
16.4.2 Laser Requirements 464
16.5 Small-Arms Laser Sight 464
16.6 Power Beaming 464
16.6.1 Introduction 464
16.6.2 Military Utility 465
16.6.3 Principal Components 466
16.6.4 Concept of Operation 467
16.6.5 Laser Requirements 468
16.6.6 Summary 469
16.7 Laser Propulsion 469
16.8 Optical Air-Data Systems 470
16.8.1 Laser Requirements 471
16.9 Guided Bullets 471
16.9.1 Military Utility 472
16.9.2 Laser Requirements 472
16.10 Mine Detection 472
16.10.1 Military Utility 472
16.10.2 Laser Requirements 473
16.11 Laser Hydrophone 473
16.11.1 Military Utility 475
16.11.2 Principle of Operation 475

16.11.3 Laser Requirements 476
16.11.4 Summary 476
16.12 Summary of Novel Applications 477

Part III Safe Use, Education, and Training 479

CHAPTER 17

Safe Use of Lasers and Legislation 481

17.1 Introduction 481
17.2 Effective Laser Safety 484
17.2.1 The Laser-Safety Standards 485
17.2.2 Other International Safety Standards 486
17.3 Hazard Classification of Lasers 487
17.4 Maximum Permissible Exposure 489
17.4.1 Hazard Distances 489
17.4.2 Hazard Zones 490
17.4.3 Calculation of NOHD and EOHD 492
17.4.4 Laser Hazard Area Trace 493
17.4.5 Engineering Controls 493
17.5 Laser-Safety Officer 494
17.6 Visiting/Host Nation Interactions 494
17.7 Potential Hazards to Personnel 496
17.7.1 Eye Anatomy and Hazards 497
17.8 Hazards to Skin 502
17.9 Other Potential Hazards Presented by Lasers (Nonbeam Hazards) 503
17.10 Summary of Potential Hazards Presented by High-Power Laser Beams 504
17.11 Common Causes of Laser Incidents 504
17.12 Laser-Safety Training 506
17.13 Laser Safety in Practice 509
17.13.1 Safety-Design Considerations for Laser Research and Development Laboratory 509
17.14 Laboratory Testing 511
17.14.1 Engineering Controls 511
17.15 Safe-Operating Guidelines/Procedures 513
17.16 Field Testing and Trial Planning 514
17.17 Planning of Trials 515
17.18 Range Laser-Safety Officer 517
17.19 Range-Safety Officer 519
17.20 Flight Trials 520
17.20.1 Flight Trials Involving Ground-Based Laser Trials 520
17.20.2 Aircraft-Based Laser Trials 521
17.21 Some Additional Safety Thoughts 521
17.22 Summary 522
17.23 Legislation 522
17.23.1 Protocol IV 523

17.24 Summary 524
Selected Bibliography 526

CHAPTER 18

Education and Training 529

18.1 Introduction 529
18.2 Doctrine and Concepts 530
18.3 Training 531
18.4 Synthetic Environments 533
18.5 Role of Education 534
18.6 Training Design 535
18.7 Equipment Training 539
18.8 Simulated Training 540
18.9 Laser-Safety Training 541
18.10 Equipment for Training 542
18.11 Summary 543
Selected Bibliography 543

APPENDIX A

The Development of Infrared Threat-Seeker Technology 545

A.1 Introduction 545
A.2 Generic Optical Seeker 546
A.3 Target-Tracking Techniques 546
A.4 Impact of Seeker-Technology Developments 549
A.5 Imaging-Seeker Technology 549
A.6 Summary 550

APPENDIX B

Jammer Development 551

B.1 Introduction 551
B.2 Development of Jammer Techniques 551
B.3 Incoherent Sources for Infrared Jamming 553
B.3.1 Black Bodies 553
B.3.2 Arc-Lamp Technology 553
B.4 Coherent Sources for Infrared Jamming 555
B.5 Summary 556

APPENDIX C

Missile-Warning System Technology 557

C.1 Introduction 557
C.2 Ultraviolet Missile-Warning System 560
C.3 Infrared Missile-Warning System 562
C.4 Two-Colour Missile-Warning Systems 563
C.5 Summary 564

APPENDIX D
Rules of Thumb and Guidelines 567
D.1 Laser-Directed Energy-Applications 567
D.2 Lasers, Optics and Testing 568
D.3 Scattering and Propagation 571
D.4 Systems 572
D.5 Pointing and Tracking 573
D.5.1 Analysis 574
D.5.2 Human Eye 574
D.5.3 Detectors 575
D.5.4 Miscellany 576
Selected Bibliography 579

Acronyms 581

Glossary 583

About the Author 635

Index 637

Preface

This book is intended for use by scientists, engineers, and technologists, with varying degrees of training spanning undergraduate students through the graduate level to postdoctoral students and research fellows as well as well-experienced professional engineers. In the case of the more junior or less experienced people, this book aims to provide an insight to the technology or its applications, whereas for the more experienced engineers, the book should provide a guide to requirements and the derivation thereof, as well as an indication of alternative or emerging applications of the military laser technology.

The critical aim of this book is to provide an insight into the physical principles of the device technology that underpins many laser-based military systems in one form or another. From this knowledge a deeper understanding of the fundamental requirements and the potential performance, as well as limitations of such systems, may be assessed, given the appropriate operational parameters.

There is a vast range of such military systems that continues to expand, much like our universe. This book concentrates on military laser-based systems that are either in relatively common usage or close to deployment; there is a chapter giving a brief description of some emerging applications, which are expected to become operational in the near term.

The information presented in this book makes the assumption that the reader has a basic appreciation of quantum theory (i.e., the existence of the photon and energy levels within atoms and molecules) and the associated physical principles. The use of complex concepts has been kept to a minimum, as has the use of equations in one form or another, which is not connected with advice given to Professor Stephen Hawking when writing *A Brief History of Time* that "every equation included in the text would halve the potential sales"; in this case, it is to ease the understanding of particular phenomena in system terms especially in qualitative terms. A selected bibliography is given at the end of each chapter, so that the more interested or curious reader can access more detailed information.

This book is not intended to compete with the many excellent books that have been published describing the basis of the operation of a laser and the fundamentals of laser design. Careful investigation has indicated, as far as the author is aware, that there is a dearth of books concentrating on the military applications of such technology and, in particular, military laser systems. Hence, it is not intended to be yet another laser book.

Over the last five decades, laser technology has developed a lot of mystique, along with an extensive and complex lexicon of its own. Therefore, a secondary aim of this book is to steer the interested or less experienced engineer through these complexities by attempting to define the common jargon using simple scientific terms; this aspect of the technology is covered in a glossary. Moreover, some rules of thumb that the author has found useful over the last four decades or so have also been included.

The author is extremely appreciative of the assistance and encouragement provided by very many colleagues around the world. Needless to say, that an undertaking of this scale has involved many discussions and debates so the list is almost endless; however, the following have made significant contributions which I am pleased to recognise and thank:

Dr Chris Howle, Dstl;
Professor Mark Richardson, Defence Academy U.K.;
Professor Lars Sjöqvist, FOI (Sweden);
Dr Ove Steinvall, FOI (Sweden);
Dr Ken McEwan, Dstl;
Dr Richard Hollins, Dstl;
Dr Tim Kendall, Dstl;
Mr Kevin Burrett, Dstl;
Dr Philip Milsom, Dstl;
Mr Mark Elson MBE, Dstl;
Mr Grant Perkins, QinetiQ;
Dr Andrew Scott, QinetiQ;
Professor Keith Lewis, Scovis (formerly with QinetiQ);
Dr Nick Wood, QinetiQ;
Dr Mark Gruneison, AFRL (USA);
Dr Tim Newell, AFRL (USA);
Dr Ron Kaspi, AFRL (USA);
Professor Jas Sanghera, NRL (USA);
Mr Jeff Hecht;
Mr John Albertine (formerly of U.S. DoD);
Dr Rita Peterson, AFRL (USA);
Dr Espern Lippert, FFI (Norway);
Professor Miloslav Dušek, Palacký University (Czech Republic);
Dr Christian Seidel, Cassadian (Germany);
Dr Peter Kightley, QinetiQ;
Professor See Leang Chin, Laval University (Canada);
Mr Joe Grobmyer, AMRDEC (USA);
Dr Marc Eichhorn, Institute Saint Louise (France/Germany);

Mr Matt Flower, UK MoD;
Dr Eric Liggins, QinetiQ;
Dr Bruce Stuck, (formerly with Brooks AFB, US);
Mr Peter Nunn, (formerly with BAe Systems);
Mr David Dwyer, GE (formerly with OCTEC);
Dr Ian Elder, Selex Galileo;
Professor John Barr, Selex Galileo;
Dr Laurence Cox, Thales Optronics (formerly with Dstl);
Prof Johan Nilsson, ORC University of Southampton;
Professor Andy Clarkson, ORC University of Southampton;
Dr Bob Grasso, Northrop Grumman Corporation (US);
Professor Gerry Buller, Herriot-Watt University;
Professor Geoff Nash, University of Exeter (formerly with QinetiQ);
And many others!

Of course, despite all of these excellent contributions, suggestions and help, there will be errors which are entirely mine.

Part I
Fundamentals

CHAPTER 1

Military Laser Technology and Systems

1.1 Introduction

A laser is only a sophisticated light source with many special optical properties; however, for many years the laser was considered to be a technical solution looking for a problem, a phrase probably coined by Professor Theodore Maiman's assistant Irnee J. D'Haenens: how times have changed. In the decades since 1960, the developments with laser technology have revolutionised the performance of many military systems, providing their operators with significant enhancements in performance particularly in sensing, soft-kill (countermeasure) and hard-kill (laser damage) options, guidance and navigation. Consequently, there have been many military projects aimed at exploiting one or more of these special optical properties of the laser, or its fundamental characteristics, since the first reported practical demonstration of laser action in May 1960 by Theodore Maiman.

The first reported application of laser technology concerned sensing and targeting, and in particular, the use of a pulsed source to measure the distance to distant or remote objects. This provided unprecedented measurements of accuracy to a target from the remote spotter's position. More recent examples of the use of the laser in the sensing function have included gated-imaging and laser-radar applications. There are a number of projects that exploit the high radiant-intensity characteristics of the laser beam for directed-energy applications, particularly for countermeasure systems used for platform protection or anti-air applications.

Optical systems, techniques and technologies have been used on the battlefield for many thousands of years to provide information to the military to gain an advantage over an enemy. However, just as in the case of Newton's third law of motion, there is an optical countermeasure equivalent: to every optical system, there is the potential for a countermeasure effect to negate its performance. This may be considered the first law of optical countermeasures. It is usually possible to create an effect that will negate or disrupt the performance of a sensor, or system, denying information to an observer; in some cases, the techniques can be remarkably simple, such as the use of a bright light, as discussed in this book.

The direct application of photons has been used as an optical-countermeasure technique, in one form or another, for several millennia to provide a defensive capability to defeat sensors or destroy components. The initial active-countermeasure approaches using photons concerned the use of focussed sunlight, either to destroy objects or for dazzling an opponent's soldiers. Smokescreens offer a very simple countermeasure method through attenuation and scatter; smoke in one form or another has been used to deny information to a range of sensors, for example, enabling manoeuvres to be undertaken without an enemy knowing exactly what is under way. The use of a laser beam to dazzle a sensor system may be considered a modern application of obscurant techniques. Countermeasure techniques have developed significantly, as have the use of active systems to aid a range of military functions.

More recent optical-countermeasure applications have concentrated on the use of optical techniques to disrupt the normal operation of a sensor or to prevent observation of a scene or designed operation of a target-tracking system, thereby causing a weapon to miss its intended target. Laser technology is not necessary to provide the photons required to disrupt the normal or designed operational performance of a sensor; many undesirable effects in the normal operation of a sensor may be achieved using any intense optical source. For example, the initial infrared jammer systems used heated carbon rods to generate the infrared energy that was chopped into a modulated waveform needed to defeat the target-tracking system in a heat-seeking missile. Later developments have seen the use of arc-lamp technology for this application and it is only recently that laser-based sources have entered operational service. The second law of countermeasures may be: don't use a laser unless absolutely necessary. Generally, the use of laser technology in these applications leads to a more efficient and effective countermeasure system.

The high spectral purity of the laser emission may be exploited for many sensing applications. For example, the laser may be used in a system that acts as a vibrometer, which senses the vibratory motion of a target from the changes in the photon energy reflected by a target. The sensed Doppler shift in the reflected energy may be used to classify or identify a target. This form of sensing may also be used to determine the speed of an aircraft, instead of using a pitot tube. Other sensing methods rely on the absorption of specific laser lines to identify particular species in the atmosphere.

The application of the special characteristics of the photons emitted from a laser has led to the development of many very effective military systems, not only the countermeasure devices described briefly above. The original motivation for the development of the laser as an intense source of optical energy was allegedly as a death ray, comparable with the original vision for radar, and was featured as such, in the James Bond film, *Goldfinger*. In fact, the use of the laser as a source of photons for a compact tactical directed-energy weapon, which vaporises its targets, is still to be demonstrated; this scale of destruction is unlikely to be realised in the foreseeable future.

The use of laser-source technology and related techniques in guidance and navigation systems has developed rapidly since the original demonstration of laser action and, indeed, probably represents the most frequent use of laser technology in military laser systems. Cruise missiles exploit these characteristics to provide pinpoint accuracy, during the navigation, mid-course and terminal guidance phases of flight, to hit a previously selected target: in this case, the laser-based system may be used to profile the ground and compare the observed profile with that in a stored map of the topography along the intended flight path to establish a missile's position accurately. Similarly, the integration of a simple quadrant-detector on a dumb bomb, to form a seeker to collect laser photons reflected from a target, along with a flight-path control system creates a laser-guided bomb, which has enabled the hit-point accuracy to be measured in metres rather than miles.

Many of the modern military applications have only been possible thanks to the availability of laser light and its special characteristics. The unique characteristics of the photons in a laser beam are usually cited as:

- Monochromatic emission, although certain classes of lasers, such as dyes and vibronic devices, prove to be an exception to this rule and recent investigation of very short pulses in the femtosecond regime has demonstrated very broadband (super continuum) emission in the filaments;
- Spatial coherence, so that the shape of a wavefront can be predicted into the future;
- Temporal coherence, which means the colour of the laser beam is monochromatic and the phase can be predicted at a time in the future;
- Polarisation, a defined state;
- Highly directional, providing the potential for high radiant intensity.

However, there are many other features that are associated with a laser beam that follow from these fundamental characteristics, and add value, so that they become the source of choice for use in a wide range of optical systems. These features may be interpreted as follows:

- High radiant intensity;
- Short-pulse generation (originally known as giant-pulse generation);
- Variable repetition rate to continuous wave;
- Speed-of-light velocity of the beam propagation;
- High spectral purity;
- Excellent modulation control.

These features of the laser output have been exploited in one form or another to provide the various military functions described in Part II. Some of these applications are quite mature, others are still developing and new applications are being devised and researched. Currently, there is investigation and development of a number of soft-kill weapon systems used to enhance platform survivability and defeat optical seekers used in anti-air missile systems. In this case, it is the realisation of compact, efficient and reliable laser technology that has made the use of soft-kill weapons viable. Similarly, with sensing applications, the use of laser technology has enabled obstacle avoidance systems to become a reality and new applications, such as the use of this type of system for the detection of trip wires, are being developed.

1.2 Laser Development

The initial development of the laser as a mere photon source, based on the amplification of stimulated emission, followed from the development of the microwave analogue, known as the maser (microwave amplification of stimulated emission). Academicians Nickoli Basov and Alexander Prohorov and Professor Charles Townes shared the 1964 Nobel Prize for physics for their pioneering work that led to the first demonstration of laser action by the late Theodore Maiman at Hughes Research Laboratory in Malibu, California, in May 1960.

There was a lot of competition to demonstrate the first laser source amongst various research laboratories throughout the world. There was intense competition between Hughes Research Laboratory and Bell Labs, a West-East competition within the United States. The first team to declare success was led by Professor Maiman; the famous paper was published in the journal *Nature* in 1960, rapidly followed by another solid-state laser (U^{3+} in CaF_2). The first demonstration of a gas laser (HeNe) at the Bell Labs followed soon after the realisation of these solid-state devices in 1960. It is with some irony that many research teams considered that a gas, such as a helium-neon mixture, offered the ideal lasing medium and would lead to the first successful demonstration of laser action. The excellent frequency stability of this class of laser has resulted in it being the source most commonly used in ring-laser gyroscopes.

Theodore Maiman's laser was a flashlamp-pumped ruby rod (chromium-doped aluminate, where the absorption of shorter wavelength photons gives rise to its red colour). The laser action took place between two states in the chromium (Cr^{3+}) ion, the dopant in the aluminium oxide that forms the ruby crystal. Soon after this demonstration, there were many others; some of the key developments are:

- Neodymium [1] and later Nd:YAG [2];
- Gas laser: HeNe [3] and later the ubiquitous red line [4];
- Alkali [5];
- Semiconductor: GaAs [6];
- Diode-pumped solid-state laser [7];
- Ceramic lasers: Dy^{2+}:CaF_2 [8];
- Fibre [9];
- Noble gas: argon ion [10];
- Gas lasers: CO_2 [11];
- Gas dynamic laser: carbon dioxide [12];
- Pulsed chemical: HCl [13];
- Chemical: HCl [14];
- Dye [15];
- Copper vapour [16];
- Vibronic solid-state lasers: a nickel doped magnesium fluoride (Ni: MgF_2) [17];
- Free-electron laser [18];

- Chemical oxygen iodine laser [19];
- Vertical-cavity surface-emitting laser: GaInAsP/InP [20];
- Thin disk laser: Yb: YAG [21];
- Quantum-cascade laser [22];
- Filamentation/ultra-short pulse emission [23];
- Vertical external-cavity surface-emitting laser [24];
- Super-continuum generation [25].

Some of these laser sources are the critical device components of many laser-based military systems used today. The technologies are considered further in Chapter 2 and the use in military systems is considered in Part II.

The decade of the 1960s was also important for the investigation of nonlinear optical phenomena; the availability of laser beams enabled intense optical fields to be formed in crystals so the nonlinear aspects of the susceptibility (χ) in the polarisation vector could be investigated and then exploited. The first demonstration of nonlinear conversion was in 1961 [26] and was frequency doubling of photons from a ruby laser using a quartz crystal. Third-harmonic generation followed a year later [27] and the same year saw sum-frequency generation demonstrated [28].

This activity was followed by the investigation of optical parametric techniques [29] (optical parametric amplification) and [30] (optical parametric oscillator). Parametric techniques essentially equate to photon splitting, so that two photons are created with a different (longer) wavelength from the photon incident on the nonlinear crystal that undertakes the wavelength-conversion process. This technique of frequency conversion is very important for generation of those wavelengths where direct generation of a given wavelength is difficult, as in the case of solid-state lasers generating infrared wavelength photons in the 4- to 5-μm band. However, it should be noted that the use of nonlinear techniques tends to add complexity and are usually less efficient than direct generation devices. Wherever possible, direct-generation techniques are usually invoked, as they are generally less complex and more efficient.

Many of the early lasers had their gain medium, which undertakes the laser action, excited by flashlamp sources that emit a broad spectrum of light. Other techniques were demonstrated, such as use of radio frequencies to excite a state in the laser ion, atom or molecule, or, as is now common, the use of another laser, to create the population inversion. This latter approach has been perfected with diode-pumped solid-state lasers and has resulted in highly efficient and compact sources. Currently, wall-plug efficiencies of some diode-pumped solid-state lasers can be well in excess of 50% and, in the case of fibre lasers, this has exceeded 60%. Wall-plug efficiency (i.e., the total efficiency of the system) is a critical consideration for most if not all laser-based military systems. Other techniques for generating the population inversion may involve:

- Chemical reactions;
- Electrical discharges;
- Electron beams;
- Ion beams;

- Gas expansion (gas dynamical processes);
- Radio frequency waves;
- Other techniques, including explosives.

The first decade or so of laser development also saw many important technique demonstrations that were to be developed and perfected in the following decades. The application of nonlinear techniques has been noted above, and further important examples are the creation of giant pulses [31] and the use of a laser cavity that forced the resonant beam to make multiple zig-zag bounces through the gain medium to reduce the impact of thermo-optical distortion of the beam; these were investigated and demonstrated in the late 1960s and early 1970s. The Q-switched laser with zig-zag slab gain medium architecture is very common in many solid-state lasers.

1.3 Laser Systems Development

The first reported military system was a range finder that measured the time of flight of a pulse of light from a flashlamp-pumped ruby laser to a target (and back to a detector), soon followed by a device based on a carbon dioxide source. This development was important for providing accurate range estimates to targets for artillery operations. Laser-guided bombs were another significant military application that enabled the accuracy of bombs delivered from large stand-off ranges to be revolutionised.

This first decade of the laser also saw many important technology demonstrations such as the use of a ruby laser to destroy a visible-band camera, using a modest power flashlamp-pumped ruby laser over a range of several hundred metres. This technique evolved into the laser directed energy weapon studies of the 1970s, which used gas lasers in one form or another, such as the very powerful deuterium fluoride chemical lasers or the carbon dioxide devices. The 1970s also saw the first use of laser technology to defeat heat-seeking missiles with an early laser-countermeasure demonstration system.

Recent developments with laser technology have made this device the source of choice for many military applications although other optical sources may be used, as highlighted by optical jamming systems. A further example concerns damage induced by laser beams; however, any intense optical source can be used to destroy various components through focusing the rays from the source, which can burn a hole in any material given sufficient power density, but a laser can do it more effectively. It is worth noting that this was the same in-band defeat mechanism that destroyed a camera taken to the moon by the Apollo 12 mission and occurred when astronaut Alan L. Bean was positioning a camera and a specular reflection of the Sun, from part of the structure of the lunar module, was incident on the camera.

Modern laser technology has enabled laser-based sensing systems such as laser radar (Ladar and lidar) and active imaging to be realised. The basis of this approach is to use a laser beam to illuminate a scene or object and then undertake an analysis of the scattered and reflected photons.

Laser-radar systems, often termed Lidar or Ladar systems depending on their function, enable many sensing functions to be undertaken depending on the complexity of the system, such as:

- Target detection and recognition;
- Obstacle avoidance;
- Three-dimensional terrain mapping and bathymetric sounding;
- Surveillance and reconnaissance, including location of obscured or hidden targets, such as those positioned in amongst trees and under camouflage, as well as detection of buried mines;
- Targeting and fire control, including automatic target recognition (ATR), automatic target identification (ATI) to a high confidence level; additionally these techniques can identify the operational status such as operating or non-operating;
- Improved mission planning from mapping;
- Guidance, including aim-point selection from stored databases to enable optimum engagement of fixed, mobile and moving targets in various domains;
- Navigation, including terrain following and obstacle avoidance for manned and unmanned platforms;
- Non-cooperative target identification (NCID) against air and surface targets at extended ranges;
- Battle damage indication (BDI) and battle damage assessment (BDA) in real time;
- Detecting and sensing of hazards and hazardous agents;
- Improved aim-point selection for many types of munitions;

Many of these functions are not feasible without the use of photons generated by laser action; the development of practical and efficient laser sources have made remote sensing a reality. The specific advantage that the laser-based system has for remote sensing arises from the coherent nature of laser light emitted by the transmitter. It is possible to get a great deal of information regarding the characteristics and behaviour of the objects that are detected, given appropriate processing, as well as knowledge of the various processes that are occurring in the atmosphere.

Recent developments with active imaging, or gated imaging, permit sensing systems to detect distant targets in a cluttered environment at long ranges. In this case, short pulses of laser light are used to illuminate a chosen scene and the detector array is opened after a specific time delay to capture the illuminated scene. The delay in opening the detector may be varied so that a series of images may be captured, enabling additional information to be captured throughout the scene, regardless of clutter.

The development of optical gyroscopes, such as the ring-laser gyroscope, enabled the strapdown inertial navigation system to be realised in the 1980s. This type of navigation system, using laser-based sensors, has been in operational service for some time providing highly accurate navigation performance, which is entirely self-contained and free from external interference, such as jamming.

However, one of the critical issues concerns the development of low-cost and reliable devices and this is the research objective of many research projects. Other research interests include:

- Enhanced efficiency;
- Increased tuneability (where feasible);
- New wavebands in the electromagnetic spectrum;
- Greater power (radiant intensity);
- Simultaneous multiple-line emission;
- Reduced size and weight.

Significant device-development thrusts concern semiconductor and fibre-laser technologies. The utility of these lasers has become appreciated and many new projects are actively investigating the application of this versatile and efficient technology for a broad range of military systems as discussed in this book. It may be worth remembering that the best solution laser system for any specific application is the one that offers the least complexity, whilst delivering the performance required to fulfil the entire task. In the military context, the performance aspect is multifaceted and will usually involve many considerations beyond the power, shape and colour of the laser beam, which will be discussed in some detail in the following parts of this book.

1.4 Aim of the Book

An important objective of this book is to discuss the principles of the use of lasers in military applications, as this technology provides a critical subsystem for many military systems and not merely an experimental curiosity of the research laboratory. A further objective is to remove some of the mystique that still surrounds lasers. Other major objectives include:

- The current state-of-the-art of laser technology;
- How the device technology is currently exploited;
- Some consideration of the fundamental requirements and considerations that are required to realise an efficient and effective system;
- Safe use and operation of the laser-based systems;
- Consideration of the current safe-operation requirements and standards;
- Requirements for training and other lines of development.

1.5 Organisation of the Book

The book is divided into three principal parts. Each part is essentially self-contained to aid the busy reader establish a rapid understanding of the critical issues. An important objective is to provide an insight and background into the fundamental aspects of the laser technology and its application to a range of systems for both graduate and undergraduate students, as well as the experienced engineer requiring underpinning information or fundamental knowledge. Moreover, it aims to be much more than just another laser book, as there are many outstanding books available describing the fundamentals of just about every aspect of laser action; some of these works are referred to in this text.

This book does give a description of the fundamentals of laser technology, as well as its use in a range of systems designed for a variety of military applications. Equations and mathematical descriptions have been kept to a minimum; references and bibliographies are provided to direct those students and engineers who require more in-depth analysis and detailed information about specific aspects of the technology and the various databases.

1.5.1 Part I: Lasers and Fundamentals

The objective of this segment is to provide a qualitative description of the fundamentals associated with optics, radiometry and the operation of a laser source, the various types of laser devices and the impact of the atmosphere on the beam. The technical aim here is to remove some of the mystic surrounding laser technology. Consequently, the topics covered in this introductory part are:

- Laser action and a definition of many of the terms associated with laser technology;
- Laser devices: the features and characteristics of the various classes of laser device, such as the relevant solid-state, liquid-phase and gaseous lasers;
- Beam director technology for achieving the energy-on-target objectives;
- Laser beam propagation and how this phenomenon can modify and limit the laser flux that propagates with time towards a designated target;
- Radiometry and its role in assessing the performance of laser systems, along with the range of terms used in this branch of optics, also considering a range of techniques that may be used to characterise a laser beam.

1.5.2 Part II: Military Laser Systems

The objective of this section of the book is to provide qualitative information about the function and operation of the various systems that use laser technology to undertake their function to provide military capability. The current status of development of each application is estimated along with possible future system performance requirements, such as undertaking as many functions as possible at the eye-safer wavelengths. This segment will consider and describe the various principal subsystems and their function that normally are integrated to form the particular system. For each application, there is description of the typical laser beam performance requirements needed to fulfil the task.

1.5.3 Part III: Safe Use, Education, and Training

This section describes the fundamental aspects of laser safety and in particular considerations that are important to the safe operation of lasers and laser-based systems. This is a critical aspect of the development of laser systems and covers guidelines for safe operation of laser technology in the laboratory, as well as on an evaluation range and in operational service.

Consideration is given to procedures that may reduce hazards during the development of the technology, in addition to testing the performance to ensure that it is achieving its operational performance. An outline of, and reference to, the laser-safety standards in the United States and Europe is provided in this part of the book, along with a description of the common hazards, such as eye and skin damage. Additionally, this part of this book provides some information about the Protocol IV of the Vienna Convention of the International Committee of the Red Cross, which governs the use of laser weapons.

This section also covers an outline of the principal aspects of training personnel to operate the equipment successfully and safely, so that the laser-based system operates as designed; more detailed aspects need to be devised for each specific system to ensure that the critical aspects of the operation of the laser-based system become familiar to the operators. Training personnel with the laser in a safe mode is absolutely critical, so that tactics and methods for operational use of any laser-based systems can be perfected; particularly as in many cases the laser-based system represents novel technology. As a consequence, new tactics and concepts of use are required to achieve optimum performance.

The glossary covers common abbreviations and definitions of acronyms used when describing laser devices and their operation. The appendices cover the development of infrared seekers, the development of infrared jammers and missile and threat warning, which have supported development of platform-protection systems, and a selection of rules of thumb often used to aid estimates of potential performance of systems or the validity of a measurement to a first approximation.

References

[1] Johnson, L. F., and K. Nassau, "Infrared fluorescence and stimulated emission of neodymium in calcium tungstate," *Proceedings of the Institute of Radio Engineers,* Vol. 49, No.11, 1961, p. 1704.

[2] Geusic, J. E., et al., "Laser oscillations in Nd-doped yttrium aluminium, yttrium gallium and gadolinium garnets," *Applied Physics Letters,* Vol. 4, No. 10, 1964, p. 182.

[3] Javan, A., et al., "Population inversion and continuous optical maser oscillation in a gas discharge containing a He-Ne mixture," *Physical Review Letters,* Vol. 6, 1961, p. 106.

[4] White, A. D., and J. D. Ridgen, "Continuous gas maser operation in the visible," *Proceedings of the Institute of Radio Engineers,* Vol. 50, No. 7, 1962, p. 1697.

[5] Jacobs, S., et al., "FA15 optical pumping of cesium vapor," *Journal of the Optical Society of America,* Vol. 51, 1961, p. 477.

[6] Hall, R., et al., "Coherent light emission from Ga-As junctions," *Physical Review Letters,* Vol. 9, No. 9, 1962, pp. 366–368.

[7] Keyes, R. J., and T. M. Quist, "Injection luminescent pumping of $CaF_2:U^{3+}$ with GaAs diode lasers," *Applied Physics Letters,* Vol. 4, No. 3, 1964, p. 50.

[8] Hatch, S. E., et al., "Hot-pressed polycrystalline $CaF_2:Dy^{2+}$ laser," *Applied Physics Letters,* Vol. 5, No. 8, 1964, pp. 153–154.

[9] Koester, C. J., and E. Snitzer, "Amplification in a fiber laser," *Applied Optics,* Vol. 3, No. 10, 1964, p. 1182.

[10] Bridges, W. B., "Erratum: Laser oscillation in singly ionized argon in the visible spectrum," *Applied Physics Letters,* Vol. 5, No. 2, 1964, p. 39.

[11] Patel, C.K.N., "Continuous-wave laser action on vibrational-rotational transitions of CO_2," *Physics Review,* Vol. 136, No. 5A, 1964, p. A1187.

[12] Gerry, E., and A. Kantrowitz, *Avco Everett,* 1966.

[13] Kasper, J. V. V., and G. C. Pimentel, "HCl chemical laser," *Physical Review Letters,* Vol. 14, No. 10, 1965, p. 352.

[14] Kasper, J. V. V., and G. C. Pimentel, "HCl chemical laser," *Physical Review Letters,* Vol. 14, 1965, No. 10, p. 352.

[15] Sorokin, P. P., and J. R. Lankard, "Stimulated emission observed from an organic dye, chloro-aluminum phthalocyanine," *IBM Journal of Research & Development,* Vol. 10, No. 2, 1966, p. 162.

[16] Walter, W. T., et al., "6C3 - Efficient pulsed gas discharge lasers," *IEEE Journal of Quantum Electronics,* Vol. 2, No. 9, 1966, p. 474.

[17] Johnson, L. F., R. E. Dietz, and H. J. Guggenheim, "Optical maser oscillation from Ni2+ in MgF2 involving simultaneous emission of phonons," *Physical Review Letters,* Vol. 11, No. 7, 1963, pp. 318–320.

[18] Eliass, L. R., et al., "Observation of stimulated emission of radiation by relativistic electrons in a spatially periodic transverse magnetic field," *Physical Review Letters,* Vol. 36, No. 13, 1976, p. 717.

[19] McDermott, W. E., et al., "An electronic transition chemical laser," *Applied Physics Letters,* Vol. 32, No. 8, 1978, p. 469.

[20] Soda, H., et al., "GaInAsP/InP surface emitting injection lasers," *Japanese Journal of Applied Physics,* Vol. 18, No. 12, 1979, pp. 2329–2330.

[21] Giesen, A., et al., "Scalable concept for diode-pumped high-power solid-state lasers," *Applied Physics B,* Vol. 58, No. 5, 1994, p. 365.

[22] Faist, J., et al., "Quantum cascade laser," *Science,* Vol. 264, 1994, pp. 553–556.

[23] Braun, A., et al., "Self-channeling of high-peak-power femtosecond laser pulses in air," *Optics Letters,* Vol. 20, No.1, 1995, p. 73.

[24] Kuznetsov, M., et al., "High-power (>0.5-W CW) diode-pumped vertical-external-cavity surface-emitting semiconductor lasers with circular TEM/sub 00/ beams," *IEEE Photonics Technology Letters,* Vol. 9, No. 8, 1997, pp. 1063–1065.

[25] Ranka, J. K., et al., "Visible continuum generation in air-silica microstructure optical fibers with anomalous dispersion at 800 nm," *Optics Letters,* Vol. 25, No. 1, 2000, pp.25–27.

[26] Franken, P. A., et al., "Generation of optical harmonics," *Physical Review Letters,* Vol. 7, No. 4, 1961, p. 118.

[27] Terhune, R. W., et al., "Optical harmonic generation in calcite," *Physical Review Letters,* Vol. 8, No. 10, 1962, p. 404.

[28] Bass, M., et al., "Optical mixing," *Physical Review Letters,* Vol. 8, No. 1, 1962, p. 18.

[29] Wang, C. C., and G. W. Racette, "Measurement of parametric gain accompanying optical difference frequency generation," *Applied Physics Letters,* Vol. 6, No. 8, 1965, p. 169.

[30] Giordmaine, J. A., and R. C. Miller, "Tunable coherent parametric oscillation in LiNbO3 at optical frequencies," *Physical Review Letters,* Vol. 14, No. 24, 1965, p. 973.

[31] McClung, F. J., and R. W. Hellwarth, "Giant optical pulsations from Ruby," *Journal of Applied Physics,* Vol. 33, No. 3, 1962, p. 828.

Selected Bibliography

Hecht, J., *Beam: The Race to Make the Laser*, New York: Oxford University Press, 2005.

CHAPTER 2

Laser Action, Components and Definition of Terms

2.1 Introduction

A laser beam has many unique characteristics that have been used in a variety of applications, military as well as civil. For some applications, the important characteristics of laser light are the spatial and temporal coherence and its monochromicity. Another important characteristic of the beam is its high directionality that leads to high radiant intensity, which is an important parameter for most directed-energy applications. As an example, a modest power laser pointer may have a radiant intensity comparable with that of the Sun, given a suitably small divergence.

The generation of photons in a laser follows the same basic physical process as those occurring in any other light source, through a change in the energy state of the body that is acting as the source; however, the critical difference with a laser source, compared with other sources, is the stimulation of the energy transition by another photon; the energy transition for incoherent sources is spontaneous. The photons liberated in a laser-energy transition, and the stimulating photon, have identical characteristics, which leads to the unique properties of laser light. The term given to the laser light generation is stimulated emission, owing to the chain of events required to cause a suitable event.

The emission process is a direct consequence of conservation of energy and the wavelengths are determined by quantum theory. For example, when an electron in an atom moves from a higher energy level to a lower energy level the difference in energy between these two energy levels has to be given up by the atom as the electron moves to the lower level to conserve the energy of the system. The energy difference between the two levels involved determines the wavelength or frequency of the emission. The relationship between the energy difference (ΔE) and the wavelength (λ) is given by:

$$E_2 - E_1 = \Delta E = hc/\lambda \tag{2.1}$$

where E_2 is the higher energy level, E_1 is the lower energy level in the energy transition, h is Planck's constant and c is the velocity of light. Hence, the shorter wavelengths correspond to photons that are more energetic.

The special properties exhibited by a laser beam are a result of the precise mechanisms used in the laser cavity, or resonator, to generate the photons. The laser source has to be quite a sophisticated device to exploit the attributes of laser action. This sophistication may be in the design and construction of the device, or in the growth processes used to make the laser source in the case of semiconductor lasers. Very sophisticated techniques may be applied during the photon generation process, either within the laser device, or externally, to exploit fully some feature or characteristic of the laser action.

This chapter considers and describes the operation of a laser in terms of the fundamental physical processes from a qualitative viewpoint; many of the detailed explanations of laser physics, or quantum electronics as it is formally described, are complex and are dealt with in many excellent texts and so will not be described here. This chapter also introduces some of the terms used to describe the laser technology, as well as those used to indicate the characteristics and performance of its operation and the generated beam.

2.2 Laser Classification

It has already been noted that there are various types of laser devices, often characterised by the form of the gain medium and then qualified by the pumping scheme used to create the population inversion, a fundamental requirement for stimulated emission and discussed further below. The most common broad classifications for laser devices are solid-state lasers and gas lasers, although liquid-phase lasers have special properties well aligned with some military applications. The basic operation of these lasers is discussed in some detail in Chapter 3.

For many military applications, it is largely accepted that the preferred approach is a laser system based on solid-state laser technology. The perceived advantages of solid-state technology are:

- Compact configurations;
- Long shelf-life;
- Ruggedness;
- Electrical excitation;
- Potential for high efficiency;
- Nontoxic;
- Potential to operate over extreme temperature ranges;
- Flexibility of output characteristics;
- Potential for multifunctional capability;
- Minimal logistics footprint.

Solid-state technology also faces some challenges such as its complexity and the consequential impact on reliability and cost. Very high-power operation for laser directed-energy applications is still a challenge for solid-state lasers, both in terms of achieving the emitted power and maintaining the requisite beam quality, because of the thermal management requirements. Careful thermal management is required to avoid thermally induced optical effects, particularly thermal lenses in the gain medium that are difficult to compensate, and stress-induced fracture of the gain medium. The use of fibre-laser technology may mitigate the majority of these drawbacks for high-power applications, as the emission characteristics are largely defined by the active core region, although they have other limitations, such as limited emitted power emission from a single fibre; however, beam combination techniques using multiple fibre amplifiers are being developed to produce high-power sources.

The principal attributes of gas lasers with well-conditioned gas flow may be summarised as:

- Ruggedness;
- No thermally induced birefringence in the gain medium;
- Potential for very high power emission (scaleable performance);
- Good or high beam quality, although some very high-power devices do exhibit thermally induced effects that reduce the beam quality;
- High efficiency;
- Potential to operate over extreme temperature ranges.

The greatest drawback of the gas laser is generally its size along with the required logistical support, including supply and disposal of dangerous chemicals for some devices; however, these types of device have generated extremely high powers with good beam quality and demonstrated high reliability. Generally, gas lasers are good reliable work horses for research studies to support development activities.

2.3 Laser Action and Stimulated Emission

Laser action is dependent on a change in the energy status of a laser specie, such as an atom or a molecule in the gain medium, which gives rise to photon emission. There are three fundamental types of energy level encountered in laser devices and these transitions lead to photon generation:

- Electronic levels concern the configuration of the electronic structure in a specie. The electronic transitions normally involve the outer electronic shells of the laser species with energy differences of the order of an electron volt (eV), corresponding to an emission wavelength around 1 μm.
- Vibrational levels result from the quantisation of the various degrees of freedom of a molecule. A vibrational transition occurs when a molecule shifts from one vibrational state to another, such as occurs when there is a change in the vibrational pattern of a molecule. These energy-level changes are much smaller and usually correspond to an emission in the mid-wave infrared.

- Rotational levels are energy levels that are a consequence of the quantisation of the rotational states of molecules. These energy differences are much smaller and result in emission in the far infrared.

In general, it may be assumed that a transition between a single set of energy levels, such as a pair of electronic levels, involves a very precise change in energy state and thus leads to a small spread of wavelengths in the beam. However, other transitions, or combinations thereof, may be less sharp, as occurs in a dye laser, leading to a broader wavelength emission, which may be exploited for some applications, as is discussed later in Chapter 3. It is also possible to have combinations of these effects such as a vibronic laser, also discussed in Chapter 3. In this case, the combination of vibrational and electronic transitions leads to tuneable emission. Additionally, other phenomena including impurities and environmental effects can result in broadening of the gain profile (i.e., resulting in a spectrally broadened emission).

Complex quantum mechanical rules determine the probability of a transition occurring. A successful laser will use the transitions that are most likely to occur, although meta-stable states may also be used as the transitions are slower and thus can provide energy storage in an upper energy level. This feature can be important for Q-switched generation of energetic pulses of photons.

The term laser action is used to describe the photon generation process in a laser device, through some form of change in the energy state of the source. It is designed to provide a device that will allow a direct exploitation of stimulated emission, predicted by Albert Einstein. A fundamental requirement of successful laser action is that the gain achieved during the exploitation of the stimulated emission process must exceed the losses occurring during that process; this positive gain-loss balance is crucial to the success of any amplification process. A laser is essentially a photon amplifier and the laser acronym (light amplification by stimulated emission of radiation) has been attributed to Gordon Gould.

The practical realisation of laser action takes place in a resonant cavity, consisting of at least two mirrors and a gain medium placed between them. The mirrors, which may be the facets of a semiconductor chip, are arranged to enable the photons generated during the energy transition in the gain medium to pass back and forth. As the photon beam makes additional passes through the cavity, the number of photons increases and consequently, so does the optical power. Generally, one mirror is usually highly reflecting and the other partially transparent, enabling a small fraction of the resonated photon flux to be emitted to form the laser beam. A schematic diagram of a very simple laser is shown in Figure 2.1.

The photons are generated during laser action as a result of stimulated emission from the energy transitions in the laser specie contained in the gain medium. Stimulated emission arises when an ion, atom or molecule (laser specie) is in an excited state and absorbs a photon of identical wavelength (as well as other physical properties) to that which the excited specie will emit in reverting to its ground state or lower laser level. Both photons are emitted simultaneously giving rise to optical gain for that specific event. For there to be a high probability of the stimulated emission event occurring, there is a need for more species to be in the excited state than those in the lower laser level. This condition is a reversal of normal population of energy states in materials, determined by Boltzmann statistics, and is called

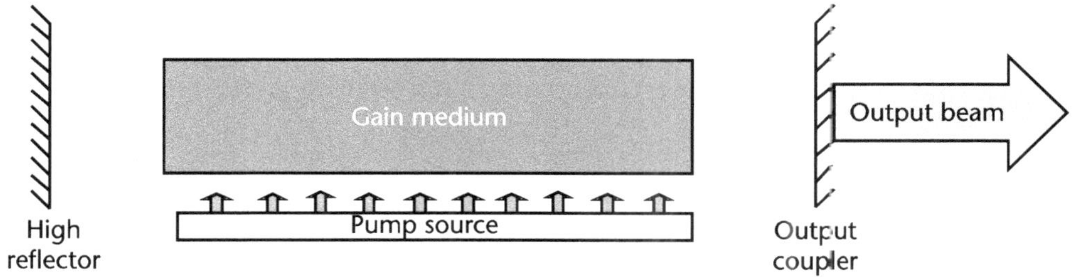

Figure 2.1 Basic laser schematic.

a population inversion. A simple illustration of stimulated emission is shown in Figure 2.2.

The laser emission process is nonlinear, as the laser device has to absorb energy to achieve the initial conditions for laser action and so that the gain in the number of photons created by stimulated emission exceeds the losses. Once this condition has been achieved, the laser beam can be created. As the applied energy to the device (gain medium) increases, so does the output power, usually in a linear fashion. Eventually, a point will be reached where the linearity ceases, owing to other mechanisms contributing to the build up of losses and reduction in efficiency, such as thermal effects or emission limitations in the pump source at higher operating powers, and the rollover condition appears. At this point the efficiency decreases. This effect is illustrated in Figure 2.3.

2.3.1 Gain Medium, Pumping and Population Inversion

The gain medium contains the laser species that will be excited through an excitation process, known as pumping, to create the population inversion necessary for the stimulated emission. The excitation may be applied to the gain medium by a variety of mechanisms. In the case of an optical technique, this may be applied to the end of the gain medium, known as end pumping, or laterally (side pumping).

The gain medium used to create the photons can be solid, liquid or gaseous and the attributes of these broad types of laser devices are considered below. The gain medium is positioned in the resonant cavity and contains the species that undergo the various energy transitions needed to create stimulated emission of photons,

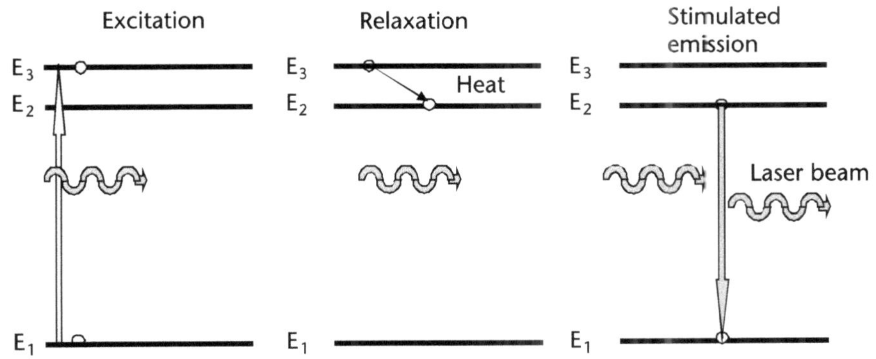

Figure 2.2 Stimulated emission process.

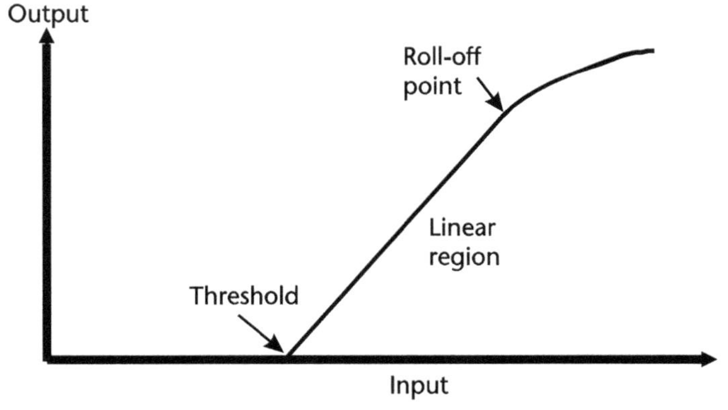

Figure 2.3 Laser-output characteristics.

determined by the energy conservation equation in (2.1). Needless to say, the actual specification and design of the gain medium are usually quite complex. For example, some considerations of the various gain media are:

- A solid-state gain medium must be of high optical quality to avoid a distortion of the generated beam, the wavelength transparency must cover the operational emission range and the concentration of laser species must be consistent with the pumping regime to allow optimum use of the applied excitation.
- Fluid gain media have somewhat different design considerations usually based on achieving uniform flow to avoid inducing distortions into the beam. In the case of liquid lasers, the suppression of the formation of even microscopic bubbles is a critical requirement for the generation of a good quality laser beam (i.e., one with minimal distortion of its shape).

The gain medium is excited to create population inversion by some technique that involves absorption of energy, illustrated by the pump energy arrows in Figure 2.4; on the right side of this figure the quantum defect is illustrated, which gives rise to heating of the gain medium. The pumping or excitation process takes a laser specie from a lower energy state to a higher one; this is a crucial prerequisite for

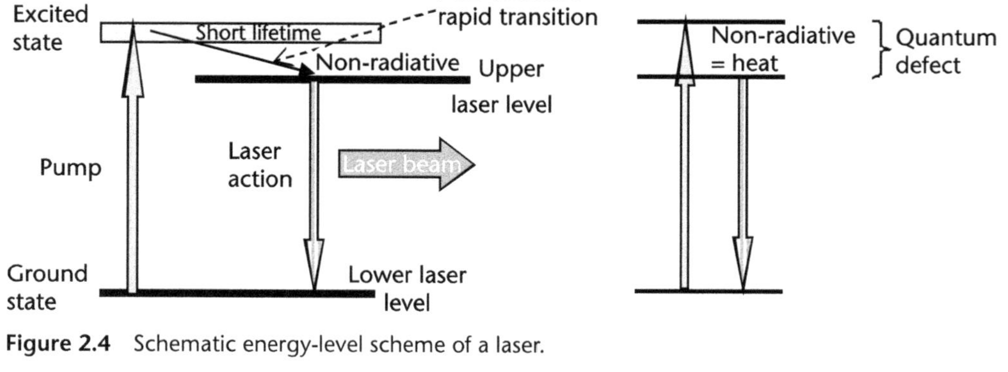

Figure 2.4 Schematic energy-level scheme of a laser.

creating stimulated emission. Several mechanisms may be invoked to pump the medium. Some common techniques include:

- Absorption of photons from another source, such as another laser, for example, a diode, or a flashlamp, but an unusual example is the explosively pumped iodine laser;
- Absorption of radio-frequency waves;
- Chemical reactions resulting in species being elevated to an excited state;
- Collisions amongst atoms and molecules in the gain medium, or collisions between electrons and the laser species;
- Forward biasing of a semiconductor to displace the holes and electrons in a semiconductor junction;
- Interaction of free electrons with ionised atoms;
- Acceleration of electrons.

The likelihood of this process happening is referred to as an absorption cross-section, dependent on many factors as the absorption process is often a multistep process.

A population inversion occurring naturally in any specie is uncommon because of the thermodynamic processes that encourage any excited specie to attempt to revert to its lowest energy state. With some irony, this is also the process that gives rise to spontaneous emission. The consequence is that as the energy of a state increases then the population in that state throughout the material decreases with respect to lower-energy states. The pumping process aims to provide a selective increase in the population of a particular energy level (the upper laser level) with respect to the lower laser energy level, as shown in Figure 2.5.

Achieving a population inversion is a crucial requirement for stimulated emission to occur. This is because the laser gain is proportional to the difference in the probability of the occurrence of stimulated emission and the absorption of energy by whatever mechanism, or combination of mechanisms. Therefore, if laser action

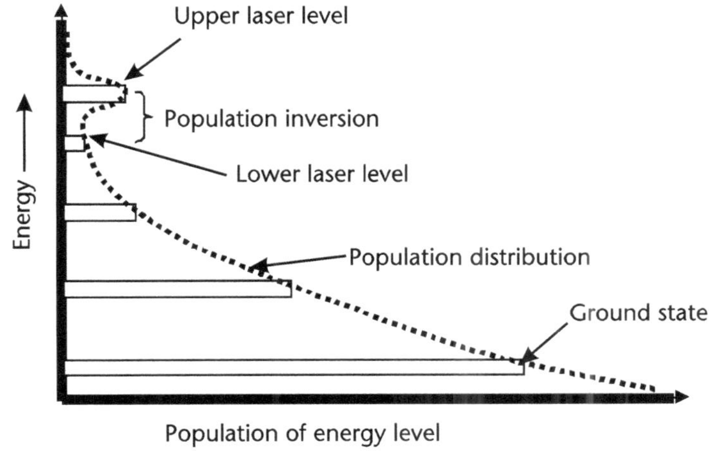

Figure 2.5 Population inversion schematic.

needs to be sustained, and it normally is, the lower level must be de-populated at a rate at least as fast as the upper level is excited, otherwise the laser action will cease, as described for a three-level pulsed laser below, a process often termed bottle-necking.

It is quite common to classify a laser by the type of the pumping scheme and the form of the gain medium (i.e., the diode-pumped solid-state laser). In this example, the laser has a solid-state gain medium excited by a diode laser. This design of laser aims to produce a compact and efficient laser source and was a successor to the flashlamp-pumped solid-state laser; these and other types of lasers are described in Chapter 3.

2.3.2 Laser Transitions

The dynamics of the laser transition are normally very complex and take account of a range of probabilistic statistics concerning the behaviour of the laser specie in its various energy states, including its upper and lower transition levels. Some of these statistics include the time that an atom or molecule spends in its various states and the probability of undertaking particular transitions to and from the various energy levels within that laser specie. To enable a population inversion to be achieved efficiently and maintained during laser action, it is normal to use a number of energy levels during the excitation and emission process. In the case of a three-level laser, as used in the original demonstration with a ruby rod, the absorbed energy takes the atom to a higher energy level; there is then an initial transition to the actual upper laser level, known as nonradiative decay, and usually results in heating of the gain medium, which reduces efficiency. This nonradiating transition is the source of heat being deposited within the body of the gain medium and is discussed further in Section 2.8.9. The laser emission occurs in the transition to the lower energy, usually called the ground state. This process is illustrated in Figure 2.6.

One critical requirement of this process is that the time taken for the excitation created by the absorption of energy, in whatever form, must be rapid compared with the lifetime of the specie in the upper laser level. Moreover, the probability of the transition from the upper laser level, which is normally not populated, to the lower laser level must be high for the stimulated emission transition to occur readily.

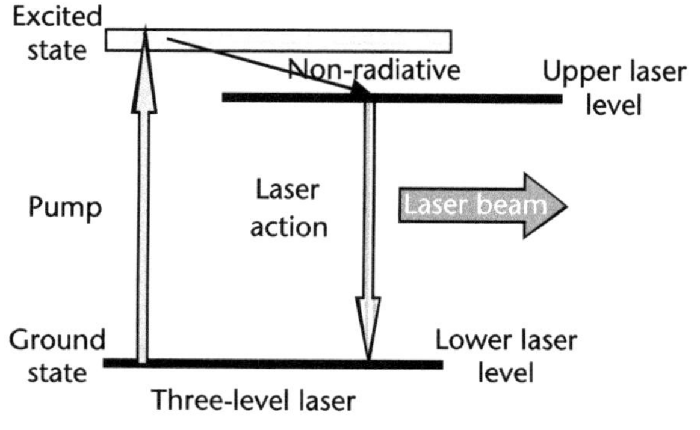

Figure 2.6 Three-level laser scheme.

One of the major drawbacks of the three-level laser is the problem of maintaining the population inversion during laser action, particularly as the radiation-less transfer usually results in an increased population of the lower energy level through internal heating from the radiation-less process. This mechanism leads to a high threshold for the onset of laser action and increases the effect of energy upconversion. As the laser action progresses, it becomes increasing difficult to maintain the population inversion needed for the laser-photon generation and it is quite common for the laser action to cease unless special measures are taken.

A four-level scheme is exploited by use of the neodymium ion in a suitable host, which has an additional energy level involved in the laser-action process. As shown in Figure 2.7, the lower laser level in this scheme is just above the normal ground state for the ion or laser specie. An important requirement in this case, is that the transition from this energy level to the ground state must be rapid to ensure efficiency of the excitation scheme for maintaining a population inversion.

2.3.3 Laser Modes

The dimensions of the resonant cavity where the laser action takes place are such that many resonant oscillations are possible, which have an influence on the beam quality of the emission. These oscillations lead to transverse and longitudinal modes, which determine the spatial characteristics and absolute frequency (bandwidth) of the emission. These modes are described using the identical terminology to the microwave practice, that is, transverse electromagnetic (TEM_{00}), where the subscripts represent the number of interacting modes in orthogonal directions to the cavity's propagation axis.

Figure 2.8 shows some examples of the lower-order modes. It is only the lowest-order mode TEM_{00} that can be characterised unambiguously. The diameter (D) of such a beam, with wavelength λ and beam waist radius ω_0, may be estimated as a function of range I according to the equation:

$$D \approx \omega_0^2 \left\{1 + \left(4\lambda R / \pi \omega_0^2\right)^2\right\} \tag{2.2}$$

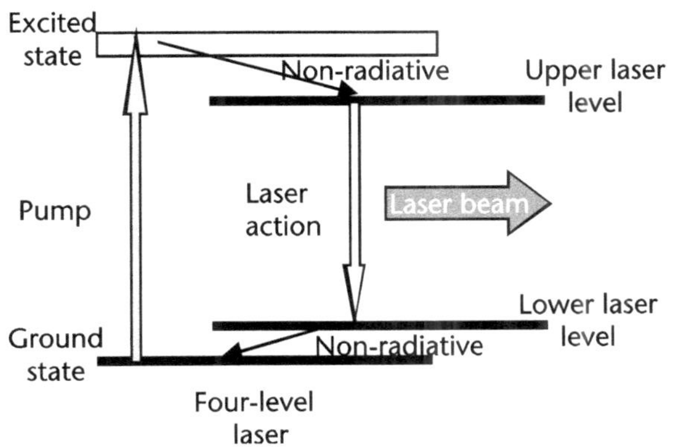

Figure 2.7 Four-level laser scheme.

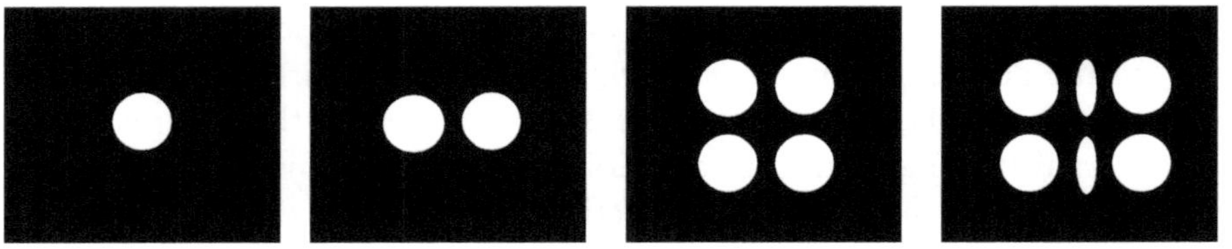

Figure 2.8 Examples of transverse laser modes.

This equation is only strictly accurate for TEM_{00} beam propagation in a vacuum.

2.3.4 Near- and Far-Field Beam Profiles

Careful analysis of (2.2) will show that for some considerable distance from the exit aperture the beam diameter is largely invariant for a fixed set of initial conditions; moreover, it approximates to the beam waist for this TEM_{00} condition, as the second term in (2.2) is small. Where this approximation holds is known as the near-field range (R_{nf}) and may be calculated from:

$$R_{nf} = \pi d^2/(4\lambda) \tag{2.3}$$

The extent of the near field may also be defined in terms of the Rayleigh range (R_R), where near field $\leq \pi\omega_o{}^2/\lambda$ or $R \leq R_R$ close to the region around the beam-waist radius (ω_0) and R_R is the Rayleigh range, which indicates the distance where the beam radius increases by $\sqrt{2}.\ \omega_0$, as shown in Figure 2.9 and expressed as

$$R_R = \pi\omega_0^2/(M^2\lambda) \tag{2.4}$$

where M^2 is the propagation factor and is discussed in Section 2.6.

When the beam appears to diverge, and hence, the diameter starts to show an angular dependency (θ_{ff}) as a function of range, then the second term in (2.2) is dominant. This condition is known as the far-field range and the diameter of the beam may be estimated from:

$$D \approx \theta_{ff} R = 2\pi R/(\pi \cdot d) \tag{2.5}$$

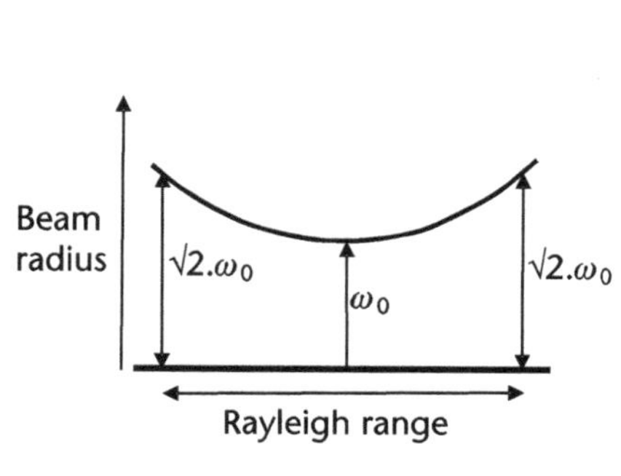

Figure 2.9 Rayleigh range.

as

$$\theta_{ff} \approx 2\lambda/(\pi \cdot d)$$

Similarly, the far field may be described in terms of that propagation distance far exceeding the Rayleigh range; in this case, the far field is defined as the distance R fulfilling the following condition:

Far field $\geq \pi\omega_0^2/\lambda$ or $R \gg R_R$ (this condition shows the divergence of the beam)

Figure 2.10 shows a schematic of the geometrical considerations.

A consequence of these features of the propagating profile of the beam is that care is required in recording the intensity profile very close to a laser in some circumstances and assuming that it will be what is projected over a significant distance (i.e., in the far field). A classic example is when the device uses a positive-branch unstable resonator, as discussed below. In the near field, a classical annulus shape should be recorded, but in the far field a number of effects will fill the void in the middle of the beam intensity pattern.

Some applications demand the generation of the lowest-order mode (TEM_{00}) to achieve the most monochromatic and highest-intensity output in a circular Gaussian profile; however, many of the directed-energy weapon applications do not require such fidelity. For example, most welding applications use a beam with a top-hat profile. All of the directed-energy applications are likely to require a good beam quality to ensure a high probability on time on target for the laser energy, whilst delivering a high radiant intensity.

2.4 Oscillators and Amplifiers

The small divergence of a typical laser beam results from the multiple passes that a photon beam makes as it moves along the optical axis between the mirrors that form the laser cavity, but there is also the wavelength-dependent spreading of the beam as it emerges from the gain medium. Those photons that are created in directions away from the optical axis are lost to the system in a conventional laser, as most lasers do not have reflectors around the sides of a resonator; the classical exceptions are the zig-zag slab and fibres, which use total internal reflection. Hence, if the oscillation is concentrated along the optical axis, the stimulated emission process now becomes a more efficient process, as a larger fraction of the transitions

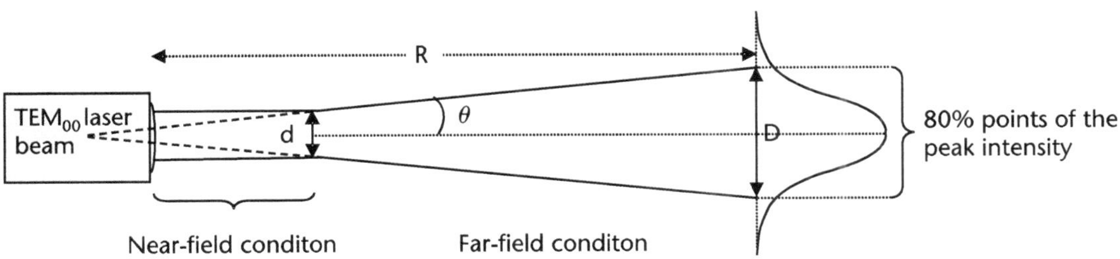

Figure 2.10 Illustration of the near- and far-field profiles of a TEM_{00} beam profile.

contribute to the emission, and furthermore, the divergence of the beam is constrained to a small value.

The mirrors define the resonant cavity that forms the oscillator and they provide the positive feedback to create the oscillation, to give an analogy with an electronic system; the mirror reflectivity characteristics, as a function of wavelength, influence the operating wavelength and allow a certain amount of bandwidth control. Normally, one of the mirrors is not fully reflective, so a specific fraction of the beam within the oscillator is allowed to escape to create the laser beam. The fraction escaping is related to the gain-loss balance of the resonant cavity in the device; the higher the gain the greater the amount of power that can escape and still maintain laser action by the source. This condition arises because the circulating power in the cavity is sufficiently large in a high-gain system to ensure that the total losses in the cavity, including the fraction allowed to escape from the cavity to form the beam, do not exceed the gain from the stimulated emission process.

The function of a laser amplifier is to boost the power of the beam created in an oscillator. In this case, the device does not form a beam but takes its photons from a separate source and increases the intensity through further stimulated emission as the beam passes through the gain medium. Some amplifiers do not have mirrors if the photon beam makes a single pass, others may have a mirror arrangement to give a small number of passes through the gain medium, in order to achieve efficient extraction of the stored energy. In this case, the mirrors are arranged to avoid creating a resonant cavity.

High-power systems often use a master-oscillator, power-amplifier configuration. The master oscillator generates the fundamental beam with good beam quality, as demanded by many applications. The amplifier then increases the optical power whilst maintaining the beam quality (beam divergence), and thus, increasing the beam brightness, provided the amplifier does not introduce any aberrations into the beam during the amplification process.

Many different types of resonant cavity have been designed and demonstrated for a range of particular applications; however, most are variations on a few specific designs and some examples are shown in Figure 2.10. Resonators may be classed as stable, where a ray of light generated parallel to the optical axis is contained continuously, whereas in an unstable resonator the rays diverge from the resonator's optical axis, and so escape from the cavity without the need to use a partially reflecting mirror. Some examples of resonators classed as stable include:

- Plane parallel, or Fabry-Perot cavity, which has two plane mirrors aligned parallel to each other. Its inherent simplicity is somewhat offset by the challenges in maintaining this critical alignment throughout the operational environment, but as it is on the edge of stability, as discussed later, the cavity lengths are usually quite short.
- Confocal resonator, which has two concave mirrors with identical radii of curvature separated by double their focal length (i.e., their focal points coincide).
- Hemi-confocal resonator, which is essentially a variation on the above with one mirror replaced with a plane mirror.

- Hemispherical resonator in which the separation between the curved mirror and the plane mirror is equal to the radius of curvature of the concave element.
- Spherical cavity, sometimes called a concentric resonator, where the two identical mirrors are separated by twice their radii of curvature (i.e., coincidence of their centres of curvature).

In general, a stable resonator requires some focussing power within the cavity to achieve the threshold condition for a ray generated with a small misalignment with respect to the cavity's optical axis. In the case of the plane-parallel cavity, this is usually provided by the thermal lens in the gain medium, originating from the thermo-optical properties of a solid-state gain medium. A plane-parallel cavity is on the edge of stability, so the cavity length dimensions are usually short and frequently used in an extra-cavity optical parametric oscillator, discussed later in this chapter for nonlinear frequency conversion. A particular characteristic of a stable resonator is that the optical energy is concentrated along the optical axis of the resonator, so the most efficient extraction of energy from the gain medium occurs in this part of the resonator. This leads to a laser beam with an almost Gaussian profile with the highest intensity along the direction of propagation and is typically used with lower-gain devices.

A positive-branch unstable resonator is configured with a concave mirror and a smaller convex mirror aligned with it at the other end of the resonant cavity, so that the rays diverge from the optical axis of the resonator, generating a beam with a doughnut shape in the near field. The mirrors are arranged so that the rays appear to diverge from a point outside of the cavity; the configuration is often designed to have a modest magnification of around two. The near-field beam profile, a doughnut shape, is shown in Figure 2.11.

The unstable resonators spread the photon energy within the cavity volume resulting in energy extraction from a larger fraction of the gain medium compared with the stable configurations. As a consequence, this design tends to lead to a higher energy conversion, but the extraction efficiency in the region of the cavity axis is reduced. A typical application for this configuration is in pulsed high-gain devices. The availability of a resonator with all-reflective optical elements is also

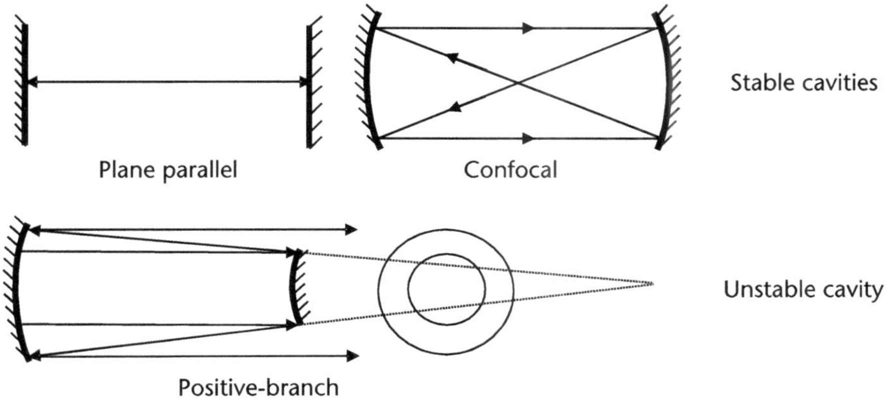

Figure 2.11 Examples of common resonant cavities.

important for use in the ultraviolet, where transparent materials may be a limiting factor in the laser design.

The condition determining the stability of a cavity is given by the relationship:

$$0 < G_1 \cdot G_2 < 1 \tag{2.6}$$

where $G_1 = (1 - d/R_1)$ and $G_2 = (1 - d/R_2)$, d is the separation of the mirrors and R_1 and R_2 are the radii of curvature of the mirrors forming the resonant cavity. The cavity stability conditions are shown for of the G_1 and G_2 parameters in Figure 2.12.

The scale of the separation of the mirrors that form the resonant cavity is many orders of magnitude greater than the optical wavelengths of any laser. This may imply that the resonant condition may be satisfied for a large number of wavelengths (λ).

$$D = n(\lambda/2) \tag{2.7}$$

where D is the cavity length and n is an integer number of half-wavelengths ($\lambda/2$).

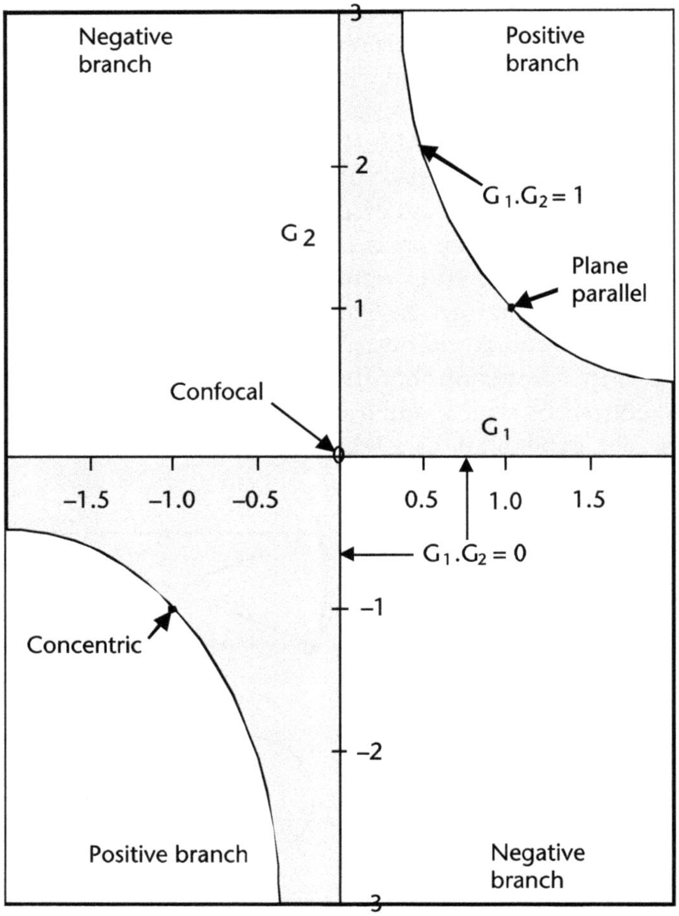

Figure 2.12 Cavity stability criterion.

In reality, an oscillation is only possible for a relatively small band of wavelengths, which are defined by the gain bandwidth of the energy transition in the gain medium.

The configuration of a resonant cavity permits two distinct types of mode:

- Transverse modes determine the cross-sectional intensity profile of the beam. These modes are described by the notation TEM_{xy}, where x and y indicate the number of intensity nulls in the emission, in the two orthogonal transverse axes. The fundamental mode (TEM_{00}) has its intensity peak on the optical axis and no intensity nulls (i.e., it has a Gaussian profile). This is the form of transverse mode required by many optical systems, as it determines the beam quality and hence divergence.
- Longitudinal modes correspond to different resonant conditions along the length of the resonator that results in slightly different frequencies; the range of the possible frequencies is governed by the gain bandwidth, as discussed above.

Most applications are more concerned about the control of transverse modes to achieve good beam quality rather than control of the precise frequency of the emission. It is quite common for the emission to change frequency, and hence wavelength, through a change in the longitudinal-mode resonance condition This frequency change is quite small as it is of the order of a few modal spacings (i.e., of the order of megahertz with a potential line width of kilohertz); generally, this shift in operating frequency is only important for some specialised applications, such as vibrometry which are considered in Part II.

A critical design consideration in the use of any oscillator in a laser system, is the inclusion of an isolator to prevent laser light being fed back externally into the oscillator and causing a catastrophic failure of the entire system. A number of approaches are well established, including use of a polarised beam and a Faraday isolator or careful system design to avoid a direct path back to the oscillator. The generation of a back reflection in high-power fibre lasers through stimulated Brillouin scattering has to be controlled and suppressed for exactly the same reasons described above.

2.5 Beam Generation and Its Control

In any system application, the laser emission is usually critically important to the successful exploitation of the laser-beam's characteristics. In some applications, the laser is required to emit a modulated waveform with a definite mark-space ratio,[1] either to disrupt the normal operation of a target or so that the receiver in the laser system can determine some characteristics or information from the received reflected signal.

1. The mark-space ratio is the ratio of the on period to the off period in a modulated waveform, compare with duty cycle.

A laser device may be operated to give either a continuous stream of photons, known as continuous-wave (CW) emission, or as a series of pulses, hence the use of the terms continuous-wave or pulsed laser. In the case of the CW emission, it is necessary to choose an energy-level scheme within the laser specie that will allow the maintenance of a population inversion throughout the emission process. For this type of emission, it is crucial to avoid undesirable losses, leading to increased threshold for the onset of laser action, and consequently, resulting in very inefficient operation. Some of the losses may be envisaged as bottle necks in the energy transfer process that limit the magnitude of the population inversion that may be sustained.

Most laser applications require a pulsed emission, the more common form of operation of many military laser devices, owing to the requirement for high peak power. The laser design may allow the normal relaxation processes of the transition from the upper to lower laser levels to form the pulsed output, known as relaxation oscillations. However, it is normal to add an additional component to the resonant cavity that controls the loss within, and hence, the emission characteristics that leads to a well-defined modulated waveform in the laser beam. In this case, the energy storage capacity and storage lifetime of the upper energy level is an important parameter in the design of the laser system.

2.5.1 Q-Switching

The device that controls the loss in the cavity is known as a Q (Quality) switch and exploits the fact that the laser is an oscillator; therefore, its operation can be described by a quality factor, which relates the energy stored per pass to the energy dissipated in each transit of the photon beam through the resonant cavity. The way that the Q-switch works is to keep the quality factor low so that the energy builds up in the gain medium, because threshold for laser action is not achieved owing to the high loss. If the loss is suddenly removed, then a large population inversion can be achieved in a high-Q cavity generating a very high number of photons rapidly. Q-switching does not increase the energy in a pulse but shortens its duration, so that the peak intensity rises.

The process of Q-switching is shown in Figure 2.13, relating the cavity losses, the generation of the population inversion and finally the pulse generation. The duration is shown as the full width of the half maximum (Δt).

Various designs of Q-switches have been designed and demonstrated. The original technique was a mechanical approach, which involved a rotating mirror to control the deflection of the beam followed by the photons through the resonant cavity, and hence, the loss in the cavity. In this case, optical alignment is only achieved for a short period, depending on the number of facets and the rotation speed of the mirror assembly. More modern approaches include:

- An active modulator, which controls the transparency in response to an external electrical signal; in this case, the device is usually used to block one of the mirrors during the high-loss phase. Acousto-optical devices form a diffraction grating in a suitable crystal to deflect the beam (in a similar fashion to the mechanical device); application of an electrical signal controls the form of the diffraction grating, and hence, the propagation direction. An

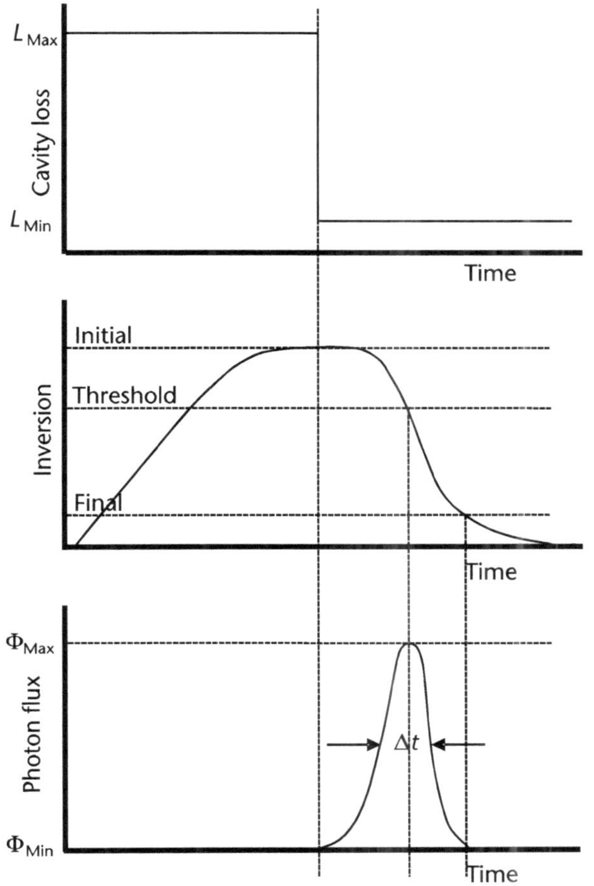

Figure 2.13 Evolution of Q-switched pulse.

electro-optical Q-switch uses the effect of an electric field on the refractive index of a nonlinear material, so that one plane of polarisation experiences a different refractive index compared with the orthogonal plane of polarisation. Consequently, the external electrical field changes the birefringence of the crystal, effectively rotating the plane of polarisation of the beam and increasing or decreasing the loss, so the basic laser cavity requires the insertion of a strongly polarising element, such as a Brewster plate.

- A passive modulator usually has a high optical loss at low incident intensities, but becomes transparent at higher intensities. They may use a photochemical element with saturable organic dyes, which bleach in the presence of a high-intensity beam and then return to an absorbing state. Another approach uses a semiconductor material that bleaches, known as a SESAM (semiconductor saturable absorber mirror) and is based on fibre Bragg-grating technology. The high-absorption state in such materials is usually a result of a single transition, which becomes saturated in the presence of an intense beam. The reaction time of such a passive device is usually of the order of a microsecond and they are reliable devices but synchronisation of pulse trains is more difficult than with active Q-switch devices. The generation of a Q-switched pulse is shown in Figure 2.14.

The great advantage of the use of the modern Q-switching approaches is that the duration of the laser pulse, or pulse length, and the pulse-repetition frequency or rate of pulse generation can be controlled very easily and provide a consistent output. The pulse duration produced by these techniques is governed by the bandwidth of the modulating device.

One of the important reasons for using a Q-switch is that the laser produces a series of characteristic pulses of photons with a very high instantaneous power, which is ideal for a range of military applications considered in Part II. Moreover, this form of output provides a route to the use of nonlinear processes, such as wavelength conversion, as discussed later.

When developing a Q-switched laser, the upper laser level must have an excited-state lifetime that is compatible with the Q-switching process. This means that this energy level must effectively be capable of energy storage, as indicated above, during the pumping process that occurs between each of the Q-switched pulses being emitted. This technique was once known as giant-pulse generation, because of

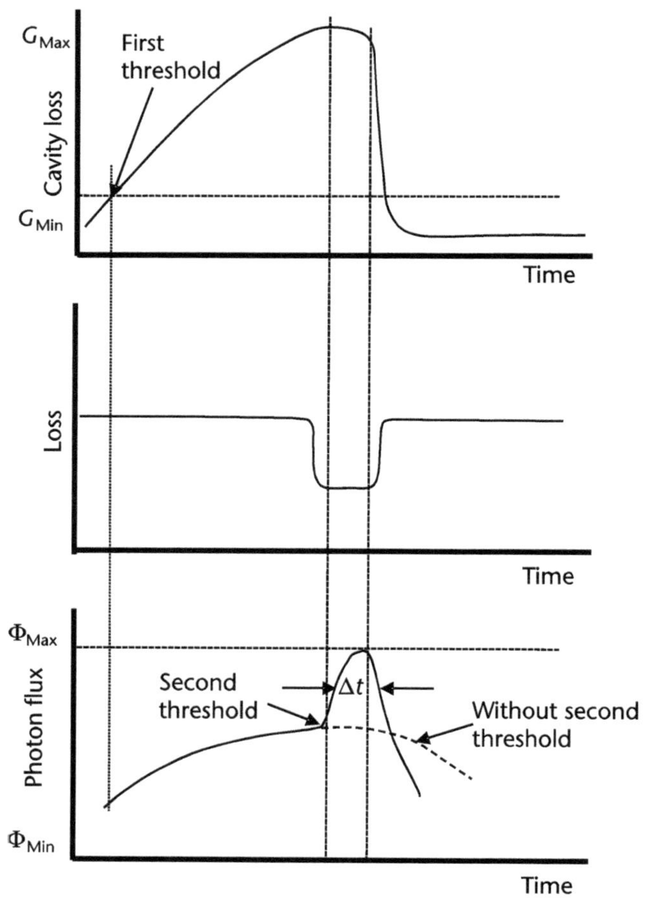

Figure 2.14 Passive Q-switch operation.

the very high instantaneous power in each pulse. The instantaneous power (P_i) is calculated from the pulse energy (E_p) and the pulse duration (τ), that is:

$$P_i = E_p / \tau \tag{2.8}$$

2.5.2 Cavity Dumping

Some laser species, such as the organic dye molecules used in liquid lasers, are not able to support an actual Q-switched mode of operation to generate short-duration pulses. This is because the upper laser levels in these molecules do not allow energy storage as their lifetime is too short. An alternative technique is used in these cases, known as cavity dumping.

The principle of cavity dumping is the exact opposite of the technique used in Q-switching, which starts with a high loss that is rapidly converted to a low-loss system; so, in a laser system using a cavity dump, a low-loss system is converted to a high-loss system, typically through the use of a controllable deflection mechanism. In this type of modulated laser system, the energy storage is in the cavity and not the gain medium. A typical device will produce a pulse-duration commensurate with the cavity round-trip time, and is governed by the length of the laser resonator.

The fundamental approach of this technique is to send the laser energy directly out of the resonant cavity without the use of a partially reflecting mirror (output coupler). A typical cavity design will use highly reflective mirrors and what amounts to a pop-up mirror to deflect the beam. Several techniques have been shown to be feasible using active modulators, described above, in this case to deflect the beam out of the resonant cavity. For example, an acousto-optical device may be used to deflect the beam on to a polarising beam splitter. A schematic of a cavity-dumping scheme is shown in Figure 2.15.

2.5.3 Mode Locking

Mode locking is another technique that may be applied to control the emission from a laser cavity where groups of photons are clumped together and aligned in phase, as they oscillate through the cavity. Each time they arrive at the output coupler, part of the light escapes as an ultrashort pulse. The bunch of photons then makes another round trip and a fraction of the flux is again emitted. The pulses are

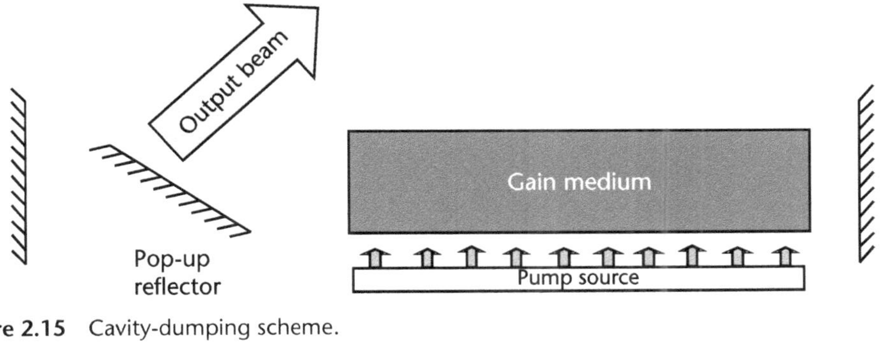

Figure 2.15 Cavity-dumping scheme.

separated by the cavity-round-trip time (2L/c, where L is the length of the resonant cavity and c is the velocity of light). Very high repetition rates are feasible, usually in the megahertz regime.

The fundamental idea of the mode-locking technique is that the modulation of the phase of the photons in a cavity, by an optical element, causes the different longitudinal modes to oscillate together in phase. A critical requirement is that the modulation frequency has to correspond with the cavity round-trip time, and requires a laser design that oscillates in many longitudinal modes; consequently, the cavity length is usually long, and these devices are prone to become misaligned during use. The phenomenon of mode-locking occurs when the modulation frequency is equal to the cavity round-trip time and the laser emission consists of a superposition of closely and uniformly spaced laser wavelengths, which have zero phase difference at the laser aperture.

The pulse length of the emitted beam is inversely proportional to the laser's oscillating bandwidth, hence, the need for a source with broad gain bandwidth such as a dye laser, but mode-locking of other lasers is now common place. This technique generates pulses in the picosecond regime and typically with a repetition rate measured in megahertz. The minimum pulse duration is set by the Fourier transform of the spectral width of the emission.

2.5.4 Gain Switching

This form of modulation occurs when the gain in the lasing medium is rapidly increased and is usually achieved through pulse pumping of the gain medium, which may cause inhomogeneities in the gain medium. These inhomogeneities may result in a reduction in the frequency stability and the coherence length of the emission, which may be important for some sensing applications. The pulse duration generated by this technique is of the order of the relaxation of the excited stated of the lasing species.

2.5.5 Other Techniques

External modulation techniques may be used to impose a modulated waveform on a CW beam. Normally, this technique is used when it is not practical to include optical elements in the laser cavity, so extracavity techniques are used, such as:

- *Mechanical modulators:* examples include a rotating chopper with transparent and opaque segments, or an oscillating shutter;
- *Spatial-light modulators:* devices that vary their transmission according to an external electrical current;
- *Electro-optical devices:* Kerr cell or Pockels' cell, which uses a current to change the polarisation of a crystal (aligns the polarisation vector to that of a polarising filter).

Q-switching may be used in conjunction with cavity dumping methods. In this case, the gain medium is initially excited to a very high level of population inversion whilst in its high-loss mode. The cavity may then be switched into a low-loss condition, which results in a rapid increase in the circulating optical power through stimulated emission. At a point where the intracavity flux is close to its peak the cavity flux may be dumped resulting in a higher peak power than occurs through use of either technique individually. Again, the pulse duration is determined by the physical length of the resonator.

2.6 Other Laser Components

A laser can be considered to be a relatively simple optical source with a complex quantum-mechanical process to generate the beam with the special properties. A vast number of different components have been used in a laser device in order to achieve or exploit a particular effect. Some of the more common specialist components have already been described above and other fundamental components of most laser devices are considered next.

2.6.1 Mirrors

Mirrors are required to form the resonant cavity of the device, with the gain medium contained between them; their influence on the laser-emission characteristics were discussed in more detail in Section 2.4. Generally, one of the mirrors (the output coupler) acts as the interface for the free-space propagation, but, in a cavity dumping system both mirrors are fully reflective, and consequently, a separate technique is used to form the output beam, which was discussed further earlier.

Bragg gratings are used in fibre lasers to form the resonant cavity. Modern techniques using very short laser pulses (ultrafast laser pulses) enable the Bragg gratings to be written directly into the fibre, thus avoiding free-space optical components or the need for splicing components on to the active region. This latter technique enables robust monolithic structures to be created.

2.6.2 Coatings

Optical coatings can have a major impact on the performance of a laser device. Most mirrors have sophisticated optical coatings to achieve the desired optical performance, usually in terms of reflection and transmission. Other components in the resonant cavity, such as a solid-state gain medium, will have antireflection coatings to control stray reflections, such as Fresnel reflections caused by the difference in refractive indices between media. Optical coatings are also used for range of other purposes, such as:

- Spectral filters to select specific wavelengths for transmission or reflection;
- Beam-splitting to provide more than one beam from a single input beam;
- Neutral density attenuation to reduce transmission over the entire operating range;
- Polarisation control to transmit one plane of polarisation selectively compared with the orthogonal plane of polarisation;

- Protective surfaces to provide some protection of soft optical components from the local environment.

The challenge of designing appropriate optical coating to ensure that a laser achieves its optimum optical performance throughout the operational environment is a serious challenge that should not be underestimated. Many laser-based systems have experienced serious reliability, and therefore, availability issues, because of failure of optical coating technology when being operated in the field.

2.6.3 Windows

A window may be defined simply as a piece of flat, optically transparent material, with the primary function of separating one environment from another. Window materials have an important function: to seal the laser device from the environment, this function is particularly important for high-power systems and is technically very challenging.

One of the most important characteristics of a laser window is to allow the beam to emerge from the resonant cavity and have no effect on the characteristics of the laser light beam passing through it. Some of the major challenges in the design of an optical window are to:

- Minimise optical absorption;
- Minimise Fresnel reflections;
- Avoid deviation of the beam;
- Avoid changing the spatial characteristics of the beam;
- Eliminate any outgassing effects that would contaminate the laser cavity.

There are many well-characterised materials available for optical windows that may be used for low-power applications, such as sapphire and silicon. However, there are significant challenges to find suitable windows for high-power applications; a strong candidate is spinel.

2.7 Direct and Indirect Wavelength Generation

In broad terms, the wavelength emitted by a laser is given by two precisely defined energy levels in the laser gain medium. In most cases, this is a very simplistic model, as the actual transition may involve simultaneous changes in a number of quantum states which may lead to:

- An emission on a series of closely spaced transitions giving a number of wavelengths over the gain bandwidth of the gain medium;
- A continuum of transitions over which many wavelengths may be generated;
- An emission on a family of transitions involving the sharing of some energy levels, such as occurs when a single excited state can decay to several lower levels, or vice versa, during the pumping to create the population inversion.

There are a number of line-broadening mechanisms that lead to the extended shape of the gain curve for a laser-gain medium, which determines the range of wavelengths that may be emitted; examples of these energy-transfer mechanisms include:

- Collision broadening;
- Pressure broadening;
- Doppler broadening;
- Atomic collisions;
- Lattice vibrations.

It is the width of this gain curve that determines the tuneability of a device. Some gas lasers have a gain curve with an appreciable width, sometimes of a micrometre or more, so the emission wavelength may be tuned across most, if not all, of this bandwidth. Broad gain curves are less common in solid-state lasers, owing to the applicability of fewer of the broadening mechanisms above. Special optical elements, such as a Fabry-Perot etalon, may be used to select a narrow wavelength or tune the emission.

The process of photon generation described earlier, with a given wavelength, may be termed direct generation. If there are no constraints on the use of different gain media and the laser system allows the use of multiple laser devices for any application, then direct-generation techniques may provide an adequate solution to many requirements. However, as noted earlier, for many military applications there is usually a requirement to use solid-state technology, which has an impact on the range of wavelengths that may be generated by a direct transition process, owing to the transparency of the host.

The wavelength emitted from a transition is determined by the difference in energy between the upper and lower energy levels involved in the stimulated emission process. If the energy difference (ΔE) is expressed in electron volts, then the wavelength (in nanometres) of the emission is given by:

$$\lambda = 1{,}239/\Delta E \tag{2.9}$$

This equation is useful for determining the prospects for direct generation of laser light with a given wavelength from various types of transitions. This equation indicates that an energy difference of about 0.3 eV is required to allow an emission wavelength of around 4 μm in the mid-wave infrared. This scale of energy difference occurs between vibrational energy levels in gas molecules but is not consistent with electronic transitions, which are typically about a factor of four greater; electronic transitions are consistent with solid-state lasers emitting in the near infrared waveband around 1 μm.

Figure 2.16 shows some of the wavelength achieved by direct generation from the visible to the far infrared. In this figure, arrows are used to indicate monochromatic transitions (i.e., those normally associated with laser action between well-defined energy levels and boxes indicate a band of wavelengths that occur with some of the gas lasers, such as deuterium fluoride and various compositions of semiconductor lasers).

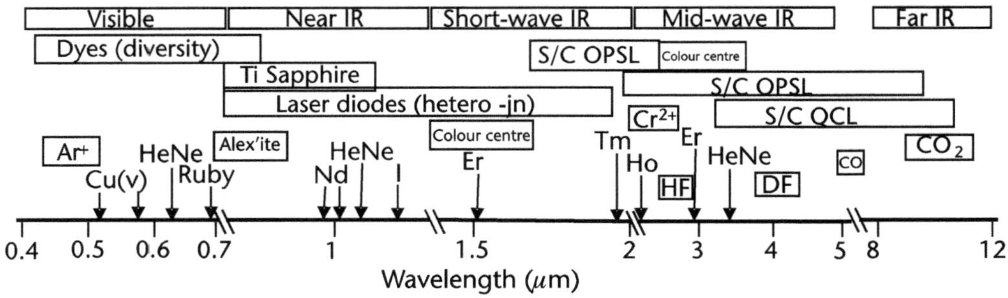

Figure 2.16 Range of direct emission wavelengths.

Consequently, a different approach is required to generate photons of the desired wavelength(s) in an inaccessible waveband from the conventional solid-state laser. Other reasons for using alternative approaches, often called indirect generation, include the need for a tuneable emission or multiple wavelengths from a single source. For many military applications, it is not practical, nor operationally feasible, to use multiple lasers in a laser-based system for a single application.

One of the important considerations concerning indirect generation is the transparency range of the nonlinear crystal. Figure 2.17 shows the transparency range of some of the common nonlinear crystals used in optical parametric oscillators or for harmonic generation. Some crystals have a very broad transparency range ascribed to them, but only part of this range may be useful for high-power generation.

Figure 2.18 shows the transmission range of three common nonlinear materials used in optical parametric oscillators, including gallium arsenide, which in its optical patterned form, is becoming a very effective crystal. Lithium niobate may be used in bulk or periodically poled forms; currently this crystal is usually used in the periodic form [periodically poled lithium niobate (PPLN)].

The route to indirect generation is by the use of nonlinear optical processes, often called $\chi^{(2)}$ processes, owing to the use of the electric susceptibility (χ) of the electric field. There are many techniques that may be used to change the input wavelength to some other defined value(s); in general, the modified emission bands are governed by the transmission band of the nonlinear crystals used to create the new wavelength(s). Some of the more common $\chi^{(2)}$ processes are:

- Optical parametric generation including optical parametric oscillators and amplifiers;
- Harmonic generation, such as frequency doubling;
- Sum and difference frequency mixing.

Other techniques that are used include:

- Raman scattering: a $\chi(3)$ processes;
- Super continuum: a chromatic dispersion process associated with self-phase modulation.

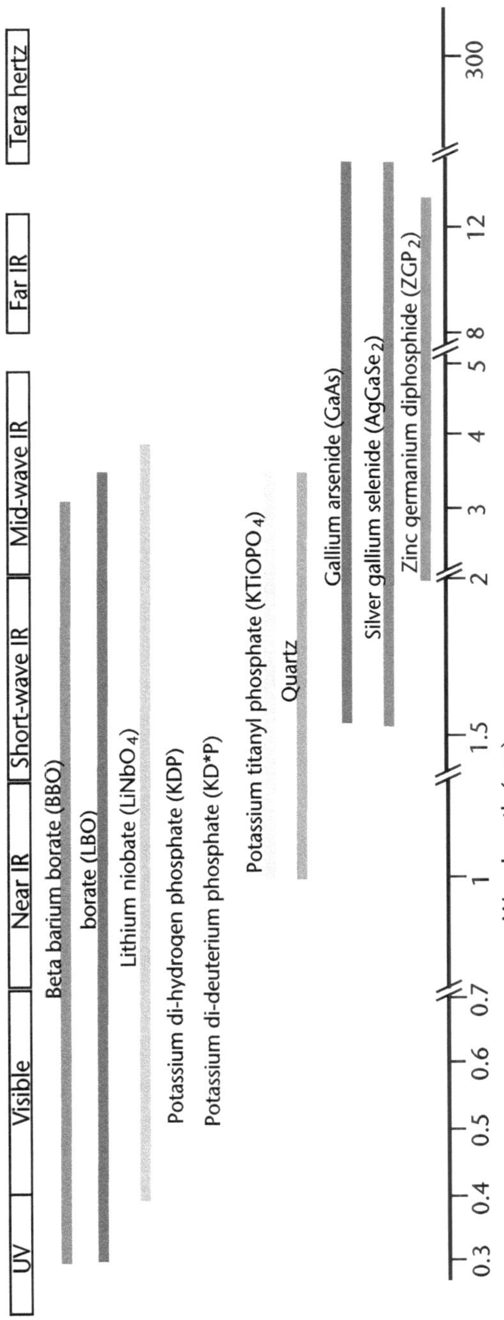

Figure 2.17 Useful transparency range of some common nonlinear materials.

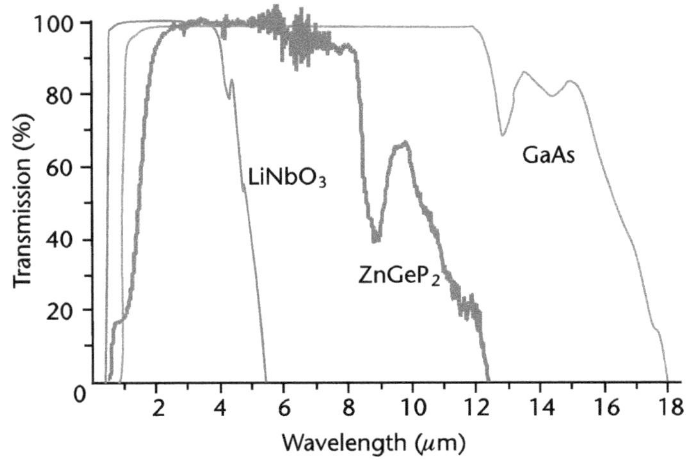

Figure 2.18 Transmission characteristics of three common nonlinear materials. (Courtesy of Dr. P. G. Schunemann, BAE Systems, USA.)

These nonlinear crystals are often grown through a relatively complex crystal-growth process, which results in a single-crystal boule being produced, from which the crystal is cut. Care is required in the cutting and polishing process to avoid laser-induced damage occurring close to the surfaces of the processed crystal, owing to the high instantaneous power being used with short-pulse pump lasers being brought to a focus within the crystal. Typically, lattice defect-induced damage may occur during the cutting process, unless a suitable technique is used. The probability of any damage occurring may be mitigated through polishing the surfaces to a suitable depth beyond the cut, so that these defects are polished out.

2.7.1 Optical Parametric Processes

The optical parametric process is relatively elementary, as it is a three-photon interaction, where a short-wavelength (high frequency) photon is annihilated creating two longer-wavelength (lower frequency) photons. The origins of this process are in the nonlinear polarisation of the interaction medium used to create the interaction, often referred to as a nonlinear element or crystal.

There are various forms of the parametric process; the most common form is the optical parametric oscillator and is usually associated with an intense pump beam as it passes through a suitable medium, such as a birefringent crystal in a resonant cavity. In this case, the spontaneous parametric emission, where two lower-frequency photons are created, initiates the process and these photons resonate within a cavity; hence, the term optical parametric oscillator. However, if photons exist in the cavity with appropriate frequencies, then stimulated emission takes place and this is through the parametric amplification process.

2.7.2 Optical Parametric Oscillators

The use of the optical parametric technique equates to spitting of the energy of the incident photon (λp) into two other photons with lower energy, and hence, longer wavelengths, so that energy is conserved. Complex interactions within the nonlinear crystal provide the mechanism for the transfer of energy from the pump to the signal beam. Simultaneously, a third beam, with a frequency equal to the difference between the pump and signal frequencies is generated; this is known as the idler beam and has a wavelength (λi) longer than the signal (λs) beam. A schematic of the typical optical arrangement of an extra-cavity optical parametric oscillator and a pump laser is shown in Figure 2.19.

In practice, it is only necessary to apply a pump beam to the nonlinear crystal, as parametric noise within this crystal provides a weak signal beam, as indicated above for the optical parametric oscillator, which increases in power. The crystal may be in the laser resonator if it is an intracavity device. It is more common to have the nonlinear crystal in a separate cavity, although this is not a separate laser and not a wavelength conversion using laser action, so no stimulated emission occurs. The input mirror has to be reflective at the signal wavelength (and possibly the idler wavelength as well) but transparent at the pump wavelength. As a minimum, the mirrors of this resonator need to allow a single resonance of the signal or idler wavelengths, but the cavity may also be doubly resonant, so that both new wavelengths are resonant.

Short-duration pulses, from a Q-switched laser, are used to provide the incident light for this process to exploit the high instantaneous power of such pulses, which generate the extremely high electric fields in the nonlinear crystal that are required to initiate the nonlinear conversion. This is because parametric gain, used in this technique, is an intensity-dependent mechanism. Solid-state devices often use techniques based on parametric gain to generate infrared wavelengths. One of the attractions of this nonlinear method is that the generated wavelengths are tuneable. Devices have been demonstrated that will emit over a broad range of wavelengths, in some of the more extreme examples from the visible to the far infrared. A more typical tuning curve is shown in Figure 2.20.

There are basically two ways of operating an optical parametric oscillator either by using a bulk crystal such as potassium titanyl phosphate mentioned above, or by using a periodically poled crystal, such as periodically poled lithium niobate. In both cases, the nonlinear crystal may be positioned inside the resonant cavity (intracavity) or outside the cavity beyond the output coupler (extracavity), as indicated above.

The tuneability of the optical parametric oscillator results from the conservation of energy during the conversion process, but the conversion from a high-frequency pump photon to the two lower-frequency photons is not unique. The

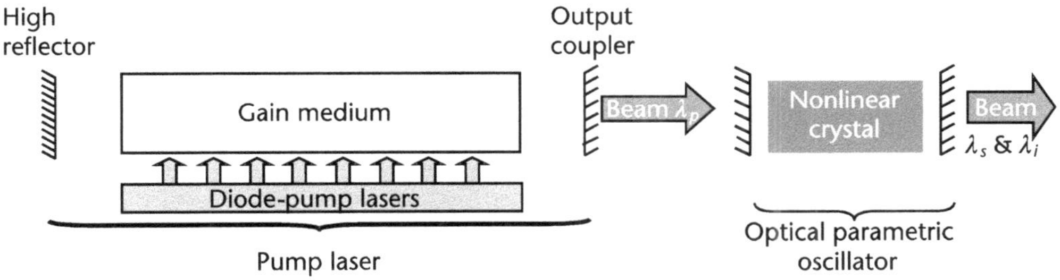

Figure 2.19 Optical arrangement of a pump laser and optical parametric oscillator.

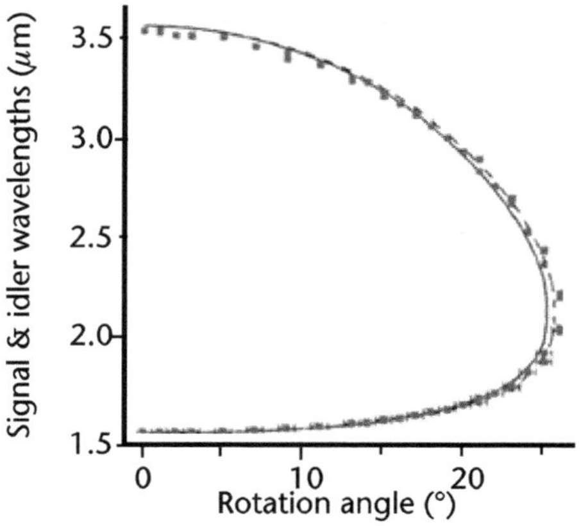

Figure 2.20 A tuning curve for a potassium titanyl phosphate nonlinear crystal.

frequencies of the signal and idler may vary in a linked fashion provided that the sum of their respective energies is equal to the incident photon, that is,

$$E_p = E_s + E_i \tag{2.10}$$

where E refers to the energy of the photons defined by their respective subscripts for the pump (p), signal (s) and idler (i) components, in terms of angular frequency is:

$$\omega_p = \omega_s + \omega_i \tag{2.11}$$

or

$$1/\lambda_p = 1/\lambda_s + 1/\lambda_i \tag{2.12}$$

2.7.3 Phase Matching

The need for phase matching arises owing to the dispersion in the refractive index of nonlinear crystals. Phase matching is one of the most restrictive requirements placed on a crystal and reduces the number of potential crystals available for nonlinear conversion to a few hundred out of the many thousands of known crystals. This number of crystals is reduced further to a handful by the need for other requirements to be satisfied simultaneously. These restrictions led to the development of the quasi-phase-matching technique, which is discussed further later.

Not only is there a need for energy to be conserved in the parametric interaction, but the momentum of the interacting photons must be conserved, which may be considered in terms of the wave vectors of the three interacting photons, that is:

$$k_p = k_s + k_i \tag{2.13}$$

where $k = n_\lambda\, \omega/c$ or $k = 2\pi\, n_\lambda/\lambda$ and, n_λ is the wavelength-dependent refractive index of the material, ω is the angular frequency and λ is the wavelength. Again the subscripts refer to the components of the parametric event.

The three refractive indices associated with the pump, signal and idler beams can be changed by:

- Angle tuning, known as critical phase matching, so waves follow different paths though the crystal with respect to the optic axis of the crystal.
- Temperature tuning, known as noncritical phase matching, where a change in temperature alters the indices of the crystal. This technique offers the advantage of relative simplicity, as realignment of the cavity is not necessary during tuning, but the tuning range is generally relatively narrow compared with the angular-rotation method. This is the only practical approach to tuning with noncritically phase matched devices, considered next.

The conservation of momentum condition defined in (2.13) is satisfied when the three waves travel along paths within the crystal that present the appropriate refractive index to each wave, and thus theoretically no momentum is lost (often expressed as $\Delta k = 0$, that is, $k_p - k_s - k_i = 0$). In this case, there is cumulative growth of the signal and idler beams at the expense of the pump beam power.

For all practical materials, the phenomenon of dispersion occurs as the beam passes through the nonlinear material; hence, the energy and momentum conservation conditions cannot be satisfied simultaneously, owing to dispersion of the light in the medium. In general, the refractive index increases with photon frequency; thus, the magnitude of the **k**-vector of the pump photon is too long to achieve the phase-matching condition. The common solution is to use the birefringence in an anisotropic crystal to compensate for the material dispersion.

When an intense beam is incident on a birefringent bulk crystal that has its faces cut at a specific angle with respect to its crystallographic axes, the crystal decomposes the incident ray into two rays (ordinary and extraordinary rays). The phase-matching condition may be achieved, for example, in a negative uniaxial crystal by making the pump beam an extraordinary wave and the signal and idler beams ordinary waves. If the birefringence is larger than the dispersion, there will be a set of specific frequencies where the phase matching condition is satisfied for a specific angle of incidence, with respect to the optical axis of the crystal.

The angle of incidence of the pump light is chosen so that these rays overlap within the crystal for a sufficient period to enable the nonlinear process to occur, despite the fact that the two rays travel at different velocities within the crystal. The signal and idler wavelengths may then be tuned by rotating the crystal relative to the direction of propagation of the pump beam defined by the appropriate $\mathbf{k}_p$ vector.

Under appropriate phase-matching conditions, there is a flow of photon energy from the pump beam into the signal and idler beams, so that the intensities of these beams increase rapidly. The trick required for successful operation of the optical parametric oscillator, is to ensure the magnitude of the pump intensity above threshold for parametric operation and interaction lengths do not permit back conversion, resulting in the return of energy to the pump beam. A factor of five above threshold for parametric conversion is a common rule-of-thumb value.

There are essentially two types of phase matching in common usage with bulk nonlinear crystals, which are dependent on the orientation of the polarisation vector of the two beams generated in the optical parametric oscillator:

- Type-I phase matching occurs when the signal and idler beams have the same polarisation, but are orthogonal to that of the pump beam.
- Type-II phase matching has the polarisation vector of the two beams orthogonal and that of the signal beam is orthogonal to the polarisation of the pump beam.

Type-III phase matching has been investigated but is not common; in this case, the two generated beams have orthogonal polarisation vectors but it is the idler that has its polarisation vector orthogonal to that of the pump beam.

An optical parametric oscillator may be designed in many different ways; a common design is to have the mirror reflectivities arranged so that one of the downconverted wavelengths is reflected. This is known as a singly-resonant oscillator and it is usually designed that the signal wavelength is resonated within the cavity; in this case, there is no feedback of the idler.

As the name implies, a doubly resonant oscillator has both of the downconverted wavelengths resonated within the Fabry-Perot cavity, so both wavelengths are reflected. This usually leads to a lower threshold for operation, but greater demands for optical alignment to achieve operational stability. The optical emission from a doubly resonant device is often characterised by spikes, usually attributed to small changed in the length of the resonant cavity, which is often associated with a clustering effect as two resonant conditions need to be satisfied simultaneously.

Owing to a condition known as walk-off, the effective parametric interaction length of nonlinear crystal may be less than the actual crystal length, with the interaction length being defined as that length where the parametric conversion process takes place. The walk-off condition arises because of the use of extraordinary and ordinary waves in the crystal to achieve the phase matching condition; in the case of the ordinary wave, the phase velocity of the wave vector is in the same direction as the Poynting vector, which is generally not the case for the extraordinary wave, except along the principal axes of the crystal. Hence, if the phase-velocities of these two waves differ, the envelopes of these two waves deviate from each other, limiting their overlap and thus limiting the conversion efficiency.

2.7.4 Quasi Phase Matching

Quasi phase matching is an alternative method of undertaking nonlinear conversion. It is often seen as a simpler way of providing the conditions for effective nonlinear conversion, through application of an alternative technique of compensating for the phase-velocity dispersion, resulting from the wavelength-dependent refractive indices, discussed above.

A periodically poled crystal material needs to be birefringent to allow nonlinear conversion to take place, but the materials do not need to be centro-symmetrical, a fundamental requirement for phase-matching techniques. The fundamental requirement for the creation of a suitable nonlinear material for quasi phase matching is the ability to form domains within the crystal, where the crystallographic axes are reversed between each adjacent domain. The scale of the domain size is compatible with the coherence length (l_c) of the crystal for the incident wavelength, to ensure that the two rays formed by the action of the birefringence remain in phase. Beyond the distance of half the coherence length ($l_c/2$), back conversion starts, and so the nonlinear process declines. However, if the domain orientation is reversed after this distance of half the coherence length ($l_c/2$), then the nonlinear conversion progresses without back conversion. As a result of this quasi-phase-matching condition, the intensities of the signal and idler beam increase, as they propagate through the reversed domains along the crystal length.

Figure 2.21 indicates the cumulative growth of the signal and idler waves in a domain-reversal crystal, which may be periodically poles or optically patterned. For qualitative comparative purposes, the plot also shows the growth in the nonlinear emission from a phase-matched crystal.

2.7.5 Harmonic Generation

Harmonic generation concerns the interaction of light with matter to produce multiple harmonics of the incident frequency in the pump beam. One of the attractions of this technique is the generation of wavelengths not readily accessible with conventional laser action, especially with solid-state lasers. A common example is the frequency doubling of the output from a neodymium laser to give green light.

The mechanism that results in the frequency doubling is a consequence of the interaction of the intense pump beam in a nonlinear crystal in a process that is the opposite of the parametric oscillator operating at degeneracy; in this case the emission is half the wavelength (i.e., all of the photons have the same frequency). Many

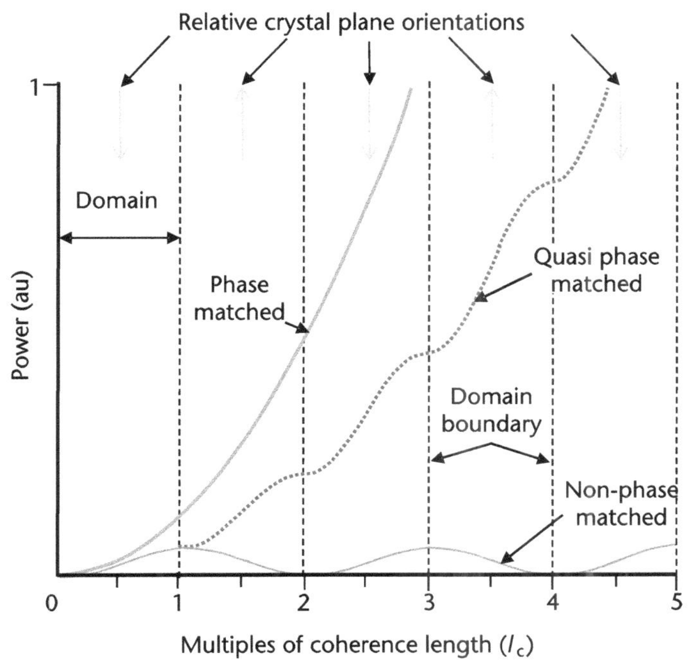

Figure 2.21 Quasi-phase matching. (Courtesy of Dr. E. Lallier, Thales, France.)

Table 2.1 Typical Nonlinear Crystal Characteristics

Material	*Transmission Range (μm)*	*Figure of Merit**	*Damage Threshold*	*Thermal Condition/ Thermo-Optic Coefficient*
PPLN	0.2–4.1	1	1	1
KTP	0.4–3.5	0.42	1.5	1.6
PPKTP	0.4–3.5	0.95	1.5	1.6
ZGP	2–11	8.8	0.5	2.3
GaSe	1–20	6.8	0.15	—
OPGaAs	0.9–18	6.2	0.5	2.3

Courtesy Dr. E. Lippert, FFI, Norway. *The figure of merit (FOM) is given by: FOM = $d_{eff}^2/(n^3 . \lambda^2)$ and d_{eff} is the nonlinear coupling coefficient, n is the refractive index of the material and λ the wavelength. Moreover, when the pump intensity is I_p, the gain (G) is given by: $G \alpha \sqrt{(I_p . \text{FOM})}$.

nonlinear crystals are available for this technique. The magnitude of the conversion effect is related to the incident pump-power density in the nonlinear crystal, which may be in the laser's resonant cavity or external in a separate cavity, although a single-pass through the nonlinear crystal is sometimes used (i.e., without any oscillation). It is usual to focus the incident beam to a small spot within the nonlinear crystal when using Q-switched pulses in an extracavity configuration, although efficient CW frequency-doubled beams are now quite common, using intracavity techniques.

Higher-order harmonic generation is normally a multistep process. For example, the use of third-harmonic generation used to create ultraviolet light from a neodymium laser involves frequency doubling of the fundamental and mixing both of these photon beams (both frequencies) in another nonlinear crystal to produce the third harmonic. In the case of fourth harmonic generation, the frequency-doubled beam is passed through a second frequency-doubling crystal. These higher-order harmonic generation mechanisms are dependent on very high power beams incident on the nonlinear crystal and tend to be very inefficient and, hence, rarely even contemplated for military applications.

2.7.6 Raman Conversion

Stimulated Raman scattering is a $\chi(3)$ nonlinear technique and is a consequence of wave mixing occurring between the incident electromagnetic wave and a mechanical vibration (phonon) in a suitable medium, which may be a solid, liquid or gas. This is an inelastic scattering process that changes the energy of the photons and this technique may be used to generate a number of wavelengths around the incident wavelength, some of which will be at a shorter wavelength, whilst other lines will have a longer value. In fact, the longer wavelength process is generally more efficient, owing to the higher probability of this type of transition, occurring, and hence, tend to be more intense; however, it should be noted that Raman conversion is a relatively weak process, requiring an intense beam to invoke the nonlinear process.

These two types of emission lines are known as anti-Stokes (the phonon energy is added to the incident beam) and Stokes (energy subtracted from the incident pump) lines or wavelengths, respectively; the actual wavelength shift is determined by the conversion medium and is usually quoted in wave numbers (cm^{-1}). The Stokes lines tend to be more intense, as noted earlier.

The actual value of the frequency shift is dependent on the host material (and is constant for a given material) but varies between materials. The wavelength shifts are often quite modest and so are not used to access wavebands significantly different from the incident light.

One of the attractions of this technique is that the incident light can generate a number of laser lines simultaneously once the threshold condition for the elastic-scattering process has been achieved. This is a consequence of the various rotational and vibrational states of the nonlinear material involved in the conversion process.

In the mid-wave infrared this technique is not very attractive, as the Raman coefficient varies inversely with the wavelength (λ), so at 4 μm it is about 1/8 of the value for visible light; hence, it is more sensitive to losses and thereby needs greater optical intensities, which can lead to air breakdown and plasma formation.

2.7.7 Sum and Difference Frequency Mixing

The sum and difference frequency mixing technique has many similarities with harmonic generation described above; in fact, harmonic generation is a specific case of this nonlinear technique. As the name implies, when two or more photon beams with different frequencies (say ω_1 and ω_2) are passed through a nonlinear crystal, additional beam frequencies are generated with a higher frequency ($\omega_1 + \omega_2$) and a lower value ($\omega_1 - \omega_2$). Solid-state and gaseous media have been used to achieve sum and difference generation.

2.7.8 Super-Continuum Generation

Super-continuum generation exploits the nonlinear component of the refractive index of a propagating medium and results in self-phase modulation. The requirement is for very short pulses, typically in the femtosecond regime. A schematic of a super-continuum spectrum is shown in Figure 2.22, which shows the emission from the titanium-sapphire laser close to 800 nm, along with the super continuum extending from the ultraviolet to the near infrared.

Another super-continuum spectrum is shown in Figure 2.23; in this case, the super-continuum is generated in various optical fibres that are transparent in the mid-wave infrared. These lasers were pumped with short-pulses in the mid-wave infrared close to 2.5 μm.

2.8 Performance Characteristic Terms and Parameters

There is a vast array of terms used to describe the operation characteristics of a laser beam. Some of the more important ones are defined here and considered specifically in terms of relevance to some military applications.

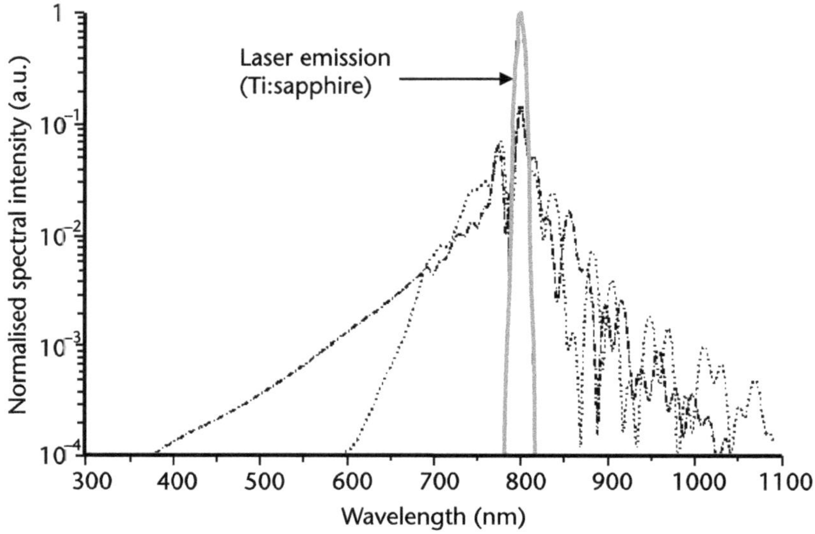

Figure 2.22 Super-continuum spectrum. (Courtesy of Professor S. L. Chin, Laval University, Canada.)

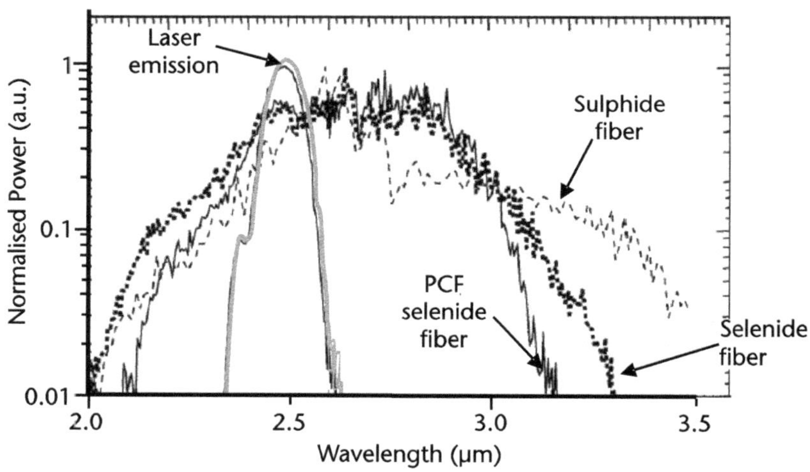

Figure 2.23 Super-continuum spectrum in the mid-wave infrared. (Courtesy of Dr. L. B. Shaw, NRL, USA.)

2.8.1 Beam Quality and Divergence

The size and length of the gain medium broadly determines the modal quality of the beam, because a resonator can sustain transverse and longitudinal oscillations. The Fresnel number (F_N) provides an indication of the potential or otherwise for good beam quality from a resonator. For example, long thin resonators, such as fibre lasers, are likely to generate an output beam with low divergence and, hence, have potential for high radiant intensity. The Fresnel number is defined as

$$F_N = a^2/\lambda L \qquad (2.14)$$

where a is the radius of the gain medium, λ is the wavelength of operation and L is the cavity length.

The existence of multiple transverse modes in the resonant cavity has a dramatic impact on the appearance, and particularly the intensity characteristics of a laser beam, which is commonly known as the beam quality. There are a number of ways of expressing the beam quality of an emission; a common parameter is known as M^2 (M squared) propagation ratio, others relate to the diameter and divergence product of the beam or Etendue.

$$\begin{aligned}&\text{Brightness} = \text{Power}/(\text{Area} \times \text{Solid Divergence Angle}) \text{ or}\\ &B = P/\left(\lambda^2 M_x^2 M_y^2\right)\end{aligned} \tag{2.15}$$

where M_x^2 is propagation factor in the x direction and M_y^2 is propagation factor in the orthogonal direction.

The M^2 parameter expresses how much the far-field beam profile spreads with respect to the ideal or diffraction-limited beam. The propagation ratio (M^2) is defined as:

$$M^2 \approx z\lambda/\pi\left(\omega_o \cdot \omega_z\right) \tag{2.16}$$

where z is the propagation distance, ω_o is the spot size (radius) at the beam waist and ω_z is the spot size (radius) at distance z.

There is a rule of thumb that the M^2 approximately equates to the square root of the number of transverse modes for a small number of higher-order modes.

The importance of the beam quality of a laser emission is often a much overlooked parameter in many laser-based applications; however, it can have a critical impact on the performance of many laser-based systems. The beam quality has a very strong influence on the radiant intensity of the beam, critical for many high-power directed-energy and countermeasure applications. It is the matching of the beam divergence to the pointing stability of the beam director, which determines the probability that a laser beam is incident on its designated target. For many military applications, it is this requirement of the energy being on its target along with the requisite radiant intensity that determines the effectiveness of a laser-based system. Consequently, if the energy is not on target then the laser-based system will be ineffective.

There are a number of ways of expressing the divergence of a laser beam; the most common ones are the angular extent of the beam when its intensity has decreased to either $1/e$ of the peak intensity, or to the $1/e^2$ value of the peak value. Care needs to be exercised as to whether the quoted divergence of the beam is the half or full-angle value; in the case of laser-spot sizes, the usual convention is to use radii. It is recommended to choose a convention, such as half-angle to the $1/e^2$ points and base all laser-based system requirements and definitions on such a convention, to avoid confusion and the inevitable failure to meet a specification.

2.8.2 Power

The output power of a laser is often the performance parameter that many people take as the most important indication of performance of a laser-based system; however, for many military applications this performance parameter may well be a secondary consideration, as the fundamental requirement may really be for a very high radiant intensity, or achieving a very short, but energetic, pulse with a precise timing.

The output power, expressed in watts, of a pulsed laser device may be expressed in many forms:

- Average power is a product of the energy per pulse and the pulse-repetition frequency; this value has to be factored by the duty cycle of the laser if it is being operated in an intermittent mode, for example, when used as a source for a laser-based jammer.
- Peak power is a similar calculation to that used to express the average power, but ignores the period where the device is quiescent (i.e., ignores the duty cycle).
- Instantaneous power, defined above, indicates the equivalent power in a single pulse.

The power of a CW beam is far simpler and expresses the number of joules (or fractions thereof) per second in the beam (i.e., watts).

2.8.3 Radiant Intensity

This is an important measure of potential performance for many countermeasure applications and is defined by the quotient of the power and the solid angle of the beam divergence, giving units of watts per steradians (W/sr). In the case of a pulsed laser, it is normal to use the average power in this calculation. This term may also be used to express the brightness of a beam.

2.8.4 Efficiency

There are many ways of expressing the efficiency of the operational performance of a laser. The wall-plug efficiency expresses the fraction of the total input power to the entire laser system that is actually converted into the optical emission. This expression of efficiency has a very significant impact on the installed-system requirements, such as the prime power needed to operate the laser and the management of the waste heat. The wall-plug efficiency of many modern military systems is often of the order of 10%. Clearly, inefficient lasers are to be avoided and not used for high-power generation.

Another common expression of the efficiency of the laser device is the slope efficiency, which is the gradient of the relationship between the input power (or energy), used to create the population inversion, to the power (or energy) in the output beam, as shown in Figure 2.24. This measure of efficiency is often used to optimise the operational performance of a laser, but ignores the impact of other subsystems, such as cooling. It also ignores the fact that a laser is a nonlinear device

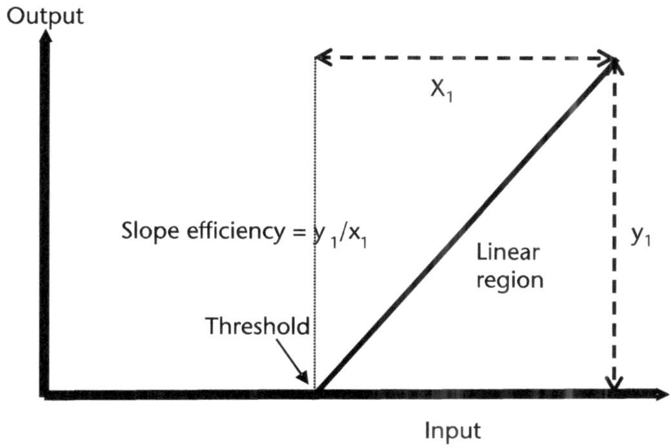

Figure 2.24 Slope efficiency.

and it takes a specific amount of input power to reach a threshold for laser action, usually called the lasing threshold. Another expression of slope efficiency is optical-to-optical conversion efficiency for an optically pumped device.

2.8.5 Time to Full Brightness

Many solid-state lasers use a gain medium with thermally induced optical distortion characteristics that results in a distortion of the optical path through the resonator, which may be crucial to the successful operation of a stable resonant cavity. This phenomenon results in the formation of a thermal lens within the gain medium, which causes a fluctuation in the brightness of the beam, owing to changes in the power and fluctuations in the divergence of the beam. A resonant cavity is often optimised to give its peak output after a short duration, sometimes measured in tens of milliseconds. This delay in achieving full radiant intensity from the laser aperture is called the time to full brightness.

Any thermal lens formed in a gain medium, by definition, is dependent on the temperature profile within the device so the focal length changes with temperature, resulting in a change in the optical alignment of the beam within the cavity. The mechanisms responsible for the creation of a thermal lens are:

- The change in refractive index with temperature (dn/dt);
- Mechanical stresses induced by the thermal profile;
- Thermal expansion leading to deformation of the gain medium, for example, leading to bulging of the end faces of a slab or rod resulting from the greatest intensity being along the central axis of the gain medium with a beam with a Gaussian profile.

These effects are influenced by the operational duty cycle of the device, because of the varying thermal load on the laser cavity, but the effects can be mitigated by the use of a zig-zag slab configuration, discussed in Chapter 3. The impact on potential performance in the delay in achieving full brightness of a laser-based system can be catastrophic, as some military systems are reliant on bright beams being available almost instantaneously to be directed on to a target and remaining there for the duration of the intended interaction.

This performance parameter does not apply to gas laser technology. In general, the full-power emission from a gas laser is usually virtually instantaneous, similarly with liquid lasers and many semiconductor devices.

2.8.6 Pulse Length

Pulse length, pulse width or pulse duration is one of the fundamental characteristics of a pulsed laser and describes the time elapsed from the onset of the emission to its cessation in any pulse. Different applications may demand different pulse shapes, such as a particular shape or very steep leading edge. Care is required in the choice of pulse length as this can result in the formation of a plasma if the instantaneous power exceeds the breakdown potential of the atmosphere, resulting in the designated target being shielded from the laser. Moreover, evaluation of the potential for laser-induced damage of optical components, especially optical coatings and transparent elements, in a laser-based system by the very short pulses with high instantaneous power is an important design consideration to avoid self-destruction of the laser system.

A new phenomenon of ultrashort pulses generation is currently of interest for some military applications, which leads to self-phase-modulation and results in white-light filament formation.

2.8.7 Pulse-Repetition Frequency

Pulse-repetition frequency (PRF) of a laser describes the number of complete cycles of the modulations in the output beam that occur in one second. Some applications require a small number of pulses per second with a very precise pulse-repetition interval (i.e., a consistent period between pulses). Other applications demand a very high pulse-repetition frequency, so that the beam appears to be a virtually CW emission. A related term is the pulse-repetition interval, which is the period between pulses and is important for determining the unambiguous ranges of time-of-flight systems.

2.8.8 Duty Cycle and Duty Factor

This performance characteristic reflects how often the laser is emitting photons in a given period. For example, a device may be operated on a 25% duty cycle so that it is only emitting for a total of 250 ms in any 1-second period.

The duty factor expresses how often this emission may be required in any given period such as in any 1-hour period. The duty cycle and, to a slightly lesser extent, the duty factor have an important impact on the cooling requirements for a laser system, as it determines the energy balance within the laser system, and hence, its overall performance, as well as it size and weight.

2.8.9 Quantum Defect

This term relates to the difference in energy between the excitation process and the emission photons in an optical pumping scheme, primarily related to the excitation of solid-state lasers; it is sometimes called the heating ratio. The quantum defect indicates the energy that is lost during the excitation process and thus leads to reduced performance, resulting from internal heating of the gain medium. The temperature rise (ΔT) may be estimated from:

$$\Delta T \; \alpha \left(P_{heat} \cdot t \right) / K_c \cdot A \tag{2.17}$$

where P_{heat} is the heat loading from the quantum defect, t is time, K_c is the thermal conductivity of the gain medium and A is its cross-sectional area.

To achieve high performance, as well as high efficiency, from a laser it is usual to attempt to minimise the quantum defect. There are several approaches that may be exploited such as:

- Choice of a set of energy levels in the laser specie that allow the use of a pump wavelength close to the emission wavelength;
- Use of a number of energy levels within the laser specie that effectively generate two pump photons for each excitation photon, in the two-for-one pumping scheme;
- Cryogenic cooling of the gain medium to exploit energy levels close to the pump wavelength, through control of the population of the lower energy levels in the gain medium, where the normal thermal induced population is reduced through a reduction in kT, where k is Boltzmann's constant.

Thermal management is a critical consideration for power scaling of stimulated output for any type or class of laser. Figure 2.25 indicates how the heat generated from the quantum defect and other mechanisms originates in the gain medium; other mechanisms include excited-state absorption, cross relaxation and impurities in the gain medium.

It is normal to remove the heat from the gain medium; in fact, some laser designs, such as the thin-disk and fibre devices, offer a relatively easy approach to heat removal. It is often necessary to undertake positive action to remove the heat resulting from laser action to avoid damage to the gain medium. A consequence of the removal of heat is the creation of thermal gradients with the gain medium, which can have deleterious effects on the beam quality of the emission. A common effect is thermal lensing within the gain medium, discussed earlier, but another effect is thermally induced stress, particularly if the gain medium is bonded from smaller components.

It is important to minimise all mechanisms that will generate heat in the gain medium, not only because of the potential impact on the gain medium and subsequently on the emission characteristics, but on the efficiency of the laser system. Inefficient lasers tend to be large lasers.

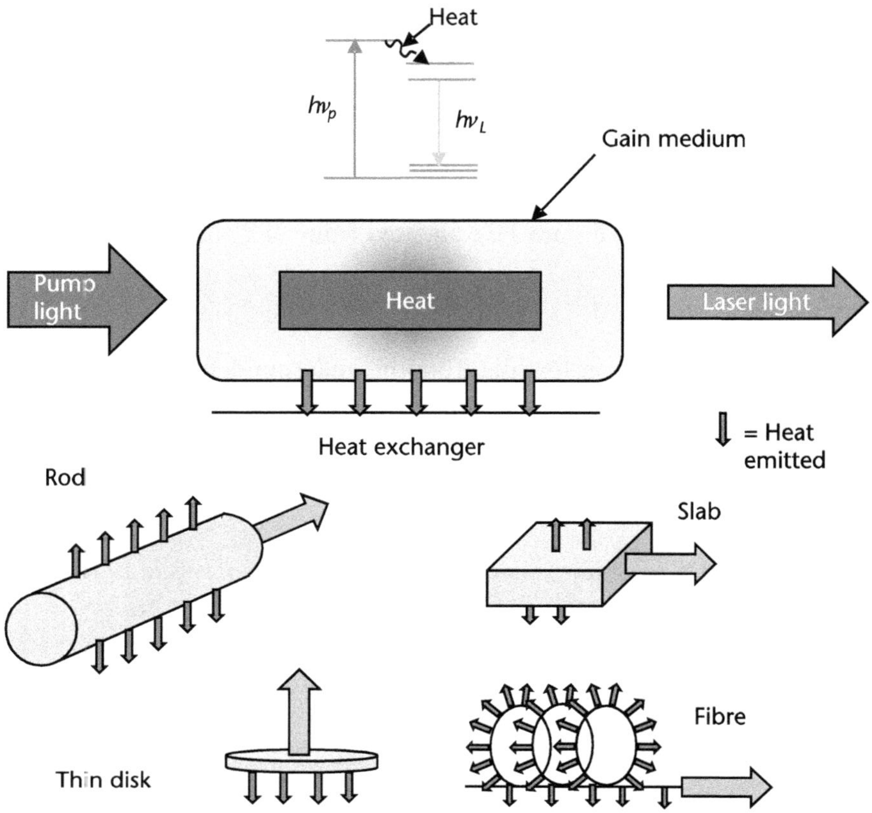

Figure 2.25 A schematic of some of the thermal-management issues in laser devices.

2.8.10 Coherency and Speckle

Some applications, such as laser-radar systems, require very high coherency to achieve the desired detection performance, so careful design with additional components are required to achieve the desired spatial and temporal coherence. Similarly, laser-based systems requiring very narrow spectral emission tend to have increased complexity, particularly in the case of solid-state devices, owing to the need to control the emission transition, as described earlier. Increased complexity generally leads to reduced reliability and increased cost.

A rough indication of the coherence length (l_c) of a device is given by:

$$l_c = c/\Delta\nu \tag{2.18}$$

where c is the velocity of light and $\Delta\nu$ is the frequency bandwidth of the laser source.

It is the coherency of a laser that gives rise to speckle, which occurs as a result of interference when laser light is scattered from a diffuse surface. For some sensing applications, the suppression, or mitigation of speckle effects, is important to ensure good target detection performance.

2.8.11 Beam Stability

Beam stability is the motion of the centroid of the emitted beam profile about a nominal propagation axis, which results from instabilities within the resonant cavity, such as a varying thermal lens within the gain medium. This performance characteristic is often overlooked in the design of a laser, but it can have a significant impact on the energy-on-target characteristics of any laser-based system, owing to introducing additional jitter into the pointing system and also complicating the optical alignment process.

2.8.12 Line Width

For most military applications, the natural spectral extent of an emission, laser line width, is not a major consideration. However, there are some sensing applications that require the frequency bandwidth of the emission to be measured in the kilohertz regime, to be compatible with a particular detection technique, or sensing of a particular event.

It is usual to consider the laser emission to be monochromatic, but as discussed earlier, there are a number of mechanisms that lead to either line broadening or simultaneous line emission.

In the case of a requirement for a very narrow laser emission, it is usual to include an additional optical element into the resonant cavity such as a Fabry-Perot etalon. This is a typical frequency-selective element, which will constrain the laser emission to a single longitudinal mode. Frequency selection may also be achieved using a ring-resonator configuration. Moreover, a dispersive component may be used as a tuning element by changing its orientation within the resonant cavity, as discussed next.

2.8.13 Tuneability and Diversity

Some military sensing applications require the laser emission to be tuneable to detect the presence of certain species. By varying the spectral content of the emission it is possible to tune into the absorption band of particular warfare agents, for example.

Sources such as a dye laser or a carbon dioxide laser have the physical capability of emitting over a relatively broad spectral band or over range of wavelengths. Insertion of a frequency-selective element, such as an etalon, a diffraction grating or a prism in to the cavity, may be oriented to select a particular wavelength for resonance. The tuning range may extend from a few tens of nanometres to several micrometres. An alternative approach is to use an optical parametric oscillator, which can offer up to several micrometres in a tuning range.

Wavelength diversity provides an additional dimension to the wavelengths that may be emitted from a laser cavity. For example, in the case of a dye laser, it is possible to remove one laser dye and replace it with another and thereby move the gain curve by up to several hundred nanometres. This scheme is dependent on the cavity optical elements and coating being compatible with the range of operational frequencies. In this case, the laser dye may be in a solution or in a solid-state host.

2.8.14 Polarisation

Some laser designs, such as a gas laser using a sealed cell with windows angled at Brewster's angle will generate a beam with a polarised output. Other devices may be designed to generate beams without oscillations in any specific plane (i.e., without any plane of polarisation, such as horizontal or vertical planes of polarisation). The inclusion of optical elements such as Brewster plates or polarising beam splitters may be used to create a beam with a specific polarised output. The use of a wave plate in the laser cavity may be used to modify the polarisation characteristics; for example, a half-wave plate will rotate the plane of polarisation, whereas a quarter-wave plate will change the nature of the polarisation, such as linear to circular.

A typical application is the modification of a beam's temporal characteristics through the use of a Pockels' cell as an extracavity device.

2.9 Summary

A laser exhibits many unique characteristics that result from the complex method used for the generation of the photons through stimulated emission. These unique properties may be exploited for a range of military applications, considered more fully in Part II, but including:

- Range finding;
- Target designation and marking;
- Sensing, including detection of chemical or biological agents;
- Situational awareness, such as obstacles along a flight path;
- Target identification and recognition;
- Navigation, including mapping and terrain profiling;
- Platform protection;
- Directed energy.

In principle, a laser is a relatively simple optical device that needs the application of engineering precision and careful choice of materials to realise a fully functional and effective device. Attention to detail in the design of the device is crucial to enable the full exploitation of the exploitation of the unique characteristics of laser action.

A number of sophisticated optical techniques may be applied to the device to modify the emission characteristics. Some of these optical subsystems may be integrated into the resonant cavity, whilst others have to be positioned outside of the resonant cavity where the laser action takes place; in this case, they are known as extracavity devices. These optical techniques enable the spatial and temporal characteristics to be changed along with the wavelength of the photons.

Selected Bibliography

Alfano, R. R., *The Super-Continuum Laser Source*, 2nd ed., New York: Springer, 2006.

Boudrioua, A., *Photonic Waveguides,* New York: John Wiley & Sons and ISTE, 2009.

Brignon, A., and J. P. Huignard, (eds.), *Phase Conjugate Laser Optics,* New York: Wiley Interscience, 2003.

Diels, J. C., and L. Arissian, *Lasers: The Power and Precision of Light,* New York: Wiley VCH, 2011.

Dimitriev, N. G., et al., *Handbook of Nonlinear Optical Crystals, Vol. 64*, 2nd ed., Heidelberg, Germany: Springer, 2006.

Ferreia, M. F. S., *Nonlinear Effects in Optical Fibres,* New York: John Wiley & Sons and OSA, 2011.

Finlayson, D. M., and B. D. Sinclair, (eds.), "Advances in Lasers and Applications," *Proceedings of the Fifty Second Scottish Universities Summer School in Physic,* Bristol, U.K.: Institute of Physics Publishing, 1999.

Hall, D. R., and P. E. Jackson, *The Physics and Technology of Laser of Resonators*, Bristol, U.K.: Adam Hilger, 1989.

He, G. S., and S. H. Liu, *The Physics of Nonlinear Optics,* Singapore: World Scientific Publishing, 1999.

Hecht, J., *The Laser Guide Book,* Blue Ridge Summit, PA: TAB Books (McGraw-Hill), 1992.

Hitz, C. B., et al., *Introduction to Laser Technology*, 4th ed., New York: John Wiley & Sons and IEEE, 2012.

Hooker, S., and C. E. Webb, *Laser Physics,* Oxford, U.K.: Oxford University Press, 2010.

Kailsiky, Y., *The Physics and Engineering of Solid State Lasers*, Bellingham, WA: SPIE Optical Engineering Press, 2006.

Koechner, W., *Solid-State Laser Engineering,* Berlin, Germany: Springer, 1999.

Koechner, W., *Solid-State Laser Engineering, Vol. 1,* 2nd ed., Berlin, Germany: Springer-Verlag, 1985.

Little, C. E., *Metal Vapour Lasers, Physics, Engineering and Applications,* New York: John Wiley & Sons, 1999.

Injeyam, H., and G. D. Goodno, *Laser Handbook,* New York: McGraw-Hill, 2012.

Marhic, M. E., *Fiber Optical Parametric Amplifiers, Oscillators and Related Devices,* Cambridge, U.K.: Cambridge University Press, 2007.

Meyers, R. A., *Encyclopaedia of Lasers and Optical Technology,* San Diego, CA: Academic Press, 1991.

Mills, D. L., *Nonlinear Optics*, Berlin, Germany: Springer-Verlag, 1991.

Milonni, P. W., and J. H. Eberly, *Lasers,* New York: John Wiley & Sons, 1988.

Milonni, P. W., and J. H. Eberly, *Laser Physics,* 2nd ed., New York: John Wiley & Sons, 2010.

Nambiar, K. R., *Lasers: Principles, Types and Applications,* New Delhi, India: New Age International Publishers, 2005.

Nikogosyan, D. N., *Properties of Optical and Laser Related Materials: A Handbook,* New York: John Wiley & Sons, 1997.

Powers, P. E., *Field Guide to Nonlinear Optics*, Bellingham, WA: SPIE Optical Engineering Press, 2013.

Powers, P. E., *Fundamentals of Nonlinear Optics,* Boca Raton, FL: CRC Press/Taylor & Francis Group, 2011.

Prather, D. W., et al., *Photonic Crystals: Theory and Application,* New York: John Wiley & Sons, 2009.

Scheps, R., *Introduction to Laser-Diode Pumping of Solid-State Lasers,* Bellingham, WA: SPIE Optical Engineering Press, 2007.

Shen, Y. R., *The Principles of Nonlinear Optics,* New York: John Wiley & Sons, 2002.

Siegman, A. E., *Lasers,* Sausalito, CA: University Science Books, 1986.

Silfvart, W. T., *Laser Fundamentals,* Cambridge, U.K.: Cambridge University Press, 2004.

Stegeman, G. I., and R. A. Stegeman, *Nonlinear Optics: Phenomena, Materials and Devices,* New York: John Wiley & Sons, 2012.

Sutherland, R. L., *Handbook of Nonlinear Optics*, 2nd ed., New York: Marcel Dekker, 2003.

Svelto, O., and D. C. Hanna, *Principles of Lasers,* 2nd ed., New York: Plenum Press, 1982.

Svelto, O., and D. C. Hanna, *Principles of Lasers,* 4th ed., New York: Plenum Press, 1998.

Tang, C. L., and L. K. Cheng, *Fundamentals of Optical Parametric Processes and Oscillators,* Amsterdam, the Netherlands: Harwood Academic Publishers, 1995.

Thyagarajan, K., and A. Ghatak, *Lasers, Theory and Application,* New York: Springer, 2010.

Vere, A. W., *Crystal Growth: Principles and Progress,* New York: Plenum Press, 1987.

Webb, C. E., and J. D. C. Jones, *Handbook of Laser Technology and Applications,* Bristol, U.K.: Institute of Physics Publishing, 2004.

Weber, M. J., *Handbook of Lasers,* Boca Raton, FL: CRC Press, 2001.

Young, M., *Optics and Lasers,* Heidelberg, Germany: Springer-Verlag, 1992.

Yariv, A., *Quantum Electronics*, 2nd ed., New York: John Wiley & Sons, 1975.

Zernike, F., and J. E. Midwinter, *Applied Nonlinear Optics,* New York: John Wiley & Sons, 1973.

CHAPTER 3

Laser Devices for Military Applications

3.1 Introduction

Many thousands, if not hundreds of thousands, of different types of laser devices have been investigated since the first demonstration of laser action in 1960. Attempts have been made to explore just about every conceivable aspect of laser action or its emission characteristics. However, most of these devices have been experimental configurations that led to erudite publications and theses, and only a small percentage has been developed for military applications. Despite the success of some specific technologies, such as semiconductor and fibre devices, there is still a significant range of different types of laser devices used in military systems that are based on what some people would consider old technology.

Considering these devices in terms of the three fundamental classes, based on the physical state of the gain medium (i.e., solid, liquid or gas), along with the excitation technique, is useful as the fundamental characteristics have some broad similarities. Many other terms are applied to describe some laser devices, such as the form of the gain medium, its shape or the path taken by the photon beam in the gain medium. All of this information usually provides some indication of the strengths and weaknesses of the device technology, and hence, the potential suitability for any given application.

The purpose of this chapter is to consider the techniques that may be used to determine the requirements for laser devices used in military applications and to offer a specialised description of the most common types of laser technology used in military systems. It will present a brief description and simple diagram of the operation of these devices; where appropriate, there will also be a short discussion of the pros and cons of this type of laser technology, along with an indication of technical challenges and future potential. Finally, a detailed summary of the fundamental performance characteristics of the emission will be provided.

One of the great challenges of developing successful laser devices for military applications is efficient and effective packaging. This requirement is usually application specific, so it will not be discussed here in detail, other than to note the care is required to ensure that the structure is effective and does not amplify any local resonances.

3.2 Military Laser-System Considerations

One of the greatest challenges concerning the development of a laser subsystem for any military application is to provide all the desired beam-parameter characteristics needed to fulfil the function, when it is needed and in a form that can be used effectively by a nonlaser specialist. These crucial requirements are usually far more demanding than the turnkey operation of commercial systems, owing to the complex and severe environment in which the laser-based system may be required to operate without routine, or other, maintenance.

The operational environment of a laser-based system may be very complex and the laser device will be required to provide the defined operational emission parameters throughout the extremes of the entire defined environment. Each application should have specifically defined operational, storage and transportation environments. This definition should include the following:

- Operational temperature range may extend from –54°C to +70°C.
- Storage and transportation temperatures (nonoperational) may exceed the above.
- Operational altitudes are clearly specific to a military aircraft or spacecraft.
- Transportation environment (nonoperational) includes flight altitudes.
- Operational vibrational environment is usually specified in terms of a power spectral density (g^2/Hz) over a specified frequency bandwidth.
- Transportation vibrational environment (nonoperational) has similar terminology.
- Operational acceleration characteristics are both lateral and longitudinal, particularly important for aircraft applications.
- Shock environment is both operational and nonoperational whilst in storage or being transported, expressed as an acceleration (g) over a short duration.
- Operational electromagnetic environment is particularly important for maritime and aircraft applications; additionally, the device must not disrupt the function of other systems on the platform, often expressed as a relevant field strength.
- The presence of abrasives, corrosive liquids, fuels, oils sprays and other fluids may have an influence on the design and construction of the device.

There are many other environmental considerations in each specific application that can impact on the success or failure of an acceptance test. The first six points above are usually termed the shake-and-bake testing; ironically, it is often more demanding to achieve the requisite performance at the lowest temperatures, as some systems may appear to be more sluggish when very cold.

In summary, the operational environment is usually very challenging. With the device being severely shaken in some instances, thermally cycled in the extreme, dropped or bashed, and still expected to work on demand. These requirements are clearly a challenge for all designers, but many have been very successful.

Many military applications require the simultaneous presence of a demanding combination of specific operational parameters. For example, the combination of a high power beam with very good beam quality maintained during a variable duty cycle and an emission in specific parts of the spectral wavebands during each shot. In many instances, meeting just some of the requisite beam parameters may be considered an achievement, but meeting all of them simultaneously and on demand throughout the operational environment is usually a significant challenge on the design philosophy.

Another critical consideration is how to operate and maintain the laser-based system successfully in the field so that the completed system can achieve its full potential to support military operations. Some desirable characteristics and challenges in terms of the operational considerations are:

- Achieving high wall-plug efficiency to reduce the waste-heat-management problem and minimising the power that has to be found on the battle field;
- Developing a compact and inert configuration;
- Achieving a robust and simple device that is safe to operate.

An important consideration in the introduction of a new laser system, whether a weapon or a sensor, is its impact on the support organisations and structure, known in the United Kingdom as Defence Lines of Development (DLODS), defined as:

- *Training:* all aspects of training provisions needed to support a new capability.
- *Equipment:* development of the associated primary support equipment needed to achieve a particular defence capability.
- *Personnel:* provision of operators, maintainers, trainers, planners and other support personnel to operate the developed equipment.
- *Infrastructure:* storage, transportation, training and production facility requirements.
- *Doctrine:* development of the military tactics required to utilise the equipment to its full capability and to define the user requirements (operational performance requirements).
- *Organisation:* the supporting contractors and defence organisations required to manage the defence capability throughout the life of the weapon.
- *Information:* supporting information systems and interfaces.
- *Logistics:* provision of the system support strategy associated to sustain the defence capability through the life of the system.

A thorough analysis of a new technology's impact on all of the support activities and organisations is a critical factor in determining the total value of this technology and its through-life cost.

One of the challenges particularly associated with high-power laser systems is to develop a training mode that is safe to use in all senses of the term. It is absolutely vital, as is discussed later, that the military personnel are able to be trained in the use of the new system and develop the appropriate skills and tactics to achieve the optimum operational value from the equipment and, additionally, to be able to adapt and refine the operational tactics and procedures to counter a rapidly changing threat environment.

3.3 Laser-System Specifications and Requirements Definition

A systematic approach towards defining laser-based system specifications is absolutely essential to justify the demands made on the laser-system design team. The starting point is normally a system study, a crucial step in defining an accurate laser-system specification. From this, the performance requirements of the subsystems that support and interact with the laser device may be derived.

Accurate specifications for every laser-beam characteristic is vital if a cost-effective solution is to be achieved; this is normally realised from a detailed consideration of the subsystems within the device as well as those that interact within the entire system, and beyond, at the platform integration level. Overall, it is the installed performance that ultimately defines the effectiveness of any operational system, whether using lasers or other effectors.

It is customary to tailor the scope and complexity of the system study to the precise military application; however, the studies will follow this pattern to a varying degree. A system study will usually consider all the aspects of the projected operational use of the new laser system; this includes:

- A threat study that investigates the potential spectrum of targets, which may include weapons, and a realistic projection of future performance, their operational characteristics and their potential vulnerabilities;
- A definition of conceptual laser-based systems to provide a suitable technique to engage and/or defeat the spectrum of identified threats;
- A definition of scenarios to allow the investigation of potential performance of the various concepts;
- An investigation of the operational environment, which includes the dynamics of the host platform, as well as the extremes of thermal cycling (storage, transport and operational use) and vibratory motion, especially appropriate to candidate installation positions within or on a platform;
- An identification of candidate architectures, their potential subsystems and functions, along with an analysis of their interfaces, including the definition and refinement of error budgets to ensure a harmonised system so that any one operational parameter is not excessively difficult to achieve;
- An analysis of potential perturbing elements, such as the range of atmospheric propagation phenomena, which may cause modifications to a laser beam as it propagates;
- An indication of techniques and methods for testing the performance of the subsystems and then assessing the effectiveness of the installed system;
- An identification of training methods and support requirements;
- An identification of potential logistical requirements.

This generic outline of a system study may be considered appropriate for a laser-based directed-energy system. A similar approach would be recommended for any other laser-based technique, but should be clearly tailored to that operational use of the laser system, with particular emphasis on specific challenges for the operation of the new system.

One of the most vital aspects of a system study is the performance specification or requirement for each of the subsystems, as well as for the complete system. These data may then be used to undertake potential performance estimates and to define a concept of operation. Each performance specification normally defines the mandatory performance value for each defined parameter and a desired value that would lead to either enhanced system performance or a relaxation in some other stressing aspects of the performance requirement. In general, all systems are a compromise between the conflicting demands of each component and subsystem, which leads to a realistic and ideally effective solution.

A summary of a process for the definition of subsystem specification (SSS) and projected performance estimation is shown in Figure 3.1. The starting point for the definition of subsystem specifications and requirements flowed down throughout a proposed system is the cardinal points specification (CPS). These parameters may be used in a system-performance model to determine the defeat probability for a given target set, which may then be used to estimate system performance. Data from this model may be used to define the electro-optical system and its pointing-error budget. The comparison of these parameters with current state-of-the art and technical reality is vital if a cost-effective solution is to be achieved. Several design iterations are usually needed to achieve the desired technical goal, particularly to achieve the required installed performance.

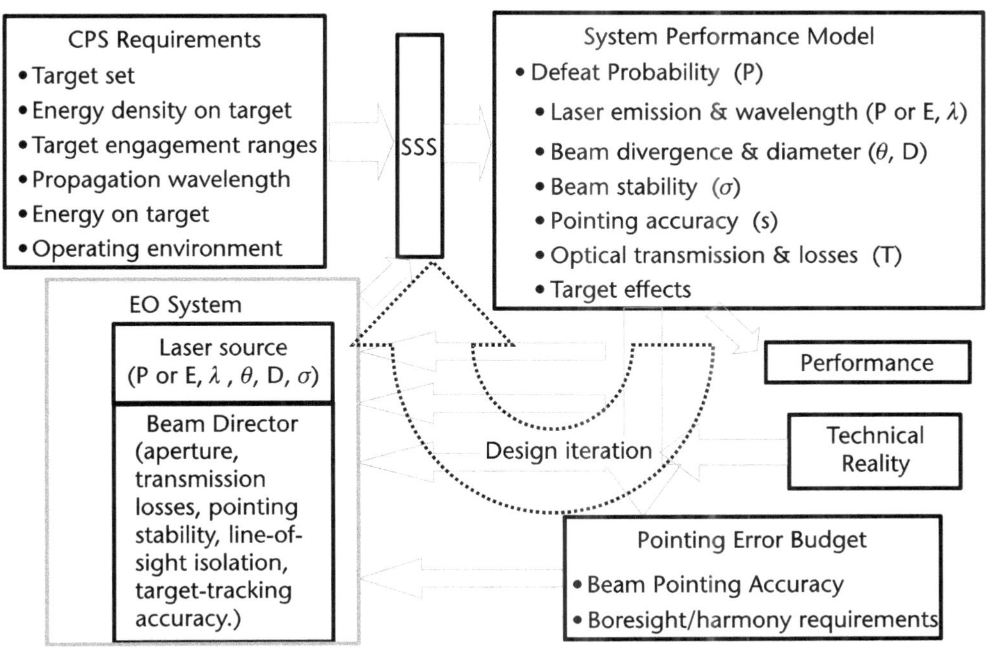

Figure 3.1 Requirements' flow for a laser-based system.

During the definition of the CPS and SSS, it is vital to consider how the equipment will be tested. An equally important consideration is how the equipment will be used during the training of its operators. Equipment that cannot be used safely or realistically during training is of little value.

3.4 Solid-State Lasers

A solid-state laser is an optical device that generates its photons in a nonconducting solid material. The solid gain medium may be crystalline, glass or ceramic with a dopant that has the appropriate energy-level structure to allow light generation by stimulated emission. The physical characteristics of the gain medium mean that the only realistic pumping scheme is through use of other photons. Originally, this was with flashlamps, but for many applications, diode pumping is now the technique of choice. It is much more efficient, robust, reliable and generally more compact. Modern laser-diode technology offers very long operational lifetimes of many thousands of hours. One drawback is the initial cost of the diodes, but through-life costs can be significantly reduced.

The crystalline, glassy or ceramic material is known as the host. A range of crystals has been developed for this application; synthetic garnets are an example. Yttrium aluminium garnet is one of the class that is often used and is usually referred to as "yag." Similarly, various glass hosts have been developed; some common examples are silicate, phosphate and fluorophosphate glasses, along with ceramics. The main operational performance requirements for the host materials are:

- Optical transparency at the pumping and emission frequencies;
- Good heat-transfer characteristics and/or a negligible change in optical parameters with temperature;
- Suitable elastic modulii to give high resistance to fracture and shock, particularly important for high-power systems;
- Ease of growth without distortion or twinning, or any other aberrations;
- High laser-induced damage threshold;
- Low thermal-expansion coefficient;
- Hardness;
- High and stable optical quality.

One of the more exciting recent developments has been the production of laser gain media from ceramic materials using a sintering process. This process offers a route to large gain media in various shapes with very good optical quality at a modest price. The principal challenge has been to achieve good optical quality with the larger samples. This has been largely solved with carefully controlled purification of the constituents, followed by precision heating and pressing processes in a carefully controlled atmosphere.

The active species in any solid-state laser is an ion that is embedded in the host matrix, which is introduced as an impurity during the growth process. The level of doping of the laser species varies from ion to ion and from host to host. It is generally of the order of 1%, although some species may be a factor of three or four higher, and of course, others may have a far lower concentration.

As noted above, the only practical pumping scheme involves optical techniques. The original approach was the use of flashlamps but this was inefficient for the following reasons:

- The pumping is a multistage process with electricity being used to excite the flash tube, which is then used to pump the laser species.
- The flashlamp emission is broadband and has similarities with the emission from a black body and is multidirectional.
- The absorption feature in the laser ion tends to be over a relatively narrow wavelength band, so only a small fraction of the flashlamp emission is in-band to, or coincident with, the absorption feature, but at least there are no concerns to ensure that this broadband emission wavelength needs to be stabilised to coincide with the desired absorption feature.

The efficiency of such an optical pumping scheme is the product of the efficiencies of all of the stages. As the first step of generating the black-body emission is very inefficient, the final efficiency will be very low. Additionally, careful design is required to realise robust operation of the system using flashlamps. Moreover, the need for the generation of very high potential and rapid-rise currents is eliminated when using laser diodes. This has many advantages, particularly for efficiency and especially for safety.

The successful development of compact laser-diode devices for optical pumping has dramatically improved the performance of solid-state lasers in many respects. The great advantages of diode pumping are:

- Efficient light generation, theoretically as high as 80% or beyond;
- Spectral emission matched to the absorption feature, so no light is wasted;
- The emission from a diode can be well matched to the appropriate, or specific, absorption feature (this minimises the quantum defect and thus the internal heating of the laser crystal);
- Well-defined emission direction that can be matched to the gain medium dimensions.

Diode pumping has, therefore, made a dramatic improvement in the efficiency of solid-state technology; typically, in some specific devices the optical-to-optical efficiency can easily exceed 60%. Additionally, laser diodes are robust and reliable, and help make laser devices more compact. However, the major drawback is still the cost of the diodes, but at least many of the technical challenges concerning the emission lifetime have been resolved.

The use of diode-pumped solid-state laser technology in military laser systems is now usually the preferred technological solution. The usual reasons cited for this preference may be summarised as follows:

- Compact configurations, giving system flexibility and reduced logistics burden;
- Long shelf life, reducing the logistics burden;
- Ruggedness, which more compatibility with the military environment, and again reducing the logistics burden;
- High efficiency, reducing power demands and reducing the heat-loading, including asymmetrical thermal gradients, in the crystal leading to better beam quality;
- More stable output-power stability, leading to better beam stability and greater flexibility of operation;
- Long operation lifetime, reducing the logistics burden;
- Nontoxic;
- Potential to operate over extreme temperature ranges;
- Flexibility of output characteristics, offering the prospects of multifunctional operation;
- Elimination of the need for kilovolt potential to drive the pump source and the prospect of an exploding flashlamp.

Laser diodes also offer the flexibility of different pumping configurations (i.e., longitudinal or transverse, which will influence the output power, lasing threshold power requirements and beam quality of the output beam). For example, longitudinal pumping allows better overlap between the pump beam and the generated mode volume. This leads to a lower lasing threshold and higher gain, and thus potentially to the emission of a higher-power beam with better beam quality.

It is now common to fibre-couple the pump light into the resonator for longitudinal pumping. This scheme has several advantages:

- Beam shaping of the diode beam to match the mode volume without additional optical elements;
- Removal of an external heat source from the laser cavity and gain medium, reducing thermal gradients in the laser;
- Potential reduction in size of the laser subsystem;
- Design flexibility and ease of replacement.

Transverse pumping is appropriate when the gain medium has a relatively short absorption depth, as occurs in many neodymium-doped gain media. Additionally, many diode bars or stacks can be arranged around the gain medium to give a very uniform pump intensity profile, which is important for very high-power applications, especially if good beam quality is to be maintained throughout sustained operation.

The major challenges facing the teams developing solid-state laser devices for any military application is getting the requisite laser power at a cost that is affordable whilst maintaining performance throughout the operational environment. A particular challenge is controlling the chirp of the emission from a diode as the operating temperature of the system changes, especially if the lasing specie has a narrow absorption feature. Relatively simple techniques have now been developed to compensate for this effect, usually involving judicious control of the local thermal environment.

Solid-state lasers are normally associated with wavelength emissions from about 0.7 μm to about 3 μm. Owing to the nature of the electronic transitions, the wavelength is broadly dependent on the laser ion involved in the stimulated emission. Some of the common laser ions used in military applications and their corresponding emission wavelengths are given in Table 3.1. Approximate values are given, owing to the dependence of the host environment on the actual emission wavelength; for example, the neodymium ion can emit from about 1.04 to about 1.07 μm from the same transition depending on the type of glass or crystal used in the host gain medium. Moreover, neodymium has a number of other relatively strong transitions at wavelengths shorter and longer than the dominant one at around 1 micron; although these emission lines have less gain they are very useful for some specialist applications. A suitable choice of mirror coatings or use of intracavity specific filters, to control the losses at particular wavelengths, enables the desired emission wavelength to be achieved.

Specific lasing ions used in vibronic lasers have not been included as they are discussed in detail in Section 3.7. Only a generic illustration is given in Table 3.1. More details of the actual tuning range of some selected crystals are provided next.

3.4.1 Future Challenges for Solid-State Devices

The application of laser diodes for pumping solid-state has transformed the optical performance of these devices. However, there are still many performance aspects that could be improved. Some of the major challenges for this technology are:

- Reducing cost;
- Reducing complexity;
- Enhancing efficiency;
- Increasing robustness;
- Enhancing the emission bands to the mid-wave infrared without the use of nonlinear techniques.

Table 3.1 Characteristic Emission Wavelengths for Some Common Devices

Laser Ion	*Wavelength (μm)*
Neodymium (Nd)	~1
Ytterbium (Yb)	~1
Erbium (Er)	~1.5 and 2.9
Thulium (Tm)	~1.9
Holmium (Ho)	~2, ~3 and ~4
Vibronic (various)	~0.7 to ~2

3.4.2 Rod Lasers

This class of solid-state laser was the first type of device demonstrated in 1960 and the generic term describes the shape of the gain medium. The host may be crystalline or glass and doped with a vast variety of laser ions, usually the metal or rare-earth species; currently the most common laser specie is probably the rare-earth neodymium ion. A relatively simple configuration is possible where the gain medium is placed between the reflectors in the resonant cavity, as shown in Figure 3.2. An intracavity Q-switch may be used to produce a carefully controlled pulsed output.

One potential configuration for higher-power devices is to have a ring of flashlamps arranged around the laser rod. The output power can be boosted further with the use of amplifiers, which may be single pass or have multiple paths through the gain medium in the amplification stage.

A major challenge with this shape of gain medium is to design an efficient technique to get the pump light into the gain medium. The early devices used a helical flashlamp, but a more modern approach is to use a linear flashlamp and an elliptical reflecting cavity with the laser rod at one focus and the flashlamp at the other. This latter approach is more robust but quite bulky and, consequently, heavy.

A further drawback of this class of device is the fact that many of the host materials used in this type of gain medium are susceptible to the creation of a thermal lens during sustained operation. This leads to distortion of the beam and in some extreme cases, extinction of the beam. Techniques were developed to mitigate the effects of the thermal lens, but this usually resulted in a delay in the laser achieving full brightness. Additionally, the thermal gradient across the rod led to other aberrations and a varying beam quality and hence beam divergence.

In general, with a rod geometry, direct diode pumping is only practical using longitudinal pumping, so efficient operation is difficult. Hence, the use of a flashlamp is still common with conventional rod-laser devices. When a linear flashlamp is used, as shown in Figure 3.2, it is usually with a water jacket to minimise the heating of the laser rod; moreover, this coolant may use fluorescence converters, such as laser dyes in solution, in an effort to increase efficiency.

Large glass rods in excess of a metre long have been produced with high optical quality and demonstrated the capability of generating very energetic pulses with

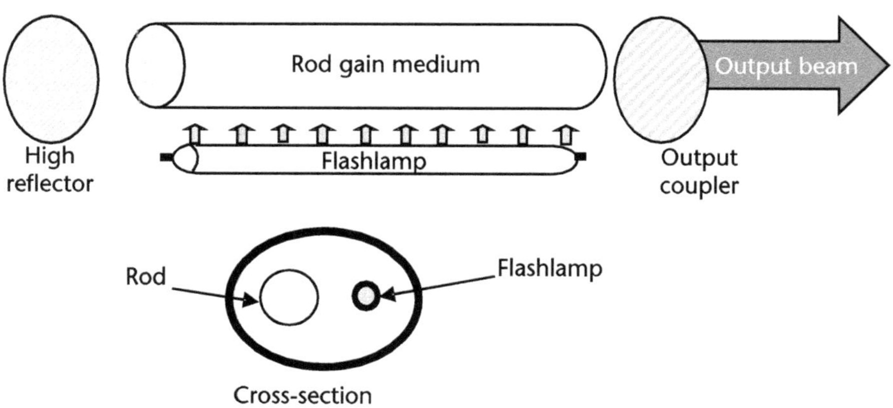

Figure 3.2 A rod laser schematic.

a good beam quality, albeit at a very low pulse-repetition rate. The production of similar sized crystalline laser rods remains a challenge for the crystal growers, which is one of the reasons for developing ceramic host materials.

Qualitative performance of rod lasers is outlined in Table 3.2.

These are essentially quite simple laser devices that are somewhat flexible in use, but have a range of performance limitations not compatible with many modern military applications. One distinct advantage is that they are relatively cheap to produce, but through-life costs may be significant, owing to the predominant use of flashlamp technology. The modern fibre laser may be considered to be a very long, thin rod laser; moreover, a disk laser may be considered to be a short, fat rod laser. Both of these types of laser are considered further below and are capable of far higher performance operation than is routinely demonstrated by most conventional rod lasers.

3.4.3 Slab Lasers

The next development with solid-state laser devices was the investigation of slab-shaped gain media to overcome many of the perceived short comings of rod lasers. This development was driven by many requirements; from the military viewpoint, the benefit is that the geometry is ideal for pumping with a laser diode, leading to compact, rugged and relatively efficient sources.

This form of laser may be operated in a similar fashion to the rod devices, as shown in Figure 3.2, with the slab replacing the rod in the resonant cavity. This type of laser can be excited with linear flashlamps or laser diodes. In the case of

Table 3.2 Qualitative Performance of Rod-Laser Devices

Performance Parameter	*Characteristic*
Wavelength range	Visible to mid-wave infrared depending on laser ion and host
Tuneability	Not usually, but see vibronic lasers
Output	Pulsed or CW
Power	Can have a high average power for big pulsed glass lasers (but see PRF); energetic pulses in excess of 1 kJ demonstrated; typical average powers of the order of a few hundred watts from some crystalline hosts
Pulse-repetition frequency (PRF)	Depends on hosts; high for crystal hosts of many tens of kilohertz, but low for glass hosts (less than 1 Hz); mode-locked lasers have megahertz PRF using a crystalline host
Beam quality	Usually poor for sustained operation at high PRF, unless time to full brightness is compromised
Time to full brightness	Can be fast but with a degraded beam quality in sustained operation
Efficiency	Usually low, often less than 1% (with flashlamps) but 5%–10% with diode pumping (optical to optical)
Size	Bulky
Reliability/maintenance	Low because of the obsolete technology (flashlamps), higher with diode pumping
Complexity	Low because of the simplicity (flashlamps); higher with diode pumping
Cost	Low, but through-life costs may be high (flashlamps), higher with diode pumping

flashlamp pumping, this may be in a close-coupled configuration, that is, with the linear flashlamp(s) immediately adjacent to the slab side(s) and enclosed by a faceted reflector. Like the rod laser, this is known as side pumping, as illustrated with diodes in Figure 3.3. Some of the technological issues of the rod device may have been overcome, but it is still an inefficient and bulky device, unless laser diodes are used to create the population inversion.

The principal advantage of this gain-medium geometry is the ability to pump it with a laser diode, or more often, an array of diodes. The use of this pumping technique has many advantages compared to flashlamp-pumping methods, including:

- Inherent increased efficiency;
- Enhanced pumping intensity;
- More uniform pumping of the gain medium;
- Increased simplicity;
- Design flexibility and use of longitudinal pumping or side pumping;
- Reduced size;
- Increased robustness;
- High-power and/or high repetition rate operation.

One of the challenges facing the manufacturers of slab technology has been the production of the very large slabs of crystals, such as yttrium aluminium garnet, required for the very high-power applications with the requisite optical quality and homogeneity. A number of techniques have been investigated for assembling several small slabs into one large single slab. One of the more successful approaches is known as diffusion bonding; it uses a combination of pressure and heat to glue the components together. The bonding mechanism uses the van der Waals forces between the atomic layers of the adjacent surfaces. These have to be polished flat with high precision (small error tolerance). The development of the high-quality, loss-loss ceramic-crystal hosts has solved some of these technical challenges

A further refinement of this type of laser is the use of the zig-zag slab resonator geometry, which describes the path taken by the generated beam as it passes through the gain medium. The geometry of the slab is designed so that the generated photon beam is internally reflected against the sides of the slab an odd number of times in total as it passes through. The principal advantage of this technique is that any inhomogeneities and thermally induced aberrations are averaged out in the transit of the slab by the beam, leading to better and more stable overall beam quality. The zig-zag optical arrangement in the gain medium is shown schematically in Figure 3.4.

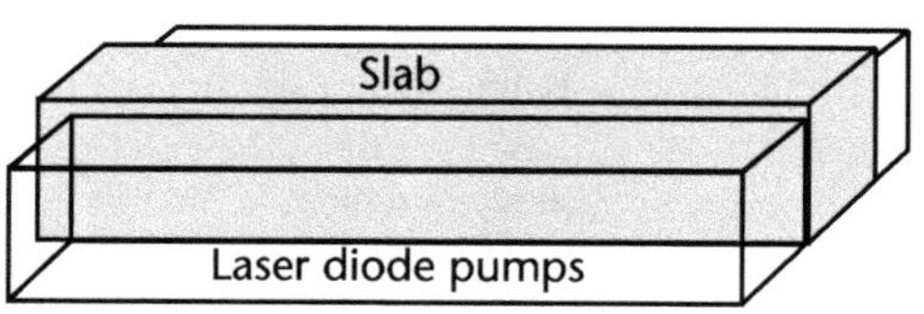

Figure 3.3 A slab laser schematic with diode side-pumping.

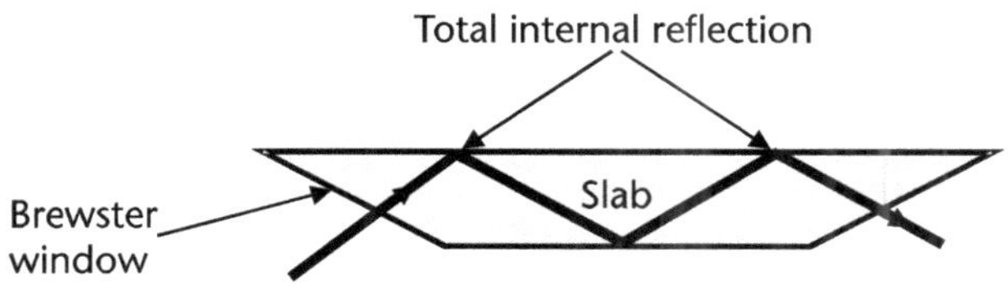

Figure 3.4 Zig-zag slab schematic.

The qualitative performance, given in Table 3.3, is only considered in the context of a diode-pumped device. The flashlamp-pumped slab laser, which is not really appropriate use of this geometry, will be similar to the performance of a flashlamp-pumped rod laser outlined earlier.

The diode-pumped solid-state lasers may be used as high-power sources, particularly when using the neodymium-doped devices. A recent demonstration by the Northrop Grumman Corporation proved it was possible to generate in excess of 100 kW from such a laser using a master oscillator power amplifier configuration and combination of several beams; the Textron Company has also demonstrated a similar performance.

A common use of diode-pumped solid-state lasers is the pumping of nonlinear crystals to generate wavelengths not readily available from solid-state lasers. The common examples are:

- Frequency doubling of a neodymium laser to generate green light;
- Shifting to longer wavelengths to gain access to other wavebands, such as shifting the wavelength of a 1-micron laser into the eye-safe wavelengths (~1.5 μm) or into the mid-wave infrared.

Table 3.3 Qualitative Performance of Diode-Pumped Slab-Laser Devices

Performance Parameter	*Characteristic*
Wavelength range	Visible to mid-wave infrared depending on laser ion and host
Tuneability	Not usually, but see vibronic lasers
Output	Pulsed or CW
Power	Can have a high average-power output from crystalline hosts, particularly when used in an oscillator and amplifier configuration: tens of kilowatts
Pulse-repetition frequency	Depends on hosts: high for crystalline hosts of many tens of kilohertz, but low for glass hosts (less than 1 Hz); mode-locked lasers have megahertz PRF using a crystalline host
Beam quality	Usually good and stable when using a zig-zag configuration, but usually a trade-off with time to full brightness
Time to full brightness	Can be fast but with a slightly degraded beam quality in sustained operation
Efficiency	Usually far better than 25% (optical to optical conversion)
Size	Compact
Reliability/maintenance	High
Complexity	Increased, owing to the use of another laser (the diode)
Cost	Higher, but through-life costs may be potentially lower

The solid-state slab laser offers significant performance advantages when compared with a rod laser for a small increase in complexity in the case of the diode-pumped solid-state device. The diode technology is significantly more expensive than the flashlamp technology, but this may be mitigated somewhat by the increased reliability and reduced through-life costs. This technology has demonstrated very high power beams suitable for laser-weapon applications, as noted earlier.

The use of the diode-pumping schemes makes a compact and efficient device. The flexibility of the technology means if offers photons suitable for wavelength conversion, either through harmonic generation or exploitation of parametric processes.

3.4.4 Slod Lasers

The term “slod” is derived from a cross between slab and rod and refers to the shape of the gain medium; it is sometimes referred to as a D-section rod, owing to its cross-sectional shape. The design aim was to apply modern excitation technology to the rod laser’s well-understood and favourable characteristics whilst improving its relatively poor performance. One development was to produce a high-power pump source in a rugged and compact package to pump an optical parametric oscillator, for subsequent conversion of this pump light to other wavelengths.

The engineering solution was to modify the shape of a laser rod, so that it could be pumped efficiently by a laser-diode array using a transverse pumping arrangement. A flat was created along the length of an otherwise circular laser rod as shown in Figure 3.5. The curved surface was optically coated to reflect any unused pump light back into the gain medium, to ensure an optimum excitation configuration.

The uneven pumping of the gain medium resulted in several inadequate aspects of performance. The most serious defect was probably the poor beam quality, but other shortfalls in performance were the significant delay before full brightness was achieved and a very lengthy time to achieve the optimum emission characteristics.

The qualitative performance of a diode-pumped slod laser is considered in Table 3.4.

Although this type of laser design retained some of the advantages of the laser rod’s simple technology, the uneven pump-intensity distribution and consequent poor beam quality, amongst other limitations, make it a less-than-optimum solution.

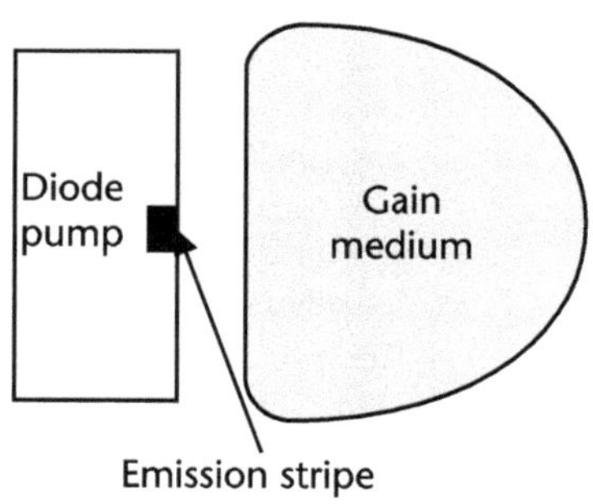

Figure 3.5 Cross-section of a slod and its pump diode.

Table 3.4 Qualitative Performance of Slod Lasers

Performance Parameter	*Characteristic*
Wavelength range	Visible to mid-wave infrared depending on laser ion and host
Tuneability	Not usually, but see vibronic lasers
Output	Pulsed or CW
Power	Modest, owing to the uneven pumping regime
Pulse-repetition frequency	Depends on hosts: high for crystal hosts of many tens of kilohertz, but low for glass hosts (less than 1 Hz); mode-locked lasers have megahertz pulse repetition frequency using a crystalline host
Beam quality	Poor for sustained operation at high PRF, owing to uneven pumping intensity distribution, even with a trade-off with the time-to-full-brightness parameter
Time to full brightness	Can be fast but with a degraded beam quality in sustained operation, usually slow
Efficiency	High, potentially better than 10% (optical to optical conversion)
Size	Compact
Reliability/maintenance	High because of the application of modern technology
Complexity	Low because of the simplicity
Cost	Moderate owing to relatively simple concept, with the potential for modest through-life costs

3.4.5 Plastic Lasers

This branch of solid-state laser technology refers to the polymer host material used to hold the laser species. The plastic host was developed to enable a solid-state dye laser to be created, to simplify the dye laser configurations used with the liquid phase, considered below (Section 3.8).

The use of conventional solid-glass materials doped with laser dyes is quite attractive, as they have low optical scatter, low distortion, are hard and stable, and can be polished smooth and flat. However, it is usually impossible to use such a host directly as the constituent materials need heating to temperatures in excess of 1,000°C during production, thereby destroying the laser dye. Consequently, alternative host materials and techniques for fabrication at a lower temperature have been sought.

There are two major approaches: synthesising polymer or glassy materials at low temperature or producing a glass with a porous structure. The former technique allows a laser dye to be added to the constituent chemicals and is known as pre-doping. A porous structure allows a dye solution to be introduced into the structure to impregnate it via diffusion and is termed post-doping of the solid matrix.

Predoped hosts usually produce solid materials with good optical quality and uniform dye concentration. These dye-doped solids can have properties approaching those of glass, but tend to be somewhat softer, which can have an adverse impact on laser performance, particularly if high-energy operation is required. A major breakthrough in the optical quality of dye-doped materials, particularly laser rods, shown in Figure 3.6, has been a refinement of the production method, involving purification of precursors and careful control of the curing process. The new and improved process has led to a significant reduction in optical scatter in the laser rods and discs

Figure 3.6 Plastic laser rods.

and control of the refractive index profile, which in the past had led to strong thermal lensing within a rod.

In contrast, post-doped materials are usually more glass-like, with many of the excellent bulk properties of glass, but the material can be susceptible to losses owing to excessive optical scatter from the pores within the host matrix. Clearly, there is a trade-off between ensuring that the average pore size is large enough to absorb the dye solution and achieving a good dye-doping concentration, and still have very good optical quality with low losses, particularly optical scatter.

These laser materials may be used in any conventional laser configuration described above. Optical pumping is the only feasible option and is usually achieved through laser pumping or with a flashlamp in a close-coupled configuration. There have also been successful demonstrations of laser-diode pumping, but that was for the generation of a near infrared beam.

These lasers operate in the same way as their more mature liquid-phase cousins and so the fundamentals of laser action in a dye laser will not be discussed here; the details are in Section 3.8. A summary of the qualitative performance of plastic dye solid-state lasers is given in Table 3.5.

Plastic dye lasers offer the potential of a low-cost laser source suitable for some sensing and countermeasure applications. The devices do not have the durability or the average power of the flashlamp-pumped liquid-phase devices, but are far simpler and significantly more compact. Plastic dye materials may be used for wavelength conversion in wavelength-diversity modules, which may be added to other laser devices.

The big challenges for dye-doped polymer materials are to identify fabrication process that will make the host more glass-like and enhance the service life time of the material. Making the host more glass-like, with very low optical scatter, would bring many advantages to the beam quality.

One solution to the service lifetime issue was to use a dye-doped disc and spin it a relatively slow frequency. This was known as the spinning disc laser, which has also been adopted in other solid-state devices.

Table 3.5 Qualitative Performance of Plastic Dye Lasers

Performance Parameter	*Characteristic*
Wavelength range	Visible to near infrared depending on laser dye
Tuneability	Yes, usually over many tens of nanometers; these devices also offer wavelength diversity by changing the laser gain medium to one containing a different dye
Output	Pulsed, determined by excitation source
Power	Modest average power
Pulse-repetition frequency	Depends on pump source
Beam quality	Poor for sustained operation at high pulse repetition frequency, owing to poor thermal conductivity (of the host)
Time to full brightness	Fast
Efficiency	Low, less than 1% (optical to optical)
Size	Compact
Reliability/maintenance	High because of the application of modern technology
Complexity	Low because of the simplicity
Cost	Very cheap

3.4.6 Ceramic Lasers

This term refers to the host used in the laser gain medium. There has been an interest in the development of alternative host materials made by simple processes, such as sintering, to enable greater flexibility in the doping concentration and profile of the host, as well as the size of the crystal.

One of the greatest challenges in the development of ceramic hosts has been to achieve the optical clarity that is achieved with more conventional hosts. New techniques have been developed to control the size of the grains in the sintering process, but the greatest advance has been in the refinement and purification of the precursor materials.

The development of ceramic materials as an alternative to crystalline or glass hosts has been pursued as a potentially viable route to high-power devices. Ceramic materials have some considerable advantages over other hosts, such as:

- Higher thermal conductivity to enable higher power operation and reduced aberrations such as minimising susceptibility to thermal lensing;
- Higher thermal shock resistance linked to the higher thermal conductivity;
- Greater shock resistance/strength, in which grain boundaries are more accommodating to additional lattice strains and resist crack growth (the strength of the ceramic host is inversely proportional to the square root of the grain size);
- Ability of produce gain media with specific nonuniform doping profiles;
- Potential for higher doping concentrations linked to the strain tolerance of the grain boundaries;
- Potential for larger sized gain media associated with the growth process, which may allow complex shapes to formed, which cannot be realised in other hosts.

The major challenge of routinely producing hosts with low optical loss has largely been achieved.

3.5 Fibre Lasers

A fibre laser is a solid-state rod laser with a long and very thin gain medium surrounded by an undoped glass, which has many advantages for the efficient extraction of waste heat. The effective dissipation of waste heat and the management of the heat loading in a solid-state laser gain medium are critical issues and usually the fundamental cause of aberrations in the emitted beam intensity profile produced by a conventional solid-state laser, as discussed in Chapter 2.

The fibre laser can be considered as a brightness converter, owing to its very high optical-to-optical conversion efficiency, as well as a wavelength changer. The device takes the highly divergent laser light from the pump diodes and usually emits light at a slightly different wavelength in a much brighter beam. Beam brightness is a fundamental requirement of all directed-energy applications; additionally, beam brightness is important for many sensing applications. Moreover, fibre laser technology can be configured to control the mode and polarisation of the emission from each fibre laser. Mode control is normally a fundamental requirement for higher-power devices intending to combine the emission from several lasers coherently, as discussed later.

The basic design of a conventional fibre laser is a doped core surrounded by a cladding region of similar material to the core but not doped and with a slightly reduced refractive index compared with the core. This optical arrangement, with the differing refractive indices, provides the confinement of the light along the length of the optical fibre core. This core region is often a few microns in diameter and the cladding several orders of magnitude greater, although large-mode area devices have been demonstrated for high-power applications; these fibres have a core diameter of the order of 10 to 50 μm.

A number of configurations of the doped core (gain medium) and the cladding region have been developed in order to achieve efficient excitation of the gain medium. The simplest approach to pumping the core is to direct light into the cladding region taking care to match the pump beam with the acceptance angle (defined by the numerical aperture) of the optical fibre. This is shown schematically in Figure 3.7, along with the definition of numerical aperture.

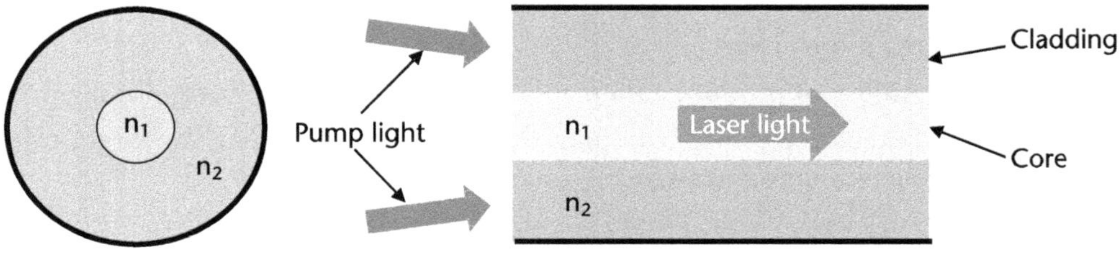

Numerical aperture = $\sqrt{(n_1^2 - n_2^2)}$

Figure 3.7 Simplified pumping of an optical fibre.

A number of specialised techniques have been developed for coupling of the pump light (from laser-diodes) into the core of the fibre. Some of these techniques are relatively complex and expensive. However, one novel technique, known as GTWave®, uses two cores for injecting the pump light into the third core, which is the gain medium, as shown in Figure 3.8. In this case, the two pump cores are in contact with the cladding region surrounding the doped gain medium in order to achieve highly efficient coupling of the pump light into the core leading to efficient laser action.

Another form of speciality fibre is known as polarisation-maintaining fibre, which is designed to maintain the polarisation characteristics of the output beam from the optical fibre. This type of optical fibre, with its relatively large stress rods, is often referred to as Panda fibre because of the appearance of its cross-section as shown in Figure 3.9. In this figure, the position of the stress rods with respect to the core can be seen. This technique is effective, relatively simple and robust, but presents some challenges as far as accurate alignment is concerned when splicing two fibres together.

The operation of a fibre laser is a classic exploitation of stimulated emission, using laser diodes to establish the population inversion in the gain medium (core). The relaxation from excited energy level to the lower energy level leads directly to the laser emission. Many different dopants have been used to give a range of wavelengths, usually in the near and short-wave infrared. A simple illustration of a double end-pumped fibre laser is shown in Figure 3.10.

Figure 3.11 shows a cross-section of a double-clad core used in some fibre lasers to enhance pumping efficiency (i.e., getting the pump light into the core). This figure indicates the various components of the optical fibre and the position of the

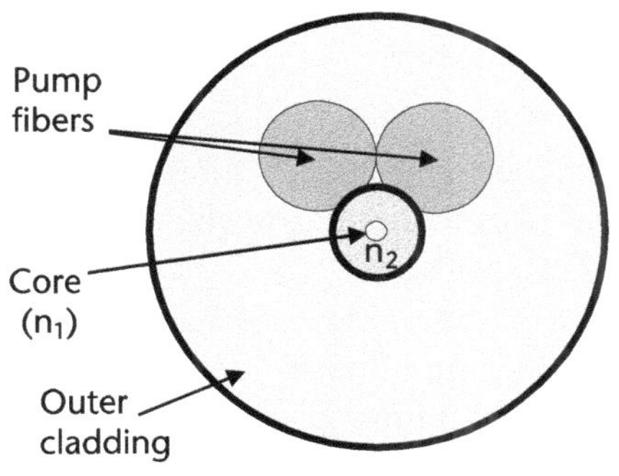

Figure 3.8 GTWave® cross-section.

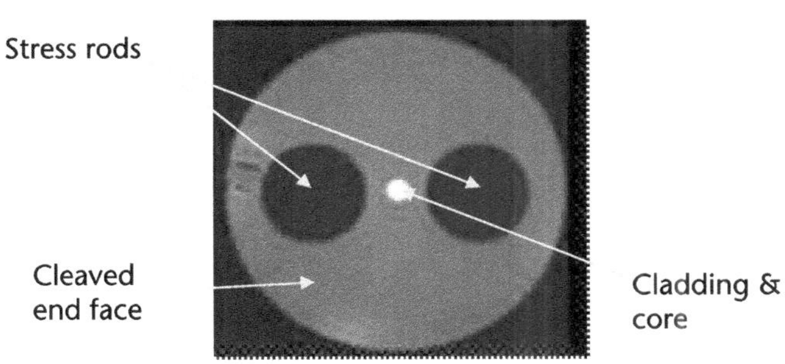

Figure 3.9 Panda fibre cross-section.

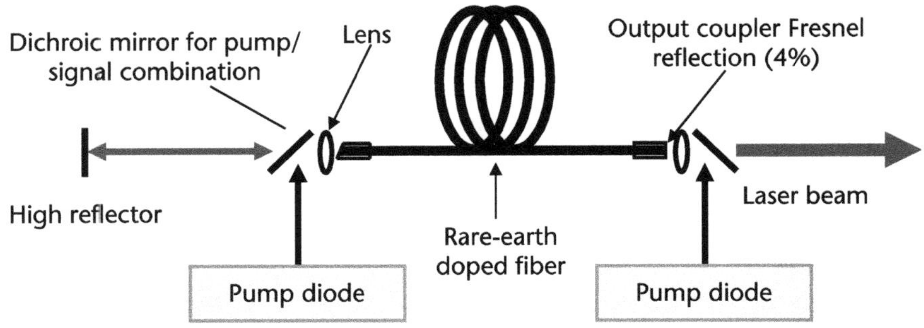

Figure 3.10 Fibre-laser schematic. (Courtesy of Professor J. Nilsson, ORC, U.K.)

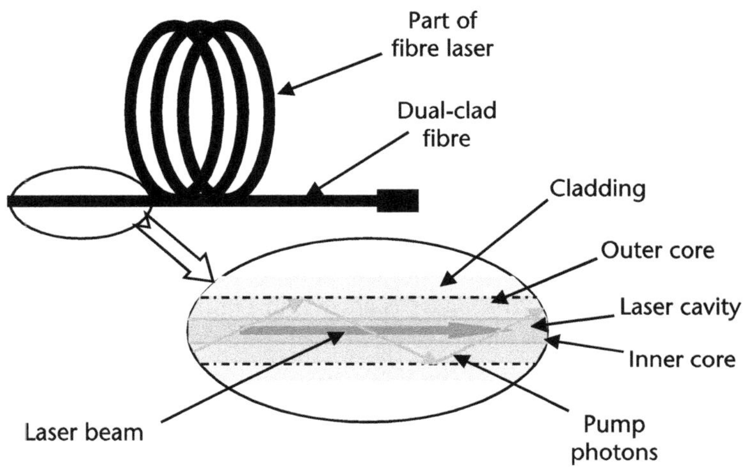

Figure 3.11 Dual-clad fibre core.

laser generation in the inner core. Sometimes the outer core is referred to as the inner cladding and is fundamental to the efficient coupling of the pump light into the gain medium (i.e., the core).

More sophisticated pumping schemes may involve the combination of multiple pump diodes to give a high intensity beam to end pump a fibre, as shown in Figure 3.12. Alternatively, side pumping may also be realised in a scheme similar to the technique also shown in this figure. The GTWave approach may be seen as a sophisticated form of side pumping.

These lasers tend to operate most effectively in either the continuous-wave regime or with pseudo continuous-wave (CW) emission. A high pulse-energy generated in a high repetition-rate operational regime may be problematical, owing to the instantaneous power density in the core of the fibre being close to the laser-induced damage threshold (greater than 100 MW/cm^2). Consequently, high-energy, short-duration pulsed operation may lead to an instantaneous peak power that is likely to exceed the laser-induced damage threshold of the core. Additionally, similar care is required when these devices are operated in a high repetition-rate, low energy regime, in order to keep the peak power below the bulk damage threshold of the core.

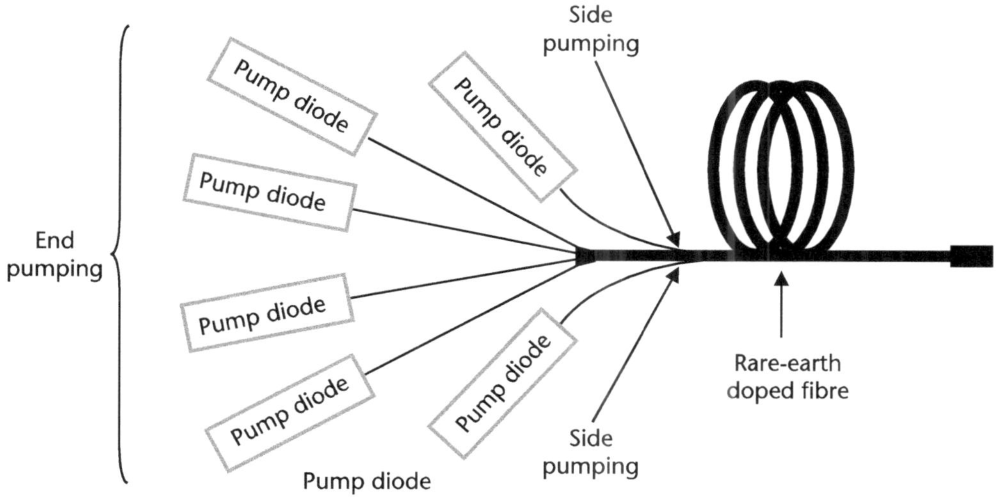

Figure 3.12 Fibre-laser pumping schemes.

There has been an explosive increase in the performance of fibre-laser technology during the last decade. The initial efforts were aimed at telecommunication applications with a particular emphasis on the eye-safe wavelengths around 1.5 μm; however, the more recent developments have been with emission around 1 micron and with a specific aim of providing an efficient and effective laser source for machining of materials, particularly cutting and welding of metals. A laser-based directed-energy weapon endeavours to achieve a target-interaction effect that may be considered to be cutting or reshaping of materials at range.

Some of the causes of the significant enhancement in performance have been largely a result of these developments:

- Effective doping with rare-earth ions;
- Development of efficient and compact pump sources;
- Control and reduction of losses.

These high-power demonstrations have used master oscillator, power amplifier configurations. The fibre laser is ideally suited for this form of operation. The fibre-laser oscillator generates the initial beam with good beam quality and is then amplified through a series of one or more fibre amplifiers to generate the high-power beam, whilst maintaining the excellent beam quality formed in the master oscillator, which is dictated by fibre design and negates the effect of thermal lensing.

A fibre-based master-oscillator power-amplifier device may be pumped in a range of configurations; these are considered in Table 3.6.

The enhancement in performance of ytterbium-doped fibres is summarised in Figures 3.13 and 3.14 shows the power-scaling achievements for a number of different types of fibre lasers over a 10-year period.

The recent and continuing development of fibre-laser technology has a firm foundation because of its attributes. These are:

Table 3.6 Pros and Cons of Fibre-Laser Pumping Methods

Pumping technique	*Advantages*	*Disadvantages*
Co-pumped	No output pig-tail required, can trade efficiency for line width ($\Delta\lambda$), reduced thermal load in active fibre	Longer effective length, large thermal load on the multimode combiner (MMC)
Counter pumped	Shortest effective length, de-couples total pump absorption from line width ($\Delta\lambda$)	Output MMC and splices at the output degrade beam quality, large thermal load on MMC, high thermal load on active fibre
Bidirectionally pumped	Lowest thermal load on MMCs, reduced effective length, partially decouples total pump absorption from line width ($\Delta\lambda$)	Output MMC and splices at the output degrade beam quality and additional components required

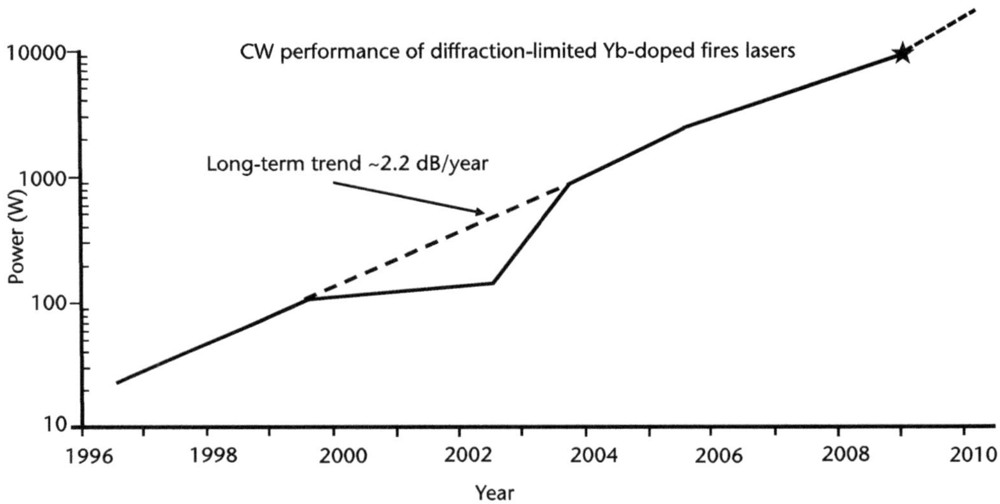

Figure 3.13 Emission enhancements from 1-micron fibre lasers. (Courtesy of Professor J. Nilsson, ORC, U.K.)

- Simple and rugged technology;
- Compact configurations;
- Easily adaptable shape;
- Good heat dissipation and easy management of waste heat, leading to emission characteristics insensitive to temperature;
- Potential for good beam quality from single-mode generation;
- High efficiency from use of high doping concentrations of ytterbium ions in silica;
- Potential for low cost;
- Potential for scaling to the megawatt-power regime (through beam combination techniques);
- Flexible architectures (multi-pass oscillators or master-oscillator, power-amplifier configurations).

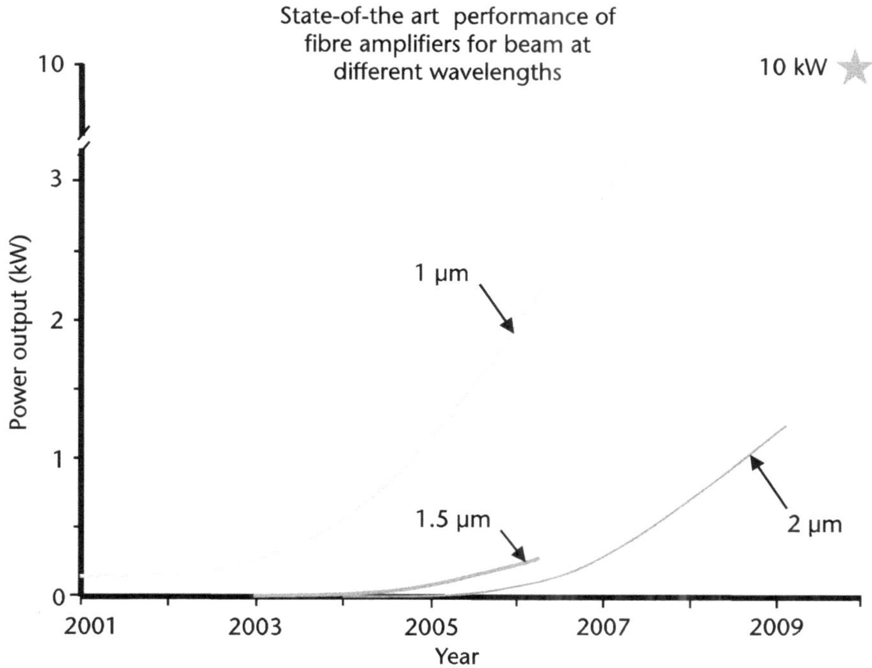

Figure 3.14 Progress with power-scaling of various fibre lasers. (Courtesy of Nufern, USA.)

The development of robust and reliable splicing techniques has led to the development of all-fibre configurations, thus avoiding the use of any free-space optical components. Moreover, the perfection of techniques to write fibre Bragg gratings directly into the optical fibres and define the resonant cavity have enabled robust, compact monolithic all-fibre structures to be produced. This type of device design has allowed for further size reductions, alongside enhanced reliability and robustness.

The qualitative performance of fibre-laser technology is summarised in Table 3.7.

Some directed-energy applications need a source of many tens of kilowatts. Given the current status of technology, the power density in the core will exceed the damage threshold of a single fibre laser. Multiple fibre lasers therefore need to be combined.

Fortunately, fibre-laser technology is ideal for combining multiple examples of devices for scaling to very high powers of the proportions projected for use in directed-energy weapons because the beam characteristics may be accurately controlled. A number of techniques are feasible:

- Incoherent methods, where the beams of separate lasers are simply overlapped; this technique is also known as geometric coupling;
- Coherent methods, using either the spectral or phase methods for generating a single beam.

Table 3.7 Qualitative Performance of Fibre-Laser Devices

Performance Parameter	*Characteristic*
Wavelength range	Typically between ~1 and 2 μm, largely governed by suitable glass hosts, but aspirations for mid-wave infrared emission
Tuneability	Dependent on the dopant, thulium fibre devices are tuneable, whereas neodymium-doped fibres are not
Output	Pulsed or CW
Power	High, multiple kilowatts per fibre (greater than 10 kW now available)
Pulse-repetition frequency	High, usually CW
Beam quality	Good to excellent, M^2 close to 1
Time to full brightness	Fast
Efficiency	High, potentially better than 70% (optical to optical)
Size	Compact.
Reliability/maintenance	High because of the application of modern technology
Complexity	Low because of the simplicity
Cost	Moderate owing to relatively simple concept, with the potential for modest through-life costs

These methods have been used by other laser sources to generate a single beam from multiple inputs. The coherent methods are more complex but provide a brightness enhancement with a carefully controlled beam intensity profile in a potentially close-to-diffraction-limited beam. Hybrid techniques are likely to be developed using combinations of all three methods, as summarised in Figure 3.15.

A summary of the various beam-combination techniques is shown in Table 3.8.

Incoherent beam combining (or geometrical coupling) is the simplest approach and is likely to be effective over several kilometres, although technical challenges exist to ensure that the various beams remain focussed on a single spot on the target during a dynamic engagement. Coherent beam combining techniques offer some distinct advantages in terms of the radiant intensity along the propagation axis, but stringent requirements are demanded on the output characteristics of each component beam in order to maintain an effective emission.

Fibre laser technology continues to develop at a very rapid rate. The many attributes of the fundamental technology make it very suitable for a broad range of military functions and applications, from sensing to directed-energy weapons.

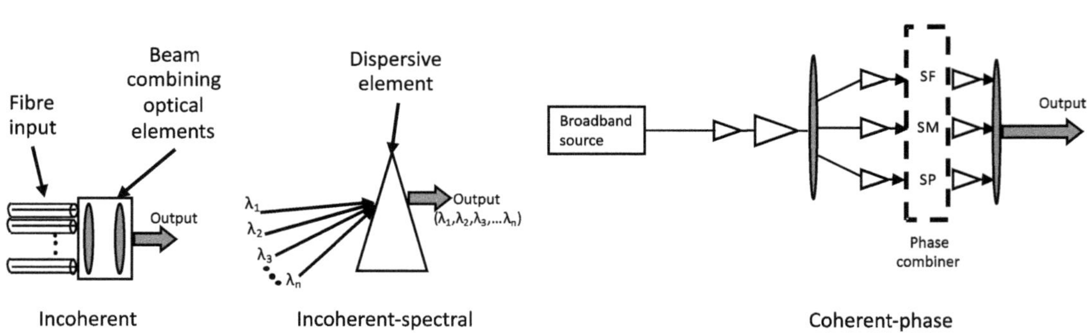

Figure 3.15 Basic beam-combination techniques.

Table 3.8 Summary of the Principal Beam-Combination Techniques

Beam combining technique	Advantages	Disadvantages	Schematic illustration
Incoherent/geometric	•Simplicity; •Higher power from a single laser.	•Lower brightness •Potentially bulky; •Need steering optical elements for each beam.	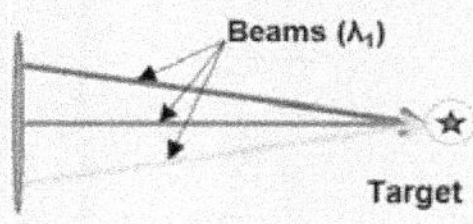
Spectral	•Potential for excellent beam quality; •Power concentrated in a single lobe; •Active phase control not required.	•Limited gain bandwidth; •Power scaling limited by quality and size of grating.	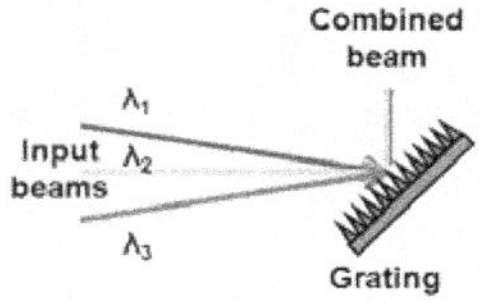
Coherent	•Highly scalable; •Graceful degradation with component failure; •Potential for adaptive optical compensation; •Fine beam steering. •Target in the loop feasible.	•Active phase control required; •Limited power in the bucket unless high fill factor achieved.	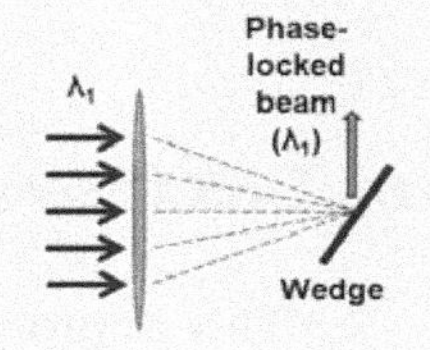
Passive	•No active phase control.	•Scalability; •Adverse impact of thermal environment; •Non-linear effects.	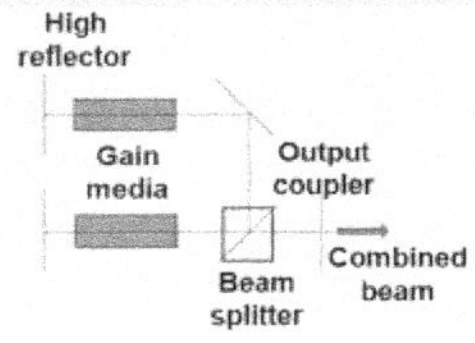

Many high-power demonstrations have been made in the recent past. A continuous-wave power in excess of 1.5 kW, with close to diffraction-limited beam quality has been reported from a master-oscillator power-amplifier configuration. The company IPG offers a high-performance fibre laser with an output of more than 10 kW with a single-mode beam, with the prospect of a 20-kW system in the near future. Both of these demonstrations were at wavelengths close to 1 micron.

Progress with the fibre output power from one micron amplifiers for use in coherent beam combination is shown in Figure 3.16; in this case, the output beam has single frequency in a single mode. Of particular importance is the reduction in size of the system, largely attributable to the enhanced efficiency of the laser device.

The major challenge for fibre lasers really lies in the high-power operation regime and particularly the control of the nonlinear phenomena that limit output in single-frequency devices. The thresholds for the nonlinear phenomena, such as stimulated Brillouin or stimulated Raman scattering are directly proportional to the core area and inversely proportional to the length of the fibre. Hence, there is a move to shorter fibres with a larger core area to achieve high-power operation without adverse nonlinear interactions. Additionally, operation of fibre lasers at a longer wavelength tends to delay the onset of these nonlinear phenomena.

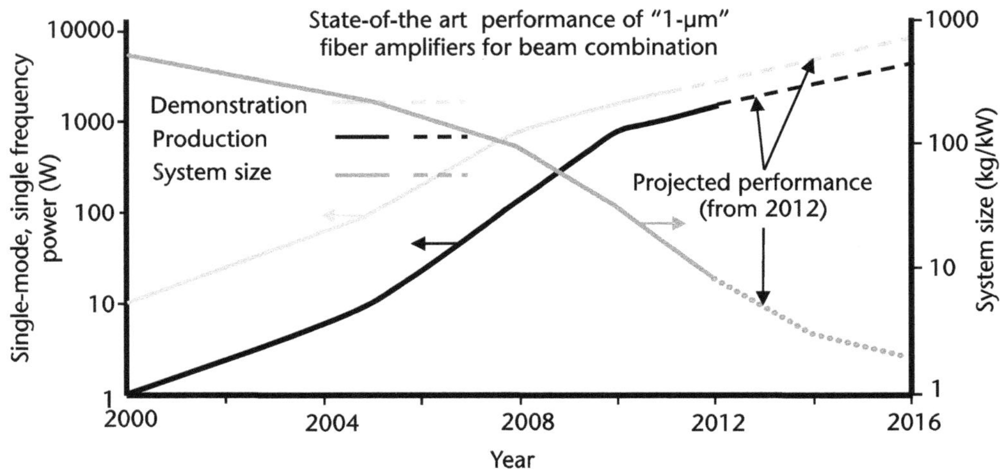

Figure 3.16 Progress with 1-μm fibre lasers for coherent beam combination. (Courtesy of Nufern, USA.)

3.6 Disk Lasers

This approach can be viewed as the exact opposite laser configuration to the long thin fibre devices, discussed above. This technique has been developed in the recent past and offers a relatively simple but effective approach to high-power generation in a compact configuration. The development of these lasers has largely been pioneered by Professor Adolf Giesen.

The fundamental principle involves using a very thin but fat disc that can be face-pumped by laser diodes, thus taking advantage of the longitudinal pumping scheme discussed above. The use of a very thin disk in close contact with a heat sink is intended to avoid thermal gradients in a gain medium that are disruptive to the intensity profile and beam stability of the emitted beam, as well as to the population inversion in any three-level laser system, as discussed in Chapter 2. Another benefit of this approach is that it is scaleable to high powers.

The gain medium is essentially a laser rod that is only about 400 μm long (height) but several centimetres in diameter. One face is in direct contact with a heat sink via a heat spreader, which also acts as a bonding agent, to keep the gain medium flat. This configuration is designed to permit efficient axial cooling of the entire gain medium and ensure a uniform thermal load. Moreover, the back surface of the disk can be coated to provide the high reflector of the resonant cavity, and additionally, can be designed to allow multiple passes of the gain medium by acting as a folding mirror; sometimes this arrangement is known as an active mirror. A schematic of the basic laser arrangement of a simple thin-disk device is shown in Figure 3.17.

Spinning disc devices, as discussed earlier, have been developed to mitigate a range of systematic errors induced by thermal effects to enhance beam quality or time to achieve full brightness.

The design is ideal for multiple, high-intensity, fibre-coupled diode pumping of the front face of the disk. This configuration permits full exploitation of longitudinal pumping and can be scaled further by:

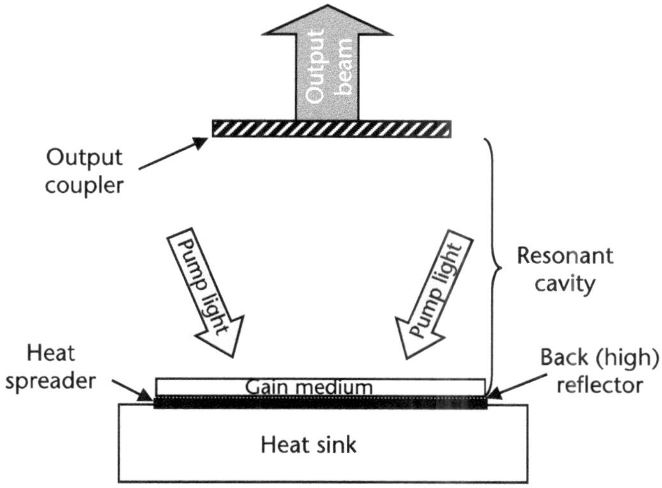

Figure 3.17 Schematic of a thin disk laser (W-cavity configuration).

- Increasing the diameter of the pump beam;
- Using multiple pump beams;
- Using multiple passes of the gain medium.

Another proposed technique for scaling involves an array of thin disks operated in a zig-zag arrangement. Several disk lasers have been coupled together to give a single high-power beam. Various other high-power configurations have been proposed, such as arrays in a cylindrical geometry and edge (transverse) pumped disks acting like a waveguide.

These devices may be operated to generate pulsed or continuous wave beams, hence offering significant design flexibility compared with fibre laser; however, fibre lasers giving pulsed emissions are progressing.

The high-profile thrust for this technology has been for generating high-power 1-micron beams, but this design has also been exploited to give visible-band emission by including an intracavity frequency doubling crystal in the resonator. Other devices have been demonstrated for generating the pump light required for activating an optically pumped semiconductor.

One of the major drawbacks of this class of laser is the beam divergence characteristic, because the Fresnel number is high. Therefore, for many potential military applications, such as laser-directed energy weapons, careful conditioning of the beam will be necessary, such as the reduction of the beam divergence through judicious use of beam-expansion telescopes, or use of intracavity beam shaping techniques.

A summary of the qualitative performance of solid-state disk lasers is given in Table 3.9.

This technology has developed significantly since the mid-1990s and has provided very valuable sources for a range of commercial applications. It is an interesting development of the rod-laser technology, which became largely obsolete when laser-diode pumping techniques for solid-state laser technology were perfected.

Table 3.9 Qualitative Performance of Disk-Laser Devices

Performance Parameter	*Characteristic*
Wavelength range	Visible to mid-wave infrared depending on laser ion and host, but usually visible and near infrared
Tuneability	Not usually, but see vibronic lasers
Output	Pulsed or CW
Power	High power demonstrated (> 7 kW per disc)
Pulse-repetition frequency	High
Beam quality	Poor, but beam size may be changed by use of suitable beam-expansion telescope (note this process reduces the divergence, but does not change M^2)
Time to full brightness	Fast
Efficiency	High, potentially better than 70% (optical to optical)
Size	Compact
Reliability/maintenance	High because of the application of modern technology
Complexity	Low because of the simplicity
Cost	Moderate owing to relatively simple concept, with the potential for modest through-life costs

The high-power, 1-micron source has a potential military application as a source for laser-based directed-energy weapons and may be a serious competitor to fibre lasers. Additionally, the compact lower-power, 2-micron source is a candidate for use with an optically pumped semiconductor disk laser.

The principal challenge for thin disk devices for military applications probably centres on displacing the lead that fibre lasers have established for the high-power regime through efficient beam combination. Improvements to the beam quality of disk lasers have been demonstrated recently for high-power application.

3.7 Vibronic Lasers

These solid-state lasers have some attractions for a number of military applications as the output wavelength is tuneable, in some cases over a considerable range. There are a number of gain media that have the appropriate characteristics suitable for giving a tuneable emission, generally they emit in the near or short-wave infrared.

These solid-state lasers have many similarities with the other devices, described earlier; for example, the active species are impurities in a glass or crystalline host. Many of the hosts are doped with chromium ions, as shown in Table 3.10, indicating qualitative performance. In fact, the alexandrite crystal is chromium-doped chrysoberyl (Cr: $BeAl_2O_4$), but other dopants such as titanium, cobalt and thulium are also used.

The tuneability of these lasers arises from the interaction of the crystal field of the host and the lasing ion to create vibronic energy sublevels, or subbands, of specific electronic energy levels of the laser ion. In this case, during a transition a species changes its electronic state as well as its vibrational state. These broad energy states, may, in theory, be either the upper energy level of the transition or the lower; in some cases, both energy levels are vibronic.

Table 3.10 Qualitative Performance of Vibronic Lasers

Performance Parameter	*Characteristic*
Wavelength range	Ultraviolet to mid-wave infrared depending on laser ion and host, but usually visible and near infrared, with examples of Ti: sapphire ~ 660–1,190 nm; alexandrite, ~700–825 nm; Cr: LiSAF, ~760–920 nm; Cr: LiCAF, ~ 720–840 nm; Cr: Emerald ~720–840 nm; Cr: Forsterite, ~1.16–1.35 μm
Tuneability	Yes
Output	Pulsed or CW for some gain media
Power	Low- to moderate-power demonstrated
Pulse-repetition frequency	Moderate, mode-locked Ti: sapphire is the choice for very high pulse repetition frequency devices
Beam quality	Moderate to good
Time to full brightness	Fast
Efficiency	Low, of the order of 1% to 5% in general (optical to optical)
Size	Bulky, but large for laser-pumped
Reliability/maintenance	High because of the application of modern technology
Complexity	Low because of the simplicity
Cost	Moderate owing to relatively simple concept, with the potential for modest through-life costs

The pumping mechanism is similar to the constraints consistent with other solid-state gain media. There is the option of flashlamp pumping or excitation with another laser and, although it has not been fully developed, the option for pumping with laser diodes has been demonstrated. The transition that creates the tuneable emission occurs between these broadened electronic energy levels, as shown in Figure 3.18. In this schematic some of the energy-level structure of alexandrite is shown for both the single fixed line and the tuneable lines.

These lasers may emit several laser lines simultaneously in an untuned cavity. If a tuning element is available, then this component will select one line in preference

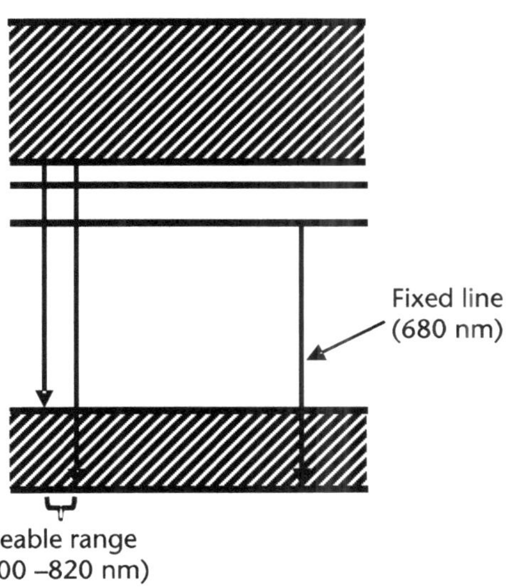

Figure 3.18 Schematic of vibrational energy levels.

to the others and that wavelength will predominate through receiving all the available gain within the cavity.

The problem with this type of solid-state laser is the trade-off between the gain bandwidth, the ability to create the population inversion and achieving a high probability of stimulated emission. This operational design issue arises because the gain bandwidth is inversely proportional to the product of the excited-state lifetime and the cross-section for the stimulated-emission event in a solid-state laser. As these lasers have broad gain bandwidths, the product of the lifetime and the cross-section is much lower compared with other solid-state lasers. Hence, special arrangements are usually needed to create effective laser action, owing to the lower excited-state lifetime or reduced gain. For example, the titanium-sapphire gain medium has a very short excited-state lifetime, of the order of 3 μs, so special flashlamps are required or pumping from another very-short pulse laser, but the gain is high; hence, in this case, it may also be pumped by a continuous wave laser. Modern approached have seen these gain media pumped with diode lasers.

The qualitative performance of a selection of vibronic solid-state lasers is given in Table 3.10.

Vibronic laser were the subject of intense research and development activity during the 1980s. Interest continues in a few of the crystals, such as alexandrite and titanium sapphire. In fact, titanium sapphire is a common source for super continuum emission, discussed later.

The tuneable output from these devices makes them suitable for some sensing, and potentially, some low-power directed-energy applications. The demonstration of compact diode-pumped devices would be an advantage.

3.8 Liquid-Phase Lasers

The only common liquid-phase laser is the dye laser, which has been demonstrated in many forms for a vast array of applications. Some of the attractions of this type of laser are its inherent simplicity as well as its versatility.

The structure and composition of a laser dye have an important influence on the spectral emission from such a laser. More than 50 laser dyes are available for use in the gain medium of a dye laser, enabling energy to be emitted from the ultraviolet to the near infrared, simply by using a different dye; this is wavelength diversity. Additionally, many dyes can be tuned over many tens of nanometres, providing extra flexibility. The spectral coverage of the dyes available from one supplier is shown in Figure 3.19.

Liquid-phase lasers may be designed to give a continuous-wave output beam or a pulsed emission, depending on the form of the optical excitation of the organic dye solution. The optical excitation can be coherent from another laser or incoherent, using the energy from a flashlamp or arc lamp. An alternative approach involves the use of light from a surface discharge, which offers the prospect of highly efficient operation. Laser diodes have also been used for this function for some specialised applications.

The optical setup of a dye laser can be extremely simple, which leads to its flexibility and general versatility. However, this does have a cost in terms of mechanical complexity if sustained output is required, along with some of its other attributes,

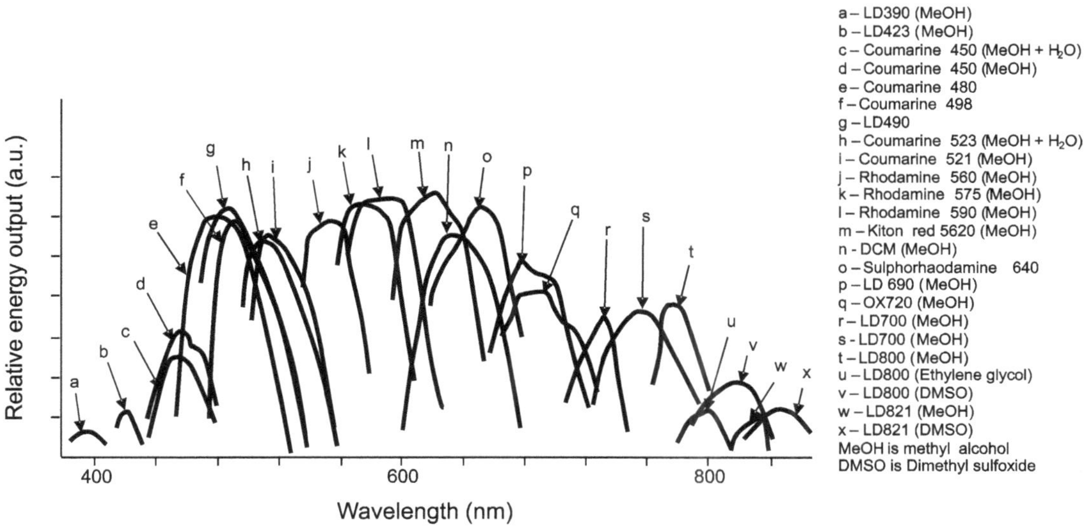

Figure 3.19 Laser-dye emission characteristics. (Courtesy of Exciton, USA.)

such as narrow line width, high repetition rate and high power. For this type of operation, the liquid laser needs a mechanical pump to circulate the dye solution from a reservoir and additional optical elements are required in the cavity to control the output. A simple dye laser cross-sectional configuration is shown in Figure 3.20.

The principal attributes of dye-laser technology may be summarised as follows:

- Simplicity, which leads to low cost and potentially high reliability;
- Versatility, which makes the liquid-phase dye laser technology attractive for many applications;
- Very broad spectral emission capability, including simultaneous multicolour emission;
- Scalability, in which the output power of a dye laser is related to the size of the laser gain medium, enabling the power generated to be scaled using the same optical configuration;

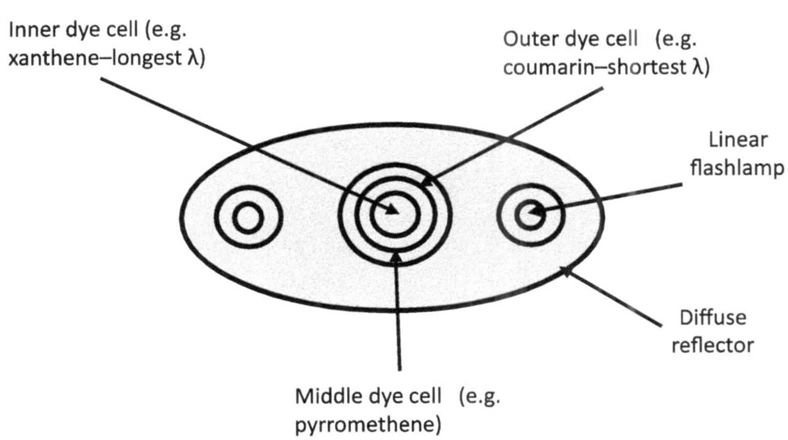

Figure 3.20 Liquid-phase dye laser (cross-section of double ellipse).

- Ease of removal of heat from the gain medium, so this laser technology can operate at a high repetition rate with high-energy pulses, which permits a high average power output.

However, care is required when designing the circulation system so as to achieve a compact arrangement that allows an easy and clean procedure for flowing the dye solution through the dye cell; otherwise, it can become a messy configuration. The flow may be longitudinal or transverse, the latter approach being more relevant for high repetition-rate operation.

For effective laser operation, a dye molecule requires a number of characteristics such as strong fluorescence, good photo-stability and low excited-state loss. Laser dyes tend to have higher purity than commercial-grade dyes used for other purposes; this leads to better lasing efficiency. Many dyes are soluble in liquid and it is often possible to dissolve a laser dye in a choice of solvents. In fact, the solvent can have a very strong influence on the emission, governing the output power and influencing the rate of degradation of the dye solution.

During laser action, optical energy from an external source is absorbed by the organic molecules, raising these molecules to an excited state. As a rough first approximation, the wavelength absorbed by the molecules is proportional to the length of the backbone of carbon atoms that form the chain structure; however, in reality, the parameters controlling the absorption band are somewhat more complex. Owing to the complex nature of the structures of these organic compounds, the energy-level structure is very complicated. However, a simplified diagram can be drawn to illustrate the principal interactions, as shown in Figure 3.21.

A laser-dye molecule essentially forms a four-level system in the gain medium, where the upper and lower energy levels have an extended span, which leads to

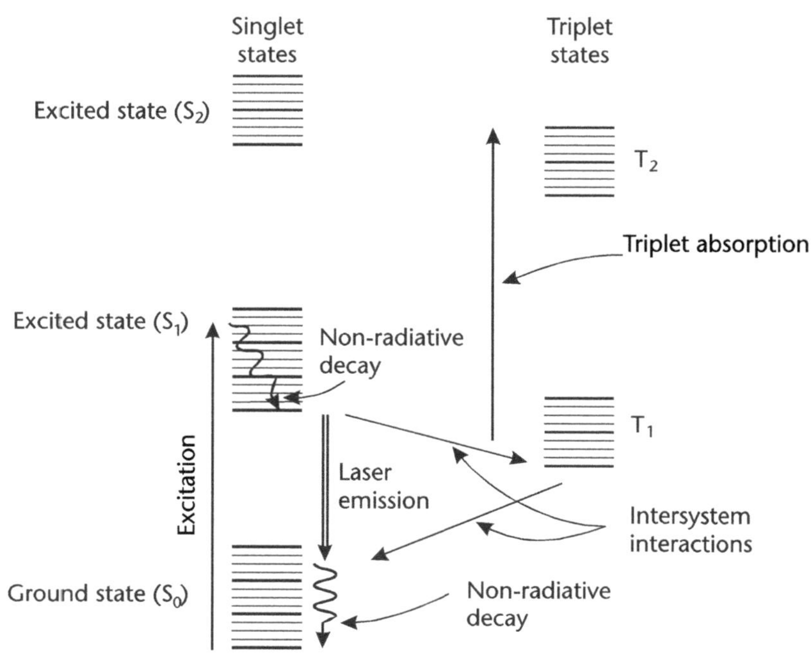

Figure 3.21 Simplified energy level structure diagram of a generic laser dye.

the broad absorption and tuneability, very similar to the energy-level structure of vibronic lasers, discussed earlier. The electronic states form the basic structure and transitions between these states lead to emission in, or close to, the visible part of the spectrum. The energy difference between the states, which leads to the visible band emission, range from ~3.1 eV to 1.65 eV.

Figure 3.21 shows these electronic levels have other substates superimposed on them, resulting from the vibrational and rotational motion within the molecule. Energy differences caused by rotational motion are small and when the other line-broadening mechanisms are included, a continuum of states emerges, both for absorption and emission. Consequently, a laser dye can absorb energy, as well as emit light over a broad range of wavelengths, particularly in comparison with other types of laser. Broad emission bands of typically 10 to 20 nm or more, are possible.

The lifetime of these states is very short, of the order of a few nanoseconds. Consequently, the temporal-emission profile of the output is closely related to that of the excitation source, particularly for short-duration excitation pulses, usually of the order of microseconds. Cavity-dumping techniques may be applied to some laser sources to concentrate the emission into a series of short-duration, high-intensity pulses of a few nanoseconds duration, as discussed in Chapter 2. Additionally, the broad bandwidth of the lasing emission does lend itself to the generation of ultrashort pulses (in the picosecond regime) via the mode-locking technique.

A laser-dye solution offers a quick and easy route to making a gain medium with high optical quality. This solution is contained in a transparent cell to form the gain medium in the laser cavity, which may be pumped transversely for high repetition-rate operation or longitudinally. The great flexibility of this class of laser is provided by the extensive range of organic compounds suitable for use in the gain medium, as illustrated in Figure 3.19. Many techniques have been demonstrated to give simultaneous multicolour operation from a single device.

As noted above, many solvents are available for creating a laser-dye solution: alcohols and ethylene glycol are commonly used. Although some laser dyes are soluble in water, a surfactant may be needed to prevent dimerisation. The solvent used to make the solution can have a significant influence on the optical properties of the laser gain medium, particularly the longevity or usefulness of the dye solution, often termed the service life of the dye solution.

The light used to excite the laser dyes can cause the dye molecules to decompose, particularly in the case of broadband pumping with a flashlamp. A number of life-enhancing techniques have been demonstrated that can ameliorate the effects of unwanted and potentially destructive shorter wavelength energy. Furthermore, the use of an additional dye solution between the pump source and the gain medium, which can absorb the harmful shorter wavelength photons as well as changing their wavelengths via fluorescence conversion, provides a better match to the absorption band of the active dye and thus enhances the efficiency of the laser.

Photo-degradation-causing mechanisms are complex and research has shown that formation of singlet oxygen in the solution is particularly detrimental; unfortunately, this is one of the critical active ingredients of COIL. Consequently, elimination of dissolved oxygen from a dye laser solution enhances the longevity of the solution, but an alternative triplet-quenching agent will be needed, such as cyclo octo tetraene (COT).

Table 3.11 Qualitative Performance of Dye Lasers

Performance Parameter	*Characteristic*
Wavelength range	Visible to near-infrared depending on laser dye
Tuneability	Yes, usually over many tens of nanometres; these devices also offer wavelength diversity by changing the laser gain medium to one containing a different dye
Output	Pulsed or CW, determined by excitation source
Power	High average power feasible (greater than 5 kW), more than 100 J/pulse demonstrated
Pulse-repetition frequency	Depends on pump source
Beam quality	Good even with sustained operation, but may have significant divergence, which can be controlled with a beam-expansion telescope
Time to full brightness	Fast
Efficiency	Generally low, but may be up to 5% (optical to optical) with careful design
Size	Relatively compact with flashlamp pumping, more bulky with laser-pumped configurations
Reliability/maintenance	Moderate because of the use of flashlamps
Complexity	Low because of the simplicity
Cost	Very cheap

Another major drawback of any dye laser technology is the presence of an additional set of energy states in the dye molecule known as triplet states, which have long lifetimes. Consequently, careful design and the use of an adequate flow rate is required to avoid loss of laser species to these meta-stable states, resulting in laser action being quenched if sustained laser operation is required. Furthermore, the triplet-state population may be controlled through the use of triplet quenchers, such as cyclo octo tetraene.

A summary of quantitative performance of dye-laser technology is given in Table 3.11.

Liquid-phase dye lasers are inherently simple and reliable devices that offer a route to efficient direct generation in the visible and near infrared. These devices are inherently tuneable and provide the option of extended wavelength coverage through diverse techniques, so a single cavity can emit over a very broad range of wavelengths. These attributes, along with the scalability to high average power, make these devices attractive for many military applications.

The challenges for dye-laser technology are mainly centred on achieving improvements in operating efficiency. Additionally, there are clear system requirements in military applications for making the whole configuration more compact and reducing the potential logistics burden. Part of the solution to this requirement may be solid-state dye laser technology.

3.9 Gas-Phase Lasers

Gas-laser technology embraces many different types of device configurations and it has been at the forefront of high-power sources for decades. For example, the first 1000-kW device was demonstrated by 1970; the gain medium used in this laser was carbon dioxide. The same gain medium was used for the demonstration of the first megawatt-class device. There have been many different types of gas laser, probably owing to subtle changes being made to the operating conditions of a laser head, such as the operating pressure in the discharge tube, enabled a range of different aspects of laser performance to be demonstrated.

The modular approach and the simple device design, with a cell to contain the gas, enabled an extensive range of potential gas mixtures to be investigated and the most promising ones optimised. Clearly, it is a quicker and more streamlined technique than is possible with a solid-state gain medium. A simple gas laser configuration is shown in Figure 3.22; the upper image is of a longitudinal excitation with a transverse scheme below.

The use of a gaseous medium has other significant advantages. In particular, it permits the use of a wide range of excitation techniques as the free movement of the gas molecules allows many energy-transfer mechanisms to be invoked. It is common to use electrons, ions and photons, as well as chemical reactions or gas expansion techniques. Hence, gas lasers are frequency categorised by the type of excitation technique.

The common excitation techniques are:

- *Discharge:* an electrical current is passed through the gas mixture either along the length or across it and the electrons create the population inversion, which may occur directly through collisions with the lasing species or through intermediary species. The discharge may be continuous or pulsed and is initiated with a voltage spike. Transverse-excitation techniques are usually applied to high repetition-rate systems.
- *Particle beams:* these are either electron or ion beams from an accelerator, and in the case of electron beams, it is a variation of the pulse-discharge

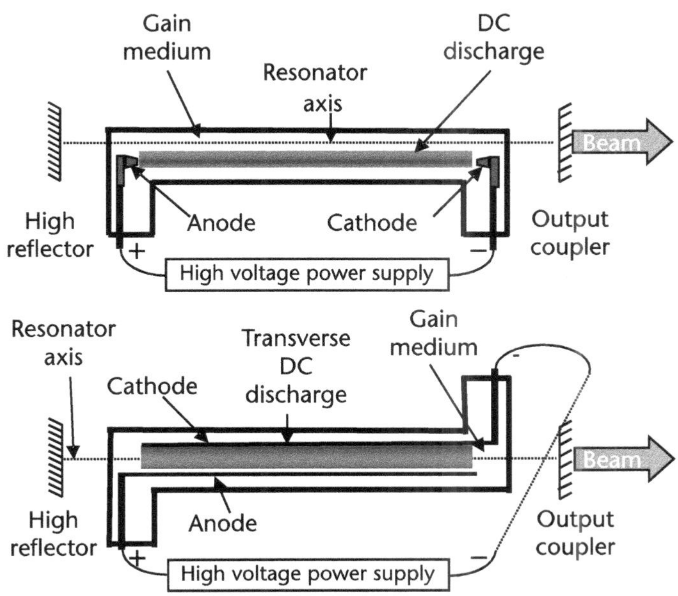

Figure 3.22 Simple gas-laser configurations.

technique. This form of excitation tends to be bulky and is not normally compatible with tactical military applications.

- *Radio-frequency radiation:* these electromagnetic waves, including microwaves, may be absorbed by the laser gas as they are directed through the gain medium, creating a population inversion directly or creating an intermediate level in the excitation process. These waves may be used for pulsed or continuous wave excitation.
- *Chemical reaction:* this makes use of the energy released in an exothermic reaction. A common example is the reaction between two species to generate a third that is in an excited vibrational state. It is crucial that this is a very rapid reaction in order to create a population inversion. Gas flow through the gain medium is also important, to avoid the laser action being quenched by spent gas. Fresh fuel must be supplied to the combustion chamber in order to ensure sustained operation. This type of gas laser usually requires an initiator, such as a spark, to start the combustion process. Again, pulsed and continuous-wave operation is feasible. This type of laser has been used in many of the very high-power laser demonstrations.
- *Optical:* this method has been discussed in detail above for solid-state and liquid lasers and similar considerations apply to the excitation of a gaseous gain medium. This technique is not very common in gas-laser devices, owing to its relative inefficiency compared with other approaches that can be more easily applied. The modern exception is with diode-pumped alkali lasers.
- *Gas dynamics:* this technique has some similarity with jet-engine operations and associated dynamics as a hot high-pressure gas is expanded rapidly into a low-pressure environment or a vacuum. The adiabatic expansion rapidly cools the gas but leaves many of its constituent molecules in an excited state, creating a population inversion. The gas flow must be rapid as the excited molecules will return quickly to the thermal equilibrium of the new ambient conditions, destroying the population inversion. This was the technique used to produce the first very high-power laser, with an output in excess of 100 kW within 10 years of the first demonstration of a laser.
- *Other techniques:* a number of other methods including use of the fission products of a nuclear decay have been investigated, but these are not considered further in the context of practical military devices.

The advantages of gas-laser technology are:

- Mature high-power technology;
- Potential for high efficiency;
- High beam quality and low thermo-optical dependencies;
- Low complexity, leading to high reliability.

The downside is that these devices are bulky and the emission-wavelength regime occurs in a less than ideal part of the spectrum for many common military applications. An import system consequence is that a large beam director is required to ensure a small close-to-diffraction-limited spot at range.

The major challenge for this technology is in reducing the size of the devices. The long-wavelength emissions have an impact on the size of the optical system required to direct the beam, as noted above, but that fact is inherent in the long wavelength infrared emission, whatever the device technology.

Recent research studies have aspired to harness the good properties of both solid-state and gas-laser technologies by investigating diode-pumping of alkali vapour, such as those with caesium vapour as the gain medium. Alkali vapour lasers are an interesting prospect for high-power operation, as they combine the advantages of a gas laser with those of a solid-state device. In some cases, the alkali vapour is in a sealed cell and is diode pumped; in other configurations the gas is flowing. These devices are discussed in more detail later.

3.9.1 Carbon-Dioxide Lasers

The ubiquitous carbon dioxide has been used in many ways for a range of military applications as a result of its versatility. Its fundamental emission is between 9.2 and 11 μm, with a dominant line around 10.6 μm. Its versatility stems from the different classes of carbon-dioxide lasers that exist. These result from subtle and not-so-subtle changes in the internal operation of the various devices and configurations that give rise to different emission characteristics.

The gain medium in a carbon-dioxide laser is a mixture of that gas along with nitrogen and helium to provide the optimum conditions for the stimulated emission from the excited carbon dioxide molecule. The nitrogen helps in the excitation and the helium helps in the decay process and acts as a buffer gas in the heat transfer. Various gas compositions are used but there is usually a similar ratio of carbon dioxide and nitrogen but with helium as the dominant species.

Common examples of carbon dioxide lasers include:

- *Waveguide devices:* a sealed tube with a small internal diameter is usually excited with a radio-frequency wave transverse to the optical axis or a direct-current discharge. Designs have been developed with a folded cavity to boost output power by increasing the gain length, but retaining a relatively small package, sometimes referred to as z-fold devices. These devices may be operated in a pulsed or continuous-wave mode.
- *Pulsed transversely-excited sources:* this class of laser uses a pulsed electrical-discharge across a relatively high-pressure gas (usually at atmospheric pressure), hence the term transversely-excited atmospheric laser (TEA laser). These lasers give a pulsed output with relatively short-duration profiles (a few nanoseconds) and may be operated as a sealed system or with a flowing gas, either longitudinal or transverse.
- *High-power flowing-gas axial configurations:* electrical discharges applied along the length of the gain medium are the traditional excitation mechanism. The use of a flowing gas, which can be recycled, assists with the heat removal. Use of a rapidly flowing gas composition along the longitudinal axis helps improve the efficiency by removal of hot gas that reduces the gain in the stimulated emission process. Again, the gas may be recycled but careful conditioning is required to avoid any reduction in beam quality.

- *High-power transverse-flow devices:* this configuration has the gases flowing across the cavity, with the electrical discharge mutually perpendicular to both the discharge and the resonator's axis. This type of laser usually has its gases recirculated and the rapid flow assists with achieving high efficiency, but the short cavity leads to reduced beam quality because of the small Fresnel number. Transverse flow is also used in the gas-dynamic laser, with the resonator formed across the nozzle used to expand the hot, high-pressure gases.
- *Sealed-tube low-power devices:* these are excited by a longitudinal electrical discharge, but measures have to be taken to avoid the breakdown of the carbon dioxide into carbon monoxide and free oxygen. Alternatively, the device may be excited by transverse radio-frequency waves, as used in the various waveguide designs. These devices are normally operated in a pulsed regime but continuous-wave devices are available. As with waveguide devices, the removal of waste heat is a potentially limiting factor for high-power operation.

A summary of the qualitative performance characteristics of a carbon dioxide lasers is given in Table 3.12.

This type of laser in one configuration or another has featured in many military applications from range-finding, designation, sensing and countermeasures, to high-power directed-energy demonstrations. The major drawback is the large size of even the more modest-power devices, but the technology has many fundamental strengths.

3.9.2 Chemical Lasers

The prospect of very high efficiency operation of a high-power laser was one of the motives for investigating the use of a chemical reaction to create the conditions for stimulated emission. Its feasibility is based on the fact that various chemical reac-

Table 3.12 Qualitative Performance of CO_2 Devices

Performance Parameter	*Characteristic*
Wavelength range	Far infrared in the 9- to 11-μm band
Tuneability	Yes, usually over the above waveband
Output	Pulsed or CW, determined by excitation source
Power	Very high-power operation feasible, or high average power (of the order of 1 MW)
Pulse-repetition frequency	Depends on pump source, but greater than 100 kHz
Beam quality	Good even with sustained high-power operation
Time to full brightness	Very fast to instantaneous
Efficiency	Moderate, can be up to 20% in a multiline emission
Size	Relatively bulky with most configurations
Reliability/maintenance	Relatively good
Complexity	Low because of the relative simplicity
Cost	Moderate

tants can store energy very efficiently and, more importantly, release it very rapidly, a fundamental requirement for the generation of a population inversion.

Historically, the most common chemical lasers initially used carbon dioxide and then diatomic hydrogen halides such as hydrogen fluoride and deuterium fluoride; however, there is now great interest in the oxygen-iodine device. As will be evident from Figures 3.23–3.25, the fundamental design of such lasers had its origins in rocket and jet-engine design.

The initial effort to generate very high power beams was based on a gas-dynamic laser, which was essentially a rocket engine. This type of gas laser was used to provide a tool for investigation of potential high-power effects, as noted above. A gas-dynamic laser uses combustion to heat a mixture of nitrogen, carbon monoxide, nitrous oxide and helium, to a high temperature and pressure.

The chemical reaction that is critical to the operation of this type of laser starts in a combustion chamber where a stable, high-temperature flame front is established by burning methane with pure oxygen. Once this condition is established, there follows an injection of the main reactants (nitrous oxide and carbon monoxide) and subsequently a further addition of methane to form water vapour. Nitrogen is finally injected so that an appropriate gaseous lasing medium can be formed. When this gas mixture passes through a convergent and then a divergent nozzle, the hot gaseous nitrogen in the mixture achieves a population inversion, which rapidly transfers its energy to the carbon-dioxide molecules that then undergoes laser action. These gases are then vented from the laser system.

This form of laser was found to be scalable, which was exploited to provide a megawatt-class source. A schematic diagram of part of this type of device is shown in Figure 3.23 and depicts the process after the formation of the carbon dioxide from the main reactants of carbon monoxide and nitrous oxide (i.e., the region where the population inversion occurs). This was the laser source used in the Airborne Laser Laboratory that established the feasibility of using a laser beam to destroy other weapons, and hence, provided the foundation for practical high-power laser weapons.

Another common form of chemical laser uses hydrogen halide molecules, which emit from transitions between the vibrational energy levels that have an energy difference for photon generation in the mid-wave infrared. The fundamental wavelengths are emitted when the excited state drops by one vibrational level.

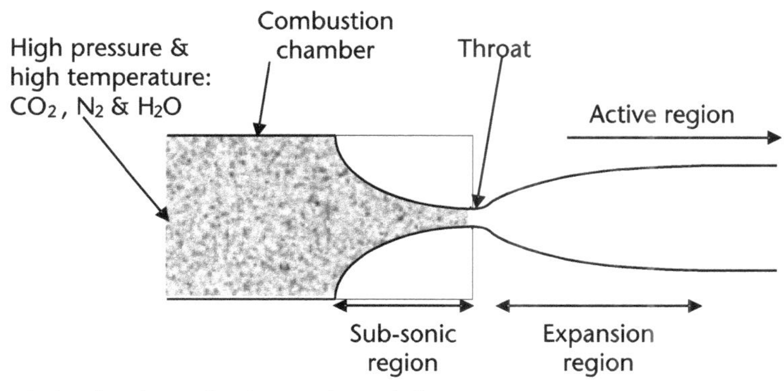

Figure 3.23 A simple schematic of a gas-dynamic laser.

These vibrational states also have rotational sublevels and so, in a transition, both vibrational and rotational transitions can occur simultaneously. This leads to a series of lines for each vibrational-state transition. There is, in general, a concentration of power at the shorter wavelength end of the emission band. Moreover, it is possible for a transition to occur over two vibrational states that give half of the fundamental wavelength for that molecule; this type of transition is known as an overtone.

The transition in an iodine device happens between electronic states, usually corresponding to a wavelength emission of 1.315 μm. This is the transition exploited in the chemical oxygen iodine laser.

The generic operation of a classical chemical laser concerns the injection of the reactant gases into a combustion chamber through a set of nozzles that encourage the fuel and oxidiser to mix and react to form excited species. The excited gaseous mixture then passes through the resonant cavity orthogonally to the resonator's axis. The used gas is removed from the resonator and collected for recycling, which permits the laser action to continue and results in the generation of a laser beam. Particular care is required in the design of the nozzles to ensure efficient mixing of the gas and to maintain the design gas pressure and flow rate. A schematic of the laser arrangement based on the use of flowing gases to form deuterium fluoride is shown in Figure 3.24.

Chemical-laser designs have also included sealed sources rather that the flowing gas approach shown schematically earlier. In these cases, a sealed source is used and is usually operated at relatively low power. An electrical discharge is used to crack sulphur hexafluoride (or nitrogen tri-fluoride) to liberate fluorine, which is then mixed with the free hydrogen. This reaction takes place in the mixer/reactor shown in Figure 3.24. It is worth noting that the reactor requires careful design to achieve optimum operating conditions and is only shown as an idealised subsystem in Figure 3.24. In this case, the third port of the reactor, which is not labelled in Figure 3.24, is likely to be used to introduce a third gas, such as oxygen.

A molecular sieve (scrubber) is positioned after the resonant cavity to absorb the spent laser species but permits any unused reactants to be recycled (not shown in the schematic above). All of these designs use additional gases such as heli-

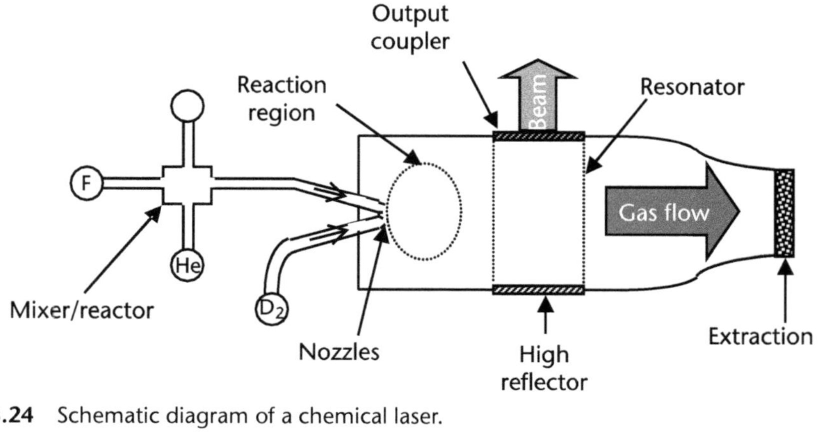

Figure 3.24 Schematic diagram of a chemical laser.

um to act as a diluent to reduce the rate of the combustion, and hence, prevent self-destruction.

These lasers may be operated as either pulsed or continuous-wave devices. Additionally, frequency-selective elements may be used to allow single-frequency operation.

A summary of the qualitative performance characteristics of hydrogen-halide lasers is given in Table 3.13.

The potentially high performance of this class of chemical laser was considered attractive for a range of sensing and countermeasure applications, particularly in the sealed configuration. However, the rapid development of solid-state devices has displaced this type of laser for many of those applications. However, there is still some interest in the very high power devices for laser-based directed-energy applications. The major challenges for chemical lasers arise from the collection and disposal of toxic materials and also the sheer size of the devices. Even the lower-power systems are substantially larger than the equivalent solid-state devices.

3.9.3 Chemical Oxygen Iodine Laser

This class of laser is at the heart of the airborne laser programme of the United States Air Force. A Boeing 747 has been modified to allow the integration of this laser and all of its subsystems to create a flying laser directed-energy system for strategic ballistic-missile defence.

The basis of this chemical laser is the use of singlet oxygen delta $\{O_2(^1\Delta)\}$ to create the excited state in the iodine gas. The excited molecular oxygen $\{O_2(^1\Delta)\}$ is very energetic and provides a very efficient and effective excitation mechanism to create the population inversion in atomic iodine. The singlet oxygen delta is created in a singlet-oxygen generator (a chemical reactor) and involves wet chemistry. Fundamentally, a mixture of hydrogen peroxide and potassium hydroxide (known as

Table 3.13 Qualitative Performance of Chemical-Laser Devices

Performance Parameter	*Characteristic*
Wavelength range	Mid-wave infrared (fundamental): HF ~2.6 to 2.9 μm and DF ~3.6 to 4.2 μm; near and short-wave infrared for the overtones
Tuneability	Yes, usually over the above waveband
Output	Pulsed or CW, determined by the mode of operation
Power	Very high-power operation feasible, or high average power, potentially an MW-class source
Pulse-repetition frequency	Depends on excitation, but greater than 100 Hz
Beam quality	Moderate
Time to full brightness	Very fast to instantaneous
Efficiency	Moderate, can be up to 20% as a chemical reaction, but ~2% in the case of an electrically excited device
Size	Relatively bulky with most configurations; it is a gas laser
Reliability/maintenance	Relatively good
Complexity	Low because of the relative simplicity
Cost	Moderate

basic hydrogen peroxide) is mixed with chlorine in the reactor to create the highly reactive gaseous molecular oxygen.

The singlet oxygen delta and the iodine gases are introduced into the reaction chamber of the laser through an array of sophisticated nozzles to dissociate the iodine and create the population inversion as it passes through the expansion nozzle into the resonant cavity. The gas flow is orthogonal to the resonator axis, and is thus a transverse-flow device, as described above for the gas dynamic laser.

The basic design of one of the modules that makes up part of the laser system is shown in Figure 3.25.

A summary of the potential qualitative performance characteristics of chemical oxygen iodine lasers is given in Table 3.14.

This device may be considered the ultimate very high-power gas laser. A single module is capable of emitting in the region of 50 kW; therefore, a single module or a few ganged together is suitable for some tactical laser directed-energy applications. More efficient modules may be feasible, which can scale to far higher powers.

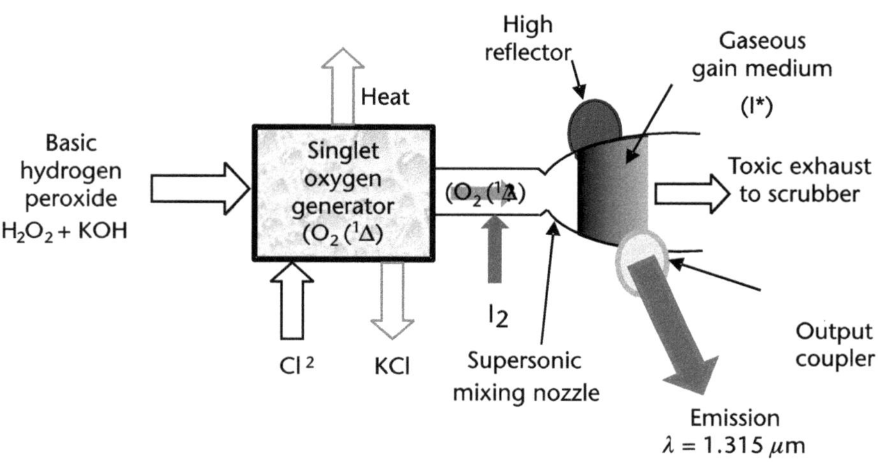

Figure 3.25 A schematic of a chemical oxygen iodine laser.

Table 3.14 Qualitative Performance of Coil Devices

Performance Parameter	*Characteristic*
Wavelength range	Near infrared (1.315 μm)
Tuneability	No
Output	Pulsed or CW, determined mode of operation
Power	Very high-power, modules in excess of 50 kW
Pulse-repetition frequency	Depends on excitation
Beam quality	Moderate
Time to full brightness	Very fast
Efficiency	Moderate
Size	Large
Reliability/maintenance	To be determined
Complexity	High
Cost	Enormous (ABL)

There has been some interest in the development of an electrically excited oxygen-iodine laser. In this case, electrical discharge techniques are used to create the singlet oxygen delta. The aim is to develop a more compact device and avoid the wet chemistry used in the usual singlet-oxygen generator.

This technology faces the same challenges as all chemical lasers, discussed earlier. The successful development of the electrical device would ease many of the logistical issues. The use of a different active lasing specie would enhance the propagation in the maritime environment.

3.9.4 Explosively Pumped Iodine

Explosive materials contain an enormous amount of stored energy that can be liberated relatively easily by simply activating a detonator. This approach has been used to excite the gain medium in an amplifier for an iodine laser.

The fundamental approach is to surround the gain medium, contained in a sealed cell, with a suitable quantity of explosive material, as is shown in Figure 3.26. A seed beam is injected from a separate master oscillator into the amplifier. When the explosive is detonated, the iodine in the gain cell dissociates and is raised into an excited state. Theoretically, it is possible to make several passes of the gain medium before the amplifier is destroyed as the speed of light is so much faster than the pressure waves of the explosion.

Clearly, this device is only capable of single-shot operation but very energetic pulses of the order of a megajoule have been demonstrated.

3.9.5 Alkali Lasers

The optical pumping of alkali vapour in a gain medium has been investigated since the early days of lasers, but a practical device was not feasible until suitable diode sources became available. The principal challenge has been to achieve efficient pumping despite the narrow absorption feature in the alkali vapours. However, alkali gases have the advantage of a small quantum defect: there are two energy levels close to the ground state that can be pumped to create a population inversion. The emission is typically in the near infrared.

These devices attempt to combine the advantages of gas and solid-state lasers, hence, the advantages of such technology can be summarised to be:

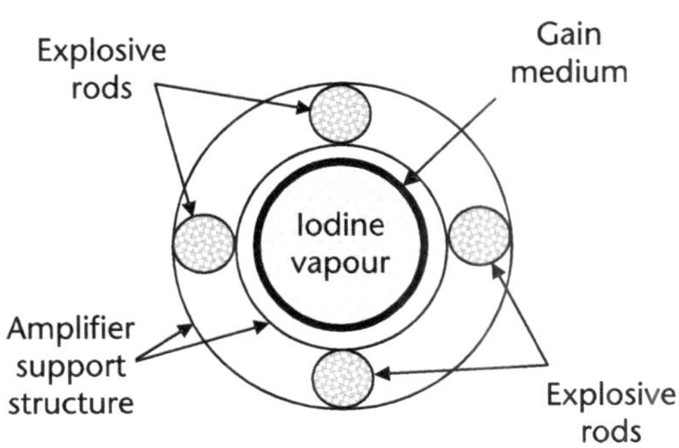

Figure 3.26 A cross-sectional view of an explosively-pumped iodine amplifier.

- Compact configurations;
- Long shelf life;
- High beam quality combined with high-power emission;
- Potential for a sealed device (for lower-power devices);
- Ruggedness;
- High efficiency;
- Potential to operate over extreme temperature ranges;
- Flexibility of output characteristics;
- Potential for multifunctional capability;
- Low complexity, leading to high reliability.

The current state of the art is that potassium, rubidium and caesium vapours have been lased and powers in excess of 1 kW have been demonstrated from a flowing-gas configuration. These lasers have the potential to scale beyond 100 kW and possibly to the megawatt class.

3.10 Semiconductor Lasers

This class of laser has been developing rapidly since the 1980s, about 20 years after the first demonstration of an injection laser diode. It is clearly a solid-state device, but it is conventionally considered as a separate branch of laser-source technology, owing to it having somewhat different operating principles for creating the population inversion and to its vast range of emission wavelengths.

Semiconductor laser technology is a substantial family of devices with many different structures and a plethora of compositions. However, a common characteristic is that these devices are very small, yet very efficient. This section will concentrate on those laser devices that have a direct application to the theme of this book, rather than the diode lasers that are, for example, used to pump a gain medium.

In a conventional laser diode used in commercial applications, electricity is used to create the population inversion and the light is formed by a recombination of holes and electrons, which works well for emission into the short-wave infrared. However, as the bandgap for this type of emission is reduced to encourage emission at longer wavelengths, nonradiative recombination processes become significant, such as Auger recombination. In this process, the energy from recombining electrons and holes is transferred to other carriers and becomes the limiting mechanism owing to the large number of thermally generated carriers present.

Several different approaches are used to achieve efficient emission in the midwave infrared range and this category of semiconductor lasers can be classed into two broad groups: those that are powered by electricity and those that rely on optical excitation for their laser action. The former group is referred to as laser diodes (or more simply as diodes) as a result of their basic operation, and the latter as optically pump semiconductor lasers (OPSLs).

The fundamental structure of a laser diode involves the formation of an interface between two subtly different semiconductor material compositions, known as a pn junction. The term pn junction (or, more commonly, junction) arises from the fact that one side of the junction is doped with donor elements to give an excess of electrons (n type semiconductor for negative carriers) and the other with excess holes (positive carriers – p type).

The holes and the electrons are the current carriers in a semiconductor. The electrons and holes are free to move within the structure, although the hole mobility, a measure of how quickly they move under an applied electric field, is much smaller than the electron mobility. Provided that there is a direct bandgap, it is the recombination of the holes and electrons that lead to the creation of photons. When a diode is operated with a small forward bias and no optical feedback, the result is usually spontaneous emission, and the device acts as a light-emitting diode. High forward-bias drive currents are required to create a population inversion, which results in stimulated emission.

Forward biasing of the diode, which is a two-terminal device, involves applying a positive voltage to the p side of the junction and the corresponding negative drive to the n side. The purpose of forward biasing is to change the internal potential barriers, resulting in carrier injection from one side of the junction to the other, as shown in Figure 3.27.

Within the junction region, the excess electrons decay to the lower energy created by the holes and thus recombination occurs. To achieve sustained laser action, a resonator is needed to provide the optical feedback. This can be formed directly on the semiconductor chip by creating facets on either end of the junction and applying suitable coatings. In some cases, where a high gain is available, the Fresnel reflection may be adequate to form the output coupler.

As with all lasers, the wavelength of the emission is determined by the energy difference between the upper and lower energy levels involved in the decay process. In the case of semiconductor lasers, the energy difference is largely determined by the band gap between the conduction band with the free electrons, and the valence band with bound electrons. This bandgap is dependent on the lattice structure created by the chemical composition of the semiconductor. Consequently, a multitude of binary, ternary and quaternary compositions have been investigated and

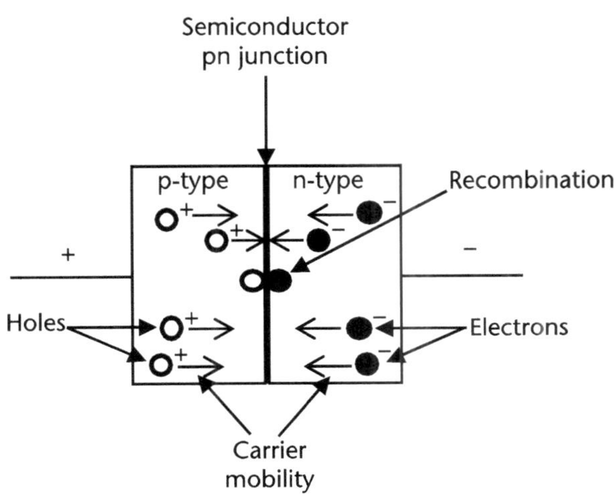

Figure 3.27 A forward-biased pn junction.

developed to provide the desired wavelengths. Some of the resulting compositions have been made up of many different elements and referred to as an alphabet soup. However, there are certain rules that govern which elements can be combined with others to form a suitable structure. This has led to the creation of terms such as III-V (three-five) devices, a reference to the columns in the periodic table where the constituent elements originate.

There are two broad classes of emitters. The structures that emit from the edge of the chip are known as edge emitters and can be formed into monolithic arrays called diode bars. The ones that provide light from the surface of the structure are called surface emitters and may also be formed into arrays to give a high-power emission. The physical dimensions of the active region are very small in one direction and not very much larger in the orthogonal direction, leading to an asymmetric beam shape; the more divergent beam direction is termed the fast axis and other direction, which spreads more slowly, is known as the slow axis. It is worth noting that the M^2 values of both directions may be quite close to their diffraction limited values, despite the beams being highly divergent. Waveguide-like structures have been perfected to concentrate light into a specific direction, which is often used during the pumping of solid-state lasers.

The edge emitters are configured to form an oscillator extending over hundreds of micrometres or more along the plane of the pn junction. This results in a beam with high divergence emerging from a small area on the cleaved surface of the chip. Vertical-cavity devices oscillate perpendicular to the plane of the pn junction; hence, the resonator contains only a thin layer of the gain medium, and need high reflectivity mirrors to sustain the oscillation. These devices tend to have low threshold currents for laser action and generate a better quality, lower divergence beam.

The actual structure of the diode is usually very complex and made up of much more than just a block of n-type material grown next to a block of p-type material. There are usually many different layers added on either side of the junction and at the interface of the junction between the n and p materials to enhance performance; for example, layers that aid the stimulated emission process, control the current flow or guide the photons. All of this structure is grown on a simple substrate using epitaxial techniques.

As noted earlier, there is a transition from spontaneous emission to stimulated emission, this occurs at a specific value of the forward bias current and is known as the threshold current. It is often used as a figure of merit for diode devices. The relationship between the diode drive current and the emitted power indicates laser efficiency; hence, the desire is for a steep curve, as shown in Figure 3.28, with as low a threshold current as is possible.

Since the positive demonstration of laser action in a semiconductor device, increasingly sophisticated structures have been designed with the aim of confining the carriers within the active region. Some of the important pn junction designs are:

- *Homo-junction:* these structures comprise a single semiconductor composition (e.g., gallium arsenide), with each side having an appropriate doping to provide the desired n and p characteristics. This structure provides very little vertical confinement of the generated photons and is therefore very inefficient and only suitable for pulsed operation.

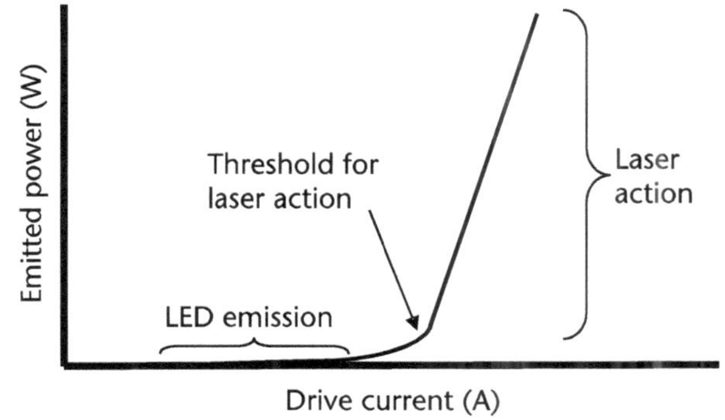

Figure 3.28 Emission characteristic of a diode.

- *Single hetero-junction:* as the term implies, the active region is sandwiched between materials with a different composition, and with different band gaps. These materials are usually selected so that they have different refractive indices to aid the confinement of the generated light to the active zone. These have largely been superseded by the double hetero-junction.
- *Double hetero-junction:* in this case, the active regions are sandwiched between two layers with different compositions, which further enhances the confinement of the photons on the top as well as the bottom of the active region. The enhanced confinement led to greater operational efficiency and the capability for continuous-wave operations.

Significant research activity has been invested into the reduction of the stripe width (active zone on the surface) in hetero-structures to give increased current density in the gain region, thus lowering the threshold current and improving the efficiency. Greater confinement also limits the laser modes that can oscillate, often resulting in a single transverse mode nominally in one direction, which has a dramatic (positive) impact on beam quality.

A further enhancement of diode lasers came from the development of a waveguide structure in the active layer. Two fundamental approaches have been followed:

- *Gain guiding:* this approach is from control of the drive current in the junction, through restricting it to a narrow region of the active layer. This technique is frequently used in high-power devices.
- *Index guiding:* this approach, as used in hetero-junctions, uses the properties of refractive-index differences to confine the laser light. Examples include:
    - Buried hetero-structures, where the active layer is embedded within a semiconductor with a lower refractive index, resulting in excellent confinement and high efficiency;
    - Double-channel planar buried hetero-structures, in this case, the active stripe is buried in a mesa structure, formed as a result of a semiconductor grown in a way that allows two channels to be cut around the active region to create the isolation of the stripe;

    - Ridge-waveguides, in which a ridge is formed above the active region, so that the edges of this ridge reflect the photons being guided in the active region and insulating layers may be applied to assist with the current confinement.

The design of the resonator has a significant impact on the emission characteristics and relatively sophisticated approaches have been developed to overcome some of the perceived drawbacks of the simple faceted technique, sometimes called the Fabry-Perot resonator. Examples devised for single-frequency operation include:

- *Distributed-Bragg reflectors:* this method uses a periodic grating in a passive part of the cavity (i.e., where there is no gain). This technique is used to control the spectral line width.
- *Distributed feedback:* this device uses a grating immediately adjacent to the gain region, whose function is to scatter light in the active layer to give optical feedback. This is independent of the resonator mirror (such as the cleaved facets). This grating behaves as a frequency-selective element and so determines the wavelength of the oscillation in the resonator.

External cavity optical arrangements may also be used to give a tuneable output if a diffraction grating is used in the optical arrangement. Rotation of the grating will determine the emitted wavelength within the operating bandwidth of the device. Moreover, the use of very long external cavities can result in very narrow laser line widths.

Additional flexibility in the emission properties of these devices may be achieved through the design of semiconductor devices by creating structures with quantum wells or with alternating layers of different semiconductors (super lattices). The quantum properties of such structures can be exploited to allow different wavelengths to be generated or enhancement to other emission characteristics.

The development of modern growth techniques, which allow a mismatch between the laser structure and the substrate of one percent or more, has removed many of the emission-band restrictions, and allows more wavelengths to be emitted through application of strain within the structure. The technique involves the growth of thin layers that can accommodate the strain in the mismatches of the two lattices without disrupting the growth of the thicker layers. The strained layer forms a quantum well in the active layer. The use of strain provides an extra degree of freedom in the design of the semiconductor, which may be exploited to give enhanced optical performance through band-gap engineering.

Some of the other semiconductor laser devices developed include:

- *Edge-emitting arrays:* these are a collection of many parallel laser stripes closely spaced in a linear array, with the spacing between the stripes of a few multiples of the stripe width, some optical coupling between the individual stripes is feasible, leading to a partially coherent beam.
- *Surface emitters:* there are two fundamental types:
    - Planar cavity surface-emitting lasers have a laser cavity in the plane of the active zone and uses cleaved facets to direct the laser light out of the

surface. An alternative and more elegant scheme, it uses a Bragg grating to scatter the generated beam through the surface.

- Vertical-cavity surface-emitting devices have a very short cavity formed by reflectors above and below the active region, to give a resonator in the vertical plane and orthogonal to the plane of the junction. The active region is usually formed from quantum wells, which are very thin and result in low threshold currents. A close-packed array of individual lasers can be formed.

These diode structures are ideal for laser action in the visible, near infrared and short-wave infrared spectra. Many of the military applications require emission in the mid-wave infrared, which presents a number of significant challenges for the structures presented above. Consequently, a number of alternative approaches have been investigated to modify the fundamental bandgap transition that is required for energy differences of the order of 0.3 eV for mid-wave infrared generation.

It is not adequate merely to change the material composition so that the bandgap is reduced; this is because the loss mechanisms become dominant, owing to the very small energy difference between the states. Many techniques have been investigated to suppress these nonradiative losses, but even operation at very low cryogenic temperatures (e.g., 4K) has not suppressed the losses sufficiently to give sustained operation. One of the dominant nonradiative losses is Auger recombination, as previously mentioned; a three-body interaction, where the energy released by holes and electrons recombining is transferred to other carriers rather than generating photons. One of the techniques investigated to reduce this limiting factor has been to modify the alignment of the quantum-well band. In the case of the Type-II alignment, the minimum of the conduction band occurs at a different position from the maximum of the valence band; some refinements in this structure, known as the W laser, have enabled a strong overlap between the electron and hole wave functions to be maintained in their Type-II structure, leading to maintained high efficiency.

A number of different approaches to bandgap engineering have been investigated with the goal of manipulating the bandgap position, usually through the quantum confinement of the carriers (i.e., through the use of quantum wells). This confinement is achieved by the creation of a structure where sandwiches of semiconductor materials with differing band gaps are used to form quantum wells. Material compositions have been devised to enable the relative position of the valence and conduction bands to be displaced and achieve the desired energy difference; these are shown in Figure 3.29 and used to categorise the quantum well types, as I, II or III.

Some examples of the modern approaches to generation of wavelengths in the mid-wave infrared for countermeasure and remote sensing applications are considered next.

3.10.1 Optically Pumped Semiconductor Lasers

There has been a continuing interest in the development of an electrically driven, long wavelength emitting semiconductor laser. This configuration has the potential of being both highly efficient and very compact but, until recently, the loss

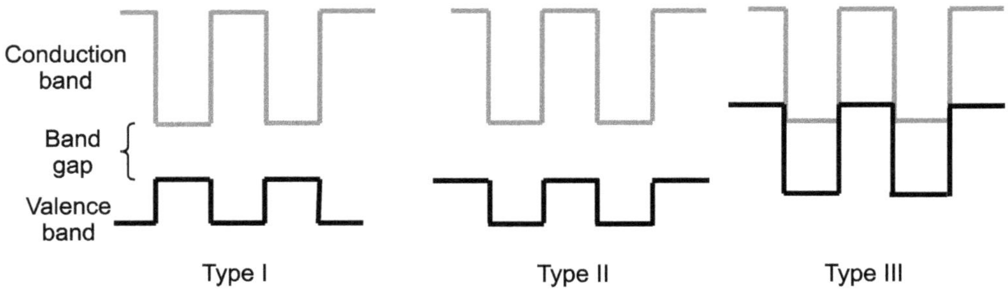

Figure 3.29 Illustration of various semiconductor band alignments. (Courtesy of Professor G. Nash, University of Exeter, U.K.)

mechanisms were dominant. Consequently, optical pumping techniques were investigated. In this configuration, a shorter-wavelength beam is used to create the holes and electrons required for laser action. This approach has several attractions:

- The absence of intentional dopants minimises the losses resulting from free-carrier absorption.
- A conducting path across the gain medium is not required, so the laser chip is easier to grow and fabricate.
- Separation of the pump source and gain medium has many heat-management advantages.
- There is potential for high wall-plug efficiency.

The original approach to this technology used complex compositions to form a suitable band gap for mid-wave infrared emission from interband transitions in a double hetero-structure junction. The use of an optical pumping scheme was selected in an attempt to maintain a lower junction temperature so that the numbers of carriers were reduced and the Auger recombination losses were minimised.

However, even with the use of optical-pumping techniques and cooling of the gain medium to about 80K, this type of structure still failed to produce a device with emission characteristics suitable for military applications; although, this class of laser was able to demonstrate the potential value of semiconductor laser technology for defeating heat-seeking missiles during some initial practical application studies.

This form of mid-wave infrared source has been transformed by the successful development of multiple quantum well structures in a Type-II structure. A quantum well can be formed in a thin layer of semiconductor material when it is sandwiched between two other layers of semiconductor with larger band gaps. In this case, the free electrons are trapped in a very thin zone with dimensions of around 15 to 20 nm. Under these conditions, the quantum-mechanical properties of the electrons become significant and the energy levels of the valance and conduction bands become quantised. Multiple quantum-well structures may be produced by forming a structure with alternating layers with high and low bandgaps. Care is required in the design of the high bandgap barrier to prevent tunnelling through them and destroying the confinement. A schematic of a multiple quantum-well structure is shown in Figure 3.30, which is a form of W structure.

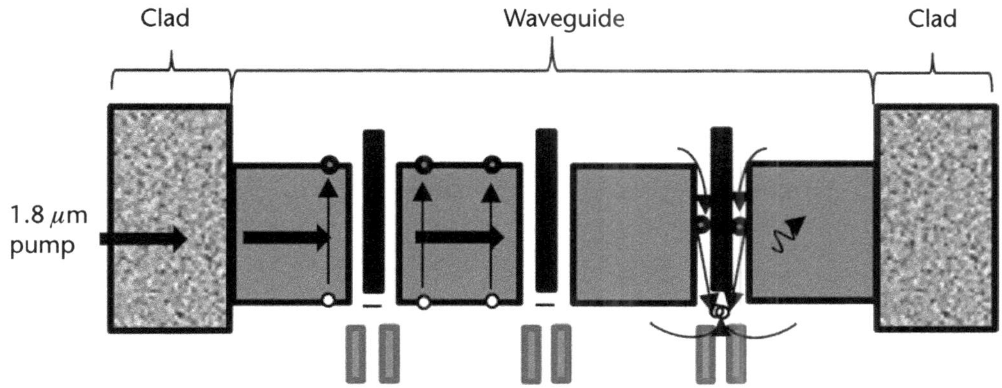

Figure 3.30 A schematic of a multiple quantum-well structure. (Courtesy of Dr. R. Kaspi, AFRL, USA.)

The principle of operation of an optically pumped semiconductor device is that photons are used to excite the confined electrons within the quantum wells. The light from a shorter-wavelength diode, such as a 2-micron emitter, is focused on to the active zone on the chip containing the multiple quantum-well structure. These pump photons are absorbed in the waveguide to create electron-hole pairs. The population inversion is created by the normal absorption process to change the energy state of the electrons. These carriers are transferred into the quantum wells and recombine to generate the photon beam. The resonant cavity is formed by the cleaved facets on the chip and the suitable application of reflective coatings.

In the particular configuration shown in Figure 3.30, two indium arsenide electron wells surround a thin indium gallium antimonide layer where the holes are confined. There is a layer of indium gallium arsenic antimonide surrounding these quantum wells, which absorbs the incident pump light and where the electrons and holes are created. These carriers become trapped in their respective quantum wells and the recombination occurs with transitions between the lowest electron energy level and the highest hole energy level. The gallium-antimonide cladding layer provides good confinement for the electrical carriers. There is also the added advantage of some confinement in the vertical direction, owing to a small refractive index difference between the gallium antimonide and the indium gallium arsenic antimonide. Moreover, there is the potential for reduced beam divergence resulting from the extended vertical and lateral sizes of the active region of these devices compared with diode structures.

Several important developments have aided the performance of this type of laser. One of these was the development of long-wavelength pump diodes and a corresponding substrate that is transparent to the pump wavelength. These features have enabled the chip to be mounted with its active layers next to the heat sink, which allowed the gain medium to be pumped through the substrate. Another important consideration is that the pump source is somewhat removed from the quantum-well structure, thus reducing the heat load on the device.

This arrangement has brought many advantages to the operation of this class of laser, including sustained operation and near total elimination of the pulse droop (a reduction in energy with time) caused by increased heating from non-radiative effects. A schematic of this configuration is shown in Figure 3.31, with the epi-layer down configuration.

The optical arrangement of these devices allows for the pump laser to be separated from the active region. This design flexibility has a number of advantages:

- The waste heat of the pump source does not have an impact on the optically pumped region.
- The mid-wave infrared stage can be made extremely small and mounted directly on to many other subsystems (e.g., within a beam director).

A summary of the qualitative performance characteristics of optically pumped semiconductor lasers is given in Table 3.15.

This form of laser has many potential military applications in remote sensing as well as for platform protection using infrared-countermeasure techniques. The flexibility of the technology, along with the multiline and multiband emission in a compact package offers the possibility of additional applications, as well as robustness against other technological developments.

3.10.2 Quantum-Cascade Laser

A quantum-cascade laser is essentially an electrical device; it is a modern approach to achieving laser action in semiconductor materials through direct generation. This type of laser source is capable of an emission in the mid-wave infrared and beyond (i.e., with an even smaller band gap than quoted above for mid-wave infrared).

A quantum-cascade laser is a unipolar device. The design of the laser involves a complex periodic structure in the conduction band, achieved using periodic sandwiching of different semiconductors with alternative active and injector/relaxation regions.

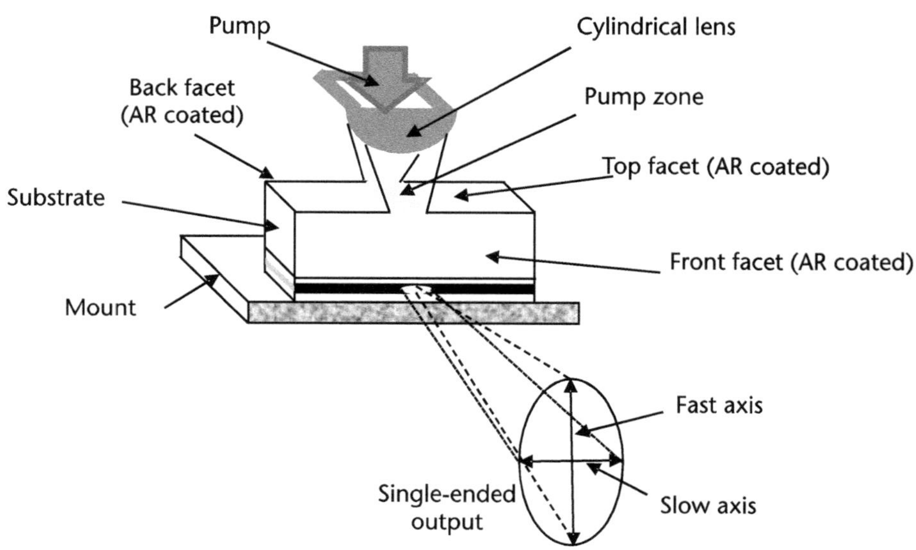

Figure 3.31 Optical configuration of an optically pumped semiconductor. (Courtesy of Dr. R. Kaspi, AFRL, USA.)

Table 3.15 Qualitative Performance of OPSL Devices

Performance Parameter	*Characteristic*
Wavelength range	Mid-wave infrared to far infrared, 3 to 10 μm
Tuneability	Yes, usually over a narrow range and simultaneous multiline and multiband feasible
Output	Pulsed or CW, determined mode of operation (at depressed temperature)
Power	Moderate power (watt class)
Pulse-repetition frequency	Depends on excitation, but greater than 10 kHz
Beam quality	Moderate to good
Time to full brightness	Very fast, once cooled
Efficiency	Moderate, potential to approach 20%
Size	Very compact with most configurations; system flexibility with separation of pump laser if required
Reliability/maintenance	Relatively good
Complexity	Low because of the relative simplicity
Cost	Moderate

The laser action takes place in the active region and a mini-gap is engineered into the structure to prevent the electrons from tunnelling out of the upper laser level into an adjacent quantum well without first making a relaxation transition, and ensure that they are channelled into the next upper laser level in the following active region. A typical structure has of the order of 25 to 40 active periods in the active region; a few periods of an active region are shown schematically in Figure 3.32.

In this periodic structure, the radiative transitions needed for laser action occur between electron subbands confined within the conduction band quantum wells. This technique for generating photons contrasts with the more conventional radiative transition that occurs between hole and electron states in the interband semiconductor devices considered above. In this case, only one type of carrier is

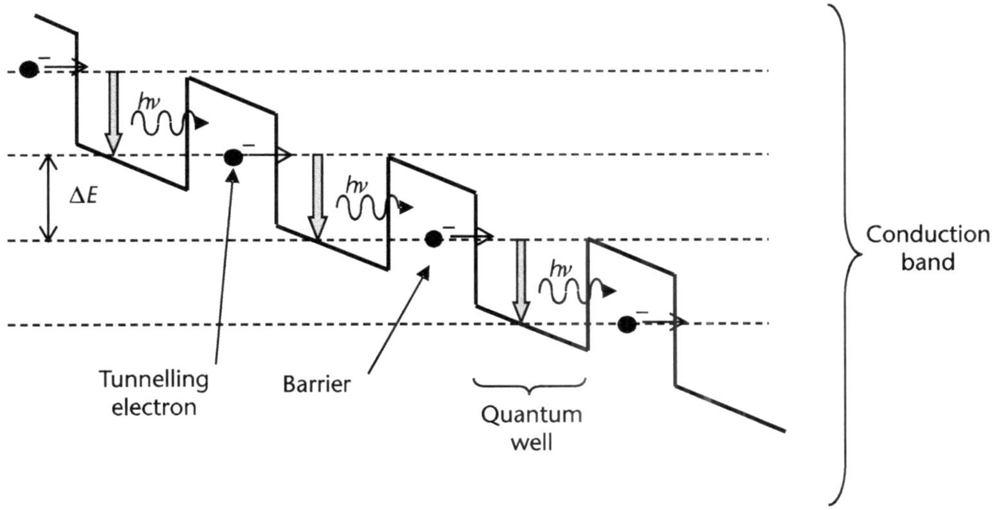

Figure 3.32 The periodic structure of a QCL. (Courtesy of Prof. G. Nash, University of Exeter, U.K.)

involved, usually electrons, to avoid some of the radiative-loss mechanisms, as discussed further later.

The physical process involved in the laser action starts with the injection of electrons, from the applied current, into the first quantum well of the active region via a tunnelling process, which then undergoes a change of state to a lower energy level, emitting a photon. This change of state within the conduction band of the structure, and the avoidance of recombination, means that the free electrons are not lost in the radiative transition. Hence, as the energy states in each stage of an active region are slightly displaced with respect to each other throughout the active region, an electron having undergone a transition in one quantum well may tunnel into the next well and be in an excited state to then repeat the emission process.

This cascade process can continue throughout the periodic structure in the active region and electrons will be recycled many times by subsequent injections into the successive active regions located beyond the initial injection region. Hence, each electron may generate as many photons as there are quantum-well stages, leading to the potential for high optical power with high efficiency. However, a potential drawback of this technology is that the threshold voltage is proportional to the product of the number of stages and the potential drop per period, leading to a large threshold voltage.

The intersubband nature of the fundamental laser action provides these devices with a few advantages compared with the interband approaches. For example, Auger recombination processes, which can dominate the interband devices at higher operating temperatures, are eliminated. The dominant nonradiative loss is electron-phonon scattering, which may cause the electrons to relax in the subbands without emitting a photon, leading to low radiative efficiency.

Careful design is required to avoid increases in operating temperature and ensure continuous-wave emission at room temperature. Moreover, the emission wavelength can be varied over a relatively wide range within a single combination of materials, such as the material system of indium gallium arsenide/aluminium indium arsenide on an indium phosphide substrate (InGaAs/AlInAs/InP).

The big challenge concerning this laser technology remains the net gain that may be achieved across the structure. Significant losses can result from the formation of the waveguide used to channel the laser action in the structure, which currently limits the available gain. Significant enhancements continue to be made to the design of the active gain region to reduce these and other losses, and thus, increase net gain, leading to higher average-power emission.

The original design of quantum-cascade lasers had emission wavelengths beyond the mid-wave infrared, typically between 5 and 12 μm. Refinements made to the complete structure have led to efficient emission shorter than 4 μm. The challenge for achieving even shorter wavelength emission is in engineering quantum wells in the active region that are not only deep enough to accommodate the desired laser transition, but also sufficiently robust to prevent escape of the free electrons from the upper laser level into states beyond the barrier. Such losses are detrimental to population inversion at higher temperatures or in the presence of higher electrical fields. The need for narrower active regions is required to achieve the higher energy states, resulting in a requirement for the application of significantly higher electrical fields for operation at the shorter mid-wave infrared wavelengths.

A summary of the qualitative performance characteristics of quantum-cascade lasers is given in Table 3.16.

Quantum-cascade laser technology continues to advance rapidly with output power of the order of watts with very good beam quality being available in the mid-wave infrared. The prospects for even better performance are very good. The emission characteristics of this type of laser are well matched to a range of military applications such as sensing, countermeasures and communications.

3.10.3 Interband-Cascade Lasers

This class of laser is subtly different from the quantum-cascade device described above. In this case, the transition does involve inter-band transitions, as the title implies. The structure of the device comprises multiple periods of a basic unit of an electron quantum well in the conduction band coupled with a hole quantum-well structure in the valence band. The band structure of the device is sometimes referred to as a Type-III alignment but is also termed a hybrid Type-II/III; for simplicity, it will be referred to as Type-III in this book.

These devices are designed so that when a voltage is applied to the diode, the electron ground level in the first quantum well is above the valence-band edge of the second quantum well; moreover, the hole in the ground level of the second quantum well is above the conduction-band edge of the first quantum well. This arrangement of the respective energy levels is shown schematically in Figure 3.33 for a single period in the structure.

During the operation of this type of laser, electrons resonantly tunnel into the first quantum well via an injection process. These electrons are blocked by a barrier created by the conduction band of the second quantum well, and consequently, make an indirect transition to the valence band of the second quantum well, resulting in photon generation, as shown in the figure. These electrons then undergo a further tunnelling process to the next pair of quantum wells and the decay process is repeated. The electrons traverse a series of active regions, with optical transitions occurring at each stage through interband transitions.

Table 3.16 Qualitative Performance of Quantum-Cascade Lasers

Performance Parameter	*Characteristic*
Wavelength range	Mid-wave infrared to far infrared 3.5 to ~ 11 μm
Tuneability	Yes, usually over a narrow range and diversity feasible
Output	Pulsed or CW, determined mode of operation (at room temperature)
Power	Moderate-power operation feasible (watt class)
Pulse-repetition frequency	Depends on excitation, but greater than 10 kHz
Beam quality	Moderate
Time to full brightness	Very fast
Efficiency	Moderate, potential to approach 20%
Size	Very compact with most configurations
Reliability/maintenance	Relatively good
Complexity	Low because of the relative simplicity
Cost	Moderate

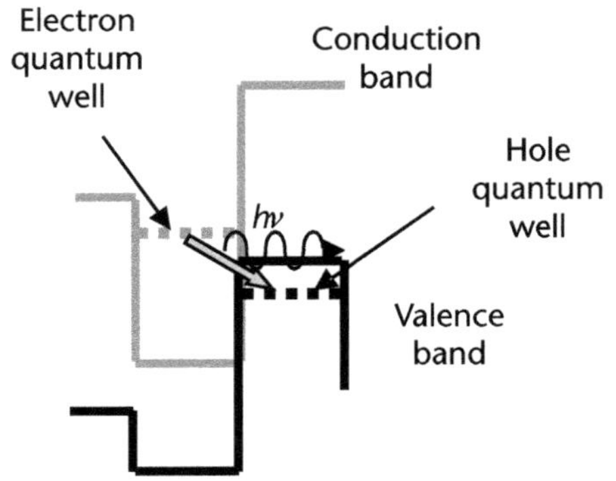

Figure 3.33 Energy-level arrangements in an inter-band cascade laser. (Courtesy of Professor G. Nash, University of Exeter, U.K.)

This type of cascade laser configuration combines the advantages of a cascade structure with those of a Type-III bandgap arrangement. Moreover, these devices are designed to reduce the adverse impact of Auger recombination effects and other internal losses by suppression of resonances between the bandgap and the Auger transition energies. However, the spatial separation of the electrons and holes in this Type-III configuration, compared with other bandgap configurations, means that there is reduced overlap of their respective wave functions, resulting in reduced optical efficiency and a higher current density being needed to drive the device. The advantage of this configuration, with respect to the other cascade devices, is a lower drive potential and reduced phonon-induced nonradiative losses.

A summary of the qualitative performance characteristics of interband cascade lasers is given in Table 3.17.

The emission characteristics of this design are compatible with a number of military applications that require modest power beams of only a few watts, such as remote sensing and infrared countermeasures.

3.10.4 W Structures

These are also electrically driven structures. They get their name from the displacement of the bandgap between the valence and conduction bands, so that the transition has the appropriate energy difference for mid-wave infrared emission. W-structure devices are a very successful refinement of the interband cascade lasers.

This type of structure has been very successful in suppressing the Auger recombination loss mechanism. Moreover, the design has enabled a strong overlap between the electron and hole wave functions more typical of Type-I structures. This is a critical requirement for high efficiency operation.

3.10.5 Strain-Engineered Diodes

This structure is based on a Type-I bandgap, which is the simplest arrangement where the holes and electrons occur in the same place. The radiative emission is from recombination via a direct transition. In this case, there is the potential for high efficiency from good overlap of the respective wave functions. Additionally,

Table 3.17 Qualitative Performance of Inter-Band Cascade Lasers

Performance Parameter	*Characteristic*
Wavelength range	Mid-wave infrared
Tuneability	Yes, usually over a narrow range and diversity feasible
Output	Pulsed or CW, determined mode of operation (at room temperature)
Power	Moderate power (watt class)
Pulse-repetition frequency	Depends on excitation, but greater than 10 kHz
Beam quality	Moderate
Time to full brightness	Very fast
Efficiency	Moderate, potential to approach 20%
Size	Very compact with most configurations
Reliability/maintenance	Relatively good
Complexity	Low because of the relative simplicity
Cost	Moderate

this relatively simple structure can lead to good carrier confinement, through the use of waveguide and cladding layers.

The structure of the diode comprises compressively strained quantum wells within cladding waveguide barriers, offering a suitable compromise between the need for good electronic and optical confinement and the characteristics of low series resistance needed for low threshold operation. This design is aimed at providing increased optical gain with correspondingly lower threshold current density. This in turn should reduce nonradiative Auger recombination losses. A schematic diagram of this arrangement is shown in Figure 3.34.

The strain being introduced into the quantum well structure is achieved through the use of barriers with different lattice constants, as described above, from those of the quantum-well layer. Compressive strain affects the valence-band structure within the quantum well, leading to a better matching of the hole and electron density of states. The consequence of this is a potential increase in gain and a reduction in carrier density, which in turn reduces the impact of Auger recombination losses.

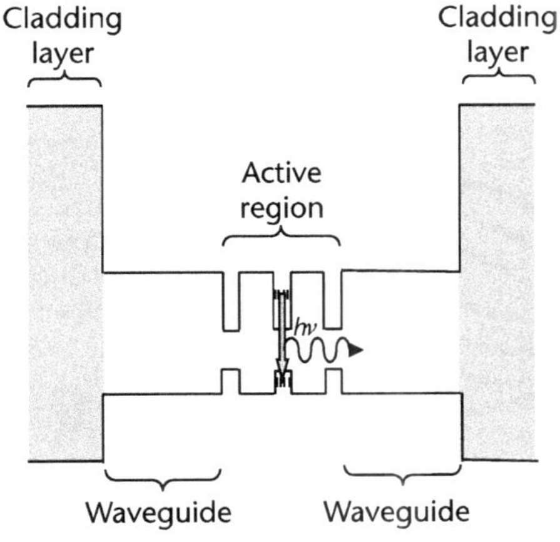

Figure 3.34 Schematic of a strain-engineered diode. (Courtesy of Professor G. Nash, University of Exeter, U.K.)

It is anticipated that this simple design could be adapted for multiline emission from the same device by incorporating different quantum-well structures, designed for different wavelength emission, within the same structure.

A summary of the potential qualitative performance characteristics of strain-engineered diode lasers is given in Table 3.18.

This type of laser is in active development, as are the above mid-wave infrared devices; its design offers a relatively simple configuration suitable for applications requiring modest power mid-infrared sources.

3.10.6 Optically Pumped Semiconductor-Disk Lasers

An optically pumped semiconductor-disk laser has a structure similar to a vertical-cavity surface-emitting laser with a semiconductor region containing the quantum wells for the carrier confinement, which leads to laser action. The active region is in the form of a disk with a thin absorbing region similar to the solid-state disk laser described above. Figure 3.35 shows a typical optical arrangement for this type of laser and Figure 3.36 shows some detail of the semiconductor disk in a schematic

Table 3.18 Qualitative Performance of Strain-Engineered Diodes

Performance Parameter	*Characteristic*
Wavelength range	Mid-wave infrared
Tuneability	Yes, usually over a narrow range and diversity feasible
Output	Pulsed or CW, determined mode of operation (at room temperature)
Power	Moderate power (potentially watt class)
Pulse-repetition frequency	Depends on excitation, but greater than 10 kHz
Beam quality	Moderate
Time to full brightness	Very fast
Efficiency	Moderate, potential to approach 20%
Size	Very compact with most configurations
Reliability/maintenance	Relatively good
Complexity	Low because of the relative simplicity
Cost	Moderate

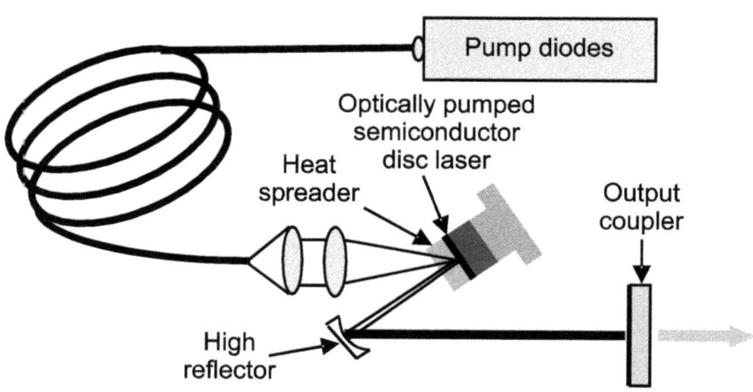

Figure 3.35 Optically pumped semiconductor disc laser.

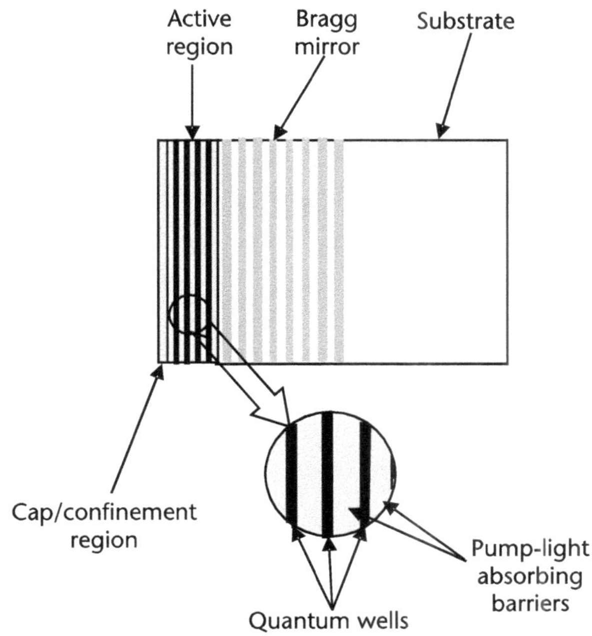

Figure 3.36 Schematic diagram of the semiconductor-disk structure.

fashion. These devices may be operated to give a continuous wave or pulsed output beam.

The use of semiconductor materials and device engineering enables the structures to be designed for a specific wavelength emission. However, the use of different semiconductor material systems in the active region allows great flexibility and versatility in the emission wavelengths. Additionally, with the use of second-harmonic generation, the emission wavelengths can extend into the ultraviolet, as shown by double circles in Figure 3.37.

Table 3.19 shows the typical wavelength emission bands for direct generation from various semiconductor compositions.

This technology is developing rapidly for many low to modest power applications. The laser configuration using semiconductor techniques helps overcome some of the design constraints of diode-pumped solid-state lasers and helps tailor the emission wavelength to match the application. The principal drawbacks involve the need to use an external cavity and the requirements to manage the waste heat in the chip, as the emission characteristics are thermally sensitive.

Table 3.19 The Wavelength Emission from Various Compositions Used in Semiconductor-Disk Lasers

Wavelength Spread (nm)	*Composition*
640 to 690	InGaP/AlGaInP
850 to 870	GaAs/AlGaAs
920 to 1,180	InGaAs/GaAs
1,190 to 1,300	GaInNAs/GaAs
1,500	InGaAs/InGaAsP
1,900 to 2,800	GaInAsSb/AlGaAsSb

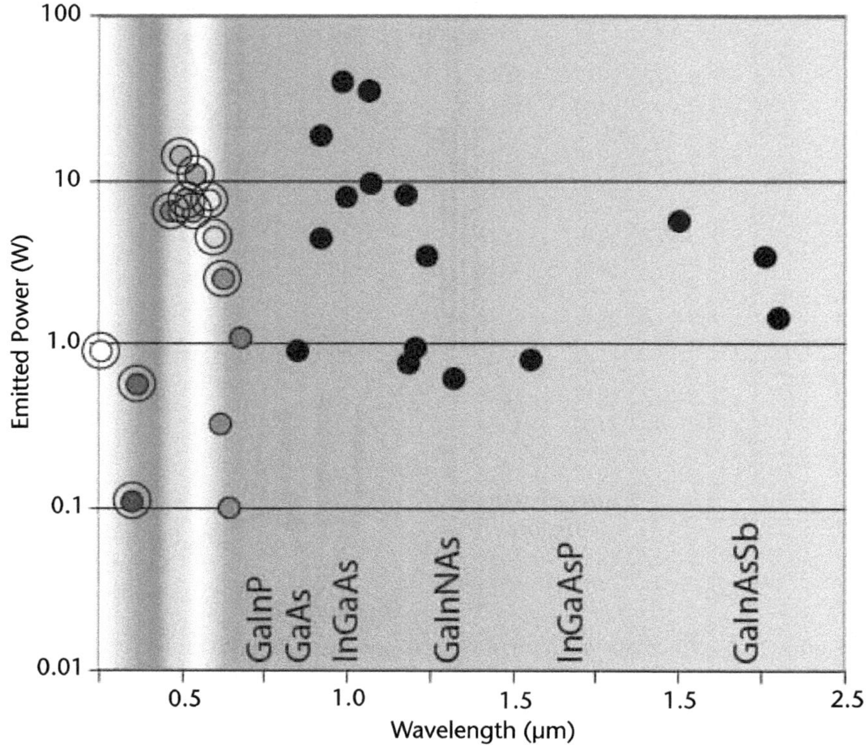

Figure 3.37 Emission wavelengths of semiconductor-disk lasers. (Courtesy of Dr. D. Burns, University of Strathclyde, U.K.)

3.11 Free-Electron Laser

The free-electron laser is a device that appears to be an ideal source for laser-directed energy applications. Not only is it capable of very high optical emission power with good beam quality, it has a significant tuning range, which may have important operational consequences. For example, the option to select an emission wavelength can allow for a very high atmospheric transmission of a high-power beam over a range of climatic or meteorological visibility conditions. This flexibility could be important for high-power applications, especially for maritime operations.

The principle of operation of this device is the extraction of light (synchrotron radiation) from high-speed electrons passing through a magnetic field with spatially periodic variations in intensity as well as direction. The synchrotron radiation is emitted from high-energy electrons when a magnetic field interacts with them to change their velocity, either by accelerating them or slowing them down. The synchrotron radiation has some similarities with spontaneous emission but as the free-electron beam does not have discrete energy levels the synchrotron radiation is not limited to discrete transitions.

In this device, the energetic free electrons are directed through a magnetic field formed by an array of permanent magnets with alternating polarity, as shown in Figure 3.38. This regular array of magnets, known as a wiggler or an undulator, forms the spatially periodic magnetic field, which interacts with the applied elec-

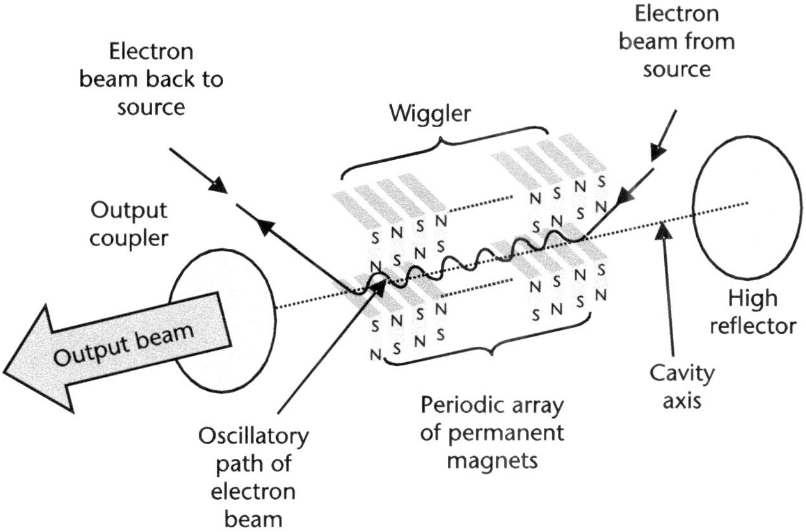

Figure 3.38 Schematic diagram of a free-electron laser.

tron beam. The electron beam is bent first one way then the other by this periodic magnetic field, hence the origin of the term wiggler for this subsystem.

The impact on the electron beam is for it to be organised into packets (or clumps) of electrons separated by the period of the magnetic field. Hence, each packet of electrons experiences the same magnetic field at the same time, separated spatially by a single period of the magnetic array. When the electrons have the same energy, they emit synchrotron radiation at the same wavelength and the photons from the various clumps add together in phase.

The basis of the light generation process is not stimulated emission; therefore, the emitted beam is not truly laser light but will have many of its conventional properties. Moreover, the recycling of the electron beams through the cavity can lead to very high power emission and a recapturing of the electron energy can lead to high-efficiency operation.

There are two fundamental types of free-electron laser distinguished by the wavelength and the interactions between the photons and the electrons. The free-electron lasers envisaged for military applications operate in the Compton regime (i.e., with a wavelength less than 20 μm and electron energies greater than 20 MeV). In this regime, where interactions between individual particles are involved, the individual electrons have high energy but the total current is low. The energy transfer is between individual electrons and the electromagnetic field. Consequently, the gain is low and only present when the light is travelling in the same direction as the magnetic field.

This configuration is usually operated as an oscillator, owing to the relatively low gain and low intrinsic efficiency. Careful design of the resonator is necessary to provide the optimum operating conditions, particularly the mirror design and permitted extraction techniques. Efficiency improvements result from the use of an electron beam, where there is little if any spread in the energy distribution.

A summary of the potential qualitative performance characteristics of free-electron lasers is given in Table 3.20.

Table 3.20 Qualitative Performance of Free-Electron Lasers

Performance Parameter	*Characteristic*
Wavelength range	Vast, ultraviolet to far infrared and beyond
Tuneability	Yes, usually over multiple bands feasible, this laser technology has the greatest tuneable of all devices
Output	Pulsed or CW
Power	Moderate to very high power (potentially MW class)
Pulse-repetition frequency	Depends on excitation
Beam quality	Moderate to good
Time to full brightness	Very fast, once the electron accelerator is operational
Efficiency	Low
Size	Enormous, only suitable in its current form for integration with the very largest platforms
Reliability/maintenance	To be determined
Complexity	High at a result of the fundamental approach
Cost	Astronomical

This device is being developed principally for laser-weapon applications. The technology is capable of generating very high optical powers in the multimegawatt regime. The technology base for this type of device continues to develop and could be enhanced significantly with research into the development of super-conducting materials operating at room temperatures. However, the technology is on a large scale and likely to be suitable for only very large ships.

3.12 Cryogenic Lasers

This is a general term given to any laser device when the gain medium is cooled to 77K or colder. At these low temperatures, a number of detrimental effects concerning laser action are suppressed. Examples include:

- The thermal conductivity of the laser gain medium is increased significantly, because of the increase in the mean free path of the phonons; this results in a reduction in thermal gradients and a reduction in thermal-lensing effects.
- The thermal-expansion coefficient is reduced significantly, which reduces the distortion of the gain medium during high-power operation; this feature also reduces thermal-lensing effects and results in enhanced beam quality.
- The thermo-optic coefficient is reduced, again beneficial to beam-quality enhancement.
- Enhanced laser and absorption cross-sections of rare-earth ions, because of reduced thermal-broadening mechanisms; the consequence is enhanced gain and reduced threshold, so that shorter Q-switched pulses can be generated, for example.
- The thermally induced population of the lower laser level is reduced in quasi three-level laser systems, leading to enhanced performance; an example is the ytterbium-YAG emission at 1,030 nm, which behaves as a quasi-three-level

system at room temperature, but at cryogenic temperatures this becomes a genuine four-level system.

It is recognised that cooling the gain medium is an added complexity, but it does lead to enhanced performance and provides access to some transitions that are problematical at room temperature and above. Moreover, provided that the quantum defect can be minimised or made negligible, the amount of cooling needed to maintain laser action at the cryogenic temperature can be very small.

3.13 Cannon Lasers

There has been occasional interest in designing a laser system for use in cannons or guns as a normal ordnance round. The military objective was to provide a simple optical countermeasure to supplement the existing defensive systems. The basis of most of these concepts has been to use the energy released in an explosion to generate the population inversion, as discussed for the explosively pumped iodine amplifier.

Laser action can be generated in the barrel of the cannon and the gun mount may be used to direct the laser beam at the target. A round may be designed to originate from a barrel that includes a back reflector and possibly some conditioning optical devices to ensure that the beam has a relatively recognisable spatial intensity profile. The energy supplied by the explosion can be very large and lead to a substantial pulse of light.

Another concept concerns achieving laser action from a projectile fired from a gun, so that the laser is in free flight and ignited after leaving the gun barrel. There are at least two significant challenges for this concept, beyond a laser design that will survive all of the forces involved in a cannon launch and maintain optical alignment, namely:

- Generating light over a very broad solid angle so the beam can be directed at its designated target;
- Generating a normal low-divergence laser beam, whilst having the independent means of directing the countermeasure at the target.

This approach is technologically weaker but remains an option to generate an isotropic radiator akin to an aircraft flare from a projectile.

Potential techniques for generation of visible and near-infrared laser light include:

- Explosively pumped dye laser (solid or liquid phase);
- Iodine laser for 1.315-μm operation.

Potential techniques for generating mid-wave infrared radiation may be via a deuterium-fluoride device and use of an explosive to initiate the exothermic chemical reaction (i.e., exploiting the chemical-laser technique and the stored energy in an explosive material).

Potential techniques for generating photons in the far infrared are via carbon dioxide laser technology. There are two possible excitation routes:

- Exploiting the excited state of deuterium-fluoride species to transfer energy into a ground state carbon-dioxide molecule, through a form of optical pumping;
- A mini gas-dynamic system, as described above, using the explosion in the barrel to fracture compressed carbon dioxide pellets, which then expand and form the population inversion. It may be possible extend the emission by the sequential fracture carbon dioxide pellets; with appropriate interlacing techniques, temporally modulated output may also be feasible.

It may also be feasible to generate a multiband output in a single gun barrel. For example, for the generation of a far-infrared emission as described above, the process will involve simultaneous mid-wave and far-infrared photons. However, techniques for the simultaneous emission with combinations of the various other wavebands are less obvious.

The advantages of explosively driven schemes are that they allow a potentially easy integration with conventional munitions to extend the possible defensive capability of an existing system. The use of explosive-pumping techniques has a number of advantages but the important aspect is the scale of the pumping energy that is possible virtually instantaneously, leading to very energetic pulses in each waveband. Successful development of the technique would allow a countermeasure channel to be available on any platform with a gun. Additionally, it is possible to generate a multiband output or range of different wave-band emissions, either simultaneously or sequentially.

There are many fundamental obstacles to be overcome before this relatively simple idea could become a reality for military applications. A few of the major challenges are:

- The cannon barrel must be pointing at the target sensor for the duration of the engagement, possibly a few hundred nanoseconds up to several hundred microseconds.
- Achieve the desired pointing stability and sightline isolation on a moving platform.
- Achieve the necessary match between the beam divergence and the pointing stability to ensure optimum, or close to optimum, energy-on-target characteristics.

A summary of the potential qualitative performance characteristics of explosively-pumped lasers is given in Table 3.21.

This interesting idea has a lot of potential, but some significant technical challenges need to be overcome. The major challenge is devising robust techniques for directing the photon energy on to the target and keeping it on that target for an adequate period before being used by the military.

Table 3.21 Qualitative Performance of Explosively-Pumped Lasers

Performance Parameter	*Characteristic*
Wavelength range	Vast, ultraviolet to far infrared feasible
Tuneability	Not feasible, but wavelength diversity possible and probably a reality
Output	Pulsed emission, the only realistic option
Power	Moderate to high pulse energy
Pulse-repetition frequency	Not very realistic
Beam quality	Poor even with special design features
Time to full brightness	Very fast
Efficiency	High
Size	Small
Reliability/maintenance	To be determined, dispensable by definition
Complexity	Very simple
Cost	Very low

3.14 Super-Continuum Lasers

There has recently been significant interest in the exploitation of some propagation features of very short pulses in the femtosecond regime. The generation of these very short pulses with even modest pulse energies results in an instantaneous optical power in, or very close to, the terawatt power scale. The propagation through air of this type of pulse leads to filament formation. Interestingly, this filament has very broadband emission, and so, has many potential applications including remote sensing and countermeasures.

The source most commonly used for super-continuum generation is titanium sapphire emitting close to 800 nm, discussed briefly in Section 3.7. Laser pumping of the titanium-sapphire crystal, with a frequency doubled Nd:YAG laser, generates the stimulated emission. The pulse is compressed through careful conditioning of the beam by emitting the blue part of the emission spectrum before the longer-wavelength red component. This emission scheme leads directly to shortening of the pulse as the red component travels faster than the blue wavelengths.

The beam self-focuses via the nonlinear Kerr effect and as the intensity increases, this leads to plasma formation, which in turn defocuses the beam. Currently, there appears to be an upper limit of a few millijoules of energy contained within each filament. By increasing the output of the femtosecond laser, more filaments are produced. The divergence is effectively zero (maybe of the order of 0.2 to 0.3 mrad) and over the short propagation paths of tens of metres that have been used thus far, the spatial profiles of the filaments have remained constant as they propagate.

Figure 3.39 shows an example of a filament generated from a femtosecond emission from a titanium-sapphire laser (45 fs, 50 mJ, 10 Hz) captured in a laboratory. The image shows the natural colour of the filament.

Laser schemes are being developed to produce a multiterawatt system that has a volume of about (60 cm^3).

Not surprisingly, some of the limiting factors for these lasers are the optical coatings and optic elements within the cavity. To prevent damage within the laser resonators, the beam sizes are usually increased. For the femtosecond systems that

Figure 3.39 An image of a filament generated by ultrashort pulses. (Courtesy of Professor S. L. Chin, Laval University, Canada.)

operate with peak powers in excess of 5 TW, there is a need to generate the laser light under vacuum, because the ionisation potential of the air (nitrogen) is lower than the field strengths associated with these power levels.

3.14.1 Super-Continuum Generation Using Photonic-Crystal Fibre and Planar Waveguides

Photonic-crystal fibre may be used to create ultrabroadband emissions by exploiting the dispersive properties of these particular fibres. The practical advantages of using current compact femtosecond laser technology, combined with photonic-crystal fibre or planar waveguide technology are being investigated. The source lasers emitting ultrashort pulses are readily available and these photons are injected into a photonic crystal fibre with a core diameter of about 2 μm or smaller. In this case, the super-continuum spectrum is limited by the material in the microstructured optical fibre. For a normal photonic-crystal fibre formed in silica, the emission extends from approximately 380 nm to about 1.6 μm. Alternative materials have been used to form the microstructured optical fibre and emission wavelengths beyond 4 μm have been demonstrated.

There are a number of mechanisms that contribute to the super-continuum generation as the pulse propagates along these fibres, including:

- Self-phase modulation;
- Self-steepening of the pulse;
- Soliton splitting;
- Raman shifting.

The Stokes detuning of lower-order solitons could be used for tuning the soliton emission to a particular emission waveband, especially in the mid-wave infrared where atmospheric filamentation has so far shown inefficient conversion when using conventional materials.

3.14.2 Filament Ultraviolet Generation

The peak-pulse irradiance (power density) of the near-infrared laser pulse from a very short pulse titanium-sapphire laser (emission centred at approximately 800 nm) reaches well above 10^{13} W/cm^2. This is sufficiently intense to generate higher harmonics in a nonlinear medium. The conversion efficiency of the third-harmonic energy is typically of the order of 0.5%, which could generate ultraviolet femtosecond pulses with very high peak powers from such an intense terawatt source; the third-harmonic peak power could range from megawatts to gigawatts.

This phenomenon raises the prospect of generating an intense ultraviolet pulse at a remote location, which is important for many sensing applications. Ultraviolet radiation in general and short-wavelength ultraviolet radiation in particular do not propagate well in the atmosphere because of strong Rayleigh scattering and molecular absorption by species such as ozone.

The efficient generation of the third-harmonic photons in a pulse during the filamentation process has been attributed to a nonlinear phase-locking mechanism between the pump and third-harmonic pulses. Moreover, the third-harmonic pulse also undergoes super-continuum generation during the filamentation process. The result is a further extension of the continuum into the ultraviolet waveband. Consequently, with the use of high pump power, the two separate super-continuum spectra will merge, creating a super-continuum extending from around 230 nm into the infrared. A typical example of a super-continuum spectrum in the infrared is shown in Figure 3.40.

An important aspect of the mechanism involved in this form of third-harmonic generation in the filamentation of the beam from a high peak-power laser is that the harmonic is generated along the entire filament channel and does not suffer any de-phasing effects. This, in turn, will generate a long interaction length.

The generation of the third-harmonic radiation pulses from the fundamental around 800 nm is of particular importance, as its wavelength (267 nm) overlaps with the absorption peaks of many large biological warfare agent molecules such as tryptophan and tyrosine; this technique could be important in biological-warfare-agent detection.

Some of the consequences of filamentation may be summarised as follows:

- The laser pulse can self-transport to a significant distance (many kilometres) in the atmosphere at high intensity; this is not feasible with longer laser-pulse emissions.

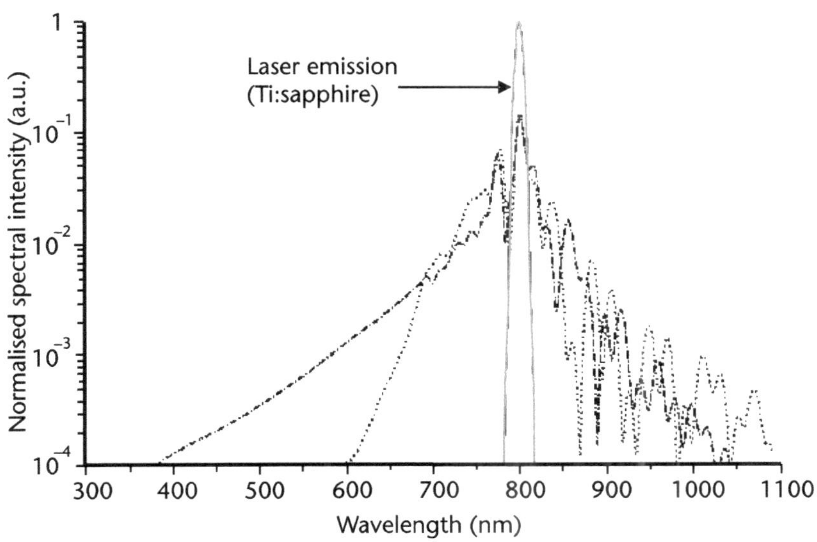

Figure 3.40 Super-continuum spectrum from the ultraviolet to the near infrared. (Courtesy of Professor S. L. Chin, Laval University, Canada.)

- The laser pulse can generate ultraviolet radiation (third harmonic of the fundamental) efficiently at a distance point in the atmosphere wherever filamentation occurs; this is not possible with other more conventional techniques for generating ultraviolet radiation because of absorption by ozone and Rayleigh scattering;
- The laser pulse can self-transform into a white-light laser pulse (super continuum), generating a spectrum that extends from 350 nm to 9 μm, together with a broadened third harmonic of the fundamental laser line.
- The white-light source and the filamentation constitute a super-broadband radiation source anywhere in the atmosphere where filamentation occurs, ranging in wavelength from 230 nm to 9 μm.
- Generation of electromagnetic pulses and terahertz radiation can happen anywhere in the atmosphere where filamentation occurs.
- Inside the filament, all molecules will be multiphoton/tunnel ionised and fragmented or excited, resulting in the fingerprint fluorescence of each species encountered by a filament.
- The fluorescence generated in a filament will most probably have an amplified spontaneous emission type of gain, which has been established for nitrogen molecules.

3.15 Waveguide Lasers

Many different types of laser may be designed to operate as a waveguide device. This is achieved by combining the optical confinement, typical of adjacent structures with different refractive indices as occurs in fibre lasers, with the energy storage of bulk lasers in a geometry that supports integrated optical devices. As such, they have received significant interest for decades and have a growing repertoire of fabrication methods, especially inscription by femtosecond lasers, leading to an increasing variety of waveguide-based gain media. Although semiconductor lasers are based on waveguide techniques, gas and dye waveguide lasers have been demonstrated, this text will focus solely on solid-state waveguide lasers. The example of fibre lasers will only be used here to illustrate the fundamental principles.

The definition of a waveguide laser at its most fundamental level is simply a laser in which the gain medium is a guided-wave structure. The guidance, or confinement, of the light is accomplished by surrounding the gain-medium core with a cladding that has a lower refractive index, and a thickness greater than the penetration depth of the evanescent field. Light that enters the waveguide experiences total internal reflection at the core/cladding interface, and thus remains confined to the core where it can propagate over long distances without the divergence it would experience in a bulk solid-state medium.

In a structure such as the core of the fibre laser, light will be confined to the laser gain medium, or core, if its angle of incidence falls within the waveguide's acceptance angle or numerical aperture (NA). This acceptance angle increases with the difference between the refractive index of the core and the cladding. A large numerical aperture will provide a higher coupling efficiency but will also allow a larger number of modes to propagate within the waveguide structure. The number of modes sustained within the waveguide is also proportional to the waveguide thickness; however, it is possible to reduce their number, and thereby improve the beam quality, either by making the waveguide smaller or by reducing the core/cladding index difference (i.e., creating a smaller numerical aperture). A single-mode waveguide laser can be constructed by controlling these parameters so that only the fundamental mode is supported and the remaining modes are eliminated.

Solid-state waveguide lasers have been manufactured using glass, polycrystalline, and single-crystal gain media in a variety of geometries. Planar or slab waveguide lasers confine light in only one direction within a thin slab or layer of the gain medium sandwiched between the cladding layers. Such devices can be either side-pumped or end-pumped. Channel waveguides confine light in both transverse directions. These include: embedded channel designs in which the guiding region is below the surface of the material; and ridge designs in which material is etched away to leave a ridge or strip of either gain material or cladding. Channel waveguides are more likely to be end-pumped.

Waveguide lasers can be operated with discrete cavity mirrors but offer additional advantages when incorporated into a monolithic resonator design. In this instance, the laser would either have mirror coatings applied to the end faces, or distributed Bragg reflectors fabricated into the material, a common practice with fibre lasers. A monolithic resonator makes the laser even more compact and robust. It also reduces the amount of free-space coupling in the system, improving alignment stability, enhancing environmental tolerance, and reducing losses.

The main advantage of a waveguide laser over its bulk counterpart is its high intensity, made possible by optical confinement of the light to a relatively narrow core region. This confinement of the light permits the use of longer gain lengths and results in better overlap of the pump and resonator modes, leading to a lower pump threshold. This is especially helpful in three-level and quasi-three-level systems where pump absorption is an important consideration. Finally, the small dimensions of a waveguide facilitate the rapid removal of waste heat to avoid harmful thermal effects such as lifetime quenching, thermal lensing or bulk fracture.

These operational benefits are offset by higher losses that originate from the fabrication processes used and the nature of the waveguide structure itself. As with its bulk counterpart, a waveguide gain medium will have optical absorption and scatter, as well as coupling losses at the input and output faces. The higher intensities in a waveguide will exacerbate non-linear (multi-phonon) absorption. The waveguide will incur additional losses from scattering at the core/cladding interface and light leakage into the cladding region. The latter is usually caused by coupling to higher-order, unguided modes as a result of stress or irregularities in the structure. Most of these loss mechanisms can be minimised through careful attention to detail during the fabrication processes.

An additional benefit of waveguide lasers is that some interesting and useful gain media can be fabricated more easily in waveguide form than as an optical fibre. More broadly, a waveguide laser can have a much smaller footprint than a fibre laser (or bulk laser), small enough to be integrated with other optical devices such as modulators or switches on to a single chip in a compact and robust package. Waveguide lasers are thus particularly well-suited for use in integrated optics, as seed lasers, and in other applications where compact, high-gain coherent sources are required.

3.16 Laser Performance Summary

Table 3.22 provides an indication of the current highest performance achieved by various laser devices, along with an indication of some important beam parameters; the value in parenthesise indicates the wavelength where the peak output performance was achieved.

Figure 3.41 indicates graphically the performance of the strategically important solid-state sources for directed-energy applications as well as sensing.

3.17 Summary

Many different types of laser technology have been developed for a wide range of military applications. These laser devices have used solid, liquid and gaseous gain media; the devices used have traditionally been based on well-established device techniques and methods.

Meeting the laser performance requirements needed to fulfil many of the military applications are a significant challenge. Most of the military applications require photons on demand, so there is little opportunity for the various subsystems to stabilise or reach a steady-state condition. Moreover, many applications require

Table 3.22 Summary of the Highest Performances Reported for Various Laser Devices

Type of Laser	*Current Peak Output*	*Beam Quality (M^2)*	*Wavelength (μm)*
Solid-state slab	15 kW	~2	1–3 (1)
Rod	3 kJ	4^+	1–3 (1)
Ceramic	15 kW	~2	1–3 (1)
Plastic	Few J class	5+	1–3 (1)
Fibre	10 kW	<2	1–3 (1)
Thin disk	7 kW	>3	1–2 (1)
Liquid dye	>100J	>0.5	0.4–0.8 (0.6)
Gas CO_2	>500 kW	~2	9–11 (10.6)
Chemical DF	>2 MW	>2	~4 (4)
COIL(per module)	>100 kW	~2	1.15 (1.15)
Semiconductor OPSL	10W	~2	$2–10^+$
QCL	10W	~2	3–14 + THz (8)
Semiconductor Disk	10W	~2	Broadly tuneable to 2 μm
Free-electron	100 kW–3 MW (projected)	~1.5	Broadly tuneable
Ultrashort pulse	TW	~1	Broadband

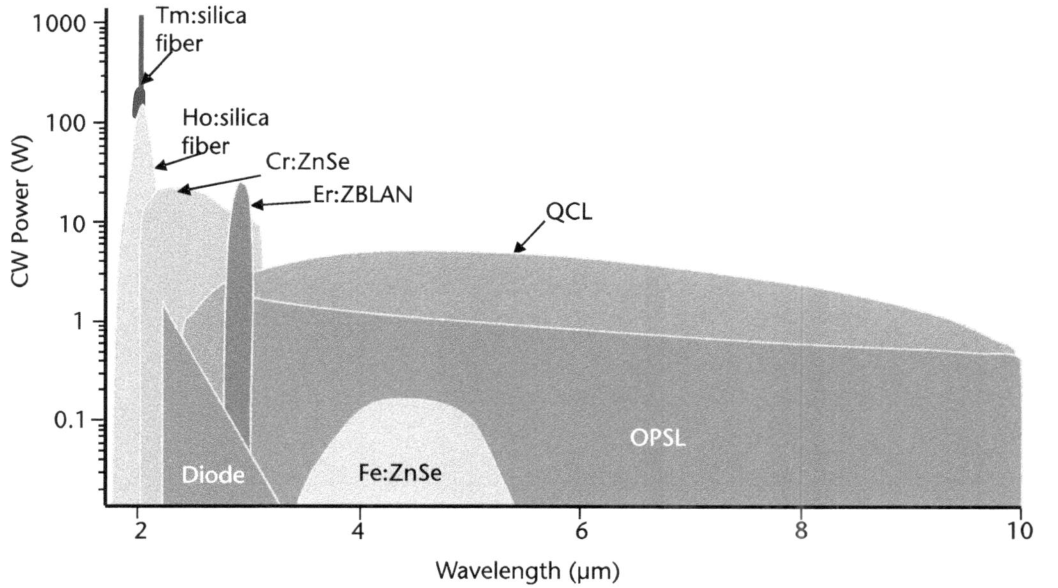

Figure 3.41 State-of-the art performance of modern solid-state sources. (Courtesy of Dr. C. T. Newell, AFRL, USA.)

combinations of beam parameters that are difficult to achieve individually, let alone in combination and on demand.

Despite the development of laser technology based on well-established techniques, there has been a persistent demand from the military operators for better technology, more capable and flexible devices and in particular, laser-based systems that are smaller, cheaper and lighter. Recently there has been a particular emphasis on the cost aspect, both with initial purchase and through-life costs.

One development that has had an enormous impact on laser technology for military applications is the diode laser. Initially, it was investigated as a laser device in its own right, but the diode laser was then developed for pumping solid-state lasers, which transformed the performance of the device's efficiency and size amongst many other aspects. More recent developments have seen the diode laser provide output efficiently in various parts of the electromagnetic spectrum.

One of the most interesting and exciting developments in semiconductor technology is the quantum-cascade laser. This very compact and efficient device technology now offers emission in the mid-wave as well as the long-wave infrared with the prospect of terahertz emission.

An equally important development has been with fibre lasers. The original development of low-loss efficient fibre amplifiers for the telecommunications industry has been translated into efficient and powerful devices for many military applications; these applications range from the low-power applications of sensing and countermeasures to laser-damage weapons.

It would appear that gas-laser technology is currently less popular for military applications. However, for sheer power, the chemical oxygen-iodine laser is still the champion. In general, gas-laser technology offers a low-risk approach to high-power generation with very good beam quality, but the scale of the device technology is large, which inhibits its integration with compact systems.

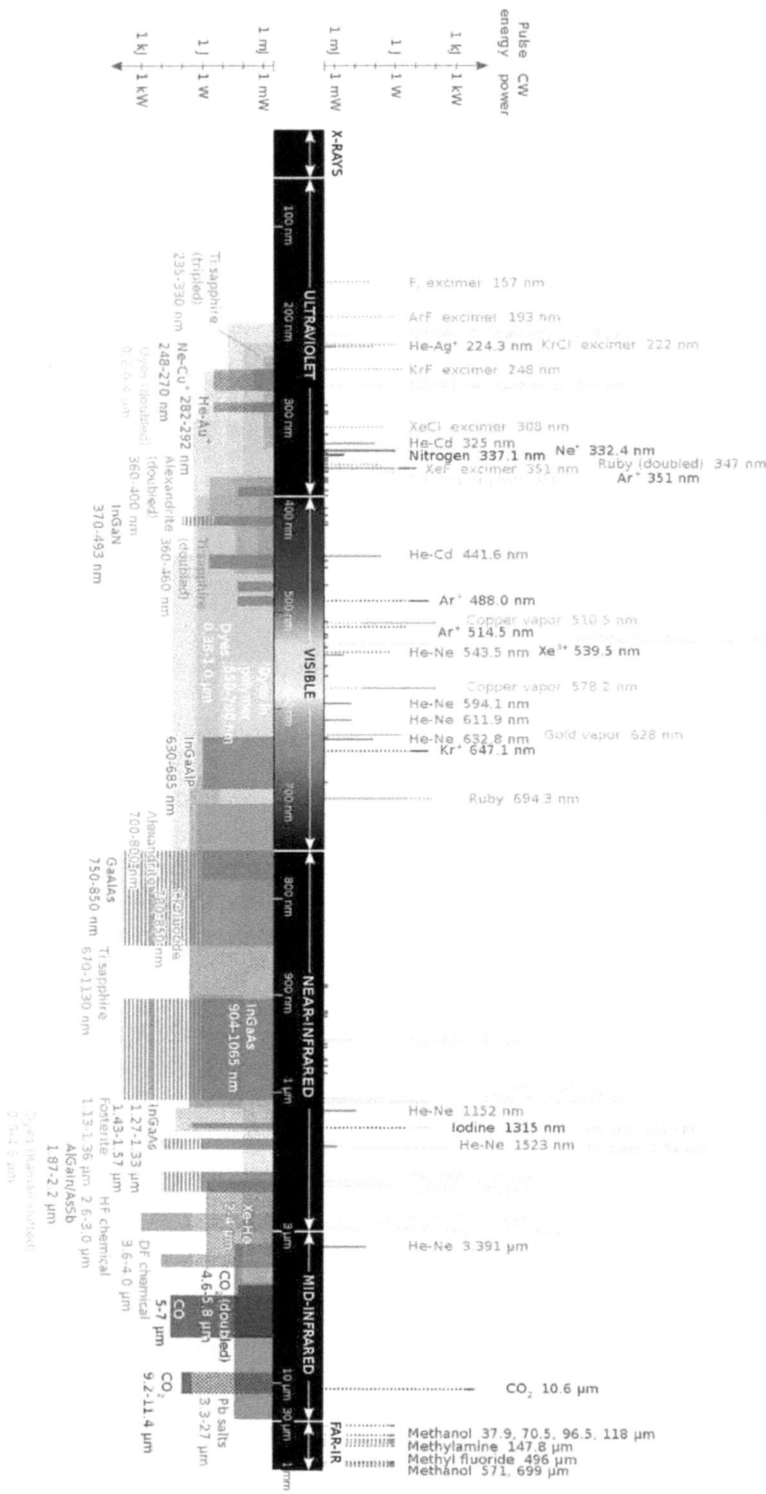

Figure 3.42 Laser emissions across the spectrum from commercial devices. (Courtesy of Dr. Espen Lippert, FFI, Norway.)

Liquid lasers offer a niche capability for a range of applications requiring energetic pulses in the visible part of the spectrum, usually for countermeasure applications. However, they tend to be somewhat bulky; moreover, the successful development of solid-state dye-laser technology offers a more robust source for some military applications.

Figure 3.42 is included to provide an indication of the emission lines available from some of the more common laser devices. This figure shows common laser lines from commercial lasers, but with the continuing move towards use of commercial-off-the-shelf technology into military systems, more of these devices could be used in military systems in the future.

Selected Bibliography

Agrawal, G. P., *Nonlinear Fiber Optics*, 3rd ed., San Diego, CA: Academic Press, 2001.

Agrawal, G. P., *Applications of Nonlinear Fiber Optics,* Burlington, MA: Academic Press, 2008.

Alfano, R. R., *The Super-Continuum Laser Source*, 2nd ed., New York: Springer, 2006.

Anderson, J. D., *Gas Dynamic Lasers: An Introduction*, New York: Academic Press, 1976.

Becker, P. C., N. A. Olsson, and J. R. Simpson, *Erbium-Doped Fiber Amplifiers: Fundamentals and Technology,* San Diego, CA: Academic Press, 1999.

Bloom, A. L., *Gas Lasers,* New York: John Wiley & Sons, 1968.

Boudrioua, A., *Photonic Waveguides,* New York: John Wiley & Sons, 2009.

Boyd, R. W., *Nonlinear Optics,* San Diego, CA: Academic Press, 2003.

Brau, C. A., *Free-Electron Lasers,* Boston, MA: Academic Press, 1990.

Butcher, P. N., and D. Cotter, *Elements of Nonlinear Optics,* Cambridge, U.K.: Cambridge University Press, 1990.

Chin, S. L., *The Fundamentals of Laser Optoelectronics,* Singapore: World Scientific Publishing, 1989.

Chin, S. L., *Femtosecond Laser Filamentation*, New York: Springer, 2010.

Davis, S. J., and M. C. Heaven, (eds.), "25th Anniversary SPIE Conference: Gas and Chemical Lasers and Intense Beam Applications III," *SPIE Proceedings*, Vol. 4631, 2002.

Diels, J. C., and L. Arissian, *Lasers: The Power and Precision of Light,* New York: Wiley VCH, 2011.

Digonnet, M. J. F., (ed.), *Rare-Earth Doped Fiber Lasers and Amplifiers,* New York: Marcel Dekker, 1993.

Finlayson, D. M., and B. D. Sinclair, (eds.), "Advances in Lasers and Applications," *Proceedings of the 52nd Scottish Universities Summer School in Physic,* Institute of Physics Publishing, Bristol, U.K., 1999.

Endo, M., and R. F. Walter, (eds.), *Gas Lasers*, Boca Raton, FL: Taylor & Francis/CRC Press, 2006.

Freund, H. P., and T. M. Antonsen, Jr., *Principles of Free Electron Lasers*, 2nd ed., London, U.K.: Chapman & Hall, 1999.

Ghatak, A. K., and K. Thyagarajan, *Introduction to Fiber Optics,* Cambridge, U.K.: Cambridge University Press, 1998.

Hall, D. R., and P. E. Jackson, *The Physics and Technology of Laser Resonators,* Bristol, U.K.: Taylor and Francis/IoP, 1990.

Hecht, J., *City of Light: The Story of Fiber Optics,* New York: Oxford University Press, 1999.

Hecht, J., *The Laser Guide Book,* Blue Ridge Summit, PA: TAB Books (McGraw-Hill), 1992.

Hitz, C. B., et al., *Introduction to Laser Technology*, 4th ed., New York: John Wiley & Sons/IEEE, 2012.

Hooker, S., and C. E. Webb, *Laser Physics,* Oxford, U.K.: Oxford University Press, 2010.

Joannopoulos, J. D., et al., *Photonic Crystals: Molding the Flow of Light,* 2nd ed., Princeton, NJ: Princeton University Press, 2008.

Kailsiky, Y., *The Physics and Engineering of Solid State Lasers*, Bellingham, WA: SPIE Optical Engineering Press, 2006.

Kaminskii, A. A., (ed.), *Laser Crystals,* Berlin, Germany: Springer-Verlag, 1981.

Kaminskii, A. A., (ed.), *Laser Crystals: Their Physics and Properties,* Berlin, Germany: Springer-Verlag, 1990.

Kaminskii, A. A., *Crystalline Laser: Physical Processes and Operating Schemes,* Boca Raton, FL: CRC Press, 1996.

Kapany, N. S., *Fiber Optics: Principles & Applications,* New York: Academic Press, 1967.

Kapany, N. S., *Military Applications of Fiber Optics,* Fiber Optics reprint series, Vol. 14, Brighton, MA: Gate Keepers Publishers, 1994.

Koechner, W., *Solid-State Laser Engineering,* Berlin, Germany: Springer, 1999.

Koechner, W., *Solid-State Laser Engineering, Vol. 1*, 2nd ed., Berlin, Germany: Springer-Verlag, 1985.

Koechner, W., and M. Bass, *Solid-State Lasers: A Graduate Text*, New York: Springer-Verlag, 2003.

Little, C. E., *Metal Vapour Lasers, Physics, Engineering and Applications,* New York: John Wiley & Sons, 1999.

Injeyam, H., and G. D. Goodno, *Laser Handbook,* New York: McGraw-Hill, 2012.

Marhic, M. E., *Fiber Optical Parametric Amplifiers, Oscillators and Related Devices,* Cambridge, U.K.: Cambridge University Press, 2007.

Marshall, T. C., *Free-Electron Lasers,* New York: Macmillan Publishing Company, New York, 1985.

Meyers, R. A., *Encyclopaedia of Lasers and Optical Technology,* San Diego, CA: Academic Press, 1991.

Milonni, P. W., and J. H. Eberly, *Lasers*, New York: John Wiley & Sons, 1988.

Milonni, P. W., and J. H. Eberly, *Laser Physics,* 2nd ed., New York: John Wiley & Sons, 2010.

Nambiar, K. R., *Lasers: Principles, Types and Applications,* New Delhi, India: New Age International Publishers, 2005.

Nikogosyan, D. N., *Properties of Optical and Laser Related Materials: A Handbook,* New York: John Wiley & Sons, 1997.

Okhotnikov, O. G., (ed.), *Fiber Lasers,* Weinheim, Germany: Wiley-VCH, 2012.

Okhotnikov, O. G., (ed.), *Semiconductor Disk Lasers,* Weinheim, Germany: Wiley-VCH, 2010.

Powers, P. E., *Field Guide to Nonlinear Optics*, Bellingham, WA: SPIE Optical Engineering Press, 2013.

Powers, P. E., *Fundamentals of Nonlinear Optics,* Boca Raton, FL: CRC Press, Taylor & Francis Group, 2011.

Rulliére, C., (ed.), *Femtosecond Laser Pulses: Principles and Experiments*, 2nd ed., New York: Springer-Verlag, 2004.

Saldin, E. L., et al., *The Physics of Free Electron Lasers,* Berlin, Germany: Springer-Verlag, 2000.

Sanghera, J. S., and I. D. Aggarwal, *Infrared Fiber Optics,* Boca Raton, FL: CRC Press, 1998.

Scheps, R., *Introduction to Laser-Diode Pumping of Solid-State Lasers,* Bellingham, WA: SPIE Optical Engineering Press, 2007.

Schubert, E. F., *Light Emitting Diodes*, 2nd ed., Cambridge, U.K.: Cambridge University Press, 2007.
Shen, Y. R., *The Principles of Nonlinear Optics,* New York: John Wiley & Sons, 1984.
Siegman, A. E., *Lasers,* Sausalito, CA: University Science Books, 1986.
Silfvart, W. T., *Laser Fundamentals,* Cambridge, U.K.: Cambridge University Press, 2004.
Svelto, O., and D. C. Hanna, *Principles of Lasers,* 2nd ed., New York: Plenum Press, 1982.
Svelto, O., and D. C. Hanna, *Principles of Lasers,* 4th ed., New York: Plenum Press, 1998.
Thyagarajan, K., and A. Ghatak, *Lasers, Theory and Application,* Springer, New York, 2010.
Uhrin, R., *Laser Crystal Growth,* Bellingham, WA: SPIE Optical Engineering Press, 2000.
Webb, C. E., and J. D. C. Jones, *Handbook of Laser Technology and Applications,* Bristol, U.K.: Institute of Physics Publishing, 2004.
Weber, M. J., *Handbook of Lasers,* Boca Raton, FL: CRC Press, 2001.
Young, M., *Optics and Lasers,* Heidelberg, Germany: Springer-Verlag, 1992.

CHAPTER 4

Beam-Director Technology

4.1 Introduction

Laser systems are generally only effective when the beam from a source is incident on a target; this may be a designated target cued from a separate sensor, such as a radar system or a thermal camera. Alternatively, the target may be sought through a systematic autonomous search of a volume of space around a platform or defended area.

The energy-on-target requirement is fundamental to the success of the vast majority of, if not all, laser-based military systems, but especially for countermeasure, weapon, sensing and targeting tasks. This is because of the highly directional properties of a laser beam; hence, a method is required to direct this narrow beam at a target, whether the application is searching, scanning or precision pointing at an object. The pointing device that provides this function is usually called a beam director and its technical performance is critical to the success of any lased-based system, owing to the crucial need to get the photons on to or into their target, there is very little margin for error with many applications involving laser-based systems, because of the small divergence of the beam.

The specific pointing requirements of the various military applications may require all or some combination of the fundamental pointing functions, outlined above. In most cases, some form of beam director will be used, although it is possible to undertake some of these functions by suitable configuration of the laser cavity by, for example, configuring a conjugate laser cavity to permit limited fine-beam steering via a programmable aperture within the cavity. The discussion of beam-director technology in this chapter will consider target-tracking, pointing and scanning functions to illustrate the various fundamental requirements of the use of laser beams.

The beam-director system is often one of the most technically challenging and critical of all of the subsystems in any laser-based system, particularly for the more demanding military applications. In the case of the simplest systems where less pointing precision is required, the beam director may be a sighting device such as a telescope, which has to be aligned accurately with the propagation axis of the laser beam, as may be used for a land-based range finder. However, the majority of applications require a more sophisticated beam-direction system, especially if sightline stabilisation is necessary to achieve the requisite scanning or pointing function from a moving platform. Moreover, a target-tracking function may be required to ensure that the energy-on-target requirements are met in the case of an engagement with a moving target.

The purpose of the beam-director system is to point the laser beam at a designated target and keep it there for a sufficient period, which may range from a fraction of a second to many seconds. This involves the detection, acquisition and location of targets, selection of the appropriate target and then tracking it; furthermore, for some laser-damage systems it is necessary to identify susceptible zones on the target, particularly for very large targets. The latter activity is known as aim-point selection and refinement; this means that the laser beam can be concentrated on to the desired location to meet the necessary energy-on-target requirements needed.

In the case of moving targets, it is crucial to be able to predict accurately where the target will be when the laser is fired, which implies that relatively sophisticated target-tracking algorithms need to be used in the target-tracking system. Additionally, if the laser-based system is mounted on a moving platform, the beam-directing element must be moved in a manner to counter the angular motion of the vehicle (i.e., so that it is space stabilised to eliminate own-ship motion).

When a laser is used for a sensing or targeting application, the requirement on the beam director is usually somewhat different from the directed-energy case outlined above, as the beam is normally required to find and interrogate a target zone in space to find, classify or identify an object. For example, with a laser-radar (Lidar) system, the laser beam may be required to scan a zone of space and record the reflected laser light so that an image may be formed and refreshed during the scanning process. This image may be in a single dimension or a multidimensional rendering of the scanned scene. In this case, the sensing system is used in conjunction with an inertial-reference system to provide the acquired image with a position-fix with respect to a reference bearing, axis frame or geographic location. The reflected data may be superimposed on an image of the scene, or a map depending on the application.

The beam-directing element, which is often a mirror, requires angular freedom to achieve the pointing or scanning functions, as well as stabilisation in space, if this feature is required. The simplest and most mature approach is angular motion about two orthogonal axes of metal frames (gimbals) (i.e., giving two degrees-of-freedom of angular rotation). This approach can lead to a very compact pointing device but has the disadvantage of a specific angular position where there is a pointing singularity, known as the nadir position, where it is impossible to point accurately at any moving target or provide accurate sightline stabilisation, owing to a loss of pointing authority (or precision-pointing ability). This condition, which is analogous to gimbal lock in a free gyroscope, usually occurs within about 2° of the singularity position when the two axes of rotation are very close to alignment.

In a dynamic engagement, the occurrence of the pointing singularity is usually only momentary; moreover, nadir avoidance methods have been devised to minimise the impact on the installed system performance. Moreover, techniques have been identified to sense the onset of this condition and undertake various strategies that apply techniques to mitigate the effect on the pointing ability of a two-axis system, such as organising the orientation of the position of this singularity so that it rarely occurs. Alternatively, if the subsystem senses the onset of this condition, the tracking algorithm and control logic may change mode to enable the target to be tracked without the beam being emitted. However, this limitation is mainly a concern for directed-energy applications and is less of an issue for sensing systems.

A beam-director design with three or more degrees-of-freedom of rotational motion has been used to avoid the nadir condition. The availability of the third axis enables a small displacement to be added to the angular position vector of the gimbal positions so that the axes of rotation do not become aligned. The use of additional degrees of freedom may be required if the beam director has a moving environmental cover, so that the effects of variable frictional forces on the pointing vector are avoided.

The use of the additional axis, or axes, of rotation adds additional size and complexity to the beam-director's design. The third axis is usually on the inner axis of rotation (inner gimbal) and may have as little as 5° or 10° of angular displacement, whereas this inner gimbal may have up to 180° angular freedom. However, the angular freedom of the inner gimbal is usually a little more than 90°, and the outer axis of rotation could provide continuous rotational capability. In the case of a 360° rotation or more, slip rings are needed to convey power to the sensors and torque motors on the inner gimbals and collect feedback signals from the sensors.

The angular motion of the target-tracking, sensing and beam-pointing systems have traditionally been achieved through the use of mechanical gimballed-arrangements; however, there is modern research into solid-state beam-steering techniques that offer enhanced precision, as well as speed of response. Currently, the solid-state or nonmechanical techniques have a limited field of regard; therefore, hyper-hemispherical fields of regard would require some significant additional complexity in the optical design if a purely nonmechanical technique was to be implemented.

For the more accurate and demanding beam-director applications, such as many of the directed-energy and countermeasure applications, it is normal to have a target-tracking system embodied directly within the beam-pointing system. Examples of beam directors for this demanding application are shown in Figure 4.1. This is usually an electro-optical device, such as a narrow field-of-view imaging system (camera), which is integrated within the pointing system and shares a common aperture, so that the tracking and pointing axes are co-axial to avoid cross-eye effects. Alternative approaches are possible, which may lead to a simpler system but these tend not to be compatible with very accurate pointing and tracking requirements, often of the order of microradians or less.

The other common application of laser-based directed energy is the directed infrared countermeasure systems used for defeating heat-seeking missiles. These platform-protection systems also require precision beam-director systems, but in this case the pointing stability is normally measured in many tens of microradians, rather than in a handful of microradians. Examples of two such compact systems are shown in Figure 4.2.

4.2 Design Considerations for Beam Directors

The steering requirements of a laser beam through a beam director sub-system normally fall into one or more of the following classifications of pointing:

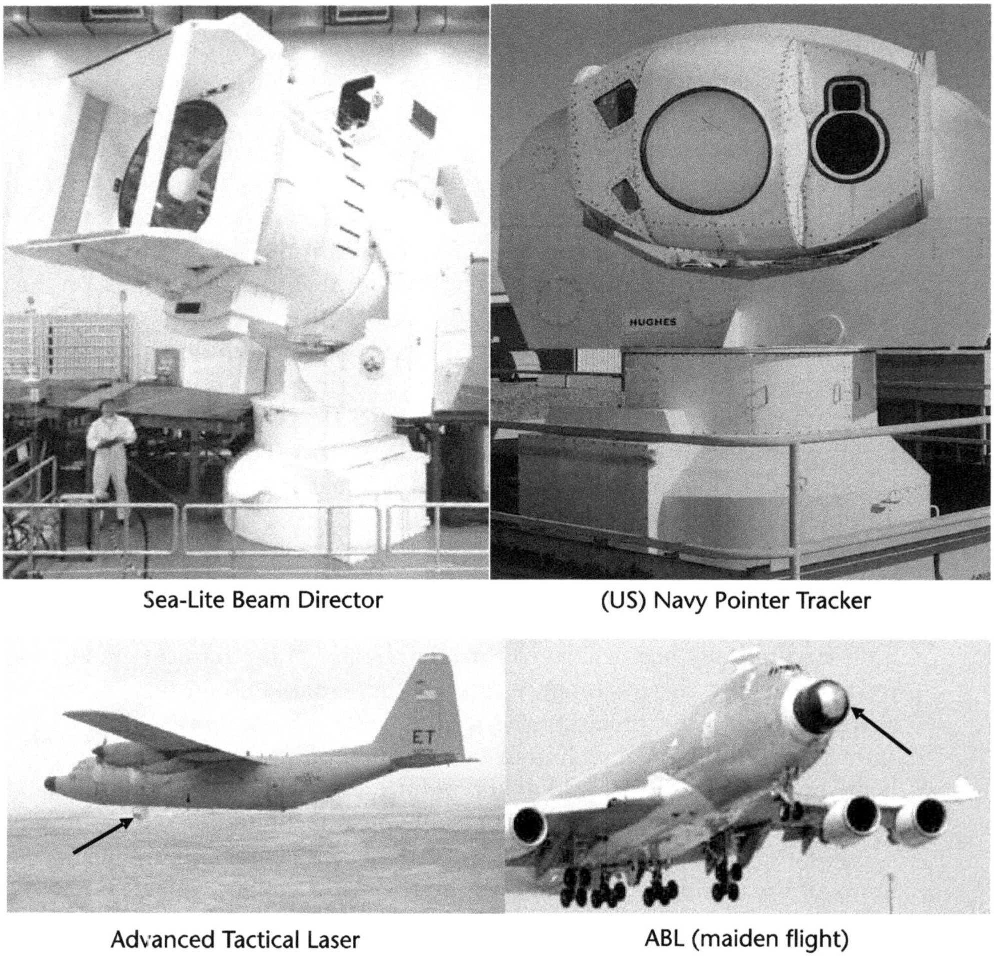

Figure 4.1 Beam directors for high-energy laser applications. (Courtesy of the U.S. DoD.)

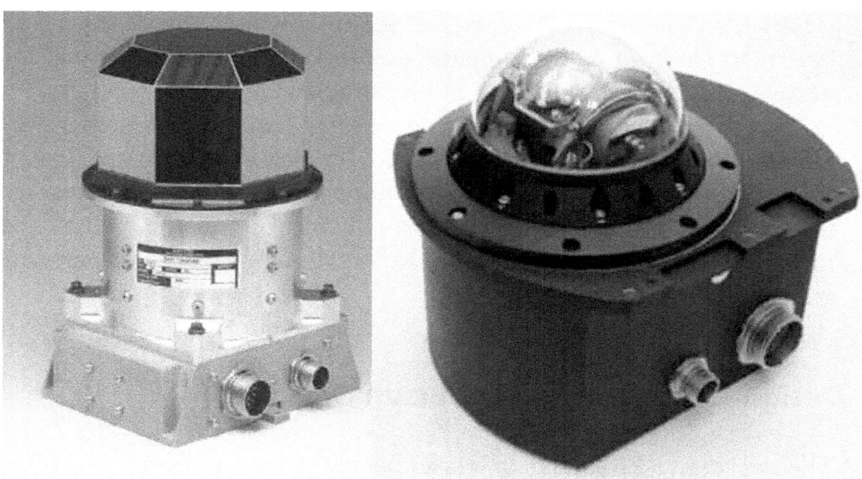

Figure 4.2 Beam directors for directed infrared countermeasure systems. (Courtesy BAE Systems (NA), USA and Selex E S, Edinburgh, U.K.)

- *Scanning:* stepping a laser beam in small increments from one angular position to another, or moving the beam continuously, so that the angular steps are infinitesimally small. This form of beam directing is often required in sensing systems and may demand a scan angle of 90° or more with a scan time of the order of a second or faster depending on the application;
- *Pointing:* directing a beam to specific bearings within the field of regard of the system in response to an external cue. This is a mode used during the first stage of target acquisition and tracking with a high- or low-power directed-energy system. The slewing demand for some applications may be many hundreds of degrees per second over a field-of-regard in excess of 2π steradians.
- *Dwelling:* the mode where a beam spends a specific time at a given bearing or on a specific feature of a tracked target. This may be required during a sensing task or when undertaking a directed-energy interaction. The dwell period may extend from small fractions of a second to many seconds for a directed-energy interaction.

When designing a beam director, it is essential to undertake a system study to gain a full understanding of how the beam director is to be used and the modes of operation that it is required to perform. The design trade-offs can then be undertaken with respect to the type of laser beam that is to be steered and controlled. It is important to embody the performance characteristics of the laser beam into this system study, as the beam can have a dramatic impact on the pointing performance of a beam director, and thus, the overall performance of the entire system.

The functional parameters of a beam director that are usually covered in such studies include:

- Field-of-regard or steering angle in one or two planes;
- Degrees-of-freedom and managing the nadir problem in a two-axis system;
- Angular agility;
- Laser-beam divergence and imaging requirements to support a sensing function or target-tracking requirement, for example;
- Aperture size and minimising vignetting effects;
- Spectral bands of operation and control of dispersion;
- Transmission in the various channels;
- Control of the steering angle.

4.3 Generic Beam-Director/Pointer System for Directed-Energy Systems

Beam directors face many challenges, such as finding a designated target, provided by an external cue, as well as maintaining the energy on that target. Moreover, it may be required to undertake additional functions, so that the progress of the laser-target interaction can be monitored and the effectiveness of the interaction between the beam and the target can be assessed. The system may be considered to undertake beam-steering and pointing functions, along with a target-tracking capability. A schematic of a beam director system is shown in Figure 4.3.

For this application, a generic beam-direction system may be considered to be a mechanically gimballed structure that has or is integrated with:

- A situational awareness or target-alerting system, to provide the targeting cues;
- A target-tracking system, to establish the position of the designated target and aid the pointing function;
- A system processor and gimbal-control system to control the steering angle including beam stability;
- A pointing element and motor-drive mechanism, to achieve the energy-on-target requirements through a precision pointing and stabilisation functions;
- Inertial sensors to provide a definition of the reference axes, to provide the angular feedback for control and stabilisation functions;
- A power supply to support all of the electronics and motor control subsystems;
- An environmental cover to protect the delicate components from the external environment.

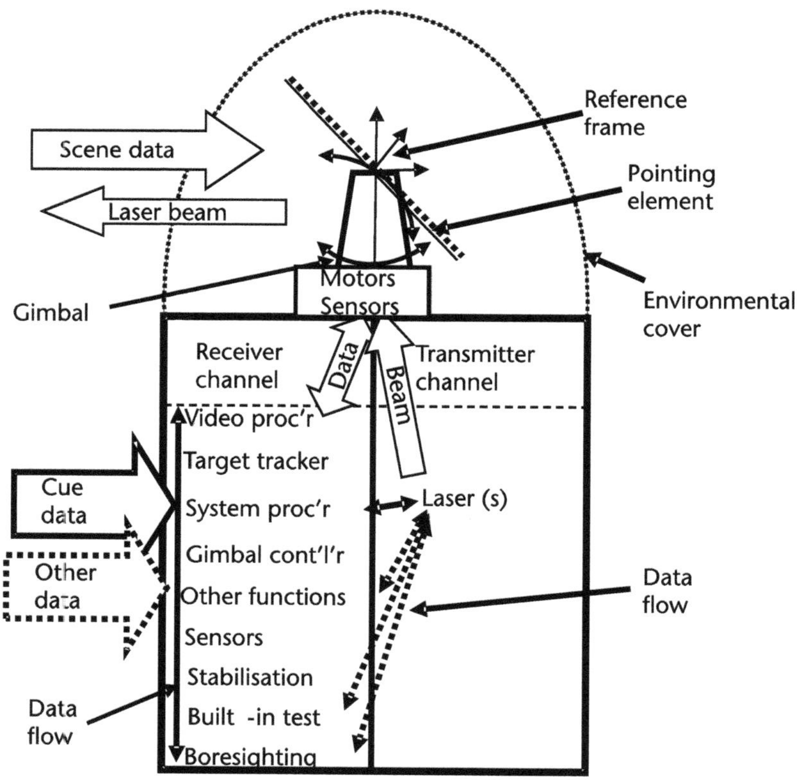

Figure 4.3 Generic beam director.

4.3.1 Gimbal

The gimballed structure is a mechanical arrangement of supporting framework mounted on precision bearings (with a high-accuracy preload to provide a very smooth motion) that has a minimum of two degrees of angular freedom. It is a rigid frame supporting the beam directing element (mirror) and allowing rotation of the mirror about one axis, whilst the frame itself can be rotated about an orthogonal axis, and by which, it is attached to the rest of the system. The structure may have optical, and usually, electrical, slip rings.

The beam director will have a number of degrees-of-freedom of rotation to enable a target to be tracked and engaged with precision over a wide field of regard, which may be greater than a hemisphere (hyper hemispherical) depending on the specific system requirements; this requirement is quite common for laser-based countermeasure applications. As noted above, a system with only two degrees of freedom is simple and very compact, but has a pointing singularity at the nadir position when the gimbal's axes of rotation line up.

Techniques have been developed to minimise the impact of the nadir condition on both the pointing and tracking functions. One of the common techniques is to monitor the angular position of the inner gimbal and when it approaches the nadir position make a prediction about the future position of the designated target and initiate a commanded slew of the gimbals to intercept it a few milliseconds later beyond the singularity. With a suitable field of view of the target-tracking (or fine-tracking) sensor, the target can be tracked throughout by use of this nadir-avoidance routine. In general, the time spent by the gimbal in the nadir position is very short owing to the motion of the target and the platform; moreover suitable orientation of the beam director can minimise its occurrence.

4.3.2 Gimbal Functions

An actual beam-director system is much more than the generic pointer described above, because it is usually designed to undertake a number of vital functions to ensure that a laser directed-energy weapon system, of whatever emitted power class, meets the required energy-on-target characteristics. These additional functions are possible because of the sophisticated sensors integrated within the beam-director subsystem. Some examples of beam director functions include:

- Target tracking, usually employing electro-optical techniques to provide the requisite low-noise target-tracking data, such as monitoring the retro-return from the target in one or more wavebands;
- Precision-pointing routines, linked to the precision aim-point monitoring system, to keep the energy on a target during an engagement (this is often of the order of a few microradians of accuracy for those applications requiring very high pointing accuracy);
- Maintenance of harmony of optical axes through a bore-sighting function (i.e., avoiding cross-eye effects).

In addition, for the more demanding and sophisticated military applications, the beam director will undertake additional functions to ensure that the propagating beam has the optimum performance characteristics. These are likely to include:

- Sensing of the atmospheric-propagation conditions along the path to the target;
- Target illumination and probing;
- Target identification and recognition techniques;
- Aberration measurement and correction, using methods such as adaptive-optical techniques;
- Periodic compensation of any drift in the harmonisation between pointing and tracking vectors (i.e., re-harmonisation);
- Techniques for aim-point refinement (on the target);
- Countermeasure effectiveness assessment and battle-damage identification.

These processes are often termed beam control, which are critical for ensuring optimum delivery of the launched beam to the target-interaction zone.

4.3.3 Blind Arcs (Blockages)

One critically important consideration with any gimballed system is the determination of those sightlines, or bearing within a given reference frame, where parts of the platform structure present an obstruction. These blind arcs are determined through an investigation to determine the bearings and the angular extent of these blocked sightlines around a platform to measure the angular extent of the blockages. The bearings of these obstructions are measured in a defined reference frame and stored in the system processor. This process is often called a platform survey.

4.3.4 Optical-Train Design

The design of the optical train of the beam director is one of the great challenges of this subsystem for any application, but particularly for high-power, directed-energy systems. The principal requirements include:

- Minimising scatter and loss from the optical surfaces, by use of precision shaping and polishing of the optical elements and application of very robust optical coatings (however, care is required in the optical design of the entire train to ensure these optical coating are not destroyed by the incident laser flux);
- Control of any scattered or stray light;
- Control of the power density on all optical elements;
- Minimising absorption by the optical elements, consideration of the use of thermal control (cooling) of optical elements;
- Avoidance of optical aberrations and distortion of the beam profile;

- Control and, where possible, elimination of optical back scatter of the weapon/countermeasure beam into other optical sensors that would, for example, blind or destroy those sensors, which may be used for aberration control or target tracking.

The optical train and its supporting structure must be athermalised to ensure that it remains mechanically stable throughout the operational thermal environment. Equally important for high-power applications is the maintenance of cleanliness of all optical surfaces, through hermetic sealing of the entire system, to ensure correct operation and avoidance of self-destruction through high-power interaction with foreign objects. It is normal to use a fully reflective optical train wherever possible for the directed-energy beam path, in order to minimise the compensation issues associated with refractive optical elements.

4.4 Threat-Alerting System

The operational requirement for many laser DEW (directed-energy weapon) or countermeasure systems requires a field of regard of at least 2π steradians to provide target-engagement opportunities over more than a hemisphere. Consequently, it is essential that the system is given cues on the position or location and trajectory of threats in a specific reference frame. Although the threat-alerting system is not actually considered to be an integral part of the beam-director system, it is critical that it provides the cues to the target-tracking device in the beam director, or to the operator of the pointing element in the case of the simpler systems. Moreover, the target cues need to be supplied in a timely and ordered fashion; all components of data latency have to be minimised, particularly for high-speed/dynamic interactions.

A number of different types or classes of sensors may be used to provide the threat-alerting function. For the more sophisticated laser directed-energy weapon systems the alert may come from a surveillance radar, possibly used in conjunction with other sensors, to provide an accurate location of the threat targets. The use of staring, wide-angle camera-like sensors is quite common, as they should maintain a fixed alignment with respect to each other and the beam director(s); these devices may operate in the electro-optical or infrared part of the electromagnetic spectrum. Staring sensor systems are normally configured into a spatially distributed array to provide the appropriate coverage of anticipated threat approach directions, which for air platforms could be a whole sphere (i.e., 4π steradians). In the simpler systems, the search for threats could be undertaken using the human eye with an optical aid such as a telescope, a thermal camera, low-light television or charge-coupled device (CCD) array.

One of the crucial requirements for a high probability of success in the handover process, which should be seamless, is that the angle-of-arrival accuracy of the threat, provided by the threat-alerting sensors, should be significantly less than the field of view of the tracking sensor. In that case, and given a suitable interpretation of the handover data, there is a high probability that the tracking sensor will find the target when provided with the target cue. A further, and equally important requirement, concerns the freshness of the threat-cue data, thus minimising data latency. This is particularly important in the case of engagement of moving targets from a moving platform, especially if the engagement involves high-speed crossing targets.

The handover requirements from the threat sensor to the tracking sensor are normally established through an error-budget analysis. It is normal to aim for a handover probability close to 100% as the design requirement. This error-budget analysis will usually cover:

- Accuracy of the cue bearing estimate (from the threat-warning sensor);
- Alignment or harmony of the threat-warning sensor-array reference system and that of the beam director within a common reference frame;
- Any systematic errors, such as those generated by thermally induced effects in the body structure;
- Any random effects, such as aero-flexure or distortion of the platform, particularly lever-arm effects between any angular reference sensor subsystem and an effector, such as a beam director;
- Data latency effects and target motion rates;
- Field of view of the tracking sensor.

In the case of a threat-sensing/alerting system being installed on a platform, such as a ship or an aircraft, it is absolutely critical that a platform survey is undertaken to establish where the platform structure obscures parts of the field of view/sightlines of each sensor; this is the identical installation problem discussed above for the steering angles across the field of regard of the beam director. This information only has to be established once for a given installation on a specific class of platform; the information is usually stored in the system processor or similar computer. Equally important, is a complete understanding of the overlap region between the fields of view of staring sensors and ensuring a unique target declaration for each detected target by the alerting subsystem, which is essential to avoid ambiguity in target referencing.

4.5 Target-Tracking Subsystem

Most beam directors have a dedicated and sophisticated target-tracking device, integrated within its optical system to undertake closed-loop tracking of a designated target; its function, as its name implies, is to track a target and establish the sightline, or tracking vector, to that target. It is normal for this subsystem to have the potential, in conjunction with the video processor, to be able to track a number of threat targets simultaneously. It is common to use a mirror positioned within the beam director's body to direct energy from the observed scene into the camera unit.

Some form of imaging system with a relatively small field of view is often used to track a target; thermal cameras are usually used but other imaging sensors working in a different part of the electromagnetic spectrum are equally feasible. This camera usually has a focal-plane array with a format of at least 256 by 256 elements (pixels) to provide the necessary resolution to track a target accurately. For those high-accuracy applications requiring microradian accuracy, a more sophisticated array could be used, augmented by active-tracking techniques, using retro-reflection observations from a target.

The size of the array required in the cameras is normally determined by:

- The field of view required to ensure a high probability of handover from the cueing sensors;
- The angular resolution required to ensure that multiple targets can be resolved;
- The angular resolution and data (frame) rate needed to achieve and sustain the requisite tracking accuracy;
- The need to resolve any specific features on a target.

The detector-array device in the tracking camera may use any of the modern high-performance sensor materials; the fundamental requirement is to have adequate signal-to-noise characteristics, above the background and clutter, for the image processor in the camera to be able to distinguish the designated target from frame to frame (i.e., establishing adequate contrast with a realist integration time). The sensor used is commonly sensitive to a portion of the mid-wave infrared, so in this case there are a few binary or tertiary semiconductor alloys available, but they need to be cooled to cryogenic temperatures to achieve the requisite detectivity. The mid-waveband infrared is favoured, as it has greater operational flexibility, better transmission and reduced sensitivity to turbulence.

This tracking sensor is often called the fine-track sensor; it operates in the infrared, tracks the designated target via the thermal contrast that the target (e.g., missile) has within the thermal scene captured by the camera. The thermal contrast may be positive or negative, enabling the video processor to operate on a suitable discrimination technique. This sensor may track a target by collecting and observing photons from:

- Emission from a rocket or engine plume;
- Aerodynamic heating;
- Laser retro-reflections, as occur in the cat's eye effect from some sensors (discussed in more detail in Chapter 8);
- Thermal emission from the laser-interaction zone (for high-power applications).

The design of the optical train in the fine-track sensor is optimised to give a narrow field of view, consistent with a high probability of successful handover from the target cueing system and adequate target resolution; additionally, a narrow field of view will provide robust tracking performance in the presence of decoys, clutter and other potential false targets. The optical train of this sensor is also designed to minimise the effects of back-scatter of photons from the directed-energy source on the tracking, or sensing, performance of this sensor. Additionally, the size of the optical aperture is chosen to ensure that sensor can acquire the designated target at the design acquisition range of the laser-based system it is supporting.

A further fundamental requirement is that the optical axes of the tracking and pointing systems are harmonised, so that their axes are co-axial and co-aligned throughout the field of regard. This is discussed further below.

4.5.1 Target-Tracking Algorithms

There are a number of target-tracking algorithms, ranging from the very simple to the highly complex and very sophisticated. The simplest algorithms are generally not suitable for tracking a moving target accurately, even with a suitably stabilised beam-director system, as the tracking vector lags behind the actual sightline. This is an effect known as pointer lag, as illustrated in Figure 4.4.

The use of a Type-1 (tracking) algorithm allows a target moving with constant angular velocity to be tracked with a fixed pointer lag, the value (angular offset) being proportional to the magnitude of the angular motion; however, if the target sight-line angular rate changes (accelerates or retards), then pointer lag would continue to change in magnitude (increase or decrease according to the sign of the rate of change). In the case of the implementation of a second order, or Type-2 algorithm, the pointer-lag phenomenon would only occur during the presence of angular acceleration of the sightline angle. However, these higher-order algorithms are susceptible to noise effects, resulting from the differentiation processes used in the feed-forward loops, used to predict future position of the tracked object. Careful design of these tracking loops is essential to ensure that the gains and offsets used in these control systems are optimised to ensure that noise-induced effects are minimised.

In practice, several techniques may be used to ensure very accurate tracking and subsequent control of the directed-energy beam, such as:

- Switching between tracking algorithms depending on the target-engagement parameters;
- Use of active-tracking techniques where the hot spot formed on the target is used as the tracking point in the aim-point refinement routine.

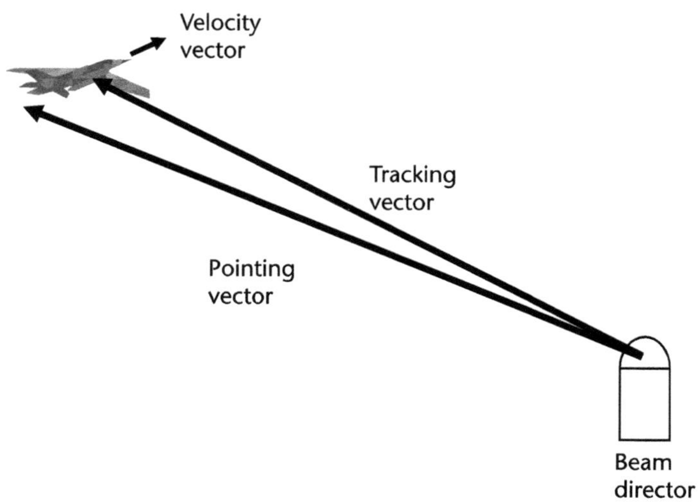

Figure 4.4 Pointer lag.

4.5.2 Aim-Point Refinement Techniques

The use of an imaging sensor in the beam director provides the fundamental target-tracking function, but this may be enhanced by the use of the retro-return of laser energy from a target, which may be interpreted by the video processor to provide an effective tracking source. The retro-return signals can take a number of forms, such as an enhanced thermal return from a laser-induced hot spot or an optical return from a sensor on the target. However, the use of a target return means the beam director's tracking system has to have a sensor, such as a camera or quadrant detector, which is sensitive to the wavelength(s) of the retro return or emission.

A single- or dual-band tracking system would be feasible, using a dichroic optical element to form the two independent optical tracking channels leading to the sensing through imaging cameras for each waveband. More than two separate channels would be a complex optical system, which, would be bulky and potentially less reliable than the single-channel system.

An alternative approach is to use the geo-location, where the target point is defined within a suitable area of the engagement zone and the position and orientation of the platform is known in the same reference frame. The latter data would have to be derived from a full inertial-navigation system on the platform. This information on the target point and platform position would enable a sightline vector from the beam director to the target to be defined and then stabilised using suitable inertial sensors and beam-directing algorithms. This approach is more likely to be allied to some of the target-suppression applications using electro-optical countermeasure techniques.

4.6 Pointing Element

The pointing element is usually a mirror mounted in a gimbal assembly, as already described for the generic pointing system, which enables the beam to be directed into space with great precision. The gimballed assembly may be designed to give a number of degrees-of-freedom of rotation to the pointing vector, usually between two and four, depending on the application. Moreover, the final target-tracking and beam-pointing optical elements, used to provide these functions, are often combined on a single mechanical element and mounted on the same structure to preserve harmony of their optical axes, whilst preserving the separate optical paths.

Some lower-power directed-energy systems use a common aperture optical system to ensure harmony of the various optical axes, although control of optical back scatter is a critical issue with this configuration, which is likely to exclude this technique from being applied to the very high power applications. Techniques for minimising the impact of the optical back scatter have been demonstrated, usually involving camera shuttering methods, often termed camera blanking or specific use of optical-filter technology and subtle differences in the emission and tracking wavebands.

The design and fabrication of the mechanical gimbal assembly is one of the critical aspects of this subsystem, as the axes of rotation have to be mutually orthogonal and maintained as such, over the operational environment and entire field of regard. Control and minimisation of random errors are essential for high-accuracy pointing and tracking, as they are impossible to compensate for in the short time scales of an engagement. Systematic errors may be eliminated, although sophisticated error-characterisation techniques may be necessary. Clearly, a high-precision pointing and tracking system requires the following:

- Precision-engineering methods ensure that design tolerances are achieved to provide smooth and precise rotational motion. Sophisticated bearing design is an essential requirement to maintain a constant load and thus stability of the axes of rotation.
- Brushless electrical motors are frequently used to provide the precision torque needed to achieve a rapid and accurate slew commanded rotation, followed by a rapid stop at the desired angle in the settle period, prior to the start of the closed-loop tracking mode of operation, with precision positioning and accurate responsive control.
- Advanced design methods to undertake analysis of modal responses to vibratory motion and avoid undesirable resonant characteristics.
- Precision instruments, such as gyroscopes and resolvers, provide the angular motion data for the operation and control of the pointing and tracking system through the sensing of the angular motion of the beam director's gimbals. It is common to use the resolvers during the slew of the system, such as a commanded motion to a given bearing, followed by use of the higher resolution of the gyroscopes for the accurate control of the beam-directing element. Separate rate sensors are used to avoid any cross-coupling effects that would compromise the tracking accuracy of the higher performance beam directors. The angular-rate feedback from the gyroscopes plays a crucial role in the sightline-stabilisation process.
- Careful choice of all materials used in the mechanical assembly ensures both compatibility of contact, without degradation or corrosion, and the avoidance distortion, mechanical relaxation or displacement over the operational conditions of the system.
- Environmental sensors allow appropriate real-time compensation of errors induced by changes to the ambient conditions.

4.7 Target Sightline Stabilisation

The gyroscopes and the target-tracking sensor within the beam director provide the vital data for the line-of-sight isolation of the tracking and pointing vectors, or inertial stabilisation, to compensate for motion of the beam-director system when installed on a moving platform. All of these sensors are operated in a closed-loop scheme to provide accurate measurement of angular rates to enable accurate sightline isolation. The closed-loop approach enables low-cost gyroscopes to be used successfully; however, well-characterised devices must be used in the system, so that the drift of these sensors does not impair performance of the stabilisation loop.

The principle of sightline stabilisation is that the angular sensors detect motion of the gimbals and supply details of this motion to the control loops. They in turn use these data as error signals to offset own-ship motion. For the lower-precision systems, the sightline-stabilisation loop deduces the body motion induced in the beam director from the motion of the tracked image on the detector array and sensed motion recorded by the gyroscopes. The required compensation is deduced from comparing the motion of the target image on the fine-track sensor with the angular motion sensed by the gyroscopes. The higher-precision systems will have at least a high-accuracy attitude and heading reference system to provide the inertial attitude reference for the space stabilisation. In both cases, the conjugate motion is applied to the sightline angle to provide the appropriate compensation of the detected platform motion.

4.8 System Processor

The system processor is a computer controlled subsystem that undertakes all of the critical functions associated with the command and control of the engagement process, operation of the laser and its beam director and the associated subsystems. The processor for a high-precision directed-energy application may contain, in addition to a system control unit, the following subsystems and functions:

- The target-tracking algorithm processes the data from the received image captured on each frame of the fine-track sensor (camera), establishes the track (trajectory) of the target, via frame-by-frame image processing, to define the target motion through correlation techniques, and provides these data to the gimbal controller.
- A threat database contains susceptibility information about a range of targets likely to be encountered on a given mission.
- An image-processing routine that identifies a potential target from the data collected from the target-tracking sensor to select an aim point on a target using the stored threat database, which gives susceptibility data for the identified target, and the known engagement conditions. In the case of an unidentified or unrecognised threat, undertake a rapid assessment of potential susceptible points and select a suitable aim point given the known engagement conditions.
- A gimbal/pointing vector controller that undertakes the pointing function of the beam director. This control system will contain the appropriate class of target-tracking algorithms, or switch between various types of tracking algorithms, so that the damage-class beam may be pointed very accurately at the selected aim point on a moving target without the pointing vector lagging behind the tracking point (i.e., maintaining energy on the selected aim point). The closed-loop operation of the target-tracker and this control system also inherently provides the system-stabilisation function, but for the most accurate systems may be assisted by local inertial-navigation units to ensure very high precision sightline stabilisation.

- A system database of the characteristics (systematic errors) of the subsystem that compensates for systematic errors and provides installation and platform interaction information.
- A control system that assesses the ambient conditions and applies the appropriate compensation to the pointing vector, via the gimbal controller.
- The adaptive-optical control system (if embodied as a subsystem) for beam-shape control, which is a fundamental part of a beam-control subsystem.
- The laser-control logic that activates the weapon laser, controls its emission parameters and then maintains the desired flux on the aim point In the case of a laser-countermeasure system, it applies the chosen modulation switching to form an appropriate jam waveforms.

The gimbal-system controller provides the command signals to a servo unit, which generates the power signals to the gimbal motors and, hence, drives the optical axis of the laser beam to the demanded position and maintains the pointing vector on the selected aim point of the target. The system controller also provides an interface for other devices to monitor the behaviour of the system.

The system processor used in lower-power and less sophisticated directed-energy systems is usually a subset of the above list of functions; however, the basic principles still apply in terms of optimising the performance of the beam-direction system. Whether it is a high-performance system or otherwise, maintenance of energy on target is a critical requirement parameter; in the case of the high-precision system that requirement may not be quite adequate, the energy may need to be focused on to a specific part of the target.

The system processor usually has the following information about its subsystems:

- Harmony, or known orientation and angular offsets, between the reference system of the threat-alerting/cueing system and the beam director(s).
- Harmony between the optical axes of the tracking system in the beam director(s) and the laser beam(s), that is, the pointing vector (s).
- Algorithms for the correction or compensation of the systematic errors (as a function of the ambient conditions and the operating environment), which will initially be undertaken during an optical boresighting process and then monitored during an engagement, so that real-time compensation may be applied; in the case of the very high precision out-of-band directed-energy systems, additional subsystems may be required to assist with this critical process.
- A map of the possible engagement zone that is placed around a platform for each beam director (if there is more than one) so that blind-arc engagement of a target, or unintentional multiple engagements of the same target are avoided.
- Knowledge of the fields regards each beam director, if more than one is fitted on the platform (particularly for platform-protection systems), so that the most appropriate beam director can engage a designated target. In the case of a laser system using more than one beam director, such as that used as a part of a platform-protection system, the processor also has the ability to switch the jamming beam from one transmitter (beam director) to another, or use

both beam directors against a designated threat, which is quite common with lamp-based systems but less so with laser-based systems.

4.8.1 Built-In Test and Boresighting Functions

The system processor usually initiates the built-in test procedures to ensure that each of the subsystems is fully functional. The system processor will also undertake and control the boresighting function used to ensure that the optical axes of the target-tracking and pointing systems are in harmony: generally, sophisticated methods are necessary to ensure minimal misalignment, particularly for the very high-power, high-precision out-of-band damage-class weapon systems.

A number of sensors are often integrated into the beam-director assembly and its subsystems to ensure that all of the subsystems are available and operating according to specification, particularly before the initiation of a target engagement. Tests of the subsystems' electronics are also undertaken and monitored before and during a target engagement. There will be algorithms within the system processor to monitor the performance of the beam director.

The boresighting function that establishes the harmony of the optical axes may be implemented in a variety of ways using the subsystems within the beam director; three approaches are quite common, but a number of others are also feasible:

- One technique involves a calibration process, undertaken during assembly of the system, where the error between the two principal optical axes of the beam director system are determined throughout the field of regard and stored in the system controller. The stored information is then subject to various algorithms to compensate for systematic errors. During operational use, sample measurements are made periodically, at previously measured cardinal points using the fine-track sensor/camera, so that the current offsets may be compared with those recorded during the calibration process and then applied as a systematic angular correction to all pointing commands.
- An alternative approach is to undertake a full multipositional measurement of the pointing vector on the fine-track sensor/camera at known orientations of the tracking vector. The position and size of the spots formed on the fine-track-sensor array enable an estimate of the systematic error between the principal axes to be extracted from this pattern, from which the systematic error-correction that needs to be applied pointing command can be determined.
- Continuous monitoring of the position of the laser beam at some point in its path from the source aperture to the beam-directing element using a beam splitter and a quadrant detector. This approach requires a stable emission from the laser to gain an assessment of the beam displacement as a result of systematic errors in the optical train. This technique is often known as continuous auto boresighting.

In each case a low-power and stable beam is required, preferably using the actual directed-energy beam, but in a highly attenuated state, as discussed next.

The techniques described above, may be considered internal techniques or self-contained methods; however, there are other boresighting approaches that use external optical elements and sensors to ensure the principal optical axes are co-axial. An example is to use an external mirror or prism at a fixed position to compare the receiver and transmitter optical axes and also use the system to view several distant objects and compare the position of the target point with the pointing axis position using a suitable detector, which could be the fine-track sensor. The pointing vector can then be adjusted, using a range of software or mechanical methods, to ensure it is co-axial with the tracking vector. This type of approach is generally suitable for the simpler beam-director systems.

In principle, the techniques and methods used for the high-precision weapon systems will be similar to those described for the generic system; however, low-power alignment lasers and real-time compensation techniques, allied to an adaptive-optical system, are necessary to ensure minimal pointer-to-tracking vector error. Moreover, it is likely that additional sensors will be used to monitor the position of the laser-target interaction zone during an engagement with a target.

4.9 Other Functions

4.9.1 Beam Control

The system processor may also be required to help direct and deliver the optimum energy to the target. The use of an adaptive-optical system is a critical technique to achieve this goal, as it attempts to minimise aberrations induced into the directed-energy beam, from whatever cause, to shape the beam and deliver the optimum flux to the target. This activity is often called active beam control. Figure 4.5 shows the

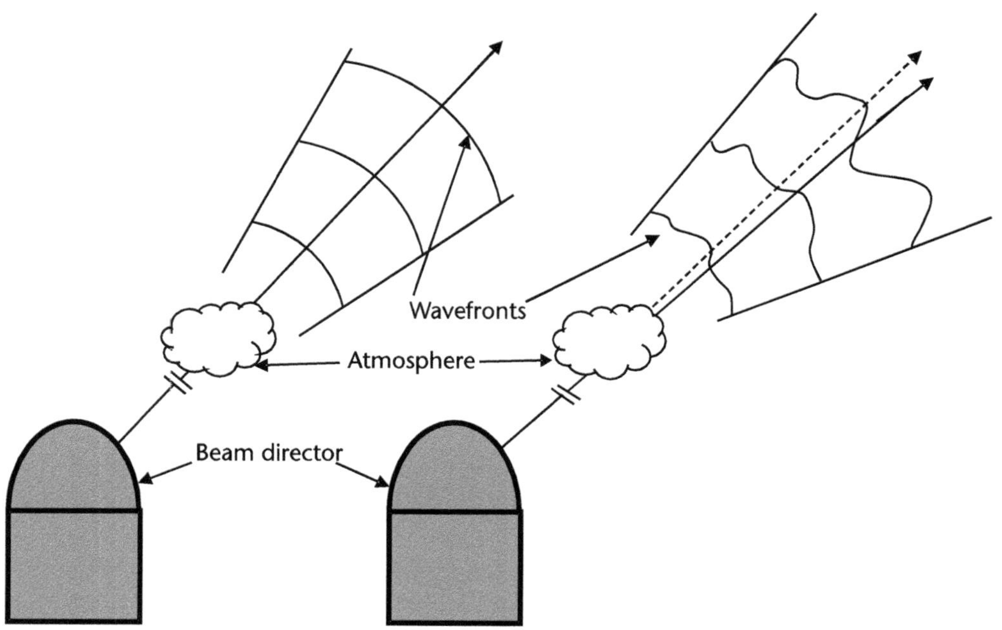

Figure 4.5 Schematic illustration of the impact of aberrations on beam propagation.

need for aberration correction, the illustration on the left shows the ideal propagation case, whilst on the right is the effect of aberrations on the propagating wavefront, showing the distortion in the wavefronts resulting in an increase in beam divergence and modification to the propagation direction.

Adaptive-optical techniques involve the use of additional sensors that establish the scale of aberrations induced into a laser beam as it propagates to a target. Algorithms are required to determine the precise value of the aberration introduced into the propagating wave and calculate the correction required to create the conjugate beam within the system, before being launched to the target. In this case, a sophisticated type of mirror is required to create the conjugate beam, which will be distorted by the aberrations along the entire optical path, resulting in the optimum beam shape at the target.

A couple of different types of mirror are available to form the conjugate beam, either a segmented element or a continuously deformable mirror. The principle of a closed-loop, adaptive-optical technique for closed-loop systems is shown in Figure 4.6, showing the incident aberrated wavefronts to the laser system and the formation of the conjugate wavefront on the launched beam.

The very high-performance systems require full correction of the aberration, which involves correction to the focus as well as the angular deflection, known as tip-tilt correction. Most moderate-power systems would require at least a tip-tilt correction, also known as first-order correction. This is shown schematically in Figure 4.7.

4.9.2 Battle-Damage Assessment

Another important function for any directed-energy system is the monitoring of the effectiveness of the laser-target interaction, often called battle-damage indication and assessment, or in the case of laser-based countermeasure systems, countermeasure effectiveness. In Figure 4.6, the camera monitoring the input beam from the target may act as the accurate target-tracking sensor, or a separate camera may be used as indicated.

A number of techniques may be used to monitor a high-power engagement, usually based on sensing a change in the thermal emission from a target; this

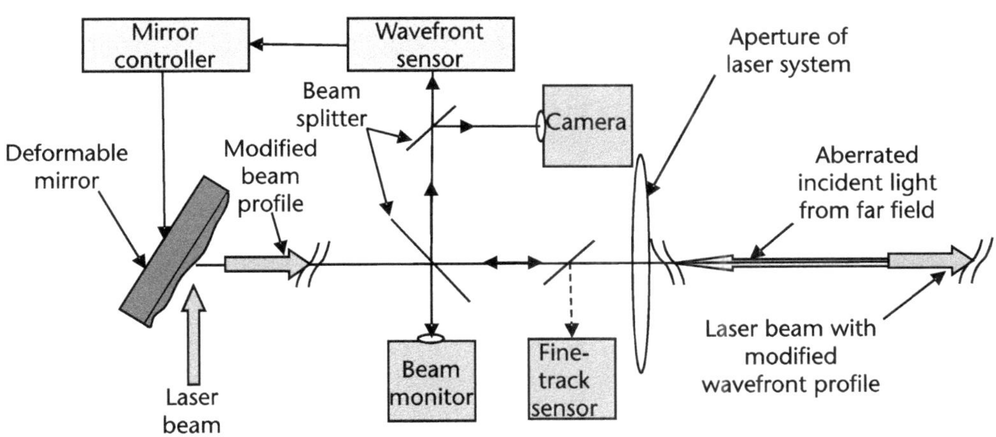

Figure 4.6 Principle of a closed-loop adaptive-optical system.

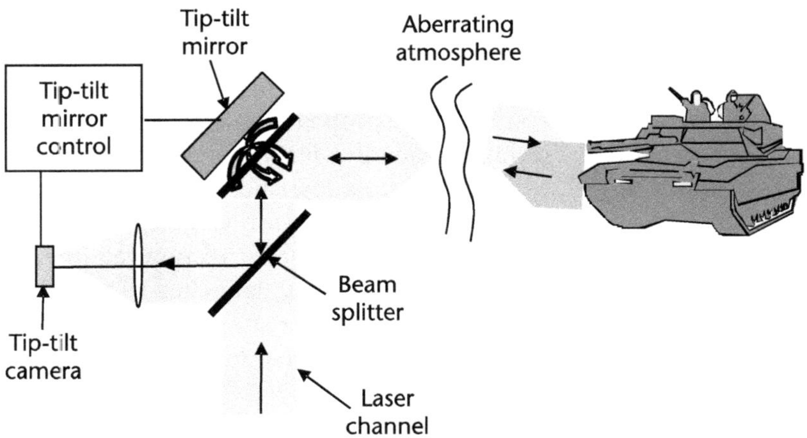

Figure 4.7 Tip-tilt correction. (Courtesy of Dr. A. M. Scott, QinetiQ, U.K.)

feedback function may be achieved using the existing high performance sensors within the beam director. However, other techniques based on the spectral analysis of an emission from a target, such as laser-induced breakdown spectroscopy, will need additional sensors. Some of these techniques are discussed further in Chapter 13, which discusses remote sensing.

Several techniques are quite common for assessing countermeasure effectiveness. These techniques are based on observation of the change in the characteristics of the retro-return from the illuminated seeker, which may be achieved using the target-tracking sensor. A typical metric concerns monitoring the reduction in the retro-return as the seeker in the approaching missile is deflected during the jamming process. Alternative techniques may involve estimation of perturbations in the approach trajectory.

4.10 Video Processor

The fundamental task of the video processor is to analyse the incoming image stream from the target-tracking sensor or camera, detect the location of the targets of interest, track them from frame to frame and pass an aim-point to the closed-loop controller. The video processor has an important role in maintaining a tracking box around the acquired target. This is a process where each target known to be within the field of view of the fine-track sensor is given a unique label and bounded within each frame. The centroid or edges of the image of the target may then be determined using contrast-based techniques within each tracking box, so that target motion can be determined on a frame-by-frame basis.

Modern digital sensors are capable of capturing images containing millions of pixels with up to 16 bits of intensity data per pixel at frame rates of up to many kilohertz. The first step in the video processing chain is often to select a region of interest to process and to reduce the bit depth to increase the processing rate. A video processor may start off processing the entire image (at a low frame rate) to detect the target of interest and subsequently process a much smaller region at a higher frame rate to lower latency and ultimately improve the stability of the closed-loop tracking.

The target detection algorithm can range from the relatively simple centroid tracker, which finds the centre of energy of hot spots within a single image, to much more complex filters. These filters attempt to match the video data with known target signatures over many frames. The results from the image processing on each frame can then be combined to create target tracks using techniques based on those originally developed for radar systems.

Combining the frame-to-frame target motion in pixels with the platform and gimbal motion as well as their orientation can allow the target motion in real-world (inertial space) to be estimated. This can often be used as a cue to differentiate between background clutter and true targets. Other techniques to reduce the effect of clutter within the image include tracking the background motion in image space to highlight targets that are moving or changing in some way.

4.11 Power Supply

The power-supply requirements are often taken for granted in this type of sophisticated system, but having the appropriate conditioned power available, on demand, is crucial for the successful operation of this type of subsystem. One of the critical requirements is that there is sufficient power available during a commanded slew, for example, so that pointing vector can be suitably oriented as quickly as possible; moreover, the supplied power must be able to support the demanded angular motion, so that the gimbal system moves as smoothly as possible enabling the pointing vector to be positioned accurately and rapidly over a range of environmental conditions.

Modern high-performance power-supply systems, which are very compact, are readily available for this type of application. One of the very important design considerations is to establish the power demands over a range of operation conditions for the various modes of operation of this system. Careful consideration is required of potential worst-case power demands on the power supply, so that an appropriately-sized power-supply system is integrated with the system. However, some careful consideration is required concerning the aggregation of worst-case conditions; otherwise, an expensive and large system will be installed but only a fraction of its capability ever used.

Most platforms have raw power available for these applications; however, it is necessary to condition this type of electrical power, as it is normally unregulated and subjected to brown outs, particularly when there is an excessive demand from any of the other installed systems. Hence, careful conditioning and smoothing of this type of power are vital if uninterrupted performance is necessary (i.e., assured performance on demand).

4.12 Environmental Cover

Several approaches have been used to protect the gimbal structure from the natural elements, particularly for airborne applications. Some systems use a plane window that moves with the gimballed structure, which is more applicable to the higher-power applications. An alternative technique is to use a fixed structure often called a dome, which is more common for lower-power applications such as laser-based countermeasure systems. Multiple plane-panes of a material with low transmission loss have been used to form a dome-like structure for lower-power applications. Some examples are shown in Figure 4.8.

In the case of a multifaceted structure, these panes are anti-reflection coated to optimise transmission and minimise Fresnel reflections. The panes are accurately flat and the faces are machined to be parallel to minimise ghost effects and other aberrations. Additionally, the panes are bevelled on all edges so that they fit together accurately and minimise split-pupil effects. Systematic errors may be offset by optical characterisation; for example, to avoid any split pupil effects as the beam passes through the joint, it is possible to displace the beam position slightly.

New developments concern the use of a spherical or curved structure, but this approach presents some design challenges as this type of optical element has optical power. The optical elements of the beam-pointing element need to be at the centre of curvature of this component, which is not a trivial design task for a compact system. Moreover, the component becomes very difficult to manufacture to the appropriate tolerances if the beam director is going to have a significant hyper-hemispherical field of regard. Another challenge is the application of an appropriate anti-reflection coating to the inner surface, to give an effective optical performance over the entire field of regard. Many of these design challenges may be overcome through very careful value-engineering techniques and some compromise in performance at the edges of the field of regard of the beam-director subsystem. This design has been developed successfully by the Israeli company Elbit Systems with their J-MUSIC system illustrated in Figure 9.9 for the installation on their Boeing 707 and in Figure 4.9.

The challenging performance requirements of high-power laser-damage weapons demands the use of very high-quality optical materials that will not distort the beam and crucially have negligible absorption at the emission wavelengths. Additionally, the optical aperture is likely to be large, so the window must not distort under its own weight and must be tough enough to provide a barrier from the environment. The requirements for negligible distortion and aberrations generally indicate the need for a plane structure, such as the window in the centre of the image shown in Figure 4.8.

Figure 4.8 Environmental covers. (Courtesy of Selex E S, U.K., NRL, USA and BAE Systems (NA), USA.)

Figure 4.9 The J-MUSIC hyper-hemispherical dome. (Courtesy of Elbit Systems, Israel.)

Developments with materials such as spinel have enabled large-aperture low-loss environmental windows to become a practical reality for high-power systems. Sapphire and silicon continue to be favoured for lower-power applications.

4.13 Laser-Integration Requirements

This is not strictly part of the beam director, but it is yet another one of the many critical design integration requirements that need very careful consideration so that unwanted errors are not introduced into the pointing vector. The techniques used in the integration have to ensure that the axis of the laser beam is co-axial and co-linear with the optical axis of the beam director, and additionally, techniques may need to be integrated into the design to align the axis of the laser beam with the pointing vector during the course of an engagement.

There are many options that will allow remote positioning of the laser source and piping the photons to the beam director using an appropriate technique, such as interfacing with optical fibres. In fact, for some applications the delivery fibres could form part of the gain medium. Use of a remote laser source eases the system-integration issues, as well as providing some system design flexibility.

Direct integration with the beam director to form a monolithic system may be necessary for the very high-power systems, where highly efficient integration techniques are critical to achieve a very stable configuration. This latter approach can lead to restrictions in the beam director's positioning options, particularly for maritime applications; however, this approach has the advantage of fewer restrictions concerning the piping of the energy into the beam director and maintaining the harmony of the axes sets of the various subsystems.

The very high precision devices will include a beam optical alignment and position-sensing system to ensure that the laser beam axis is aligned with the input axis of the beam director. It is crucial to the success of an engagement that the precise harmony of the relevant axes is maintained throughout the laser-target interaction, as well as through the entire operational and environmental envelope.

Another critical requirement for the integration of the laser beam with the beam director is to ensure that the divergence of the emitted beam exceeds the angular noise, or jitter, on the pointing vector. It is crucial that the laser beam spends a high proportion of its time on its designated target. The higher the desired probability of the beam being on its designated target, the greater the ratio needed for the divergence to exceed the pointing stability, which has an impact on irradiance at the target. For example, for a probability of time on target exceeding 90% the ratio is approximately 5:1.

4.14 Generic Beam-Director System for Laser-Based Sensing Systems

The use of a laser in a sensing system has similar challenges to those described for a generic directed-energy system, although the extreme precision-pointing requirements implicit in the high-power laser-DEW system are not usually relevant for these military applications. However, many of the beam-director design requirements and considerations outlined above, particularly for the lower-power laser-countermeasure application, are equally relevant to this class of application.

The purpose of a laser-based sensing system is to gather information about its local environment. This may involve the detection of objects or the presence of chemicals or other warfare agents. Most laser-based sensing systems have various modes of operation involving a search for potential targets, detection, acquisition, location and, in some cases, identification and recognition of detected returns through well-established processing techniques, such as spectroscopy. When the sensing system is in search mode, the function of the beam director is to allow the system to create a picture or map of the volume being searched and determine the presence of potential targets for subsequent evaluation.

The design of the beam director for this application may have some similarities with the beam-director system used for the directed-energy application. For example, a beam director with a transmitter and a receiver channel, sharing at least part of a common optical system is a classical monostatic configuration. For this type of laser-based sensing system, the receiver channel is optimised to detect the photons returned to this system through reflection or scattering along the propagation path; these signals are often in a noisy or cluttered environment so the signal-to-noise ratio is low. A major difference between these two beam-director systems is that for some sensing applications, the beam director needs to undertake a regular systematic search of space. Additionally, the size of the receiver aperture may dominate the size of the system, owing to the signal-to-noise requirements to achieve detection and subsequent accurate analysis of the probed space in the receiver subsystem.

The ability to undertake a scanning function is an integral part of a search mode; in fact, some sensing systems such as obstacle-avoidance laser radars (Lidar) operate in this mode continuously. The search for potential targets may be undertaken using a simple type of scan at a fixed angle in elevation but swept through a sector in azimuth. If the platform that the sensing system is mounted on moves, then the sensing system will search a sector of space in a push-broom style, so that a fresh segment is interrogated on each sector scan. Alternatively, a two-dimensional scan may be undertaken, so that either the elevational or azimuthal angles are changed on each scan sweep, enabling a volume of space to be investigated, or a look-back (or forward) function to be used to check a given observation on interlace scans.

A number of very simple approaches to scanning systems are well established, such as simply rotating optical wedge (Risley prisms) or rocking a flapping mirror.

One of the important requirements of this type of scanning system is to be able to correlate the received energy from each point during the scan with the position of the beam director's pointing element in that scan, so that an image can be formed from the detected returns; this process enables a point-cloud image to be formed as discussed in Chapter 13. It is normal to use optical encoders and inertial sensors or a simple inertial-reference system to provide the datum for the scan-position calculation. The angular position of the scanning element throughout its scan pattern may then be related to this datum and the received data combined to form a compound image of objects detected in each scan.

Similar components and subsystems may be used to control the scan angle and its rate in this application as applied to the pointing element in a countermeasure slew command through a given angle. These types of angular sensors will be used to monitor and control the angular rate and the angular extent of the scan.

For those sensing systems that have a mode for recognition or identification of the detected objects generating the returns in the search scan, the beam director in those systems may need to have a specific pointing-function mode. In this case, the cue for the bearing for further probing of the return will come from the search-mode data, often using image-processing techniques, as described above. In general, modest pointing performance accuracy will be necessary for this probing function.

A number of techniques are used for undertaking the analysis of the photons collected by the receiver system, which are considered in more detail in Chapter 13. However, in general this analysis will require additional sensors, and associated processing and analysis algorithms, to be embodied in the receiver system.

With the development of more advanced solid-state laser technology and high-performance two-dimensional detector arrays, the concept of flash-sensing systems has become a reality. With this type of system, it is not necessary to scan the scene of interest to the sensor, as it may be used as a staring system, which has very important system implications, including size, weight and power, as well as speed of analysis.

The discussion so far in this chapter has concentrated on mechanically configured pointing and scanning systems. When limited fields of regard are required for lower-power systems, such as the sensing application, then nonmechanical beam steering techniques may be used.

4.15 Advanced Beam-Steering Techniques

Nonmechanical beam steering methods are being developed to provide rapid and accurate rotational motion that avoids the complexity of mechanical methods. In general, these novel methods only permit small angular displacements. The advantages of nonmechanical beam steering methods are:

- Compact;
- Lightweight;

- Agile;
- Low power consumption;
- Potential for multispectral operation.

The basic principle used in nonmechanical beam steering is to create a linear phase retardation profile over the laser beam at the exit aperture of the beam-steering element; the slope of the corresponding wavefront ramp will determine the angle through which the beam can be steered. To steer a beam through a large angle, large phase or path-length differences are required together with a large aperture; this is a crucial challenge for the nonmechanical beam-steering device, especially if a compact system is required.

A range of techniques using macro-optical components are being investigated:

- *Risley prisms:* the principle of operation is to use a cascaded pair of achromatic prisms that rotate about the optical axis of the system. Rotation of the prisms displaces the beam through a conical field of regard. Complex scan patterns can be generated through appropriate differential motion of the prisms.
- *De-centred lens system:* a beam may deflected in its passage through a telescope by moving the exit lens with respect to the input lens; the lenses are separated by the sum of their focal lengths.
- *Macro-optical mirrors:* an optical arrangement similar to the de-centred lenses may be realised with curved mirrors. The use of mirrors reduces some of the optical complexity required to control dispersion and compensation for refraction.

Micro-optical technology offers a route to simple, compact, lightweight beam-steering devices with low-power consumption. To control a laser beam in two dimensions, it is necessary to fabricate two-dimensional arrays of the micro-optical elements. Examples of such micro-optical elements include the following:

- Electro-optical prisms that are used in an extracavity configuration. A series of small prisms are formed as prismatic domains in a ferro-electric material, such as lithium niobate, or in electro-optical ceramics, such as lead lanthanum zirconium tantalite (PLZT). The beam-steering function is achieved by applying an electrical field to change the domain shape.
- Dual-axis scanning micro-mirrors that are an analogue of the macro-optical system, with the gimballed micro-mirrors reflecting the laser beam through controlled angles. There are several design challenges concerning the nonlinear response of the micro-actuation system, as it generally does not scale with size and achieving a flat surface across the array presents a technical challenge, but not an insurmountable requirement.
- Blazed-grating beam steering, which use an array of micro-optical elements, such as micro-telescopes, micro-lenses or micro-prisms with a fixed pitch. Each of the micro-optical elements samples the incident laser beam and

directs a beamlet into space. These beamlets interfere coherently and form a diffraction pattern with a characteristic main lobe pattern surrounded by side lobes; the direction of the main-lobe pattern is governed by the grating equation. Active blazing of the periods of the micro-optical array is required to steer the laser beam. This has been demonstrated using a two-dimensional array of micro-optical telescopes, either in a Galilean or Keplerian arrangement, and moving one substrate containing one set of lenses of the telescopic arrangement with respect to the other, in a similar fashion to the de-centred lens technique, discussed above.

- Phased-array methods that involve the use of optical phased arrays, which impose a phase-delay ramp across a laser beam by controlling the spatial variation of the refractive index on the geometric rays across the aperture of the device. Three techniques are being investigated:
  - Spatial modulation of the refractive index using thin films of liquid crystals, which may be used in a reflective or transmissive mode to create the desired depth of optical phase modulation,
  - Electro-optical ceramic,
  - Arrays of microelectromechanical mirrors.

Figure 4.10 shows a schematic of an electro-optical prism for extracavity beam steering using micro-optical elements in the upper part of the figure. In the lower part of the figure are examples of micro-optical telescope arrays, a Galilean arrangement on the left and a Keplerian one on the right. In these cases, the beam steering is achieved through motion of one of the substrates in each beam-steering system, i.e. by decentring the lenses in the telescopes.

An example of rapid non-mechanical beam steering for a low-power system is the use of micro-opto-electromechanical systems (MOEMS) positioned in the focal plane of a macro-optical system. Minute displacement of the MOEMS device in the focal plane leads to rapid displacements of the laser beam.

Figure 4.11 shows one technique for intracavity beam steering. The basis of this technique is a confocal resonant cavity with a programmable aperture, such as a spatial light modulator using liquid-crystal technology. Beam steering is achieved by moving the low-loss segment in the light valve, so that the resonant axis is displaced within the resonant cavity, as shown in the figure.

Some consideration is being given to the development of hybrid beam-director systems to exploit the advantages of mechanical and nonmechanical beam-steering techniques. The application of the nonmechanical steering technology to the fine-beam-steering function is an obvious starting point for low-power applications.

4.16 Summary

Beam directors are an essential part of any laser-based system whether they are used for sensing, communications or directed-energy applications. The fundamental task of the beam director is to steer the beam on the target location with angular

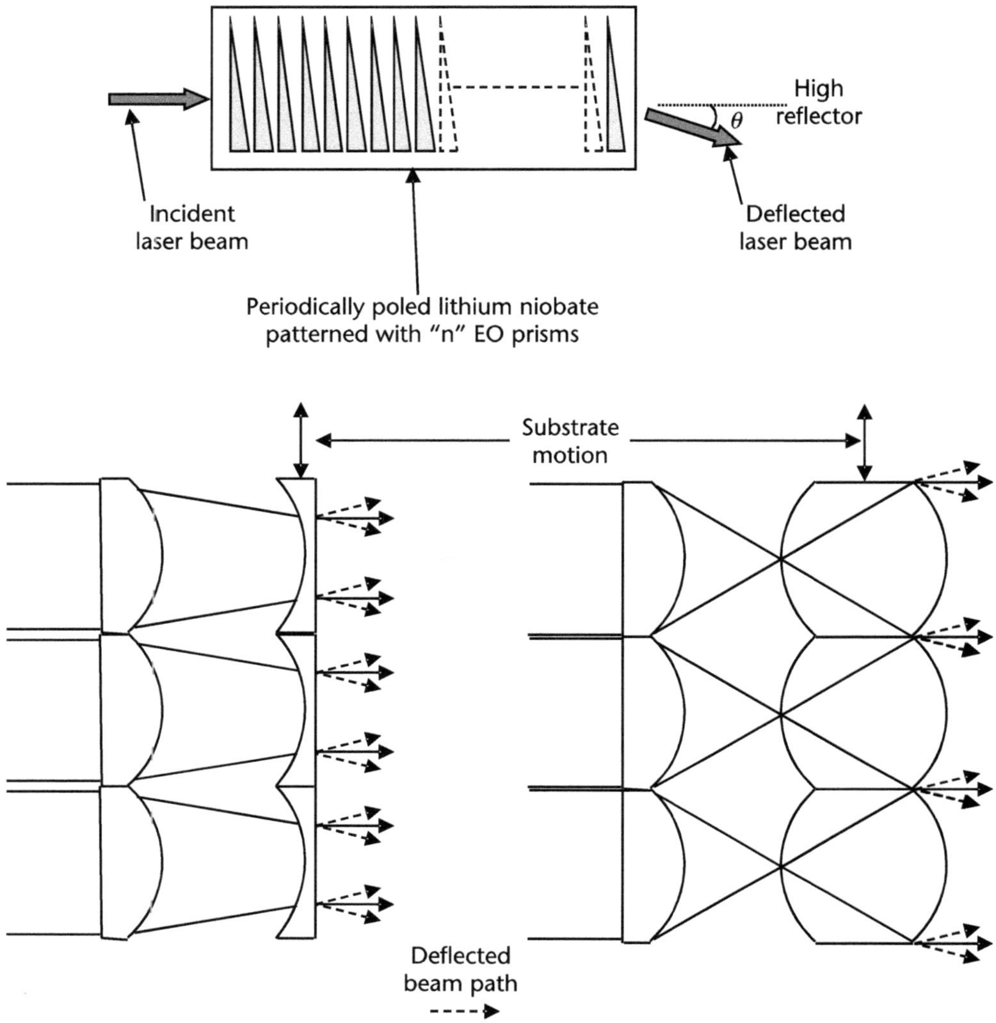

Figure 4.10 Micro-optical components for extracavity beam steering.

precision. In order to fulfil its assigned function, it may require support from other subsystems; for example, to acquire and track a target or to monitor an interaction with a volume of space.

The most demanding laser-based beam-steering requirements are associated with laser-damage weapons; the pointing accuracy of the laser beam will be in the microradian class or better. This scale of pointing accuracy requires multiple modes of operation, which are supported by many subsystems of varying complexity, to ensure the precision delivery of the photon flux to the target zone. This class of beam director will usually be made to very fine mechanical tolerances and based on the more traditional mechanical designs, using bulk optical elements and gimbal structures to steer the beam.

There have been a number of interesting developments with nonmechanical beam-steering techniques over the last decade. Currently, this technology is only suitable for low-power applications such as sensing, for example, associated with

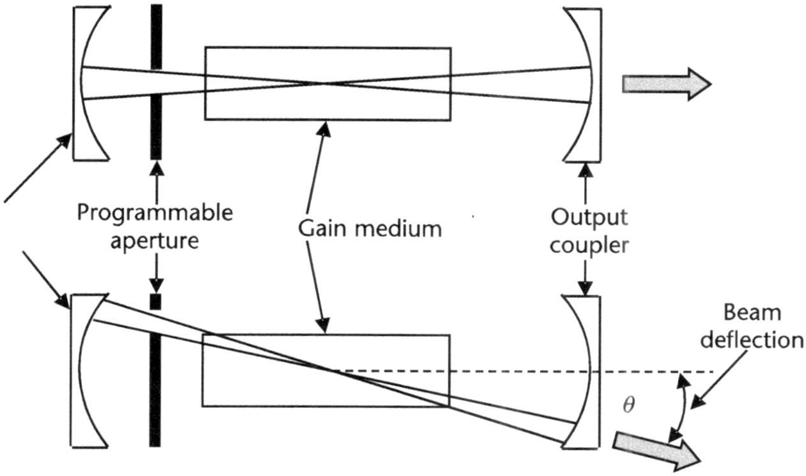

Figure 4.11 Intracavity beam steering with a programmable aperture.

laser-radar systems. These nonmechanical techniques offer a route to compact, lightweight, agile beam control with high precision. One of the major technical challenges concerns the development of techniques that will allow large fields of regard, comparable with those of a mechanical device.

Future developments in beam director technology may see hybrid pointing devices using mechanical motion integrated with nonmechanical techniques for fine beam-steering. Such a system could be realised with a phase-combined fibre laser source array, as discussed in Chapter 3. The other long-term design aim for beam-director technology is a configuration that would conform with a structure, such as an aircraft's fuselage or a ship's superstructure, whilst maintaining a large field-of-regard; use of nonmechanical techniques may offer a practical solution to this requirement challenge, but with some operational constraints.

Beam-control techniques have been developed for use with the higher-power directed-energy applications to optimise the laser flux interacting with a target. These techniques are based on adaptive-optical techniques, which may be built into the optical train of the beam director and are fundamental to the beam-control system.

Techniques have also been developed for assessing the progress of the interaction of a directed-energy beam with a target, which is known as battle-damage indication and assessment. In the case of a countermeasure system, this technique may be as simple as monitoring the apparent optical return from the seeker of an approaching missile; however, more complex techniques have been developed involving the analysis of various emissions from the laser beam interaction with the target structure.

Selected Bibliography

Merritt, P., *Beam Control for Laser Systems,* The Directed Energy Professional Society High Energy Laser Joint Technology Office, Albuquerque, New Mexico, 2012.

Seyrafi, K., and S. A., *Introduction to Electro-Optical Imaging and Tracking Systems,* Norwood, MA: Artech House, 1993.

Titterton, D. H., and J. L. Weston, *Strapdown Inertial Navigation Technology*, 2nd ed., Reston, VA: The IEE/American Institute of Aeronautics and Astronautics, 2004.

CHAPTER 5

Laser-Beam Propagation

5.1 Introduction

The Earth is surrounded by a gaseous blanket that can have a dramatic effect on the spatial and temporal properties of the laser energy in a beam as it propagates through the atmosphere. The modifications to the laser beam start as soon as the photons leave the source's exit aperture all along the path to its target.

The principal effects concern absorption, scattering and bending, or deflection, of the photons in the beam. Many of these effects are dependent on the wavelength and flux of the propagating beam, as well as the content of gases and presence of aerosols, temperature gradients and motion of the atmosphere.

In the case of low-power, laser-based applications, such as laser-based jammers used in platform-protection systems or devices used for sensing, the emitted energy is modest, typically with an average power of the order of a watt; therefore, for these types of applications, the more extreme propagation-limiting effects, such as thermal blooming and plasma formation, may be neglected. However, all of these effects, and others, are critical to high-power laser propagation, which occurs with high-power laser-based directed-energy weapons.

In general, for normal atmospheric propagation of laser beams, there are five major considerations:

- Molecular absorption, which defines the atmospheric windows, that is, the wavelength bands where the more favourable transmission is likely to occur and results from anomalous dispersion.
- Molecular and aerosol scattering (Rayleigh and Mie), which are dependent on particle size and emission wavelength.
- Atmospheric turbulence causing scintillation, beam-wander and broadening effects.
- Aerosol absorption that contributes to beam extinction and, in the case of high-power beams, is a major cause of thermal blooming, owing to local changes in the refractive index of the propagation path.
- Refraction is a bending of the light path.

All of these phenomena contribute in one way or another to reducing the energy in the beam that reaches the target. For many applications, the important consideration is the reduction in energy or power density (irradiance).

This chapter will consider the fundamental aspects of the consequences of the physical and naturally occurring phenomena on laser beams. Propagation in a vacuum or at high altitudes will be close to ideal, but this situation is not relevant to many tactical military applications.

5.2 The Earth's Atmosphere

The Earth's atmosphere has many complex properties because it provides a transition between the solid or liquid of the surface of the Earth and the vacuum of outer space. The atmosphere consists of a number of gases and suspended particles, the temperature and pressure of the atmosphere varying with latitude and altitude, as well as exhibiting seasonal, diurnal and local variations. Clearly, this is a complex problem, which can have a dramatic effect on the spatial and temporal characteristics of any propagating laser beam.

The dynamics of the seasonal and diurnal variations of the atmosphere are instrumental in the formation of weather patterns throughout the world; moreover, these dynamics have important consequences on the behaviour of electro-optical and infrared sensors and laser-based systems. Conversely, favourable weather conditions can result in enhanced propagation, but reliance on luck does not lead to a robust technique for any application.

The lowest layer of the atmosphere is known as the troposphere, which extends from sea level to about 11 km (7 miles) altitude at middle latitudes. In this layer, the temperature generally decreases with increasing altitude, at the rate of 6.5K per kilometre (1°F per 300 ft), but decreases at a higher rate in the tropical latitudes; generally, this effect is known as the lapse rate. Variation in this rate leads to some interesting refractive effects, such as the formation of mirages or ducting effects. The pressure variation ranges from one atmosphere at sea level to a few tenths of an atmosphere at the top of this layer. Beyond the troposphere is the stratosphere, where the temperature is reasonably constant at about 218K; other layers exist at far greater altitudes, but do not concern most tactical military applications of laser technology.

5.3 Beam-Attenuation Mechanisms

The fundamental physical processes that attenuate the propagating electromagnetic radiation in the Earth's atmosphere are:

- Resonant absorption by the rotational and vibrational motion of diatomic and tri-atomic molecules that form the atmosphere (i.e., anomalous dispersion);
- Rayleigh scattering by molecules in the atmosphere that are much smaller than the incident wavelength of the energy;

- Mie scattering by particles comparable in size with the incident wavelength.

These processes are considered in more detail in the following sections, along with the impact on the spatial and temporal characteristics of the propagating beam.

5.3.1 Absorption

Absorption of electromagnetic energy by gases in the tropospheric region of the atmosphere is mainly a result of the presence of di-atomic and tri-atomic molecules along the propagation path, as noted above, which has a consequential effect that is strongly wavelength dependent. These molecules, such as ozone, water vapour and carbon dioxide, have vibrational frequencies that correspond to wavelengths from the ultraviolet to the far infrared. These resonant absorption features are broadened into bands, as a result of collisions and thermal-motion effects.

A number of atmospheric windows exist, as shown in Figure 5.1, where the effects of attenuation from absorption are small. Ranges of wavelengths define these atmospheric windows; for example, the mid-wave infrared window covers the 2- to 5-μm range. However, this region has some zones where the absorption is high, and therefore, it can be subdivided into three distinct zones, commonly known as bands, where the attenuation is low.

Attenuation is not completely absent from these wavelength bands or other regions defined by the atmospheric windows, such as the visible and near-infrared regions. This is also a result of the presence of the di-atomic molecules of oxygen and nitrogen along any propagation path in the atmosphere. There is a small amount of absorption, owing to collision and pressure broadening of the predominant gases that form the atmosphere.

There is also very weak absorption of visible and infrared radiation in the atmosphere, caused by the wings of the strong absorption by nitrogen and oxygen molecules, which peaks in the ultraviolet. More important effects in the infrared waveband are absorption by aerosols and the continuum absorption by water vapour (7 to 30 μm) and nitrogen (3.7 to 4.7 μm) and from the trace gases, such as methane, carbon monoxide, nitrous oxide and ozone.

5.3.2 Scattering

There are fundamentally two distinct scattering mechanisms, known as Rayleigh and Mie, which remove photons and, therefore, energy, from a discrete optical beam. Scattering describes the redistribution of energy by particles in the atmosphere. This scattering may be in the propagation direction, forward scattering, or backwards towards the source (i.e., back scatter); a common example is the scatter of light by fog from a car's headlights. Both of these effects reduce the potential effectiveness of any laser-based system, and limit the effective operational range of sensors, and active laser-based systems.

Rayleigh scattering is from particles in the atmosphere that are very much smaller than the optical wavelength. In the case of the mid-infrared wavelengths, this means particles of submicrometre size and smaller; for the visible-band photons, the diameter of the scattering centres needs to be no more than about 100 nm.

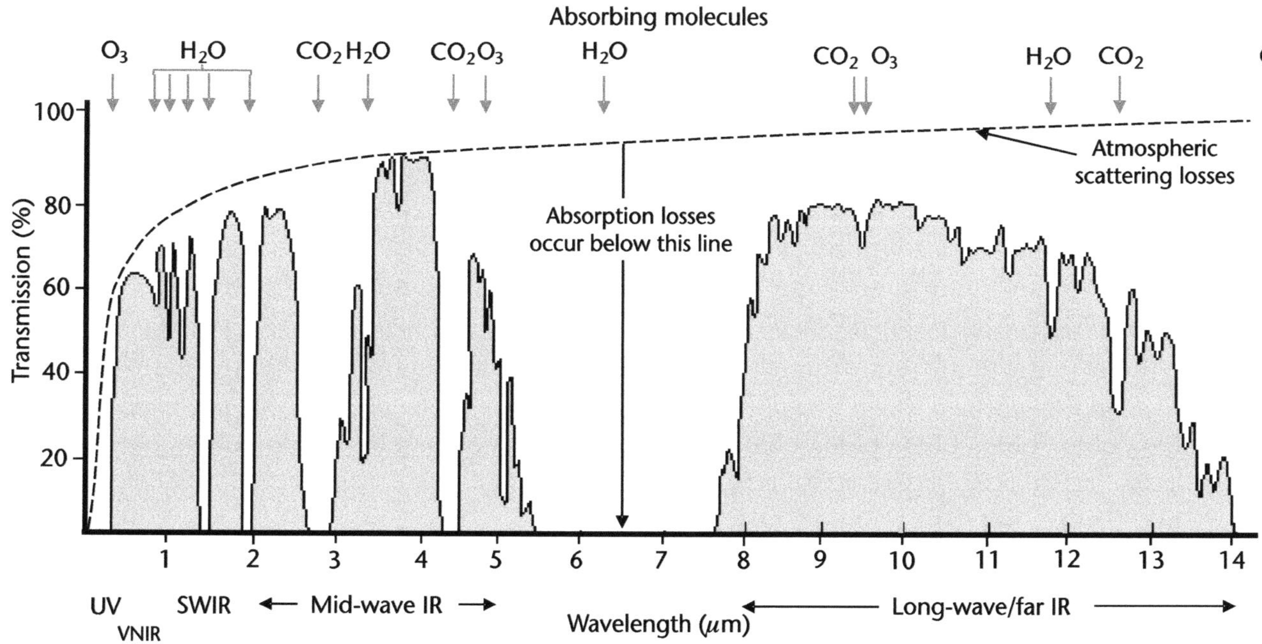

Figure 5.1 Transmission of a typical atmosphere in Northwest Europe for a multikilometre horizontal path at sea level. (After: R. D. Hudson Jr., "Infrared System Engineering," John Wiley & Sons, New York.

An example of Rayleigh scattering is molecular scattering of sunlight by the various gases in the atmosphere, resulting in a blue sky.

Mie scattering is caused by particles comparable in size with the incident wavelength. In the case of photons in a mid-infrared laser beam, this means scattering from atmospheric aerosols, such as dust or water droplets in fog; similarly for visible-band photons. Therefore, the effects of Mie scattering are very dependent on atmospheric conditions and the aerosol content of the atmosphere, which can change very dramatically over a short time scale, perhaps through a change in wind direction. Rayleigh scattering may sometimes be viewed as a small-particle limiting case of the more general Mie theory.

In general, scattering is not a significant attenuating mechanism for medium-altitude laser-beam propagation, as may occur during anti-air engagements, unless a cloud obscures the target. This may well cause a sensor to lose lock on its target. For engagements close to the ground, dust and other particles from the battlefield may create a significant attenuation of the laser beam. In the maritime environment, there can be an interesting mixture and distribution of aerosols with the larger ones closest to the sea surface.

5.3.3 Extinction

This term is used to express the reduction in the intensity of a beam of radiation as it propagates through a medium. Extinction of a beam is a result of the absorption of the photons by the species in the atmosphere and then scattering out of the beam by aerosols and other particles present in this gaseous medium. The application of this phenomenon is exploited in the denial defeat mechanism, with the use of smokes and obscurants.

Extinction can be a serious problem for laser-based directed-energy weapons, as discussed briefly later; turbulence can also be a significant factor in the reduction in the energy on target, which is also discussed later. Careful consideration is required in the choice of the wavelengths used in high-power infrared lasers as the extinction can be highly wavelength dependent and may be the dominant factor in the reduction in radiant intensity at a target; formerly the choice was between deuterium-fluoride (DF) and carbon-dioxide (CO_2) sources, but now there is the choice of high-power beams from one-micron solid-state lasers. In general, the shorter wavelength radiation experiences less extinction, but is more prone to turbulence-induced effects.

5.4 Atmospheric Transmission

In practice, a range of aerosols and particles of varying sizes may be present along the propagation path, such as fog, cloud, smoke and dust, particularly close to the ground. Consequently, the transmission of visible and infrared radiation in the atmosphere is affected in a similar manner, through absorption and scattering by molecules and by the presence of aerosols and other particles. Of the two processes, absorption usually proves to be more dominant at infrared wavelengths, whereas for visible light, scattering is normally the more dominant impact.

Atmospheric transmission loss is a summation of the combined loss effects of absorption and scattering, as well as any other effects that result from these phenomena such as plasma formation with very intense laser beams. Mathematical models exist for the prediction of the transmission of optical radiation through the atmosphere.

Figure 5.1 shows a plot of the percentage transmission as a function of wavelength measured over a horizontal path length of several kilometres at sea level for an atmosphere typical of Northwest Europe. The plot shows transmission resulting from the combined effect of molecular and aerosol scattering, as well as absorption; but note that selective absorption by water-vapour, carbon-dioxide and ozone molecules is the dominant process.

Figure 5.1 clearly shows the high transmittance of the atmospheric windows separated by regions of high absorption in the infrared, whereas in the visible band there is generally fairly good transmission across the whole waveband. There are two principal windows that may be used by infrared sensors, such as thermal imaging cameras or infrared seekers; they are known as the 3 to 5 μm (or mid-wave infrared) and 8 to 14 μm (far infrared) atmospheric windows.

The mid-infrared atmospheric window is often considered to be from 2 to 5 μm for other sensors, such as heat-seeking missiles or surveillance sensors, and has been exploited by these systems. Close inspection shows that there are several regions, known as bands, within this window suitable for exploitation by the various thermal-camera and seeker technologies:

- Band I extends from around 2 to beyond 2.5 μm and is used by sensors with uncooled detectors, such as the lead-salt devices, for example, lead sulphide (PbS).
- Band II extends from approximately 3 to 3.5 μm and is used by sensors with cooled detectors, such as lead sulphide or lead selenide (PbSe).
- Band IV extends from nearly 4 to about 4.8 μm and is used by sensors with cooled detectors, such as indium antimonide, but this band has a substantial absorption around 4.3 μm, owing to the carbon-dioxide absorption, leading to the term upper and lower Band IV, or the blue wing and red spike.

The consequence of the selective absorption and scattering on the operation and performance of infrared systems is to force the sensors systems to operate at wavelengths that fall within in the atmospheric windows to achieve successful operation over several kilometres, or beyond. However, some specific specialised systems may use some of the highly absorbing regions of the spectrum, for example, covert short-range communications or active fuzes.

Care is required in the choice of laser wavelength(s) for all laser-based directed-energy applications, particularly where a lengthy propagation path is required. This is because there is fine structure within these wavelength bands (atmospheric windows) where the absorption can be large. For example, in the case of a laser-based countermeasure system, the laser energy could be severely attenuated by the atmosphere but the signature of the target would be largely unaffected, owing to its broad-band characteristics, and thus in the extreme limit the jam-to-signal ratio (J:S) could tend to 1. It should be noted that these atmospheric effects do not have a significant impact on the performance of broadband incoherent countermeasure sources, such as flares or arc-lamps.

Extra consideration is required with the choice of wavelength band of operation for a laser weapon, as in some circumstances the Mie scattering processes for some atmospheric conditions become dominant. For example, in the maritime environments, such as the littorals, Mie scattering of photons with wavelengths around 4 μm, from a high-power deuterium-fluoride laser, can be much greater at this wavelength than occurs with 10-μm photons from a carbon-dioxide laser. Hence, a far higher-power deuterium-fluoride laser would be required to achieve the equivalent effects attained with a carbon-dioxide source, given the same irradiance or energy density required to achieve a given defeat mechanism, which could lead to other detrimental propagation effects. These effects, such as thermal blooming, are considered in some detail below.

Similar considerations apply to the wavelengths of operation used by military sensing systems. Most of these sensors use modest-power emissions, so the considerations given for the laser-countermeasure system will apply. Additionally, there are techniques such as range-gated imaging where the effects of scatter and other clutter phenomena are minimised, as discussed in Part II.

5.4.1 Weather

It is quite common for the prevailing atmospheric conditions in any season to be described in terms of the visible range (visibility) and the type of weather including precipitation, such as fog, rain or snow. Visibility is a subjective evaluation of the distance at which an object can be seen by an observer, such as a dark object above the horizon in daylight or a light during darkness. Meteorological visibility is an attempt to be less subjective by measuring the distance when the transmission is reduced to 2% of the launched optical power or energy.

Weather type is far less subjective; however, careful definition of the rate of precipitation, its type or classification of fog is necessary to make an appropriate calculation of transmission at a given wavelength. Extensive weather databases have been established for the majority of geographical locations around the world along with seasonal variations.

5.4.2 Humidity

Increasing humidity results in an increased aerosol concentration and reduces the transmission within each atmospheric window of the spectrum, owing to increased Mie scattering and continuum absorption. However, the 8- to 14-μm atmospheric window is more susceptible than the mid-infrared atmospheric window to humidity because of absorption rather than scatter. Variations in atmospheric humidity are generally greater in summer than during winter in temperate conditions. Therefore, atmospheric transmission within these windows will be less in summer in temperate and tropical regions than in the desert or during the winter season, particularly for the 8- to 14-μm window.

From an investigation of the contribution of the molecular absorption for typical summer and winter conditions, and the aerosol absorption and aerosol scattering for clear and hazy conditions for the two infrared-atmospheric windows; the following conclusions may be reached:

- Transmission in the 8- to 14-μm window is dominated by molecular absorption.
- The 3- to 5-μm atmospheric window is influenced more by aerosol scattering in summer than in winter.
- For short ranges (less than 10 km) the 8- to 14-μm window appears to offer better atmospheric transmission for most circumstances.
- Only in conditions of high humidity and extreme clarity in visual terms (i.e., in the visible band) does the 3- to 5-μm atmospheric window have a superior transmission performance, particularly for ranges in excess of 10 km.

The impact of atmospheric humidity on infrared sensors systems is to give somewhat favourable transmission of the longer infrared waveband. The impact of humidity on laser-based countermeasure systems is small, as it is unlikely to generate any significant changes in the potential performance (no notional change in J:S, that is, potential jamming capability). Humidity is likely, however, to have a more significant effect on the shorter wavelength laser-based, directed-energy weapons. The impact of humidity on active sensing systems is similar, owing to the wavelength dependence, but many of the sensing systems use wavelengths in the short-wave infrared, so the impact will be more marked.

The consequences for visible-band laser systems and sensors, is that there will be more aerosol scattering and also Rayleigh scattering (owing to its $1/\lambda^4$ dependency) will be higher than for the infrared wavebands. As a result:

- Radiation will be scattered more, so precision pointing of a laser beam at sensors, such as optical seekers, may be less critical for some lower-power applications, but the effective range of a system will be reduced, unless the optical power emitted can be increased.
- Higher humidity is often associated with a shorter visual range, because the water droplets coalesce or condense on atmospheric nuclei, which are much smaller than the visible-band wavelengths, and consequently scatter more effectively.

The impact of high humidity on the lower-power, laser-based systems will be to reduce the notional performance and, as the humidity increases, the degradation in performance will increase. For the higher-power, directed-energy applications, particularly those operating close to one micron, the consequences could be quite dramatic in the maritime environment, although this could be mitigated to some extent by having a tuneable emission wavelength.

5.4.3 Haze, Fog, Cloud and Rain

The transmission of infrared radiation is superior to that of the visible wavelengths when small aerosols are present, particularly if the meteorological conditions are very hazy or conventional smoke is present; this is the basis of the long-standing conventional countermeasure concealment technique, as mentioned above. This is because the mean particle size of about 0.5 μm is a relatively small fraction of the infrared wavelength (say, for 10 μm in the far infrared), and therefore, any scattering by haze or smoke particles will be negligible, particularly if the scattering particles are very much smaller than the infrared wavelength (i.e., Rayleigh scattering dominates).

Scattering by fog and clouds on the other hand, is intense, because the size of the droplets extends from 5 to 15 μm, i.e. comparable with the photon wavelength (a Mie-scattering processes). The presence of a high concentration of aerosols has an impact on image quality, leading to a blurring of the image, which limits the ability of a thermal system to detect, acquire and then track or identify a target.

Rain affects the performance of an infrared sensor in two ways: it reduces the thermal contrast of the scene by coating everything with a thin layer of water and it affects transmission in both atmospheric windows almost equally, particularly in heavy rain. Attenuation in these circumstances may amount to as much as 20%/km, giving rise to only 33% transmission over a 5-km path. The reduction in thermal contrast is a problem for a number of types of infrared sensors, as this effect will severely limit the acquisition range. There is a commensurate impact on the laser transmitter characteristics; however, rain is unlikely to be a significant factor for anti-air engagements with laser-based systems, because of the reciprocity of the effects of rain on both systems.

The effect of the presence of rain on the performance of laser-systems and sensors operating in the visible band is a result of changed scattering and/or reflection processes, rather than reduction in thermal contrast in the case of sensor functions.

5.4.4 Summary of Transmission Effects on Beam Propagation

Table 5.1 provides a broad indication of the impact of atmospheric conditions close to the surface of the Earth on laser-beam transmission in various wavebands. It must be stressed that this is a coarse estimate and for a precise prediction it is essential to undertake an accurate calculation, using accurate estimates of the atmospheric content and the actual laser wavelength(s), as described later. Table 5.1 also provides an indication of the laser-based systems and sensors that operate in the various wavebands.

The fraction of the optical power in a transmission along a defined path may be estimated using Beer's law:

$$T = \exp\left[-\int_0^r \int_{\lambda_1}^{\lambda_2} \alpha_d(\lambda,\ell)\,d\lambda\,d\ell\right] \tag{5.1}$$

where T is a measure of the total transmitted energy, α_d is the atmospheric extinction coefficient density function (for units, see below), r is the distance to the target from the laser, ℓ is the range integration variable, λ is the wavelength integration variable, λ_1 is the shortest wavelength of interest in the propagation band of wavelengths and λ_2 is the longest wavelength of interest in the propagation band of wavelengths.

The atmospheric extinction coefficient has units of reciprocal length and reciprocal wavelength but not reciprocal area. Hence, care is required in the definition

Table 5.1 Impact of Weather on Sensors and Systems in Various Wavebands

Weather Conditions	*Visible and Near IR*	*Short-Wave IR*	*Mid-Wave IR*	*Far IR*
Low visibility	High	Moderate	Low	Low
High humidity	Low	Low	Moderate	Moderate
Rain/snow	Moderate	Moderate	Moderate	Moderate
Fog/cloud	High	High	Moderate	Moderate to low
Dust	High	Moderate to high	Moderate	Moderate
Smoke	High	Moderate	Low	Low
Typical laser systems	EOCM, laser-DEW, laser sensing , laser targeting	EOCM, laser DEW, laser sensing, laser targeting	Laser DIRCM, laser sensing	Laser CM, laser sensing
Typical sensors	Binoculars, night-vision goggles, TV cameras, image intensifiers	Surveillance, beacons, TV cameras, image intensifiers	Thermal imagers, FLIR, seekers	Thermal imagers, FLIR

of the correct coefficient (α_d) appropriate for the atmospheric propagation conditions along the specified optical path and the exact laser wavelength. Knowledge of the precise wavelength and line width is important for this calculation, owing to the narrow line width of a typical laser line. This aspect of estimation of atmospheric transmission is discussed more fully in Section 5.4. Stronger absorption at some wavelengths can adjust the spectral composition of the transmitted light.

If the atmospheric extinction density function coefficient is constant along the propagation path and over the spectral line width of the transmission, then Beer's law may be simplified to the following expression:

$$T = \exp(-\alpha r) \tag{5.2}$$

where α is the extinction parameter (km^{-1}) and is the sum of the absorption and the scattering effects at that wavelength, and r is the length of the propagation path.

5.5 Models for Prediction of Transmission

A number of models exist to estimate the transmission of electromagnetic radiation through the atmosphere. The most popular ones have been developed by the U.S. Air Force Research Laboratory (AFRL) at Hanscom Air Force Base (Massachusetts): they are band models offering various degrees of resolution of wavelengths propagating through a defined atmosphere. The fundamental assumption applied to band models is that the atmospheric absorption lines can be represented by a simple mathematical expression. Band models compute transmittance averaged over a spectral band $\{\tau_{av}(\nu)\}$ defined by the following expression:

$$\tau_{av}(\nu) = (1/\Delta\nu)\int_b^a \tau(\nu)\,d\nu \tag{5.3}$$

where the limits of the integral are $a = +\Delta/2$ and $b = -\Delta\nu/2$ and $\Delta\nu$ is the spectral band of the model.

These computer models allow for the calculation of atmospheric transmittance and radiance based on absorption and scattering phenomena, as considered above, for a variety of propagation-path geometries. The Lowtran model offers relatively low resolution in the transmission calculation of 20 wave numbers (20 cm^{-1}), whereas Modtran is an order of magnitude finer at two wave numbers (i.e., 2 cm^{-1}). Fascode is a high-resolution model that uses the Hitran database directly, which contains details of gas composition, interactions and an extensive compendium of absorption-line parameters, and is suitable for prediction of transmission of laser lines. These models are considered in a little more detail next.

5.5.1 Lowtran Model

Lowtran is the name given to a series of models produced by the AFRL team for predicting optical transmission through the atmosphere (low-resolution atmospheric transmission): this model calculates the transmittance (or radiance) for a specified path through the atmosphere using the band-model technique, applying routines to calculate molecular continuum absorption, molecular scattering and aerosol absorption and scattering. The radiance model calculates atmospheric self-emission, solar and/or lunar radiance single scattered into the path, direct solar irradiance through a slant path to space and multiple scattered solar and/or self-emission radiance into the optical path.

The Lowtran model uses 32 layers to represent the atmosphere from sea level to 100 km. The thickness of the layers varies according to altitude: 1-km incremental layers up to 25 km, 5-km increments from 25 to 50 km (i.e., the top of the stratosphere), and the last two layers are 20 and 30 km thick, respectively. It should be noted that the detailed structure just above the ground (land or sea) is not represented in this model. An important phenomenon is the maritime boundary, which is nonstandard and can present a particularly complex propagation environment.

This model calculates attenuation and refraction effects for each layer, which are then summed along the transmission path. The physical characteristics of each layer are determined by input data and predetermined standard models of various regions of the Earth and its seasons. The option to specify a particular atmosphere exists and it should be noted that the model assumes that the atmosphere is in thermal equilibrium.

5.5.2 Modtran Model

Modtran (moderate resolution atmospheric transmission) is a development of the Lowtran model and offers superior resolution (2 cm^{-1}), as noted above, as well as having improved temperature and pressure dependency models. Other improvements included in Modtran are an increased number of molecular species included in the atmospheric representation (CH_4, CO_2, CO, HNO_3, H_2O, NH_3, N_2O, NO, O_3 and SO_2). Thus, high-altitude calculations (up to 60 km) are more accurate than those produced by Lowtran, provided the local thermal equilibrium is satisfied.

Neither of the Lowtran and Modtran models are suitable for accurate assessment of transmission of laser lines, merely estimating potential effects; their principal value is for predicting the impact on more conventional broad-band sources. The following section is directly related to the estimation of the transmission of laser beams.

5.5.3 Fascode (Laser) Model

Fascode has been developed for calculating the propagation of radiation with a very narrow optical bandwidth, such as a laser line. The characterisation of the aerosol and molecular content of the atmosphere along the optical path is very similar to the representation used in Lowtran. The difference is that the Fascode routine uses the Hitran (high resolution atmospheric transmission) database directly, to determine the band-model parameters of the transmission path over the spectral ranges defined in the input parameter list for the specified atmosphere and season. The use of this model is absolutely critical if the atmospheric transmission of a given laser line is required to any realistic accuracy, because the fine structure of the transmission function in the apparently high-transmission segments of the atmospheric windows may result in the laser wavelength coinciding with a highly absorbing region (feature), that is, the fine structure of the atmospheric transmission.

5.6 Turbulence

The phenomenon of optical turbulence occurs as a result of fluctuations in the refractive index of the atmosphere within the air mass, along the propagation path followed by an optical beam, typically generated by the convection currents within the air mass. Turbulence is responsible for a variety of effects, such as scintillation, that is, the temporal-intensity fluctuations of a beam. In the case of laser-beam propagation, owing to the coherent nature of the photons in the beam, the effects of turbulence result in beam wander and beam broadening, which is shown schematically in Figure 5.2; the instantaneous position of the centroid, as a function of time, is shown as a + in the left part of the figure and the broadened-beam profile is shown as a dotted profile in the right part of the figure in comparison with the natural-beam shape. The mean position of the beam may be deduced from this distribution. With imaging systems, the effects of turbulence result in motion of the image (image dancing), as well as blurring of the image.

Differential heating of the ground, caused by albedo variations, dominate turbulence effects in many cases over land, creating rising convective cells, which

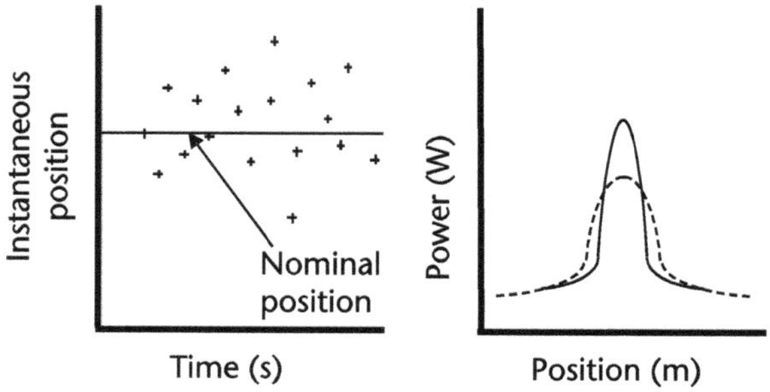

Figure 5.2 Beam wander and broadening.

decay in size producing a Kolmogorov-type cell cascades. Additionally, variations in the natural absorption of energy by the atmosphere, such as solar radiation, contribute among other effects to the motion in the atmosphere, especially in the maritime environment. The motions of these cells, with density variations from one zone to another, also have a temporal dependency. Those experiencing higher absorption will have a higher temperature and hence lower density than other parts of atmosphere experiencing less absorption.

An important physical consequence of the change of density of the various cells of air is the variation in the refractive index of air from cell to cell. Less dense (rarer) air has a smaller refractive index than denser air. Moreover, changes of air density in the cells, with respect to those around them, mean that the cells move; the less dense cells rise and the denser ones fall. This phenomenon is shown schematically in Figure 5.3, along with an illustration of terms used to describe turbulence effects, which are defined next.

The motion of these zones of differing air density leads to atmospheric turbulence and a random distribution of the index of refraction of these cells through the atmosphere, as these cells move around. Despite the fact that the magnitude of the individual fluctuations in the index refraction is very small, the cumulative effect on the spatial and temporal characteristics of a photon beam, as it travels along a path through the atmosphere, may be very significant. Basically, this is a consequence of the dependence on phase variations, as shown in:

$$\text{Phase} = 2\pi/\Delta d \cdot \lambda \tag{5.4}$$

where Δd is the path difference and λ is the wavelength, which results in a $1/\lambda$ multiplier to magnify the effect of phase changes.

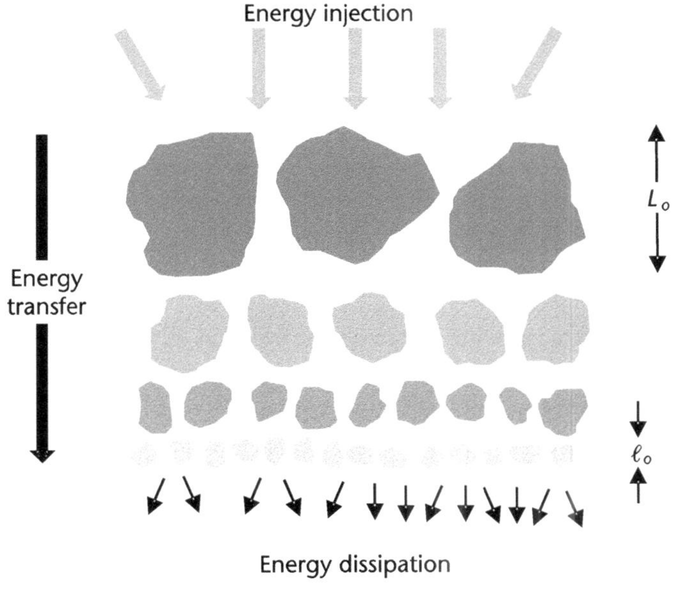

Figure 5.3 Schematic illustration of energy dissipation in a turbulent atmosphere.

The turbulence-induced fluctuations in the refractive index produce a distortion in the wavefront of the beam as it passes through the various turbulent zones; this distorted wavefront continues to propagate and is itself, distorted further, as it travels, unless corrected. This is discussed further later.

A direct consequence of the random motion of the cells, which causes the variation in refractive index along the propagation path, is to modify the spatial characteristics of a laser beam. There are three fundamental effects under normal propagation conditions:

- Modification to the local intensity of the beam across its spatial profile, owing to interference effects, results in scintillation (rapid changes of the beam's intensity).
- The centre of the laser beam moves with respect to the nominal directed (projected) optical axis of the beam (i.e., the sightline between the source and its target point). This is known as beam wander and is independent of the propagating wavelength and results from changes in the refractive index and movement of the cells in the propagating medium.
- The spatial profile of the laser beam is also extended, reducing the peak intensity. This is known as beam broadening; it is approximately inversely proportional to the propagating wavelength and results from changes in the refractive index of the air.

A laser beam propagating in the presence of severe turbulence experiences large interference and the consequences become dramatic, resulting in radical changes (break up) of the spatial profile of the beam. The beam structure now has patches of very high intensity resulting from the constructive interference of components of the beam that are in phase and other areas with negligible intensity where the components are out of phase, the high-intensity zones are sometimes called blobs, intensity spikes or surges.

Kolmogorov developed a theory to explain the effects of turbulence and allow the prediction of its impact. Originally this theory was derived to explain turbulence in fluids, but it has been developed to predict the impact of turbulent effects in the atmosphere and estimation of the consequences on laser-beam propagation. He deduced his theory from the fact that convection produces large-scale cells that decay into smaller cells, and eventually when the inner scale is reached, the decay chain is terminated by local heating.

The important parameters are:

- The refractive-index structure parameter, C_n^2, which expresses the temperature profile or distribution, considered further later;
- The inertial range defined by the inner- and outer-scale lengths, of the atmospheric cells:
    - Inner scale (l_0), which relates to the length of the dissipation of the kinetic energy in the atmospheric eddies into the atmosphere and shown schematically in Figure 5.3;

- Outer scale (L_0), which relates to the solar-heating scale creating eddies in the atmosphere, also shown schematically in Figure 5.3. Close to the ground, it approximates to half the height of the beam above the ground ($h/2$).
- The wavelength of the energy: the shorter the wavelength, the greater the turbulence induced effects for a given C_n^2 (i.e., inversely proportional to wavelength);
- The height above the ground: close to the ground, the reduction in turbulence-induced effects is proportional to $h^{-4/3}$. At some distance above the ground, typically 500 to 600m, the factor is $h^{-2/3}$. However, the actual value depends on where the inversion layer is actually located; for example, during the day in a hot desert it could be as high as 2 km returning to a few hundred metres above ground level during the night.

It may be concluded, that the effects of turbulence are more severe close to the ground, particularly on hot sunny days when using shorter wavelengths to make the observations or deliver an effect. In general, a good estimate for the value of C_n^2 in a hot desert is of the order of 10^{-13} $m^{-2/3}$, whereas a calm overcast day in temperate latitudes is likely to have a C_n^2 in the region of 10^{-15} $m^{-2/3}$ to 10^{-16} $m^{-2/3}$ during most of the day; moreover, this value may be attained briefly in desert-like conditions for a brief period close to sunrise and sunset, as shown in Figure 5.4. This is the phenomenon of thermal crossover when the ground and air temperatures become closely matched, so that there is no convective cell generation; it is known as the diurnal variation.

The refractive-index structure parameter C_n^2 is often used to describe the magnitude or the strength of the turbulence in the atmosphere. The average value of C_n^2 over a given period along a defined optical path may be measured using an active sensor known as a scintillometer, which estimates the distortion of a transmitted optical pattern over a known distance and relates it to C_n^2. A rough guide to the effects of turbulence strength, with very approximate C_n^2 values, is given here:

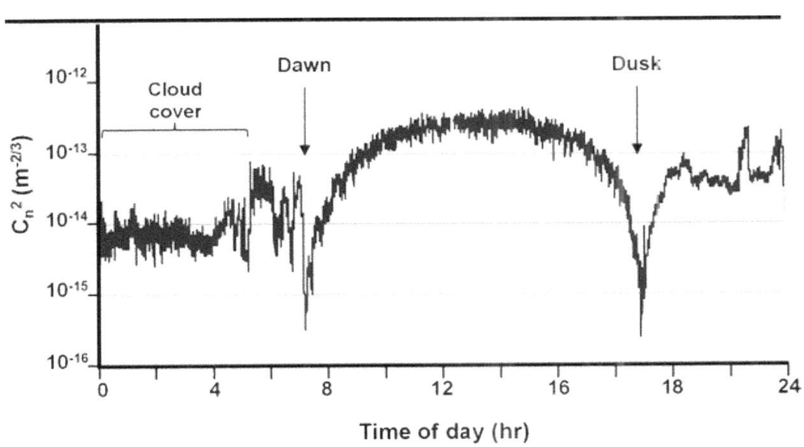

Figure 5.4 Diurnal variation of C_n^2.

- Low turbulence ($C_n^2 \approx 10^{-16} m^{-2/3}$) with some low amplitude or negligible scintillation and negligible beam-spread and beam-wander effects;
- Moderate turbulence ($C_n^2 \approx 10^{-14} m^{-2/3}$) with some moderate amplitude scintillation, but no beam break up (i.e., incomplete cancellation between various parts of the beam) and small beam-spread and beam-wander effects;
- High turbulence ($C_n^2 \approx 10^{-13} m^{-2/3}$) with severe scintillation and beam break up (patch-like zones appearing and disappearing in the beam) and significant beam-spread and beam-wander effects, which will have an impact on the apparent beam stability.

The effects of turbulence are usually more severe close to the ground, particularly on hot sunny days and the effects tend to decrease with altitude; moreover, the effects of scintillation are approximately inversely proportional to wavelength. There is a strong link in the characteristics of the turbulence to the thermal-structure parameter of the atmosphere (C_T^2) and the difference in the temperature of the ground and the surrounding air, often simply called the air-ground temperature difference; over the ocean, it is the air-sea temperature difference, which has a somewhat different time constant. Consequently, there is a diurnal variation in the value of C_n^2 throughout the entire day, with minima close to dawn and dusk, as noted above, and is clearly apparent in Figure 5.4.

Cloud cover can have a significant impact on the value of C_n^2 during the early part of the day, as is shown in Figure 5.4. Cloud cover acts as a blanket, there is less of a thermal difference between the land and the air and the turbulence is reduced and, hence, so is C_n^2.

A consequence of the inverse-wavelength dependence of some turbulence effects on the propagation wavelength (approximately λ^{-1}), means scintillation and beam spread is more of a problem for visible-band sensors than for sensors that operate in longer wavebands. Hence, the sensors operating in the mid-wave infrared and especially those operating in the far-infrared waveband, particularly under low to moderate turbulence conditions, experience less degradation in performance. Therefore, turbulence tends to be more of a problem for visible-band sensors and visible-band laser-based systems than for similar infrared systems.

5.6.1 Turbulence Mitigation (Aperture Averaging)

A simple approach for minimising the impact of turbulence on the performance for sensors is to choose an optical aperture large enough to average over the intensity fluctuations produced by the turbulence along the propagation path. A similar approach may be used to mitigate the impact of turbulence on laser-beam propagation through increasing the diameter of the laser-emission aperture, although this technique may be impractical for some system applications.

If the aperture of an optical receiver (D) is larger in diameter than the spatial scale of the atmospheric scintillations owing to atmospheric turbulence, the receiver will average the fluctuations of the received waveform over the aperture area. This averaging process leads to a reduction in the signal intensity fluctuations compared with a point receiver. Typically, scintillation at a receiver is reduced significantly, compared with that measured at a small aperture, if the aperture diameter is extended to a few tens of millimetres. Additionally, the aperture-averaged signal should contain fewer high-frequency components than recorded by a smaller detector, because the higher frequency components are associated with the small-scale elements of the scintillation pattern moving across the beam, which will be spatially averaged out by the detector. A general rule is that the aperture-averaged variance of the log-amplitude in the presence of turbulence seems to be proportional to $D^{-7/3}$, where D is the aperture diameter of the sensor.

5.6.2 Turbulence Mitigation (Adaptive Optics)

Adaptive-optical techniques have been developed to compensate for turbulence-induced effects; they are critical to the success of laser-based systems that require precision target tracking, accurate beam pointing, and minimising distortion of the propagating wavefront. A typical application of this technology is laser-based directed-energy weapons, which is considered in a system's context in Chapter 7.

Beam-spreading effects, as a result of atmospheric-turbulence, are shown in the left side of Figure 5.5. On the right of this figure is a compensated beam, resulting from the effective compensation from the application of adaptive-optical techniques under identical propagation conditions.

An adaptive-optical system can provide various levels of correction to the shape and direction of the propagating wavefront to compensate for aberrating effects introduced by the atmosphere. The simplest systems provide compensation for beam-wander effects caused by turbulence in the propagation medium, but ignore distortions caused to the wavefront of the beam. This relatively simple approach, or first-order correction, is known as tip-tilt correction, and is also discussed further in Chapters 4 and 7.

A more sophisticated system involves not only the sensing of the beam-wander effects, but also the distortion of the wavefront. The latter process is called a wavefront sensor and a schematic is shown in Figure 5.6; various wavefront-sensing

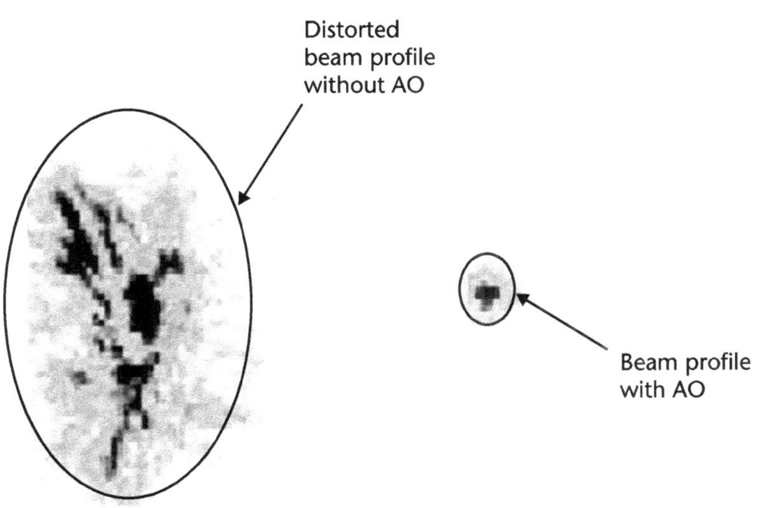

Figure 5.5 Atmospheric turbulence effects on the left and the corrected beam on the right.

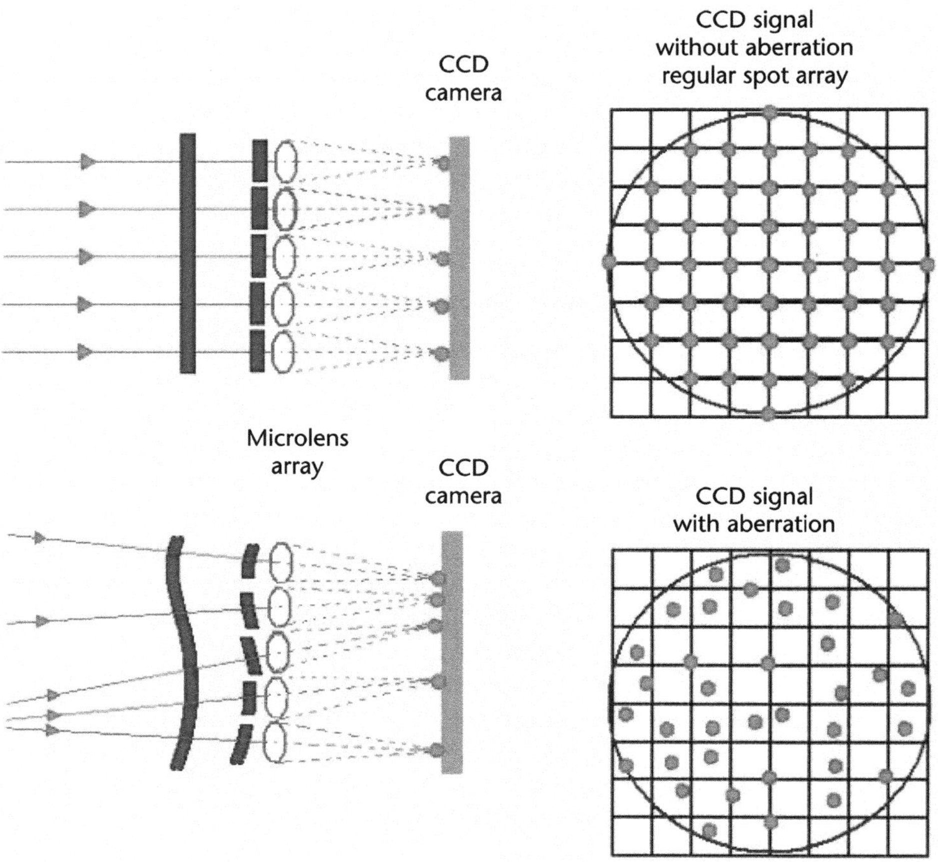

Figure 5.6 Shack-Hartmann wavefront sensor. (Courtesy of Dr M. Eichhorn, ISL, Germany.)

techniques have been developed but the most common is based on the Shack-Hartmann method, as shown in Figure 5.6.

A complete adaptive-optical system for compensating beam-wander and wavefront distortion is shown in Figure 5.7.

5.6.3 Turbulence Mitigation (Lucky Imaging)

Lucky imaging is a technique for passive correction of scenes imaged through turbulent atmospheres, which results in an improved image resolution. The technique works by collecting a sequence of short-exposure images as the atmospheric turbulence evolves. An image-quality metric is applied to the series of captured images to identify those regions in the image where the distortion is low. An alignment algorithm is used to remove the distortion from parts of the short-exposure images and then assembles a full-frame image from the various zones judged to have low distortion. A frame-stitching algorithm is used to create a single frame from the series of short-exposure frames to form an image with low distortion. The technique may be used over sequences of frames to create an apparently continuous motion picture of a turbulent scene with reduced distortion.

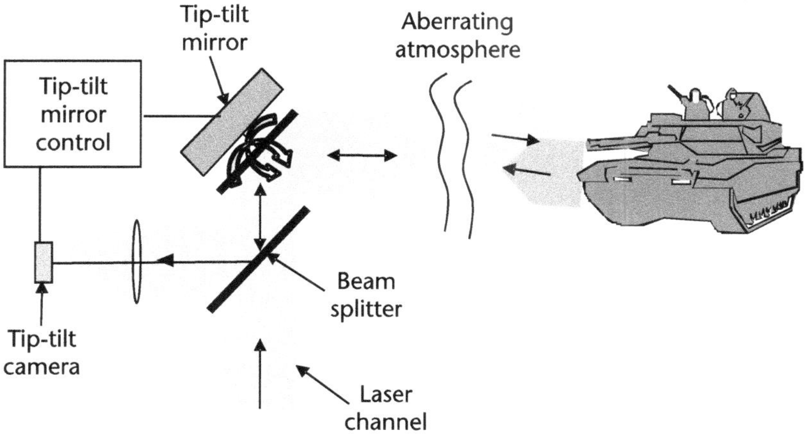

Figure 5.7 Schematic of an adaptive-optical system. (Courtesy of Dr. A. M. Scott, QinetiQ, U.K.)

Additional techniques, such as phase diversity, may be applied to improve the quality of the image, although they require additional optical components. In this case, the phase-diversity image-quality metric compares two equally defocused images either side of the true focus.

Lucky imaging has been used in conjunction with burst-illumination targeting systems, discussed in Part II. The application of this image-distortion technique has been demonstrated to improve targeting performance through enhanced recognition and identification ranges. Lucky imaging has been applied to both passive and active systems.

5.6.4 Wind Effects

The impact of wind on atmospheric turbulence has two contrasting consequences:

- The flow of air over uneven and rising ground generates turbulence-like effects to the leeward side of the perturbing structures and hills.
- The flowing air (wind) mixes with the lower atmosphere in the boundary layer and so reduces the lapse rate and, therefore, also reduces the turbulence created by the thermals.

The effects of wind on the spatial structure of a beam propagating through a moderate to highly-turbulent medium can lead to increased distortion of the beam profile in the direction of the cross-wind. Furthermore, in the maritime environment the presence of wind increases the aerosol content of the air, which may have an impact on the transmission characteristics of a laser beam. Similar changes in the distribution of aerosols occur over land with the disturbance of dust and other mineral particulates.

It is worth noting that wind also has an impact on the contrast presented by thermal targets with respect to the background. Wind flow over a target is equivalent to forced convection and reduces the surface temperature and, thus, the thermal contrast, which has implication on detection performance, especially detection range.

5.6.5 Summary of Turbulence and Wind Effects

The potentially moderate to severe impact of turbulence on the beam-propagation characteristics of laser-based systems, depending on propagation conditions, results from:

- The coherence of the beam;
- The small beam divergence used by most laser-based systems in order to have high radiant intensity, which leads to improved effectiveness.

Atmospheric turbulence leads to a reduction in the average radiant intensity. A strong turbulence will cause a reduction in the time the laser beam spends on its target or in a specific direction. However, this is far less of an issue for laser-based sensing systems. The beam spread of a laser emission leads to a reduction in the radiant intensity; scintillation causes a fluctuation in the radiant intensity, generally leading to a reduction in brightness on average, thus reducing the effective range of a system. Motion of the centroid of the beam adds to the pointing jitter from the beam director and so reduces the time the beam spends on its target; although the simultaneous increase in the beam spread may counteract the additional turbulence-induced jitter effects to a certain extent, depending on the precise magnitude of the effects.

Turbulence in the atmosphere distorts the image being used by an electro-optical or infrared tracker in a beam director, which leads to:

- Reduction in detection and acquisition ranges and/or a delay in achieving these processes;
- Reduction in the range at which a target may be classified or recognised, if the subsystem has this function;
- Distortion and shimmering of the target, leading to less accurate target tracking. The resultant effect is that the beam again spends less time on its target, which potentially reduces weapon or countermeasure effectiveness. The consequences for laser-based sensing systems are similar when compared with the weapons systems.

For the high-performance directed-energy systems, compensation techniques, such as adaptive-optical techniques, will almost certainly be required to achieve the very demanding irradiance for a successful target interaction. In general, it may be assumed that beam conditioning and beam control will be vital for successful high-power interactions with targets.

Some relatively simple passive techniques may also be applied to minimise the first-order effects of turbulence-induced perturbations, such as aperture-averaging or lucky-imaging techniques.

5.7 Thermal Blooming

This phenomenon usually arises when high-power beam propagation occurs in the lower atmosphere. This results in changes to the spatial characteristics of the propagating laser beam, as shown in Figure 5.8, with an initially unbloomed beam and the beam profile caused by thermal-blooming effects. Absorption of energy from photons in the beam by the molecules and aerosols along the propagation direction causes local heating of the atmosphere and a consequent change in the local refractive index of the atmosphere.

Generally, for a Gaussian beam, more energy is absorbed along the centre of the propagation path, as a direct result of the intensity distribution in the beam, which reduces the refractive index preferentially along the propagation axis. This change in the refractive index effectively creates a lens in the atmosphere whose focal length may be either positive or negative. Hence, the photon energy may be focussed, or more usually de-focussed, owing to the reduction in the refractive index along the beam-propagation axis. The consequence of the defocus is a reduction in the power density at the target area. This is illustrated schematically in Figure 5.9, showing the formation of a divergent lens.

Thermal blooming is a significant problem for high-power, laser-based, directed-energy-weapon systems. The effects can be less severe if there is lateral motion of the beam or a cross-wind; as these effects sweep away the heated air with the

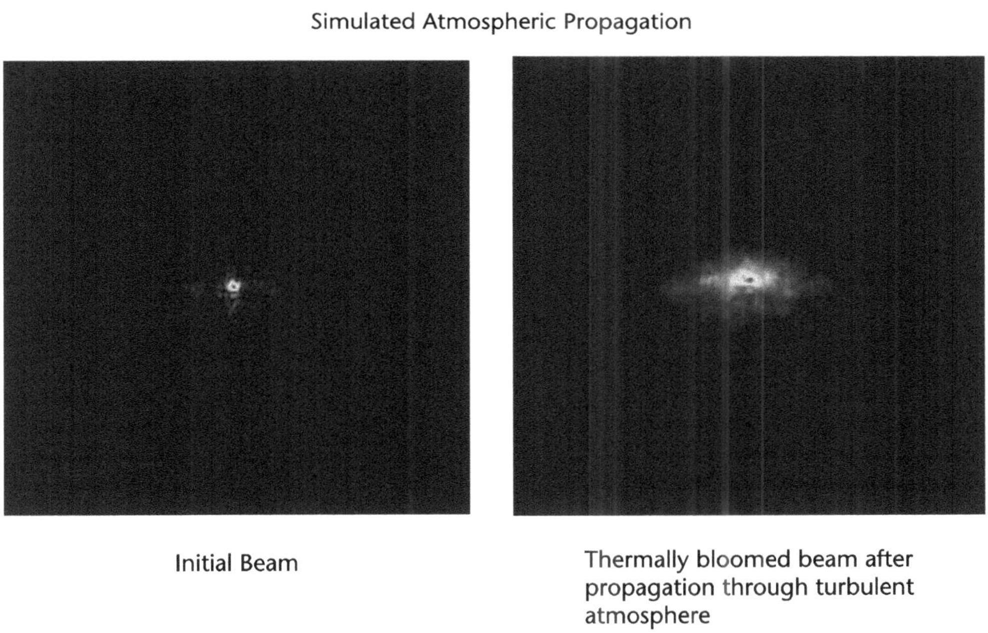

Figure 5.8 Effects of thermal blooming on the spatial profile of a laser beam. (Courtesy of Mr. Don Seeley, DoD HEL/JTO, USA)

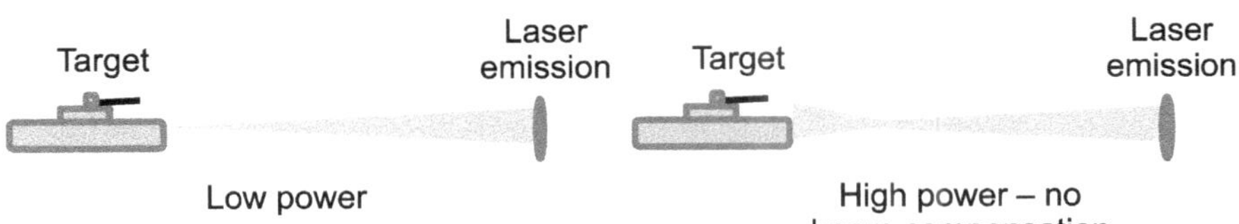

Figure 5.9 Schematic illustration of thermal-blooming effects.

reduced refractive index from the propagation direction and continually replace it with unaberrated air. This means that despite the high-power photons heating the atmosphere from absorption of the beam, the beam continually passes through fresh air without any significant defocusing properties. As a consequence of this phenomenon, thermal blooming is a potential limiting feature of some high-power laser systems, as discussed in Part II.

As noted above, atmospheric absorption is dependent on the wavelength of the propagating beam and the atmospheric content; therefore, great care is required in selecting the wavelength of the high-intensity laser beam to achieve high transmission, as well as reducing the impact of absorption to minimise thermal-blooming effects. The ability to tune the wavelength is one of the reasons for interest in the free-electron laser for maritime, laser-based, directed-energy applications, as discussed above in Chapter 3.

Consequently, some natural compensation of thermal blooming can be achieved, provided that the target-engagement geometry can be chosen so that the interaction sightline between the beam director and the target has some lateral component to its motion, causing the laser beam to pan across a sector during the engagement; in this case, there is less likelihood of thermal blooming reducing the fluence reaching the target, as discussed above. The onset and impact of thermal blooming also reduce with altitude, owing to the reduction in absorption processes encountered by the beam.

Figure 5.10 shows the impact of a potentially limiting feature created by thermal blooming at low and high altitude for a powerful beam, in terms of the range dependency of achieving a given moderate to high irradiance (for example, of the order of several hundred W/cm^2) at any given down-range position. The

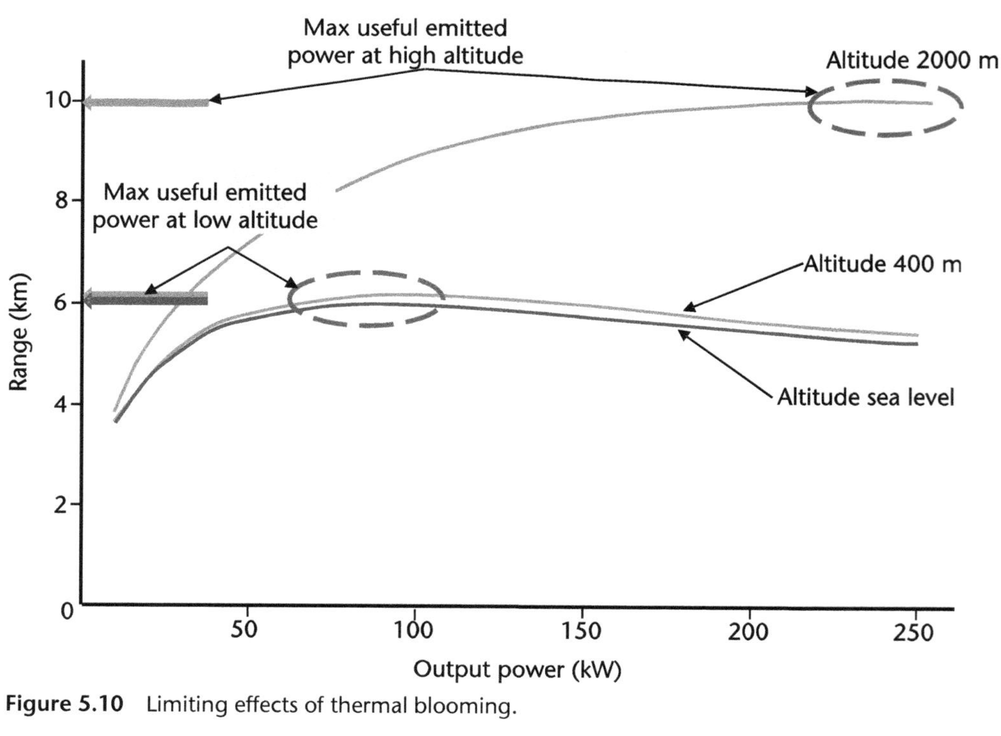

Figure 5.10 Limiting effects of thermal blooming.

consequences of the enhanced absorption at lower levels in the atmosphere, and the resultant spread of the beam are readily apparent, so in this instance of severe thermal blooming at low level, the maximum projected-power that is likely to be useful for achieving a desired irradiance, as stated above, maybe of the order of 100 kW. Increasing the launched power will not improve the attainment of the desired irradiance at ranges beyond about 6 km in this hypothetical (and challenging) case.

Absorption of the propagating laser beam by the atmosphere is wavelength dependent, as illustrated in Figure 5.11. This plot shows the maximum range where a given launch power may be effective as a function of wavelength in a maritime environment in terms of a desired fluence at a target. This diagram makes it very clear that brute force approaches of launching ever-increasing power may be self-defeating and judicious choice of operating wavelength is crucial to minimise distortion of the beam profile.

Analyses of these effects presented in this theoretical plot infer some interesting conclusions, for example:

- The iodine wavelength (1.315 μm) has limited utility in this particular environment out to about 3.5 km and increasing the launched power reduces capability.
- Other wavelengths have a less dramatic limiting function.
- Careful choice of wavelength could mean a difference in effective range increasing by a factor of two for the same launched power.

It should be noted that this plot only concerns the maximum range that of a given irradiance for a given launch power. It does not infer any effectiveness at that range in the laser-target interaction dynamics. Estimates of potential performance of laser directed-energy systems are complex activities, which require application techniques well beyond the intended scope of this book.

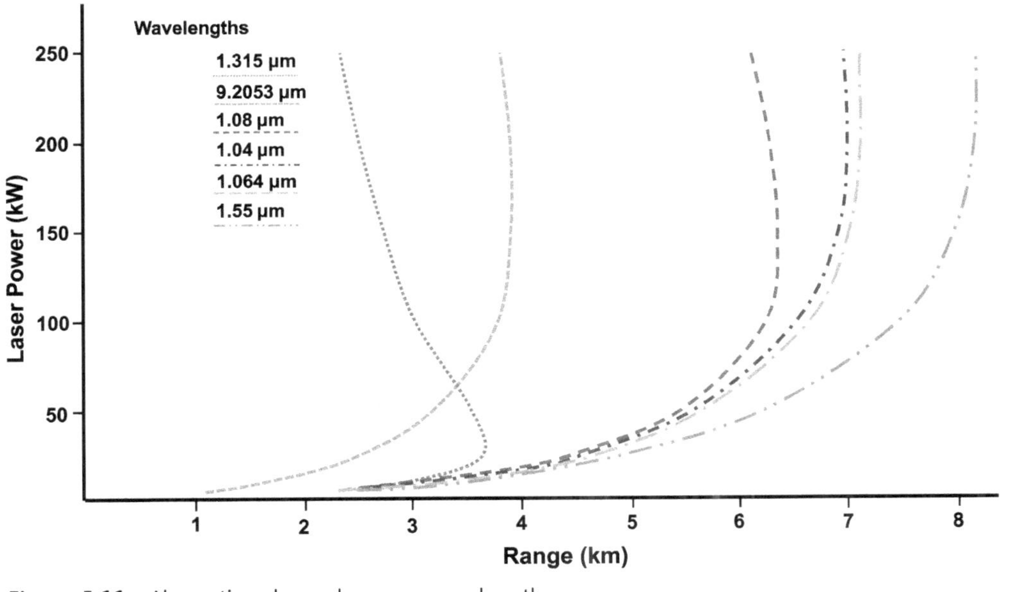

Figure 5.11 Absorption dependency on wavelength.

5.8 Ionisation

When high-intensity laser beams are brought to a focus, the ionisation threshold of any molecular species at, or close to, that focus may be exceeded by multiphoton processes, liberating electrons and creating a plasma. The presence of particles and aerosols along the laser path may reduce the threshold power-density required for plasma formation; additionally, the ablation of solid material from the surface of a target may provide the mechanism needed for the creation of a plasma close to the target. In general, multiphoton processes increase with irradiance and very intense beams of radiation are required to ionise clean or pure air. This is especially true in the infrared, owing to the lower photon energy, but occurrences of pure air are rare during most military operations.

The formation of a plasma along the propagation path of a laser beam is detrimental to the intensity of the propagating beam at the target, as the beam becomes decoupled from the target's surface, and the plasma cloud moves away from the target's surface. However, if the plasma can be constrained to the immediate vicinity of the target's surface, enhanced coupling may occur, but this condition is very difficult to maintain in a dynamic engagement environment. Similarly, this phenomenon may be used as a protection measure for blocking intense beams by seeding the formation of a plasma from material being ablated from a target's surface.

5.9 Nonlinear Propagation (Filaments)

This is a relatively new phenomenon, which may be observed when very short laser pulses in the femtosecond regime, with extremely high instantaneous peak intensities, propagate in air.

The propagation of intense femtosecond laser pulses in air and other transparent bulk media does not follow the conventional physics of normal propagation of light, as it occurs in the nonlinear regime. These self-focusing laser pulses propagate over significant distances overcoming diffraction effects both in space and time through the formation of filaments. During the propagation of these filaments through air, the photon pulse self-transforms into a white light laser pulse; its bandwidth spans continuously from about 350 nm in the ultraviolet to around 9 μm in the far infrared from a very short-pulse laser, such as a titanium-sapphire source. Additionally, this laser pulse contains a broadband third-harmonic pulse around 267 nm of the original peak wavelength of the short-pulse device (titanium-sapphire wavelength). Consequently, the total bandwidth of the twin pulses in this white-light continuum spans the ultraviolet from 230 nm continuously across the spectrum to around 9 μm in the infrared; hence, the use of the term super continuum.

A long filament formed during the propagation in air is made up of multiple self-focusing events. A proposed physical reason for the long distance propagation of these filaments follows a moving focus model, in which successive slices of the input laser pulse focuses at different positions. Inside a self-focused zone, the intensity is so high that air molecules are tunnel ionised, resulting in a weak plasma. The balance of the self-focusing and plasma defocusing of the slice leads to an apparent intensity clamping, or limiting, as the peak intensity at the self-focus cannot be increased indefinitely with increasing pulse energy.

Other slices of the short pulse will self-focus in a similar manner successively along the propagation axis. Therefore, a self-focus of a pulse at any position along the propagation axis is surrounded by the larger background of the laser-beam's electromagnetic field, which provides a type of background reservoir for sustaining the filamentation effect. The result is that the energy in the self-focus is fed from this background reservoir; it is not the self-focus zone itself that propagates. Moreover, it is the series of self-focal spots that forms a filament, as observed during very short-pulse propagation.

The photon emission wavelengths from within the self-focused region result from the nitrogen molecules fluorescing after ionisation and recombination, which is then amplified, so this light has the characteristics of amplified spontaneous nitrogen fluorescence. These filaments may extend to several kilometres in air.

The formation of multiple filaments is a consequence of distortion in the wavefronts, owing to a variety of physical effects such as:

- The strength of the focusing (Kerr lens) and defocusing (plasma) nonlinear phenomena;
- Intrinsic wave front irregularity of the laser pulse;
- Inhomogeneity of the propagation medium, such as the atmospheric turbulence in air;
- Aberrations in the optical system and in the atmosphere.

The very high power of such laser pulses is generally much greater than the critical power required for self-focusing, so the distorted (or warm zones, as they are called) regions across the aberrated wavefront lead to multiple self-focusing zones around the beam-propagation axis. The propagation of these foci results in multiple-filamentation formation along the propagation axis. These foci are sustained by a common background reservoir, described above, but as each of these hot spots derives its energy from the common background reservoir, there is a competition for energy to sustain them; this is similar to mode competition in a laser cavity.

There is interaction between the various focal zones, which depends on the relative distance between hot spots in the distorted wavefront and the interference of the fields around these hot spots that could then generate a new array of warm zones. These could in turn self-focus further along the propagation path; provided the field in the pulse has enough power to form a new self-focus. This interaction between the array of multiple filaments plays an important part in determining the number of self-foci needed to ionise the air molecules sufficiently. Consequently, the closer the initial warm zones are in the aberrated wavefront, the more new filaments will be created along the propagation path, leading to a more efficient super-continuum generation and enhanced nitrogen fluorescence.

An additional benefit of nonlinear propagation effects is the ability to direct an intense laser pulse through clouds; femtosecond filaments can survive an encounter with water droplets as large as 95 μm, and they can be transmitted through atmospheric clouds that are optically dense, which would normally result in a transmission of about 5% of the incident beam. This phenomenon is attributed to a dynamic energy balance between the quasi-solitonic structure in the filament and the surrounding laser photon flux, which acts as an energy reservoir.

5.10 Beam Propagation Through Wakes and Plumes

This phenomenon provides an example of the potential impact on installed performance of a laser-based system, as a result of the local propagation conditions close to an aircraft, where there are zones of extreme turbulence. The exhaust is typically a mixture of gases with a temperature of the order of 800K, but the plume contains a mixture of hot and cold zones with a complex thermal profile.

The efflux from a jet engine has hot exhaust gases that are mixed rapidly with the cooler ambient atmosphere, which produces a mass of disturbed or turbulent air close to the engine nozzle and for some hundreds of nozzle diameters downstream. Figure 5.12 shows a region of highly-turbulent gas stretching out behind an aircraft. This zone behaves as a region with a very large refractive-index structure parameter (C_n^2). Close to the engine nozzle, the magnitude of the turbulence has been estimated to be four or five orders of magnitude worse than the most severe effects experienced by natural atmospheric turbulence, as may occur in a hot desert close to the ground, that is, C_n^2 of the order of 10^{-9} m$^{-2/3}$ to 10^{-8} m$^{-2/3}$.[1]

Figure 5.12 The problem of propagation through highly-turbulent media. (Courtesy of Swedish Armed Forces.)

1. It should be noted that although this C_n^2 value is very large, the parameter C_n^2L is often cited as the important parameter (where L is the interaction length) and the relatively short interaction length in the plume reduces the impact somewhat, although this zone of turbulence will create very strong lensing effects close to the laser aperture, which enhances defocusing and thus break up of the beam profile.

5.10.1 Propagation Characteristics

The consequence on a laser beam propagating through this region is intensive beam wander and spreading, as well as break-up of the spatial profile of the beam, as shown in Figure 5.12, for both fixed- and rotary-wing aircraft. Therefore, the direction of a laser beam through an efflux should be avoided if degraded performance cannot be tolerated. For some laser-based systems, the presence of the efflux creates a blind arc in the system's target engagement zone.

An additional concern relates to the possibility that the tracking system in a laser-based directed-energy weapon, such as a laser-based countermeasure system, cannot detect or track a target in the vicinity of the efflux. The thermal contrast required to detect a target in the background clutter could be degraded, for example, as the tracking feature being used by the beam director, such as the plume from the target's rocket engine, starts to be obscured by the aircraft's efflux. In this case, the track on the approaching threat could be lost by the tracking system. Equally important are the potential situations where either the approaching threat is not detected by threat warning sensors or the hand-off to the beam director cannot be achieved, as a target cannot be detected, acquired or tracked.

5.10.2 Impact on Installed Performance

A major concern is the impact that this phenomenon will have on the installed performance of any laser-based systems, as highly-turbulent media have a greater impact on coherent radiation than on incoherent light, causing beam break up, as well as beam spreading and motion of its centroid. Hence, the changes to the spatial characteristics are greater on a laser emission than on the emission from an incoherent source, such as an arc lamp. Moreover, the divergence of the laser beam is much less than that from an arc lamp; hence, the reduced pointing stability may have an impact on the efficacy of laser-based countermeasure systems. All of these deviations from the ideal, even in good propagation conditions, could have important consequences on countermeasure effectiveness in some specific cases.

The impact of wakes and plumes from engines on the installed performance of platform-protection systems, such as laser-based directed infrared countermeasure systems, is of particular importance for transport aircraft and similar aircraft configurations, with engines mounted under the high wings and the beam directors (of DIRCM transmitters) positioned on the fuselage. It is anticipated that this phenomenon of propagation through the engines' wakes could reduce the expected platform protection offered by the laser-based countermeasure when engaging near co-altitude threats in particular quarters, especially about the beam.

The consequences of severe atmospheric turbulence on the operational performance of laser-based sensing systems, operated from airborne platforms, are likely to be less significant than for the airborne laser-based, directed-energy systems. This qualitative assessment is based on the fact that the sensing systems are normally anticipated to be operated in the forward or rear arcs of an airborne platform. If the laser-based sensing systems are to be used to probe the scene around the lateral positions of an aircraft, then similar consequences to the performance are likely to occur, as described above for the laser-based, directed-energy systems.

5.11 Aero-Optical Effects

This phenomenon is a consequence of the flow of air over a structure, which results in the compression or expansion of that air close to an aperture in a structure or a system protruding from an aircraft such as a beam director in an air flow. This variation in pressure leads to changes in air density, causing variations in the refractive index of the propagation medium; consequently the beam of radiation will bend as it passes through the zone of varying refractive index. This bending occurs because the air flow has the optical properties of a thin prism or a wedge. The aero-optical effects may also result in:

- Dynamic wave-front distortion;
- Pseudo-steady aerodynamic lensing effects with a slowly varying mean value;
- Aero-buffet-induced line-of-sight jitter;
- Beam-train coupling;
- Shock-wave effects.

The consequence of these changes in the beam-propagation direction and other induced aberrations on the transmission and receiver channels is a reduction in the tracking accuracy or the pointing precision of the laser-based system in the more severe cases. The phenomenon of the aero-optical effect is shown schematically in Figure 5.13.

There is a deflection of a beam each time it passes through the wedge. Compensation techniques may include an estimation of the optical wedge using probe beams and then applying a suitable aiming correction to the pointing direction of the laser beam.

Careful positioning of the beam directors for laser-based systems can mitigate many of the effects caused by this phenomenon. For example, positioning a beam director where the boundary layer is very thin, such as on the nose of an aircraft or at the front of a pod, so that it projects beyond the boundary layer flowing over a structure. Other techniques involve shaping of the air flow around the beam director to minimise the optical wedges on or adjacent to the transmission aperture.

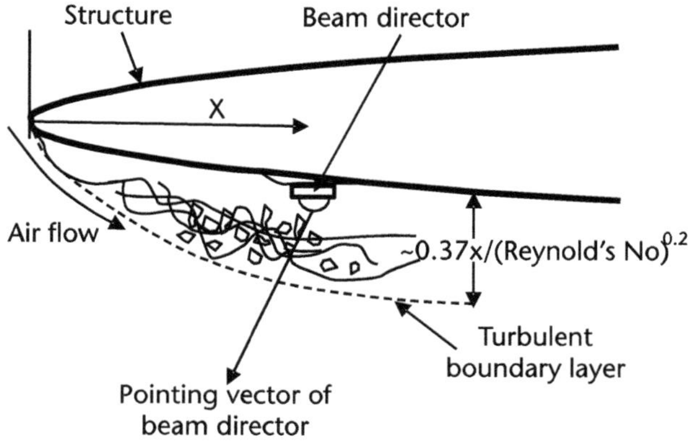

Figure 5.13 Aero-optical effect on laser-beam propagation.

As noted above, airborne sensing is often undertaken with the system looking forwards or backwards. Hence, provided that transmitter and receiver are suitably positioned the impact on a laser-based sensing system is likely to be minimal. Moreover, as extreme precision in the pointing vector is not mandatory in this case, the impact on the optical performance with the system in other positions may not be detrimental either.

The impact of aero-optical effects is considered in terms of the potential degradation in performance of a laser-based countermeasure system in the following section.

5.11.1 Aero-Optical Effects Around a Laser-DIRCM Transmitter

The impact of the air flow around the beam director results in three anticipated effects:

- Turbulence in the boundary layer around the aircraft, which creates a random phase screen in front of the transmitter;
- Air pressure is increased in front of the transmitter and reduced behind it, leading to refraction and lens-focusing effects;
- Velocity of the airflow over the transmitter is increased above the free-stream airflow velocity, leading to shock waves where the local velocity exceeds the speed of sound.

The effects may be modelled mathematically using a suitable representation of the air flow and the protrusion of the optical system, to estimate the impact of these effects on the Strehl ratio of the beam profile and its displacement. The effects of boundary-layer turbulence, refraction and focusing by shock waves are anticipated to be small for a laser-based countermeasure system operating in the mid-wave infrared, providing that the laser-beam diameter is of the order of 10 mm and the transmitter body diameter is at least an order of magnitude larger, typically more.

The imaging system used when tracking the beam director (fine-track sensor or camera) may well need aberration correction from an adaptive-optical system to achieve diffraction-limited operation to support high-precision laser-based systems. In the case of high-power laser-based systems, the aero-optical effects may pose a limitation on the lethal field of regard.

5.12 Summary

The atmospheric propagation conditions have the potential to alter radically the operational or designed performance of any laser-based system.

The composition of the propagation medium (the Earth's atmosphere) has important consequences on the transmission of any laser beam, owing to the presence of the various di-atomic and tri-atomic molecules and an ever varying distribution and concentration of aerosols. These gases and aerosols lead to some extinction of the beam as a result of absorption and scattering effects, which, in turn, can lead to other phenomena that can interfere dramatically with the laser-beam and target-interaction effects, particularly for high-power beams.

In some circumstances, the dynamics of the atmosphere and its weather can seriously degrade the anticipated operational performance of laser-based systems and sensors. The effects of severe atmospheric turbulence may lead to a significant reduction in pointing accuracy, radiant intensity and irradiance, unless compensation techniques are applied. The more extreme atmospheric turbulence effects lead to very dramatic scintillation effects and eventual break up of the beam.

High-performance, laser-based systems will require the application of compensation techniques, such as adaptive-optical techniques to correct for the aberrations, and thus, maintain operational performance in the presence of moderate to severe turbulence along the propagation path. It is anticipated that such sophisticated compensation techniques will not be required for the lower-power, laser-based, directed-energy systems, and indeed, may not be feasible for these applications in a highly dynamic interaction.

Atmospheric absorption of the beam can lead to very significant effects on the power that can pass through the atmosphere, particularly for high-power systems. The phenomenon of thermal blooming can lead to defocusing of a high-power beam and modification to its spatial profile. Careful choice of the operating wavelength and appropriate engagement geometry can mitigate some of the more extreme operational effects of this beam-irradiance limiting phenomenon, but in extreme conditions, with very intense beams, thermal blooming can limit or clamp the optical irradiance that can be delivered to a target. Under these conditions, increasing the power of the system may only decrease the optical power transmitted to the tracked target.

Another potentially limiting effect of high-power propagation is atmospheric break down, resulting in plasma formation. The creation of a plasma is generally to be avoided, because it will decouple the photon beam from a target; however, in some instances the presence of a plasma can enhance the coupling into a target, provided it is in the immediate vicinity of a target. The major challenge in the exploitation of this potential enhancement effect is to ensure that the plasma remains in the immediate vicinity of the target, especially if that coupling technique is critical to the defeat mechanism is being invoked.

A recent development in ultrashort pulse devices has led to the demonstration of filament formation in the atmosphere resulting from the self-focusing of the very high intensity beams. These filaments exhibit many potentially valuable properties for both sensing and modest-power, directed-energy (e.g. countermeasure) applications.

Exhaust plumes from aircraft engines can have a significant impact on the propagation characteristics of a laser beam being transmitted from an onboard system. The complex thermal profile of the plume, resulting from the mixing of the hot exhaust with the ambient air, leads to an extreme example of beam propagation in turbulent media.

An additional phenomenon that can complicate beam propagation from an aircraft is a result of air flow around and over a protruding aperture, such as a beam director, which can result in the formation of optical wedges close to, or over, an aperture. This phenomenon is known as the aero-optical effect and leads to deflection of the transmitted and received optical paths and other undesirable effects.

The use of compensation techniques is crucial to maintaining the performance of the very accurate systems, especially directed-energy systems.

Selected Bibliography

Andrews, L. C., *Field Guide to Atmospheric Optics,* Bellingham, WA: SPIE Optical Engineering Press, 2004.

Andrews, L. C., and R. L. Phillips, *Laser Beam Propagation Through Random Media,* Bellingham, WA: SPIE Optical Engineering Press, 2005.

Andrew, L. C., R. L. Phillips, and C. Y. Hopen, *Laser Beam Scintillation with Applications,* Bellingham, WA: SPIE Optical Engineering Press, 2001.

Bohren, C. F., and D. R. Hoffman, *Absorption and Scattering of Light by Small Particles,* New York: John Wiley & Sons, 1983.

Chu, B., *Laser Light Scattering,* San Diego, CA: Academic Press, 1991.

Geary, J. M., *Introduction to Wavefront Sensors,* Bellingham, WA: SPIE Optical Engineering Press, 2006.

Houghton, J. T., *The Physics of Atmospheres,* Cambridge, U.K.: Cambridge University Press, 1977.

Hudson, R. D., Jr, *Infrared System Engineering,* New York: John Wiley & Sons, 1969.

Jakeman, E., and K. D. Ridley, *Modeling of Fluctuations in Scattered Waves,* Boca Raton, FL, Taylor & Francis, 2006.

Meteorological Office, *A Course in Elementary Meteorology,* London, UK: Her Majesty's Stationary Office, 1975.

McCartney, E. J., *Optics of the Atmosphere,* New York: John Wiley & Sons, 1976.

Middleton, W. E. K., *Vision Through the Atmosphere,* Toronto, Canada: University of Toronto Press, 1963.

Perram, G. P., et al., *An Introduction to Laser Weapon Systems,* Albuquerque, NM: The Directed Energy Professional Society High Energy Laser Joint Technology Office, 2010.

Sasiela R. J., *Electromagnetic Wave Propagation in Turbulence: Evaluation and Application of Mellin Transforms,* 2nd ed., Bellingham, WA: SPIE Publishing, 2007.

Smith, F. G., *The Infrared and Electro-Optical Systems Handbook, Vol. 2, (Atmospheric Propagation of Radiation),* Ann Arbor, MI: ERIM and Bellingham, WA: SPIE Optical Engineering Press, 1993.

Strohbehn, J. W., (ed.), *Laser Beam Propagation in the Atmosphere,* Berlin, Germany: Springer-Verlag, 1978.

Tyson, R. K., and B. W. Frazier, *Field Guide to Adaptive Optics,* Bellingham, WA: SPIE Optical Engineering Press, 2004.

Tyson, R. K., *The Lighter Side of Adaptive Optics,* Bellingham, WA: SPIE Optical Engineering Press, 2009.

Veichel, H., *Laser Beam Propagation in the Atmosphere,* Bellingham, WA: SPIE Optical Engineering Press, 1995.

CHAPTER 6

Radiometry, Testing and Evaluation of Laser Systems

6.1 Introduction

Radiometry is the starting point for testing active optical systems. With any optical system, it is easy to generate many megabytes of data, but reducing the recorded data to a form that enables an understanding of the military importance and then assessing the value may be a separate challenge. The aim of this chapter is to provide an introduction to the principles of radiometry, which forms the fundamental basis for the measurement of optical radiation, and some techniques that may be applied to evaluate laser-based systems.

The assessment and the evaluation of a laser system, or any optical system, is much more than counting the photons that emerge from the aperture and deciding on the size of the bucket needed to collect them. The characterisation of the beam is important, and moreover, if the laser emission is short of the mark, there is little chance of the laser system providing the military capability defined by the requirements or capability studies.

The most important characteristic of any military system whether it uses a laser or not, is the performance that it provides when it is installed on a platform: the installed performance. This represents how the device and its subsystems behave in an operational environment when integrated to form a complete system and provide a specified military capability.

However, the indication of performance acquired during the evaluation and testing phase is far more important than deciding whether a supplier or contractor has achieved the requisite spot-point performance or not. The assessment phase is about identifying the conditions where the system works as designed and where the shortfalls are; this phase of testing should determine where, within the performance envelope, the installed performance is good, marginal or inadequate.

Suspected defects in the required performance may be evaluated and resolved if there has been a systematic, rigorous and progressive scheme for the testing of the subsystems, as well as the integrated system. Such a methodical approach may provide an indication of marginal performance, with respect to the specification in a component or subsystem prior to integration with a platform for the installed-performance testing, which may allow any of the problems identified later to be resolved rapidly. This approach is also valuable if there is a need to extend the performance of a system at a subsequent date.

It is crucially important that the military operators appreciate and understand fully the implications of any shortfall in system performance. Full knowledge of the performance of any system in its operational environment enables the military planners and operators to make value judgements. For example, it may be possible for part of the problem to be resolved through changes to tactics, training and procedures.

Surprises in the lack of performance of equipment during front-line operations must be avoided. This starts at the component level and progresses through stages of increasing system and threat environment complexity throughout development test and evaluation activities. This evaluation programme ends with a comprehensive operational test and evaluation programme to provide an assessment of the installed performance.

This chapter offers systematic approaches to a comprehensive test programme. Its cost should never be underestimated, particularly if it involves aircraft, ships, military vehicles or missiles. Hence, the need for test plans that are efficient and give rigour, as well as careful and methodical techniques for data collection.

A general rule of thumb is that the costs, in terms of time and money, at least double at every stage of the evaluation process. Similar comments apply to the time taken to test the kit, analyse the responses and interpret the potential military value. Moreover, there is no real substitute for a complete data set that is recorded accurately and timely.

6.2 Radiometry

Radiometry is the measurement of optical radiation, which for these purposes is electromagnetic radiation in the frequency range between 3×10^{11} Hz and 3×10^{16} Hz. This range of frequencies covers a vast part of the spectrum and corresponds to wavelengths between 10 nm and 1,000 mm and so includes the regions commonly called the ultraviolet, the visible and the various bands across the infrared to the terahertz region and beyond. In fact, there is an overlap between that part of the spectrum called the far infrared and radio frequencies; in the latter case, it is normal to refer to electromagnetic radiation as belonging to the radio part of the spectrum when it is detected by measuring currents generated in an antenna, whereas infrared radiation is detected by measuring its heating effect.

In the consideration of radiometry, there are essentially two fundamental cases:

- The extended source involves a surface radiating into a solid angle, where the radiance is a function of the viewing angle.
- The other case is radiation from a point source and has no angular dependency, which is independent of viewing angle as it is an isotropic radiator.

The observational dependencies of an extended source and a point source are shown in Figure 6.1.

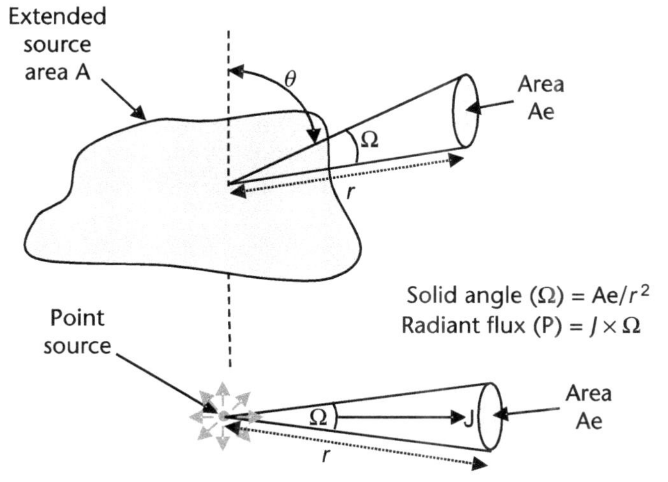

Figure 6.1 Geometrical considerations of extended and point sources.

The base units of radiometry are the watt and joule; other radiometric units in common usage, along with their physical meaning, are:

- Watt (radiant flux);
- Watt per square metre (irradiance), commonly referred to as power density;
- Watt per steradian (radiant intensity);
- Watt per square metre per steradian (radiance).

The other crucial fundamental term to understand is photometry. This term refers to the measurement of light, which is defined as electromagnetic radiation detectable by the human eye. Consequently, photometry and its associated units are restricted to the visible region. The internationally agreed wavelength range covering the response of the human eye is from 360 to 830 nm. Moreover, all of the photometric quantities are weighted by the spectral response of the eye, which peaks in the green and decays to practically zero in the ultraviolet and in the deep red and beyond.

There are two basic approaches to making photometric measurements:

- Use of optical radiation detectors constructed to mimic the spectral response of the eye, through, for example, use of spectral filters;
- Spectro-radiometry coupled with appropriate calculations for weighting by the spectral response of the eye.

The usual photometric units include:

- Lumen (luminous flux);
- Candela (luminous intensity);
- Lux (illuminance);
- Candela per square metre (luminance).

The difference between radiometry and photometry may be summarised as follows. Radiometry includes the entire optical radiation spectrum (and often involves spectrally resolved measurements), whereas photometry concerns optical measurements within the visible spectrum, but most importantly, the measurements are modified (or weighted) to take account of the response of the human eye. This section provides some guidance to the fundamental science and terms used in photometry and radiometry.

6.2.1 Photometry

The primary aim of photometry is to measure and characterise visible optical radiation, in such a way that the results correlate with the visual sensation experienced by a normal human observer exposed to that light. Modern photometric practice involves measurements made with photo-detectors, to remove the subjective nature of visual comparison techniques of measurements. This technique is now referred to as physical photometry and provides a basis for absolute measurements.

Clearly, to achieve the aim of physical photometry it is necessary to have an internationally agreed model of the response characteristics of human vision. This model, shown in Figure 6.2 has under gone a number of refinements since the relative spectral responsivity of the human eye was first defined by the Commission Internationale de l'Éclairage (CIE) in the 1920s. It was redefined as part of colorimetric standard observers a few years later in 1931.

This function is now known as the spectral luminous efficiency function for photopic vision, or more commonly as the $V(\lambda)$ function, is defined in the spectral range from 360 to 830 nm. This response function for the human eye is normalised to unity at its peak, defined as 555 nm in the green part of the spectrum, as shown

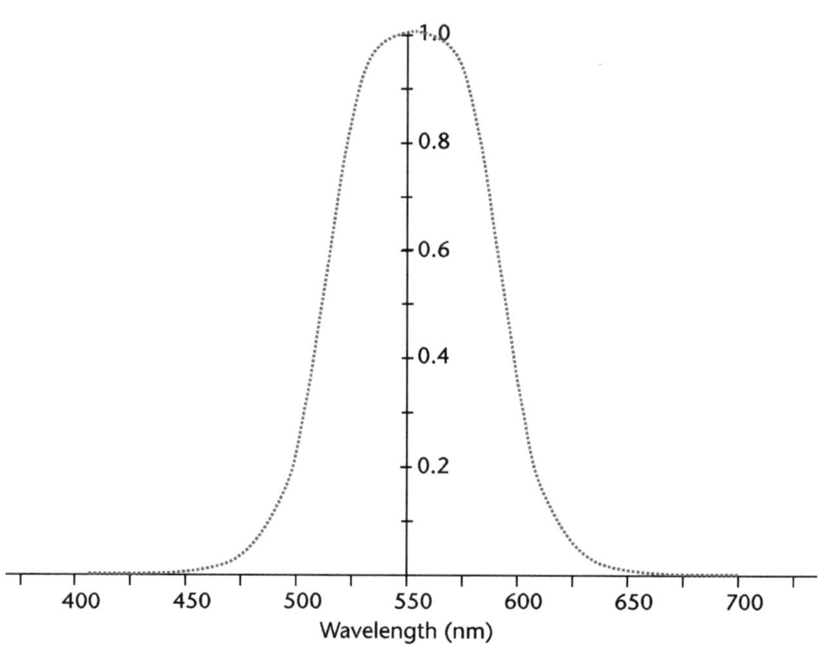

Figure 6.2 International agreed spectral response of the human eye (CIE V(λ) function).

in Figure 6.2 is for photopic vision. Consequently, any sensor aiming to mimic the behaviour of the human eye will be required to have this spectral response.

This model is commonly used to define the response of the human eye, owing to the wide acceptance of this model by the optics community. The values were republished by the Commission Internationale de l'Éclairage and also by the Comité International des Poids et Mesures (CIPM), following the redefinition of the candela in 1979. Tabulated values of the $V(\lambda)$ function at 1-nm increments are available in the literature, but the specific details are beyond the intended scope of this book.

For most practical purposes, the spectral band from 380 to 780 nm is more than adequate for physical photometric purposes, as any characterisation using this range of wavelengths results in negligible error, because the value of the $V(\lambda)$ function response falls to less than 10^{-4} outside this zone. Thus, the use of a photodetector having a spectral responsivity matched to the $V(\lambda)$ response function replaces the role of human eyes in photometry.

In contrast, radiometry concerns physical measurement of optical radiation as a function of its wavelength, as noted above. It is possible to relate a photometric quantity X_v to the corresponding radiometric quantity X_e, by the equation:

$$X_v = K_m \int X_e V(\lambda)\, d\lambda$$

The constant, K_m, is taken to be equal to 683 lm/W at the peak response of the human eye (555 nm), in the above equation. It relates the photometric quantities and radiometric quantities, and is called the maximum spectral luminous efficacy (of radiation) for photopic vision. The value of K_m is defined by an international body to be to be 683 lm/W at a specific frequency (equivalent to a wavelength of 555.016 nm in standard air), that is, not quite at the peak of the response of the human eye, leading to an actual value of K_m slightly larger than 683 lm/W. However, this simplifying assumption incurs negligible error in moving from radiometric to photometric quantities.

These definitions relate to a specific response of the human eye (i.e., photopic vision). Photopic vision, in general terms, relates to a relatively narrow field of view and in relatively high luminance levels (higher than approximately 1 cd/m^2) so the cones of the retina are the predominant sensors in operation in the human eye.

At lower light levels, the spectral responsivity of human vision deviates significantly from this model defined by $V(\lambda)$. This type of vision is called scotopic vision and occurs at very low levels of luminance (less than approximately 10^{-3} cd/m^2). Additionally, this type of human vision is characterised by use of the rod sensors on the retina, the spectral responsivity in these condition peaks at 507 nm, and is designated by the $V'(\lambda)$ function. When the human eye is operating in the scotopic regime then the perception of colour is largely lost, so all animals, not just cats, appear black in the dark. The response of the rods and cones is shown in Figure 6.3, showing the respective peak responses at 507 and 555 nm.

The accepted term for human vision in the region between photopic vision and scotopic vision is called mesopic vision. Currently, there is no internationally accepted spectral luminous efficiency function for the mesopic region. It is common practice for almost all photometric quantities to be stated in terms of photopic human vision, even for low light levels well beyond the photopic regime.

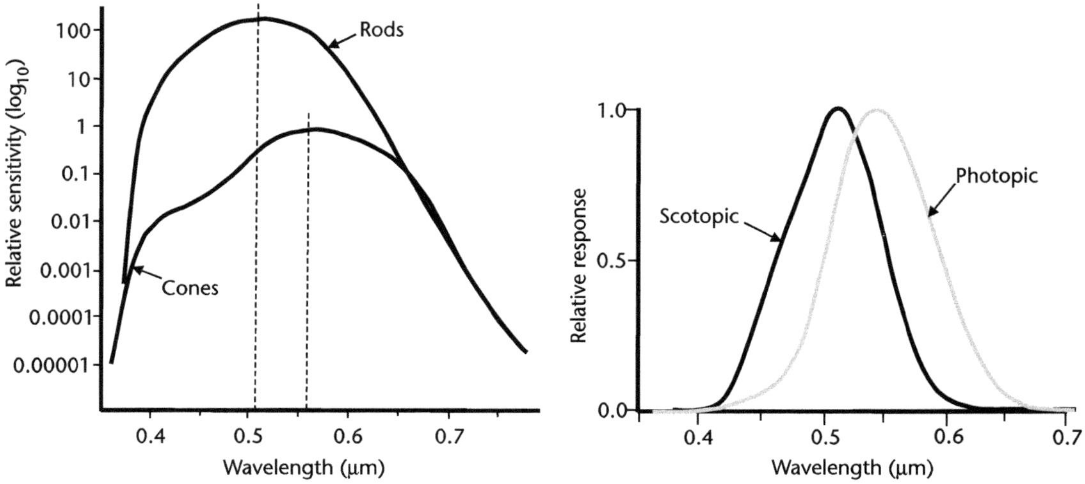

Figure 6.3 Response of rods and cones in a human eye.

6.2.2 Candela and Luminous Intensity

The candela is a fundamental Système International d'unités (SI) base unit and is the unit of luminous intensity. Historically, the intensity of light sources was measured by comparison with the intensity of a flame from a standard candle, using visual-bar photometers. Hence, the unit of luminous intensity was formally known as the candle. Standard candles were gradually superseded by flame standards of oil lamps, and later the luminous intensity was referenced to the freezing point of platinum black bodies, which became known as the international candle, then the candela.

However, difficulties in fabricating the black bodies and developments leading to improved measurement techniques and detector accuracy led to the definition of the candela in relation to the watt, so simplifying the standard. The current definition is: "The candela is the luminous intensity, in a given direction, of a source that emits monochromatic radiation of frequency 540×10^{12} Hz and that has a radiant intensity in that direction of (1/683) W/sr."

The candela may well be the SI base unit, but the luminous flux (unit, the lumen – lm) is perhaps the most fundamental photometric quantity. The rationale for this view is that the other photometric quantities are defined in terms of the lumen with an appropriate geometric factor.

In the following sections, a definition will be provided of each radiometric parameter and the equivalent photometric parameters will be quoted; the conversion to the photometric parameter is undertaken for each parameter, as described above, to take account of the spectral response of the human eye.

6.2.3 Radiant Flux and Luminous Flux

Radiant flux (P, also called optical power or radiant power) is the energy (U) in joules, radiated by a source per unit time. The unit of radiant flux is the watt (1W = 1 J/s):

$$P = dU/dt$$

Luminous flux is the rate of flow of light, as modified by $V(\lambda)$. The unit of luminous flux is the lumen (lm). Sometimes the term luminous flux is used to express the total luminous flux in photometry.

6.2.4 Radiant Intensity and Luminous Intensity

In this section the photometric terms are shown in parentheses. Radiant intensity (J) or luminous intensity is the radiant flux (luminous flux) from a point source, as shown in Figure 6.1, emitted per unit solid angle in a given direction, as defined by:

$$J = dP/d\Omega$$

where dP is the radiant flux (luminous flux) leaving the source and propagating in an element of solid angle $d\Omega$ containing the given direction. The unit of radiant intensity is W/sr, and that of luminous intensity is the candela ($cd = lm/sr$).

6.2.5 Irradiance and Illuminance

Irradiance (H) or (illuminance) is the density of incident radiant flux or luminous flux at a point on a surface, and is defined as radiant flux (luminous flux) per unit area, as shown in Figure 6.4, and is given by

$$H = dP/dA$$

where dP is the radiant flux (luminous flux) incident on an element dA of the surface containing the point. The unit of irradiance is W/m^2, and that of illuminance is lux ($lx = lm/m^2$).

6.2.6 Radiance and Luminance

Radiance (N) or (luminance) is the radiant flux (luminous flux) per unit solid angle emitted from a surface element in a given direction, per unit projected area of the surface element perpendicular to the direction θ, as shown in Figure 6.5.

$$N = d^2P/\left(d\Omega\ dA\cos(\theta)\right)$$

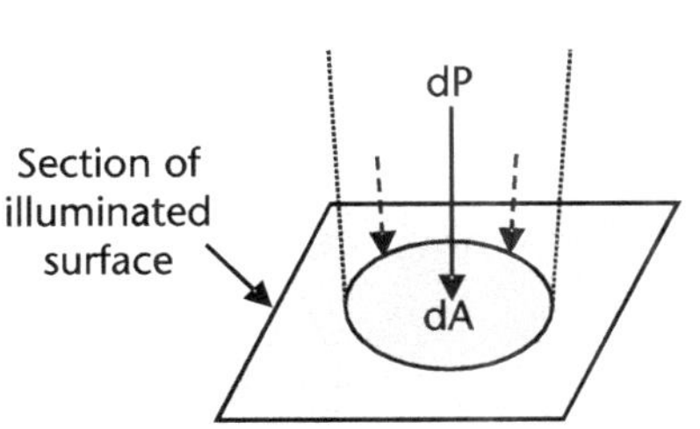

Figure 6.4 Irradiance.

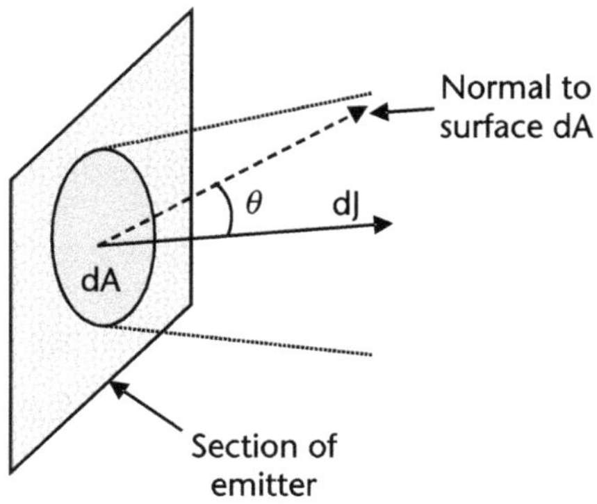

Figure 6.5 Radiance.

The unit of radiance is W/sr/m^2, and that of (luminance) is cd/m^2, where dP is the radiant flux (luminous flux) emitted (reflected or transmitted) from the surface element and propagating in the solid angle $d\Omega$ containing the given direction θ, dA is the area of the surface element, and is the angle between the normal to the surface element and the direction of the beam. The term $dA\cos(\theta)$ gives the projected area of the surface element perpendicular to the direction of measurement.

6.2.7 Radiant Exitance and Luminous Exitance

Radiant exitance (RE) or (luminous exitance) is defined to be the density of radiant flux (luminous flux) leaving a surface at a point. The unit of radiant exitance is W/m^2 and that of (luminous exitance) is lm/m^2 (i.e., it has the same dimensions as lux, but strictly this is the unit of illuminance). These quantities are illustrated in Figure 6.6 and defined by:

$$RE = dP/dA$$

where dP is the radiant flux (luminous flux) leaving the surface element. Luminous exitance is rarely used in the general practice of photometry.

6.2.8 Radiant Exposure and Luminous Exposure

Radiant exposure or (luminous exposure) is the integral of irradiance with respect to time or (illuminance) over a given duration t, as defined by

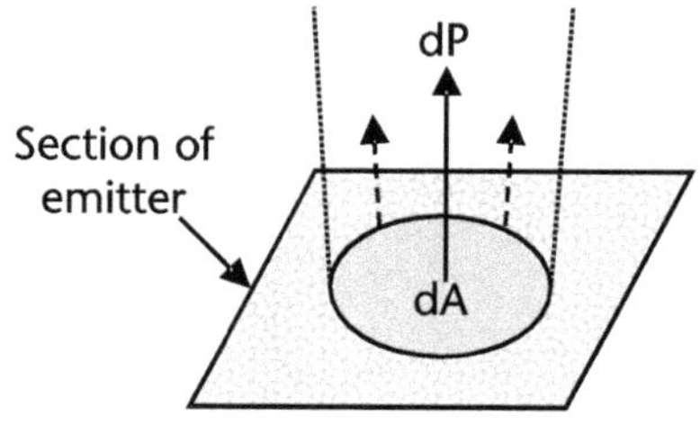

Figure 6.6 Radiant exitance.

$$\text{Radiant exposure} = \int_0^t H dt$$

The unit of radiant exposure is J/m^2 or W s/m^2, and that of (luminous exposure) is lux second (lx s).

6.2.9 Radiant Energy and Luminous Energy

Radiant energy (U) or (luminous energy) is the time integral of the radiant flux or (luminous flux) over a given duration t, as defined by:

$$\text{Radiant energy} = \int_0^t P(t) dt$$

The unit of radiant energy is the joule (J), and that of (luminous energy) is lumen second (lm s).

6.2.10 Total Radiant Flux and Total Luminous Flux

Total radiant flux or (total luminous flux) is the geometrical total radiant (luminous) flux of a light source. It is defined as either:

$$\text{Total radiant flux} = \int_0^\Omega J(\Omega) d\Omega$$

where $J(\Omega)$ is the radiant (luminous) intensity distribution of the light source over the solid angle Ω or

$$\text{Total radiant flux} = \int_0^A H(A) dA$$

where $H(A)$ is the irradiance distribution, over a given closed surface A, surrounding the light source.

6.2.11 Radiance Temperature and Colour Temperature

Radiance or brightness temperature (unit: kelvin) is the temperature of the blackbody radiator (Planckian radiator) for which the radiance output at the specified wavelength has the same spectral concentration as for the actual thermal radiator considered (at the same temperature). It is sometimes referred to as the monochromatic temperature. The brightness (radiance) temperature is measured by making a visual comparison between the brightness of a tungsten filament and the unknown radiator, such as a grey body.

Colour temperature (unit: kelvin) of a grey-body radiator (i.e., a radiator whose emissivity is independent of wavelength) is the actual temperature of a black body that has the same colour as the grey body, and thus provides an estimate of the true temperature of the grey body. Alternatively, it may be defined as that temperature of a black body that has the same co-ordinates on the chromaticity diagram as the body itself. This term is commonly used to specify the colours of incandescent lamps even though the chromaticity characteristics of real incandescent lamps are not exactly on the black-body profile.

Distribution temperature (unit: kelvin) is the temperature of a black body with a spectral power distribution closest to that of the light source in question, and is used for quasi-black body sources such as incandescent lamps.

Correlated colour temperature (unit: kelvin) is the temperature of the Planckian radiator whose perceived colour most closely resembles that of the light source in question. Correlated colour temperature is used for sources with a spectral power distribution significantly different from that of Planckian radiation (e.g., discharge lamps).

A summary of the photometric and radiometric terms are given in Table 6.1, along with the symbols commonly associated with the radiometric quantities.

6.3 System Considerations for Testing and Evaluation

When designing a test and evaluation programme, it is worth considering how the requirements for the system and the technical specification for the system, its subsystems and their components are derived. From a system engineering viewpoint, this is often undertaken in a structured fashion, involving a decomposition process; the way system requirements have been defined provide an indication how to test the performance of a system and verify that the system is doing what it has been designed to accomplish. It is important to remember that any system requirements that cannot be defined cannot be tested.

This processes starts with the user requirement stating a required military capability and flows through many stages to the component level, as shown in Figure 6.7, which leads to the system and subsystem design. A military requirement is an unambiguous statement of the military capability that the designed system must deliver in military operations (i.e., what the system will do on the battlefield during combat). It does not indicate systems solutions, but it will indicate how it is to be

Table 6.1 Summary of Photometric and Radiometric Terms

Photometric Parameter	*Units*	*Equivalent Radiometric Parameter*	*Units*
Luminous flux	Lumen (lm)	Radiant flux (P)	Watt (W) – 1W = 673 lm (at a specific wavelength)
Total luminous flux	Lumen	Total radiant flux	Watt
Luminous intensity	Candela (cd) or (lm/sr)	Radiant intensity (J)	Watt per steradian (W/sr)
Illuminance	Lux (lx) or (lm/m^2)	Irradiance (H)	Watt per square metre (W/m^2)
Luminance	Candela per sq metre or (lm/sr/m^2)	Radiance (N)	Watt per steradian per sq metre (W/sr/m^2)
Luminous exitance	Lumen per sq metre (lm /m^2)	Radiant exitance (RE)	Watt per square metre (W/m^2)
Luminous exposure	Lux seconds (lx s)	Radiant exposure	Watt per sq metre second (W/m^2 s)
Luminous energy	Lumen seconds (lm s)	Radiant energy (U)	Joule (J)
Colour temperature	Kelvin (K)	Radiance temperature	Kelvin (K)

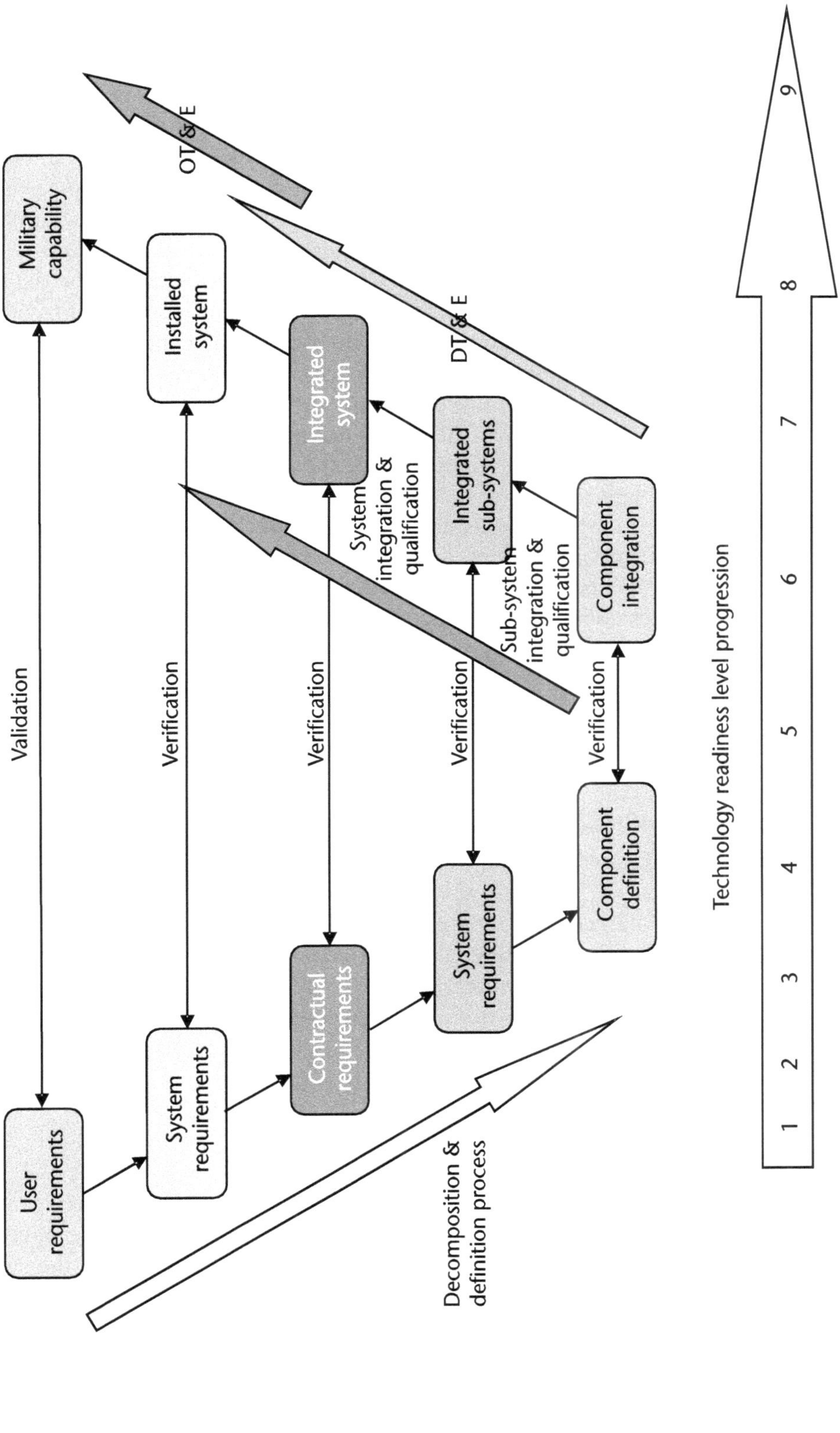

Figure 6.7 A Vee diagram. (Courtesy of Dr. N. Law, Dstl, U.K.)

tested, especially in operational terms. The user requirement is then interpreted by the system engineers expressing what it will do in terms of a functional definition, not how it will be accomplished. The technical solution follows from the system requirements, along with the testing philosophy and test definitions required to achieve the requirements.

The right side of this diagram also shows how the integration of a system progresses from the component level and the relevant testing stages culminating with the operational test and evaluation (OT&E) activity. This diagram also shows the stages of verification of component, subsystem and system performance during the subsequent integration stages. The final stage being the validation of military capability, as demanded in the defined user requirement.

The Vee diagram shown in Figure 6.7 also shows schematically how the technical maturity of the subsystems and the integrated system increases through the various stages of testing. This diagram also gives a very coarse indication of the advance in the technology readiness levels of the system; however, it is worth noting that the values on the left part of this arrow are only indicative of component and, hence, system immaturity, whereas the values on the right side are more indicative of the relative advance in system maturity. The definitions of technology readiness levels are discussed later.

Testing of a component or a system is used to determine whether a test article fulfils its intended function defined by the military requirement. In contrast, the evaluation process may be defined as a testing scheme that leads to a determination of what a system, or subsystem, can actually achieve, Therefore, test and evaluation may involve:

- Demonstration;
- Measurement;
- Analysis of performance;
- Assessment of results;
- Prediction of potential installed performance.

A well-structured test and evaluation programme will provide confidence whether or not the military requirements have been met, it is fit for purpose and, most importantly, safe to use. Undertaking a well-designed test and evaluation programme may also lead to system design improvements, as well as the development of effective tactics, training and procedures. Alternatively, it may indicate reasons for short-falls in anticipated performance.

6.3.1 Development Test and Evaluation (DT&E)

During the integration stages of the system development, various levels of testing are undertaken to build up a picture of how the subsystems and then the fully integrated system will perform. These tests on the subsystems followed by the system tests are often described as design and verification tests, which are used to establish that the system has been built as intended. The design and verification testing is a critical subset of the development test and evaluation phase; moreover, it is used in the analysis of the performance of the system to establish whether or not the performance specifications have been met.

The development test and evaluation process is a comprehensive series of tests, which provides the data to support the verification process indicated in Figure 6.7. This phase of testing helps to answer the fundamental question of whether the system has been built correctly, and in particular, has the integration process been effective. This aspect of system development and maturation is fundamental.

6.3.2 Operational Test and Evaluation (OT&E)

These tests are normally the final stage in the test and evaluation process prior to release into military service; the purpose of the operational test and evaluation is the validation that the system meets the stated user requirement. Consequently, the design of these tests and trials is aimed at establishing credible performance estimates in a realistic environment across the performance envelope of the platform. In the case of equipment for aircraft, the final stage is Military Aircraft Release (MAR) certification, discussed briefly later.

Each of the armed forces of any nation will have their own carefully specified processes and procedures for the operational test and evaluation of a limited number of examples of each military system, which is in an operationally ready state. These test items will either be fabricated using advanced prototype designs or examples from the initial (or pre-) production stage of the project, which is discussed further in Chapter 18. The objective will, however, be identical to establish the installed performance of the system in an operational environment using military personnel.

The operational test and evaluation procedure is normally the final test stage before equipment reaches initial operational capability, which will use production equipment. In the case of equipment for use on aircraft, it will normally require military aircraft release certification, a carefully documented and rigorous procedure, designed for specific aircraft and defined applications. Military aircraft release certification may be staggered with limited operations permitted initially, until full confidence is established, or design modifications are completed and fully evaluated. The critical requirement being to ensure the equipment is safe to use on the aircraft.

A number of test systems have had to be fully refurbished prior to a flight trial, as many aspects of the prototype system were simply worn out, owing to the consequences of extensive testing or through frequent demonstration.

The route to establishing installed performance is shown somewhat simplistically in Figure 6.8. It should be noted that the scales lack units, but are definitely nonlinear and possibly logarithmic.

6.3.3 Technology-Readiness Levels

A systematic approach has been developed for assessing the degree of development of a component based on a range of criteria.

This process has been developed from an assessment procedure developed by the National Aeronautics and Space Administration (NASA) in the United States of America. It has nine levels of readiness, where technology-readiness level nine is

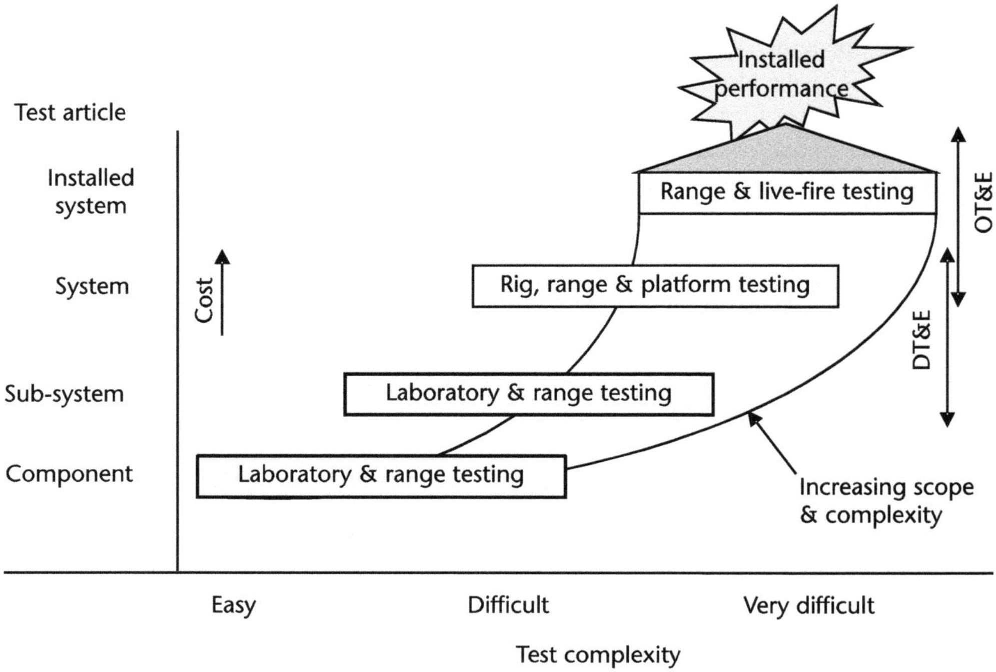

Figure 6.8 The complex path to establishing installed performance.

the most advanced and represents equipment that is mature and has been tested in its intended operational environment.

As in all cases involving a general systematic process, care is required in the actual interpretation of the meaning and intention of a descriptor of any such process. It is quite common for a device to have completed all of the criteria associated with various levels up to a given point (X), but then may still have to complete perhaps one or more in the next level, and additionally, a system may have achieved a few others in a more advanced stage. Therefore, it is quite common for a device or systems to claim a technology-readiness level $X/(X + 1)$.

Care is also required in estimating or projecting a technology-readiness level for a system: just because a series of subsystems are considered to be at a given technology-readiness level, it does not mean that the system in its entirety is at the equivalent technology-readiness level. However, in such cases it is likely, but not certain, that the system would mature quite rapidly. As a rule of thumb, the technology-readiness level of a system cannot exceed the technology-readiness level of the least mature subsystem or component,

The technology-readiness levels are defined in Figure 6.9.

6.4 Approaches to Testing and Evaluation

6.4.1 Introduction

A crucial requirement at the start of any military system development programme is the need to have a test and evaluation strategy from which, a comprehensive test and evaluation plan can be devised.

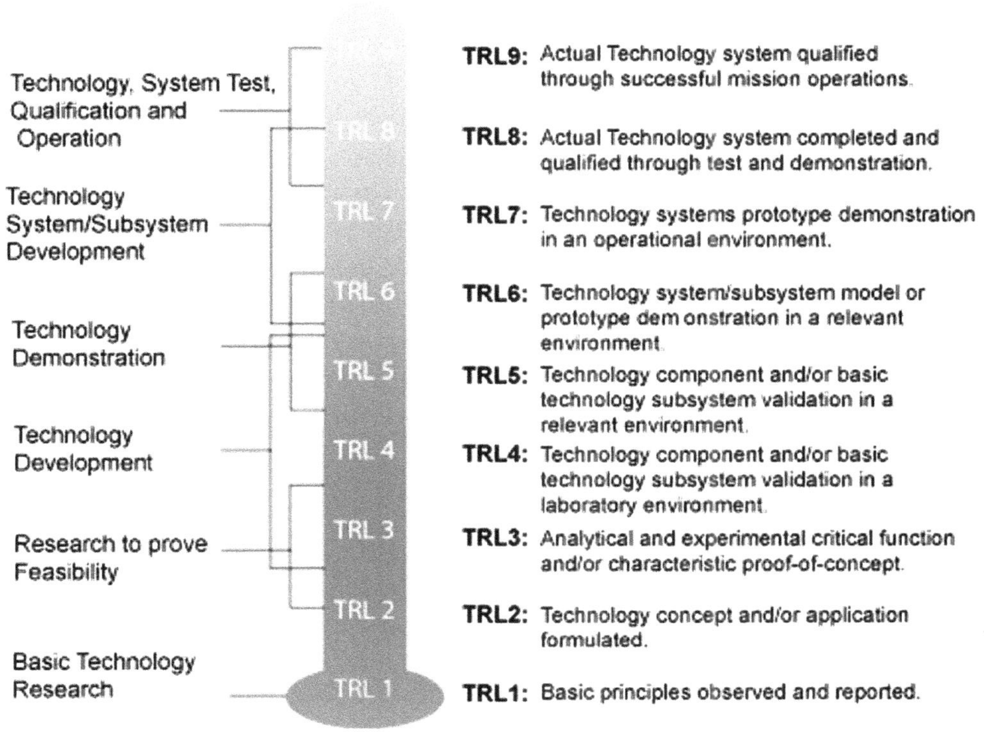

Figure 6.9 Summary of technology-readiness level definitions. (Courtesy of MoD DE&S, U.K.)

The initial stages of the test and evaluation processes will enable the level of maturity of the components in the subsystems, as well as the technical maturity of the complete system, to be understood. This understanding then enables the development and test programme to be designed to raise the technology readiness level and then the system-readiness level. There is little prospect of success if a subsystem has only been tested in the benign environment of an optical table and its next test stage is a flight test as part of an operational test and evaluation programme, particularly if the subsystem concerned is a complex device.

6.4.2 Test and Evaluation Strategy

The test and evaluation strategy may be a very brief document, which will provide the foundation for all of the test phases and will usually define:

- The objectives for the proposed equipment development in relation to a cardinal-point specification for the system, the scale of the required testing and crucial time linkages (if applicable to other tests scheduled in the programme);
- Test evolution strategy and philosophy, covering hardware testing and the role of modelling and simulation;
- Defined break points in the development programme and the technical linkages between the project phases, for example, stipulating that a particular

development subsystem must attain a given maturity/capability level before a subsequent phase, such as flight trials, can commence:

- Test performance objectives and goals linked to criteria for success for each test phase in the proposed evaluation programme;
- The linkages between the test phases and the range of success criteria for transition between phases, such as achieving given technology-readiness levels or system-readiness levels.

- Definition of any requirements for use of specialist techniques or test approaches needed to fulfil project objectives;
- Definition of any specific test directives, which could include operation on particular test ranges or evaluation in a particular environment;
- Other critical test requirements specific to the test article and the military objective.

6.4.3 Test Design

Given an agreed test and evaluation strategy, the design of the experimental evaluation process can be undertaken. For a complex programme, there is likely to be a hierarchy of test documentation and procedures. The top-level design of the experiment will be derived directly from the test strategy to provide a route to acceptance for the project. A series of documents will be compiled providing a test description for each test required supporting this objective.

This process of generating a series of successive requirements to define test plans, as discussed above, from a top-level document that is linked to the test and evaluation strategy is crucial when decisions have to be made about transitions between phases and leading ultimately to qualification. It is recommended that each test, however minor, is linked to a specific aspect of the overall requirement. This process provides a potentially very powerful audit trail, and moreover, the data collected during each stage of the evaluation may be used in the analysis of any identified shortfall in performance in subsequent testing phases. The latter aspect also enables potential routes to solutions of encountered problems to be identified. For example, if a lack of performance is identified during system testing, then it may be possible to trace this problem back to component effects, reduced component output or to an issue with its integration, rather than just demanding more power from the laser.

6.4.4 Test Plan

A test plan defines all aspects of a specific test, whether it is a component test or a flight test, although the latter will embody far more detail, as indicated later. It will be noted that the following description of a test plan has many similarities with the trial's planning discussed from the laser-safety viewpoint for range testing in Chapter 17. Clearly, a single document is required; however, here the emphasis is on the mechanisms for testing and evaluation and the plan in Chapter 17 is with the emphasis on the laser-safety issues on a range; from these joint considerations, a single document will emerge. During many of the laser-evaluation activities, the laser-safety aspects may not be relevant either because of the class of the laser device under test or the fact that the laser beam is contained fully within an enclosure; therefore, for this test the laser is a Class-1 device (the definition of classes of laser emission from a safety standpoint are defined in Chapter 17), as there is no optical hazard presented by the beam, owing to the presence of the enclosure to contain the potential hazard.

Generically, it should include the why, how, where, when, by whom and with what statements in the description, covered by:

- Objectives and scope of the test or trial's plan and a definition of this particular phase of the evaluation programme include:
    - Equipment or component to be evaluated;
    - Parameters to be investigated and the methods to be used;
    - Ancillary/support equipment;
    - Support-equipment requirements;
    - Programme timescales;
    - Data collection and storage;
    - Target control (if dynamic);
    - Logistics.
- A comprehensive description of each test in the programme schedule, along with the processes and procedures for its safe execution, it is also important to have a statement concerning any known limitation(s) of the equipment being tested. This test plan will usually define which subsystems are needed to be fully functional (or at a defined acceptable level of performance) for each specific test in a more complex programme. A prediction of the anticipated outcome of each step is usually included, so that it is clear when the testing can move on to the next stage.
- A description of the test site, such as the laboratory or a range, and any operational limitation imposed by the site, for whatever reason.
- A description of where the item under test will be positioned and how it will be operated (such as a flight plan), along with nominated operators, in the case of field trials, a description of the range and precisely where the testing will be undertaken, along with any specific requirements and procedures.
- If military platforms are involved, a description of that platform and any special requirements, such as refuelling, servicing, secure storage and repair procedures and processes.
- Instrumentation and data-recording requirements describing, so it reads calibration, storage, backup, including any particular handling, protection procedures and processes.
- Nominated staff that are involved in the trial, along with responsibilities and their competencies;
- Control and chain of command providing a clear statement of hierarchy of control, so that it is clear who can stop or start the testing. This part of the plan will define functional responsibilities and authority of each person

involved in the test programme. It is recommended that there is a test director/trial's officer who has the responsibility for determining that the equipment is suitable and ready for test at each stage (this is where the buck stops as far as the scientific and technological decisions are made regarding the test programme). The plan will include reference to the laser-safety officer and the trial's safety officer, the latter is responsible for permitting the test to proceed or ordering a premature termination; it is absolutely critical that there is no ambiguity in this role. In the case of more complex tests, there will normally be reserves or deputies for each critical position. For simpler tests of components, these functions may well be vested in the same person who is undertaking the test (i.e., the technician).

- The safety plan, which includes a comprehensive risk assessment, mitigation of risk, techniques and safe-operating procedures for all of the equipment, so that compliance with the local rules can be established, coupled with an assessment and analysis to explain the scope of the evaluation and indicate the mitigation processes.
- Emergency and contingency planning that are linked to the health and safety requirements and risk assessments and the risk mitigation strategy, including the emergency shutdown procedure.
- Authorisation for the trial is the audit-trail log, which is also the approval/permission and certification process, which includes any external review requirements and licensing for the testing.
- Any specific occupational health requirements are listed, including all protection measures.
- A list of references is given for the supporting information, including related documents and procedures.
- Test/trial preparation activity, which is a definition of the activities that must be completed before this testing or evaluation can begin, so that the relevance and importance of this testing can be understood in the wider context of the entire programme.
- Management of change test/trial authorisation and processes are crucial for any trial, despite very detailed meticulous and rigorous planning trials are rarely executed according to the original plan, hence, a record of changes is vital for the post-trial analysis.

There are a few other activities that will be required in a comprehensive trial's plan for an evaluation on a range, such as:

- *Communications:* definition of the radio bands to be used;
- *Environmental impact analysis:* an in-depth investigation of the potential impact on the natural environment and procedures required to minimise it and will also include any post-trial clean-up phase;
- Additional issues that may pertain to this specific trial at this range, perhaps covering access or hours of operation;

- *Completion of trial requirements:* a list of the tasks required to be completed prior to vacating the trial's range.

6.4.5 Flight Trials

A flight trial's plan involves additional requirements and processes concerning the safety of flight and associated regulations, particularly involving operation of equipment on an aircraft and use of suitably qualified and trained staff. Many of these requirements are specific to the platforms involved and the operational test squadrons undertaking the testing, for example. Therefore, specific guidelines are difficult to provide, other than a need to engage personnel from these organisations in the trial's planning discussions at the earliest possible opportunity. Moreover, normally the equipment may only be installed on an aircraft/platform by a suitably qualified engineer; all of whom will be fully qualified and trained for the task and follow carefully defined procedures, which have to be adhered to prior to flight in order for the installation to be certified fit for flight.

6.4.6 Documentation and Compliance

For relatively simple laboratory tests, these requirements may well be covered by a simple document. In the case of a complex test, this may develop into quite a tome. Moreover, the test programme should be configured so that results can be used to refine and validate the simulation tools, which should be reflected in the test plan.

A critical step in the design and development of any test and evaluation plan, particularly if it involves laser technology, is to involve the laser-safety officer or equivalent, at an early stage of the planning process. It is also worth ensuring that the current laser-safety regulations are understood fully. It is normally far easier to get approval for any test programme when using a range already approved for laser operations rather than at a new range.

It is worth remembering an old guideline for system testing during trials: "never undertake a trial without already knowing the outcome to a first order." Additionally, it is often stated that any field trial usually measures gross system effects, so it should not be designed to tease out details of subtle or obscure scientific phenomena. The recommended place to undertake such evaluations of fundamental studies is the laboratory, where conditions can be easily measured and controlled.

6.4.7 Simulation and Mathematical Modelling

An important consideration in the test and evaluation programme is the use and role of mathematical modelling and simulation during testing; this can be a valuable tool for understanding a range of critical interactions and the impact of changes to design parameters.

There are many different sorts of modelling tools and techniques, far too many to be reviewed in detail in this text. However, some of the more valuable ones for providing an understanding and insight into the technical issues that are likely to influence system behaviour during tests are the following:

- An overall system model enables realistic performance predictions to be made, as well as understanding the impact of changes to subsystems and interfaces. This type of model is normally formulated at a relatively high level with some occasional simplifying assumptions, but will aim to provide a tool suitable for understanding the performance of the complete system in its operational environment. Even with today's advanced computer technology and sophisticated simulation techniques, creating a very high-fidelity representation of every component in a system will be time consuming. The resolution or fidelity in terms of performance prediction for the complete system is likely to be ±5% or worse.
- Models of the individual subsystems, which are likely to be in far more detail and have a relatively higher-fidelity representation of the components. This form of mathematical model enables a range of design iterations to be evaluated and the impact of compromises on subsystem performance to be assessed. Moreover, sophisticated models of control systems allow the performance of these subsystems to be optimised, as well as potentially making very accurate prediction of their performance. These models may be used to provide the input data files for the system model.
- An error-budget model that attempts to assess the impact that all of the tolerances in component performance and variances in measurements will have on the overall accuracy of a system.

The verification and validation of these models are crucial, so that the accuracy of the predictions are appreciated fully and understood. It is important to remember the following:

- A model that has not been verified and validated has little value and there can be little confidence in its predictions until this critical step has been completed.
- It is the system that will be integrated with a platform and will fly; the model, no matter how sophisticated, will not fly. It is only a tool to provide an indication of potential routes to the best performance from the system.

6.4.8 Scale of Testing

As noted above, the scale of testing spans a vast range of options from simple laboratory testing that may take a few minutes, but more likely a number of hours to complete with a single component with one technician, to the full-scale live-fire trial that may take many months with an army of test engineers and support staff. The following table provides some indication of the potential fidelity of such trials and tests. It is worth noting that even with the very expensive and complex live-fire trials the fidelity is not usually perfect, owing to the use of a simulated target, for example; however, unmanned vehicles or instrumented drones may be used in some circumstances.

Table 6.2 Summary of the Fidelity of Types of Evaluation Trial

Test Condition	*Threat Representation*	*Target Representation*	*Engagement Geometry*
Live fire	Real	Simulated	Real
Flight trial	Captive assets	Real	Simulated (launch geometry real)
Field trial	Captive assets	Simulated	Far-field simulation
Ground/laboratory	Captive assets	Simulated	Near field/far field
HWIL	Captive assets	Simulated	Near field/far field
Rigs (sensors/effectors)	Simulated	Simulated	Near field/far field
Modelling and simulation	Simulated	Simulated	Simulated

The requirements when defining the test strategy is to ensure that the various trials can be linked together in a coherent fashion to provide a realistic estimate of installed performance. A summary of the fidelity of typical types of trial is shown in Table 6.2.

6.4.9 Technology Insertion

During the development, or lifetime of any military system, there will be scientific and technological developments that will enhance the performance of the system if appropriately integrated with the existing system. Figure 6.10 attempts to indicate a possible route for such a process, which by its very nature, is likely to be lengthy and possibly expensive, but cheaper than a new product. This figure also illustrates the potential exploitation routes for research studies. As the figure shows, a full test and evaluation programme will normally be required to assess the level of compliant performance achieved by the technology insertion.

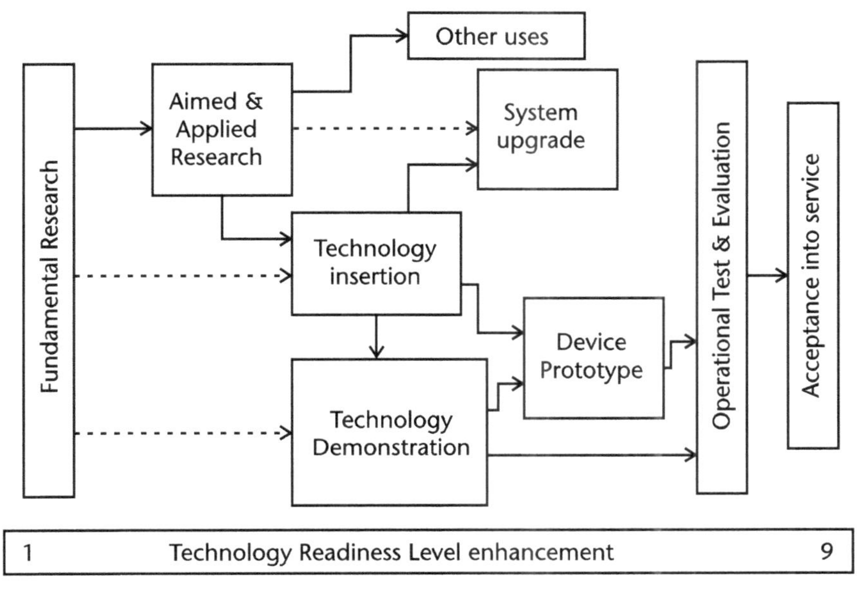

Figure 6.10 Technology-insertion opportunities.

6.4.10 Test Equipment

6.4.10.1 Power and Energy Meters

The output from any optical source can be measured using a power or energy meter. These sensors may be combined into a single sensor called a power and energy meter.

The design of the sensor determines whether the device measures power or energy; a power meter measures the received optical power in a continuous wave beam or a repetitively pulsed beam. Typically an energy meter measures the output from a pulsed laser.

The three basic types of detector used in these sensors are the following:

- Thermal sensors are very well-established devices for detecting and measuring radiation. They work because the absorbed incident radiation causes a rise in temperature of the sensor, which in turn causes some physical change in a physical property of the detecting element. Thermal detectors do not require cooling, and theoretically, respond to all wavelengths and are often used in radiometers, as these sensors provide a relatively flat response over a relatively broad spectral regime from the deep ultraviolet to the far infrared (i.e., the response is not wavelength dependent). These sensors tend to be available with a relatively large detector area and are suitable for measuring the optical output of high-power devices, well into the kilowatt regime and beyond. The response time of these detectors is typically measured in milliseconds, which limits their use for applications with short duration pulses. Probably the most common and simple thermal detector is a thermocouple, other examples include the following:
  - A thermopile sensor exploits the Seebeck effect, so a thermopile detector comprises several thermocouples connected in series, which gives enhanced responsivity over a single thermocouple, but the time constant of the response is of the order of a second unless very small junctions are used in the thermocouple array, which can reduce the time constant to tens of milliseconds.
  - A thermistor bolometer detectors exploit the change in resistance of a material as their temperature changes. Arrays of the resistive elements can be formed to make an inexpensive thermal camera.
  - A calorimeter is a simple device for direct measurement, from observation of the temperature rise of a given mass of material of known specific heat.
  - The bimetallic sensor is another simple device that has a cantilever, which bends when heated. These devices measure the deflection of the cantilever through a change in capacitance, which may be estimated easily using technology similar to that used in complimentary metal organic semiconductor (CMOS) arrays.
- Photon detectors, or quantum detectors, which usually have a detectivity that is several orders of magnitude greater than that of thermal detectors, but this is at the expense of the need to cool the devices, often to cryogenic temperatures. Consequently, these sensors tend to be very sensitive and provide

a very rapid response to the incident light, because of the direct interaction between the photons and the electrons in the semiconductor detector material. Different sensors tend to be required for different incident wavelength bands. For example, a photo-diode usually refers to photo-voltaic detectors, which have significantly greater detectivity than photo-conductive sensors. These devices operate with intrinsic semiconductor materials such as:

  - Silicon for wavelengths less than 1 μm;
  - Indium gallium arsenide for wavelengths in the short-wave infrared (typically between about 1 and 2 μm;
  - Indium antimonide for wavelengths in the mid-wave infrared,
  - Cadmium mercury telluride (CMT/MCT) for wavelengths in the mid and far infrared;
  - Germanium for the short-wave infrared wavelengths.

- Pyro-electric devices are similar to a thermal sensor because their response is proportional to the heat absorbed rather than the incident photon flux. These devices use crystals, such as triglycine sulphate or lithium tantalite, that develop an electrical charge on their surface when the temperature changes. These sensors are used for measuring pulsed emissions and operate by converting the energy in a light pulse into a voltage spike. These sensors have a broad spectral response, but tend not to have such a flat response as exhibited by thermal sensors.

All three classes of detector devices are suitable for measuring pulsed output from lasers, or any other optical source. Thermal detectors are ideal for pulse durations in the millisecond regime, as well as for highly energetic pulses. Pyro-electric devices are effective in measuring shorter-duration pulses from the femtosecond regime up to milliseconds, over a broad range of energy up to joule-class pulses. Semiconductor devices provide an accurate measure of optical energy for short-duration pulses up to microseconds, but with a very fast response time, often with a frequency response in the tens of kilohertz regime or beyond.

Photo-diodes tend to offer higher thermal stability than the other detectors. Modern thermal detectors may include ambient-temperature stabilisation devices to minimise thermal-drift errors in the output.

When using these sensors, it is important to ensure that the incident flux does not exceed the damage threshold of the detector. The incident irradiance (power or energy density) must be established from an estimate of the spot size and the energy distribution in the beam, bearing in mind that a Gaussian beam has its peak intensity in the centre of the beam. Special care has to be taken with pulsed lasers regarding the pulse duration, as the peak intensity may be very high and exceed the bulk damage threshold of a detector material.

Many trade-off options in performance and accuracy are feasible, for example, a small detector with a high damage threshold will not be very sensitive. Hence, different probes will be required to make accurate measurements of high- and low-power beams. Similarly, with detectors suitable for highly energetic pulses, care is required concerning the trade off with repetition rate; usually these devices will be effective when the incident flux has a low repetition rate and thus modest average power. When using this type of sensor with a high repetition rate the energy in the pulse will need to be attenuated.

6.4.10.2 Calibration

A further point for consideration is the use of calibrated and indicative devices and sensors during testing. Calibrated equipment is required for accurate measurement, such as the confirmation that a power-supply unit is operating within a defined range to a defined tolerance. If it is merely a case of checking that the same power-supply unit is supplying power to a component, the use of a meter that is not calibrated will be adequate. Similar comments apply regarding the measurement or observation of the optical power in a laser beam.

It is good practice to have some calibrated instruments to ensure that the non-calibrated ones are still providing a reliable output that is within their nominal range of operation, as defined or set during their manufacture. Furthermore, a few calibrated instruments may be used as transfer standards between instruments to ensure realistic operation of the noncalibrated devices.

For acceptance or validation of performance of equipment, it is imperative that suitably calibrated equipment is used throughout this process; however, during normal laboratory development, such as the optimisation of a component, prior to acceptance, then the use of equipment that is not within calibration may be acceptable, especially if its performance is known to be stable. In this case, a check of the stability of operation of noncalibrated sensor using the transfer standard method is valuable, as well as being necessary in some instances, as discussed above.

Careful consideration should be given to the precision of the instrumentation, as the measuring instruments need to be designed to have a higher accuracy than the measurement that is being made, so that the errors of the measuring device may be ignored. There may appear to be little point in measuring a piece of wood with a micrometer, marking the cutting mark with a stick of chalk and then cutting it with an axe. However, the hierarchy of precision is correct, if somewhat perverse use of instruments.

6.4.10.3 Integrating Sphere

An integrating sphere is a sophisticated type of optical component with a detector to reduce the effects of nonuniformity in sources; it is used to measure the total power of a beam whose irradiance distribution is arbitrary or a source with an arbitrary radiance distribution. It is a hollow sphere with a small aperture and a coated internal surface that forms an approximate Lambertian reflector. A detector is positioned within the hollow sphere; the detector's area must only form a small fraction of the sphere's surface area, comparable with the requirements for the entrance aperture. Furthermore, the source (emitter) must also be small compared with the area of the sphere, so that the stop that separates the source from the detector will not obscure too large a solid angle.

6.4.10.4 Radiometer

A radiometer is used for the measurement of radiant flux over a broad spectral band, thus it provides broadband spectral measurements and is useful for making measurements of laser emissions in the context of a total emission from a platform, such as a ship or an aircraft. A spectro-radiometer is a similar type of device but measures the spectral radiant flux within a small spectral band (i.e., a narrowband measurement). The spectral response band of a radiometer may be controlled using optical filters.

A radiometer has a collecting telescope to capture some of the radiant flux emitted by a source, which is focused on to a detector. The detector generates an electrical signal in response to the incident photon flux. As the radiometer is normally some distance from the source, this device measures the irradiance of a source: the other quantities, such as the radiant intensity and radiance, are calculated from the measured irradiance of the source.

It is important that the absorption by the optical train of the radiometer is minimised, which is normally achieved by use of mirrors to form the optical train. Moreover, by use of suitable reflective materials and coatings, the response of the radiometer will be determined solely by the detector.

The basic design of a radiometer is shown in Figure 6.11. A spectro-radiometer has an additional component within the optical train of the collecting telescope; this is the monochromator, which selects the radiant flux over a narrow band of wavelengths. A prism may be used as the dispersive element to provide the wavelength selection, specific wavelengths being selected by rotation of this prism. This arrangement enables the spectral distribution of the radiant flux of a source to be measured.

A further development of this instrument is the background discriminating radiometer. This device has two fields of view, so that the background radiant intensity may be estimated in one channel and the radiant intensity of the object of interest in its environment, such as a laser beam, may be measured in the other. Simple subtraction enables the radiant intensity of the source to be established.

6.4.10.5 Beam-Profiling Techniques

The spatial profile of a beam may be determined using an opaque plate with a pin hole, a single-element detector and a suitable translation stage to move the pin hole across the beam and record the detector's response. This is a laborious technique and is of little value if the spatial profile of the beam varies with time, such as from

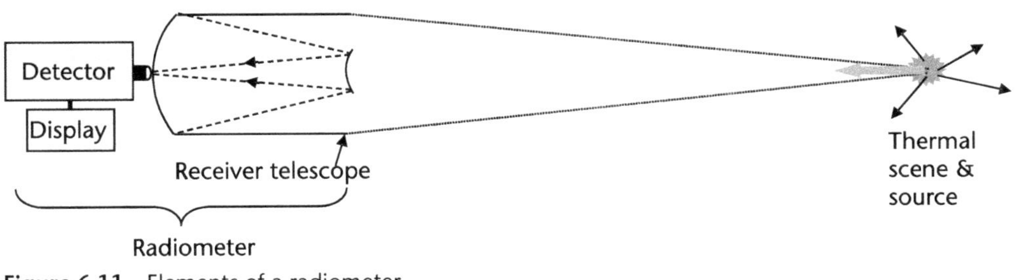

Figure 6.11 Elements of a radiometer.

pulse to pulse; moreover, if the variation in spatial profile is required as a function of repetition rate, then an alternative technique is required.

A beam-profiling camera has a focal-plane array that enables an image of an incident beam to be captured, stored in a computer and displayed in a number of easy-to-view formats. The image of the beam profile is displayed virtually instantaneously after the image has been captured. Additionally, beam-profiling cameras may be operated at a relatively high repetition rate, synchronised to the repetition rate of a laser, so that any variations of beam profile, and hence spatial profile, over time may be quantified.

6.4.10.6 Knife-Edge Measurements

This is an alternative approach for determining the intensity profile of a beam, comparable with pin-hole measurements. Hence, at a given point in the beam profile, the sharp edge is moved into the beam in conjunction with a detector and the intensity recorded as they traverse the beam profile.

6.4.11 Laboratory Testing

This is usually the first critical stage in the evaluation of a component or a system, which can take many forms. One of the attractions of laboratory testing is the fact that beam-propagation conditions can be controlled carefully and characterised accurately. Moreover, it is normally possible to undertake subscale tests rapidly to understand performance issues and investigate options to give optimised performance, particularly at the component and subsystem level.

Classical types of laboratory testing include the following:

- Optical-bench testing of components, subsystems and complete systems determines at least baseline performance parameters and demonstrates modes of operation. Critical subsystem interactions may be investigated and routes to optimised performance and the impact of design compromises may be investigated.
- Systems-integration laboratory testing, where the various subsystems are brought together to form a system and the initial performance assessments of the complete system can be undertaken. The systems-integration laboratory is also useful for investigating the interdependencies of subsystem performance in a known environment. This laboratory is likely to include a range of complex motion simulation tables to allow a realistic assessment of platform performance, particularly in the presence of own-ship motion. This laboratory is normally used during the important stage of the design-verification testing (i.e., if the system comes close to meeting the design specification). Maybe the systems exceed the performance requirements, which may enable some system trade-off studies to be considered. The latter situation is rare.
- Environmental testing of a system at various stages of its development is often termed the shake-and-bake process, although the more extremes of that form of testing for qualification are often one of the last stages in the environmental testing procedure. The environmental qualification and certification

are usually very demanding to ensure that the system functions are designed but also operate safely without the prospect of causing any danger to the platform, where it is intended to be used, throughout its performance envelope. This stage in the evaluation phase usually involves a complex series of tests (and test processes) to ensure that the system under test will not have an adverse impact on the operation of other systems on a platform nor on the platform itself. Additionally, it will be tested to ensure that the local environment will not cause any predictable malfunctions of any form, to itself or other systems or indeed to the installed platform.

- Platform simulation tests that investigate integration options, as well as the potential impact on installed performance. It is common to integrate a system with a flight test rig to evaluate potential issues prior to a flight test, such as during field trials; similar activities are undertaken with other platforms used in other environments.
- Factory acceptance testing normally starts with the more basic confirmation of fundamental system parameters, such as size, weight, power consumption, emission characteristics and baseline operation.
- Other forms of testing, which include highly accelerated life testing and highly accelerated stress testing, which are designed to highlight potential failure mechanisms in newly developed high-technology systems and devices.

6.4.11.1 Data Management and Certification

Accurate and detailed records must be kept of the performance and measurements made on the items under test, including the ambient conditions and applied stimuli. There are many reasons for this action, but common benefits include:

- Data for assessment of reliability;
- Indication of systematic trends in the performance;
- Evidence for systematic failure or potential failure mechanisms;
- Data for verification of mathematical models.

Data collection techniques must be systematic and rigorous, preferably to an agreed standard and format. A written record may be the only information that survives a mishap with a computer and/or a disc.

Certification of equipment for flight will be undertaken in a comprehensive series of detailed and prescriptive tests undertaken in the laboratory environment in one form or another. In general, these tests and ultimate certification are to ensure that the system will not have any adverse impact on the platform during any flight test or normal operational flight; similar considerations apply to platforms used on land or at sea. These tests have very specific safety objectives and are not part of the assessment of performance phase. Normally, these tests are undertaken by the design authority for the system; with a complex system involving the integration of multiple sub-systems, it is the responsibility of the individual suppliers to demonstrate the flight worthiness of their component or subsystem prior to integration into the total system. It is then the role of the system integrator and the design authority to certify the complete system for flight.

6.4.11.2 Test Parameters

When undertaking testing of a laser emission, it is essential to agree and understand the definitions of each of the terms in the parameter set defined in the requirements, which are derived from the cardinal points specification, as shown in the left side of the Vee diagram in Figure 6.7. For example, there are at least seven definitions of laser beamwidth and probably several more techniques for making these measurements; this is often a source of confusion and dispute when it comes to the evaluation and acceptance stage in a laser programme.

Any testing that involves the evaluation of high-power lasers must consider the techniques that may be applicable for the measurement of the power in the beam. This is because the output is often multimodal and so the task of establishing the beam width, even to an agreed definition such as $1/e^2$, is a challenge. Hence, some consideration should be given to an appropriate metric such as the power-in-a-bucket value at a given range, with an agreed definition of the bucket (i.e., the diameter of the calorimeter or detector aperture).

In general, it is worth remembering the limitations of the measurements that are made in the laboratory, despite the use of precision equipment and the application of well-established techniques. For example, in the case of measurement of beam parameters of a laser system, there are a number of important considerations:

- Determination of the beam-propagation factor (M^2) is not a measure of beam quality, but an indication of how the beam diverges during propagation in comparison with a pure Gaussian beam.
- M^2 measurements are very useful for comparing the relative performance of various laser beams.
- Extreme care is required during the measurement of the beam-propagation factor using established second-moment techniques.
- In order to convey more information about beam quality, it is worth considering measurement of a near-field intensity profile with a corresponding power-in-a-bucket measurement in the far field presented on a logarithmic scale.

6.4.12 Range and Field Trials

Range and field trials are the natural progression from the detailed laboratory testing and are usually started once the system has reached a technology-readiness level of 4 or 5. These tests are important in the systematic progression of the evaluation and for providing data (in all forms) to support the verification of any associated models. Some crucial aspects of this form of evaluation of laser systems are the following:

- Investigation of far-field effects includes energy-on-target testing and characterisation of beam profiles.
- Consideration and identification of artefacts of the natural environment, which have the potential to disrupt performance, for example, potential false alarm sources and evaluation of clutter rejection algorithms on target-detection characteristics and acquisition performance (of systems).
- Realistic target-acquisition and tracking testing, which are undertaken to give an assessment of algorithm performance.
- A study of free-space propagation effects, which includes the investigation of the impact of weather, turbulence and aerosols through end-to-end testing of the system.
- Provision of data requirements needed for verifying models.

The format of these range trials can be many and varied, depending on the military application. The simplest approach may involve setting up a target at one end of a field and a laser at some remote point at the other end of the range, with an appropriate beam-directing optical system. Other arrangements may be more realistic and undertaken once the initial shake-down of the equipment has been completed in the simplest trial configuration. For example, if the laser-based system is intended for an air-to-ground function, then the laser system may be mounted on a cliff face or on a tower in an attempt to simulate the engagement geometry and investigate the specific engagement issues anticipated with this geometry. Needless to say, ground-to-air effects may be simulated by reversing the positions of the equipment; however, care is required in the latter case if the laser beam is directed above the horizon and there is no suitable backstop for Class-4 devices.

It is quite common to use a pair of towers for air-to-air studies, although most towers are limited to a few hundred metres above the local ground level. An alternative could be the propagation across a ravine or valley, as care is required in the use of towers, as they tend to sway in windy conditions, or indeed as personnel move about on the platform.

Good record-keeping is even more critical here than for laboratory studies. It is recommended that meteorological data are measured at the test site and recorded at frequent intervals. If it is anticipated that strong turbulence may be a critical factor, then measurements of the refractive index structure parameter (Cn^2) should be made at regular intervals (and recorded), because changes in cloud cover can have a dramatic effect on its value, changing it by an order of magnitude in a few minutes in some circumstances.

These tests will normally give a good indication of the potential military value of the laser system, and importantly, advance the maturity of the entire system and increment the technology-readiness level. The tests may also form an important stage in the factory acceptance test process, which normally has many stages, particularly with multiple faceted requirements in a complex system. Field trials are an essential step in the evaluation and for supplying proof that a subsystem or system has reached Level 5 on the technology-readiness scale, indicating that the kit may be used effectively outside of the laboratory.

This form of evaluation trial is a critical evaluation stage before embarking on a flight trial or range trial with the equipment integrated with a platform. This type of testing is used to reduce the technical risk involved with more advanced trials, as well as providing opportunities to develop techniques for successful operation of the laser system. It is quite common to make this form of trial quite sophisticated; for example, the use of motion tables so that sightline isolation techniques may be evaluated, whilst exposing the laser system to the real-world environment.

6.4.13 Platform Trials

In the case of technology-demonstration programmes or advanced concept technology demonstration projects, this test stage may be the final stage of evaluation of a concept or a technology. Normally, this sort of demonstrator reaches a technology-readiness level of at least 7. Provided that this demonstration has been declared successful, then the next step in the technology and system progression is usually full-scale development followed in some cases by low-rate initial production (LRIP) prior to full-scale production.

In the case of a development project, it is a further advance in the development test and evaluation programme to raise the technology-readiness level towards Level 9. Success in this test phase may be the threshold condition required to start a low-rate initial production using designs of at least advanced prototype status.

Any type of platform trial can take a multitude of forms and will generally be the most expensive to date. These trials tend not to be particularly rigorous in terms of scientific results, but are usually very demanding in terms of the design, definition and execution of the test processes in order to collect meaningful data from the complete system. The design of these trials is normally a series of incremental steps using platform rigs in the laboratory, prior to integration with a representative platform, following range trials.

It is strongly recommended that the test rigs (including the laser-based system), used in the laboratory and field tests, are designed and made suitable for the platform trials from the outset. Use of a consistent and stable fundamental design with established performance parameters and known operating characteristics is a critical stage in easing the integration issues and reducing technical risk, as well as understanding the interaction issues and other challenges. It also avoids confusion in the configuration control of the test items and most importantly, avoids the introduction of specific test-related artefacts from the test rig, as the equipment is moved from one configuration to another.

A further stage in the evaluation process following a successful evaluation with the platform trials is the possibility of Capability-Concept Demonstrator trials involving potential operators from an appropriate development and test squadron. This form of trial involves using a system from an advanced technology-demonstrator project and investigating its performance in a different role or different environment.

This aspect is particularly important if the laser-based system involves a novel approach or has the potential for the introduction of a new military capability. In this case, the operators can play a critical role in the development of tactics, as well as gathering data to support evolution of rules of engagement and concept of operation. The development of tactics and doctrine is discussed further in Chapter 18.

The other critical aspect of developing new military capability, involving novel technology, is the development of the understanding of the potential performance of the novel system so that tactics, techniques and procedures can be defined. Tactics, techniques and procedures may be evolved from the more advanced trials, including Capability-Concept Demonstrator testing and some forms of advanced simulation.

Flight trials are a specific case of platform trials and have their own demanding requirements and procedures. It is virtually impossible to modify existing equipment that is not designed for flight easily so that it will meet the safety-of-flight requirement and be cleared for use on aircraft. Generally a significant and costly redesign is required if flight-requirements are not embodied into the design, particularly for the more sophisticated equipment and that activity would be merely to meet the safety-of-flight requirements. Appropriate equipment, which is fit for purpose, enables testing and evaluation to be undertaken; however, achieving the requisite military-capability performance in the flight environment is an entirely separate issue.

Certain concessions may be allowed in the system design if the laser-based system is part of a technology demonstration involving a well-defined and limited amount of flight testing and within a carefully stated flight envelope. In these carefully defined and controlled circumstances and conditions, certification for the flight testing may be less demanding; however, suitable design techniques compatible with flight operations are still a critical and rigorous requirement.

6.4.13.1 Release to Service

The goal of all of the military research and development activities, followed by the various testing and evaluation phases of an equipment programme, is to ensure that the equipment is suitable for use in theatres of operation. Any equipment-development programme is normally a lengthy process, taking many years in the case of advanced equipment, with a number of iterations in the design, which will require assessment and validation. Similarly, the route to acceptance into military service is a systematic and rigorous process.

A reasonably realistic cycle of development of an existing system through technology advances is shown schematically in Figure 6.12, which will be driven by an identified military requirement. This new requirement may have resulted from a technological advance, which will offer greater military capability, or a change in the threat posed by an enemy. This figure illustrates the various interactions and activities that could be required to fill an identified gap in military capability, along with the various iterations that may need to occur to resolve the problem before the new or modified equipment can be released into service.

A variety of development tools are used in this systematic and iterative process needed to develop new equipment or capability, which leads to the comprehensive testing required for release of new or modified equipment into service. The military acceptance activity is centred on the evaluation trials using a military platform, or range of platforms, which are governed by the military system performance requirement.

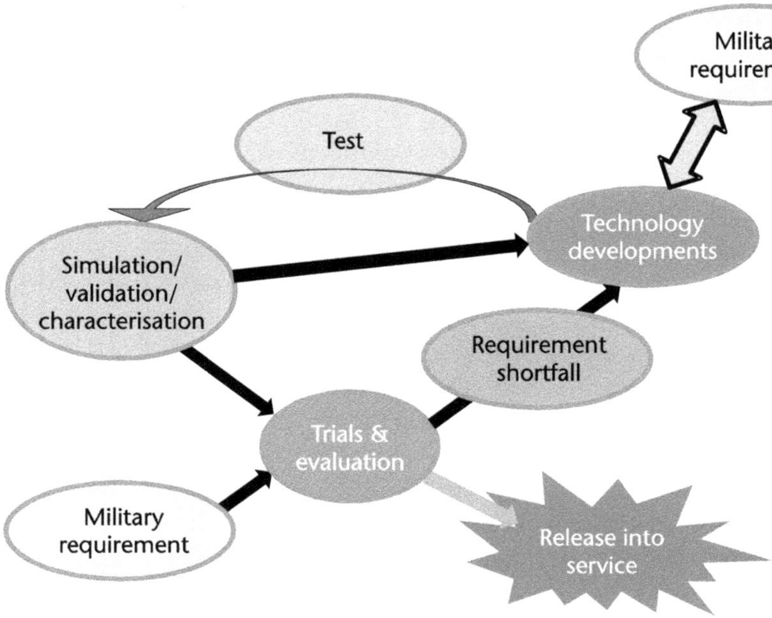

Figure 6.12 A typical development cycle leading to certification for service use.

6.5 Summary

It is critically important to understand the user requirements before a system is designed and built. This enables equipment specification, along with the interfaces for integration with platforms, to be defined. From this specification of performance, the requirements on the performance of its subsystems and components can be deduced. Test and evaluation programmes are dependent on a concise definition of performance specifications, so that the design of a system can be verified through testing processes. Sophisticated test and evaluation techniques have been developed to evaluate laser-based systems.

The radiometric and photometric quantities used to describe the various properties of a source are well established and are recommended to be followed at a base level in the evaluation of laser-based systems designed for any military application. Use of standard parameters enables the quoted performance parameters to be understood in the appropriate context. This position is merely the starting point for a test and evaluation programme.

There is a vast range of testing that can be undertaken to evaluate any laser-based system; one of the critical issues concerns the design of an efficient and effective process. All testing is expensive and as the testing becomes more complex and demanding then the time required, and consequently, the costs escalate. In general, it can be assumed that the costs of testing more than doubles at each stage. The use of sophisticated platforms, such as ships and aircraft, provide additional demands and the costs can increase exponentially. Strategic planning of the test programme is an important tool in scoping the formulation of the testing and preventing the cost of the test and evaluation spiralling out of control.

Some guidelines have been suggested for the design of a testing strategy; this starts from relatively simple laboratory tests and follows on to the complex operational test and evaluation phase of a project, leading to a full understanding of the installed performance of the system. Some of the crucial aspects in making the initial design fit for its intended purpose, such as flight testing and careful assessment, must be made to take account of the scale and duration of testing. It is crucially important that any evaluation of a system follows an integrated test plan.

An indication of the use of technology-readiness levels in the evaluation process is discussed in the context of maturation of a component and a system. As in all uses of indicative parameters, these readiness levels must be used with care; just because a particular subsystem is at a particular readiness level it does not mean that its system-readiness level is of a similar maturity. As is the case with this type of metric, caution is essential, so that credible and appropriate test evidence is supplied to achieve a realistic assessment of actual maturity.

Rigorous processes must be applied to the data collection and data logging tasks from all of the testing, so that these data may be available to all aspects of the test and evaluation programme. It is strongly recommended that data are recorded to a recognised format to enable subsequent data exchanges to be undertaken easily. It is vitally important that all tests are defined in the context of the whole test programme, and additionally, referenced to the fundamental requirements of the project.

A well-designed system usually has a well-defined relationship between its subsystems; therefore, modification of existing equipment to embody alternative subsystem technology is likely to be an expensive and time-consuming process requiring further extensive testing. Moreover, if a system requirement cannot be defined, it cannot be tested; this is axiomatic. Equipment should never go to the range for testing without knowing the answer, to at least a first approximation.

Selected Bibliography

Boyd, R. W., *Radiometry and the Detection of Optical Radiation,* New York: John Wiley & Sons, 1983.

Bukshtab, A. C., *Applied Photometry, Radiometry and Measurements of Optical Losses,* New York: Springer, 2012.

DeCusatis, C., (ed.), *OSA/AIP Handbook of Applied Photometry*, Woodbury, NY: American Institute of Physics Press, 1997.

Grant, B. G., *Field Guide to Radiometry,* Bellingham, WA: SPIE Optical Engineering Press, 2011.

Grum, F., and R. J. Becherer, *Optical Radiation Measurements, Vol. 1, Radiometry,* San Diego, CA: Academic Press, 1979.

Heard, H. G., *Laser Parameter Measurement Handbook,* New York: John Wiley & Sons, 1968.

Hudson, R. D., Jr., *Infrared System Engineering*, New York: John Wiley & Sons, 1969.

Johnson, M., *Photodetection and Measurement,* New York: McGraw-Hill, 2003.

Kaye, G. W. C., and T. H. Laby, *Tables of Physical and Chemical Constants,* Harlow, U.K.: Longman Scientific & Technical, 1992.

McCartney, E. J., *Optics of the Atmosphere,* New York: John Wiley & Sons, 1976.

McCluney, R., *Introduction to Radiometry and Photometry*, Norwood, MA: Artech House, 1994.

Ohno, Y., *OSA Handbook of Optics, Volume III: Visual Optics and Vision*, Chapter for Photometry and Radiometry, 1999.

Palmer, J. M., and B. G. Grant, *The Art of Radiometry,* Bellingham, WA: SPIE Press, 2009.

Parr, A. C., et al., (eds.), *Optical Radiometry,* San Diego, CA: Elsevier, 2005.

Williers, C. J., *Electro-Optical System Analysis and Design: A Radiometry Perspective*, Bellingham, WA: SPIE Optical Engineering Press, 2013.

Wolfe, W. L., *Introduction to Radiometry*, Bellingham, WA: SPIE Optical Engineering Press, 1998.

Zissis, G. J., (ed.), *The Infrared and Electro-Optic System Handbook, Vol. 1*, Chapter 4, "Radiometry," Bellingham, WA: SPIE Optical Engineering Press, 1993.

Part II
Military Systems

CHAPTER 7

Laser Directed-Energy-Weapon Systems

7.1 Introduction

Optical-countermeasure techniques are believed to have started several millennia ago at the battle of Syracuse, in 212 B.C.; an artist's impression of this event is shown in Figure 7.1. During this battle, Archimedes appears to have been the pioneer in the application of this technique by using an array of plane mirrors to focus sunlight on to a distant target, in an attempt to burn the sails of the ships of the attacking Roman fleet. Each mirror was silvered on both sides and had a hole in the middle to act as a sighting device, so when the Sun was imaged on the hole, the Sun's rays were also directed at the object seen in the hole. One of the reasons for the lack of success of this imaginative approach could have been the failure to keep the energy on the target for a sufficient period to achieve the desired effect. Energy on target and time on target (of the energy) are critical performance metrics for all directed-energy systems.

This class of laser-based weapon system represents the greatest technical challenge for many of the subsystems, especially the laser and, in the case of the high-power systems, aberration correction of the beam, sometimes referred to as beam control. Currently, the requisite performance has been demonstrated by the majority of components, but the major challenges are to reduce the size and complexity and then increase the reliability of the various subsystems, and, in particular, in the case of the laser source, enhance its efficiency and reduce its size, weight and cost. There are many research and development projects currently under way that are aimed at providing solutions for these technical requirements.

This chapter considers the development of laser-based directed-energy weapon systems (often referred to as Laser-DEW systems or laser-based novel-weapon systems), which create permanent damage to a target in one form or another, the technology base and the damage routes. It considers the scale of laser-based effects that can be achieved with the various classes of laser systems, which may be used as weapons. These effects are summarised in Figure 7.2. The lower-power systems, such as the dazzle and jamming systems, are considered in the following chapters. This figure indicates that there are three fundamental regimes for laser-based directed-energy systems.

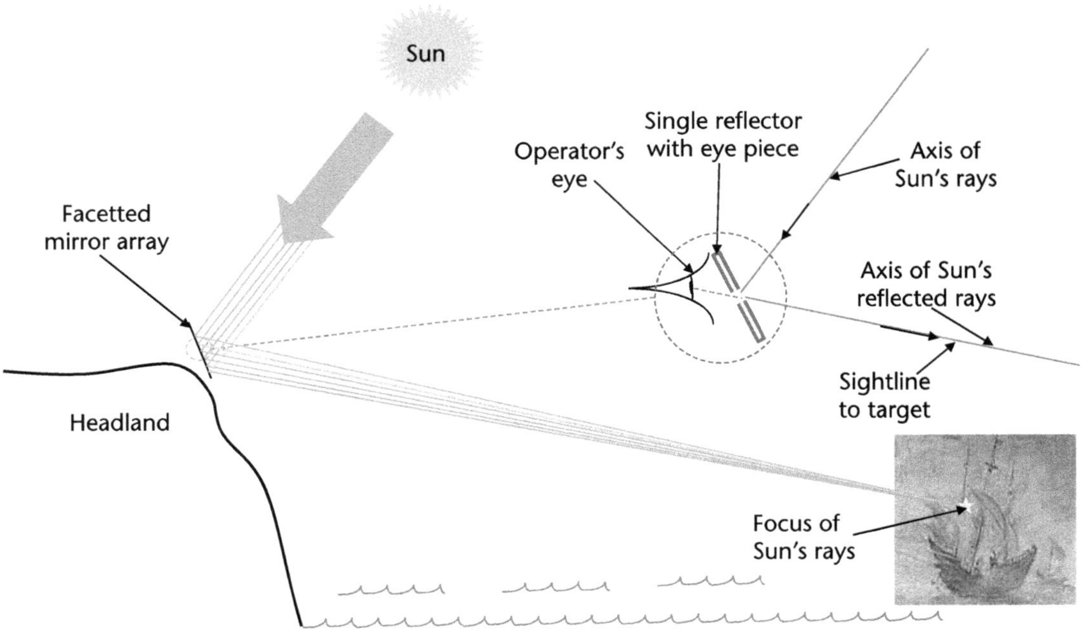

Figure 7.1 The first use of directed energy at Syracuse.

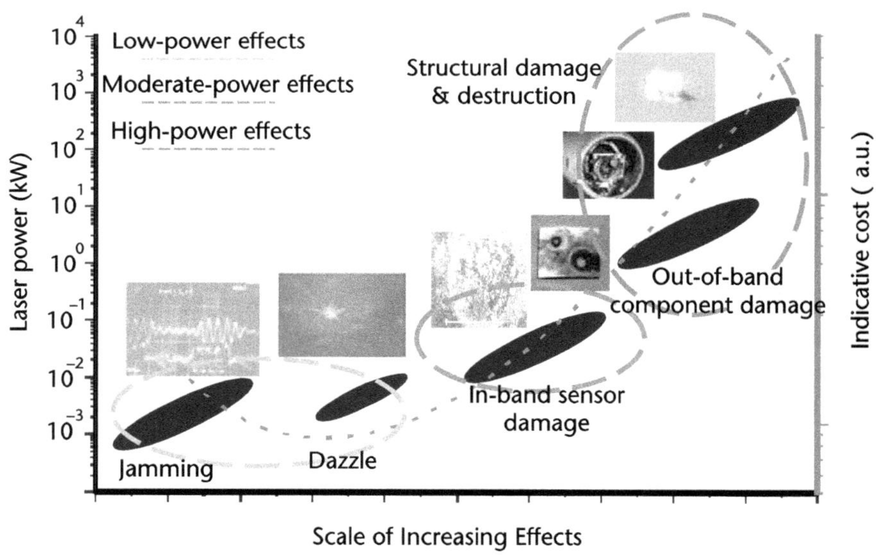

Figure 7.2 Laser-based directed-energy effects.

7.2 Historical Development

A laser-based directed-energy weapon system is often seen as an attempt to realise the death-ray concept for either tactical or strategic weapon systems; however, a better analogy may be a remote cutting and welding system. In this case, the death ray interacts with various components of a target to impede its normal operation. Hence, the modern analogy is with cutting and welding.

Very high-power laser beams were first demonstrated in the late 1960s and early 1970s using gas-laser technology, only a decade after the first demonstration of a laser source by Theodore Maiman. By 1972, the U.S. Air Force had demonstrated a 100-kW class device emitting in the far infrared, soon followed by a megawatt class gas-dynamic laser (GDL), also with a carbon-dioxide gain medium; largely based on aero-engine technology. Later in that decade, power approaching 0.5 MW was recorded from a chemical laser emitting in the mid-wave infrared.

There was continued progress with high-power gas-laser technology in the 1980s, using gas-dynamic laser and flowing-gas chemical laser technologies emitting in one part or another of the infrared waveband. A summary of the development of high-power devices and systems, along with some critical achievements, related to laser-damage weapons, are shown in Figure 7.3. This figure also shows the rise of the power emitted by solid-state lasers, including fibre lasers, in the last decade.

Very high-power laser sources were developed by a number of counties to demonstrate the potential performance of novel weapon systems for the destruction of targets. Such device demonstrations in the United States included the American Navy-ARPA chemical laser (NACL), which was integrated with an accurate pointing and tracking system to shoot down a TOW anti-tank missile in 1978 at the San Juan Capistrano test range in California. Another spectacular demonstration of potential weapon-class system performance was the destruction of a drone, known as Vandal (BQM-34), by the multimegawatt mid-infrared advanced chemical laser (MIRACL)/Navy Sealite system in 1989 at the American high-energy laser system test facility (HELSTF) in New Mexico. These were early examples of proof-of-principle tests, to establish the feasibility of a damage-class laser-weapon system.

Other significant experimental demonstrations around that time, involving the U.S. Air Force's Airborne Laser Laboratory (ALL), was a Boeing KC-135 aircraft operated with a gas-dynamic carbon-dioxide laser; this system defeated (destroyed) a Sidewinder missile (AIM 9) in flight. During the 1970s, work started on the chemical oxygen iodine laser at the USAF's Phillips Laboratory at Kirtland Air Force Base, New Mexico, now used as the laser source in the Airborne Laser system, which successfully destroyed a simulated ballistic missile in flight during 2010. However, the future operation and development of this strategic weapon system is in doubt.

Activity in the former Soviet Union included the design and demonstration of similar laser technology and the development of explosively-pumped iodine lasers. This device is essentially an explosively pumped laser amplifier; the output power is somewhat selectable, but this system has demonstrated output energy of the order of a megajoule, albeit with a low repetition rate.

March 1983 saw the start of the American Star Wars project or Strategic-Defense Initiative (SDI), which did much to end the Cold War between East and West; this project aimed to develop advanced technology for the defence of the United States and its allies and saw an emphasis on directed-energy techniques, including laser-based devices. Examples included the commissioning of the very powerful MIRACL source at the HELSTF site, as noted above, at White Sands Missile Range New Mexico. It helped to consolidate progress with laser-based directed-energy weapon technology, albeit in a ground-based proof-of-principle system. This decade also saw progress with solid-state and liquid-laser technology for anti-sensor

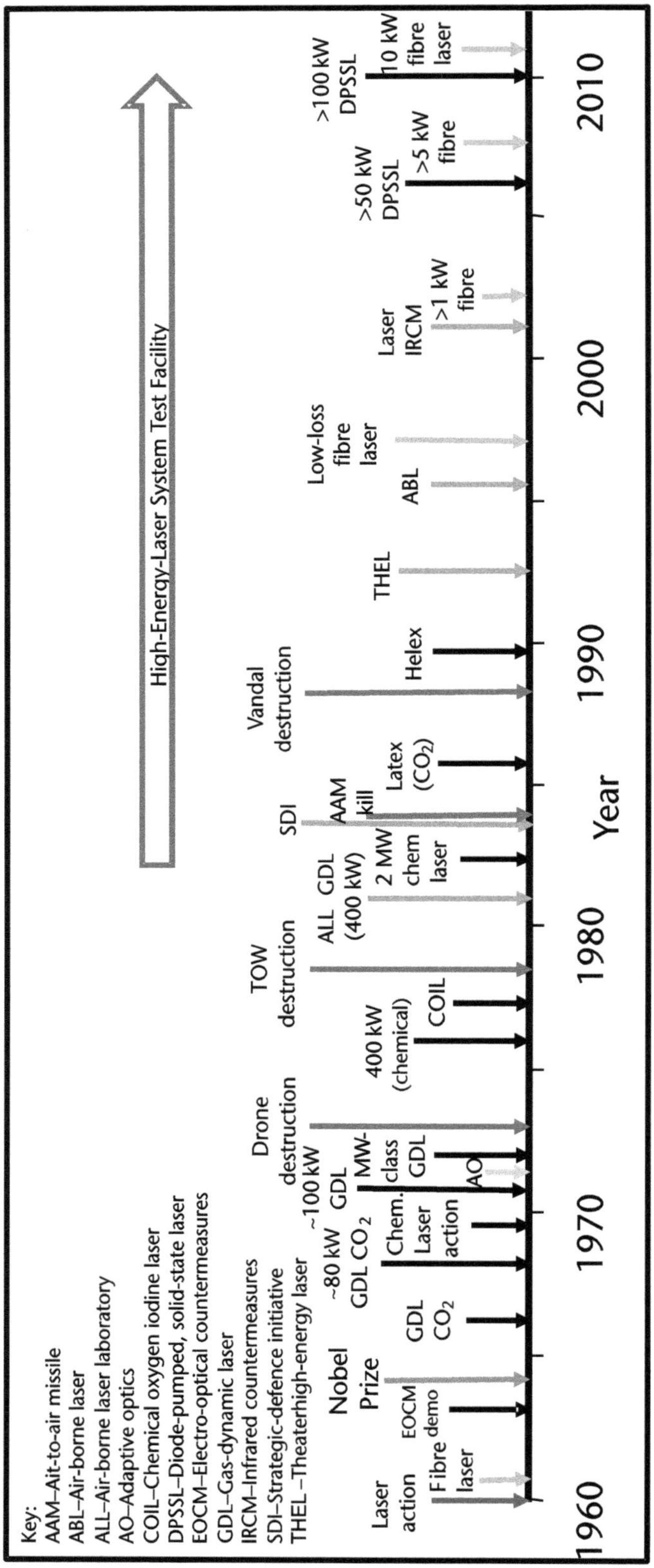

Figure 7.3 Summary of laser-based directed-energy weapon technology development.

applications, known as laser sensor-damage weapons; in general, the technical challenges for these lower-power systems were somewhat less demanding and thus not requiring state-of-the-art performance from all of the subsystems.

Very accurate pointing and tracking systems have been demonstrated during a number of these projects; these systems tend to be very complex and are often measured in terms of a few microradians. The target-tracking system within the beam director is usually based on an electro-optical camera, which forms an image of the target that is integrated with a complex target-tracking algorithm or multiple algorithms. A number of compensation subsystems have to be integrated into the beam director to ensure:

- Accurate tracking of the designated target;
- Precision boresighting of the optical axes in the beam direction system;
- Accurate compensation for systematic errors;
- Minimising of random errors.

A range of optical effects may be achieved using laser devices, as summarised in Figure 7.2. The requirements on the performance of the laser source vary widely from an average power of a few watts to megawatts, depending on the subtlety of the approach. This is a result of the various damage techniques that may be used to cause a permanent effect on a designated target. The laser damage-weapon concept is the most demanding on optical power, as shown in this figure. The laser-based systems that use the in-band damage technique, discussed later, may also be considered a damage-class weapon and these systems are also considered in some detail in this chapter.

7.3 Military Utility

Laser-based directed-energy systems offer a range of strike options whether used in the strategic or tactical roles. The technology is compatible with many military platforms for land, sea and air operations. The differences arise in the integration with the host platform and the associated demands on the sightline control between the weapon and its designated target.

The more modest-power systems, probably those weapons with the order of a 50-kW emission, would be suitable for integration with the smaller platforms, such as unmanned air vehicles or armoured fighting vehicles. The damage mechanisms used in this type of system are compatible with all types of target engagement for tactical roles, whether the designated target is on the surface, travelling close to the surface, or at higher altitudes. Additionally, this type of system may be used to attack a wide range of classes of target from ground installations and vehicles, to missiles and aircraft; consequently, the same system may be used for attack or defensive missions. Hence, this type of weapon system may be used in:

- Surface (ground)-to-surface (ground);
- Surface (ground)-to-air;

- Air-to-surface (ground);
- Air-to-air.

As noted above, virtually identical laser technology and subsystems may be used in all of these roles; hence, it may not be necessary to undertake multiple laser-weapon developmental projects to fulfil each of these roles, so the concept of joint technology and system development may be exploited. Some of the jargon used to describe the versatility of these laser-based weapon systems is target or platform agnostic.

Three examples of laser-damage weapon systems are shown in Figure 7.4; all are believed to be experimental systems to investigate and exploit novel-weapon effects. The Laser Weapon System (LaWS), which now has the weapon designation AN/SEQ-3 (XN-1), where the X designator means experimental system, is an advanced-concept demonstrator for the American Navy, the Tactical Laser System is being developed for the American Army, which is to be integrated with a heavy, extended mobility tactical truck (HEMTT), and the tactical laser system was a British technology demonstrator on a tracked vehicle (Stormer).

The technology will support engaging targets in a range of engagement geometries from head-on approaches to crossing engagements. Although head-on engagements can be especially challenging for this weapon, but not impossible, it normally takes slightly longer to cut a hole through to the susceptible zones.

[Courtesy NSWC Dahlgren, USA]

Laser Weapon System (LaWS)

[Courtesy Boeing, USA]

Tactical Laser System

[Courtesy Thales, UK]

Tactical Laser Demonstrator

Figure 7.4 Examples of prototype laser-weapon systems.

However, achieving the more demanding defeat mechanisms, such as hole burning followed by destruction, of an approaching missile can be very demanding, owing to the need for the damage front to move rapidly through many centimetres of metal and other materials. Destruction of a range of sensors on such missiles can be achieved very rapidly, especially if the wavelength is in band, as discussed later.

Hard kill of crossing targets, whilst still demanding on the optical-mechanical performance of the laser-based weapon system, is usually much easier, and hence more rapid, than the destruction of a head-on approaching target. This type of engagement has the following advantages:

- Potential for better beam propagation conditions, owing to a continuously changing sightline, which leads to reduced probability of occurrence of the thermal-blooming phenomenon (discussed in Chapter 5);
- Less metal to cut through to attack susceptible components, leading to enhanced lethality and a higher single-shot kill probability.

This type of weapon system should not be seen as a direct replacement for conventional hard-kill kinetic weapon systems when operated in a defensive role; however, when used in conjunction with these weapons, it can lead to a very effective system for a range of operations. This is a consequence of exploiting the complementary aspects of the two types of weapons. For example, a guided weapon may be used to defeat targets at long range and the laser-based weapon system may be used to destroy those threats that manage to penetrate this shield, the leakers, by exploitation of the speed-of-light projection of the laser beam's energy to give an effect with a negligible reaction time. This is the principle of layered defence.

A further potential advantage of laser-weapon systems, concerns the possibility of undertaking a small-scale or low-level damage strike, followed by an escalation of military intent using the same laser-weapon system but with more optical power if required. This is a consequence of the feasibility to generate tuneable effects from this technology (i.e., being able to select or tune the emission to generate a given damaging effect from the same weapon system).

An additional advantage is that this class of weapon buys a commander time, as the engagement and interaction time is shorter than the engagement time of conventional weapons. Hence, the laser weapon can get inside the conventional weapon system engagement cycle for many engagements by beating the conventional weapon to the punch, especially for the in-band mechanisms.

7.4 Summary of Potential Applications

Table 7.1 illustrates some of the likely applications for high-power, laser-damage weapons in the four major platform environments; in general, the system-performance parameters increase significantly for those applications further down each column in the table. The out-of-band damage route technique is assumed to provide the most appropriate interaction method for these applications, as discussed later.

Table 7.2 indicates comparisons between some requirement parameters for three probable military uses of modest power laser-damage weapons.

Table 7.1 Potential Out-of-Band Applications of Laser-Damage Weapons

Air	*Land*	*Sea*	*Space*
Self-protection (asymmetric threats)	Negating unexploded ordnance	Counter asymmetric threats (e.g., small boats)	Self-defence
Air-to-air warfare	Counterintelligence, surveillance, reconnaissance vehicles	Self defence	Anti-satellite
Ground attack	Self-defence	Support/area air defence	Global ballistic defence
Ballistic-missile defence	Air defence and anti-satellite operations	—	—

Table 7.2 Comparison of Engagement Requirements

	Air Defence	*Counter-Rockets, Mortars and Artillery*	*Unexploded Ordnance Disposal*
Targets	Cruise missiles, UAVs, precision guided missiles, aircraft (fixed and rotary wing)	Rockets, mortars and artillery	Ammunition, mines and improvised explosive devices (IEDs)
Spot size	100 mm	60 mm	40 mm
Engagement range	1.5 to 6 km	0.5 to 3 km	<300m
Time on target	<6 seconds	<20 seconds	N/A
Power	20 kW	100 kW	2 kW
Turbulence compensation	Yes	Yes/no (depends!)	No
Tracking requirement	High (μrad)	High (μrad)	Low

Anti-sensor systems that apply the in-band damage technique are also feasible. In this case, the key performance parameters are somewhat more relaxed, as discussed later in this chapter.

Figure 7.5 provides a summary of potential targets for high-energy laser weapons and Figure 7.6 indicates typical launched power requirements to achieve successful interactions with these targets.

7.5 Benefits of Laser Technology for Novel Systems

The attraction of a laser-based technology in defensive applications is that it offers a robust defeat mechanism; additionally, the laser-weapon system can make optimum use of the advantages of laser-weapon technology, as well as employing a potentially lower-cost option for tactical and strategic systems. The advantages of laser technology are:

- Potentially a very deep magazine (from the exceedingly high number of shots that may be generated from a single weapon system), given an adequate supply of prime power;

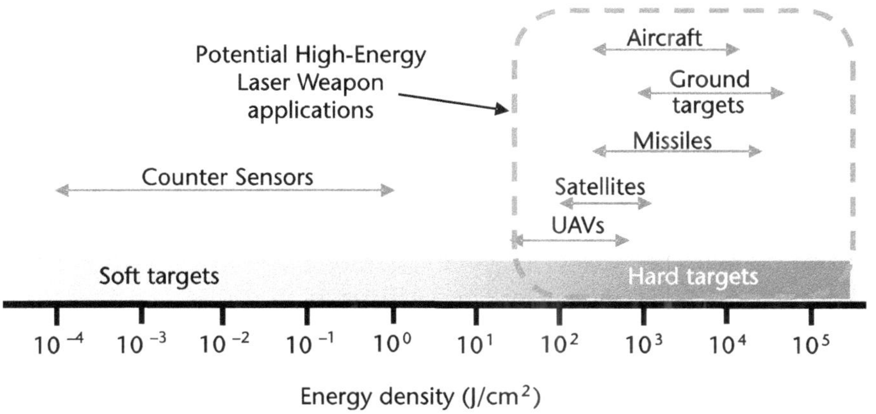

Figure 7.5 Summary of potential targets for high-energy laser weapons.

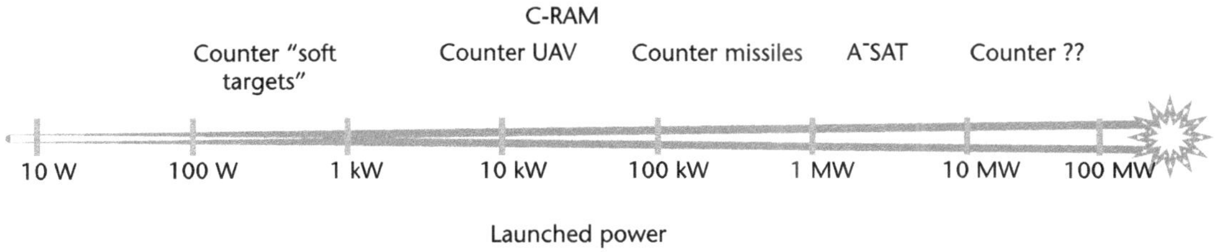

Figure 7.6 Typical launched power requirements.

- Scaleable effects from a single source from less-than-lethal interaction to full-scale counter-materiel effects;
- Speed-of-light interaction;
- Line-of-sight engagement, so it is not necessary to track and predict the motion of a moving target accurately over an extended period to estimate a future interception point (however, very accurate target-tracking is still necessary for the period of the laser-target interaction with this type of application, energy on target is a critical requirement, as Archimedes found out);
- Very short reaction time lines, so targets that unmask at very-short ranges can be engaged;
- Rapid re-engagement capability;
- Multiple-target engagement capability;
- This approach does not leave any system debris at the scene of the attack, only the destroyed or malfunctioning optical system on the engaged target and hence may be considered a green weapon;
- Highly resistant to catastrophic consequences of a hit on the weapon by the enemy, as it does not have a weapons magazine containing high explosive and electrically driven lasers are particularly insensitive, but some chemical lasers less so;
- High single-shot cost-effectiveness;
- Surgical strike with negligible collateral damage;

- Agnostic technology, so that it is applicable in multiple scenarios, air or surface;
- Low cost of engagement, estimated to be about 15 pence or 20 cents.

The major drawbacks (the cons) of laser-based directed-energy systems are usually quoted as:

- Finite interaction time with the target for some defeat mechanisms, particularly using the out-of-band approach, to realise or achieve permanent damage effects;
- Need for an accurate beam director, to ensure the laser beam remains on the target during the period required needed to complete the damage (i.e., during the finite interaction-time), which is particularly challenging for some engagements involving the high-performance laser-weapon systems;
- Complexity of the entire laser-based system, including high technology and sophisticated techniques, particularly in the beam-director subsystem;
- Lack of any direct battle-damage indication capability, however, this aspect is not unique to laser-based directed-energy systems and, in fact, the problem is often much less of an issue for directed-energy systems, compared with kinetic weapons, owing to the monitoring of the interaction to keep the beam on the target.

7.6 Laser-Beam Target-Interaction Effects

There are two principal approaches to defeating optical components, optical systems and damaging the complete weapon system. They are commonly known as in-band and out-of-band techniques. The in-band technique is usually associated with damaging a sensor component, whereas the out-of-band technique aims to defeat a complete subsystem; the principles of these approaches are shown schematically in Figure 7.7. These techniques are somewhat different in approach and, consequently, have different requirement specifications for the laser sources used. The critical issues with this type of system concern laser-target interaction effects, propagation and aberrations induced by the atmosphere for endo-atmospheric engagements.

7.7 In-Band Damage Route

The principle of the in-band damage approach is to use the optical gain of the sensor being attacked to focus the received incident energy on to a susceptible component in a focal plane of the optical sensor.[1] This enables a modest-average-power laser to be used to create permanent damage to a sensor; permanent damage occurs when

1. An unintentional demonstration was given to the world during the Apollo 12 mission on the moon when astronaut Alan L. Bean was positioning a colour Vidicon camera and it was caught by a specular reflection of the Sun from part of the structure of the lunar module. The consequence was a loss of televised pictures

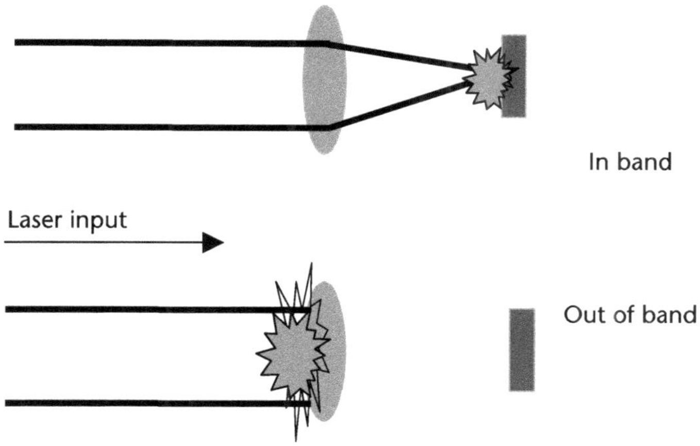

Figure 7.7 In-band and out-of-band techniques.

the laser-induced damage threshold is exceeded on the susceptible component, realised through the focussing action of the optical train in the target. The incident photon energy needs to have a wavelength that is not attenuated by the optical train of the sensor, as well as coinciding with an atmospheric-propagation window, as discussed in Chapter 5. This is the basis of a laser sensor-damage-weapon system and may operate in any of the wavebands from the visible to the far infrared and beyond, provided the laser technology can support the emission requirements.

A benefit of this approach is that a very short interaction time is required for destroying a sensor, and in theory, could be achieved from a single pulse of energy; hence, a pulsed source may be used with a low pulse-repetition frequency. In practice, a burst of pulses is used over a very short period in order to increase the hit chance and ameliorating the stochastic processes occurring along the propagation path in the atmosphere, such as turbulence-induced, beam-pointing effects. As a result, a relatively simple, low-cost yet effective beam-director device can be used, as the interaction time with the target is relatively short.

As can be seen from Figure 7.2, the power requirements of the laser emission for a laser-sensor-damage weapon are still quite modest; therefore, it is unlikely that the use of adaptive-optical techniques will be necessary, to ensure that optimum power is received by the target. However, if this class of laser-weapon system is likely to be used in a highly turbulent atmosphere, then adaptive-optical techniques would be necessary for:

- Stabilisation of the image used by the target-tracking system;
- Compensation of the directed beam to form a plane wavefront at the target.

The major drawback of this approach is that it is reliant on some co-operation from the threat system (i.e., the threat system needs to be looking at the asset, such as a military vehicle, being protected by the laser-based countermeasure system). Moreover, the operational wavebands of the threat and the emitted laser-line need

---

from this Moon landing. A somewhat more successful directed-energy demonstration than occurred at Syracuse a few millennia earlier.

to coincide, as noted earlier; this is a critical system requirement. However, if the threat system is observing the asset on a battlefield, then it is likely to have hostile intent; in other scenarios, particularly at sea, this may be a less straight forward indicator of intent. Thus, provided that the threat optical sensor can be located with sufficient accuracy, perhaps using thermal cues supported by retro-returns from the target, it can be engaged, provided that this type of interaction with an enemy is allowed by the rules of engagement.

7.8 Out-of-Band Damage Route

The principle of the out-of-band system approach is to provide a very robust technique that eliminates the dependencies and engagement constraints inherent in the laser-based novel weapon systems using in-band techniques, considered earlier. Consequently, this damage route may be considered a brute-force approach that equates to remote cutting and welding. The defeat mechanism involves the rapid heating of a surface, so that it either melts or factures.

In this case, the photon-flux intensity (irradiance) is designed to be of such a magnitude that any surface intercepted by the photon beam is damaged; however, this technique does demand very high power beams with appropriate irradiance for the task. The advantage to the weapon system is that the targets for this class of weapon may be engaged from any angle, these engagements do not require any co-operation from the threat, as is necessary in the case of the in-band techniques. Moreover, any class of target may be engaged by this photon beam such as radio-frequency guided weapons, inertially-guided missiles as well as optically-guided weapon systems.

7.9 Target-Defeat Techniques and Effects

The fundamental concept behind any laser-based directed-energy weapon system is to deliver the photons from the laser source and damage a susceptible component on, or indeed if possible, to the structure of, the designated target. Many teams have striven to produce the ultimate device that will vaporise a structure virtually instantaneously, such as a missile or aircraft, but its successful realisation remains difficult and is unlikely to be cost-effective. Additionally, this ambition is unlikely to be realised when engaging targets relatively close to the surface of the Earth, owing to the limitations imposed by the Earth's atmosphere, as discussed in Chapter 5, and the fact that the majority of defeat mechanisms require some dwell time of the photons on the target, for the photons in the laser beam to interact effectively with the material.

7.9.1 In-Band Defeat

The in-band approach to laser damage-weapons is to create a permanent effect that will destroy a sensor, a subsystem or a structural component. The defeat mechanism for this type of system is through rapid energy deposition and thus subsequent absorption of optical energy, which destroys the integrity of its target material. The permanent damage for an in-band damage system may be used for:

- Destruction of a sensor, usually a detector element or system array, resulting in the permanent loss of signal from the sensor. This technique is sometimes referred to as blinding of a sensor, such as a camera, so that the guidance signal is lost in the case of a sensor in a seeker, but a more politically correct term is jamming of the sensor.
- Soft-kill through either a laser-jamming technique or a laser-dazzle method; both of these techniques are discussed in the following chapters.

7.9.2 Out-of-Band Defeat

In the case of an out-of-band system, the application of the high-power photons may be used for a range of effects, for example:

- Destruction of a target by initially melting and then vaporising a substantial part of it, resulting in cutting a hole in a metal or composite structure and destruction of components beneath the surface and, alternatively, holes may be burned in a glass reinforced plastic radome;
- Fracturing the body of a vehicle from hydraulic shock or spallation of the inner surface, resulting in internal damage or rupture sealed vessels and loss of the structural integrity;
- Destruction of dielectric covers of guidance systems, resulting in the cracking or shattering of radomes, optical domes or other environmental covers.

The out-of-band technique is an attractive laser-based weapon approach, as permanent damage can be achieved during a relatively short-duration interaction with a target, enabling the short-range, pop-up target (late unmasking condition) to be engaged and defeated. This form of engagement opportunity offers a useful adjunct to the conventional surface-to-air weapon system, which may have a significant inner-range boundary.

The use of a damage-class (permanent) defeat mechanism may be considered to offer a defender an escalation in response to a threat beyond dazzling a target sensor or distracting an operator, which are temporary effects induced by illumination of sensors with low-power laser sources, as discussed in the next chapter. However, the use of any form of permanent damage route is generally more demanding on optical power for defeating targets, but it does usually generate a more robust defeat mechanism.

7.10 Basic Laser-Requirement Characteristics of In-Band Novel Weapon Systems

The laser requirements for this damage-class of directed-energy system may be summarised as follows:

- Wavelength is matched to the optical passband of the sensor and in an atmospheric window, but there are not any special requirements on spectral purity and the need to avoid chirping of the emitted laser wavelength.
- Pulse energy needs to be such that when focused by the optical train of the threat sensor under attack, the fluence incident on the focal plane exceeds the laser-induced damage threshold of the component positioned in the focal plane. The components most likely to be susceptible to damage are usually detectors of one form or another, but it may be any one of a number of other optical components. Typical pulse-energy requirements of the source are of the order of a joule. The use of continuous-wave emission is also possible; again the requirement is for the focused fluence to exceed the laser-induced damage threshold. However, pulsed lasers are normally used for this approach.
- Pulse-repetition rate may be low, as a single pulse of energy is all that is required or a short burst of continuous-wave emission. However, owing to the dynamic behaviour of the atmosphere, as described in Chapter 5, there are many stochastic processes that may reduce the brightness of the laser beam and also modify its direction of propagation. Consequently, a short burst of pulses is often used in a macro pulse, of duration of the order of a millisecond, as this duration is compatible with the time constant of the atmosphere. Individual pulses may be many orders of magnitude shorter, to take advantage of the effects shown for short-pulse energy deposition described by the Bartoli damage criterion; some examples are shown in Figure 7.8.
- Beam divergence is usually a compromise between the efficient use of the laser-resonator configuration that enables the maximum amount of energy to be extracted from the gain medium and the pointing stability of the beam director. A beam-expansion telescope may be used to modify the divergence of the beam to match it to the characteristics of the beam-pointing system. The actual requirement is beam brightness, as that parameter is a fundamental aspect of achieving the requisite energy density on the optical aperture of any system under attack. Typical requirements for the launched beam-divergence are of the order of a milliradian.

This damage route is often considered a candidate technique for defeating the emerging imaging-infrared threat seeker-technology or other imaging systems, such as surveillance and target-acquisition sensors. Alternative techniques are also being investigated to defeat these emerging threat technologies.

Turbulence in the atmosphere can have a significant impact on the effectiveness of the in-band laser-weapon system, owing to distortion of the optical image used to track and identify the threat, as well as the modification of the spatial characteristics of the laser beam. The beam jitter induced by turbulent motion of the atmosphere can cause the beam to miss its target. In these operational conditions, adaptive-optical techniques are needed. Thermal blooming effects are unlikely to be significant except for very high-power in-band laser-based systems (i.e., with emitted power in the 100-kW class and beyond).

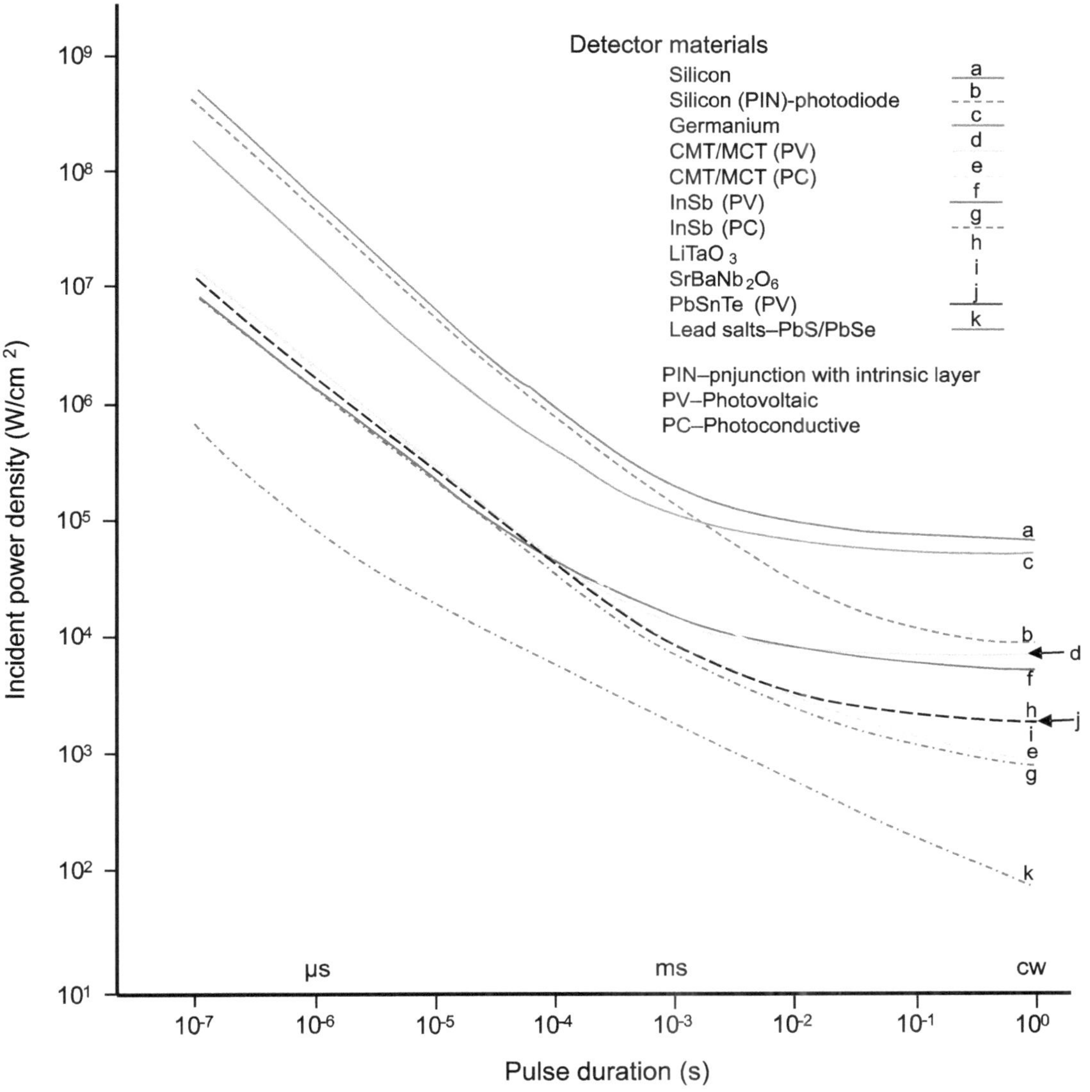

Figure 7.8 Bartoli damage criterion for some detector materials. (Courtesy of Professor F. J. Bartoli, Lehigh University, United States, Appl. Opt. 16, 11, Nov 1977.)

7.11 Basic Laser Characteristics for Out-of-Band Laser-Weapon Systems

The laser performance requirements for this class of countermeasure system may be summarised as follows:

- Wavelength does not need to be matched to the optical passband of the sensor, but it should be chosen to minimise absorption and scattering in the atmosphere, and thus, the aberrations that follow from nonoptimal transmission through the atmosphere leading to local heating of the propagation path are also minimised.

- There are no special requirements on spectral purity and avoidance of chirping of the emitted wavelength, providing that it does not incur any detrimental propagation effects.
- Pulsed or continuous-wave emissions may be used to damage a component or cut through the thin structure of a designated threat. In the case of a pulsed system, the pulse-repetition rate is usually in the kilohertz regime for cutting a hole in a structure, whereas a low repetition-rate of high-energy pulses may be used to fracture a dielectric panel or cover.
- Power or average-power requirements are many tens or hundreds of kilowatts, depending on the proposed application; however, in this case an equally important consideration is the total efficiency of the laser, often termed the wall-plug efficiency, owing to the large demand on the electrical power required to operate the laser and the management of the consequential generation of waste heat, owing to the finite efficiency of the complete system.
- Beam divergence is the usual compromise between the competing factors, considered above, but the actual requirement is beam brightness delivered to the target (i.e., radiant intensity). A typical laser will have a very small beam divergence, possibly of the order of a 10 μrad, or smaller, which has a significant impact on the pointing stability requirement of the laser-beam director.

7.12 Summary of Laser Requirements for Laser-Weapon Systems

Laser-weapon systems are being developed for many applications from relatively simple less-than-lethal interactions with targets to those designed for various defensive and offensive missions. Needless to say, the performance parameter requirements vary over many orders of magnitude. Examples of some selected key performance parameters are listed in qualitative terms for illustrative purposes in Table 7.3, along with possible projected interaction distances.

Anti-sensor system requirements are far less demanding, with emitted powers of the order of 100W and pointing jitter at least an order of magnitude more relaxed than the values indicated in Table 7.3.

7.13 Target Engagement

A typical engagement sequence is shown in Figure 7.9, which could be of the order of a few seconds or less; the figure indicates that it is a sequential process but some processes can be initiated before a prior process has been completed, as shown by the position of the data arrows (information flow).

For destruction or damage of missiles during very short-range engagements, such as a short-range pop-up threat, it is crucial that the time taken to get the photons on to the target is minimised, owing to the finite interaction time. It is important that the missile's trajectory is disrupted at a sufficiently large distance from

Table 7.3 A Summary of the Key Outline Performance Parameters for Some Laser-Weapon Systems

	Anti-Air	*Ground/Surface Attack*	*Soft Target Attack*	*Counter Intelligence, Surveillance and Reconnaissance*	*Counter Rockets, Artillery, Mortars and Missiles*	*Counter Improvised Explosive Device/Unexploded Ordnance*	*Strategic*	*Anti-Satellite*
Power (kW)	~100 to 150	~100	Up to 100	Up to 100	100+	~1+	>1,000	>1,000
Beam Quality (M^2)	<2	<<2	~2	~2	~2	~2	Close to diffraction limit	Close to diffraction limit
Jitter (μrad)	<10	<<10	<10	<10	<5	~10	Minute	Minute
Interaction range (km)	~20	Up to 100	Up to 10	Up to 10	Up to 10	<< 1	10 to >>100	10 to >>100

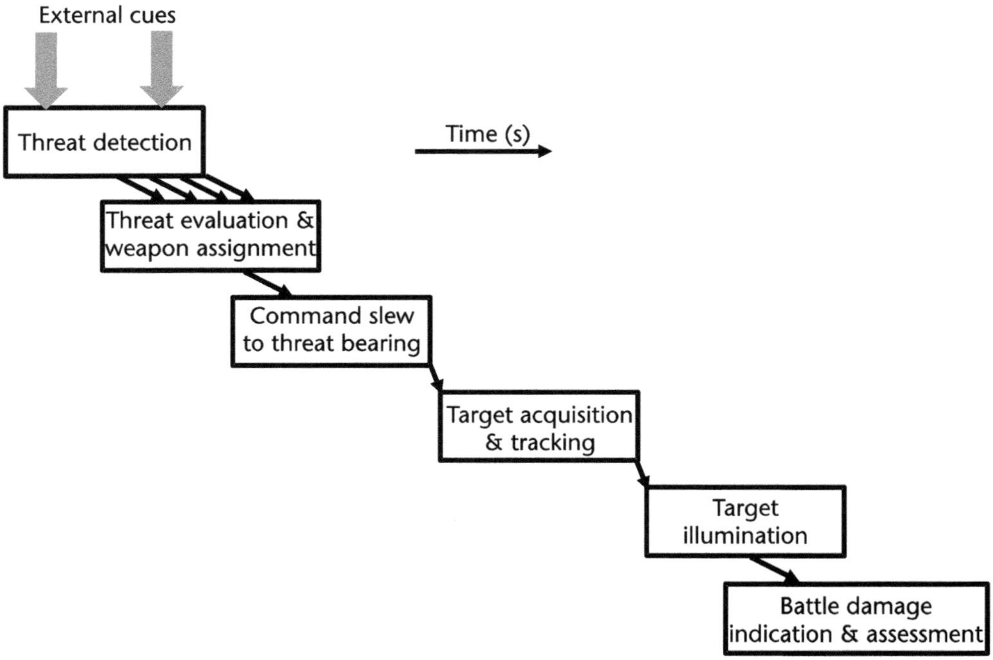

Figure 7.9 Laser-weapon engagement sequence.

its target, so that the missile misses its target (i.e., the missiles must be engaged and defeated before they reach the keep-out range[2]).

A metric often used to assess the potential performance of a laser-weapon system is known as the probability of escaping hit or $P_E(H)$. Various simulations of engagements may be undertaken with a mix of weapons defending a platform being attacked by a given selection of offensive weapons. The effect of adding a laser weapon system to the platform under attack may be assessed through changes to the $P_E(H)$.

7.14 Laser-Weapon Subsystem Definition and Operation

So far in this chapter, there has been a detailed consideration of the potential damage routes and the type of laser technology emission characteristics required to achieve these effects; however, as noted above, all of this technological endeavour will fail unless the photon energy is delivered to the designated target and interacts with a susceptible component or an appropriate part of the structure. In the case of the very high-performance systems, the laser beam may need to be directed at a particular point on the target and remain on that point for a finite time. A schematic illustration of a laser-weapon system is shown in Figure 7.10.

2. The keep-out range indicates a distance from a platform at which a threat must be destroyed/defeated to ensure that the platform is not hit (or at least to minimise that risk) by the remnants of the threat continuing to fly on a ballistic trajectory. This may be referred to as a time-dependent kill being imposed on the threat system.

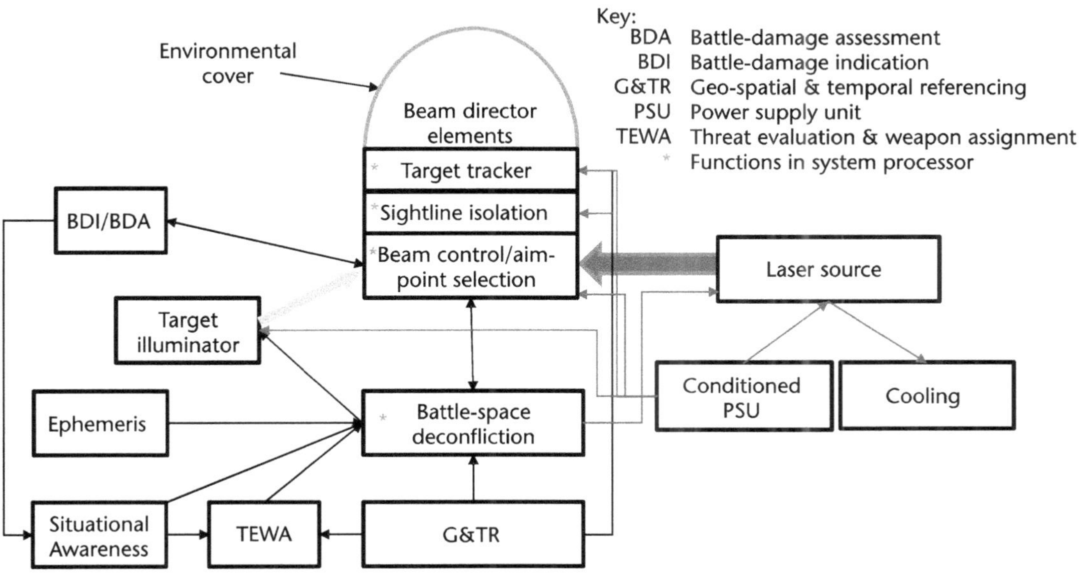

Figure 7.10 The critical components of a laser-based weapon system.

The fundamental components for a laser-weapon system, and their status of development, may be summarised as:

- The laser source, advanced and very efficient compact devices exist to generate wavelengths from the ultraviolet to the far infrared, many of them based on solid-state technology, as discussed in Part I. The in-band damage-class device is likely to have an average power in the tens of watts class, whereas the out-of-band system will demand tens of kilowatts of optical power, and beyond, for the more demanding applications.
- A beam director with a highly accurate tracking system may be referred to as an acquisition pointing and tracking (APT) system, as shown in Figure 9.5 in Chapter 9. The beam director is generally a sophisticated device and is considered in further detail in Chapter 4. The pointing accuracy requirement for the in-band damage system may be of the order of a milliradian, whereas the high-performance laser-weapon systems could well be closer to a microradian pointing accuracy. The beam director will also contain the beam-control system required to correct for aberrations from whatever source to ensure the beam delivered to the target has the optimum characteristics for its intended purpose.
- An alerting system that cues the beam-director system with its target-acquisition subsystems to help it find its allocated target rapidly and accurately. This alerting and cueing function may be from a radar-based system, a thermal-surveillance system or another optical system, using techniques such as retro-reflection methods. A critical consideration is the precision accuracy of the handover from one tracking system to another, so that this function can be achieved with a very high probability of success; therefore, the accuracy of

the cue must be significantly less than the field of view of the tracking system that receives the cueing data.

- A boresighting system to ensure harmony of the optical axes of the entire weapon system; this task may be undertaken within the beam director and forms a most important interface function. The accuracy achieved in the harmonisation of the various axes must be capable of sustaining the accuracy of the complete pointing system; additionally, this function will have to be undertaken periodically, if not continuously, to maintain the pointing-accuracy performance.
- A system processor, which undertakes the analysis of the threat data, establishes the track being followed by the threat, if it is moving, assigns an appropriate beam director, arranges the co-ordinates for the hand-off to a beam director with the appropriate field of regard for engaging the threat and then either monitors the interaction or undertakes an effectiveness assessment.
- A power source, which provides the electrical energy to power the system; the laser is likely to be the most demanding. In some instances, the demand may be around (or a significant fraction of) a megawatt for a short period of two or three seconds. This may involve novel methods such as homo-polar generators and modern battery technology to store the energy in readiness for the rapid demand. The beam director is likely to have its own power-conditioning system to allow the precision pointing required to fulfil the energy-on-target requirements over a sustained period.
- A heat management system to maintain the thermal energy balance within the whole weapon system, so that the precision alignment (harmony) of the optical axes of the various subsystems, as well as allowing the laser to operate as designed. This system is likely to have complex heat exchangers and exhaust system to ensure a rapid dispersal of the waste heat, whilst attempting to minimise any change in thermal signature of the platform installation.
- Built-in-test equipment to monitor the health and availability of the entire system.

During the design of a laser-based directed-energy system it is crucial that careful consideration is given to:

- The propagation constraints of the atmosphere, which is particularly important for high- and very-high-power laser systems;
- The target interaction effects, for example, small changes in the laser-pulse profile can make a significant difference to the rate of ablation of a target's surface;
- The operational environment.

7.15 Beam Control

Correction of laser-beam aberrations caused by an optical train or propagation through the atmosphere will be required by a high-performance laser-pointing subsystem to achieve the requisite energy-on-target characteristics. Adaptive-optical techniques have been developed significantly over the last decade, as discussed in some detail in Chapter 5.

The general principle is discussed below; a sensor is used to measure the distortion in the wavefront of an optical beam; a conjugate beam is then generated that leads to a plane wavefront being formed as the conjugate beam passes through the aberrating medium. In the case of an aberration being induced in the shape of the propagating wavefront by the atmosphere, it is normal to use a probe beam to sample the distortions of the atmosphere by forming a guide star, this allows the distortion to be assessed and the conjugate beam formed through the use of a segmented or deformable mirror. A critical requirement is that the control system used for the creation of the conjugate beam must have a reaction time and bandwidth, which is much faster than the time constant of the mechanisms creating the distortion.

Various levels of compensation may be applied to the optical system to correct the aberration, such as:

- Tip/tilt correction, sometimes called first-order correction;
- Focus, which is a far higher order and generally relevant to longer range engagements.

Figure 7.11 shows schematically the first-order correction approach of tip/tilt compensation, which is in effect a fast-steering mirror with two degrees of freedom.

The target-tracking system is also likely to have an adaptive system, using complex tracking algorithms and switching between them, so that the target-interaction zone is tracked and engaged with microradian precision; moreover, for the highest performance systems, the target-engagement interaction will be monitored throughout. If the engagement interaction is not achieved within a couple of seconds then the engagement may fail totally.

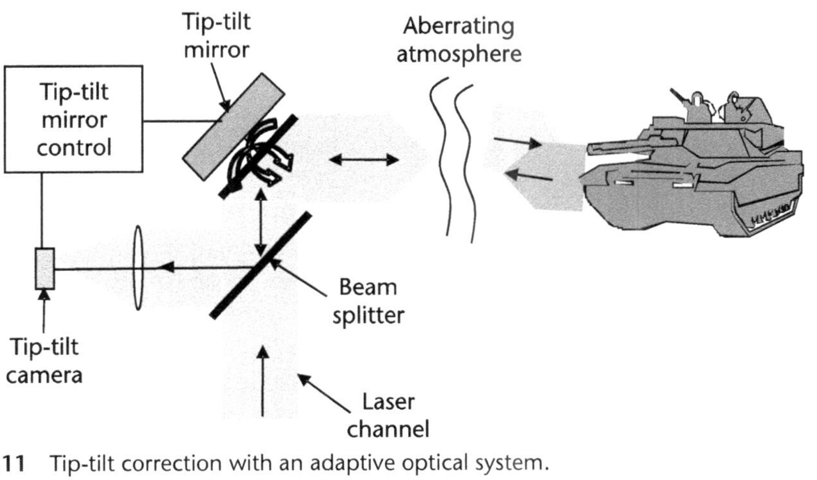

Figure 7.11 Tip-tilt correction with an adaptive optical system.

The high-power systems that demand high-performance pointing and tracking would use the full capability of the adaptive-optical system in both the transmitter and receiver channels.

7.16 Potential Laser Technology for an In-Band Novel Weapon System

7.16.1 Visible Band

There are very few technologies that will readily generate energetic pulses in this optical waveband, as such pulses really need to be produced by direct generation. The classical laser for this requirement is the liquid-phase flashlamp-pumped dye laser. This device is capable of very energetic pulses of tens of joules and more.

7.16.2 Mid-Wave Infrared

There are few solid-state sources that will give direct generation in the 3- to 5-μm band where many of the threat sensors operate. One example is a laser-pumped holmium barium yttrium fluoride device, which has generated several hundreds of millijoules in a single macro-pulse.

Electrically excited deuterium-fluoride lasers have generated multiple-joule emissions from relatively compact sources. However, the sealed systems have a relatively limited service life, without if not refreshed.

7.16.3 Far-Infrared Waveband

Energetic emissions in this band suitable for creating in-band damage come from carbon-dioxide lasers.

7.17 Potential Laser Technology for an Out-of-Band Novel Weapon System

In this case, it is not necessary to consider the passband of a sensor that is to be attacked, as a primary requirement; it is the efficient generation of the photon beam, which will propagate with minimal loss through the atmosphere and then be absorbed by the target that are the critical requirements. Hence, in-band laser wavelength considerations are not relevant here; the requirements concentrate on getting as many photons in as bright a beam as is possible to the target, with minimal distortion of the beam.

Many of the current investigations for laser sources that will fulfil the performance requirements for out-of-band damage systems have focussed on solid-state technology, and in particular, emissions close to one micron. The use of the ytterbium ion in the gain medium has a number of attractions, particularly with regard to its emission wavelength and potential overall efficiency. One of the exciting developments with this technology is the demonstration of fibre-bundle lasers, where a number of high-power, high-brightness fibres are combined to give an even higher-power beam. A number of techniques have been demonstrated, which are illustrated in Figure 7.12.

The phase-locked scheme is shown in some detail in Figure 7.13; this technique has some interesting characteristics, as discussed in Chapter 3. A couple of the attractions for laser-weapon applications are the enhanced brightness on axis and the fact that the beam can be steered by altering the phases of the component lasers.

Other laser technologies that offer high-power beams are devices such as the chemical oxygen iodine laser, but this has a number of practical issues, such as waste-efflux control for tactical applications. Moreover, the wavelength is not well matched to the atmospheric window in the near infrared, particularly in humid or damp conditions, as shown in Figure 5.1. Other chemical/gas lasers that have demonstrated high-power emission are hydrogen-fluoride and deuterium fluoride lasers; their emission is in the atmospheric windows of the mid-wave infrared, but the high-power emission is dependent on the chemical interaction forming excited hydrogen fluoride or deuterium fluoride directly during rapid gas flow to give the required performance. Hence, there is a significant issue with efflux management as sealed operation is not believed to be possible at the very high rates of gas-flow required to achieve high power emission; moreover, these devices are very large.

Carbon-dioxide lasers are well established as high-power devices for out-of-band applications; however, these devices are not small and require a relatively large optical aperture in the beam director to achieve the requisite beam size at the target. It was a gas-dynamic laser that first achieved the 100-kW milestone and demonstrated the potential utility of laser-weapon systems.

7.18 Tactical High-Power Laser-Weapon Mass and Volume Estimates

There has been some analysis of the mass and volume of high-power laser systems. This analysis suggests that these state-of-the-art devices have a power to mass factor of about 0.042 kW/kg, sometimes referred to a specific emission factor, and relative to volume the figure is 0.015 kW/litre. The equivalent values for the power supply are 2 kW/kg and 1.4 kW/litre. Hence, a 100-kW laser system will have a mass of about 2.5 tonnes and a volume of 6700 litres for the source and the power supply could add a further 500 kg and have an additional volume of about 700 litres, assuming an efficiency of 10%. Consequently, excluding the beam-director system, the mass of the system would be about 3 tonnes and occupy 7,400 litres.

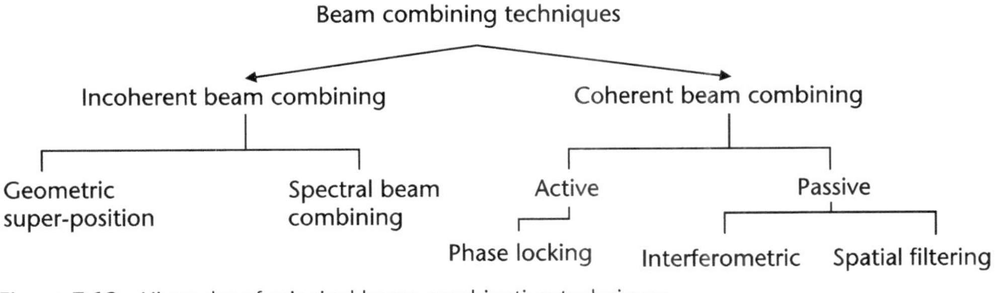

Figure 7.12 Hierarchy of principal beam-combination techniques.

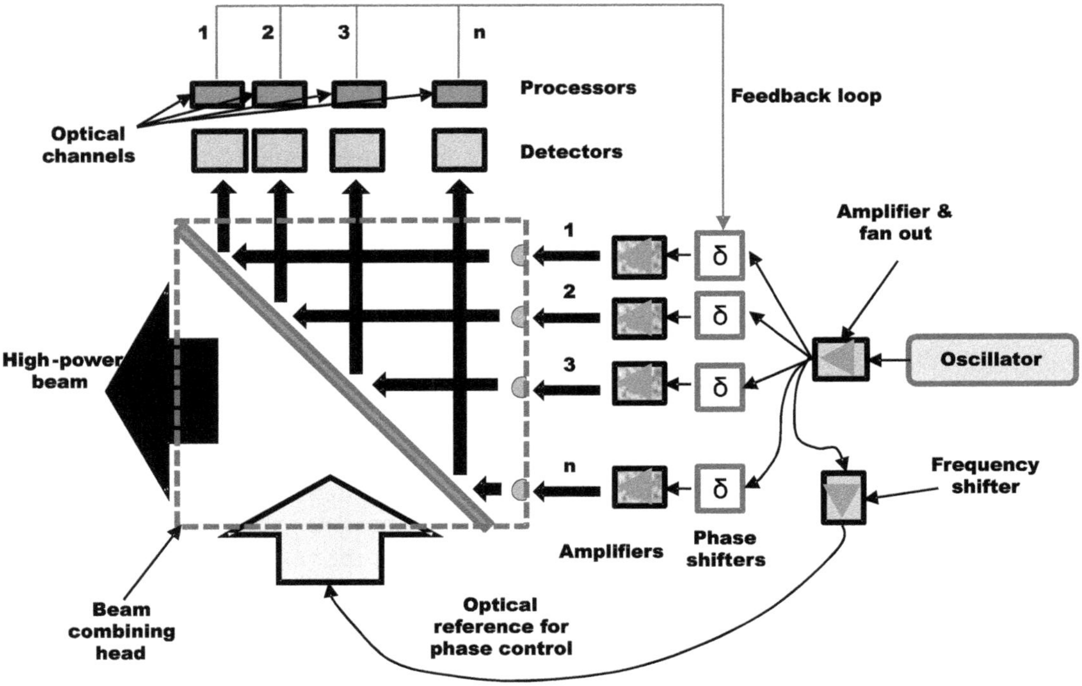

Figure 7.13 Schematic of a fibre-bundle laser concept. (Courtesy of QinetiQ U.K.)

There is an ambitious American Defense Advanced Research Project Agency (DARPA) research project to investigate techniques that will increase the laser power emitted from a kilogram of source device by a factor of five; the aim is to demonstrate 0.2 kW/kg and the volume factor target is 0.2 kW/litre, an enhancement of an order of magnitude. Moreover, there is a further programme to increase the efficiency of the laser-generation process through the design and demonstration of highly-efficient diodes. Therefore, the laser-weapon system of the future emitting about 100 kW could have a mass of 500 kg and occupy about 700 litres and have a significantly reduced waste-heat management problem, possibly with as little as three times less waste heat.

In the future wall-plug efficiency in excess of 30% is likely to be quite feasible for a high-power laser-weapon system.

7.19 Tactical Laser-Weapon-System Considerations

The major challenges facing the teams developing solid-state laser devices for this application concern getting the requisite power delivered to the target from a system that is affordable in terms of cost and weight.

The use of shorter wavelengths has a number of attractions, including the possibility of greater transmission through the atmosphere. However, one of the major system implications is that smaller optical elements can be used, compared with those required for longer wavelengths, to achieve the equivalent far-field performance, owing to the Airy diffraction relationship for a circular aperture:

$$\theta = 1.22\lambda/d$$

where θ is the angle or the first diffraction minimum, λ is the wavelength and d is the diameter of the optical aperture causing the diffraction pattern. Hence, a system using a 10-μm laser source requires an optical train an order of magnitude larger than a system using a 1-μm source to get equivalent far-field performance.

7.20 Beam-Director Technology

The beam director system is a critical aspect of the whole system, as mentioned earlier. It undertakes a number of vital functions to ensure the photon energy is not only directed at the designated target, but is kept on the target for the duration of the interaction: the critical factors of energy on target and time on target. The function of the beam director is described in detail in Chapter 4. Fundamentally, the beam director is a closed-loop system used to track a target and direct the laser photons at a designated (tracked) target. This device has to undertake a number of operations to ensure harmony between the various optical axes of its subsystems.

The beam-director technology for in-band damage-class systems is well developed, but for high-performance out-of-band systems, it tends to be bespoke using state-of-the-art subsystems. The design of the optical train of the beam director is another one of the great technical challenges of this subsystem for all laser-based directed-energy systems, but especially for the high-power systems. The principal requirements for the latter application concern ensuring that the optical transmission is as high as possible, with the reflectivity of each mirror, for example, exceeding 99.999%. Small fractions of absorption of very-high-power beams can lead to significant distortion and aberration of the beam, leading to non-optimum beam propagation; moreover, if the absorption is too great, self-destruction will result.

7.21 Laser-Integration Requirements

This is one of the many critical requirements, described in detail in Chapter 4, which needs very careful consideration so that unwanted errors are not introduced into the pointing vector. The laser-integration requirements are relevant to any of the approaches of fitting the laser to the beam director, whether directly with, or positioned remotely from, the beam director interface. Fibre-laser technology offers some interesting opportunities in this respect, as the final amplification may take place in the feed to the beam director for some applications.

Matching the laser-divergence characteristics to the pointing performance is absolutely critical to ensure that the energy-on-target requirements are achieved. This photon energy has to remain on a target for the designed required time period so that the susceptible components may be attacked successfully.

In the case of all precision laser-weapon systems, there will be an optical alignment of the sensing subsystem with the laser-beam axis for the very high precision devices, to ensure that the laser beam axis is accurately aligned to the input axis of the beam director. Moreover, there will need to be periodic checks to ensure that the harmony is maintained; in some systems, this function will be undertaken continuously, as noted earlier. These considerations have to be an integral part of the decision about positioning of the laser source and the impact on system integration options.

7.22 System Processor

The system processor is a computer controlled subsystem of the laser-weapon system, whose function is to control the laser and the beam director during the interaction with the target; the functionality of this subsystem is discussed in some detail in Chapter 4. The very-high-performance laser-weapon systems may have a number of lasers to provide sensing data, which generates critical information for the adaptive-optical (beam-control) system. In this example, the system processor receives these data, undertakes the analysis, manages the information flow and controls all of these functions to provide optimum energy on target.

There are several critical challenges for any laser-based weapon system, whether it is a simple directed-energy system or a strategic-defence system, using megawatt-class emissions, namely:

- Knowing where the target is and where it is susceptible to attack;
- Getting the radiation on to the target and keeping it there;
- Knowing that the interaction has been successful, as some laser-based weapon systems use soft-kill techniques;
- Finding the next target.

The system processor for the high-performance system is likely to have a threat database that will contain information, to varying levels of detail and confidence, about the weapons likely to be engaged by that laser-weapon system. This database will be either within the system processor or managed by it, its function is to enable:

- Identification of target-interaction aspects;
- Identification of target-engagement zones;
- Weapon assignment;
- Identification of engagement parameters and target-interaction requirements;
- Determination of the optimum laser-emission parameters;
- Management of the battle space and engagements so that other assets are not accidentally engaged, often referred to as de-confliction of the battle space.

7.23 Battle-Damage Indication and Assessment

It is vital to know when an interaction between a weapon system and a target has been successful, so that the use of weapons can be optimised and assets either used to defeat another attacker or prepared for another engagement. This philosophy applies to all laser-based weapon systems and, in particular, the release of the beam director to engage another target, because a single beam director is currently only capable of engaging a single target. However, modern laser-based source techniques could enable multiple beams to be formed, similar in principle to a multifunctional-radar system.

A number of techniques may be used to monitor the progression of the laser-induced damage front through a target. Some examples are:

- Analysis of the emission from the target as it is penetrated, such as laser-induced breakdown spectroscopy or a rapid increase in thermal signature;
- Monitoring the motion of the target, such as changes in the sightline angular rate;
- Observation of any changes in the shape or optical characteristics of the target.

The application of battle-damage assessment techniques may require the use of additional sensors, application of new algorithms and take an additional (small) period of time during a laser-beam-target interaction; however, the potential benefit to the utility of the system will be very significant.

7.24 Concept of Operation

The concept of operation needs careful consideration as these are novel weapons and should be deployed and operated in new ways in order to achieve the best results and, therefore, gain most military value from them. In the case of high-power systems, it may be better to attack a threat missile in a forward quarter aspect, as this could minimise thermal-blooming effects and have the shortest path for the laser to burn through to destroy a vital component or subsystem.

For in-band defeat mechanisms, the main application is likely to be point-defence scenarios, although there could be much to gain from careful positioning of the defence assets.

A further critical consideration in the development of concepts of operation of such weapon systems is the tuneability of response, or the fact that the same system can invoke different defeat mechanisms at different engagement ranges. This feature is shown in Figures 7.5 and 7.6 and is summarised for a maritime deployment in Figure 7.14; it is worth noting that the system has greater engagement ranges as the elevation angle increases, owing to reduced extinction of the beam.

7.25 Future Requirements

7.25.1 Out-of-Band Systems

The fundamental requirement for any future laser-based weapon system is to reduce the complexity of the laser source, to enhance reliability and reduce cost. In the case of the high-power, high-precision out-of-band laser-weapon system, there are many

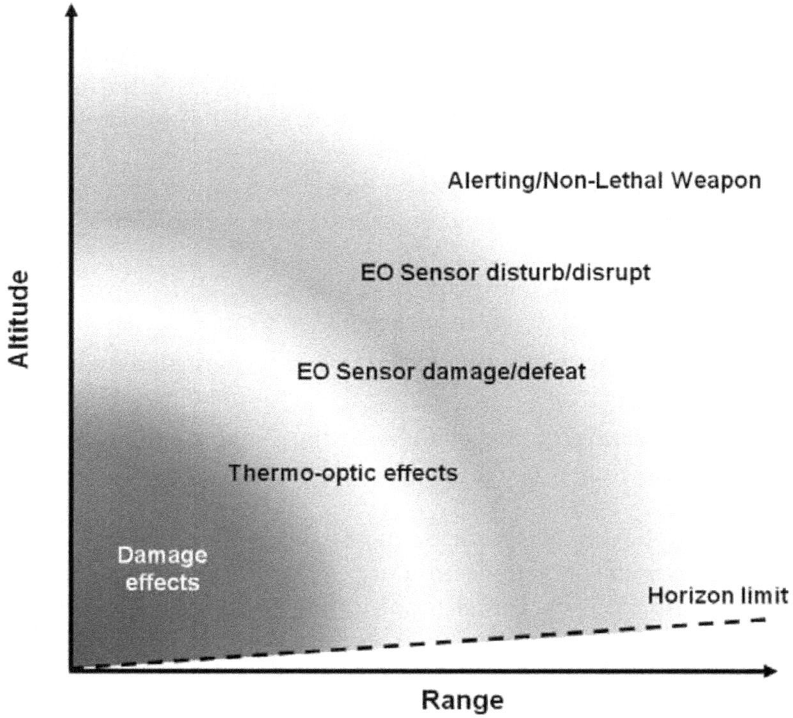

Figure 7.14 Summary of potential maritime uses of high-energy laser weapons.

technical challenges involved in design and development. Some of these challenges are:

- Simplification of the beam director, to achieve lower cost and thus enable greater use of this class of laser system, without any reduction in the pointing and tracking performance (i.e., no degradation of the energy-on-target/time-on-target characteristics);
- Reduction in size and mass of the system without compromising performance, again allowing greater utility and system integration flexibility (basically, the objective in the long term is to move from kilograms per watt of output power to grams per kilowatt from the laser source);
- Significant enhancement to the overall efficiency of the complete system and the high-power laser in particular, along with miniaturisation of the heat exchanger and better dispersal of the waste heat;
- Investigation and development of novel techniques, perhaps with beam profiling and wavelength chirping, to allow the system to operate out to greater engagement ranges without the normal limitations of the atmosphere capping the potential engagement ranges;
- Development of battle-damage indication and assessment techniques.

7.25.2 In-Band Systems

In the case of the simpler in-band technique, the future requirements mainly centre on the development of the laser technology to provide very efficient and compact solid-state devices. For the visible-band source, there is a well-defined need for an efficient solid-state dye-laser source that moves away from flashlamp-pumping of the liquid gain medium to give greater robustness and enhanced efficiency from a compact source.

There is a clear need for significant high-energy source development in the other wavebands as well, such as laser devices that provide efficient and effective direct generation from a compact system in many wavebands of the electro-magnetic spectrum. Again, there is a desire for solid-state sources, so that the benefits of this type of technology may be exploited in the system configuration.

As noted above for the out-of-band system, there is a clear and vital need for rapid and effective recognition that the photon beam has successfully reached its target. This will allow greater confidence in the use of this technique and allow a greater rate of engagement of threats, provided that the battle-damage assessment shows system defeat of engaged targets; otherwise, the dwell times may well increase. However, confidence in the system performance increases in both cases, leading to more robust performance.

7.26 Myths

Any form of revolutionary or novel weapon technology raises a range of emotions and laser-damage weapons are no exception. Many myths have been established over the last four or five decades and are usually promulgated by the detractors of this technology. Some of the more common examples are:

- Instant vaporisation of a target is the only defeat mechanism.
- A hand-held ray gun will vaporise a distant target.
- Megawatts are needed for any viable system.
- They are weapons of mass destruction capable of creating a nuclear catastrophe when the beam interacts with a target.
- Space-based lasers are omnipotent; there is no hiding place from a laser in orbit around the Earth.
- Analogous to the Death Star in science fiction, the most powerful lasers are so powerful that if they went out of control they would vaporise the Earth.
- Lasers are complex. (The physical science is complex, but understood. Avoid white coats; the design does need to be robust.)
- Lasers cannot defeat head-on targets. (What about in-band damage of seekers?)
- They are only remote welding and cutting systems.
- Lasers will not fly. (However, the laser system certainly can fly.)
- Lasers will not work in the rain or over the sea. (That depends on the wavelength of the beam, but additionally, there are many advanced sensors that

have their performance limited by adverse weather conditions, including microwave devices.)

- Beam is visible from the side. (In some circumstances that is true, but not in the infrared.)
- Plasmas are bad. (This is true sometimes, but not always.)
- There are no battle-damage assessment opportunities. (This feature is not limited to directed-energy weapons; in fact, there are techniques for assessing battle-damage effects induced by lasers.)
- It is not permitted by the Vienna/Geneva protocol. (The Vienna protocol is a specific case of designing a laser whose sole purpose is to blind.)
- Countermeasures (counter directed energy) can be implemented simply by positioning a mirror to redirect the beam, polishing the skin of the missile/target or wrapping the missile/target in a shiny film. (They would all be very interesting systems with very significant technical challenges.)
- Radio-frequency countermeasures predate optical countermeasures. (The RF version only lags behind by about two millennia.)

However, the enthusiasts for this advanced weapon technology would argue that directed-energy systems have no barriers.

7.27 Summary

The high-power laser-weapon system is the most technically demanding of all the laser-based weapon applications. The out-of-band approach demands:

- Very efficient, high-power laser technology;
- Extremely high precision beam-pointing and target-tracking from the beam director in the microradian class;
- Very high optical performance from all of the elements in the optical train to give extremely high transmission;
- Complex optical and beam-control techniques to ensure that the optimum energy-on-target characteristics are achieved.

All laser-based directed-energy weapon systems offer many advantages over conventional strike (kinetic) weapons in certain engagements, as this type of weapon can be used in a number of roles so the technology is agnostic to the application; moreover, it offers the potential for clinical strike and also demonstration of the escalation of intent using the same system. Additionally, as a laser weapon does not leave any additional debris on the battlefield, it may be considered a green weapon.

Laser-based directed-energy weapon systems have some operational characteristics that are complementary to conventional strike weapons and therefore offer the potential for a very robust defence when used as an adjunct to an anti-air system. However, these systems do have particular limitations.

Laser-based directed-energy weapon systems offer the potential for some significant advances in military operational flexibility. The same system may be used for both defence and attack operations, and for engaging all aspects of a threat target, whether stationary, on the move or flying and undertaking evasive manoeuvres.

Generally, targets that pass across the field of view of a laser-weapon system are easier to engage efficiently and destroy than those approaching head on for out-of-band systems.

In the case of an in-band approach, this technique requires the sensor under attack to be looking at the laser-based directed-energy weapon systems for a sufficient period to enable this sensor to be engaged, which could be of the order of a fraction of a second.

Much of the technology required to support the in-band directed-energy systems is well advanced; the technology for the high-performance systems is developing rapidly. The recent developments with high-power fibre lasers offer the prospect for a very compact and efficient laser source for laser-based directed-energy systems.

Selected Bibliography

Anderberg, B., and M. L. Wolbarscht, *Laser Weapons: The Dawn of a New Military Age,* New York: Plenum Press, 1992.

Basov, N. G., et al., *Chemical Lasers,* New York: Springer-Verlag, 1990.

Beason, J. D., *The E-Bomb: How America's New Directed Energy Weapon Will Change the Way Future Wars Will Be Fought,* Cambridge, MA: Da Capo Press, 2006.

Duffner, R. W., *Airborne Laser: Bullets of Light,* New York: Plenum Press, 1997.

Grunwald, E., D. F. Dever, and P. M. Keehn, *Megawatt Infrared Laser Chemistry,* New York: John Wiley & Sons, 1978.

Hecht, J., *Beam Weapons: The Next Arms Race,* New York: Plenum Press, 1984.

McAulay, A. D., *Military Laser Technology for Defense: Technology for Revolutionizing 21st Century Warfare,* New York: John Wiley & Sons, 2011.

Merritt, P., *Beam Control for Laser Systems,* The Directed Energy Professional Society High Energy Laser Joint Technology Office, Albuquerque, NM, 2012.

Perram, G. P., et al., *An Introduction to Laser Weapon Systems,* The Directed Energy Professional Society High Energy Laser Joint Technology Office, Albuquerque, NM, 2010.

Robinson, S. R., *The Infrared and Electro-Optical Systems Handbook, Vol. 8*, Chapter 5, "Directed Energy Systems," Fox, C. S., (ed.), Bellingham, WA: SPIE Optical Engineering Press, 1993.

CHAPTER 8

Laser-Based Electro-Optical Countermeasures

This type of countermeasure system may be considered to be a targeting denial system and is sometimes known as laser jamming, but is not to be confused with target-deception jamming discussed in Chapter 9. This technique may provide an opportunity to bridge the gap between shouting and shooting during military operations through inducing nonlethal effects to deter an aggressor; hence, an additional term for this use of laser dazzle is laser warning. The term laser warning should not be confused with a laser-warning system, which informs a platform or individual that they are being targeted by a laser beam. In the context of this electro-optical countermeasure system, the laser-warning-system operator is informing an individual, or group of individuals, that they have been detected through the application of a visual stimulus (i.e., a visible-band emission); moreover, the operator of the laser warner will normally use this visual stimulus to determine intent of the targeted individuals or group.

8.1 Introduction

Optical-countermeasure techniques offer a range of methods for defeating a variety of optical sensor systems, including human eyes, by exploiting temporary effects, so the laser-beam fluence is less than the maximum permissible exposure at the target (discussed in detail in Chapter 17). Consequently, these classes of laser-based countermeasure systems are lower-power, directed-energy systems and in this case, laser dazzle is the defeat mechanism in any one of its various forms, as illustrated schematically in Figure 8.1; note that if this figure illustrates the conditions for night-time conditions (i.e., assuming a 7-mm pupil), then during brighter conditions some of these distances will shrink, owing to the smaller pupil size as well as adaptation to the higher background illumination.

Modest-power lasers can disrupt various sensors and optical sights by dazzling the device that receives the energy, even from something as common and simple as a laser pointer. There was a report that during a recent conflict some allied peacekeeping helicopters were not permitted to fly, owing to the perceived threat posed by simple laser pointers being used in conjunction with an aiming device. This class of countermeasure systems may also be operated by directing the laser beam on to

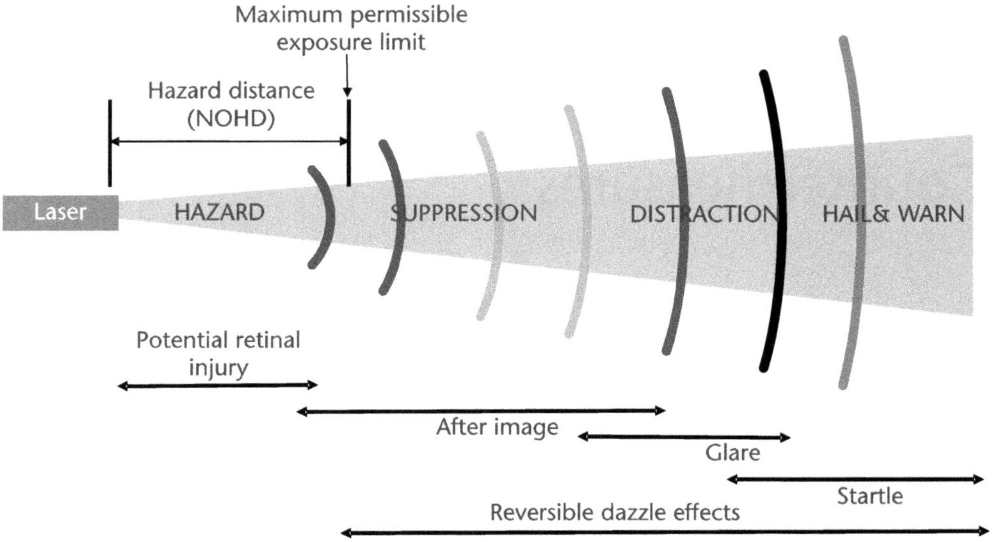

Figure 8.1 Nonlethal effects. (Courtesy of Dr. O. Steinvall, FOI Sweden.)

a target in a manner similar to the immensely powerful laser-based directed-energy weapon systems designed for destroying in-flight missiles, or using much simpler methods; however, that is where the similarity ceases.

For this class of laser-based directed-energy system, the defeat mechanism may be as simple as the distraction of a tracking or guidance system, or in the case of a platform-protection system, the obscuration of a fitted platform through the formation of an extensive dazzle patch within the sensor (i.e., intense stray-light effects). A summary of the potential effects, associated terminology and their potential military utility is shown in Table 8.1, where MPE is the maximum permissible exposure; with some observers discomfort may also be experienced with irradiances close to the maximum permissible exposure, especially in dark conditions.

There have been several well-documented uses of solar energy to distract or deny information to an enemy following the early attempt at a solar directed-energy weapon by Archimedes, described in Chapter 7. Some examples of the optimum

Table 8.1 Examples of Reversible Defeat Mechanisms

Mechanism	*Effects*	*Laser Flux Relative to MPE*
Startle	Discomfort, irritation, but after initial shock little impact on performance	Significantly less than the maximum permissible exposure, well to the right of Figure 8.1
Disability glare	Victim loses contrast in scene, visual acuity is degraded and tracking task may be impeded	Far less than the maximum permissible exposure
Veiling glare	The dazzle patch in the eye or sensor obscures at least part of the target, loss of visual acuity	A significant fraction of the maximum permissible exposure
Flash blindness	After image on retina after the laser is turned off, temporary loss of all visual acuity	Very significant fraction of, or close to, the maximum permissible exposure

use of solar energy, through positioning of attacking forces with respect to the Sun, has had a decisive influence in the result of battles are:

- The battle of Hastings (United Kingdom 1066);
- The battle at Agincourt (France 1415).

During air combat, the positioning of aircraft for an attack direction, with respect to the Sun has been an important tactic for many decades. For example, this aspect has been particularly productive when applied to selecting an approach direction of opponents, to spring a surprise attack by flying out of the Sun. Similarly, anti-aircraft batteries have used search lights to track and distract attacking aircraft.

The general principle is to use a beam of light, with a very high radiant intensity, to cause intensive scatter of photons out of the incident beam at each optical surface, resulting in dazzling of the optical sensors illuminated by the bright beam. The early use of this form of optical countermeasure involved the sensors being dazzled by sunlight to deny information to an opponent: generally, the sensors being dazzled were human eyes. The effect was temporary, as the irradiance of the Sun is of the order of 1,353 W/m^2 at the Earth's surface and when focused on to the retina is below the damage threshold of the eye for short-duration exposures.[1] Moreover, the Sun has a very broad optical spectrum, so it is capable of dazzling a wide range of other optical sensors.

The modern approach is to use a low-power laser with a fairly broad (divergent) beam, which has a wavelength that is in-band to the optical train of the sensor being targeted for attack by dazzling. This form of relatively simple optical countermeasure can be used to deny tactical information to any type of sensor operating in any optical waveband; typical targets for this countermeasure system are battlefield-surveillance and target acquisition sensors, and are relatively easy to defeat because of the in-band scattering. However, in contrast to the use of the Sun, or any other broadband emitting source, a different laser would be required for each waveband of operational interest used by threat sensors, as lasers are currently not broadband emitters, but this may change with the development of ultrashort pulse devices (super-continuum lasers).

The use of the dazzle-countermeasure technique may be considered as a variant of the more conventional denial method, usually associated with smokescreens. In this case, it should be noted that an important difference is that the effect is within the optical train of the sensor, rather than free-space obscuration in the actual scene being observed. Consequently, this approach is not affected by winds or other air-flow phenomena, but severe atmospheric-turbulence effects can lead to some undesired consequences through scintillation and beam wander. The dazzle patch moves with the sight as it moves, provided the laser beam continues to illuminate the sensor.

1. Native Americans allegedly punished their enemies by sewing open their eyelids and leaving them in the Sun all day; the damage mechanisms are likely to be corneal dehydration and keratitis. Also, when observing a solar eclipse, it is recommended not to stare too long at the Sun even with the use of suitable blocking filters/attenuators.

A fundamental requirement of this type of optical countermeasure is that the photon energy has to be maintained on the target sensor for the entire period that the denial effect is required, as the effect is only intended to be temporary and is completely reversible; for some electronic sensors the effects may be instantly reversible once the beam is off, but the human eye may take some time to recover. Consequently, unless a very widely divergent beam is used, which is atypical of laser action, only a single target may be engaged by a beam director at any one time is a robust defeat mechanism is to be achieved.

An early example of an electro-optical countermeasure system was called the Laser Countermeasure System, which used a flash lamp-pumped alexandrite source; however, after a successful prototype was demonstrated, development was suspended owing to concerns about compliance with the Vienna Protocol. Another system was developed in China and marketed as ZM 57.

Recent developments with miniature electronic systems, high-performance, low-cost sensors and high-speed computer systems have fuelled the opportunities for the rapid development of the asymmetric threat by rogue nations and terrorist groups. Some of the asymmetric threats have been developed into very sophisticated and effective systems. For example, an unmanned air vehicle, or remotely controlled aircraft, may be equipped with an array of sensors, which may be used to disrupt or inhibit a range of military activities. There are many other examples of very low-cost, simple systems that are difficult to engage and counter by conventional means.

The use of hard-kill weapons, such as guided weapons, may be impractical, and is certainly not an economical proposition for countering many asymmetric threats. Soft-kill directed-energy weapons offer a possible cost-effective solution. By the very nature of asymmetric warfare, it is necessary to have a reactive system to defeat a threat by frustrating its mission, as a minimum response to an attack. This type of soft-kill weapon system offers a flexible and controlled response option by denying an opponent sensor data and thus situational awareness, whether it is a more conventional threat or of the asymmetrical type.

More sophisticated systems may be developed that will find the threat before it is used, but that is not the subject of this chapter, which will concentrate on the techniques and methods associated with laser-based electro-optical countermeasure systems.

8.2 Military Utility

An interest in the development and application of laser-based electro-optical countermeasures has occurred as a response to the growth of asymmetric threats.

A primary military function of laser-based electro-optical countermeasures is to support peace-keeping activities by, for example, suppression of hostile fire in urban environments and other similar scenarios where innocent civilians may be present. Other examples include the protection of control posts and similar check points by the use of less-than-lethal force and a range of options for countering asymmetric threats. As noted above, counters to asymmetric threats may involve dazzling sensors in systems, such as surveillance sensors or in guidance systems.

Owing to the temporary nature of the induced effects from these low-power systems, the application of this technique may be used as a simple deterrent, or with the purpose of determining the intent or motive of people approaching an exclusion or controlled zone, by illuminating them with a bright light. For example, it may be assumed that if the intrusion is accidental or an innocent mistake, then it is likely that the intruder will leave the area quickly; those with a hostile intent may react differently.

These low-power, laser-based countermeasure systems may be used in a similar way to other obscurants, through the formation of a dazzle patch in the optical system of an observer. The dazzle patch may then be used to mask the platform fitted with the system, which may prevent the accurate tracking of a moving platform, conceal a manoeuvre or change of tactics (i.e., provide additional military capability or a force multiplier). Alternatively, the formation of a dazzle patch within an optical system denies a gunner target-tracking information. More importantly, this type of system buys the commander time enabling tactics to be modified according to the current circumstances, as well as offering a rapid and measured response.

8.3 System Operation

Dazzle-countermeasure techniques may be applied to defeat electro-optical and thermal-band sensors used for surveillance, reconnaissance, target acquisition or target tracking. These functions are often undertaken with very small target signals in a noisy background. Highly sensitive detection systems are necessary and can be susceptible to attack by optical energy, especially with moderate or high radiant-intensity beams, directed into their optical apertures.

The photon flux is usually well below the damage threshold of the susceptible elements in the optical system. Therefore, after a very short recovery period, a sensor system will revert to its normal operational performance characteristics once the intense optical source is extinguished or is redirected to another target. Consequently, it can be considered to be a temporary effect and may be attractive for a number of roles for denying information to an adversary.

A dazzle-countermeasure system directs a very bright beam of photon energy at the objective lens or optical aperture of a sensing system. It is essential that the optical train of the sensor system being dazzled is transparent to the wavelength being used. An example of a warning system is shown in Figure 8.2. The image on the left is the GLOW system unit and on the right it is shown integrated with a gun.

Figure 8.2 The GLOW laser warner. (Courtesy of Thales, U.K.)

The radiant intensity required to be incident on the target, from the countermeasure source, to cause the desired effect varies according to the ambient lighting conditions and the type of sensor. However, a general rule of thumb is that the power density needs to be close to the maximum permissible exposure for the human eye for this countermeasure technique to be effective against man-in-the-loop systems, so that veiling glare or flash-blindness effects are achieved, as indicated in Figure 8.1. This often equates to a 5 to 10-W source with a small divergent beam at range, where the maximum permissible exposure of the eye in the visible band is about 2×10^{-3} J m^{-2} for Q-switched pulses and approximately 25 Wm^{-2} for continuous-wave radiation.[2] In the latter case, the aversion response of the human observer limits the exposure to laser radiation to the case of a single pulse and a duration not exceeding 0.25 second. The envisaged applications comply with the Vienna protocol because the induced effects are temporary.[3]

8.4 Defeat Mechanism

The basis of the defeat mechanism is an in-band, target-interaction technique, described in Chapter 7, and summarised in Figure 8.3 for the dazzle-induced obscuration. The principle of the dazzle defeat mechanism is to create intense optical scatter from each optical surface within the train of an optical system, so that stray light enters the detector array and the optical picture is obscured; this effect is often termed veiling glare. In the case of the human eye, scattered light also originates within the lens, iris and the fluids, rather than principally on the surface of the optical elements. This figure shows the principal refracted rays have reduced intensity and the remainder of the incident light, which has not been absorbed by the optical system or scattered by the front surface, has gone into the scattered light paths that form the dazzle patch.

The development of compact and efficient visible- and infrared-laser technology has created an efficient means for dazzling sensors that operate in those wavebands. These sources offer the potential to be effective because of the inherently large radiant intensity of the beams, which occurs even with low-power emission. This type of beam readily creates intense optical scatter from minor imperfections on the surfaces of any optical component within the optical train of the system and any transparent object encountered along the beam's path. The positioning of the

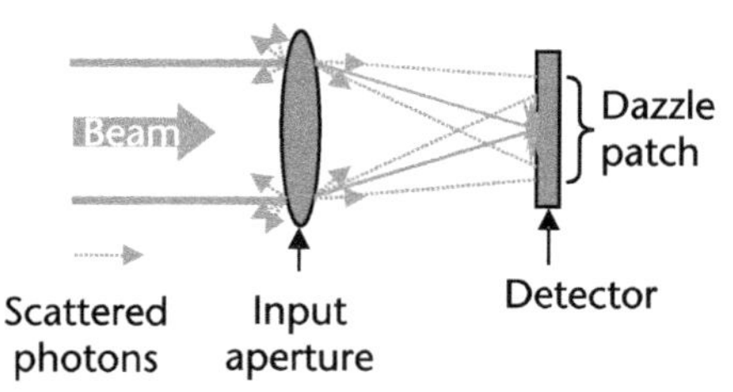

Figure 8.3 In-band dazzle defeat mechanism.

2. These values were expected to be revised in 2014.
3. Defined in Protocol IV, this is the international protocol on blinding weapons and is discussed in Chapter 17.

optical elements in the system, and the supporting structures for these optical components, can create ghost images and striking diffraction-induced effects. These combined effects can result in extensive confusion in the scene (visible or thermal) band being observed, and in the case of a target-tracking system, loss of the tracked object. Some characteristic effects are shown in Figure 8.4, as can be seen from these images, there is a loss of contrast between the target and the background; in each case an identical target is being viewed by a different system.

A subsidiary phenomenon that helps in this defeat mechanism is an effect known as blooming, which occurs when an intense beam is focussed on to a detector array. The detectors within the focused spot fill their wells in the read-out circuits and there is an overspill of electrons into the adjacent wells associated with the corresponding adjacent detectors. Hence, the image of the incident beam appears to be much larger than is actually incident on the detector and denies information about a larger zone of the visible or thermal scene. Note this phenomenon of blooming should not be confused with thermal blooming, described in Chapter 5.

A further aspect of this approach is the thermal heating and subsequent re-radiation from optical components into the optical train. The re-radiation leads to a general reduction of thermal contrast in the scene being observed, as the re-radiated energy act as a noise source. This effect occurs when:

- The optical wavelength of the incident beam is not well matched to the transparency of an optical element, or elements, in a refractive optical train.
- The optical system being engaged has a low transmission coefficient.

This phenomenon has some similarities with kinetic heating of infrared domes on the front of high-speed, infrared-homing missiles. Again, there is a loss of thermal contrast.

8.5 Pros and Cons of Laser-Based Dazzle Systems

The potential attraction of the use of low-power laser-based optical countermeasure systems may be summarised as follows:

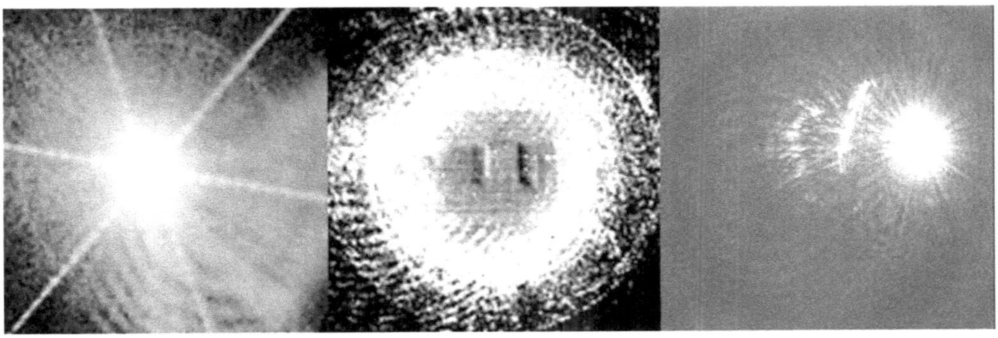

Figure 8.4 Examples of optical dazzle effects.

- High radiant intensity may be controlled to achieve the requisite induced temporary effects, as well as offering an option to match the effect to the ambient lighting conditions.
- High directionality, but the beam divergence may adjusted to ensure that a given radiant intensity is achieved on a designated target within the maximum permissible exposure criterion.
- There is the possibility of a potential for a compact source with very efficient photon generation.
- There is covert emission, owing to the high directionality of the beam, particularly if the emission is in the infrared part of the electromagnetic spectrum.
- There is the potential for tuneability to avoid blocking filters, but this adds complexity, so wavelength diversity may be a better option.
- There is the potential for low cost and high reliability.
- There is potentially an infinite supply of photons as long as the electrical supply operates.
- Simple compact systems are available for a range of operations.

The disadvantages of the laser-based approaches include:

- Usually a single wavelength emission, which may easily be blocked by a filter, provided that the laser-emission wavelength is readily predictable and does not change. The use of a tuneable or diverse emission makes the blocking mechanism more challenging. Super-continuum source could resolve this potentially limiting issue or chirping the laser emission.
- It is not a front-line repair item; the more complex a system is, the further it has to travel for repair, but cheap systems can usually be replaced easily.
- A separate laser wavelength is required for each optical waveband owing to the reliance on an in-band defeat mechanism; multiline emissions from a single source may be able to alleviate this operational restriction but then some of the simplicity is lost, additionally, the source is unlikely to be compact.

8.6 System-Configuration Options

A range of system configurations or architectures are possible from the very simple to the relatively sophisticated, which are comparable with a laser-based directed-infrared countermeasure system described in Chapter 9. Examples are:

- A handheld laser device with a laser and a sighting device with an aiming mark for very short-range engagements, similar to the system described for the laser range finder, or as shown in Figure 8.2;
- A tripod-mounted laser source that has the simple features of the hand-held approach or additional optical aids, such as a co-boresighted telescope,

which would help the manual tracking function and stabilisation of the beam on the target;

- A laser source integrated with a beam director that is linked to a cueing system which is similar to the architecture used in a directed-infrared countermeasure system, considered in the next chapter. The laser may be integrated with the beam director, either internally or externally, through a mechanical interface, or remotely through an optical-fibre link. It is important for either approach to integration, that the laser beam is injected along, and parallel with, the defined optical input axis of the beam director and maintains this harmony of optical axes throughout the operational environment; otherwise, the pointing vector will generate an unwanted random error as the pointing vector moves over the field of regard and negate the boresighting process. The beam director may also have a laser range finder to get a precise measurement of the engagement range to ensure that targets are engaged at a safe distance, (i.e., only induce temporary effects, or to modify the size of the beam to achieve a given fraction of the maximum permissible exposure, which is discussed further in Chapter 17).

8.7 Subsystem Description and Functions

The fundamental subsystems required to form a laser-dazzle system are similar to all other laser-based countermeasure systems, discussed in more detail in Chapter 7, briefly:

- A target-location and cueing device;
- The laser device emitting in the right waveband;
- An optical boresighting system to maintain optical alignment or harmony;
- A beam director;
- A system processor;
- Power supply to activate the laser and any subsidiary and ancillary equipment, each of the subsystems may have their own conditioned power-supply system.

The outline architecture of a relatively sophisticated but potentially autonomous laser-based dazzle-class countermeasure system is illustrated in Figure 8.5 and shows its principal components. In this case, the architecture implies that it is designed for an aircraft application, so there will need to be a target-tracking system, as well as a stabilisation technique applied to the beam director's pointing element, shown here as an acquisition pointing and tracking system (APT). The target-cueing information will be from the threat-approach-warning (TAW) sensor subsystem, which may be an array of staring or scanning electro-optical sensors.

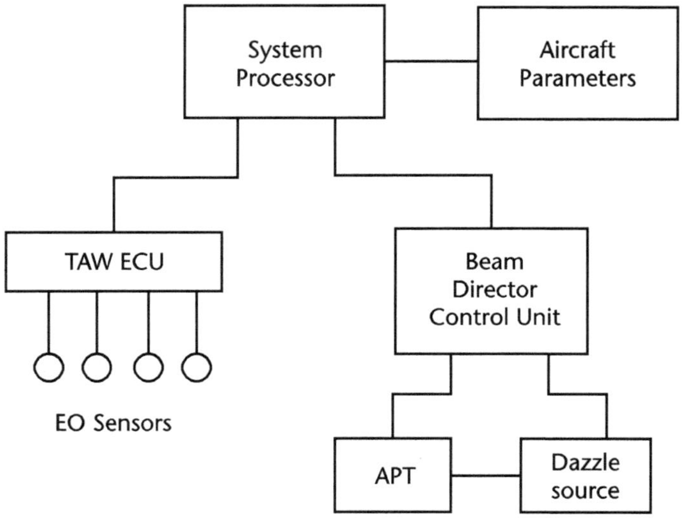

Figure 8.5 Outline system architecture.

8.8 Target Cue

A range of techniques may be used to provide details of the position of the threat to be countered using the dazzle-class laser emission. With the simplest system, the cueing technique would most likely to be through normal situation-awareness methods, using well-established observation techniques, such as use of binoculars for a sector search. Alternatively, a more sophisticated technique may involve a thermal camera linked to a scanner that searches and then interrogates each of the potential targets to establish whether there is an optical sight, using the cat's eye effect (or retro reflection), discussed below. The most sophisticated approach would, involve using a mechanical beam director, probably having a dedicated threat-warning system, such as an electro-optical staring sensor array. This is likely to be configured in a fashion almost identical to the missile-warning sensor system, described in Appendix C.

The critical requirement is to ensure that the target cue provides accurate target-bearing information to achieve a rapid handover from the cueing sensor to the target-tracking system. This function needs a careful match between the cueing accuracy and the field of view of the tracking sensor to ensure a high probability of success as discussed in Chapter 7.

8.9 Laser-Source Characteristics for Electro-Optical Countermeasure (Dazzle) Systems

The fundamental requirements of the laser emission for this countermeasure approach are:

- The countermeasure source is wavelength-matched to the optical passband of the threat sensor, but there are not any special requirements on spectral purity and avoiding chirping of the wavelength during the emission. The use of a source that changes its wavelength (frequency) could be advantageous, as it makes it more difficult to block with protection measures (as noted earlier).

- Temporal characteristics should at least be matched to the frame rate of the sensor, so that it is not necessary to use a continuous-wave laser emission unless the frame rate of the sensor is extremely high. In practice, it is normal for the pulse-repetition rate to be somewhat larger than the frame-rate of the sensor. Furthermore, the pulse duration should be much larger than the integration time of the sensor, so that the photons enter the sensor during the period it is observing a scene. The use of continuous-wave emissions has some advantages for laser-safety and also for mitigation of atmospheric turbulence effects.
- Spatial characteristics are not normally demanding, as the intention is to direct a bright beam at the target area, but not to create permanent damage to the sensor. Relaxing the beam-divergence parameters means that the laser beam is easier to maintain on a target, although this can increase the laser average-power requirements, which has an impact on the distance from the aperture where the beam is eye safe and the size of the power supply.

Average laser-power requirements are in the region of about 5 to 10W for successful dazzle of visible-band and infrared-band electro-optical systems, with a beam divergence of around a few milliradians for longer-range engagements. However, it is important that the beam divergence is matched to the pointing stability of the beam director, to optimise energy on target (i.e., time and energy-on-target) characteristics, which is vital for this countermeasure technique to achieve its intended function.

8.10 Summary of Laser-Source Requirements

The emission characteristics are probably the least demanding of all the countermeasure-application requirements, as outlined earlier. A compact, efficient and reliable source emitting either a continuous-wave or a pseudo-continuous-wave beam that is in-band to the sensor system being countered is all that is required.

8.10.1 Laser Source Options: Visible Band

A number of source technologies are available that would fulfil this requirement:

- Solid-state devices with direct generation are possible in the visible band from chromium-doped hosts such as lanthanum strontium aluminium fluoride (LiSAF) as well as ruby; it is more common to use frequency-doubled emissions from lasers such as neodymium-doped yttrium aluminium garnet (Nd:YAG) or titanium sapphire, which give green and blue emissions, respectively. It is also quite common to use the frequency-doubled emission from Nd:YAG with a wavelength-conversion device, such as a dye-doped polymer, or Raman cell, to produce other colours; these device approaches offer tuneability and wavelength diversity of the colour of the output beam. A rapidly evolving technique is to use semiconductor-diode devices, which provide direct generation in the visible band. Another rapidly emerging solid-state technology is the semiconductor disk-laser devices, usually based on multiple quantum-well structures, in a vertical external-cavity surface-emitting laser (VECSEL).

- Liquid-phase sources include devices such as the flashlamp-pumped dye laser, which will give direct generation in the visible and an output that is tuneable, as well as the option of wavelength diversity.
- Gas-laser technology, such as those devices that use ionised noble gases, for example, krypton-ion or argon-ion sources. Other options are metal-vapour lasers, such as the copper-vapour device. The major drawback with these devices is their lack of efficiency, and their bulk.

Examples of the performance of a range of small solid-state sources freely available through the Internet are shown in Figure 8.6. These devices are often termed handheld devices, owing to their size; more powerful sources are readily available with emitted power in the region of 20W, but currently these lasers would not be categorised as handheld sources.

8.10.2 Laser Source Options: Near-Infrared and Short-Wave Infrared Band

There are a few solid-state laser options that have the appropriate emission characteristics along with direct generation, such as the neodymium, erbium and thulium devices, including fibre lasers. Several laser diodes also emit in these wavebands,

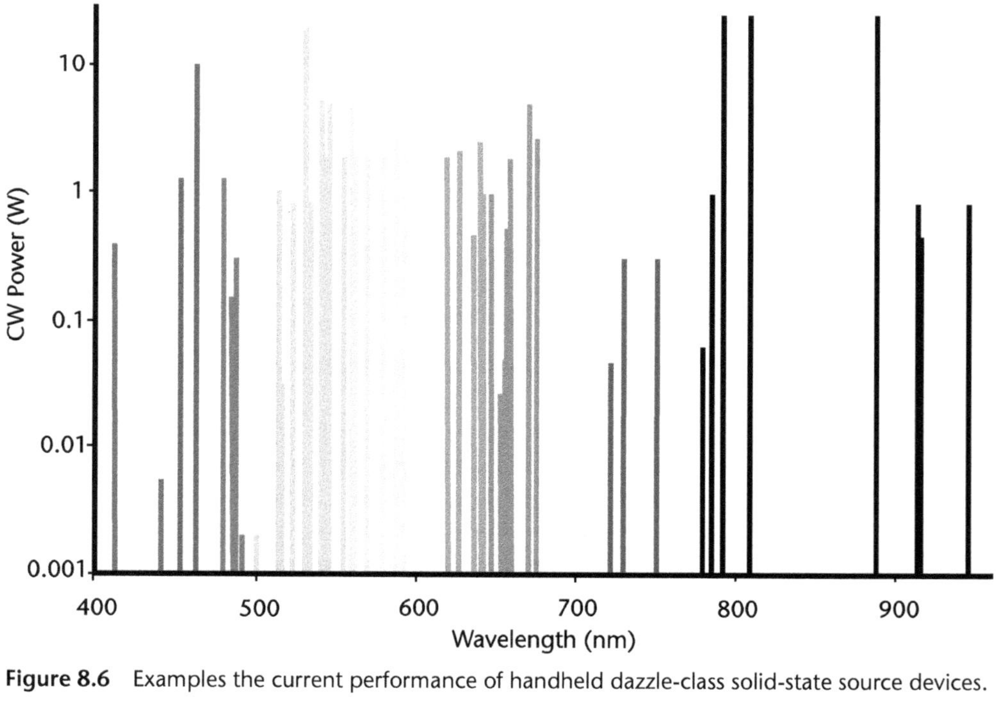

Figure 8.6 Examples the current performance of handheld dazzle-class solid-state source devices.

and moreover, so do some of the semiconductor-disk lasers. Examples of the performance characteristics of such small devices are included in Figure 8.6.

8.10.3 Laser Source Options: Mid-Wave Infrared Band

There are a limited number of sources that emit directly in this band:

- Conventional solid-state sources generally require wavelength conversion to the longer wavelengths of the mid-wave infrared region. A number of hosts, using the holmium or thulium ions or co-doping with holmium and thulium, provide an emission close to 2 μm, and erbium has an emission close to 3 μm. This emission can be shifted to longer wavelengths with a variety of nonlinear crystals such as the periodically poled materials or the bulk crystals, such as potassium titanyl phosphate in an optical parametric oscillator. Fibre lasers using similar rare-earth laser ions have generated efficient output around 2 μm, suitable for wavelength conversion. Additionally, direct generation from semiconductor diodes has been demonstrated at around 2 μm and this output may be converted to longer wavelengths using similar materials containing multiple quantum wells, or optical-parametric techniques. Recent developments with semiconductor technology have seen the demonstration of quantum-cascade lasers providing direct generation in various parts of the mid-wave infrared waveband with a wavelength greater than about 3.5 μm. Vertical external-cavity, surface-emitting laser (VECSEL) technology also provides an in-band emission across this part of the spectrum.
- Gas-laser technology offers several approaches, such as hydrogen fluoride or deuterium fluoride in the gain medium to generate broad-band emissions around 2.7 μm or 4 μm, respectively. These devices may be operated as electrically-excited sources that are sealed, despite the fact that they were originally designed as flowing gas chemical lasers. Other options for the gain medium are barium vapour or a mixture of helium and xenon. Frequency-doubled carbon-dioxide devices are candidates for the longer-wavelength end of this waveband. However, these laser devices are not small.

8.10.4 Laser Source Options: Far-Infrared Band

There is far less choice for direct emission around 10 μm; laser-beam generation in this waveband is limited generally to gas-laser technology and, in particular, to carbon-dioxide technology, which with special formulations, can cover a band from about 9 to 11 μm. There are other gases such as carbonyl sulphide that emit closer to 8 μm. Relatively efficient, mature and compact low-power carbon-dioxide lasers are readily available for this type of application.

Solid-state devices are dependent on non-linear techniques such as optical parametric generation; a nonlinear crystal well matched to this sort of emission is the optically patterned gallium arsenide material. Semiconductor lasers may be a practical source in the near future.

8.11 Beam Director

A range of beam-director technology approaches may be used to provide the target-tracking function and the pointing function to ensure energy-on-target requirements are achieved, as indicated in the system-configuration section above. The simplest approach is a sighting device co-aligned with the optical axis of the laser, such as an iron sight or telescope co-boresighted to the laser beam, as used in Figure 8.2. More sophisticated systems, particularly those fitted to moving platforms and/or intended for operation over a kilometre or more will require a sightline-stabilisation system, as well as a target-tracking device to keep the energy on target.

8.12 Laser-Integration Requirements

As usual, the axes of the laser and the optical pointing system must be coaxial and colinear for this application, although the absolute requirements are generally somewhat relaxed compared with the high-performance laser-based directed-energy weapon systems considered in Chapter 7. The critical requirement to match the beam divergence to the pointing stability of the beam director remains valid to achieve the requisite energy-on-target and time-on-target characteristics. Consequently, the simplest unstabilised systems may require a very wide beam to achieve the optimum energy-on-target requirements, which will then demand a more powerful laser emission to achieve the necessary radiant intensity to generate a dazzle patch or effect with suitable physical extent. There is a need for a system trade-off study and then system optimisation during development for each application.

This type of system may be configured in a number of ways, for example:

- The monolithic approach with the laser source mounted directly on to the beam director, or integrated within it, for the lower-power systems;
- A distributed configuration with the laser positioned remotely from the beam director, and linked to it, by an optical fibre or similar light-guiding component.

The monolithic approach is more likely to be used with the simpler configurations, although the remote positioning of the laser source offers a number of advantages in system design flexibility and integration options with platforms. Generally, a distributed system is easier to integrate with an existing platform and may have some advantages in maintainability and operation.

8.13 System Processor

The simple systems will use the human operator to provide the system-processor functions, such as the target-tracking processes and pointing the laser beam at the target. In this case, the operator will have full control of the engagement process from finding the target, tracking it, forming a fire-control solution, pointing the laser beam and assessing the effectiveness. The laser system may be mounted on a simple stabilised platform to provide some additional stabilisation of the beam, which may also be under the control of the human operator.

The more sophisticated laser-based electro-optical countermeasure system that is integrated with a beam director will require a system controller and will have a similar architecture to the laser-countermeasure system described in Chapter 9. Its principal function will be to undertake the command and control functions of the target interaction process and, where necessary, control the laser emission to ensure that the laser flux at the target does not exceed the maximum permissible exposure if a human eye is believed to be a part of the sensing system. The latter activity ensures that any effects are of a temporary nature and fully compliant with the usual interpretation of Protocol IV, but this protocol has not been universally accepted by all nations. This system may have sensors that assess the effectiveness of the interaction with the target's optical system, for example, by monitoring the retro-return signals.

In the case of the more sophisticated system, there will be at least a rudimentary built-in test system, similar to the one described for the more powerful directed-energy system.

8.14 Target Tracking and Stabilisation

The use of the retro return from a target may be used to provide a tracking source, but this means that the beam director has to have a tracking sensor, such as a camera, which is sensitive to the wavelength of the retro return. The cat's eye effect, or retro reflection, occurs when an optical system is illuminated with a well-collimated beam and the sensor has its optical axis approximately aligned with the propagating direction of the beam.

Provided that the incident beam is in-band to the optical train of the system being illuminated and this system has at least one optical element in a focal plane, then some photons will be scattered in the direction of the illuminating beam (i.e., back scattered to the observer). Most optical systems have optical elements in the focal plane and components, such as graticules, which give rise to a significant optical cross-section (i.e., a large retro return). The principle is illustrated in Figure 8.7 and the retro return from binoculars is shown in Figure 8.8.

A single- or dual-band tracking system for an effective countermeasure system would be quite feasible to support the retro-technique, using a dichroic optical

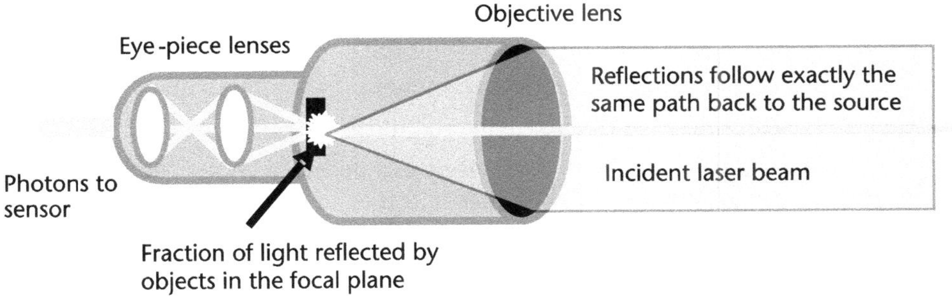

Figure 8.7 Retro principle and optical return.

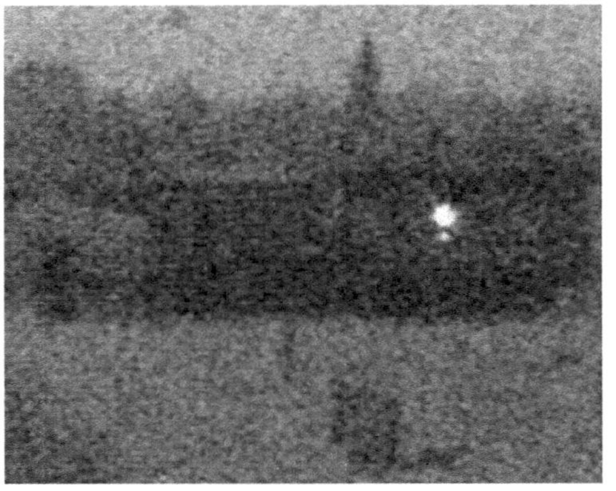

Figure 8.8 Retro return from binoculars.

element to form the two independent optical tracking channels, to provide photons to suitable imaging cameras for each waveband. More than two separate channels would lead to quite a complex optical system, which would be bulky and largely incompatible with the fundamental ethos of this type of countermeasure approach, based on simplicity and low cost.

An alternative targeting technique is to use a geo-location approach, where the target point is defined on a suitable map of the engagement zone and the position and orientation of the platform is known in the same reference frame. The latter data would have to be derived from a full inertial-navigation system on the platform. This information on the target point and platform position would enable a sightline vector to be defined and the dazzle beam stabilised to that defined bearing, using suitable inertial sensors, target-tracking algorithms and sightline-stabilisation elements, such as fast-acting (fast-steering) mirrors.

8.15 Future Requirements

The future requirements for laser-based electro-optical countermeasure system used to create temporary effects, such as veiling glare or flash blindness, largely depend on development of the characteristics of the laser source to achieve the desired effects, and thus, provide enhanced system effectiveness, as well as improved system utility. The typical enhancements that would be beneficial in addition to a more cost-effective source technology are:

- More compact and simpler technology;
- Greater electrical-to-optical efficiency;
- Easier control of fluence delivered to the target;
- Wavelength diversity or tuneability to defeat blocking filters and other counter-countermeasure techniques;
- Simultaneous multiple-wavelength emissions;
- Device technology that is more robust to the operational environment;

- Device and system technology that is easily integrated with other systems and operational platforms.

8.16 Summary

Laser-based electro-optical countermeasure systems are an example of a laser-based directed-energy weapon system. In this case, through the use of modest-power lasers of the order of a few watts, a dazzle patch is formed in an optical system being illuminated by a laser beam, which obscures at least part of the field of view. The defeat route is based on the in-band technique, and causes a temporary effect; the dazzle patch disappears when the laser is turned off or when the optical system moves out of the beam, although some sensors do have an after-image which decays as a function of elapsed time after the illumination ends. The physical mechanism for the formation of the dazzle patch is optical scatter of the photons out of the beam from imperfections on the various surfaces of the optical elements within the targeted system.

A number of system configurations are feasible ranging from a very simple handheld device, to ones involving a relatively sophisticated mechanical beam director in an autonomous system, which may be used in many applications including platform protection. In the former case, the beam director may involve a very simple sighting device, arranged so that the operator can track a target and point the laser beam at a target. A more sophisticated approach is likely to be operated in a similar fashion to a laser-based directed-infrared countermeasure system, discussed further in Chapter 9.

The laser-integration requirements are consistent with other laser-based systems; the divergence of the beam and the pointing-stability characteristics must be matched and the axes of the laser beam and the pointing system need to be harmonised and remain so, over the operational environment. Several architectures are feasible involving monolithic integration of the laser with the beam director, or a distributed system. The latter approach usually offers greater flexibility in the system integration options.

This type of system is likely to be used for a number of military roles, to deter an intruder without the need for the use of more lethal weapons; it is sometimes considered to fill the gap between shouting and shooting. This technique may have particular value in urban operations as the effects are temporary. Laser-based, electro-optical countermeasures may also be used as an obscurant by using the dazzle patch to mask a target. This technique has some similarities with the use of smokes and obscurants, but is not blown away by the wind. Alternatively, the formation of a dazzle patch may deny a tracking system the feedback required to achieve an effective interception or to create a fire-control solution.

This class of low-power, directed-energy system is designed to generate temporary effects (i.e., fully reversible). Such countermeasure systems are not normally restricted by Protocol IV of the Vienna Convention concerning the control of blinding weapons.

Selected Bibliography

Cornsweet, T. N., *Visual Perception,* New York: Academic Press, 1970.

Dando, M., *A New Form of Warfare: The Rise of Non-Lethal Weapons,* London, U.K.: Brassey's, 1996.

Driggers, R. G., and M. H. Friedman, *Introduction to Infrared and Electro-Optical Systems,* 2nd ed., Norwood, MA: Artech House, 2012.

Lewer, N., and S. Schofield, *Non-Lethal Weapons: A Fatal Attraction?,* London, U.K.: Zed Books, 1997.

Titterton, D. H., "The Development of Infrared Countermeasure Technology and Systems," in *Mid-Infrared Semiconductor Optoelectronics,* A. Krier, (ed.), New York: Springer-Verlag, 2006.

Titterton, D. H., and J. L. Weston, *Strapdown Inertial Navigation Technology,* 2nd Ed., Reston, VA: The IEE/American Institute of Aeronautics and Astronautics, 2004.

CHAPTER 9

Laser Directed-Infrared Countermeasures (Laser Jammers)

Laser-based infrared countermeasures (IRCM) have been in active service for some years now and represent the most common application of laser directed-energy-weapon technology. The primary function of this technology is platform protection through robust jamming and defeat of heat-seeking anti-air missiles. The interaction between any jammer and the target-tracking system in a missile is a critical consideration for the success of this technique; the historical development of infrared-seekers and the related enhancement in performance of jammer technology and countermeasure techniques is given in Appendices A and B.

9.1 Introduction

The requirement for the development and perfection of infrared countermeasures is a direct consequence of the extensive proliferation of infrared sensors, and particularly optical seekers used in missile systems, which may be used to attack many different types of military and civil platforms, such as aircraft. These sensors and systems have been shown to be very effective and have provided additional military capability to regular, as well as irregular, forces. One particular example is the continued development of heat-seeking missiles, which may be considered to be the ultimate asymmetric threat weapon, as some types are relatively cheap and have been shown to be extremely capable by destroying sophisticated modern, and therefore, very expensive aircraft.

The effectiveness of heat-seeking missiles is now without question. The man-portable air-defence weapon system (MANPADS) is easy to assemble from its constituent parts and convenient to transport covertly and can be made ready to fire within a very short period, in some cases under a minute. The target-engagement process has been developed so that it is relatively simple; it consists of finding a target, tracking it and then launching the weapon. The guided missile is autonomous after it leaves its launcher. Consequently, the potential covert nature of these weapons, which are passive, their inherent mobility and transportability mean that tactics of flying around the threat (i.e., avoiding the missile-engagement zone) is often not feasible other than flying at high altitude to avoid the missile-engagement zone.

However, that latter alternative may simply lead to engagement by a different type of weapon system, especially when encountering an integrated air-defense system.

There are many statistics quoted informally concerning the number of aircraft losses that may be attributed to this class of weapon since its introduction into service during the 1960s. Some statistics suggest that heat-seeking missiles have been responsible for more than 80% of all combat aircraft losses over the last 40 years. Figure 9.1 shows a summary of the documented incidents over a 35-year period to the 1990s. Recent conflicts will only add to the percentage losses attributed to the asymmetric weapons, although the most recent conflicts have seen an increase in successful engagements by unguided threat weapons. An efficient and effective countermeasure system is required to assure adequate platform protection.

The rise of world-wide terrorism and the proliferation of these simple weapons have only served to reinforce the need for more robust and effective infrared countermeasures. The requirement for effective countermeasure systems has promoted the research into the science underpinning the behaviour of infrared photons in the atmosphere, their interaction with materials and the efficient generation of these photons. The effectiveness of countermeasure techniques has been boosted by many technological developments, but in particular, by the design and demonstration of compact laser sources emitting in the mid-wave infrared part of the spectrum, which has been considered in Part I.

An effective countermeasure system may be a reactive system, which responds to the detected, or known, presence of a threat. In this case, it would undertake some action such as the emission of photons, with a wavelength in the infrared, to disrupt the performance of the threat system or weapon. In this case, the platform-protection system will need a subsystem, such as a missile (approach) warning system to provide an alert about an approaching threat.

The countermeasure system may also be used in a preemptive (or proactive) mode, so that the threat weapon system is unable to acquire a suitable target, although the target (victim aircraft) may have been detected by the threat system. The latter approach is very attractive, but some additional information is required about the target's position, and an indication of operational status, to achieve this goal with a directed-jammer system, which inevitably leads to a more complex countermeasure system. Owing to the short engagement timelines, the protection system may not have the requisite reaction time to achieve a suitable suppression of

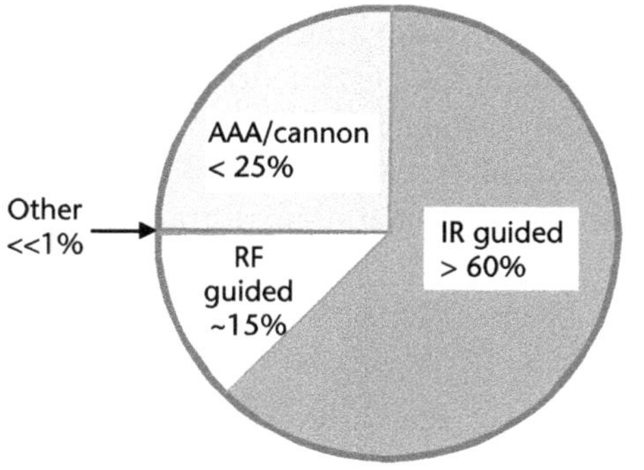

Figure 9.1 Aircraft losses over a 35-year period.

the threat. It is worth noting the flare-based systems (thermal decoys) may be used in a preemptive role quite easily.

In general, the classes of infrared-countermeasure systems discussed in this chapter are considered to be soft-kill weapons, because of the interference or disruption of the normal operation occurring within some system or subsystem in the target. A common target for a jamming system is the target-tracking system in an anti-air weapon causing it to appear to malfunction by tracking a more attractive target and thus failing to intercept its intended victim. This is a consequence of the generation of false guidance commands being sent to the control system via the jamming beam, because of the receipt of the modulated jam waveform by the missile's seeker.

This in-band approach offers a highly effective and cost-effective alternative to the use of bullets, high explosive and/or fragments from a warhead to destroy an in-flight missile. The principle underpinning any successful infrared-countermeasure system is the application of any technique that degrades the notional performance and normal mode of operation of sensors or a system (i.e., a system malfunction).

An alternative approach to defeating heat-seeking missiles is the use of pyrotechnic devices such as flares. These off-board stores are used to decoy a seeker by providing a bright infrared source for the heat-seeking missiles to track, as an alternative to the thermal emission from the aircraft. This chapter will only consider on-board systems (jammers) and the development of laser-based directed infrared countermeasure systems, which has led to the availability of robust platform-protection techniques. The general development of optical seekers and jammers is considered in more detail in Appendices A and B.

Figure 9.2 shows a schematic of a range of countermeasure techniques used to provide platform protection, supported by subsystems providing situational awareness, including approaching threats.

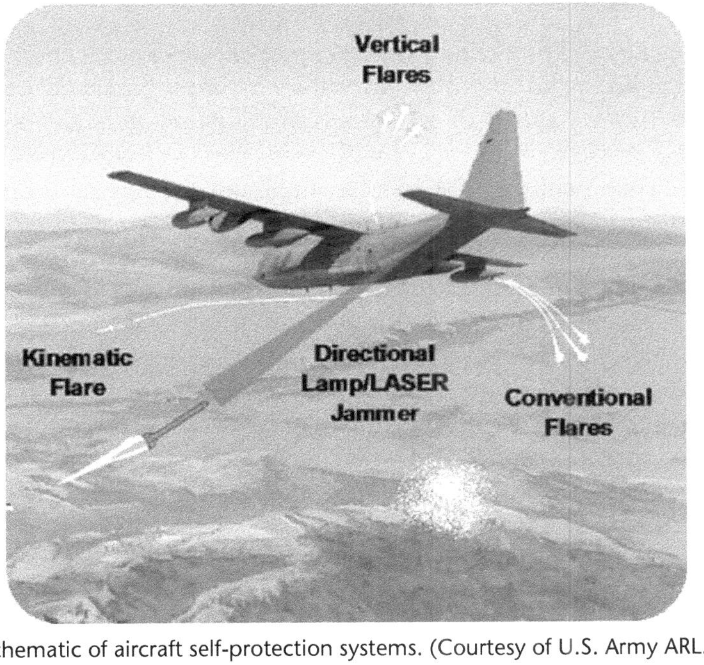

Figure 9.2 Schematic of aircraft self-protection systems. (Courtesy of U.S. Army ARL.)

9.2 Military Utility

This type of system has been designed to provide robust protection when a platform is attacked by heat-seeking missiles. This is normally a reactive system that responds to a cue from a threat-warning system, which forms a situational-awareness system on the protected platform; threat-warning technology is discussed in Appendix C.

The successful development of laser-based directed-infrared countermeasure (DIRCM) techniques has led to the ability of this type of system to defeat a wide range of heat-seeking threats very rapidly; the high radiant intensity of the laser beam provides a robust defeat mechanism. The development of advanced techniques and analytical methods means that a single beam director may defeat many threats in a very short period.

This type of countermeasure system may exploit a number of high-performance sensors to provide an installed platform with additional functionality. For example, the high-performance target-tracking camera may be used for providing additional situational-awareness information. Other devices such as a dazzle-class laser source could be integrated with the beam director to defeat other sensors on the battlefield.

9.3 Historical development

9.3.1 Jammer Operation and Development

Jamming systems are fitted to platforms to provide protection when attacked by heat-seeking missiles. These systems emit a modulated waveform of infrared energy that is in-band to the optical train of the seeker to be defeated. The function of the emitted pulsed waveform is to disrupt the missile's target-tracking system.

The development and operation of reticle-based seekers are described in Appendix B. The fundamental principle of the target-tracking system within the heat-seeking missile is to generate a time-referenced signal from a single detector in the seeker. This signal may be analysed by a simple signal processor within the electronics of the weapon to provide a direct indication of the target's position within the field of view of the seeker with respect to the optical axis of the seeker. Hence, a closed-loop control system is realised and the missile is steered to its acquired target.

The emission of the pulsed waveform, of greater optical brightness than the thermal signal of the platform fitted with the jammer, disrupts the signal-processor logic in the threat seeker. This interaction leads to a change in the amplitude, phase or frequency of the modulation of the waveform generated by the spinning reticle. This disruption leads to errors in the estimation in the tracked-target's position, confusing the target-tracking system in the seeker. The confusion results in the missile's control system demanding erroneous manoeuvres from the weapon to correct the perceived errors in the path to intercept, and consequently, it fails to collide with the target.

The original jammers used incoherent sources, operating with a classical black-body emission, to generate the appropriate spectrum of infrared wavelengths. These hot bodies were usually modulated using mechanical devices to provide the modulated jam waveform. The development of arc lamps enabled more effective pulsed (jam) waveforms to be generated, through rapid switching of the electrical supply to the discharge. Moreover, these jammer devices were omni-directional and so did not require a beam director, but their performance has become limited, owing to the modern developments in heat-seeking device technology.

The basis of the original directed infrared countermeasure systems is to use a collecting optical system enabling the broadly omni-directional emission from an arc lamp to be concentrated into a narrow beam. This focusing of the beam was used to increase the radiant intensity of the jammer beam. An increase in complexity of the countermeasure system was required, owing to the need to be able to locate the threat seeker with a directional beam and keep it on the target during the countermeasure event. The development of incoherent jammers, along with more details on the fundamental technology, is given in Appendix B.

The use of laser sources represents the most recent development of these highly efficient directed-infrared countermeasure systems. The inherent emission characteristics of laser beams, with their high brightness and ease of modulation, make them ideal candidate sources for this application.

9.3.2 Laser-Source Technology

Compact solid-state laser technology, emitting in the mid-wave infrared part of the electromagnetic spectrum, has been developed for directed-infrared countermeasure systems. A number of different techniques have been demonstrated for photon emission into the various seeker-operating bands, which is discussed in detail in Part I.

The great attraction of laser technology is that even very modest power emissions can have a very high radiant intensity, owing to the inherent small divergence of the emitted beam. Additionally, the ability of a laser to be switched to produce waveforms that have complex mark-to-space ratios and pulses of energy that are either off or on (i.e., a very high depth of modulation can lead to a very effective jam waveform capable of defeating even the most sophisticated infrared seekers). This ability to modulate a laser emission, especially its rapid turn-on and turn-off characteristics, means that the modulation ratio is effectively ideal (purely 1 or 0), so the "self S" problem does not exist (i.e., all of the jam energy goes into defeating the seeker and does not add to the signature of the platform).

Many different types of laser technology have been investigated, with an aim of providing the infrared source in a jammer system with highly efficient emission characteristics in the various seeker-operating bands. Ideally, the source device would provide the laser energy into the desired wavebands by direct generation, using a solid-state architecture. Direct generation is normally very attractive because conversion stages, such as optical parametric oscillators, add losses and complexity. This additional loss is a consequence of a variety of reasons, but primarily, because of insertion of extra optical elements into the optical train, combined with the inefficiency of the conversion process, as well as a potential adverse effect on the cost and reliability.

Solid-state laser technology has the characteristics demanded by a countermeasure system, such as being compact, efficient, inert and the potential for a long shelf life, as discussed in some detail in Part I. However, the major deficiency with this technology today is that very few options for sufficiently powerful sources with the appropriate emission characteristics exist. Direct generation is the foundation of many of these requirements, particularly high-efficiency operation, but few source technologies are able to generate the mid-wave infrared energy, around 4 microns and beyond, without using nonlinear conversion methods (i.e., they rely on indirect generation). Nonlinear conversion is not desirable, as the conversion process adds complexity and generally results in loss of efficiency in the operation of the laser source; thus, many of the advantages of using solid-state laser technology are reduced, if not lost.

Several laser-emission lines are usually required to counter the various generations of heat-seeking missiles that could pose a threat. In general, the minimum requirement is to have an emission line in the shorter wavelength part of the mid-infrared atmospheric window and one in longer wavelength part of the mid-wave infrared, as those are the parts of the spectrum where the vast majority of the infrared threat seekers operate. For more robust platform protection, other wavelengths may need to be generated by a laser-based system.

A number of laser technologies suitable for laser-based infrared countermeasure systems are discussed in Part I, including an alternative approach to conversion techniques, which involves semiconductor technology to provide direct generation of the longer-wavelength part of the mid-wave infrared. These rapidly emerging techniques concern semiconductor devices, such as the quantum-cascade technique or the optically pumped semiconductor laser, which are the subject of continuing research and development studies.

9.3.3 Defeat Mechanism

The basis of the defeat mechanism used by any jammer system is to emit photons in the thermal waveband that are modulated to create a series of pulses of infrared energy, which are in-band to the seeker being attacked. This emission is carefully controlled, so that this transmitted waveform of pulses modifies the modulation created by the reticle in the seeker of the heat-seeking missile that is subsequently passed as a modified signal to the detector. In effect, the jammer creates a false target position within the field of view of the seeker, which the target-tracking system attempts to correct immediately, by moving the seeker to position the target back into the centre of the seeker's field of view. This movement of the seeker results in the missile making a specific manoeuvre. Provided the jammer's modulated waveform has the appropriate frequency characteristics, duration and brightness, the threat missile will be prevented from intercepting its target.

As is necessary with other countermeasure techniques, the brightness (or radiant intensity) of the thermal jamming source needs to get the attention of the homing head in the heat-seeking missile (i.e., present a more attractive source of thermal energy to the seeker than the platform that is being attacked). This is achieved by ensuring the radiant intensity of the jammer exceeds the radiant intensity of the platform it is protecting by a significant factor, usually by at least 10, often an order of magnitude more. This relationship is known as the jam-to-signal ratio (or J:S).

Most modern nonimaging seekers can be defeated using a pulsed-modulation emission, known as a spin-by-carrier waveform. This form of modulation has two waveforms superimposed on the jammer emission: a spin envelope closely matched to, but usually not identical to, the spin frequency of the reticle in the threat seeker and a carrier frequency induced by the pairs of transparent and opaque blades of the reticle. An idealised jamming effect is shown in Figure 9.3; in reality, the result is more complex, owing to the motion of the infrared seeker. As shown schematically in this figure, the jam waveform get progressively out of step with the reticle modulation of the real target, leading to false target demands being sent to the seeker and the missile control system by the target-tracking system and the real target disappears from the seeker's field of view as the seeker attempts to centre the apparent target position indicated by the jam waveform. This figure shows the situation as the target moves away from the optical axis of the seeker.

There are two fundamental responses that a seeker may exhibit when illuminated by an infrared jammer, which is largely related to the J:S. In the case of the smaller J:S values or a jam waveform poorly matched to the natural modulation of the seeker's reticle, the response of the seeker's control system is to cause small perturbations of the optical axis of the seeker about the actual position of the target. This type of seeker response is a consequence of the jammer power being inadequate to create a false-target position continuously, over a sufficient period, for the false target to move the seeker over a large enough angle for the optical sightline, from the seeker to its target, to be lost. Instead, the false-target position is lost and the seeker reverts to its actual target, only for the process to be repeated, provided that the jammer is still emitting.

These oscillations of the seeker position are normally chosen to ensure an adverse interaction occurs with the missile's target-tracking algorithm, resulting in the missile making progressively larger excursions about its nominal flight path to its target. This process is often termed guidance-gain degradation.

When the J:S is larger, then the apparent motion of the false-target positions can be sustained across the field-of view of the seeker. As a result, the seeker in the missile continues to move during the illumination by the jammer and rapidly the real target of the heat-seeking missile disappears from the field of view of the infrared seeker, as the jammer drives the seeker's optical axis over its field of regard.

The large value of the J:S enables the jammer's photons to continue to enter the seeker, and interact with its detector, even when the optical axis of the seeker is no longer pointing directly at its target. There is an additional mechanism in operation during this particular type of countermeasure event, which is the optical scatter and reflection of the jammer's photons within the seeker's optical train. In general, J:S significantly greater than 10:1 is required to generate sufficient optical interaction within the seeker and thereby achieve this particularly robust defeat

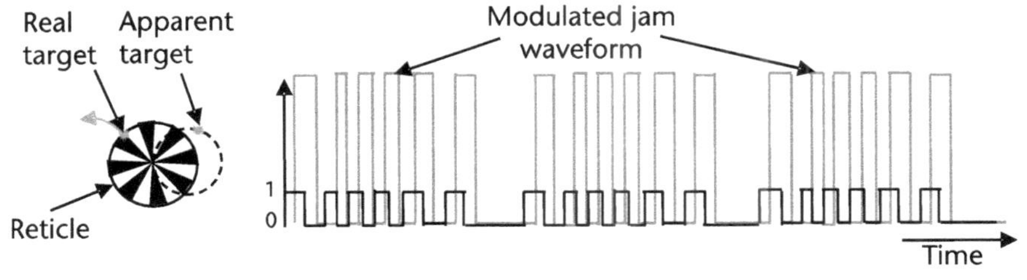

Figure 9.3 Onset of jamming effects.

mechanism, which is known as optical break lock. This is the design aim for a robust platform-protection system.

9.3.4 Jam-to-Signal Ratio

This parameter expresses the advantage that the jammer emission has over the radiated infrared signature of the platform it is protecting, as defined in Figure 9.4. This may be expressed as the ratio of the radiant intensity or the modulation amplitude of the respective emissions, provided identical quantities are used. This is shown in Figure 9.4; the jammer "DC" component is the self-S of the jam source, common with incoherent jam-source technology. As shown in this figure, it originates from the jammer system but adds to the thermal signature of the platform that the jammer is protecting.

Laser technology provides the means to generate a very high J:S and also the ability to switch between modulated waveforms rapidly without generating any self-S. Additionally, the laser technology can be designed to generate emission lines precisely where they are required in the various seeker operating wavebands.

9.3.5 Concept of Operation

A laser-based, directed-infrared countermeasure system has a characteristic narrow (low divergence) beam of modulated jam energy, which is directed towards a threat seeker; this beam may be as narrow as a few milliradians, which is some 25 times smaller than the value for the older directed-infrared countermeasure systems that used arc-lamp sources. Clearly, a simple static infrared countermeasure system architecture is no longer feasible, as it is now necessary to aim, or direct, this narrow beam into the infrared seeker of the approaching missile and maintain it on the seeker throughout an engagement. Hence, this class of protection system needs to be more sophisticated, and more complex, as the platform (and the protection system) needs to know:

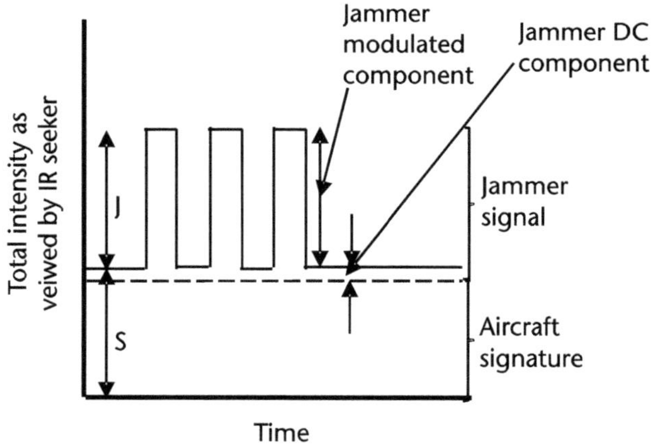

$$\frac{J}{S} = \frac{\text{Amplitude of jammer modulation}}{\text{Aircraft signature + Jammer DC signal}}$$

Figure 9.4 J:S definition.

- Whether it is under attack from a missile;
- Whether it is a heat-seeking missile;
- Where the attack is coming from (i.e., the bearing in a known/specified inertial reference frame);
- What countermeasure to deploy (if there are several systems);
- When to deploy the countermeasure (usually as rapidly as possible, but timing may be important in some circumstances);
- Where to direct the jammer beam in order to defeat the approaching threat;
- When the threat has been defeated.

These are the critical processes and fundamental functions of a defensive-aids suite (DAS) for platform protection; any type of directed-infrared countermeasure system is a fundamental contributor to such a platform-protection system. The common jargon or term for this part of a defensive-aids system is an effector.

The architecture of a typical dedicated laser-based directed-infrared countermeasure system is shown in Figure 9.5. The fundamental objective is to ensure that all of the subsystems operate as a complete system to ensure a high probability of survivability for the fitted platform. The timelines of the engagement of anti-air weapons are very short, so an autonomous system is required to ensure that the countermeasure process from the initial threat alert to the defeat of the missile takes place in a few seconds or less.

All directed-infrared countermeasure systems need an alerting system to warn the platform (and the defensive-aids suite, if it is linked to such a system on the platform) that a threat missile is approaching on a given bearing. This alerting system, known as a missile-approach warning system (MWS or MAWS), or more generally a threat-warning system, acts as a cue for the beam director used to point the jam energy into the approaching missile's seeker; this technology is discussed in Appendix C. The devices that provide the alert to the system are usually an array of electro-optical staring sensors, linked through an electronics control unit to the

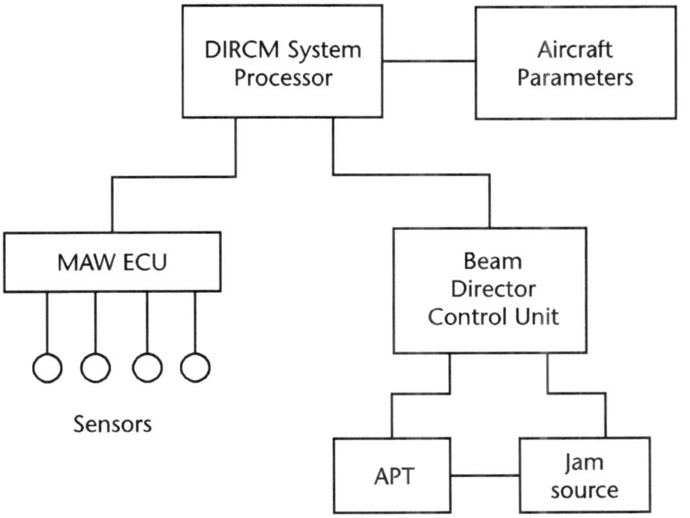

Figure 9.5 A generic directed-infrared countermeasure system architecture.

system processor. The derived threat-target information is then used by the beam-director controller to command the acquisition, pointing and tracking (APT) unit to search for, acquire and track the selected target. The laser jam source may then be given a command to generate the desired jam waveform through the directed-infrared countermeasure processor and the beam-director control unit.

The beam director (APT) has a camera to track an approaching missile; normally this is a thermal camera, although the CELEOS DIRCM system, the first directed-infrared countermeasure system, did not need a tracking camera integrated within the beam director. Figure 9.6 shows a typical engagement sequence, with the various phases of the jamming process as the missile approaches its target:

- The wide-angle field-of-view sensors of the threat-warning system detect an emission from the threat and the subsystem determines the approach bearing.
- The beam director is cued to the bearing of the approaching threat and looks for a thermal signature around that cued-bearing location.
- The beam director acquires and tracks the detected threat, as shown in the middle image of Figure 9.6; in this instance, the fine-track sensor in the beam director is tracking the missile's plume.
- The directed-infrared countermeasure processor commands the jam source to emit a suitable modulated waveform to defeat the approaching missile. In this example, the directed-infrared countermeasure system has been successful and achieved optical break lock, because the seeker is no longer looking at the helicopter and the missile has started to move away from its proportional-navigation course to intercept the helicopter.
- The threat-warning system continues to look for other potential threat targets.

The whole process shown in Figure 9.6 could take as little as about 1 second from the initial detection of an emission from a threat missile to optical break lock and the missile moving away.

A classical engagement sequence for any directed-energy system is shown in Figure 9.7, which is typical of any other laser-based directed-energy system, as discussed above and in Chapters 7 and 8.

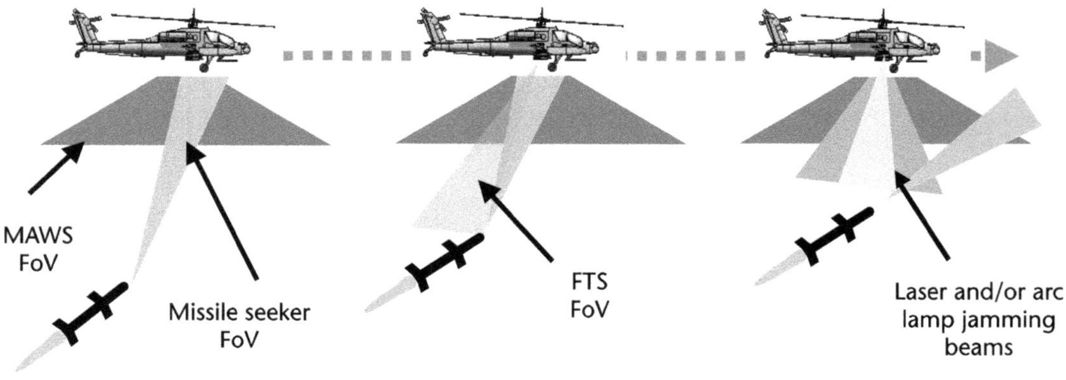

Figure 9.6 An illustration of a directed-infrared countermeasure engagement sequence.

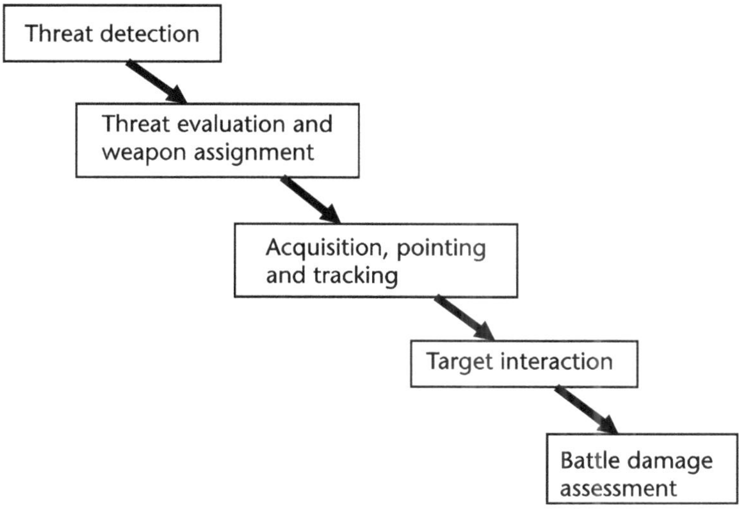

Figure 9.7 Directed-infrared countermeasure system engagement sequence.

9.3.6 Principal Directed-Infrared Countermeasure Subsystems

The component subsystems of a complete directed-infrared countermeasure system are shown in Figure 9.8 with a single transmitter; this is the small Nemesis system, which has the option of use with incoherent or coherent jammer sources. In the photograph, the various major components are shown and some are labelled, and the other subsystems are the directed-infrared countermeasure processor and power supply for the xenon arc lamps, which are the large boxes shown at the back of the picture, and the aircraft display unit, which is at the front of the photograph. Figure 9.9 shows the J-MUSIC system installed on a Boeing 707 aircraft of the Israeli Air Force.

9.3.7 Threat-Warning Subsystem

The function of the threat-warning system is to alert the directed-infrared countermeasure system that a missile is on a trajectory, which is approaching the platform fitted with the platform-protection system; it has a function similar to that of the threat-warning system, described in Chapter 8. The use of this sensor system is an additional necessary complication, compared with a staring infrared-countermeasure

Figure 9.8 A classical directed-infrared countermeasure system. (Courtesy of Northrop Grumman.)

Figure 9.9 A modern laser DIRCM system installed on a B-707. (Courtesy of Elbit Systems, Israel.)

system, owing to the inherent small beam divergence of the laser emission, as discussed above, and thus needing to know where the targets are, with respect to a defined reference frame.

Given information about the approach direction of the designated target in a defined inertial-reference frame, the beam director can be commanded to this bearing to find the designated target or threat-weapon system. Following a successful handover to its tracking sensor (camera) to allow target tracking, the narrow beam of jam energy can then be directed precisely at the approaching threat weapon or sensor system.

The technology used in this subsystem is discussed in more detail in Appendix C.

9.3.8 Beam Director

The beam-director technology used for this type of application has been described in detail in Chapter 4. In summary, the major functions it has to undertake are:

- Detect and acquire a target declared by the threat-warning subsystem (completion of the handover process).
- Track an acquired target (i.e., establish its trajectory and determine the bearing for engagement with the laser beam).
- Stabilise the optical sightline between the target and the jam source.
- Direct jam energy (modulated infrared emission) at the tracked target (i.e., along the stabilised sightline into the seeker).

- Undertake any other functions required by the directed-infrared countermeasure system, such as collection and capture of retro-reflected energy from the seeker for closed-loop infrared countermeasure functions, active tracking (of the infrared seeker) or countermeasure-effectiveness assessment. As the sophistication of the fine-track sensor and processing increases, other functionality will become feasible.

It is necessary for the beam-director system to undertake and complete several system functions before the beam director is ready to undertake its pointing functions. These concern the built-in test and boresighting functions, to establish the orientation of the optical axis of the jam source with respect to the fine-track sensor's optical axis [i.e., the tracking vector with respect to the jamming (pointing) vector].

9.3.9 Directed-Infrared Countermeasure Processor

As its name suggests, the directed-infrared countermeasure processor undertakes the command and control functions, vital to the successful operation of any directed-infrared countermeasure system. The functions and operation of the system processor described more fully in Chapter 4.

9.3.10 Infrared-Laser Requirements

The laser requirements and performance characteristics for a laser jam source used in onboard systems are summarised here:

- The emitted brightness must exceed the thermal signature of the platform that it is protecting by a substantial margin, as this is a key parameter in determining countermeasure performance and thus system effectiveness (i.e., the radiant intensity of the source determines J:S).
- There are no special requirements on spectral purity and the need to avoid chirping of the wavelength, providing that it does not incur any detrimental propagation effects; in fact, changes in frequency may have operational benefits and provide additional functionality.
- The pulse-repetition frequency must be compatible with the modulations required to form the jam waveform.
- The ratio between the on and off states is required to be of the order of 1,000:1 to avoid enhancing the thermal signature of the platform being protected (avoiding the self-S problem), ideally a modulation depth of 1.
- The average-power requirements are modest, of the order of a few watts.
- The divergence must be compatible with the pointing stability of the beam director. This is required to ensure that the laser beam spends the majority of its time on its designated target (energy-on-target requirement).
- The angular and positional stability of the beam, with respect to the input aperture of the beam director, must allow the boresighting functions to be completed, including any calibration and auto-boresighting functions.

- The stability of the beam must be significantly better than the pointing stability of the beam director, as any angular or positional motion will appear as a degraded pointing capability.
- The generated wavelengths of the laser system need to span the wavelength passbands of the full spectrum of threat seekers, as this type of laser-countermeasure system is reliant on an in-band defeat technique.

The key laser requirements are summarised in Table 9.1.

9.3.11 Candidate Laser Technologies

The preference for this application is to use solid-state laser technology, because of the well-cited characteristics of this technological approach. However, the classical solid-state lasers do not emit directly into the 4- to 5-micron band, as discussed in Chapters 2 and 3. However, recent developments with semiconductor laser technologies, such as optically pumped multiple-quantum-well devices or quantum-cascade lasers, may provide an efficient and compact source with the appropriate emission characteristics.

Many of the laser technologies considerations discussed in the previous chapters are directly relevant to this application, and thus will not be discussed again here.

9.3.12 Laser Integration

The normal requirements for the integration of the laser-based jam source with the beam director are discussed in detail in Chapter 4. The energy-on-target requirements are just as critical here as they are for any other directed-energy application. In fact, it may be even more crucial to the successful application of the technique as, dependent on design of the modulated waveform, only a particular portion of some jam waveforms may result in a specific class of seeker being defeated; if the beam is not interacting with that type of seeker at the requisite instance of time, then the opportunity to defeat that seeker may be lost.

The laser system may be fixed directly to the beam director or positioned remotely and the photon energy piped in. Direct connection can provide a much easier interface and reduce the harmony compensation and correction issues; however, remote positioning of the laser, with a suitable connector, can ease the system-integration requirements and lead to greater installation flexibility for the complete system.

Table 9.1 A Summary of Jam-Source Requirements

Laser Parameter	*Value*
Power (W)	Few (1 to 10)
Wavelength	Multiple lines in mid-infrared, between ~2 and 5 μm
Divergence (mrad)	Few so matched to beam director stability
PRF (kHz)	10 to 100
Duty cycle (%)	Up to 100
Modulation depth	High (0 or 1) ~1,000:1 or better contrast

The use of a remote location has many attractions for many of the smaller platforms, such as helicopters, requiring platform protection. Furthermore, with some consideration of using such beam directors for potential multifunctional applications, then the remote positioning of the various laser sources may be essential.

Modern development of low-loss optical fibres is expected to provide a suitable technique for piping the optical energy into the beam director. The fibre technology could be a conventional stepped index material or a passive micro-structured fibre, shown in Figure 9.10. The micro-structured optical fibres appear to have the appropriate wide optical bandwidth as well as other properties suitable for interfacing a number of laser emissions in different wavebands with a beam director.

9.4 Open- and Closed-Loop Infrared-Countermeasure Systems

When it comes to the application of the countermeasure technique, there are two fundamental approaches, open-loop techniques and the more sophisticated closed-loop method. The open-loop technique refers only to the lack of knowledge about the seeker dynamics of the approaching threat, in all other respects it is a closed-loop technique, as discussed next.

9.4.1 Open-Loop Systems

The basis of this countermeasure-system approach is that the jammer does not have any real indication of the specific weapon that is approaching the platform. Consequently, the jam waveform must be designed to defeat the spectrum of optical seekers, which are expected to be launched at the platform in its particular theatre of operation. A number of approaches have been investigated:

- One approach is sometimes called the spliced or composite jam waveform, which is illustrated schematically in Figure 9.11. In this case, the waveform

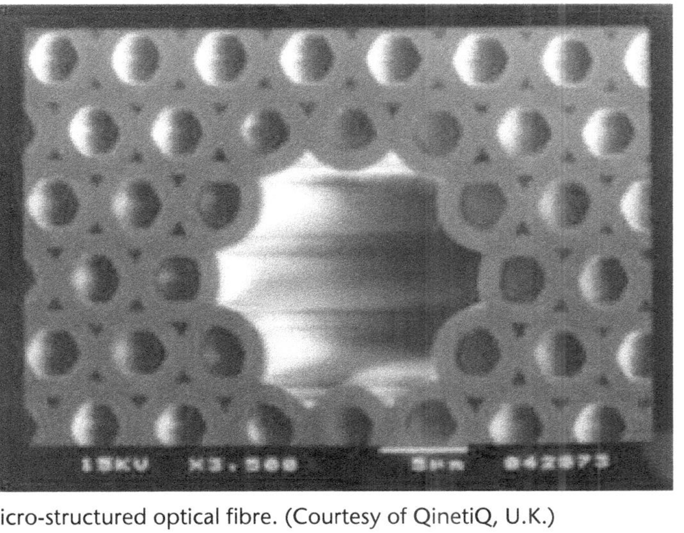

Figure 9.10 Micro-structured optical fibre. (Courtesy of QinetiQ, U.K.)

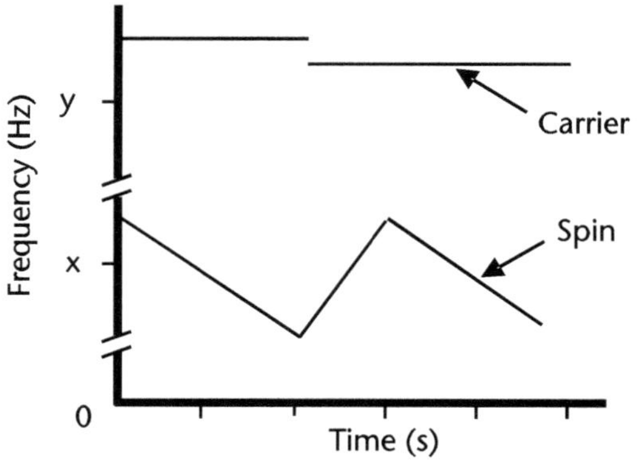

Figure 9.11 Generic swept spin by carrier (x*y) waveform.

is designed to have a number of phases, which may be aimed at confusing or defeating various types of optical-seeker technology; this is analogous to frequency hopping in a radio-frequency jammer. It is possible to arrange the modulations in the waveform to defeat a particular class of seeker at a specific phase during the jam sequence, if this is assessed to help enhance the protection provided by the jammer on any platform in defined operations. Moreover, the modulation details may be designed to defeat several different classes of seeker technology simultaneously.

- Another approach is a generic waveform that relies on a modulated waveform with a very high J:S to defeat all classes of seeker technology, the application of very large jam powers tends to reduce the sensitivity of a seeker's tracking system to the approximate match between the seeker's spin frequency (x) and the spin modulation in the waveform produced by the reticle (y); however, the jam frequency must not be blocked by the band-pass filter in the infrared seeker's detector system. Generally, there is less of a requirement to match the carrier-frequency modulations with very high J:S with those generated naturally by the reticle. This approach may be ideal, but has to be developed further before it is an accepted technique, to provide a reliable and robust protection method.

It is not possible to establish directly that a seeker has been defeated with a true open-loop technique, as no returns from the seeker are observed; it is usual with this countermeasure technique to assume that the jamming process has been successful after application of the jam waveform for a given short period. However, it is possible to capture the retro-return signal from the seeker and monitor any changes as a result of the fact that seeker's telescope (gyroscopic assembly) is being deflected during the optical-break-lock process.

Another indication of defeat of the infrared seeker may be deduced from changes in the sightline rate between the jammer and the approaching threat, as the missile changes its trajectory during the jamming process. However, this particular passive technique has some delays associated with it, as it is necessary to observe the trajectory of the missile over a short period, followed by some processing, to enable the information about any modifications to the missile's trajectory to be deduced. Data from the threat-warning sensors may also be used to help in this analysis using observations about changes in approach geometry made by the threat.

Observation and use of the retro return from the seeker, to monitor the progress of the laser-beam and infrared-seeker interaction, is sometimes referred to as the active-tracking technique. This is a simple and direct application of the same principle exploited by laser-based electro-optical countermeasure systems to locate threat optical sights. The fine-track (target-tracking) sensor may be used to monitor the optical return, provided that it has the appropriate sensitivity and spatial-resolution characteristics.

The active-tracking technique provides a virtually instantaneous indication of changes in the behaviour of the optical seeker being countered. This approach may be used to shift the pointing vector initially used by an open-loop, laser-based directed-infrared countermeasure system to coincide with the point of the retro return, and refine the aim-point of the pointing system to ensure that there is optimum energy on the designated target. Analysis of this collected retro signal may also be used for countermeasure-effectiveness assessment. Closed-loop jamming takes this process of active tracking a stage further, as considered next.

In summary, an open-loop system has the following fundamental characteristics:

- The jammer is not directly fully responsive to the specific characteristics of the possible range seeker technologies that may be used to guide a weapon to intercept a platform; therefore, it emits a jam waveform with modulations that have been optimised to defeat a range of seeker technologies defined on a threat list.
- The jamming technique consists of modifying the frequency components in the emitted modulated waveform to disrupt the designed response of the target-tracking systems in the anticipated classes of optical-seeker technology.
- Trade-off options exist for optimising performance between the various components of the jam waveform, the J:S and the time anticipated to be required to ensure a robust defeat mechanism has been achieved.

9.4.2 Closed-Loop IRCM (CLIRCM)

This is a more sophisticated and responsive technique that undertakes an interrogation of the approaching threat to establish the identity of this specific threat weapon. This technique has an extra step in the engagement process that involves an initial illumination (interrogation) of the attacking weapon's seeker with energy from the jam-class laser. The retro-reflection from the seeker's optical elements and, particularly, the modulation induced by the seeker's reticle enable the closed-loop infrared countermeasure processor system to establish the class of the seeker and its current operating condition. A fine-track sensor in the beam director may be used to capture the retro-return from the seeker, or a separate detector may be used, depending on the anticipated signal-to-noise ratio and optical resolution provided by the optical sensor in the beam director. The process is represented schematically in Figure 9.12.

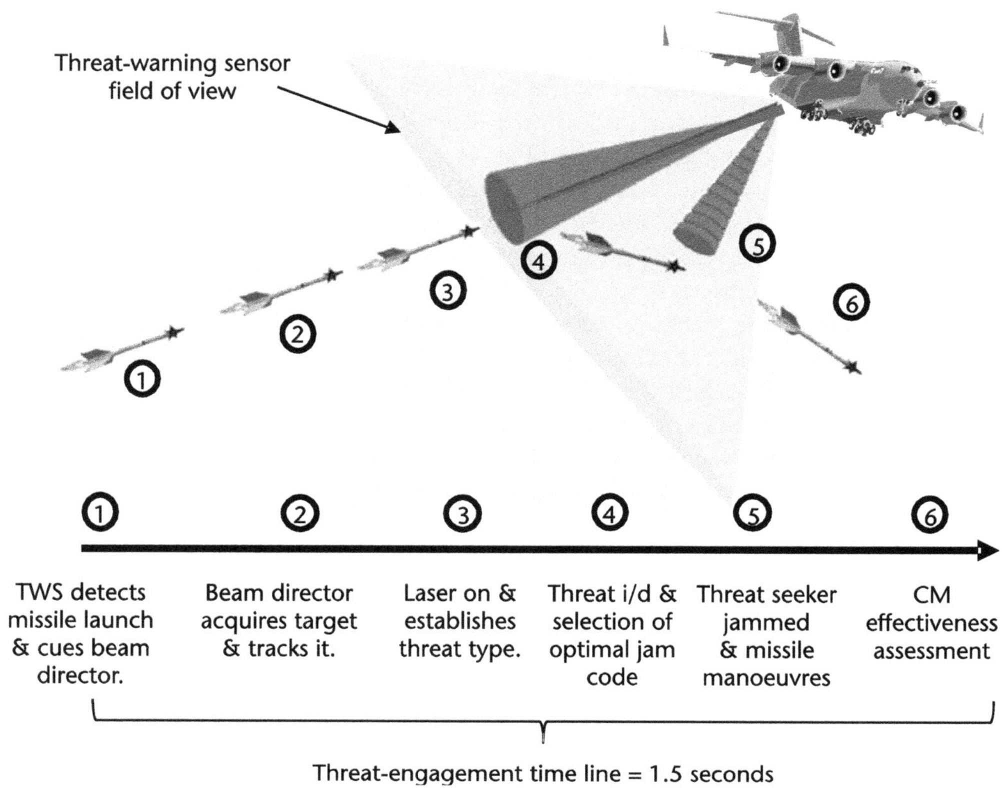

Figure 9.12 The closed-loop infrared countermeasure process and operational sequence.

The closed-loop infrared countermeasure processor has a database of information about the various optical seekers that may be employed against a platform. Comparison of the spectral characteristics embodied in the retro-return signals from the seeker, with those in the stored database, enables the seeker in the approaching weapon to be identified and, moreover, the appropriate optimum jam waveform to be applied.

An alternative approach is to calculate the appropriate jam waveform required to defeat this particular seeker from the spectral components in the retro-return signal received from the seeker. Analysis of the captured retro return, usually via fast-Fourier transform techniques, and similar spectral analytical processes, of the captured retro-return, allows a suitable jam-waveform modulation to be deduced. This deduced waveform can then be applied with these principal components inferred from the spectral analysis and then sweeping the spin frequency through a band of frequencies close to the spin component identified in the retro-return should result in a rapid defeat of the particular threat seeker.

The result of the application of the optimum waveform can be most dramatic, as seen by the image from a live-fire test using a closed-loop system, as shown in Figure 9.13. The photograph shows the smoke trail from a guided weapon, which is relatively smooth until the jamming is applied, resulting in a very dramatic turn being induced in the threat's trajectory away from its intended victim. This image shows another successful jamming event.

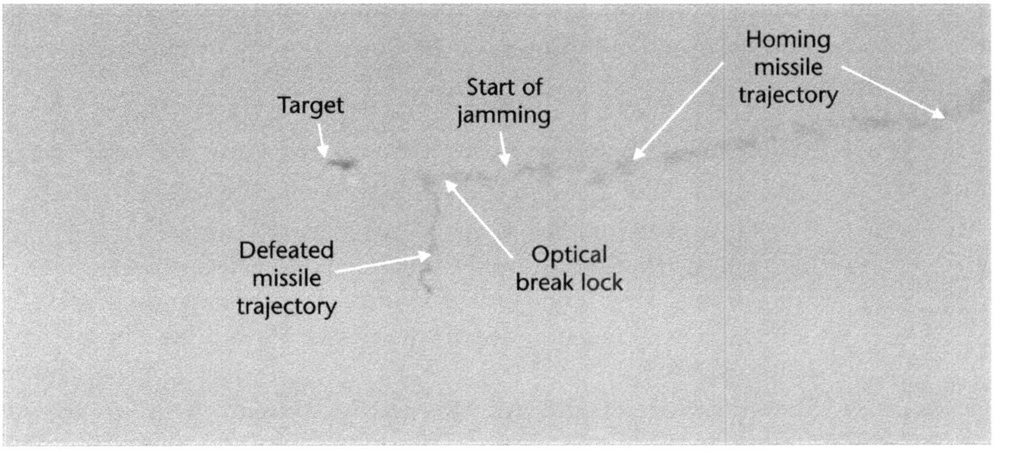

Figure 9.13 Effectiveness of CLIRCM jamming. (Courtesy of AFRL.)

The retro-signal may be used for other functions to enhance the performance of the countermeasure process, such as:

- Moving the tracking point from the plume or the body of the missile to the infrared seeker (i.e., to the front of the missile to increase the J:S, as discussed earlier);
- Monitoring the countermeasure defeat process and so determine when the missile's seeker has been defeated and know that it is safe to engage another seeker.

In summary, the use of a closed-loop infrared countermeasure technique allows:

- The jam source to be modulated with a code that is the optimum for a rapid defeat of the identified threat;
- An optimum jam-modulation frequency to be deduced for a new (or unknown) weapon and then directed into the seeker to achieve a rapid (or immediate) optical break lock (i.e., with an appropriate frequency very close to the modulation rate in the optical seeker);
- Monitoring of the countermeasure process that enables a minimum period to be spent on each threat, ensuring multiple threats can defeated efficiently and effectively.

9.5 Future Requirements

A laser-based jammer faces a number of challenges to provide robust platform protection by defeating the whole spectrum of heat-seeking missile threats. Currently, the systems have been demonstrated to be very effective but those in service tend to be large and expensive. To make this form of platform-protection technology more readily available and affordable, particularly for those platforms where available space and additional lift-capacity is at a minimum, it is crucial that the technology is developed in a number of areas. For example, the immediate requirements for the laser-based directed-infrared countermeasure systems may be summarised as follows:

- Reduction in size (and weight) of the beam director and laser source;
- Simplification of the source technology to enhance reliability and reduce cost (including through-life costs);
- Increase in the time to apply the countermeasure process, thus allowing longer-range interactions to be realised, which could be achieved through development of enhanced-performance threat-warning sensors (i.e., longer-range detection);
- Simplification of the warning-sensor system, without compromising performance;
- Reduction of false-alarm rate and sensitivity to potential false-alarm sources;
- Enhancement of the source efficiency.

Other requirements for the near term, involve the desire for the development of generic jam codes that will defeat the whole spectrum of threats using optical-seeker technology instantaneously. In this case, the effective jam code could have a form of modulation that may contain multiple simple frequency components, which would interact adversely with the approaching seeker's detector and tracking loop. The availability of this class of jam code would enhance further the performance of the simpler open-loop countermeasure systems and eliminate the need to refine the codes, although robust testing will still be required.

In the longer term, there is a clear need for the development of robust techniques for defeating the latest generation of optical seekers, such as the imaging devices. The need for this development in countermeasure techniques and methods is likely to be particularly demanding on the performance of the laser. Alternative techniques of laser operation may be necessary, such as devices that are capable of both jamming and generating in-band damage-class pulses, perhaps leading to the concept of "jamage" techniques.

It is clear that the presence of the asymmetric anti-air threat is likely to continue for the foreseeable future; therefore, it is very important that any technique that is developed to defeat imaging infrared seekers must be backwards compatible with the well-established robust techniques for defeating the whole spectrum of reticle-based seekers. The use of in-band damage techniques is only likely to have limited capability against existing reticle-based seekers. This is a consequence of the fact that the reticle is in the focal plane of the seeker's optical train and this is a relatively hard component and relatively resistant to laser-induced damage.

The other longer-term objectives and requirements will almost certainly demand greater functionality from the laser-based jamming system. Examples include:

- Greater use of the fine-track (target-tracking) sensor for other sensing functions;
- Development of the algorithms used with the warning sensors to give additional information about the approaching targets;

- Development of the current type of missile-warning sensor system into a general threat-warning (and classification/identification) system;
- Use of the laser subsystem for other functions, such a covert communications, range finding or target interrogation;
- Development of a system architecture to share a multifunctional laser source;
- Develop techniques that would allow the laser source to be positioned remotely from the beam director or use of a common source to service several beam directors.

Greater development of the functionality of this very capable system would make it a more cost-effective system and probably an indispensable asset on many more platforms.

9.6 Summary

Laser-based, directed-infrared countermeasure systems have become well established for the protection of platforms being attacked by reticle-based, heat-seeking missiles and represent a very successful implementation of directed energy. The use of a laser in this type of system has enabled the platform-protection techniques to become more robust because of the very high radiant intensity of the collimated laser beam. The beam is capable of inducing a rapid optical break lock in a threat seeker, as a result of a carefully designed adverse interaction that it imposes on the infrared seeker's target-tracking loop.

The use of a laser to generate the jam waveform has many advantages over the more conventional incoherent sources, not only in terms of higher radiant intensity, but also for efficiency and in particular generating the emission lines at the requisite wavelengths. Moreover, the use of a laser-jam source enables a more compact system to be fabricated and thus be compatible with small platforms.

Two fundamental countermeasure architectural approaches have been developed and demonstrated to defeat the target-tracking loops in heat-seeking missiles. The open-loop technique, which is so described as it does not have any information to determine the identity of the class of the approaching missile; therefore, this particular approach requires quite a sophisticated jam waveform to ensure a successful defeat of all infrared (heat-seeking) threats. The threat-tracking techniques and pointing of the modulated emission from the jammer are under closed-loop control. The closed-loop infrared-countermeasure technique determines the class of missile, by analysing the retro-return signal generated from the seeker during the illumination by the jam-class laser, and then applying an appropriate and ideal jam waveform. The latter approach enables a potentially very rapid defeat of a seeker system, despite a finite time being involved in the interrogation phase needed to identify the approaching threat system.

Techniques such as active tracking are being developed to enhance the utility of open-loop systems to monitor the retro-return signal, to enable the progress of the laser-jammer-seeker interaction to be observed. Use of this technique and analogous ones enable the laser-based countermeasure system to assess the effectiveness of the applied countermeasure technique, virtually instantaneously.

The emission characteristics of semiconductor laser technology appear to be well matched to laser requirements necessary for robust platform protection using directed-infrared countermeasure systems. Additionally, this technology offers a potential route to a reduction in the through-life costs of such systems, which is one of the major challenges for the future.

Selected Bibliography

Ball, R. E., *The Fundamentals of Aircraft Combat Survivability Analysis and Design,* 2nd ed., AIAA Educational Series, AIAA, 2003.

Biberman, L. M., *Reticles in Electro-Optical Devices,* Oxford, U.K.: Pergamon Press, 1966.

Caplan, W., *Modern Infrared Countermeasure Systems,* Norwood, MA: Artech House, (in the press).

Douda, B. E., *Genesis of Infrared Decoy Flares: The Early Years from 1950 to the 1970s.* 1st ed., NSWC Crane Division Report ADA 495417, Crane, IN, 2009.

Ellern, H., *Modern Pyrotechnics, Fundamentals of Applied Physical Pyrochemistry,* New York: Chemical Publishing Company, 1961.

Pollock, D. H., (ed.), *The Infrared and Electro-Optical Systems Handbook, Vol. 7*, Chapter 3 "Active Infrared Countermeasures," Bellingham, WA: SPIE Optical Engineering Press, 1993.

Titterton, D. H., "The Development of Infrared Countermeasure Technology and Systems," in *Mid-Infrared Semiconductor Optoelectronics,* A. Krier, (ed.), New York: Springer-Verlag, 2006.

CHAPTER 10

Laser Range-Finders

The emission from a laser, with its inherent low divergence and potential for short-duration pulses, and consequently, very high peak power, makes it an attractive option for estimating the range to a target provided the time-of-flight of a laser pulse to and from a target can be measured accurately. This is probably the simplest of the remote-sensing applications using laser-based techniques, although it has undergone many developments over the years since its first introduction; indeed, it was the first application of laser technology to a military function. The use of laser technology revolutionised the range-to-target measurement accuracy that could be achieved on the battlefield.

Measuring the range to a target has been and continues to be an important requirement in warfare, as is the ability to determine one's position rapidly and accurately, if it is not already known. The military capability requirement is very simple; acquiring knowledge of the distance from A to B to an adequate accuracy for it to be used for other military functions. This knowledge could be required as part of a triangulation activity to determine one's position, or the distance to a target, particularly for indirect-fire operations.

It has always been critically important to have an accurate range estimate to an enemy's position, especially range from one's own location, to make the best use of ammunition. As very accurate ranging techniques became available, speed of the estimation of the ranging process became even more important, as the first shot would be most likely to win an exchange; moreover, in current conflicts, accuracy is paramount to operate in complex environments, whilst minimising collateral damage.

Laser range-finding techniques form the basis for many targeting methods used on the battlefield. The fundamental technique may be adapted and extended to allow some limited additional functions to be undertaken, such as surveillance. This chapter will concentrate on the range-finding task, but some other tasks will be mentioned, which may be interpreted as laser-radar functionality. Laser-radar techniques usually involve applying multiple dimensions to the observation, often through scanning to generate a picture or map, whereas range finding is limited to a single dimension. Laser-radar systems are discussed in Chapter 13.

10.1 Introduction

The original approach applied to range finding was to use passive optical methods. These devices exploited the parallax phenomenon by observing a target from two different positions. Two images of the same target were presented to the operator, who then adjusted their relative horizontal positions to achieve coincidence. With the object of interest matched in both images, the operator simply read off the range to the target from a calibrated scale. These systems achieved reasonable accuracy depending on the size of the system; range accuracies of about 1% of range could be achieved with a device with a baseline of approximately 5 metres separating the two optical ports, over distances up to about 10 km. Needless to say, a range-to-target estimate could only be made with this type of system when the target was visible, so operation in darkness or poor visibility conditions would not be possible.

The availability of suitable pulsed-laser technology revolutionised the approach for determining range to an object; in this case, the accuracy is far better than 0.1% of range. The availability of a pulsed emission exploits the fact that, if the time taken for a beam of light to travel between two points can be determined to sufficient accuracy, then the distance can be calculated by definition; the speed of light is a well-known constant and

$$\text{Distance} = \text{Speed} \times \text{Time}$$

The simplest system approach is to emit a single pulse from the laser and measure the delay between the pulse leaving the laser aperture and the receipt of a reflected pulse from the object of interest in the scene. It must be remembered that if the observer is counting the time elapsed from the launch of a pulse to the receipt of its return from the object, then those photons have made a return journey of twice the separation.

Practical laser range-finding systems tend to use an emission with a modest repetition rate to mitigate some of the vagaries of beam propagation, so that a number of pulses are emitted over a short period; however, this leads to the potential for other complications or confusion in the range estimation, which have to be solved to make this relatively simple concept into a practical system. Examples include multiple objects in the illuminated scene, returns from clutter in the scene and range ambiguity, which is discussed in more detail later.

This type of remote-sensing system can be used to survey a scene, as though this laser system was a simple laser-radar system (discussed in Chapter 13). This application of the laser range-finder involves the scene being scanned, usually manually, and the returns provide a one-dimensional representation of the distances to intercepted targets within the scanned scene. The intensity fluctuations may provide some indication of the size and disposition of targets in the surveyed scene. Additionally, potential targets can be detected as a result of the retro return from sensors that are in band to the illumination wavelength through the cat's eye effect, discussed further in Chapter 8.

Advanced techniques have now been developed for the application of more sophisticated techniques, including covert methods involving low-probability of intercept methods, which may be applicable to targeting and sensing applications.

10.2 Historical Development

The original military application of laser technology to range finding involved the use of flashlamp-pumped ruby laser technology, entering service with the U.S. Army in the late 1960s. These ruby lasers were usually operated at a few hertz repetition rate with a few hundred millijoules in each pulse; this device clearly presented an eye-safety hazard close to the aperture! The non-ambiguous range was reasonable, but the use of a ruby laser presented a number of operational restrictions, coupled with the fact that the system was quite large and heavy, because of the necessary power supplies and a cooling systems. Consequently, it was really only suitable for use on a platform with suitable power supplies; it was certainly not practical for the dismounted soldier.

The application of a carbon-dioxide laser to this task eased the eye-safety issues; however, this was still a bulky device and normally only used on the larger military platforms. Again, powering the device was a challenge and there were some technical issues regarding the detector subsystem in the receiver channel.

Some simplifications to the system design were possible when neodymium laser technology was introduced as the illumination source, which was a 10-mJ class system (AN/GVS 5). This solid-state laser was still flashlamp pumped and the eye-safety issue returned, but as this is a four-level laser device many of the restriction imposed on the operation by the three-level ruby laser were avoided, and furthermore, mature silicon detector devices could be used. Indeed, with the advances in support technology such as the power-supply unit, it became feasible for comparatively compact systems to be produced, suitable for use by the dismounted soldier.

The use of the neodymium lasers enabled probably the first multifunctional laser system to be brought into military service; this laser system could undertake the range-finding role and designate a target for another weapon system to home on to the reflected energy from the target and then complete an interception very accurately. These multifunctional laser systems were quite powerful and thus care was required along with extensive training, owing to the potential eye-safety issue. In this case, the typical laser is a Q-switched, low repetition-rate device operating at a few tens of hertz and it is a 100-mJ class emission laser source. The original devices were still flashlamp pumped, but latterly, diode pumping has been applied, which resulted in more compact, rugged and efficient systems far more compatible for universal operational deployment.

The most modern development has been to move to the eye-safe wavelengths again, this time using a solid-state laser to generate wavelengths around 1.5 μm. Initially, wavelength conversion stages were used, but more efficient photon generation is feasible with direct generation by, for example, using transitions in the erbium ion. The move to this wavelength region has caused a change in the detector technology to a semiconductor material such as indium gallium arsenide. The use of fibre-laser technology has led to very compact and efficient systems, still compatible with the requirements for use throughout the spectrum of military operations.

The latest developments with this relatively simple sensing system have been to make it more covert, not only through the use of wavelengths beyond the visible waveband, but through single photon techniques, which are low-probability of interception detection systems, by definition. The more recent developments in advanced sensors have integrated range-finding capability into these subsystems, which are exploited in burst-illumination techniques for target identification.

10.3 Military Utility

The fundamental use of this type of system is to provide a rapid accurate estimation of the range from the system to an object or potential target of interest. This fact was noted in the introductory remarks, and is especially important, for providing range-to-target data to support indirect-fire operations. However, this relatively simple system has the potential to provide additional military capability or enhance many aspects of current military capability.

A laser range-finder may be considered to be a simple form of remote sensor with a specific function, although additional functionality is feasible. The principal military value of laser-based range finders is that the distance to a remote object can be determined easily and rapidly to a fraction of a percentage of total range separation. Hence, as more information can be gleaned from the photons received from the objects on the battlefield various tactical decisions can be supported.

Modern laser technology enables the range-finding system to be used, or adapted to undertake a series of valuable military functions; these are in addition to its basic function. Examples include:

- Surveillance of the battle space for detection and location of potential targets;
- Geo-location and navigation support through the use of triangulation methods;
- Simple range profiling of remote targets to assist classification, as discussed in Chapter 13.

In the surveillance mode, a laser range-finder may be scanned over a particular volume of space and a one-dimensional picture, for each probed bearing, can be formed from the returns detected and recorded. These may be simple back-scattered returns from objects in the path of the diverging beam or retro returns from optical devices in the propagation direction, if they are looking at the transmitter. A composite picture may be formed from the various inputs and then interpreted to identify targets or regions for further investigation.

The range-finder may be used in conjunction with a navigation system or GPS to support other operations. An example is support for surveying the current location of a firing position and locating potential targets, which may be for support of shoot and scoot tactics used with indirect-fire operations, discussed further in Chapter 15.

10.4 System Operation

As noted above, the range-finding technique is based on the accurate measurement of the elapsed time for photons to travel from a launch point to an object and return through back scatter or reflection from an object in the path of the propagating beam. The device is normally configured as a monostatic system, but bistatic approaches are possible over a limited separation angle.

The transmitter of a laser range-finding system is used in conjunction with a passive sensor, such as an optical sight/telescope, thermal camera or a television camera to direct the laser beam at objects of interest in the scene. This system may be mounted on a servo-controlled and stabilised platform for some applications. In some systems, the sighting system may be used to control the divergence of the laser beam and the field of view of the receiver system.

A laser range-finding system operates as follows: the range-finder unit is directed/aimed at a selected target, using the optical sighting system or similar pointing system. On command from the operator, a single pulse or a series of pulses are transmitted towards the target of interest and the time of flight of each pulse from the transmitter to the selected target and its return path to the detector is measured.

The transmitter is connected to a timing-detector module, which is triggered by each pulse emitted by the laser. The receiver system (or timing-detector module) has a narrow-band filter to remove as much of the background signal as possible, leaving only the monochromatic signal returned from the target. This energy is usually detected with a fast photo-detector, such as a semiconductor photo-diode or a photo-multiplier tube, which creates the electrical signal. This signal is then amplified and filtered electronically to suppress some aspects of high-frequency noise components introduced in the detection and preamplification processes to enhance signal-to-noise ratio; however, care is required with the filtering process, as the detected pulse will be of short duration with quite high-frequency components. This signal is then compared with a preset/predetermined threshold detection level; when the received signal is above this threshold value, the clock is stopped and the range to the object/target calculated.

Most tactical laser-range-finding systems provide a multiple-target indication function through a display module, which indicates the first or last indication of received pulses. The first-pulse mode is usually employed when the target object does not fill the field of view of the receiver, which avoids ranging on the clutter in the background. The last-pulse mode is normally used in the presence of fog or obscurants to avoid ranging on the array of aerosols in the field of view, or foreground clutter. This mode selection feature may also be used to resolve ambiguities when there are multiple targets in the field of view or the return signal is being confused by returns from the local terrain.

A schematic diagram of a laser range finder is shown in Figure 10.1. Generic terms are used for the subsystem components. For example, modern detector technology is called Geiger-mode avalanche photodiodes.

The use of very short duration pulses, of a few nanoseconds, enables a one-dimensional intensity profile to be recorded (fully digitised return) by the use of a suitable fast detection of each target return along the propagation path, as noted above. This type of profile information about the target, which is dependent on suitable temporal resolution within the detected pulses, may be used to classify potential targets before engaging with a weapon system. This use of laser range-finder technology for this type of sensing is discussed further in Chapter 13.

Emission of a series of laser pulses means that there is the potential for confusion when the pulsed returns are from other objects in the scene. An unambiguous range may be defined by the pulse-repetition rate of the laser, or more precisely, the

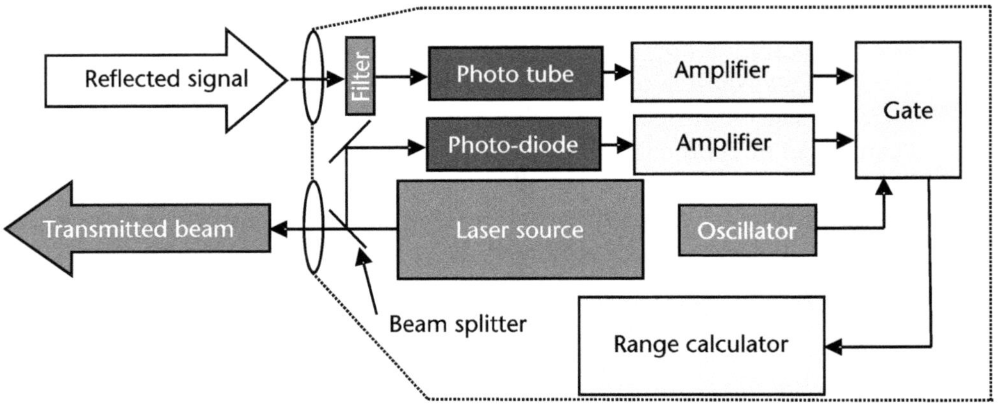

Figure 10.1 Schematic diagram of a laser range finder system.

pulse-repetition interval between pulses. The ambiguity in range estimation arises when energy is received from an object at a range at a greater distance than defined by the interpulse time; for example, a return from the first pulse in the train may be received from a more distant object during the period when the system is expecting a return from a subsequent pulse in the emission train.

To avoid completely the possibility of ambiguity in the estimation of range to the detected target with a simple system emitting a series of pulses in a short burst, the pulse-repetition frequency of the device must be set so that only one pulse is in transit at one time. For example, with a target 1 km away, the round-trip transit time is 6.67 μs, restricting the laser pulse frequency to less than 150 kHz. In the case of targets 5 km away, the pulse-repetition frequency cannot exceed 60 kHz. The ambiguity problem may be mitigated through appropriate choice of pulse-repetition frequency or application of pulse-coding techniques, including the application of a pseudo-random code, as used in some of the emerging techniques such as single-photon surveillance.

It is quite common for the range-finding and designation function to be combined in a single laser system. The pulsed emission may be exploited to provide the range estimate to a potential target and the returned energy may also be used by a seeker to direct a weapon to the point of illumination on the selected target. Laser-based designation will be discussed in Chapter 11.

10.5 Laser-Performance Requirements

An illustration of a simple laser module used in a laser range-finding system is shown in Figure 10.2; modern laser modules use laser diodes to pump the gain medium, which enhances the efficiency to about 25%. This older-style module, shown in Figure 10.2, is flashlamp pumped with a passive Q-switch to provide the giant pulses. The duration of the Q-switched pulse requirement is of the order of 10 to 100 ns, depending on the application and governed by the configuration of the Q-switch module. The pulse repetition frequency is of the order of 20 Hz with pulse energy 10 to 200 mJ, selectable for the application.

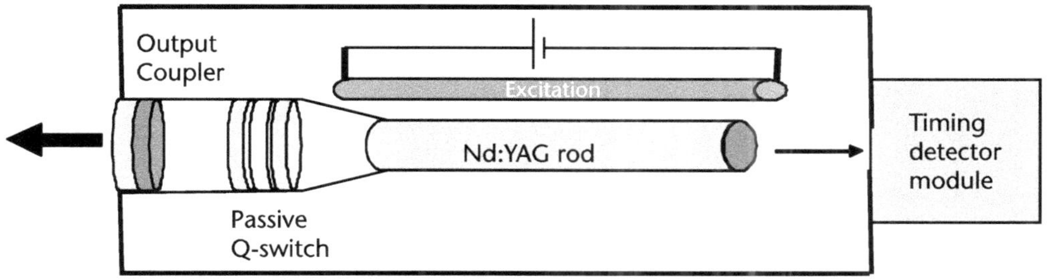

Figure 10.2 Schematic of a range-finder module.

The modern requirement is for operation in the eye-safer wavelengths, as noted earlier. To maintain a minimal size and maximum flexibility, a wavelength in the region of 1.5 μm is used and the gain medium is diode pumped for optimum efficiency.

Typical laser-performance requirements for some applications are shown in Table 10.1.

The characteristics of the laser source used in some of the more common laser range finders are shown in Table 10.2.

The laser requirements may be summarised as compact, reliable and efficient devices with a moderately energetic pulsed emission and capable of a modest pulse-repetition frequency. This points to solid-state laser technology and the use of erbium ions to provide the eye-safer operation; however, if multifunctional operation is required including designation, then neodymium laser technology will be necessary, as is discussed in more detail in Chapter 12. Full consideration of other candidate laser technologies is discussed in Chapters 7, 8, and 9.

One of the critical considerations for most applications is the size, weight and power consumed by the system, which is clearly inextricably linked with wall-plug efficiency.

Table 10.1 Typical Laser-Performance Parameters for Some Applications

Source Parameter	*Man-Portable*	*Land Vehicle/Ground, Anti-Ground, Anti-Air*		*Helicopter*	*Fixed Wing*	*Ship*
Wavelength	Eye-safe range (1.5 μm)	Eye-safe range (1.5 μm)	Eye-safe range (1.5 μm)	Eye-safe range (1.5 μm)	Eye-safe range (1.5 μm)	Eye-safe range (1.5 μm)
Pulse energy	~10 mJ	~40 mJ	~150 mJ	~40 mJ	~150 mJ	~150 mJ
Divergence	~1 mrad	~1 mrad	~2 mrad	~1 mrad	<1 mrad	<1 mrad
Pulse duration	~10 ns	~10 ns	~10 ns	~10 ns	~10 ns	~10 ns
PRF	Low/single shot	Low/1 Hz	~20 Hz	~5 Hz	~20 Hz	~20 Hz
Duty factor/ use	A shot every few seconds	Burst mode of up to 100 shots	High (up to 100%)	High (up to 100%)	High (up to 100%)	High (up to 100%)
Source/system efficiency	Critical	Important	Important	Important	Important	Less of an issue
Packaging	Critical size and weight constraints	Important size and weight constraints	Important size and weight constraints	Very important size and weight constraints	Some important size and weight constraints	Less of an issue

Table 10.2 Typical Characteristics of Layer Devices used in Some Range-Finding Applications

Parameter	*Er:glass*	*Nd:YAG*	*Shifted:YAG*	CO_2
Wavelength	1.54 μm	1.064 μm	1.54 or 1.57 μm	10.6 μm
Eye safer*	Yes	No (Class 4)	Yes	Yes
Pulse energy	>10 mJ	>100 mJ	>100 mJ	>100 mJ
Pulse duration	10 to 20 ns	5 to 20 ns	5 to 20 ns	20 to 100 ns
PRF	>10 Hz	>100 Hz	>100 Hz	>100 Hz
Pump/excitation	Diode or flashlamp	Diode or flashlamp	Diode or flashlamp	Discharge
Laser type	Solid state	Solid state	Solid state	Gas

*Note for a single pulse the Class-1 limit is 8 mJ for a single pulse.

10.6 Subsystem Description and Function

The laser range-finder module can come in many forms, which can include sightline stabilisation for use on a moving platform. When used by a dismounted soldier, the laser range-finder is usually mounted on a tripod, but for simpler applications the module may be handheld, including in a binocular format or weapon mounted. Modern range finders in binocular format may have sightline-stabilised optics.

In the case of the ground use by a soldier, the target is normally located via a telescopic sight, or a simple iron sight. The module contains the laser transmitter, timing module, detector and display as an integrated system with a carefully harmonised sighting device for selecting targets.

Harmony between the optical axis of the sighting system and that of the integrated optical module is a critical requirement, so a boresighting mechanism is usually included. This function may be achieved using a prism and observing the retro reflection from a short distance.

10.7 Single-Photon Range-Finding and Surveillance

A number of techniques have been investigated to make the range-finding process more covert as well as eye safe. One technique that has been investigated and indicates some promise is based on single-photon techniques. Prototype systems exploiting single-photon methods have also been demonstrated to expand the capability to create maps or images of a target, although this type of system could be considered to be a laser-radar system.

10.7.1 Introduction

Time-correlated single-photon counting has been used for many decades to measure fast, low-level light signals in many scientific applications. This technique measures the time difference (or micro-time) between an optical input pulse, typically in a repetitively pulsed laser beam and a returned signal recorded by a single-photon detector (i.e., through a direct-detection approach). Over many laser pulses, a histogram of the number of photon counts versus micro-time can result in a statistically accurate representation of the actual optical transient signal being measured. The estimate of range may then be deduced, as discussed earlier.

This type of technique can be used to create an image of a target through scanning the beam and application of the time-of-flight approach. Such a laser-radar type of system uses a pulsed diode-laser source and semiconductor-based, single-photon avalanche photo-diode for direct detection. A three-dimensional system scans a scene in the horizontal and vertical directions to build up a three-dimensional image of the depth profile of a target. The time-gating and high sensitivity of the time-correlated single-photon counting approach offers the potential for low-light level intensity and eye-safe operation, which may be very valuable for covert operations.

A time-of-flight prototype system uses a pair of galvanometer mirrors to scan a scene and build up a three-dimensional image of the depth of the detected object by recording the returns from the scene. The mirrors are used for:

- Scanning the laser output over the scene;
- Directing the collected return scattered light to a single optimised single-photon counting detector.

This is in contrast to the alternative approach of imaging a scene on to an array of single-photon detectors, where parallel read-out remains a major technological obstacle. This is similar to a flash or staring system, which is currently an emerging capability.

As noted earlier with more conventional approaches, the only classical way to avoid the possibility of ambiguity completely is to choose a repetition frequency such that only one pulse is in transit at any one time period. However, use of relatively low repetition-rate emissions will greatly reduce the speed of range-data collection and the effectiveness of the overall photon-counting time-of-flight technique. Consequently, alternative techniques have been investigated to solve this range-ambiguity issue.

An alternative technique, using a much higher repetition rate than allowed by the range-ambiguity rules, exploits the characteristics of a non-periodic pulse train, such as a pseudo-random sequence of laser pulses emitted at a very high clock frequency, such as 1 GHz. The emission parameters may be chosen, so that on average, only a fraction of the available time window is occupied by a laser pulse. The received pattern may then be correlated with the transmitted pattern to identify the unique target time-of-flight period.

10.7.2 System Operation

Single-photon systems may be realised using a vertical-cavity surface emitting laser diode (VCSEL) source emitting a gigahertz pulse train in a random pattern repeated over a specific period, as illustrated schematically in Figure 10.3. A silicon single-photon avalanche diode (SPAD) detector (commonly known as Geiger-mode APDs) may be used in the detector, along with a recorder with a time-stamping facility, to allow correlation with the emitted pseudo-random pulse train. A single reference pulse may be used to synchronise the time-stamping process, so that each detector event is time-stamped with respect to this pulse.

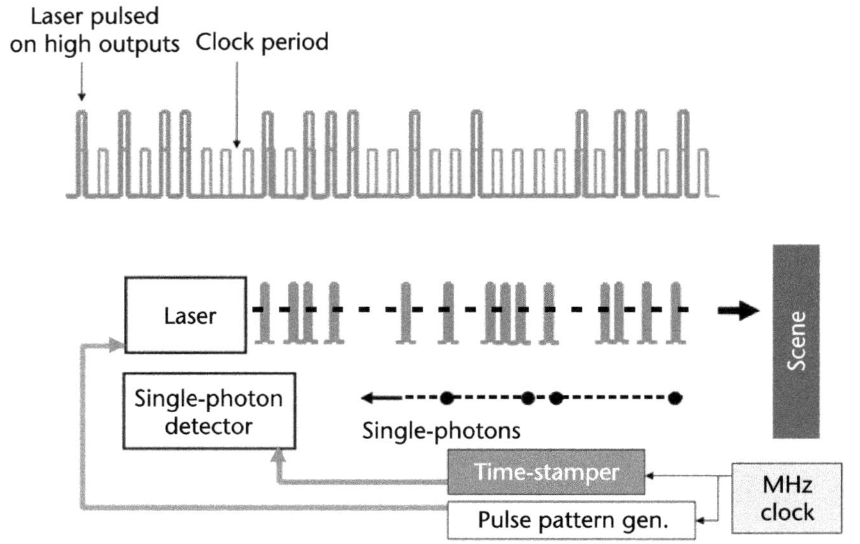

Figure 10.3 Schematic of a covert range-finder using a pseudo-random emission code. (Courtesy of Professor G. S. Buller, Herriot-Watt University, U.K.)

To achieve accurate correlation between the emission and the detection systems, it is very important to maintain synchronisation between the reference pulse and the time-stamping process. This may be achieved through phase-locking the laser-driver subsystem and the time-stamping device, using an external clock. With a suitable choice of parameters, it is possible to achieve an unambiguous range of many kilometres with depth resolution of several centimetres.

10.7.3 Subsystem Description and Function

The laser transmitter may be integrated with a single-photon avalanche detector as shown in Figure 10.4. As this is designed to be a covert system, the emission wavelength will normally be beyond 800 nm up to about 1 micrometre, owing to the availability of highly efficient detectors operating in this part of the spectrum. The use of single-photon methods is synonymous with very low-power devices which are inherently eye safe. Moreover, the emission must be compatible with high-frequency modulation.

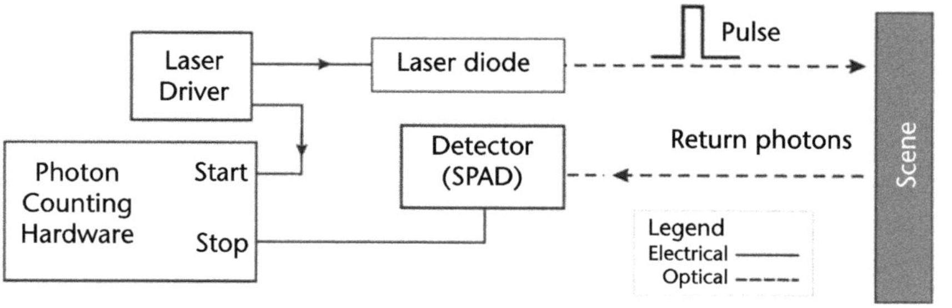

Figure 10.4 Schematic of a covert laser range-finder system. (Courtesy of Professor G. S. Buller, Herriot-Watt University, U.K.)

The laser emission may be scanned over a selected field of view in a systematic fashion, using a preprogrammed scan pattern via scanning mirrors, such as galvanometer mirrors. Such a scanning time-of-flight system, using time-correlated, single-photon counting technique, in conjunction with suitable data-analysis algorithms, can generate three-dimensional depth images of the scanned scene using very low laser-power levels. A schematic of a three-dimensional scanning system is shown in Figure 10.5.

Noise from the solar background remains a limiting factor with this type of surveillance. A shift to the short-wave infrared should reduce the solar-background clutter and noise in combination with use of improved spatial and spectral-filtering techniques.

10.7.4 Laser Requirements

The laser emission characteristics must be compatible with imposing a pseudo-random waveform.

Typical laser-performance requirements for covert systems are shown in Table 10.3. This table also includes an indication for a three-dimensional imaging system based on an extension of this single-photon technique. The three-dimensional imaging technique may be considered to a laser-radar system, as it is designed to map a target. Discussion of burst-illumination and three-dimensional laser-radar laser requirements are discussed in Chapters 12 and 13, respectively.

10.8 Future Requirements

One of the major requirements for future ground-based sensing applications concerns covert operation and the ability to integrate laser range-finding techniques with other sensors and systems, without disrupting the fundamental operation of each subsystem. For example, there is a very demanding requirement for a laser range-finder operating out to 5 km, handheld weighing less than 1 pound. Moreover, these systems are required to be compact and very efficient, particularly to aid the detection and identification of difficult targets. The size and weight

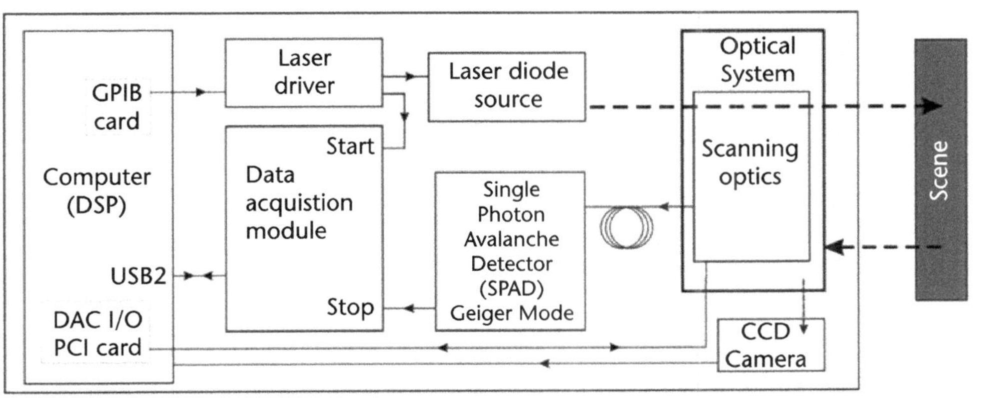

Figure 10.5 Schematic three-dimensional covert sensing system. (Courtesy of Professor G. S. Buller, Herriot-Watt University, U.K.)

Table 10.3 Laser-Performance Parameters for Covert Range Finding

Source Parameter	*Range Finding*	*3-D Imaging*
Wavelength	>800 nm, preferably ≈ 1.5 μm	>800 nm, preferably ≈ 1.5 μm
Average power	~1 μW	~50 μW
Divergence	~1 mrad	~1 mrad
Pulse duration	Few picoseconds	Few picoseconds
Pulse-repetition frequency	Megahertz range	Megahertz range

characteristics of any of these systems will need to be compatible with operation by dismounted soldiers.

Future systems anticipated for airborne applications will require greater stand-off ranges as well as the normal demands for efficient, compact, reliable and robust device technology. In general, for airborne systems there is less emphasis on covert operation, but suitable wavelength emissions are demanded for minimising eye hazards.

10.9 Summary

Laser range-finding techniques have evolved dramatically since flashlamp-pumped ruby laser technology was used to make the very earliest time-of-flight measurements between the laser and a target. Most of the relatively simple systems now operate in the eye-safe regime, which has had a very significant impact on the operational use of these systems, along with the training that has been required.

The modern laser range-finder system is very compact, rugged and efficient, which makes is suitable for use by the dismounted soldier. This type of system enables the range to a target to be determined rapidly and to an accuracy of the order of 0.1%; clearly accurate enough to support most indirect-fire operations and ensure the optimum use of ammunition. However, it is worth noting that the challenge is still to achieve an accurate estimate of angle, as digital magnetic compasses are not usually very accurate.

This technique is supported by an active laser emission, so the position of an operator may be observed by potential adversaries. However, techniques have been developed to permit covert operation, using a single-photon measurement technique.

The simple range-finder system may be used in many different ways to provide additional sensing data, as well as supporting other tasks. Examples include:

- Scene scanning to build up a picture of potential targets;
- Target scanning to aid classification of targets;
- Geo-location function aid via triangulation.

These applications will be discussed in other chapters, but it is clear that accurate estimation of range is a critical part of many sensing and some directed-energy functions. Consequently, with active-imaging systems being used for targeting functions, for example, range finding is now integrated within each pixel of the focal-plane array, which is considered in more detail in the Chapter 11, which discusses burst-illumination systems.

Selected Bibliography

Fox, C. S., (ed.), *The Infrared and Electro-Optical Systems Handbook, Vol. 6*, Chapter 2, "Laser Rangefinders," Bellingham, WA: SPIE Optical Engineering Press, 1993.

Osche, G. R., *Optical Detection Theory for Laser Applications,* New York: Wiley Interscience/ John Wiley & Sons, 2002.

CHAPTER 11

Laser-Based Targeting

The rules of engagement used by many armed forces demand that any target is identified in one way or another before being engaged, to comply with the Geneva Convention of armed conflict, especially to minimise civilian casualties. Laser-based techniques may be exploited to achieve an adequate identification of a target through:

- Use of a laser system to acquire a high magnification image of a potential target for accurate identification, often using 1.5-μm technology;
- Appropriate designation (or marking) of an identified target, which is currently undertaken with 1-μm laser beams;
- Further use of a high magnification image of a target to assess battle damage after an engagement or identification of suitable aim points.

11.1 Introduction

There are a number of military functions that may be considered under the broad heading of designation or targeting. The distinction between these two techniques may be summarised as follows:

- Targeting is just one part of the target-engagement process for countering a threat device or weapon with another weapon system.
- Designation is the process of selecting a potential target for any other military function, such as direct engagement or observation.

One subtle distinction between these two important military functions may be that targeting involves the initial stage of the target-acquisition process, including the initial search, detection and acquisition activities, whereas the designation process follows on from the acquisition of a target; designation may be defined in terms of illumination of a target, in this case with a laser. Laser-based targeting is the topic of this chapter (i.e., finding the target, which may be difficult to detect for a whole range of reasons).

The fundamental physical processes involved with these particular military functions involve the detection of the laser beam scattered from potential targets.

Initially, this part of the function involves simple detector techniques, but more modern detector-array technology has been successfully used in these systems, leading to enhanced targeting capability, as discussed in some detail later.

There are a number of forms of laser-based targeting and many terms are used to describe the methods devised to achieve this function. A common technique is range-gated imaging or burst illumination or gated viewing; these are all terms used to describe the same process. This targeting technique is a promising active-imaging technique, which may be used for target detection and recognition. This approach can use any wavelength for the illumination of the scene, provided it can be supported by the appropriate receiver technology; currently, the favoured approach is to use the eye-safe regime (around 1.5 μm), which has become a reality owing to the successful development of suitable detector devices. For example, indium gallium arsenide photo-diode technology, photo-cathode methods (image intensifier mode) extended to the 1.5-μm region and cadmium mercury telluride arrays, operated as avalanche photo-diodes.

Laser-designation systems have been designed that use a laser source that is common with a simple targeting system, such as a laser range-finder, often called laser-targeting and designation systems; the range-finding aspects were discussed in Chapter 10. Laser-designation techniques have been developed that enable a specific target to be illuminated by a laser beam after it has been detected through a targeting process. Many structures scatter laser light and thus may be designated in this way, so that a suitable seeker in a guided weapon can home on to the laser light scattered by the designated target.

Laser-based designation of targets has been very effective in achieving precision-bombing techniques. This aspect of designation is covered in far more detail in Chapter 12 on laser-based guidance. Moreover, countermeasures to this very effective guidance technique are discussed under the topic of laser decoys in Chapter 16.

11.2 Laser Targeting: Burst Illumination/Active Imaging

11.2.1 Introduction

This application of laser technology may be considered to be a form of remote sensing; in this case, the purpose is to recognise and identify an already detected distant target, usually in difficult conditions such as a heavily cluttered environment. In these circumstances, the target may not be detectable using conventional observation techniques, with or without illumination of the scene by a laser.

The continuing military problem of identifying and then classifying targets as a threat is fundamental to rules of engagement. This critically important process needs to be undertaken rapidly and at ranges beyond where the threats on the battlefield can engage you.

Passive techniques are preferred, but there are many physical processes that occur along the line of sight between a sensor and its prospective targets, which will frustrate and often prevent a target from even being detected, let alone acquired, identified and tracked. These processes that affect the transmission of photons through the atmosphere have been discussed in some detail in Chapter 5. Target clutter adds an additional complexity to this already complex task; this clutter may generate target-like returns or take the form of target-like objects in the target zone, which confuses the targeting process.

Many techniques have been investigated and some developed to provide an edge to the target-acquisition process over an enemy's defensive capability. One of the important developments in the past has been the use of thermal imaging sensor technology. This technology has provided excellent capability in many scenarios and has been extended to different parts of the infrared to realise enhanced sensing and detection performance. However, this technique in its simplest mode of operation is passive and thus has limitations in some circumstances, particularly in low thermal contrast conditions (close to the thermal cross-over condition) and when the atmosphere is in a highly turbulent state, which is well documented in many references and is discussed in some detail in Chapter 5.

Burst-illumination techniques are an attempt to enhance the targeting performance of a sensor by using a laser to illuminate a scene; the back-scattered light is viewed during the illumination by passive-imaging techniques. By operating a camera to view the illuminated scene in a particular time-gated fashion, many of the physical effects that limit the passive approach can be avoided, if not fully mitigated; however, care is required as long duration stare/integration times lead to a noisy image being captured. This technique uses wavelengths shorter than the thermal band for greater spatial resolution and enables the system to operate during the hours of darkness.

By combining active and passive imagery, it is possible to provide enhancements to the image of the scene and gain more information about the potential targets in the captured illuminated scene. This technique exploits the complementary nature of active- and passive-imaging systems.

The recent developments with avalanche-gain detectors have led to the realisation of a multifunctional focal plane array, which is compatible with both passive thermal imaging and active (two- or three-dimensional) imaging. These modern multifunctional devices may be switched from one mode to the other on a frame-by-frame basis. This development offers a clear route to enhance electro-optical systems that already contain thermal imagers and laser range-finders/designators with a multifunctional array, which offers enhanced detection capability against difficult to observe or obscured targets in complex environments. These sensor modes are extremely complementary; active-imaging systems provide their clearest images close to dawn and dusk, at the thermal cross-over between the ground and the air when turbulence is at its minimum during the diurnal cycle (discussed in Chapter 5). This time of day offers the lowest thermal contrast and poorest performance from thermal imagery.

11.2.2 Historical Development

The origins of this relatively new technique may be traced to system developments from the idea of using an incoherent infrared source to illuminate a scene and viewing that scene with a thermal detector; later developments used a thermal camera to view the scene. The development of laser technology enabled the scene to be illuminated with coherent photons to increase the ambient signal level and allowing observations of the scene to be made with image intensifiers. The latest developments have evolved from these principles, but now using a thermal-type camera and active in-band illumination.

More recent developments with both the laser technology and the receiver devices in appropriate operational wavebands have enabled an operator to capture images of the illuminated scene and identify objects of interest at a known range. Sophisticated processing has enabled these objects to be classified and identified in real time.

Many of the advances in effectiveness of this technique have arisen from the development of advanced focal-plane arrays detectors, such as the integration of the time-of-flight processing into the read-out circuitry of each pixel in the array, whilst being operated in the modern snapshot mode. Hence, each pixel may be considered to be a single laser range-finder or profiler, the latter aspect being discussed in Chapter 13.

11.2.3 Military Utility

The military value of detecting, acquiring and even identifying targets in a scene cannot be over stated. As sophisticated passive-imaging techniques have been developed, then more care has been given to the design of most military equipment, to make systems less observable across the electromagnetic spectrum, including retrofitting or modifying existing equipment. Basically, targets are becoming more difficult to find on the battlefield, let alone classify or identify.

The use of clever concealment tactics and successful application of stealth or low-observable techniques, as well as the impact of photons propagating longer distances through the atmosphere, have all made the use of passive techniques alone for long-range target detection and acquisition less effective, if not extremely challenging. Adding the dimension of active illumination to a scene of military interest with a laser has restored some of the technical edge lost by passive techniques through the successful development and implementation of low-observable techniques and technologies. The combination of active and passive images into a composite image has enabled further enhancements in targeting performance to be achieved. Additionally, a series of active images slid through the scene may be combined to generate a three-dimensional image with high precision range resolution.

11.2.4 System Operation

The fundamental principle of this target-detection technique is to use an energetic pulsed laser to illuminate a sector of interest, which is usually a cluttered scene, and use a gated detector to collect the photons scattered from the illuminated scene over a carefully defined time period. The photon beam will propagate at the speed of light through the scene, diverging as it progresses and interacting, in one way or another, with everything in its path, as it travels through the illuminated zone.

When this scene is viewed by a normal camera, such as a thermal imager, the image would be very confused, particularly if the scene contained a lot of clutter and if the target is at long range. This is shown by the image captured over a short integration period in Figure 11.1.

This confused and cluttered image may be transformed by the application of gating techniques to control the exact time when the gated camera views the

Figure 11.1 A cluttered scene view in the mid-wave infrared.

illuminated scene; effectively creating a shutter and controlling the integration period of the detector subsystem within camera. The concept of gating imaging is shown schematically in Figure 11.2. This figure shows that the gate width, if it has duration τ, then the range depth of the scene viewed by this duration of the shutter being open is given by:

$$\text{Range Depth } = (c \times \tau/2)$$

These systems tend to operate in the short-wavelength infrared, so that relatively compact systems may be produced with a good resolution for a modest aperture size and, moreover, covert to the naked eye when in operation. The use of wavelengths close to 1.5 μm has significant operational advantages, owing to the

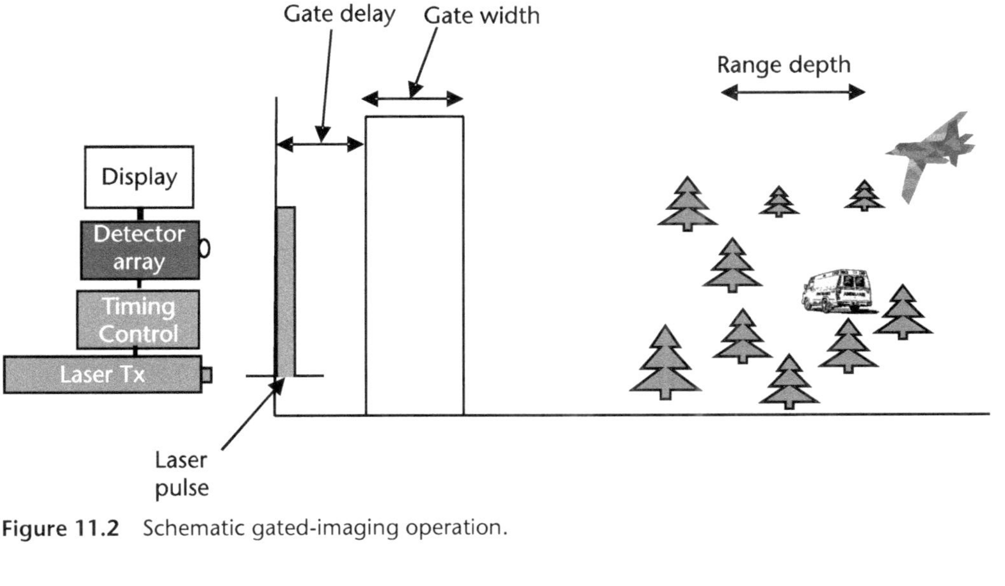

Figure 11.2 Schematic gated-imaging operation.

use of an eye-safer wavelength. This aspect of operations is discussed more fully in Chapter 17.

11.2.5 Laser Performance Requirements

Many of the demanding targeting requirements also require covert operation. The successful development of suitable imaging equipment has necessitated the use of effective laser devices, compatible with the waveband of operation of the focal-plane array in the detector system/camera. The laser technology needs to be efficient, robust and reliable, so that it can be used in any of the military environments.

Direct-generation techniques are usually preferred for the laser source, which uses an erbium laser in one form or another if the burst-illumination system is operated at 1.5 μm. Other laser ions may be used and conversion techniques can be applied for some applications, but the majority of the modern laser sources use erbium technology. If the burst-illumination system is used in conjunction with a designation system, then the source is most likely to be a Nd^{3+}:YAG source for the designation process and conversion to 1.5 μm for the targeting function.

A summary of the typical laser performance parameters for short- and long-range active-imaging systems are shown in Table 11.1, including single-detector scanning systems.

11.2.6 Subsystem Description and Function

A simple system is shown schematically in Figure 11.3; the laser provides the illumination of the selected scene with pulses of photons and the camera with its detector array is there to capture the return from the scene. The power in this technique is in the detector technology in the receiver, enabling the various spatial slices of the scene to be viewed independently, which may be stitched together to produce a three-dimensional image through use of advanced image processing. Modern sensor systems allow the range to a point on the target to be measured by each pixel within a gate period, which allows direct generation of a three-dimensional image (i.e., three-dimensional burst illumination).

The camera in the detector is controlled electronically, so that it only stares at the scene being illuminated by the laser pulse at a predetermined time after each laser pulse leaves the transmitter. Moreover, as indicated earlier, the integration time of the array is also controlled by this electronic shutter. Some images illustrating

Table 11.1 Laser-Performance Parameters for Active-Imaging Systems

Source Parameter	*Short Range*	*Long Range*	*3-D Imaging*
Wavelength	Eye safe (>1.5 μm)	Eye safe (>1.5 μm)	Eye safe (>1.5 μm)
Pulse energy	~10 mJ	>100 mJ	~10 μJ
Pulse duration	~10 ns	~10–15 ns	~3 ns
PRF	~20–30 Hz	~20–30 Hz	~1 kHz
Divergence	~10 mrad	~10 mrad	0.1 mrad
M^2	~3–4	~3–4	~1

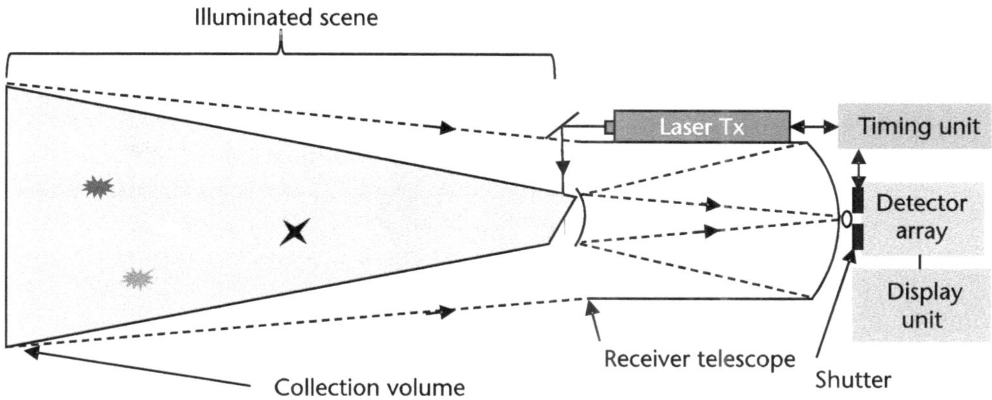

Figure 11.3 Simplified gated-imaging system.

the potential power of this technique are shown in Figure 11.4; these were collected with a focal-plane array, where each pixel is operated as an avalanche photo-diode detector and the gate position is varied.

The clutter may arise from many features in the scene, but an important mechanism is the back scatter of photons in the laser pulse from the atmosphere, which can be a critical mechanism for other sensing functions, as discussed in other chapters of this book. One of the crucially important and valuable features of gated imaging is to enable such a system to form high-contrast images over a short integration period, often of the order of a few tens of nanoseconds, at ranges that even exceed the visible range (in meteorological terms). The performance of range-gated systems is generally limited by a number of factors, including the sensor (detector) parameters, as well as the target reflectance and atmospheric-induced effects, such as speckle patterns, beam scintillation, beam wander and image dancing.

Sometimes, there is a military need to operate a gated-imaging sensor in a passive mode, because of a requirement for complete covertness. Alternatively, there may be some advantages accrued from the combination of passive and active images taken by the same sensor, particularly to improve interpretation of the scene using natural illumination. In this case, image-fusion techniques are required to be applied to the two data sets to achieve the optimum image.

The active-system technique can, for example, take a silhouette of the target against the terrain behind the target, or even against an atmospheric background, provided that a sufficiently long gate time is used in the detector and, additionally, when the natural illumination is very low. The target texture can be obtained from

1000 ns gate width — 50 ns gate centred on target — 50 ns gate behind target

Figure 11.4 Burst-illumination system images. (Courtesy of Selex, U.K.)

either the laser light and/or the natural illumination and may be superimposed on to the silhouette to improve the recognition task.

The natural illumination will affect a detected gated image differently, depending on the range to the target and duration of the period that the detector is open (gated width of the detector). If the natural illumination is high, then the integrated path radiance may reduce the contrast ratio between the target and the background. The natural illumination is incoherent, which largely reduces scintillation and target-induced speckle patterns as there is a single path for the light, rather than the double pass with active illumination.

Close to the range limit of the detection system, the shot noise in the detector is the limiting factor in the image quality. Techniques such as frame-to-frame integration are often used to reduce the scintillation and target-speckle effects; however, in this case, it is likely that the image dancing from turbulence and atmospheric-coherence time become more important. Consequently, frame-to-frame integration needs to include image registration.

This active-imaging technique allows range-resolved three-dimensional images to be formed through reconstruction of the scene from a series of sliding gated-images taken through the scene, achieved by altering the delay time for the camera shutter to open, which results in a succession of images being captured by the receiver. The use of this process of sweeping the gate through a scene allows the range to a target to be determined even in a very complex and complicated scene. The range accuracy and resolution for this imaging technique depends on the single-frame noise statistics, as well as the image-dancing and beam-wander characteristics.

The recent developments in detector technology, mentioned earlier, which enables each pixel to act as an independent range-finder, means that it is now possible to determine the range to each pixel on the target and thus form a three-dimensional (strictly 2.5-D) image of the target from a single laser pulse. This development greatly enhances imaging capability against difficult occluded targets and will also dramatically improve the ability of automatic (computer-based) extraction of information from an image. Feature extraction with two-dimensional images can be easily confused by contrast changes, such as those introduced by shadows; however, full three-dimensional information allows the ground plane to be extracted, leaving only the objects of interest. Current technology captures both the range and intensity on each pixel as illustrated in Figure 11.5, which shows a representation of a three-dimensional image of a land-rover vehicle with the intensity information draped over the three-dimensional point-cloud.

In general, the range resolution and precision achieved can be determined from knowledge of the laser signal and camera operating parameters. Use of additional data contained within each frame and application of intelligence, can lead to targeting performance that exceeds the performance expected from the system-design parameters. Examples of the enhancement to baseline performance through advanced image processing including noise-reduction and use of image-sharpening techniques. The latter technique can be applied by estimating the effect of atmospheric turbulence on the image quality, such as distortion to edges of objects captured in the image, which can also lead to image sharpening.

Recording of active images through smoke and fog has demonstrated the enhanced penetration capability achieved by illumination of the scene with beams

Figure 11.5 Three-dimensional image of land vehicle. (Courtesy of Selex, U.K.)

having a wavelength of 1.5 μm compared with the passive sensors operating in the visible or near infrared wavelengths, although it should be noted that generally passive thermal sensors operating in the far infrared have better penetration performance in these conditions compared with passive short-wave infrared devices, as discussed in Chapter 5. The use of passive imaging, using the far-infrared wavelengths, can vary significantly, as it is dependent on the particle-size distribution of the obscurant; hence, in some circumstances it may have superior penetration capability and provide a better image than those from shorter-wavelength active imaging.

The influence of target and atmospheric-speckle patterns on the quality of the image may be reduced by frame averaging, using image-motion compensation algorithms. A median filter, which averages over a block of pixels, may be applied successfully to images from a stationary sensor and target to enhance the signal-to-noise ratio.

Recent developments have shown that the active laser-imaging technique has many advantages over conventional (passive) thermal imaging, particularly now that the technology can support effective operation at around 1.5 μm, resulting from the use of the indium gallium arsenide or cadmium mercury telluride detector devices. The shorter wavelength gives improved resolution for the benefit of target identification at extended ranges, additionally the reduced imaging duration in the region of approximately 10 ns compared with an approximate 1-ms integration time for a mid-wave thermal imager, effectively freezes platform vibrations, which reduces the system requirements for stabilisation and vibration isolation.

However there are also a number of shortcomings and challenges that still have to be solved, which have restricted the full exploitation of this technology, particularly nonuniform illumination of the scene. Nonuniform illumination of the scene may result from scintillation in the atmosphere and speckle effects. These phenomena can result in images with high internal contrast being captured by the system, which are difficult for humans to interpret, as mentioned earlier.

Speckle is a coherent effect that arises when the surface roughness of a target is of the order of the coherence length of the laser emission, whereas scintillation results from beam break up as it propagates through a turbulent atmosphere, as discussed in Chapter 5. Under low turbulence conditions, the spatial noise in the image is dominated by speckle, which is generally a minor inconvenience; however, as the turbulence strength increases to moderate or higher values (as defined in Chapter 5), the optical distortion of the image caused by the atmosphere severely degrades the image quality captured by the camera. Techniques have been developed to mitigate speckle-induced effects by, for example, extending the transmission aperture.

Image-processing techniques that dramatically improve the image quality under turbulent conditions are being developed to solve this problem as illustrated in Figure 11.6. The area within the grey box has been processed to improve the image quality, using various compensation algorithms, without resorting to application of adaptive-optical techniques. A number of different techniques are being developed with the simplest based on averaging with frame and subframe registration. More advanced techniques which measure the phase distortion are also being developed and may lead to even better passive compensation of blurred images.

Lucky imaging is an example of passive compensate for image distortion caused by atmospheric aberrations. This is discussed in some detail in Chapter 5.

11.2.7 Future Requirements

Future systems need to exploit these advanced-processing techniques to generate real-time stabilised images, even in the presence of severe turbulence in real time. Such development will allow longer-range target detection and could have a significant impact upon military systems, particularly when used for ground-to-ground applications.

The normal demand for more robust and efficient low-cost laser technology applies here, especially to support dismounted military operations.

11.2.8 Summary

The principle of the technique is based on the use of a coherent beam of photons to illuminate a sector of space of interest and control the region that is observed with an imaging-sensor system. In many circumstances, this technique provides significant military advantage for targeting functions over the more classical passive thermal imaging on the battlefield.

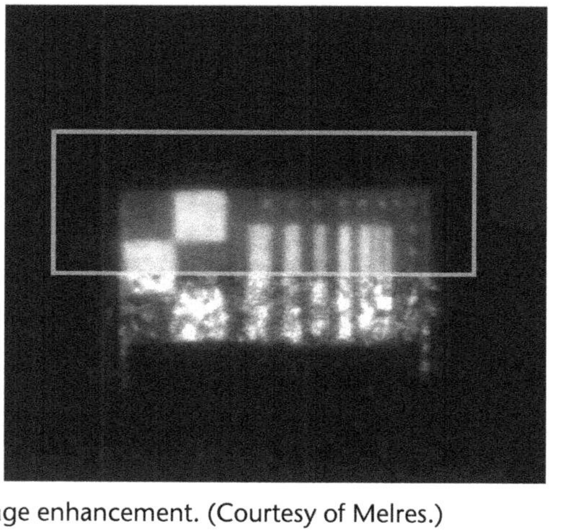

Figure 11.6 Passive-image enhancement. (Courtesy of Melres.)

The development of efficient and effective focal-plane-array detector technology, which can be supported by suitable laser-device techniques, has enabled the impact of many types of clutter in images of scenes to be reduced dramatically, if not totally eliminated; this is through the use of gated-imaging techniques. In its simplest form, the technique allows only a given (selected) portion of a laser-illuminated scene to be viewed over a very short period. This brief glimpse of the scene reduces the impact of other objects at different distances from the laser, which may provide an optical return to the detector and degrade the optical quality of the image; moreover, in many cases, these additional returns obscure the target, despite being specifically separated within the field of view of the detector system.

The use of a laser to illuminate the scene being observed enables the integration time of the detector array to be reduced significantly, in some instances this time may equate to the duration of the opening of the shutter used in the gating process. Even with very short integration times, the contrast across the scene can be maintained, so that objects may still be observed. Most of these laser devices are operated with an emission wavelength in the eye-safer regime around 1.5 μm.

Much of the power in this technique is in the processing of the large number of images that may be observed during the illumination with a pulsed laser, and even potentially, with a single pulse through movement of the gate position. The latter technique involves the detector shutter being operated with various delays after the laser is fired, enabling a series of images that appear to pass through the scene to be captured. These images may be processed to provide a vast range of information about the content of the scene; moreover, the images may be stitched together to form a three-dimensional image.

Modern developments with focal-plane-array technology, involving time-of-flight processing to a point in the scene on the chip, for each pixel in the array have been realised; each pixel can operate as an individual range-finder. Consequently, a three-dimensional image, or point cloud, of the objects in the scene can be formed, enhancing the targeting function, leading to enhanced military utility.

In some atmospheric conditions, or for operational reasons, passive imaging may be superior or necessary. The most recent detector arrays enable the operation of the same imaging camera to be switched from an active to a passive mode on a frame-to-frame basis, enabling the optimum performance to be achieved, which can have significant operational advantages.

11.3 Active Spectral Imaging

Passive multispectral and hyperspectral imaging techniques for sensing, and possibly targeting, have proved to be invaluable tools for a range of sensing applications for some period of time. These techniques have been shown to be effective for the detection and identification of targets, through the collection of spectral characteristics of targets captured in a scene. These techniques have been applied successfully to a number of military functions. The principle of passive-hyperspectral imaging is shown schematically in Figure 11.7.

Recent developments have seen the use of broadband lasers to provide illumination of a scene, instead of relying on solar illumination, which eliminates a

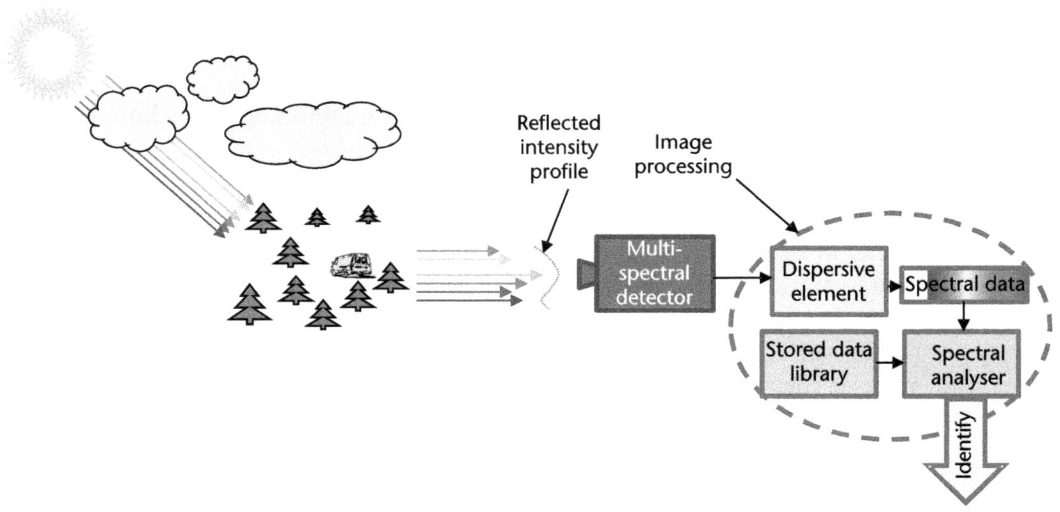

Figure 11.7 Passive hyperspectral imaging.

number of the operational constraints, which passive technique imposes on some operations, especially the limitations of use during darkness or cloudy conditions.

11.3.1 Introduction

Active-spectral imaging, which is essentially a combination of active-imaging techniques combined with passive hyperspectral-imaging methods, allows greater flexibility through complete control of the illumination of the scene. The principal problems with the more usual passive techniques concern relating the received spectral radiance to the actual spectral characteristics of a target, or its material composition, to identify it from known data. The problem of identifying a target using passive techniques, from its spectral return, arises because of the dependence of the illumination on a range of external factors, such as solar angle, clouds, shadows, atmospheric scatter and atmospheric extinction.

The passive technique is also generally limited to the visible and short-wave infrared wavebands, owing to the solar emission characteristics; active illumination will extend the spectral range to include the mid-wave and long-wave infrared parts of the electromagnetic spectrum, provided a suitable laser source is available.

This added dimension of independent illumination largely eliminates the limitations imposed by light from the sun; moreover, it provides a method for acquiring polarisation information about targets, as well as multidimensional data from the scene, as discussed in more detail in Chapter 13. The technique has progressed as a result of the advances in focal-plane-array technology, discussed earlier. In this case, spectral information is gathered in each pixel, which offers the potential for support to many military functions and capability requirements.

There has been extensive investigation of the spectral-reflectance characteristics of natural and man-made materials and objects from the passive-multispectral and hyperspectral projects over a number of wavebands. This information has helped to form the library for the discrimination and identification of returns from objects in an illuminated scene. It is interesting to note, that some objects are colourless when viewed in the visible band, but display coloured signatures in the mid-wave band.

This technique offers significant advantages for the following applications:

- Target and material identification, classification and potentially recognition;
- Stand-off detection of dangerous substances, using differential-absorption-lidar (DIAL) techniques, Raman spectroscopy and conventional spectroscopy;
- Detection of small obstacles, such as trip wires.

A range of detection methods are feasible for analysing the spectral returns, examples are:

- Fluorescence spectra;
- Differential scattering;
- Spectral back scatter.

These techniques are discussed in more detail in Chapter 13.

11.3.2 Military Utility

This new and evolving technique has similar utility to the well-established gated-imaging technique, discussed above. The added dimension of collecting spectral information or identification of polarisation characteristics of the target provides additional information to aid detection of difficult targets, possibly leading to very accurate recognition or classification of those targets. The additional spectral dimension allows some enhanced military functions to be undertaken given the appropriate supporting subsystems, examples include:

- Topography, including classification/characterisation of culture and vegetation of the terrain;
- Bathymetry, including characterisation of the sea bed;
- Stand-off detection of hazardous substances and agents;
- Surface-mine detection;
- Detection of improvised explosive devices;
- Weapon guidance: active-imaging, laser-radar guidance in three dimensions, with autonomous navigation, target search, detection, identification and precision aim-point selection.

An additional operational use of this technique is the selective illumination of a scene using an appropriate wavelength that is in-band to a specific detector in a system, such as image intensifiers, low-light television or night-vision goggles. This application of the technique enables enhanced performance from these devices for recognition and identification, even in adverse viewing conditions.

Another application is the use of wavelengths for covert illumination, so that the illumination is out of band to the normal surveillance systems. This approach has many attractions for a number of military applications.

The use of active illumination for spectral sensing has many advantages, in particular, removing the dependence on daylight for the illumination of the scene of interest. This approach also eliminates the vast majority of the other complications associated with the passive techniques, as discussed above.

11.3.3 System Configuration

The operation of this system relies on the generation of multispectral light from a laser source to illuminate the scene of interest. A number of laser configurations have been investigated, including:

- Super continuum;
- Raman lasers, such as Raman scattering in fibre lasers;
- Multiple individual lines in wavelength zones of interest (from discrete devices), and suitably combined, at a common aperture;
- Tuneable emissions from optical parametric oscillators.

A suitable detector array is used to monitor the return from the scene, triggered by the laser pulses to enable the times of flight to various potential targets be measured. Inclusion of either a tuneable filter or a dispersive element enables an estimate to be made of the spectral content of the return from the illuminated scene. Three-dimensional flash detectors have been demonstrated to give a three-dimensional video stream of data from the illuminated scene.

Detected target returns may then be analysed and assessed by comparing the captured spectral content with stored data in the library/databases. Complex algorithms may be applied to the data set to aid target identification and then recognition.

The normal mode for illumination of the scene is through flash techniques and gated viewing, as shown schematically in Figure 11.8. Scanning techniques are feasible, but at the expense of greater complexity and reduced speed of response.

11.3.4 Laser Performance Requirements

The laser-emission characteristics need to be compatible with the potential features to be investigated from the optical returns generated in the illuminated scene, such

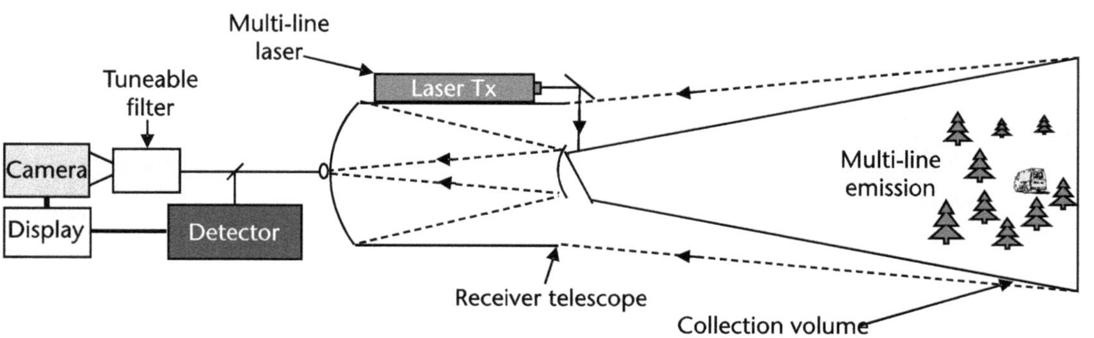

Figure 11.8 Schematic of a flash active-imaging system.

as the polarisation characteristics, other features or induced fluorescence. The combined emission must be compatible with good atmospheric propagation to the target area for these wavelengths. To exploit the different spectral features of targets, a range of laser wavelengths is normally required, preferably simultaneously, to enable rapid sensing through specific reflection characteristics generated by targets in the illuminated scene.

Generally, short-pulse, high repetition-rate emission characteristics are necessary from the laser technology used to support this application. The laser-performance parameters are summerised in Table 11.2.

An alternative approach to the use of multiple emission lines, involves a very broadband emission and use of tuning elements in the receiver channel to identify the returns from the target.

11.3.5 Subsystem Description and Function

Modern developments with solid-state laser technology offer devices with frequency agility, multiple wavelengths and short-pulse emission, which are potentially ideal for many sensing applications. Some configurations of optical parametric oscillators enable broadband-emission characteristics to be achieved from a single device. Moreover, many of these devices use short-pulse pump laser techniques, to achieve efficient nonlinear conversion, as discussed in Chapter 2. Very high repetition rates are also feasible from these devices, which are ideal for many of these applications.

Super-continuum sources are being developed, where the emission characteristics are compatible with the simultaneous wavelengths required from the illuminator for a rapid response from this type of targeting system. Recent developments with super-continuum devices have demonstrated significant emission in the mid-wave infrared, where many of the interesting spectral characteristics may be observed from many targets of interest to the military.

Tuneable filters, which are positioned in front of the detector array, are used to select the spectral pass band of the receiver system. The detector array has the same function, and can be operated in an identical manner to the gated-imaging system detector, described above. The principal difference between these two laser-based targeting systems being that each image is confined to a specific spectral range with a burst-illumination system, whereas the active spectral-imaging technique

Table 11.2 Summary of Laser-Performance Characteristics for Active Spectral Imaging

Source Parameter	*Active Spectral Imaging*
Wavelength	Eye safe (> 1.5 μm) – multilines (3 or 4) and a spectral width of ~300 nm
Pulse energy	~10 mJ per line
Pulse duration	~10 ns
PRF	~20–30 Hz
Divergence	~ 10 mrad
M^2	~3–4

provides additional spectral information, albeit at a cost of speed and complexity. These arrays are usually based on similar avalanche photo-diode array technology.

The signal-processing techniques are a critical aspect of this active spectral-imaging technology and are based on well-established techniques and methods, developed for passive multi- and hyper-spectral imaging. These image-processing techniques include a vast range of anomaly-detection algorithms, used to identify potential targets that stick out from the background, when appropriate signal-processing techniques are applied to the captured data.

The various spectral characteristics received from the illuminated target, may be used to compile a list of specific received features exhibited by this particular designated target. Reference to the databases of known target spectral characteristics may lead to a unique recognition and ultimately, its positive identification. Clearly, an efficient protocol is required to enable a rapid and efficient interrogation of the stored target data sets.

11.3.6 Future Requirements

The critical future requirements relate to enhanced tuneability and frequency agility from the illuminator. Development of enhanced image-analysis algorithms and continuing investigations to extend the database and library of effects are necessary, to match the rapid progress with active spectral-imaging techniques.

Advanced processing algorithms for switching between passive and active modes are required, along with the techniques for fusing the images from the different viewing methods, to produce the best composite image in the various parts of the spectrum.

This technique is generally limited to short ranges with the current technology; hence, application of more powerful multiband lasers would enable greater stand-off ranges to be achieved. The military utility could then expand from sensing to targeting.

11.3.7 Summary

Gated imaging has improved the targeting function significantly for both air and ground operations. This relatively new technique of active-spectral imaging uses a similar method to the gated-imaging technique, by providing active illumination of a scene and then applying multispectral and hyperspectral algorithms to detect and identify targets from the captured data. This technique is particularly important for finding targets that are normally difficult to detect, let alone identify.

The use of active illumination to support the spectral-imaging technique, removes many of the complications and limitations imposed by use of daylight with passive spectral systems. One of the most important advantages is the ability to operate during darkness, or when the natural illumination is poor. Some other advantages of active illumination are the enhancement in performance of surveillance sensors, such as night-vision goggles and image intensifiers. Covert illumination is also feasible, so that the scene can be illuminated but that illumination is not detectable by the common viewing systems.

The ability to collect spectral and polarisation characteristics from targets illuminated by the laser device, or devices, can enhance the target detection and target recognition. Application of anomaly-detection algorithms, developed to support multispectral- and hyperspectral-imaging systems, enable difficult targets to be detected and classified.

This developing technique of active spectral-imaging is demanding on laser technology, as multiwavelength emissions and frequency agility are required to exploit the potential advantages of the combination of active-gated, multispectral, imaging technology fully. Super-continuum laser technology, emitting in the appropriate wavebands, appears to offer significant advantages over other laser-based device technology.

Selected Bibliography

Eismann, M. T., *Hyperspectral Remote Sensing,* Bellingham, WA: SPIE Optical Engineering Press, 2012.

Fox, C. S., (ed.), *The Infrared and Electro-Optical Systems Handbook, Vol. 6*, Chapter 1 "Laser Radar," Ann Arbor, MI: ERIM and Bellingham, WA: SPIE Optical Engineering Press, 1993.

Osche, G. R., *Optical Detection Theory for Laser Applications,* New York: Wiley Interscience/ John Wiley & Sons, 2002.

Richmond, R. D., and S. C. Cain, *Direct-Detection Ladar Systems,* SPIE Vol. TT 85, Bellingham, WA: SPIE Optical Engineering Press, 2010.

CHAPTER 12

Laser Designation, Guidance and Fuzing

12.1 Introduction

An important, and continuing, design aim throughout the development of weapon technology has been to ensure that the payload in a projectile can achieve its design objective; one of the major reasons for any weapon failing to achieve its ultimate objective is that it does not deliver its payload to the right place at the right time. Moreover, as the need, or the desire, for the stand-off distance between opponents has increased, the problem of achieving accurate weapon guidance has only been exacerbated, in some cases quite dramatically.

As the general technology base has advanced, there have been two distinct strategic approaches to the development of conventional weapons:

- The use of very simple and therefore cheap weapons that may be used in large numbers in attempts to ensure that there is a high probability of at least one hit, taking into account dispersion of each weapon and the chances of achieving a strike on the target. These are often called dumb weapons and a major requirement is to ensure that they are cheap enough for them to be cost effective when used in large numbers. This strategy has an adverse impact on the logistics footprint, not to mention collateral damage implications.
- The investigation and design of more sophisticated weapons to ensure that they have a high probability of achieving a mission success with the use of a single weapon; these use very advanced technology and, consequently, tend to be expensive. This latter approach is probably best embodied in the development of advanced guided-weapon technology, such as the cruise missile during the twentieth century. The navigational accuracy and guidance precision achieved in these weapons enables a specific building to be attacked with minimal collateral-damage to the surrounding area.

Guided-weapon technology has faced many challenges since the first introduction of this type of weapon during World War II. If the weapon is going to be successful, then there are many technical trade-offs and compromises that have to be made if the weapon system is to be affordable, but one critical trade-off concerns guidance accuracy, the size of the warhead and sophistication of the fuze.

A range of design solutions is possible, particularly for short-range weapons. For example, if highly accurate guidance precision can be achieved, so that the random errors in the guidance system are negligible, then the guided weapon may be termed a hittile and the fuze need only sense the impact on the target during the interception, effectively the fuze becomes a contact switch. Furthermore, the warhead size may be reduced, in comparison with a lethal payload used in a missile, leading to a much smaller and lighter weapon, as it will require much less propellant, through transporting a lighter payload to its target. This type of weapon system may then be man-portable, which has critical advantages for an operator and may be used in asymmetric warfare.

The alternative approach with more conventional guidance techniques assumes that the payload will be delivered to a point close to its target. To ensure success, there will be a need for a proximity fuze to sense the most appropriate time to trigger the warhead, particularly if there are complicated engagement dynamics in the terminal phase during the interception of the target. Moreover, a large warhead may be required, necessitating the use of a larger motor. This form of technical solution is more likely to be appropriate for longer-range weapon operations.

There have been many techniques and methods investigated to realise very accurate guidance, many of these techniques involved radar systems:

- Active and semi-active radar homing systems sense the radar reflection from a distant target and provide the aim point for the interception of that target. These systems are well established and effective, but tend to be sophisticated and expensive; typically, the radar homing head will be of the order of 25% of the cost of the weapon. This entire critical subsystem is only used once in the case of the active system, whereas with a semi-active system only the receiver part of the homing head is lost once the weapon is fired.
- Radar beam-rider methods create an electromagnetic beam to define the path that the weapon is to follow to intercept its target; this radar beam also tracks the target, as well as defining the aim point. This technique requires a large radar antenna to achieve a narrow beam at a distant target, owing to the wavelength of the radiation and the physical diffraction phenomenon. However, the weapon does use relatively simple receiver technology to determine the position of the projectile within the beam. Some variants on this approach include command-to-line-of-sight methods where a radar beam tracks a target, as well as a beacon in the rear of the guided weapon; guidance commands are calculated from the offset between the beacon and the line of sight to ensure the missile manoeuvres on to an appropriate interception path with the tracked target. Again the target-tracking system is large, sophisticated and generally expensive.
- Infrared-seeker technology senses and tracks the infrared emission from a target and is considered in detail in Appendix A. This technical solution uses relatively simple technology and can result in relatively small devices, but again the homing head (infrared/optical seeker) is used just the once.

The range of trade-off options is clearly enormous, as every technology has something to contribute to the potential solution to any given military-capability requirement. In a number of cases, the successful development of laser-device technology has added an additional dimension to the options to be considered. For example, the successful development of compact and efficient laser technology has resulted in potentially exceptional guidance performance, based on variants of the radar-based techniques such as semi-active guidance or beam-rider technology outlined above. Other homing techniques using laser technology feature laser-radar techniques, such as Lidar; laser-radar techniques are discussed further in Chapter 13.

The application of laser technology to ordnance delivery has had a very significant benefit to the use and operation of many conventional weapons on the battlefield.

This chapter considers three aspects of the exploitation of laser technology in conventional weapon systems:

- Laser designators, used to provide a conventional bomb with a precision homing system resulting in a very accurate delivery of the ordnance and thus transform it into a smart weapon;
- Laser beam-riding techniques, used to provide a low-cost precision guidance method for short-range guided weapons;
- Laser-fuzing techniques, a covert approach to sensing the precise position of a threat/target during a terminal engagement.

12.2 Laser Designation

The development of the concept of the use of a laser beam to illuminate a target and then detecting the energy scattered and reflected from the target to provide a beacon for a homing sensor was one of the first airborne military applications of laser technology; it was certainly the first successful operational use of a laser-based system for weapon guidance. This application followed closely behind the use of laser devices, linked to a receiver, to determine the range to a target in a laser-rangefinder system, which led to the realisation of a multifunctional laser system combining both functions.

12.2.1 Introduction

Throughout the history of warfare there has been a fundamental requirement to achieve high aim-point/hit accuracy of all weapons, but particularly for ballistic munitions. For example, the bore of a gun is usually rifled so that a bullet spins to reduce its dispersion about the intended trajectory to its target and thus improve its accuracy. Dumb bombs have been fitted with aerodynamic surfaces to ensure that they follow a smooth and predictable path to the target zone.

The use of a laser to illuminate a spot on a target, so that a simple homing sensor could provide guidance commands to a control system on a simple iron bomb, fitted with control fins, is another example of a technique developed to reduce dispersion and improve targeting accuracy and interception precision. This technique is sometimes referred to as semi-active laser guidance and is comparable with the radar-based variant, known as semi-active (radar) homing. The target may be illuminated from a remote point on the ground or from an aircraft, so that a seeker on the weapon can track the laser light scattered by the designated target. The seeker may use a simple quadrant detector within the seeker to provide guidance of the weapon to the selected target.

The principle of semi-active guidance is that a target is illuminated from a remote position, so that the electromagnetic energy that is scattered or reflected from the target can be seen by a suitable receiver system integrated with a weapon. The weapon can be launched from a different location from the designator, although common location (or co-location) has some advantages particularly for air operations.

In the case of laser-designation using 1-micron illumination, a very simple stabilised quadrant sensor, using a silicon detector, may be used to detect the scattered energy from the target and then interprets this information in the target tracking sub-system to generate the sight-line errors for the guidance system. Thus, the weapon can home on to the beacon provided by the laser energy reflected and scattered from a designated target.

One of the attractions of this type of weapon system is that the designation equipment may be used for multiple missions; it's only the bomb, the target-sensing equipment on its nose and the control/target-tracking subsystem that become the expendables on each mission. The transformation in the precision achieved by the relatively simple weapon provides an enormous advantage to the attacking forces.

12.2.2 Historical Development

The American Air Force first tested laser-guided bombs in combat trials during 1968 for operational use in Vietnam to attack strongly defended strategic targets, such as bridges on critical supply routes along the Ho Chi Minh trail. These weapons were flown on the F-4 aircraft operating in pairs, with one aircraft providing the laser beam to illuminate (or paint) the target with a coded laser signal, and a second aircraft carrying out the bombing function. Initially, the operation used two aircraft, but the development of the Precision Avionics Vectoring Equipment (PAVE) Spike pod allowed a single aircraft to both designate and attack the designated target.

The impact of the emission requirements on the operational performance of the laser source for the early systems was quite demanding for the state-of-the art laser technology in the mid-1960s. The requirement for a relatively high energy per pulse (>80 mJ) and a moderate repetition rate (~20 Hz) in a compact system ruled out ruby-laser technology as the source device from the outset. At that time, the only practical solution for the source was the use of neodymium laser technology, and in particular, neodymium laser ions in a yttrium aluminium garnet host (Nd:YAG), which emits at 1,064 nm. These systems were flashlamp pumped, but the more modern approach is to use diode-pumping of that gain medium.

The improved optical pumping efficiency, available from laser-diode pumping, reduced both the total optical pump-power requirements for the laser system and also eased the burden on the heat removal sub-system, required to maintain a reasonable temperature within the system and thus minimise the distortion of the emitted beam and inefficiencies in the emission. Diode pumping of the gain medium, has enabled the laser system to be scaled in power for an equivalent volume or packaged into a reduced size with equivalent power, thus enhancing system flexibility without compromising capability. Developments in conduction cooling and improvements in diode-laser efficiency, as well as the overall efficiency, have increased the reliability of the laser-designation system dramatically.

Modern airborne systems embody an active-imaging system, typically together with a joule-level diode pumped Nd:YAG laser in an oscillator/amplifier configuration, to allow targeting and self-designation. Some systems use a neodymium laser to pump an optical parametric oscillator to generate an emission at an eye-safer wavelength for an adjunct active-imaging (targeting) system, with the remaining pump light (depleted pump) being used for the designation function.

Laser designation systems continue to use the 1,064-nm emission from a Nd: YAG laser, as defined in STANAG 3733 for NATO forces, but similar standards are observed for other coalition operations. The investment that has been committed to these systems and the stores is likely to rule out any shift to eye-safer laser technology for the designation function for the foreseeable future.

The original concept was for designation from aircraft to attack heavily defended targets deep inside an opponent's territory; modern development includes compact and efficient ground-based systems suitable for operation by dismounted soldiers. These types of operation may be close to the forward edge of the battle area or even within enemy territory.

12.2.3 Military Utility

The primary military function of the laser-designation is to provide an aim-point beacon for a sensor to track during its terminal phase up to the point of interception of the target. The very high optical irradiance of a laser beam means that most natural and man-made surfaces will scatter and/or reflect light over a broad range of angles, so that the position of the laser spot on a surface can be seen at medium to long stand-off ranges with a suitable receiver; these ranges may be in excess of 10 km for some favourable propagation conditions. This feature is exploited as a beacon or aim point for homing sensors in guided weapons.

The use of laser-based target designation for air-launched weapons is very well established. As noted above, the transformation of a dumb weapon such as a free-fall ballistic bomb into a precision guided weapon has had a dramatic impact on the effectiveness of bombing techniques and given a reduction in collateral damage, as well as the reduction in the logistical footprint.

Techniques for self-designation from aircraft are well established but the need to penetrate enemy territory to find, designate and then attack a specific target can present some significant risks to aircraft. The adoption of a modern approach to photon generation, through the use of compact and efficient diode-pumped laser technology has resulted in a further reduction in the size, weight and a reduced power requirement by the illuminator, leading to opportunities for more potential applications. For example, ground-based designators are now available, which are very small, and has presented an opportunity to provide additional flexibility for target designation from the ground by dismounted troops.

Figure 12.1 shows the lightweight laser designator (LF28A) with a prototype ultralightweight designator. Many modern ultralightweight systems exploit multicolour diode technology to improve laser efficiency and achieve the military operational temperature requirements without the need for electrically-driven cooling system to stabilise the temperature of the laser device. Requirements for designation from smaller platforms will drive a continuing need for smaller, more efficient laser sources into the future.

12.2.4 System Operation

The laser-guided-bomb system operates through the use of a target-designation pod on an aircraft, which usually contains at least one electro-optical sensor, such as a thermal imager and/or a low-light television device, to find an appropriate target. This targeting system may contain an optical zoom facility; a wide-angle field of view being used for the surveillance and target-acquisition mode and a narrow field of view for target identification and aim-point selection. Once an aim point has been selected by an operator, the video processor can track that point automatically and the sightline-isolation system compensates for motion of the platform carrying the pod. A fast-acting mirror, with two degrees of freedom, is usually at the heart of the gimballed system to achieve sightline stabilisation between the platform and the target.

The fire-control solution for the speed and angle of release of the bomb may be calculated when a stable tracking point has been established and the range to the target determined. As the weapon is released, the laser-designation process will be initiated with the beam directed to the selected aim point and maintained on that point by the beam-direction control system within the designation pod. A code is imposed on the beam to ensure that the seeker tracks the correct target spot, the coding of the emission enables multiple targets to be designated simultaneously by different designator systems without confusion.

Following weapon release, the initial trajectory of the weapon is ballistic; its path is determined by the dynamic forces on the body and wings of the weapon.

Figure 12.1 Ground-based designators. (Courtesy of Dr. A. D. Hayes, NVESD, USA.)

The fire-control solution will aim to direct the weapon to a position where its seeker can detect the energy reflected from the designated target. The terminal phase, under the control of the seeker, is initiated once the reflected energy from the target is detected by the seeker.

The seeker on the front of the weapon is usually a simple quadrant silicon detector at the focus of a collecting telescope. The energy received by each of the quadrants will be compared, using suitable detection and signal-processing techniques, to determine the position of the received laser spot with respect to the centre of the detector. The control system then determines the correction required to the weapon's trajectory so that the ordnance is delivered with precision to the correct interception/impact point.

The simplicity of the seeker subsystem, with its quadrant detector, means that the trajectory of the weapon in its terminal phase oscillates about the sightline to the target. The terminal-phase accuracy is largely determined by the accuracy of the weapon's control system, responding to the guidance commands from the seeker, and crucially, the maintenance of the laser beam on the selected aim point by the beam-pointing system in the pod.

Figure 12.2 provides a schematic illustration of the system operation during the attack of a bridge by a self-designated laser-guided bomb.

The development of lightweight designators, shown in Figure 12.1, means that the air-borne designation process illustrated in Figure 12.2 could have been undertaken by a dismounted soldier; such an engagement is shown in Figure 12.3. Relatively simple anti-tank-guided weapon may be operated in a similar fashion.

This is an example of co-operative designation, which has several advantages:

- The delivery aircraft can fly a profile that is optimum for weapon delivery.
- The aircraft can leave the attack area immediately after the delivery of the weapon.
- A single designator, whether on the ground or in the air, can provide the designation function for several bombers.

The disadvantages concern:

- The need for precise timing and co-ordination by all parties;

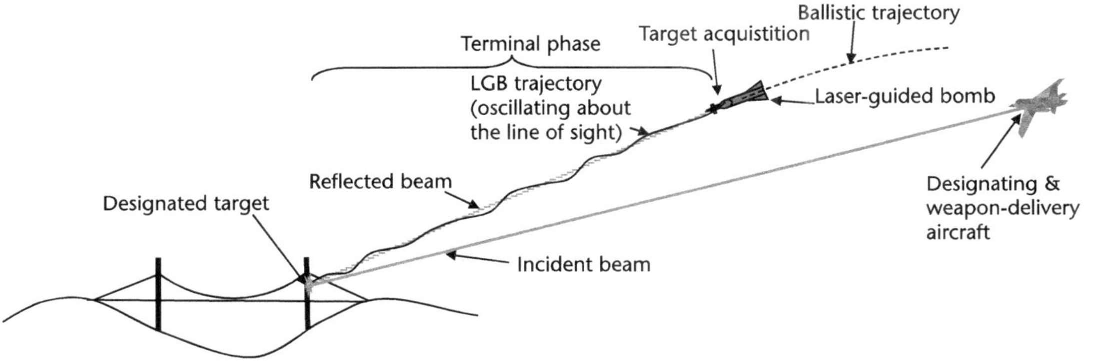

Figure 12.2 Laser-guided-bomb operation.

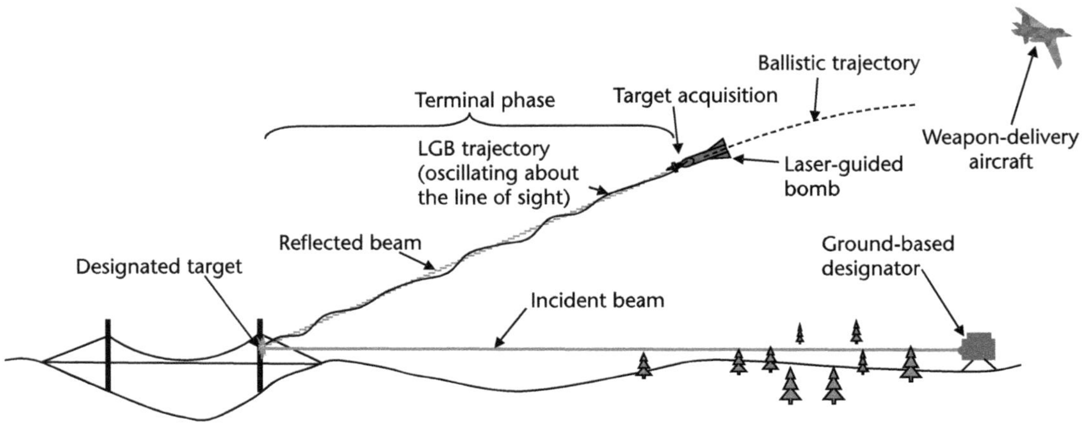

Figure 12.3 Designation by a dismounted soldier.

- Restricted chances to engage targets of opportunity.

12.2.5 Laser-Performance Requirements

The laser emission has to conform to STANAG 3733 for NATO operations, which imposes the definition of the wavelength (1.064 μm) and the pulse modulation. Consequently, a neodymium-YAG laser is used with a pulse energy in excess of 100 mJ and a modest repetition rate of about 20 Hz. The divergence requirements are such that the laser spot remains on the target throughout any engagement as the seeker approaches its target, despite any jitter in the beam director, which usually implies a half-angle divergence of the order of 1 mrad; greater divergence reduces the pointing-stability requirements, but increases the energy needed in each transmitted pulse to achieve the same range performance. Additionally, a larger beacon spot is formed on the target, which may degrade the hit accuracy on the target.

12.2.6 Subsystem Description and Function-

The basic components of the laser-designation system are the designation subsystem and the weapon.

12.2.6.1 Designation Subsystem

For airborne applications this is usually in the form of a pod that is the target-marking system, containing:

- The laser subsystem;
- The beam director, usually a two-axis ball system with a fast mirror and an electro-optical sensor for target acquisition and tracking;
- A line-of-sight isolation system;
- Target-tracking algorithms and aim-point maintenance subsystem;

- Boresight-alignment subsystem;
- Interfaces to the aircraft displays.

Figure 12.4 shows the British Thermal-Imaging and Air-borne Laser Designator (TIALD) pod system, which was used in a number of recent conflicts, but the modern system is Litening III.

A ground-based designation system is far simpler, as the beam-pointing function is undertaken by the operator, who will also undertake any target-tracking functions, as required. Similarly, the boresighting system can be undertaken manually. Line-of-sight isolation and interfaces to other major systems should not be required, other than to aid the co-ordination of an attack.

Figure 12.1 shows an image of a ground-based designator for use by a dismounted soldier.

12.2.6.2 The Weapon

The principal components of the weapon are shown in Figure 12.5, and comprise:

- The detector and seeker, which undertake the tracking of the laser spot and determine angular errors with respect to the sight line.

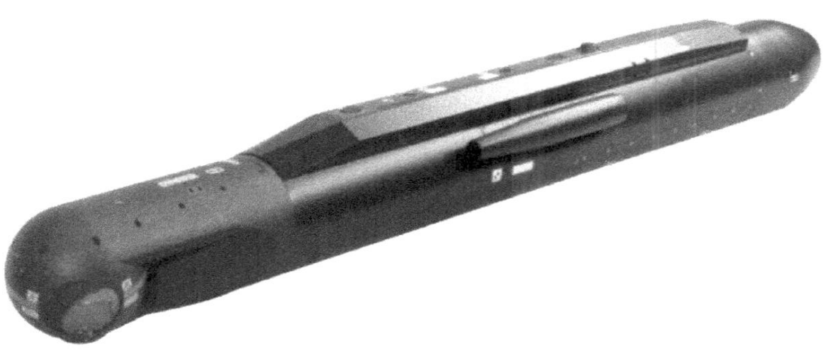

Figure 12.4 The TIALD pod. (Courtesy of Selex ES, U.K.)

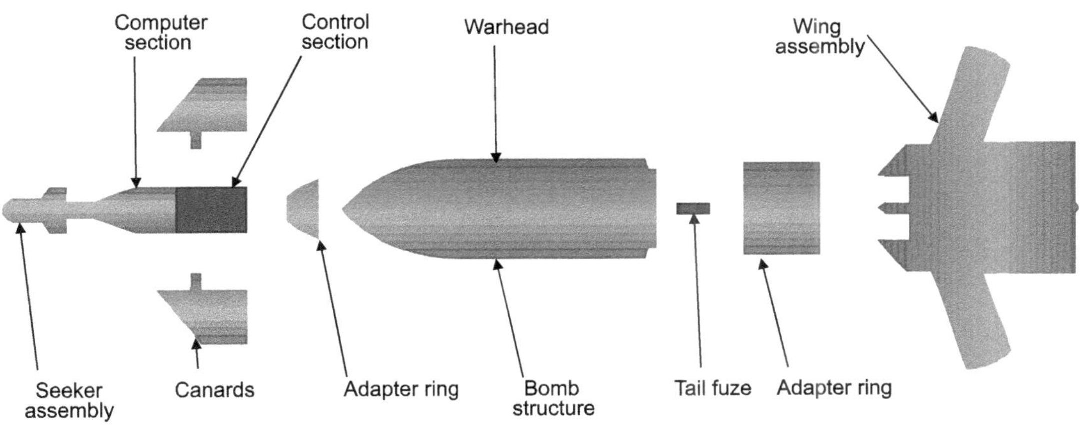

Figure 12.5 A laser-guided bomb. (Courtesy of Selex ES, U.K.)

- The computer processor and control section, which determines the trajectory corrections required from the angular errors estimated by the seeker and its target-tracking system. Canards are used to steer the weapon and are controlled directly by the control system, which gets its steering commands from the estimated error in the flight path with respect to the sightline and application of the guidance law.
- An adaptor ring that integrates the seeker and control system with the weapon.
- The general-purpose bomb.
- The fuze.
- Another adaptor ring to mate the bomb to the wing assembly.
- The wing assembly to stabilise the flight path.

An image of the seeker used in this type of application is shown Figure 12.6.

12.2.7 Future Requirements

Continued development of very reliable, high-energy, dual-mode laser systems for tactical (1,064 nm) and training (1,570 nm) airborne-designation operations will continue to be required, driving the laser technology to even smaller packages with enhanced functionality. Commercial products providing up to 300 mJ at 1,064 nm, switchable to a training mode of 90 mJ at 1,570 nm, are becoming available.

The designation role is likely to become more integrated with the advanced targeting systems described above. This is likely to lead to more capable systems with enhanced stand-off ranges and improved target recognition and identification.

The future may see a transition to eye-safer wavelengths, but current use of 1-micron optical radiation will continue into the foreseeable future.

Figure 12.6 A typical LGB seeker.

12.2.8 Summary

The application of laser technology to the guidance function of a guidance and control subsystem integrated with a ballistic weapon has resulted in the realisation of a precision-bombing system. The accurate delivery and strike by ordnance has many advantages, particularly with minimising the number of weapons required to be used to achieve a military objective and also minimising the likely collateral damage.

The laser-guided bomb concept is very simple; a dumb bomb has a couple of subsystems integrated with its external structure at the front and the back of the basic weapon. The seeker and steering unit is interfaced with the front of the bomb's structure and the wings to stabilise the flight path are at the back.

The designation subsystem concept is equally simple; a laser beam is directed on to a target, through a sightline-stabilised steering subsystem (beam director) to provide the reflected energy for the seeker to track, or use as a beacon, to determine its path to the target. The very high radiant intensity of the beam from the designating laser subsystem is scattered and reflected from the surface of the target and acts as a beacon for the seeker subsystem, which is essentially a target-tracking aid.

The air-borne designation pods continue to develop, providing ever greater sophistication and functionality to aid the targeting function in adverse operational conditions and at ever greater stand-off ranges. The designation laser continues to use the standard neodymium-doped yttrium aluminium garnet emission wavelength (1.064 μm) in a pulse-coded format defined by an international standard. There is unlikely to be a change to the eye-safer wavelengths in the near future, owing to the continued development of the standard designation systems and the existing stock-pile of such weapons.

Significant enhancements to the efficiency of diode-pumped neodymium lasers have led to highly capable miniature laser-designators being realised during the past 20 years. The availability of miniature designation systems has allowed additional flexibility for the designation function, leading to co-operative designation between air and ground forces becoming well established for a range of operations. Recent enhancements in laser performance have seen further reduction is size and additional capability from these systems.

12.3 Laser Beam-Rider Guidance

12.3.1 Introduction

Laser-based beam-riding techniques provide a relatively simple and cheap approach to short-range weapon guidance. The use of a beam-riding technique means that the weapon, the disposable part of the system, can be quite simple and, consequently, cheap and relatively lightweight. These attributes lead to a number of operational advantages, such as:

- Low cost, simple laser receiver/guidance and control computer within the missile, which requires far less sophistication compared with infrared or radio-frequency seekers (homing heads);

- Rearward-facing laser receiver, which is inherently resistant to countermeasure techniques originating from the target;
- Low-power guidance laser system, which operated at significantly lower optical power than designators/range-finders, plus the use of a zoom optical system, which reduces power density required during the early phase of the engagement;
- Operator-in-the-loop guidance, which provides additional resistance to various countermeasure techniques compared with some other guidance implementations, such as auto-trackers;
- An operator can initiate self-destruction of the missile, by switching off the guidance beam if something untoward arises, for example, if the tracked target is identified as friendly after launch or if the missile malfunctions and threatens friendly positions;
- Easily adaptable to local conditions of deployment, for example, the capability of the weapon system can be enhanced by adding new sensors to the launch platform to give day or night operation without changing the missile subsystem;
- Missiles with improved performance can be designed to fly down the same guidance beam without requiring changes to the launcher;
- Command-to-line-of-sight guidance means that a laser-beam-rider missile appears to threat-warning sensors at a fixed angle relative to the ground, which may delay threat-warning sensors declaring the approaching missile as an immediate threat.

Operator-in-the-loop guidance does have the drawback of requiring a higher standard of training for the operator than is needed for launching fire-and-forget missiles, such as the heat-seeking guided weapons; additionally, these weapons tend to have a limited-range capability.

The fundamental principle is to use a laser beam to establish the sightline from the launch point to the target to indicate the path that the missile should follow to intercept the acquired target. Several approaches may be used by the guided weapon to find its position with respect to the centre or a specific offset with respect to the propagation axis of the guidance beam, enabling salvoes of missiles to be launched and guided along the same beam.

One beam-sensing technique is based on the use of a very simple quadrant detector in the rear of the guided weapon. Another approach that allows several missiles to be fired in a closely spaced salvo is to use a spatially coded beam, so that each missile may attempt to follow a specific segment of the beam. The latter approach is sometimes referred to as laser-information-field guidance.

12.3.2 Historical Development

The development of laser-based beam-rider technology started in the 1970s with the aim of producing a relatively cheap and mobile air-defence system. The availability of suitably advanced laser technology in this decade enabled rapid developments to be achieved with this type of weapon system.

The application of a beam of laser light in the near infrared (often close to 1 μm) to form the guidance corridor, means that the beam-forming optics can be relatively small and the divergence may be controlled easily and precisely, giving a small illumination spot at the target. Clearly, the naturally small divergence of the laser beam is a significant advantage, compared with the divergence of beams generated by sources operating in other parts of the electromagnetic spectrum.

The small size of the mirrors and other moving parts means that the beam control system can be very agile, enabling highly dynamic targets, such as combat aircraft, to be tracked accurately. A relatively sophisticated target-tracking system is required, with the appropriate type of target-tracking algorithms, so that the illuminator keeps the guidance beam on the target as it manoeuvres, as discussed in some detail in Chapter 4.

This type of precision missile guidance tends to be used by relatively short-range, ground-based weapon systems. There are a number of anti-air or anti-armour missile systems, which are guided to their designated target using a laser beam-rider technique.

12.3.3 Military Utility

Missiles based on laser beam-riding technology are a significant threat to almost every air platform, from fast jets in an interdiction role, helicopters and transport aircraft in any battlefield or peace-time support role, as well as to reconnaissance aircraft. Anti-armour beam-riders are specifically designed to attack armoured ground-vehicles but they can also be used against slow-speed aircraft flying at low level. The anti-air missiles typically travel at speeds of 500 m/s, or more, and have engagement ranges of 5 to 8 km.

The key operational characteristics of these weapons are normally quoted to be:

- Precision-targeting capability, resulting from highly-accurate guidance;
- Short engagement time lines;
- Simplicity, leading to high reliability and low cost;
- Difficult (but not impossible) to counter with soft-kill (i.e. countermeasure) methods, because the beam-position sensors are in the rear of the missile.

12.3.4 Mode of Operation

The general principles of operation are similar for the various types of this class of weapon system and have some similarities with laser target-designation, described above. Targets are detected, often via an external cue, acquired and then tracked, using a cross-hair in a gyroscopically-stabilised sight (but note that tripod-mounted anti-armour beam-riders may not necessarily use gyro-stabilised sights). Once target-tracking has been achieved and the target is within missile-engagement range, the firing sequence is started by pulling a trigger (closing a switch). The trigger initiates the missile eject motor, activates the guidance-laser subsystem and this function is followed by firing the boost/sustain motors, once the missile has emerged from its launch tube.

The principle of this guidance system is to use a laser beam to illuminate the path, or guidance corridor to a selected target. Each missile has sensors in the rear of its structure, which detect the position of that missile within the laser-guidance beam, to establish its position with respect to a reference bearing within the beam, such as the centre of the beam. However, other reference bearings may be preferred.

When a missile is launched it is directed into the guidance beam, which initially has a large divergence (a conically shaped beam), in order to initiate the guidance process. Once the sensors in the rear of the missile have determined its position in the guidance beam, commands are generated in the missile's control system to direct it to a position close to the optical axis of this laser-guidance beam, or other reference within the beam, as noted above. This initial steering manoeuvre, to position the missile close to the axis of the guidance beam, is a process known as gathering and it enables the missile's trajectory to be controlled by the beam generated by the ground-based target-tracking system. This gathering process is an identical technique to that perfected for radar-guided command-to-line-of-sight missiles.

Following a successful gathering manoeuvre, the divergence of the guidance beam can be significantly reduced. The reduction of the beam divergence, typically by a factor of 30, leads to a greater operational range of the weapon system for a given emitted optical power and enhanced hit-point accuracy at the maximum interception range of the guided-weapon system.

As this is a form of command-to-line-of-sight guidance, the missile will generally have a close encounter with its target, even if the target attempts to manoeuvre to avoid the interception. Consequently, the missile will have a proximity fuze and probably a contact fuze as well, to control the warhead initiation.

Anti-air beam-riding missiles are universally used for point-defence operations and have no requirement for high target-tracking rates, except in the terminal phase. Anti-armour missiles also do not need high target-tracking rates, as they are designed to engage low-speed vehicle targets. However, target manoeuvres, as noted above, will complicate the terminal phase engagement dynamics to some extent, and consequently, present some challenges to the designers of the fuze to ensure a high probability of defeat for the weapon system.

12.3.5 The Weapon System: Firing Post

The principal subsystems of this weapon system are its launcher, where the missiles are stored ready to be fired and a target-tracking system within the laser-beam director. These subsystems may be operated in a number of ways; for example, the missiles can be launched from the shoulder of an operator, from a free-standing tripod or firing post or from a vehicle-mounted launcher, with a man-in-the-loop for both detection and tracking of a target. The operator is usually assisted by a visible-band sight, such as a telescope, although image-intensified or infrared sights are also available for situations when visible-band performance is reduced, owing to operations at night or reduced visibility typical of battlefield conditions. A complete system with the gunner is shown in Figure 12.7.

Integrated with the target-tracking sub-system is the beam director that follows the tracking point established through the target-tracking algorithm in the system

Figure 12.7 Laser beam-rider system. (Courtesy of Saab Dynamics AB, Sweden.)

processor. The laser illuminator is integrated with the beam-director component. The laser source is often quite simple technology, such as a well-collimated laser diode, emitting in the near infrared. This integrated system is often termed the firing post.

12.3.6 Guidance-Beam Generation

The simplest systems use a small divergence laser beam to define the path to the target, which is applicable to the very short-range weapon systems. In this case, the missile has a simple quadrant-detector system in the rear of the weapon. The laser transmitter is co-boresighted with a target-tracking aid, such as a telescope.

A more sophisticated system involves scanning the optical axis of a laser beam spatially over a small solid angle, to form a guidance corridor, which is co-boresighted with the target-tracking telescope's reference mark (cross hair). Accurate missile guidance is achieved by applying time and/or spatial coding to the laser beam, such that the missile is able to determine where it is spatially (lateral displacement from the optical axis) within this guidance corridor, and hence, its position relative to the cross hair positioned on the target. The missile may then change its flight-path direction to correct for any differences between the two spatial positions, according to the guidance law being used by the guided-weapon system to ensure an accurate interception with the tracked target.

The guidance law used by the missile is identical to a command to line-of-sight guidance technique. During the missile fly-out phase, a zoom optical system (initially set wide to gather the missile on to the optical axis of the target-tracking system) is used to reduce the dimensions of the laser-guidance corridor to maintain approximately constant irradiance on the guidance detector on the missile and in order to minimise the required laser power. The general scheme of operation is shown in Figure 12.8 for an anti-air engagement and the shaping of the laser beam used for the guidance of the missile to its target is shown schematically in Figure 12.9.

12.3.7 Laser Characteristics

A number of different wavelengths in the infrared have been used to provide the link between the firing post and the missile, to provide the target-tracking functions and steering commands, which may be as simple as demands to turn left/right or move up/down, through appropriate coding in the laser emission. To exploit the inherent short-range characteristics of these weapons, the beam usually has low average power and high peak power, in the case of a pulsed emission. An outline of the typical laser-emission characteristics is shown in Table 12.1.

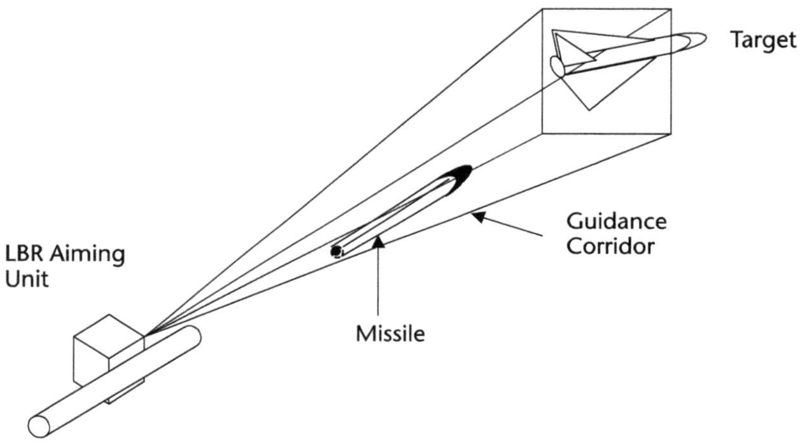

Figure 12.8 Schematic of laser-beam riding missile.

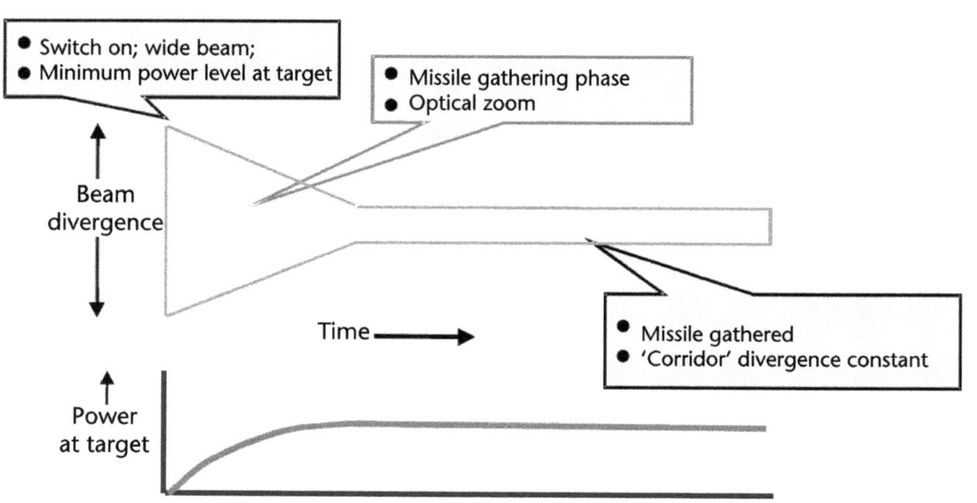

Figure 12.9 Shaping of the "beam-riding" laser emission.

Table 12.1 Outline Laser-Emission Characteristics for a Laser-Beam Rider Command Link

Laser Parameter	*Value*
Average power (W)	<<1
Peak power (W)	~10
Wavelength band	Near or far infrared
Modulation	Kilohertz PRF or chopped CW
Divergence (mrad)	~100

12.3.8 Future Requirements

Several enhancements can be envisaged for improving guidance performance and particularly eye safety. A move to the eye-safe laser emission regime could have a significant impact on the operation and training with these weapons.

The use of covert techniques, during the laser transmission from the target-tracking subsystem, could make it more difficult for laser-warning sensors (the more conventional use of the term laser warning, rather than a warning given to an intruder, as discussed in Chapter 8) on a platform to provide a warning that a weapon system is tracking it. Additionally, the use of alternative wavelengths and covert techniques may make the detection of the laser splash around the aperture of the firing post more difficult to observe from a defensive perspective and thus delay the application of countermeasure techniques by the platform under attack.

12.4 Laser Fuzing

12.4.1 Introduction

The function of any fuze is to provide the trigger to the lethality system (or warhead payload) within a guided weapon. A number of different types of fuze are used in guided weapons, the most common are known as contact and proximity fuzes. In the case of proximity fuzes, the challenge is to detect the target and send the signal to trigger the lethal payload at the optimum moment (time) during the final stages of the interception; this can be quite a challenge when the closing velocity could be in excess of mach 5 for some terminal engagements.

Matching of the spatial-detection characteristics of the proximity fuze with those of the warhead pattern along with the range of the missile-target interception dynamics are used to determine the optimum time to initiate the detonation process during an interception of a target. This is a critically important optimisation process to ensure the best use of the warhead payload and is known as the fuze-warhead matching process, which has to include the various approach directions of the interceptor and its target to achieve a fully optimised solution.

Effective fuze-warhead-matching techniques mean that the warhead effects, such as its fragmenting beam or its explosive beam, are created at the most appropriate moment, to allow a high probability of achieving the required kill/defeat mechanism; thus the missile system can realise its design objective. Optimum performance from a guided-weapon system may only be achieved when the fuze and the warhead operate as an integrated system with appropriate fuze-warhead matching.

The availability of cheap and compact laser technology has enabled an alternative transmitter to be considered for active fuzing. Laser-based devices operate in a similar way to active radar fuzes and may use similar target-sensing techniques, but have some advantages in size and weight, as well as power consumption; additionally, they may offer some advantages in spatial resolution in a similar fashion to a laser-radar system in comparison with a conventional radar system, as discussed in Chapter 13.

A laser-based fuze is generally operated as a proximity device in a relatively short-range mode. Short-range operations can be arranged through appropriate choice of emission wavelengths.

12.4.2 Mode of Operation

The primary function of any proximity fuze is to act as a situational-awareness sensor, gathering information about the arrival of a target during a close encounter in the terminal phase of the engagement. In particular, its function is to indicate the target's location, sometimes with respect to some reference frame usually defined by the missile's body axes, to the lethal payload of a weapon. When the algorithms in the fuze subsystem indicate that the target is at an appropriate point to be engaged by the warhead, the lethality system provides the signal to initiate a warhead event. Hence, in this mode the fuze forms part of a switch to activate the payload.

The terminal-phase engagement encounter with a target may be very rapid for some interceptions, as indicated above; hence, the processing must be very fast. Additionally, there is a requirement for the fuze to exhibit characteristics similar to those of a missile-warning sensor, that is:

- Reject false targets;
- Have a very high probability of target declaration;
- No false positives, which lead to premature detonation.

A classical active proximity fuze operates by forming a cone of electromagnetic radiation around the forebody of a missile and uses a sensor in the receiver channel to detect any reflections of this radiation from objects that come into contact with this beam. The projection of the cone around the missile's forebody is usually arranged to be compatible with the effective radius of operation of the warhead. This mode of operation is illustrated schematically in Figure 12.10.

When a target enters this cone of the fuze's sensing system, it gives a declaration of the presence of a target. The range of operation is usually quite short, of the order of a few metres; consequently, the fuze can use a laser transmitter of very modest power. Moreover, the use of wavelengths that are readily absorbed by the atmosphere mean that the target-sensing emission from the fuze can remain covert and largely immune to the application of countermeasures, or other confusion tactics, by the vehicle being intercepted.

The sensing cone of the fuze may be formed by arranging a number of transmitting and receiving apertures around the missile's forebody. Alternatively, scanning

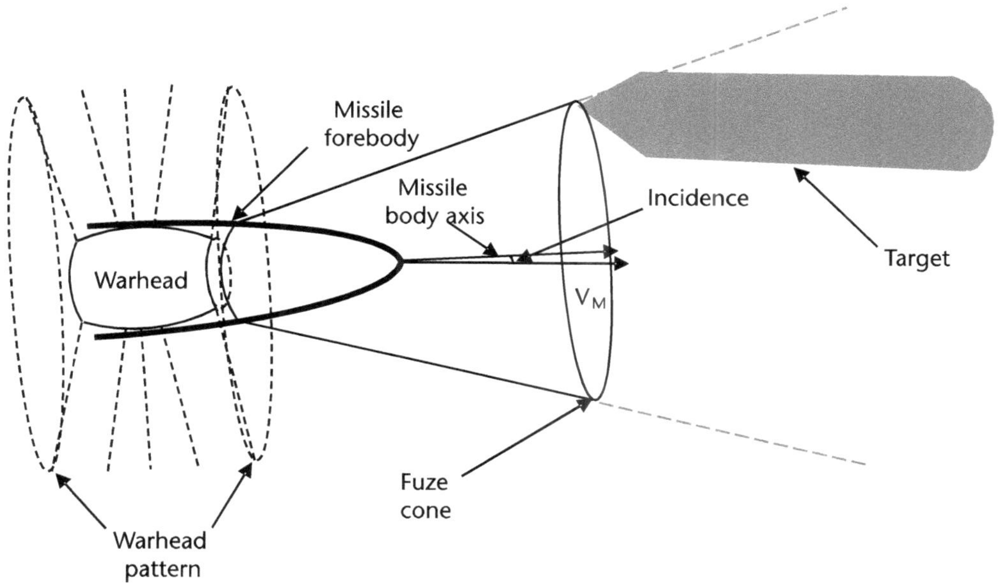

Figure 12.10 Schematic of the operation of a proximity fuze.

techniques may be used to form a cone around the missile's body. The geometry of the transmitter and receiver beams are often arranged so that the transmitter beam is directed ahead of the missile forebody and the receiver, which is positioned in front of the transmitter aperture, so that there is an overlap region, as shown schematically in Figure 12.11.

There are a number of advantages of this form of transmitter and receiver geometry:

- Optimum use of transmitter power, which permits the emitted beam's characteristics to be matched to the receiver geometry;
- Definition of a region where there is a high probability of a target being detected once it enters the overlap region;
- Defines the effective range of the fuze and somewhat simplifies the fuze-warhead matching algorithm;
- Enables a relatively covert active fuze to be designed and implemented.

Once a target enters the overlap region, or is close to it, the electromagnetic energy from the transmitter will be scattered by the target across a range of angles, with some entering the receiver aperture. This detected signal will have amplitude and temporal characteristics dependent on:

- The power of the transmitter;
- The time delay given by the separation between the target and the fuze, just like a range-finder;
- The target's reflectivity characteristics at the emission wavelength;
- The shape of the target;

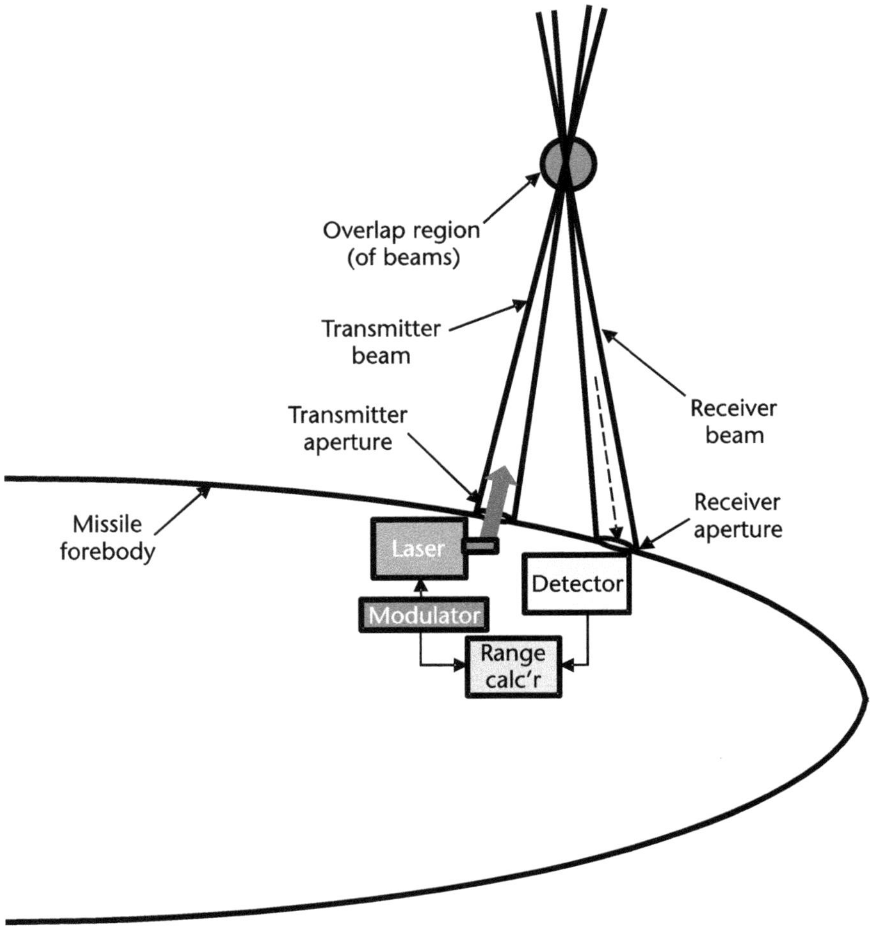

Figure 12.11 Transmitter and receiver beam overlap.

- The geometry of the interaction;
- The propagation path.

The application of a code to the transmitted beam or the use of discrete receivers with a particular instantaneous field of view allows some estimation of the angle of arrival or, more precisely, the sector of arrival of the target. The use of a coded pattern in the various segments of the transmitted beam provides similar information to the fuze/lethal system's processor. Additionally, the use of very short pulses enables the separation of the target from the missile to be estimated, from which the appropriate warhead-detonation command may be initiated, as determined from the fuze-warhead matching algorithm.

12.4.3 Principal Subsystems

The active proximity fuze comprises:

- The laser transmitter, with its wavelength being chosen for high attenuation by the atmosphere to prevent application of countermeasures by the target (i.e., jamming the receiver). There may be a single transmitter and an appropriate beam-forming system such as a scanner or multiple transmitters. The use of multiple fan-shaped beams is quite common covering somewhere in the region of a quadrant, which may be coded to provide additional information to the lethal package, through spatial and temporal patterning of each beam. This technique may be used to reduce false alarms or premature detonation of the lethal package.
- A receiver system to detect the returns generated by targets entering the sensing cone.
- A control system to process the received signals from the target and initiate the warhead.

The primary challenge for the active-fuze subsystems is to achieve an adequate signal-to-noise ratio in a cluttered environment, so that there is a very high probability of declaration without any false positives (i.e., any false alarms).

12.4.4 Laser Characteristics

Compact, modest power lasers are the primary requirement. Typical characteristics are shown in Table 12.2.

It is normal for a tactical proximity fuze to have an effective range of a few metres, depending on the application. Semiconductor laser-technology usually offers the ideal operating and system integration characteristics.

12.4.5 Future Requirements

Further exploitation of the unique characteristics of laser beams to enable the development of more intelligent fuzing techniques, such as combining the latter stages of terminal guidance and homing with the active-fuzing function. This may enable the miss distance during a close encounter to be reduced or eliminated, so that the size of the warhead may also be reduced. An additional enhancement could involve aim-point selection for use with either focused warheads or those with controllability of the direction of delivery of their lethal force.

Table 12.2 Typical Laser Parameters for Laser-Based Proximity-Fuze Applications

Laser parameter	*Typical value*
Power (W)	Average << 1, peak power ~ 50
Wavelength	Usually near infrared (poor propagation wavelength)
Divergence	Compatible with beam shape
Pulse or CW	Either (modulated CW)
Pulse length (ns)	10 to 50
Rise time (ns)	<1
Pulse-repetition frequency (kHz)	10 to 50

12.4.6 Summary

The use of electromagnetic energy to detect the close encounter of a weapon with a target has been exploited by proximity fuzes in guided weapons for decades. The common source for this critical function in a lethal system in a guided weapon is a radar transmitter, such as a magnetron.

The availability of laser emissions with the appropriate emission characteristics offers an alternative to the radar-based technique with some potentially important benefits in terms of cost, size and weight. Moreover, future laser-based fuzes may enable some additional sensing to be achieved, such as providing very accurate guidance commands in the final stages of an interception of a target.

The fundamental technique of operation of a laser-based fuze is identical to that of the well-established radar system, but with significant performance advantages.

Selected Bibliography

Garnell, P., and D. J. East, *Guided Weapon Control Systems*, Oxford, U.K.: Pergamon Press, 1977.

Garnell, P., *Guided Weapon Control Systems*, 2nd ed., London, U.K.: Brassey's Defence Publishers, 1980.

Osche, G. R., *Optical Detection Theory for Laser Applications,* New York: Wiley Interscience/John Wiley & Sons, 2002.

CHAPTER 13

Laser-Based Remote Sensing

13.1 Introduction

The special features of a laser beam lend itself to many different forms of remote sensing; this chapter will concentrate on those applications based on laser-radar technology in one form or another. The idea of using a laser beam for undertaking functions and tasks similar to those made by a radar beam have been developed over the last three or four decades (i.e., using the properties of back-scattered or reflected energy from a target, often described as returned energy).

A common civilian application of laser-radar technology is a ceilometer, which is a ground-based system used to measure the height of the cloud base. Short-duration laser pulses are fired vertically upwards and the time for the reflected pulse to return is recorded. The water molecules in the clouds scatter light more strongly than the other molecules in the atmosphere, so the return signal can be detected relatively easily. Clearly, this type of system exploits laser range-finder technology, as do many of the other types of laser radar discussed in this chapter; other laser-radar techniques concern a more detailed analysis of the characteristics of the returned energy.

The concept of a laser-radar system to observe and analyse the optical return from a distant object is relatively simple, involving a transmitter and a separate receiver; however, as is usual with any advanced technology system, there are many complications. An example is that the returned signal may have components that are caused by specular reflections and diffuse reflections, as shown in Figure 13.1. As with any sensing of this type, there is a $1/r^2$ dependency (where r is range from the transmitter to the target) on the intensity of the returned power, so long-range interactions demand a more powerful emission from the source illuminating the scene.

There are many physical processes involved when a laser beam propagates through the atmosphere, as discussed in detail in Chapter 5. Some laser-radar systems are designed to sense the photons that are returned to the transmitter, whereas other systems are configured to assess the absorption of the propagating beam, as it passes through a cloud for example, which may contain hazardous and/or toxic substances. In this latter case, the zone with the enhanced absorption may be probed by a second source to establish the reason for the absorption. The returned

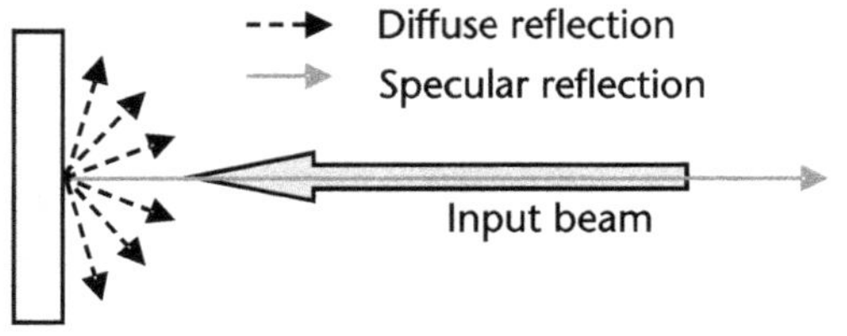

Figure 13.1 Reflected energy.

signal from the probed area may then be analysed using a number of techniques, such as spectroscopy or Doppler analysis.

The terms Lidar and Ladar describe such systems, with some subtleties in terms of their modus operandi. A number of definitions have been used to differentiate between these systems. The most common definitions are the following:

- Lidar refers to light detection and ranging measurements and is associated with soft targets, such as an aerosol cloud.
- Ladar refers to laser radar and is associated with sensing of hard targets.

The Lidar and Ladar types of systems should not be considered as simple replacements for a radar system as the characteristics of the beams used by a radar and a laser radar are radically different despite exploiting similar characteristics of electromagnetic energy. The major differences between the transmitted energy of a radar system and a laser-radar system, whether it is a Lidar or a Ladar system, concern the source wavelength, which are in different region of the spectrum, and the beamwidth of the emission.

Like radar systems, Lidar and Ladar use electromagnetic radiation to probe a volume of space. The returned energy may be analysed to provide information about objects intercepted or encountered by the transmitted electromagnetic radiation. In fact, the equations developed to understand the operation of laser-radar and radar systems, known as the laser-radar and the radar equations respectively, are identical in form; each essentially has four components directly associated with the steps in the physical process of sensing:

- Propagation of electromagnetic energy to the target;
- Reflection of electromagnetic energy by the target;
- Propagation of the scattered electromagnetic energy to the receiver;
- Collection of the scattered electromagnetic energy by the receiver.

A specific advantage that these systems have for sensing and remote sensing arises from the coherent nature of the electromagnetic energy emitted by the transmitter. It is possible to get a great deal of information regarding the characteristics and behaviour of the objects that are detected in a probed scene, given appropriate processing, as well as knowledge of the various processes that are occurring in the atmosphere. The principal advantage that a laser-based system has over a radar system is its spatial resolution for a given aperture size; additionally, in the case of a radar system, the intercepted objects generally behave as a collection of isotropic scattering objects, whereas the laser-radar beam is often smaller than the target and the full range of reflecting characteristics may occur in the received beam from the target, allowing more information about a target to be deduced from the processing of the received energy.

The shape of the transmitted beam from a laser-radar system is largely independent of the launched laser power in its characteristic narrow pencil beam shape. However, this type of beam is less than ideal for surveillance applications. In this case, if a large volume is to be probed, the laser device will need to operate at a very high repetition rate to search the volume in a suitably realistic time period. An alternative approach would involve the use of multiple beams, but this technique complicates the optical configuration, as well as the interpretation of the collected data.

A distinct advantage of the higher spectral resolution that may be achieved with a laser-radar system, compared with a conventional radar system is that it may be exploited in the recognition and identification of targets. Resolved measurements from observation of a target, such as features on parts of the target's surface, enable recognition of characteristics that are specific to a target class. Such features that would be exploited include, shape, size and motion characteristics, including velocity. The ability to measure unique features by laser-radar systems is an important characteristic for the successful application of automatic target-detection and recognition systems, as discussed below.

The shorter wavelengths used by a laser-radar system tend to be more sensitive to environmental propagation conditions, such as turbulence and natural atmospheric absorption. Radar-based systems offer greater range detection/interceptions, especially in adverse weather conditions. Hence, each of these sensing systems has a distinct role to play in any military activity.

It is quite common to group laser-radar systems according to the type of measurement made by the device, its detection technique, the modulation method, the class of laser, its function, the type of interferometer used in the detection channel (for coherent-detection systems) and probably many more descriptors. These terms, such as a 10.6-micron vibrometer, should only be used to give an insight into the function or mode of operation of the laser radar, and certainly does not indicate how well it performs; in this particular example, the label indicates it is a laser radar using a gas laser (CO_2 wavelength) to illuminate a scene and uses a coherent-detection technique to sense vibratory motion.

The term laser radar will be used as a generic term for the various types of laser-based sensing system in this chapter.

13.2 Introduction to Laser-Radar Systems

The purpose of a laser-radar system is to probe the environment around the system. There are many reasons to probe the space as environmental sensing in one form or another can play a very important role in many military operations.

Laser beams come in many forms and may be used to probe the environment in several ways. They can be used to make multidimensional measurements using the principles developed for radar systems. Some examples of the exploitation of laser-initiated effects in these sensing systems are:

- Use of the reflected energy to detect the presence of remote objects or a series of objects;
- Use of photon beams to interact and stimulate chemicals, pollutants, or organisms to emit light, from which the presence of the material may be established and subsequently identified;
- Analysis of the Doppler content of a reflected signal to measure the motion of gases, liquids and solids, as well as measurement of the characteristic vibrations of living or man-made objects;
- Evaluation of the absorption of the transmitted beam to estimate the density of particulates in the air or in liquids, such as a contaminant film on a surface;
- Estimation of the polarisation characteristics of the reflected beams to detect objects that are difficult to observe, as well as being used to detect and identify living or man-made objects, additionally use of polarimetry can be exploited to enable viewing through obscuring media (such as aerosols, foliage or camouflage).

The ability to detect and measure unique features on a target may be exploited in many military applications. Additionally, the very short wavelengths, and consequently higher carrier frequencies in a pulsed emission, used by laser-radar systems permit the detection and interrogation of very small objects, which may include a variety of objects from wires to aerosols and molecules. A further advantage of laser-radar systems for sensing, results from the fact that the laser cross-section of aerosols and rain is far larger than their equivalent radar cross-section characteristic.

The use of a laser-based sensing system to detect and identify substances in the atmosphere or on objects at a distance is an important and interesting recent development of this technology, which is usually achieved through the application of analytical-spectroscopic techniques to the observed returns. Such a system will provide capability for remote analysis and tracking of battlefield aerosols, including man-made and natural agents. In general, this will require other classes of lasers and detectors, but will build on the same advances in computing power as required by multidimensional laser-radar systems. Similar techniques are evolving for analysing liquid films for the presence, and nature of, foreign or suspicious substances.

Multidimensional measurement capability by laser-radar systems is an important technical development, but currently the full potential functionality of this technique is still limited by emitted laser power, computer processing power, and the limitations of incoherent and coherent focal-plane arrays and their signal processors. As a result, systems are limited in range and by the number of object measurements that can be made in a given amount of time. Currently, long-range sensors still require relatively bulky and inefficient beam-scanning mechanisms to provide the necessary sector coverage to probe a zone of interest. Very important progress has been made with the source technology, especially at eye-safe wavelengths, as more efficient and powerful lasers are developed, especially using fibre-based sources.

Very large staring detector arrays are being developed, which can measure time of flight or phase change in each pixel, as discussed in Chapter 11; moreover, coherent focal-plane arrays are being developed that can measure target macro (velocity) and micro (vibration) Doppler shifts in near real time. Computer processing capability continues to advance, becoming increasingly powerful with ever increasing computational speed and reduced size. Mega-pixel arrays making real-time multidimensional measurements at 10 km with frame rates of 100 frames/second or faster could be available in the near future. These systems will be smaller and more efficient than the present systems, by using flash-illumination techniques, given suitable advances in the supporting technology. Further advances in system performance are expected to continue, with the parameters of pixel-rate and effective operational range potentially increasing by orders of magnitude, with further decreases in size and weight.

13.3 Historical Development

Remote sensing with coherent laser-radar technology has been pursued since the 1970s, when it was demonstrated that Q-switched beams could be used to make observations of objects at ranges beyond 5 km. The development of diode-pumped solid-state laser technology with greater power and much higher efficiency enabled observations to be made up to 10 km, and beyond, given favourable beam-propagation conditions.

Much of the early activity was stimulated by military interest in topics such as range and velocity measurements of objects, estimates of low-level wind characteristics and vibrometry. This stimulated the relevant technologies required to support a laser-radar system, such as laser sources, detectors, modulators, optical systems, and signal-processing techniques.

There continues to be a considerable interest in certain civil applications, such as atmospheric measurements (including detection of turbulence), characterisation of aircraft-wake vortices and possibly space-borne wind laser radar (Lidar), as well as pollution monitoring. Apollo 12 carried a laser-radar system, using a flashlamp-pumped ruby laser, to map part of the moon's surface. Mapping continues to be an important function for laser radar (Lidar) systems to establish the topography of a region; pulsed beams are particularly well suited to this application, as the photons can penetrate the forest/tree canopy and generate a return from the bare Earth; signal-processing techniques allow the return from the tree canopy, if present, to be rejected/discarded. The development of efficient techniques to provide blue/green emissions from the transmitter in a laser-radar system enables bathymetry-sounding functions to be undertaken around coast lines.

The use of coherent techniques presented many challenges for the optical source in the technical thrust to make a vibrometer. The early laser-radar systems used gas-laser technology mainly because of the potential for stable operation and continuous-wave, single-frequency, photon-beam generation. Much of the initial work concentrated on the stabilisation of the emission from these lasers. This approach led to the development of slab waveguides with radio-frequency excitation, exploiting transversely-excited atmospheric pressure devices (also known as TEA lasers), including demonstration of methods for frequency stabilisation and gas catalysis. Moreover, techniques were developed from an in-depth investigation of the laser dynamics within this cavity, enabling the chirp in any transversely-excited, atmospheric-pressure, carbon-dioxide laser system to be predicted and then exploited. Other developments resulted in a carbon-dioxide waveguide laser with a programmable pulsed profile.

More recently, the emphasis has been on shorter-wavelength lasers, using solid-state technology. Initially, the development concentrated on high-power, single-frequency, diode-pumped Nd:YAG lasers with continuous-wave emission. A polarisation rotation, twisted-mode technique was applied to obtain a single-frequency TEM_{00} output emission. This type of laser technology has been used in a coherent optical-communication system, as discussed in Chapter 14. This type of twisted-mode cavity laser has exhibited an emission line width of the order of 10 kHz

Further developments have investigated laser techniques to extend the laser wavelengths to the eye-safe region, beyond 1.45 μm. Laser sources, including erbium-doped laser devices (Er: YAG and Er,Yb: Glass), have been developed, with wavelength emissions beyond 1.5 μm. Single longitudinal-mode emission characteristics have been achieved using several types of cavity design. Similarly, very stable emission from a solid-state system, using a master oscillator with a thulium laser gain medium (Tm: YAG) emitting at 2.05 μm, has been demonstrated.

The trend of using transmitters with wavelengths in the eye-safer regime, as defined above and discussed in Chapter 17, has enabled high-pulse energies to be used more safely, which has led to enhanced range performance. The use of efficient photon-counting detectors, based on avalanche photo-diode technology, in the short-wave infrared band has enhanced the performance of laser-radar systems operating in this waveband.

The most recent solid-state laser devices used in laser-radar systems have exploited the development of efficient fibre-laser technology. A critical advantage of fibre lasers is the inherent cavity stability necessary for many of the sensing applications. The development of fibre-laser techniques has involved the adaptation of 1.5-micron lasers, originally developed for telecommunications, to have the necessary emission characteristics to be compatible with laser-radar applications.

This transition of technology to solid-state lasers has enabled very compact and efficient transmitters to be demonstrated. Similarly, advances in the receiver and computer technology have enabled multidimensional sensing techniques to be developed for real-time operation. Modern computer systems, along with advanced algorithms, have enabled some of the limitations in the formation of point clouds (the array of optical returns from a scene) to be created to map an event in real time, as a scene is scanned.

13.4 Military Utility

In the general situational awareness role, a laser-based system (laser-radar) scans a zone of space and its processor forms a three-dimensional picture of the received reflections (point cloud) produced by reflections from objects in the scene. Advanced 3-D algorithms for scene analysis have increased the probability of recognising obstacles (potential hazards along the flight path) and reducing false alarms. The use of angular-motion sensors in the system enables the position of the laser beam for each position in its scan to be tagged and the co-ordinates established with respect to a known reference (such as a fixed bearing or to the velocity vector), as discussed for target recognition. In this case, the collected data, when correlated with the reference frame of the system, allow the detected obstacles to be superimposed on the image, to create a synthetic picture of the actual scene. This process applies techniques similar to those described in Chapter 11.

Image processing of the returns for the various sensing tasks generally uses stored-data characteristics of potential targets that may occur in the probed areas, so that targets of interest may be identified by comparison with the stored data (library). In the surveillance and situational awareness functions, the stored data may, for example, include reflectance characteristics of cables and wires so that these obstacles can be identified readily. The image-processing algorithm may use additional knowledge to link the various returns superimposed on the scene to indicate the positions of wires in the probed scene; correlation of the position of the measured and the stored data provides accurate navigation cues (via way marks, for example) for terrain-referenced navigation.

Another important military requirement concerns the application of remote-sensing techniques to establish the presence of any threat/toxic agent contaminant in the immediate vicinity of any military forces. Current techniques are accurate but rely on contact methods; the use of a remote-sensing method has many potential advantages, as it may be both rapid and accurate with a potentially very low false-alarm rate. Moreover, it may be viewed as a development of existing laser-based remote-sensing systems.

Laser-radar systems may also be used to detect wind shear ahead of aircraft, so that the effects of turbulence may be mitigated, if not avoided. In this case, the system detects the motion of aerosols and other small objects in the disturbed atmosphere. Again, the use of inertial sensors means that the position of the detected phenomenon may be given a location in a reference frame and the appropriate avoidance measures applied.

Modern computer technology enables the picture formed from the volume of space scanned in the searched sector to be processed rapidly. Image-processing algorithms can filter the data; the retro-returns from known objects that do not have any interest to a particular system function, such as clutter, small bushes or tree-canopy returns can be filtered out of the image, as discussed in the use of laser-radar system for mapping to provide a digital-terrain map.

The flexibility of the processing of the captured scene means that these techniques can be applied to a range of sensing applications. Some relatively common ones are considered here:

- Surveillance and reconnaissance, including location of obscured targets, such as those positioned in amongst trees and under camouflage, as well as detection of buried mines;
- Targeting and fire control, including automatic target recognition (ATR), automatic target identification (ATI) to a high confidence level (addition-

ally, these techniques can identify the operational status of a system, such as operating or nonoperating);

- Improved mission planning from mapping;
- Guidance, including aim-point selection from stored databases to enable optimum engagement of fixed, mobile and moving targets in various domains;
- Navigation, including terrain-following techniques and obstacle avoidance for manned and unmanned platforms;
- Noncooperative identification (NCID) against air and surface targets at extended ranges;
- Battle-damage indication (BDI) and battle damage assessment (BDA) in real time;
- Detecting and sensing of hazards as well as potentially hazardous agents;
- Improved aim-point selection for many types of munitions;
- Advanced approaches for current mine detection.

13.5 Basic Operation of a Laser-Radar System

The principle of operation of a laser-radar system is identical to that of a conventional radar system, involving the transmission of electromagnetic energy and sensing the light returns from the scene. In some senses, it may be considered to be a sophisticated range-finder but in general, it has enhanced functionality. For example, the radial velocity of a target may be determined from the Doppler shift in the returned signals, using a coherent system, or by making at least two simple range measurements and then calculating the rate of change of range to the target, from which velocity may be deduced.

A typical laser-radar system has an active laser transmitter, a receiver and a scanning mechanism to search, or probe, a particular field-of-regard, supported by a signal-processing system, which may undertake direct detection or coherent-processing techniques. The laser transmitter and the receiver may share a common aperture in a mono-static configuration, as shown in Figure 13.2, which is commonly used; however, the transmitter and receiver may be separated in a bistatic arrangement. The fundamental method of operation for either configuration is essentially identical but knowledge of the bistatic angle is required to ensure optimum operation in the case where the transmitter and receiver are separated.

A basic laser-radar system works by measuring the time delay between the laser-pulse emission and the detection of the returned pulse, resulting from an interaction with something (object/scattering body) encountered by the outgoing photons. Essentially, this is a sophisticated development of the well-established and fundamental laser range-finding techniques, which offers significant additional functionality.

This reflected or scattered pulse provides the absolute range to each picture element in the scene as it is observed; hence, the analogy with a sophisticated laser range-finder, as noted earlier. Consequently, by scanning the laser beam and its receiver over a field-of-regard, a three-dimensional map of the terrain or object

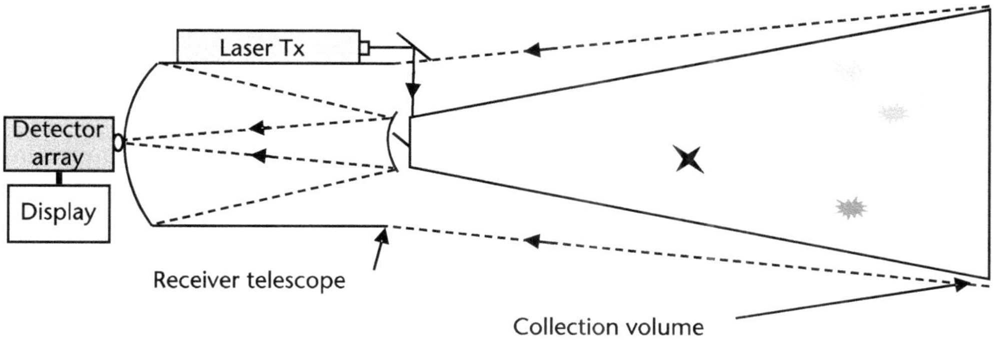

Figure 13.2 A schematic of a monostatic laser radar system.

may be generated, often referred to as a point cloud. Moreover, other objects in the scene may scatter or reflect the probing laser light; these objects may be solid targets of one form or another including various airborne aerosols, some of which may be of interest, others are merely clutter.

The point-cloud image generated by a laser-radar system from the individual detected returns contains spatial and temporal data about the targets in the captured scene, which are essentially invariant with environmental changes to a target's surface, such as water, soil on a surface or temperature. Additional information may be extracted from the received data, such as any frequency shifts induced by motion of an object or vibrations of a target, as well as variations in reflected intensity. It is worth noting that scanned laser-radar systems introduce a small range-dependent, line-of-sight error between the transmitter and the receiver in a monostatic system. This is a consequence of the finite round-trip time and the angular-scan rate, which means that the receiver collects photons from an angular position that is marginally different from the current transmitter pointing direction. The receiver must point behind the current pointing direction of the transmitter for the reflected energy to arrive on the receiver's optical axis.

A laser-radar system is capable of making a number of measurements from the interrogation of a scene with a laser beam. Some of these measurements may be made simultaneously, but the quality of these individual measurements is dependent on many factors, such as the signal-to-noise ratio, the emission wavelength and the modulation/demodulation efficiency. Examples of these measurements, which are often considered to be the dimension of a laser-radar system, are:

- Range or range profiling time-of-flight measurements, but care has to be taken to avoid ambiguity in such measurements when optical returns occur simultaneously from well-separated targets. Ambiguity arises when a repetitively pulsed laser illuminates a scene; then a detection system cannot distinguish between the receipt of a signal from a target at a particular range and the arrival of energy reflected from other targets at a greater range that were illuminated by earlier pulses in the pulse train. The unambiguous range is determined by the pulse-repetition frequency of the emission and its effects may be mitigated somewhat by selection of a pulse-repetition frequency that gives rise to an ambiguous range that is greater than the anticipated operating range of the laser-radar system. This phenomenon is discussed more fully in Chapter 10.

- Velocity, where there are two common approaches. One involves multiple range measurements at different times and then calculation of the range rate. The other technique is more sophisticated and demanding on the laser emission and receiver processing, as it involves measurement of the Doppler shift imparted on to the probing beam. This approach may also be used to measure the vibratory motion of a target, which may be exploited to detect and at least classify targets, even when camouflaged or obscured.
- Intensity, which is often used to establish the validity of the received signal.
- Bearing, which is determined from the direction of arrival of the received signal, using a suitable detector configuration.

As discussed next, other dimensions are possible and are exploited in the various forms of system applications.

The scattered and reflected photons collected from the probed scene may be used and processed in a variety of ways to provide support to an ever increasing range of military functions. These functions usually rely to some extent on the availability of supporting information from a database, some examples include:

- Wire reflectance characteristics for obstacle avoidance maps;
- Target signatures, pose and shape characteristics, as well as vibratory motion for noncooperative identification and automatic target-recognition functions;
- Topography data for navigation and obstacle avoidance;
- Bathymetry for mapping of underwater features to enable safe navigation;
- Spectroscopy characteristics, used for remote sensing of chemical or biological agents, as well as explosive materials.

The principle of laser-radar systems for remote-sensing is now well established: information is derived from the light being scattered from the target area or object being probed by the laser beam. Two regimes may be applied to this sensing technique employing either of the following:

- In the case of incoherent or direct detection, the laser is used as a powerful bright-light source. Some predetection optical filtering may be required to provide any requisite spectral information, but basically, the measurement is one of intensity only, which has some similarities with a simple pulsed-laser range finder.
- The approach of coherent light-beating detection techniques makes use of the full spatial and temporal coherence properties of laser radiation. This coherent technique mixes the scattered light field with an optical local-oscillator beam in the receiver, so that the full phase, frequency and amplitude information in the signal is established. Additionally, shot noise in the local-oscillator beam may be used to dominate thermal noise in the detector to provide quantum-limited detection of the signal beam. For effective operation, very

precise control of laser-emission parameters is absolutely vital, together with a complete understanding of the behaviour of the overall optical system, scattering mechanisms, propagation and signal processing.

Figure 13.3 shows the principles of these two approaches to the detection of the returned energy.

In theory, the coherent approach offers an advantage of up to 30-dB gain compared with the direct-detection technique.

There are many well-established ways that the photon beam from a laser may be used to probe a zone of the atmosphere for remote-sensing purposes; in fact, optical probing of the atmosphere through techniques based on elastic scattering predates the invention of the laser. The availability of pulsed-laser emissions, with high peak powers, has enabled an increased range of linear and nonlinear effects to be exploited.

Some examples of interactions between laser beams and species in the atmosphere are shown schematically in Figure 13.4, including:

- Rayleigh scattering, where the photons are scattered elastically from atoms or molecules in the atmosphere. As it is an elastic process, so there is no change in wavelength (frequency), this is the process that gives rise to the blue colour of the sky.
- Mie scattering, which is another elastic process, but in this case the scattering particles are comparable with the laser wavelength, so the scattering particles are most likely to be aerosols.
- Raman scattering, which is a $\chi^{(3)}$ process and so is inelastic with a consequent wavelength shift. In Figure 13.4 the new frequency of the scattered photons is denoted by v^* ($E = hv - hv^*$).

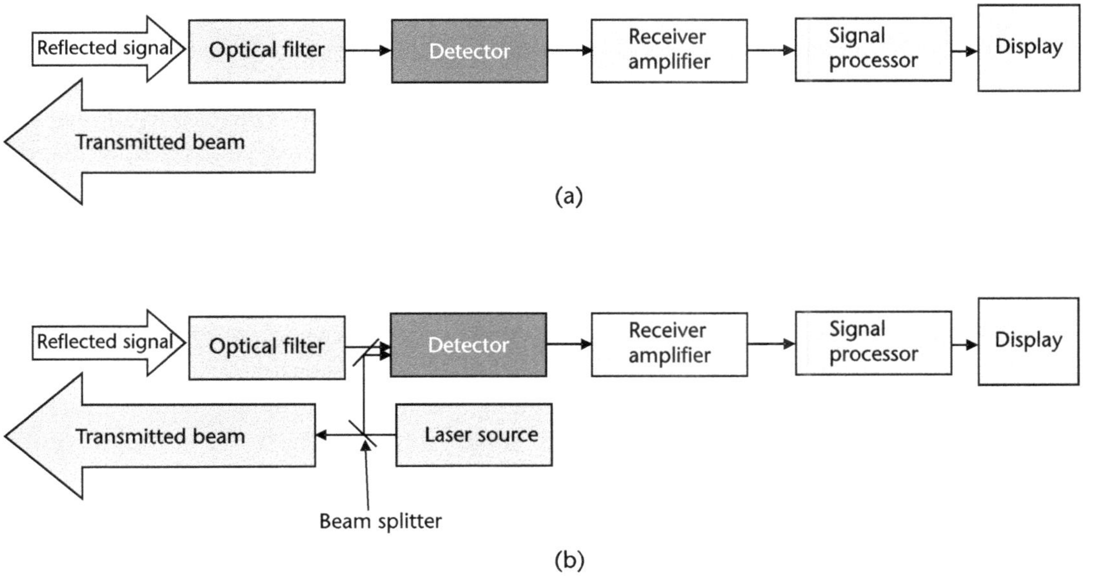

Figure 13.3 (a) Schematic of the incoherent- (direct) detection technique. (b) Schematic of the coherent-detection technique.

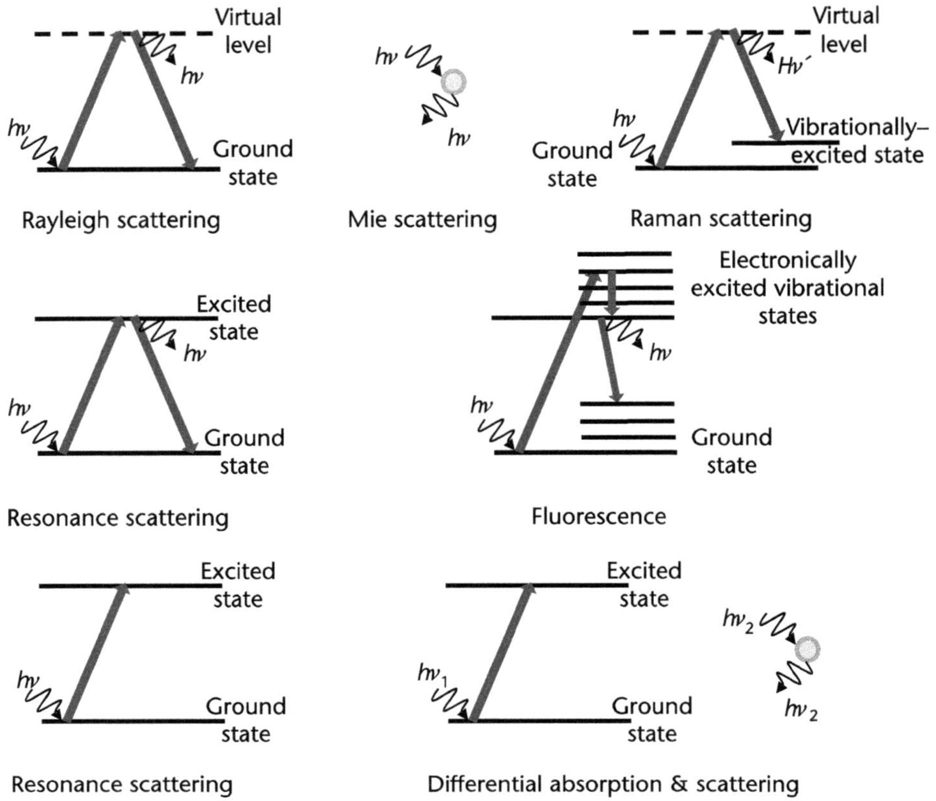

Figure 13.4 Examples of optical interactions for remote sensing.

- Resonance scattering (also known as atomic fluorescence or resonance fluorescence) is another elastic process, in which the laser-beam frequency is matched to an atomic transition with a large cross-section.
- Fluorescence is a process in which the frequency of the laser beam is matched to a specific energy transition in an atom or a molecule and is absorbed. During the subsequent relaxation process, a photon is emitted at a lower frequency. In general, the emission from molecular interactions is broadband.
- Absorption is a process in which again the frequency of the probing laser beam needs to be well matched to an energy transition in an atom or molecule, but in this case there is significant attenuation of the probe beam, as there is no re-radiation from the absorbing medium.
- Differential absorption and scattering are the principle of the differential absorption-laser radar system (DIAL). In this configuration, the differential attenuation of two laser beams is measured from the photons that are back scattered from the probed zone. The wavelength of one probe beam is closely matched to an absorption feature of a specimen of interest, whilst the other wavelength is de-tuned from that transition to provide a background measurement.

13.6 Laser-Radar Subsystem Components

A typical laser-radar system comprises four main subassemblies:

- The sensor head, containing the laser, the receiver and the power supplies.
- The beam director or scanner and its control system, this subsystem may be a relatively complicated design, particularly if it is required to undertake a number of quite different modes of operation; however, a flash illumination or staring system does not require this subsystem.
- The signal and data processor is the third subassembly.
- The display system and user interface are the fourth assembly.

For some of the applications described next, the computer system will require a database or catalogue of specific characteristics that enable the returned signal to be analysed fully through comparison with well-established information; this database is often referred to as the library. Examples include systems using automatic target recognition for identifying and classifying targets or sensing hazardous materials remotely.

Radical developments of the fundamental laser technology have enabled shorter-wavelength (e.g., short-wave infrared, around 1.5 μm) solid-state fibre-laser devices to be used in these systems, displacing the larger gas-laser technology. This technological evolution of the source devices has led to more compact and efficient systems.

The potential advantages of the use of the shorter wavelength are:

- Potential for small size and weight of the system;
- Reduced complexity and cost, because of the use of uncooled detectors and standard telecommunications components;
- Greater modulation index for anticipated returns;
- Higher target reflectivity;
- Larger Doppler shifts (proportional to λ^{-1});
- Lower divergence for a given aperture;
- Greater detector sensitivity (incoherent mode);
- Reduced power consumption (owing to the enhanced laser efficiency);
- Easier alignment (and maintenance thereof).

The identified drawbacks include greater susceptibility to atmospheric turbulence, increased noise, particularly from speckle effects, and the requirement of a higher receiver bandwidth. The laser-technology options for the various potential operating wavebands are discussed in detail in Chapters 7, 8 and 9.

The potential advantage of the application of diode laser-based coherent laser-radar systems for ranging and vibrometry has been demonstrated. The driving force for the transmitter development has been reduced cost, low observability, and compactness but also the possibility to combine this source with optical fibres to achieve a sensor-distributed system. The semiconductor-laser technology suitable for these types of applications is a strained layer, quantum-well device using distributed-feedback techniques. Range and Doppler information may be obtained by frequency shifting or chirping the laser emission. For lower-resolution applications, this can be done by tuning of the distributed-feedback laser, which is used to seed the frequency-stabilised emission. Using an external cavity and a diffraction grating, substantially larger frequency-sweep ranges may be realised, with a corresponding larger range accuracy being achieved.

The performance requirements demanded from the optical-scanning systems used in laser-radar systems vary considerably and are dictated by the mission requirements. These military missions can range from imaging, obstacle avoidance, terrain following/navigation, target detection and identification or target-tracking systems, for example. Consequently, there may be demanding requirements on the scan angle and the speed of the scan, dependent of the platform that the system is installed on, but ground-based applications may only require a very simple scanning system.

The scanning techniques used in these applications of laser-radar systems extend from fully-programmable and stabilised two-axis mirror devices to Risley prisms (or wedge pairs) for generating a variety of complex scan patterns over a fixed field of regard. Modern nonmechanical beam steering and shaping techniques are being developed for laser-radar applications, which are likely to lead to more compact and lighter laser-radar systems. These modern beam-scanning techniques are discussed in more detail in Chapter 4.

These systems scan the beam in front of the receiver optics (i.e., in the object space). Scanning can be undertaken in image space to give high-rate pixel imaging, using a galvanometer scanning in the image plane with a sophisticated laser-optical readout of the mirror position. The advantage of this approach is to provide very high accuracy angular coordinates, whilst the scanning mirror is free running. However, fast scanning introduces some fundamental problems associated with spectral broadening and lag-angle losses, leading to a severe increase of the clutter level, because of frequency broadening.

13.7 A Simple (Flash) Laser-Radar System

A relatively simple laser-radar system may be used to illuminate a scene with a flash of laser light as shown in Figure 13.5; in this case, there is no scanning system and a relatively simple pulsed-laser device is required. This schematic system uses an

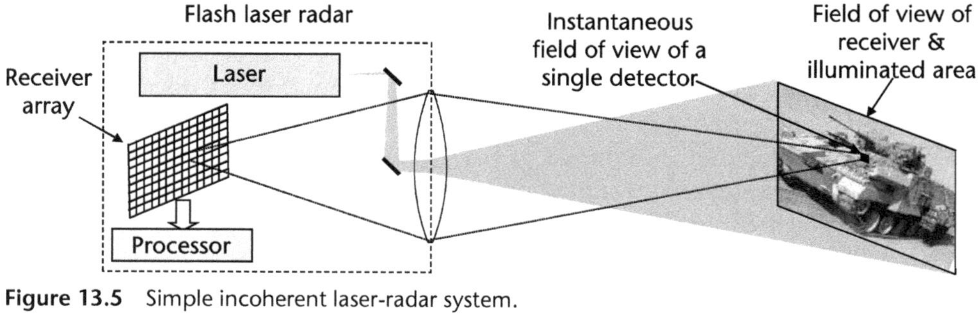

Figure 13.5 Simple incoherent laser-radar system.

incoherent approach to sensing, with the two-dimensional (lateral) resolution of the system being determined by the size of the pixels in the array and the focal length of the optical system in the receiver channel. Some estimate of the resolution of the third dimension may be made through assessing the pulse characteristics of the illuminator. The intensity of the retro return is governed by the reflection characteristics of the target, the propagation characteristics of the transmitted pulses and the range between the sensor and the target (i.e., application of Beer's Law, discussed in Chapter 5).

13.8 Dimensions of Laser-Radar Systems

A scene may be probed in many ways using a fixed or scanning laser beam and the photons scattered or reflected from the probed space detected using incoherent (intensity) measurements or coherently, using frequency space. A staring (fixed-angle) approach has become more practical with the development of more powerful transmitters and enhanced-performance detectors, as discussed above.

A single-dimensional or multidimensional image may be created using this simple approach, for example:

- *1-D profiling:* this is a single dimension of measurement and so may equate to a laser range-finder system with a single detector and plotting the return from a target as a function of time, such as an aircraft. This type of measurement contains some limited information about the target but would be of little value for identification.
- *2-D silhouette:* the second dimension may be added in several ways such as flash illumination of a scene and use of a camera (two-dimensional focal-plane array) to create an image of the objects in the scene on to a single plane (i.e., a silhouette).
- *3-D imaging:* the third dimension demands a far more sophisticated system configuration using the rapid frame rate of the imaging sensor to provide a three-dimensional image that allows shape extraction of targets in the probed scene.
- *N-D imaging:* additional dimensions based on wavelength, polarisation and other features to permit other functions to be undertaken.

These dimensions of a laser-radar system may be used with either the incoherent or coherent techniques, as shown in Figure 13.6. Many of the modern developments of laser-radar systems exploit fibre-based solid-state laser technology and coherent-detection techniques; the latter approach offering greater information about the detected targets in a scene.

13.9 Automatic Target Recognition

Techniques and methods, along with efficient algorithms, for very rapid signal and data processing have enabled real-time automatic target recognition of targets to

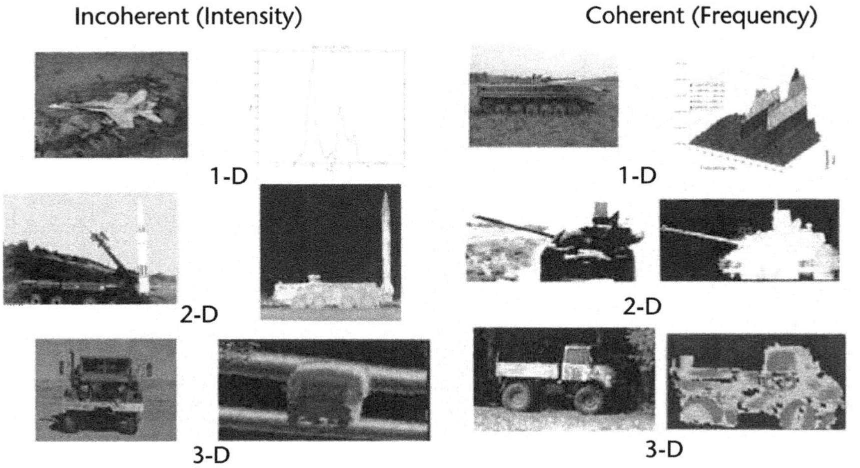

Figure 13.6 Dimensionality of different types of laser-radar systems and NATO SET Panel. (Courtesy of Fraunhofer IOSB, Germany, and NATO SET Panel.)

be achieved. The development of advanced detector technology, coupled with advanced signal-processing algorithms has enhanced the functionality of laser-radar systems. Many of the applications considered below, benefit from the use of an automatic target-recognition function in one form or another. In fact, it is the successful development of these techniques that has made many of those applications possible and new roles continue to be developed for laser-radar systems.

Automatic target recognition is now a critically important processing technique that describes a broad number of automated or assisted exploitation functions. These functions include fully and semi-automated detection, false-alarm mitigation, classification, recognition, and identification of targets in a variety of scenarios. Successful development and demonstration of this process has not only accelerated targeting functions but also led to more use of laser-radar technology in autonomous systems.

Automatic target recognition has a crucial role to play in many military functions, some common examples include:

- Time-critical targeting in any domain;
- Target tracking and cueing;
- Geo-location and registration of targets in a scene;
- Aim-point selection;
- Combat identification;
- Site monitoring;
- Force-structure analysis;
- Battle-damage assessment and identification;
- Terrain delimitation.

Automatic target-recognition systems can operate at various levels of automation that are adapted for different applications. Examples of different levels of automation may be summarised as follows:

- Interactive-decision aid that involves human intervention in the process, so that man and machine work in parallel;
- Automatic-decision aid, which involves the automatic processing of the input data and an output that is presented for a human to make the final decision;
- Autonomous system, where humans are excluded from the final decision, so that a human is not in the loop.

The time scale of the process is critically dependent on the function being undertaken; for example, with surveillance the period may cover hours of data collection and processing, whereas target acquisition with an autonomous guided weapon may need to be completed in a few seconds or less, based on real-time processing and decision making.

It is generally recognised that there are five stages or steps of discrimination performed by automatic target-recognition systems:

- *Detection:* this is the stage at which targets are distinguished from other objects and clutter in the scene, such as trees, rocks, or image-processing artefacts.
- *Classification:* this is the level at which the class or type of target is resolved, so that an object may be classified or labelled as a land vehicle or a building, for example.
- *Recognition:* this is the step where the sensed information will allow the target subclass to be determined, for example, that an object classified as a land vehicle is in fact a tracked vehicle but the data do not allow further discrimination between types of tracked vehicles.
- *Identification:* this is the stage in the data processing with the higher resolution data when the model and/or make of a recognised class of target is resolved, so that for an object recognised as a tracked vehicle it may be established that it is a tank. Additional information and analysis may enable the object to be classified as an M60 or T72, for example.
- *Characterisation:* this is the level of fidelity when very detailed physical characteristics of a target may be determined, such as a particular, or specific, equipment fit of the identified target.

This process is summarised in Figure 13.7.

A successful automatic target-recognition system has an effective sensor-exploitation algorithm in the processor, which receives information collected from

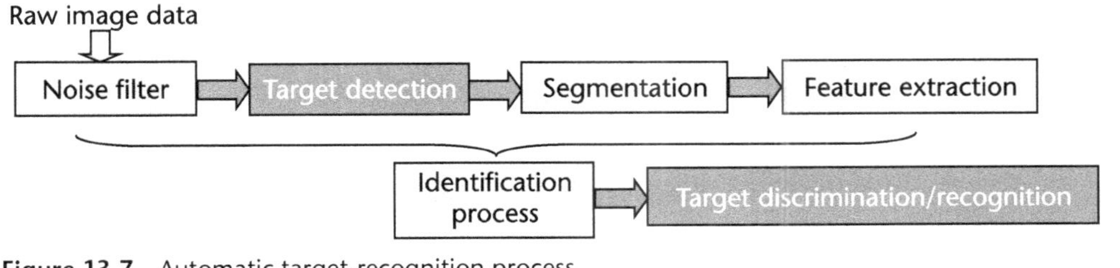

Figure 13.7 Automatic target-recognition process.

the scene by the sensor in the receiver channel. The algorithm undertakes some form of feature extraction and that information is put into a discriminator routine. The discriminator will have access to some target characteristics or knowledge derived from a variety of sources (trained features, templates and models, for example); its function is to match the sensed information with the information held in the database. It then produces some scoring metric that is evaluated in accordance with predefined decision rules. The output from the algorithm is a decision indicating the nature of the target; typically this is reported with some levels of confidence in the accuracy of that decision.

This type of approach to the application of automatic target-recognition techniques, using a laser-radar system, has many challenges, for example:

- Operating against mobile targets, which may be re-locatable (i.e., targets that are capable of moving, but are stationary at the time of engagement, or moving targets);
- The wide variety of operational conditions that is likely to occur in any military engagement, such as clutter, partial obscuration, articulations of any target that has the ability to change its shape, such as moving a turret or erecting a launcher;
- The operating environment, including adverse weather, clutter, camouflaged targets and decoys.

The sensor in the receiver channel collects information from the scene, including the target, which is then degraded by the operating characteristics of the sensor and the environment. It is this degraded signature that is then passed to the automatic target-recognition algorithm, which attempts to match this information with its stored knowledge as rapidly and accurately as possible. Clearly, the confidence level of the output from the algorithm is correlated directly with the quality of the collected data. It is normal to undertake some conditioning of the sensed data in order to reduce noise and other artefacts in the data set through a variety of filtering techniques.

The level of discrimination that may be achieved with a given degree of confidence may be assessed by the criterion of how many pixels on target are achieved by the detection system array in the receiver. This criterion is a powerful tool for evaluating data if there is enough information in the imagery to support different levels of decision making. For example, detection may be achieved with high confidence with as few as 100 pixels on target, whereas identification may require a factor of six more pixels on target in the image; based on rules derived from Johnson's criteria for detection, identification and recognition.

The challenges for automatic target-recognition algorithms currently include:

- Robust and accurate recognition performance under adverse target-detection conditions, such as when they are partially obscured either by the terrain or by poor weather conditions;
- Real-time operation under the vast range of target-engagement conditions;

- Adaptation to new target sets and modifications and other changes to existing target characteristics;
- Reduction in false alarms.

13.10 Laser-Radar Seeker

The application of laser-radar techniques to military systems has enabled autonomous weapons to be demonstrated that use automatic target-recognition methods. This involves the search of a volume of space to form a picture of the terrain in front of the laser radar and comparing it with stored data. The database may include:

- Details of the terrain in the target zone for aiding the navigation system;
- Characteristics of the designated targets, which may include spatial and temporal information.

These data points are defined with respect to the fixed reference axes of the laser-radar system and are recorded in terms of the bearing of the scanning mirror (axis) and range from the transmitter. The Cartesian co-ordinates of measured points may be calculated in terms of the system reference axes. To be able to refer to these co-ordinates, another reference frame of the position and attitude of the laser radar system is required to enable the axis transformation to be undertaken. If the host platform is moving, then line-of-sight stabilisation will be required, as discussed in Chapter 15. A slave inertial-measurement unit may need to be used if the laser-radar system is positioned remotely from the master inertial-navigation reference system and there is significant angular motion between the laser-radar system and the inertial-reference system. Additionally, frequent inertial-navigation updates are required, owing to the motion of the vehicle and it is normal to time-tag these measurements to provide suitable synchronisation of the ensemble of data.

The processed returns from the scene may be evaluated with various algorithms to undertake the detection, acquisition and tracking of objects. These may be classified according to a range of predetermined criteria, selected according to the mission objectives, usually according to size and shape. The targets may then be compared with a stored database to provide recognition of military vehicles, buildings, bridges or other man-made structures. A three-dimensional picture may be formed and the target data superimposed on the map of the area provided the laser-radar data are referenced to coordinates stored in the database for the region.

These techniques have been applied to laser-radar seeker systems for use in tactical weapons to detect, acquire and track surface targets. The seeker contains a scanning system that enables the laser-radar's beam to search its field of regard to find a target. A simple, single-plane azimuthal scan with the beam directed downwards at a shallow angle to the flight path is often adequate for a missile flying horizontally as a forward motion will provide the second dimension for the scan mechanism. This form of push-broom scanning is shown in Figure 13.8.

The processes undertaken by the laser-radar seeker are:

- A laser beam scans across the field of regard of the seeker.
- The received information is processed to create an image that includes height data.
- The captured scene is filtered to identify potential targets and eliminate background features and other objects that do not represent valid targets.
- The image objects that are potential targets are processed to estimate dimensions (length, width and height), orientation and information about any objects that have been detected on the potential targets.
- Classification of the targets in the captured scene and identification of the objects using three-dimensional target-matching algorithms.
- Guidance commands are generated for target interception.

The seeker requires stabilisation during its various modes of operation, as well as precision control of the scanning mechanism, as the transmitter and receiver are panned across the field of regard. Rate gyroscopes provide the required control function by detecting the disturbances and providing the feedback to the control loops to stabilise the sightline and ensure that a smooth and consistent scan is achieved during the target acquisition process. Sightline stabilisation and accurate beam-positioning control are critical to the success of this technique.

Strapdown techniques are also feasible, where the inertial-measurement unit in a cruise missile may provide the measurement of the angular disturbances, so an accurate angular pick-off device would be required to provide a measure of the precise position of the optical axis of the seeker at any given instant in time. Alternatively, rate gyroscopes may be used on the gimbal mechanism to record the angular disturbances and then used in conjunction with lower-precision angular pick-off devices to control the scan over the field of regard.

There are multiple trade-off options in the fundamental parameters used to define a laser-radar seeker, depending on the primary function of the seeker. An important consideration is the speed of the weapon and the desired resolution of the beam on the ground, which in turn influences the scan rate, the pulse-repetition frequency and the beam divergence. An example of some generic parameters is shown in Table 13.1.

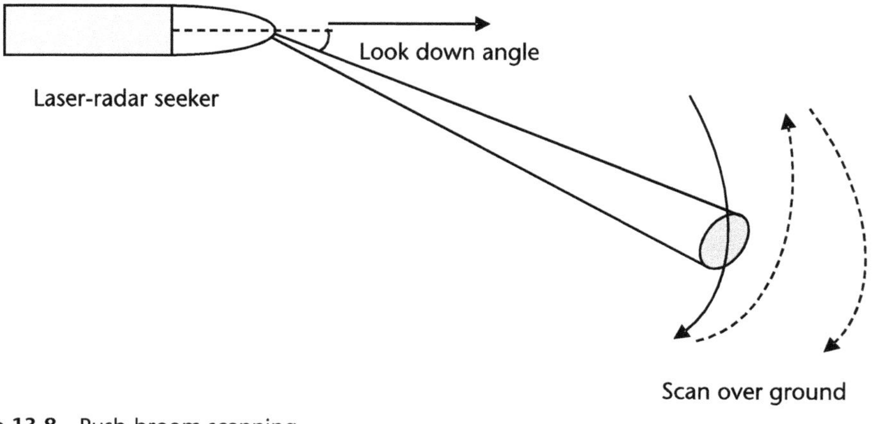

Figure 13.8 Push-broom scanning.

Table 13.1 Source Performance Characteristics for a Laser-Radar Seeker

Source Parameter	*Value*
Wavelength (μm)	1 or 1.55
Average power (W)	5 to 20 or more for long-range target acquisition
Divergence (mrad)	~1 or less for smaller targets
Pulse-repetition frequency (kHz)	10 to 100 (footprint dependent)
Line width (nm)	Noncritical (direct detection)
Scan rate (Hz)	Dependent on platform speed
Scan angle (°)	Up to 20 for target acquisition, in the terminal stages it may be a starring system

13.11 Bathymetry and Topographical Mapping

Bathymetry is a sounding technique, which is an application of laser-radar systems for three-dimensional mapping of the terrain in the presence of water, such as coastal regions, lakes and rivers. The aim of a bathymetric system is to generate a map of the solid profile of the Earth beneath the water's surface through analysis of the back-scattered laser beam to determine the absorption and scattering of the pulsed probe beam. Topographical sounding is the equivalent process over land, but the surface may be covered by vegetation.

The basic principle used in bathymetry sounding is identical to any other mapping or profiling system; these systems use very short duration laser pulses to provide a route to the measurement of the time of flight of each pulse from launch to receipt of the returned pulse by the detector and recording the various returns from the scene being illuminated. The distance to each point providing a return from an object in the transmitted beam can be determined, and given accurate knowledge of the position of the transmitter and the orientation of each beam position, the co-ordinates of each return can be calculated in space, or with respect to a suitable reference axis system. The collected data may be processed to provide a profile of the probed scene.

It is important that the transmitted wavelength is compatible with the transmission windows for the particular application. In the case of bathymetry, the wavelength needs to penetrate the surface of the water and then propagate through the column of water; blue-green wavelengths are the most appropriate for this task, as illustrated in Figure 13.9. In the case of normal topographical measurements, the near and short-wave infrared wavelengths are more effective, particularly for reduced solar noise and reduced hazards to other operators and civilians through the use of eye-safer wavelengths.

A bathymetric sounding system and topographical mapping systems are configured as classical laser-radar systems with the following subsystems:

- Laser source (blue/green wavelength for bathymetry and short-wave infrared for mapping);
- Scanning system, usually capable of various scanning patterns;
- Receiver system with a telescope collection aperture and signal processor;

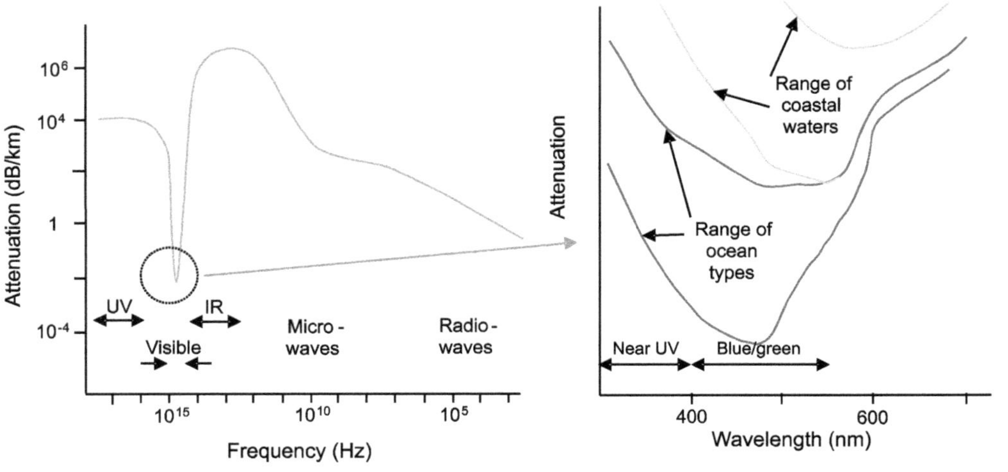

Figure 13.9 Propagation characteristics of electromagnetic waves in sea water.

- Database/library;
- Display modules.

A schematic of a bathymetric system is shown in Figure 13.10.

The bathymetric laser-radar system has significant physical challenges compared with more conventional mapping, because of the presence of the water column along the propagation path. The critical physical processes involved with this application include:

- Refraction and reflection at the water surface;
- Beam distortion in the medium;
- Extinction of the beam as it propagates, from absorption and scattering of photons out of the probe beam.

Consequently, the return beam is far more complex than the equivalent signal in a classical three-dimensional topographical system, which in general has to process discrete signals (pulses), However, as shown in Figure 13.11, a bathymetric system generates a continuous return signal, which appears like a complex waveform, comprising:

- A return signal from the surface;
- Back-scattered energy;
- Target returns.

A bathymetric system has the challenge of detecting the actual returns from the target (sea bed), which may be many orders of magnitude smaller than the reflection from the water's surface. Consequently, the receiver system has to have sufficient dynamic range to be able to detect the various components in the complex waveform of the returned signal to enable the signal processor to undertake its

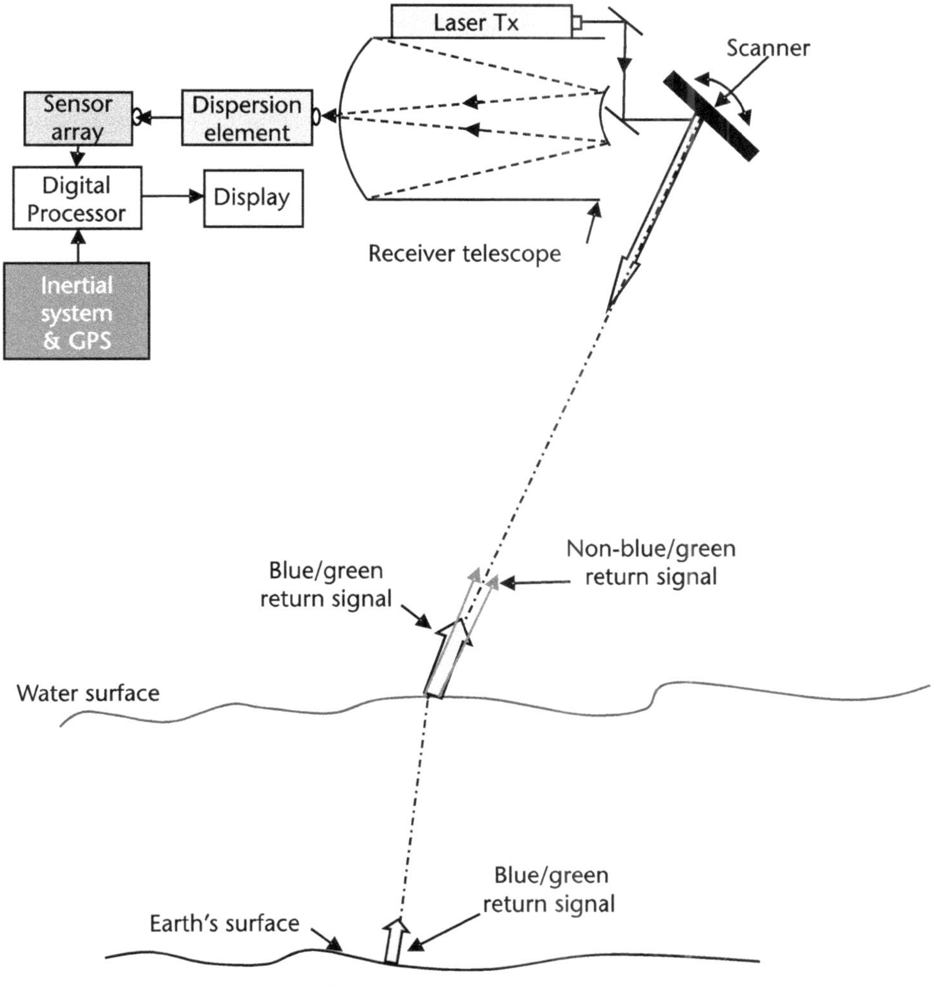

Figure 13.10 A laser-radar system for bathymetry.

functions. These systems are designed to keep the angle of incidence of the beam at the surface constant during the swath of a scan, to minimise the dynamic intensity fluctuations in the return signals. Careful selection of this angle enables the depth-bias effect (a consequence of how the laser beam propagates through the column of water) to be minimised.

The choice of a scan pattern is important, not only for minimising variations in the return waveform, but also for achieving an efficient probing of a given area and minimising shadowing effects caused by obscuration of the probing beam by other objects. The latter effect is usually more of a problem with a topographical system and is often mitigated through the selection of a scan pattern that allows the same spot in a probed area to be viewed from multiple directions.

The complexity of the signal processing and the construction of the profile beneath the water surface tend to limit the number of soundings possible per second with a bathymetric laser-radar to about 4,000, whereas a three-dimensional topographical system may be capable of up to 16 to 20,000 soundings per second.

Typical laser characteristics are shown in Table 13.2.

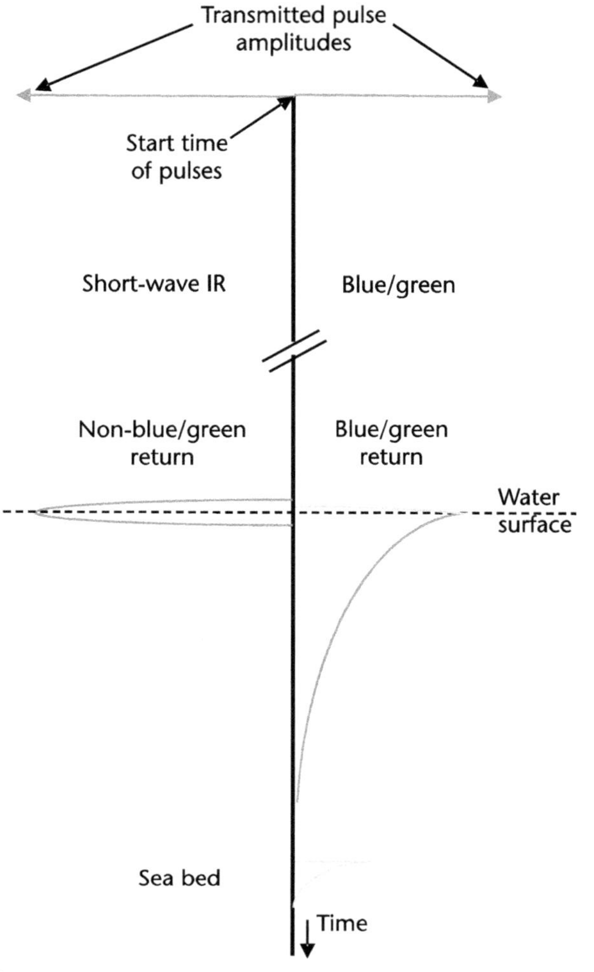

Figure 13.11 A schematic comparison of return signal from topographical and bathymetric sounding.

Table 13.2 Source Characteristics for Sounding and Mapping

Source Parameter	*Bathymetry*	*Topography*
Wavelength (nm)	532 to 550	900, 1064 or 1550
Average power (W)	1 to 10	<1 typically a factor of 10 to 100 lower than for bathymetry.
Divergence (mrad)	1 to 10 to ensure eye safety the irradiance must not exceed the maximum permissible exposure (MPE – 5 mJ/m^2).	< 1
Pulse-repetition frequency (kHz)	1 to 10	100 to 400
Line width (nm)	<1	<1

It is worth noting that photon-counting laser-radar systems may need a factor of 10 to 100 less power than traditional analogue direct-detection systems.

The depth resolution is determined by the pulse duration (τ) and the speed of light in water (c/n) and cannot exceed $(c\tau)/2n$, where n is the refractive index of water.

13.12 Obstacle Avoidance

High-performance surveillance systems, linked to air-defence units, present a significant threat to aircraft attempting to penetrate any defended air space. The effectiveness of these air-defence systems forces interdiction aircraft to fly at very low altitudes, to avoid detection by air-defence systems. These aircraft may have to fly close to the ground at very high speed, possibly in poor visibility, and probably with inadequate information about the local terrain. The possibility of a collision with a variety of hazards or terrain features is high, particularly if the aircraft relies on conventional sensors to detect potential obstacles along the flight path.

Other low-level roles, such as reconnaissance or search and rescue missions, may also require aircraft to fly close to the ground where there may be many hazards. One of the major hazards encountered during this type of mission during low-level flight over land are the man-made obstacles, which are difficult to detect, such as wires between pylons.

The functions of a typical obstacle-avoidance system installed on an aircraft cover detection, classification, and real-time display of these obstacles. The typical obstacle hazards to low-flying military aircraft include:

- Power, telecommunication, and barrage balloon cables;
- Masts;
- Guy wires;
- Tall thin buildings;
- Trees;
- Other potential mobile, as well as fixed, obstacles that may occur along a low-level flight path.

Detection of all of these potentially hazardous objects, including fixed structures, must be achieved in daylight, at night, and in adverse weather. To ensure that suitable warnings continue to be provided when the aircraft is turning, a large field of regard must be available to the sensor system, the size of which is governed by the flight envelope of the host aircraft.

Real-time and accurate situational awareness is, therefore, a crucial safety function for any low-flying aircraft. Obstacle-avoidance systems, based on laser-radar technology, offer a very valuable aid to aircraft that are required to fly close to the Earth's surface.

A laser-radar system for obstacle-avoidance applications will usually scan a beam in front of the aircraft. A detector collects the returned signals in a similar fashion to a laser range-finder for each transmitted pulse, so that a three-dimensional plot can be created indicating the position of obstacles along the flight path. Sophisticated signal processing, involving the knowledge of the beam position and bearing for each point in the scan, the delay in the return of the pulse from its transmission and the position and motion of the fitted platform in a known reference frame, enables the obstacles detected to be superimposed on a map. Clearly, real-time image-processing and display is critical for the success of this technique; moreover, use of additional algorithms to link multiple objects detected in the scanned scene, allows a simplified picture to be created for ease of use by the pilot (i.e., easing the work-load burden on a pilot).

Laser-radar systems have been developed to provide an obstacle-avoidance system for helicopters and fixed-wing aircraft. Figure 13.12 shows the system developed in the United Kingdom by QinetiQ and Figure 13.13 shows another system developed in Germany for use by the German border patrol on their EC 135 helicopters, shown in Figure 13.14.

A schematic of an obstacle-avoidance system is shown in Figure 13.15 viewing a scene with a potential obstacle in the flight path. The figure shows a camera, which can be a visible-band camera or a thermal camera, to provide a view of the scene that can have the obstacle overlaid, as appropriate.

Typical laser characteristics for an obstacle-avoidance laser-radar are given in Table 13.3.

13.13 Vibrometry

A vibrometer is an application of a coherent laser-radar system; these systems are designed to measure the Doppler characteristics of targets detected in a scene, so that remote objects can be acquired and at least classified. The technique used is based on the observation of moving or vibrating objects with a laser-radar system

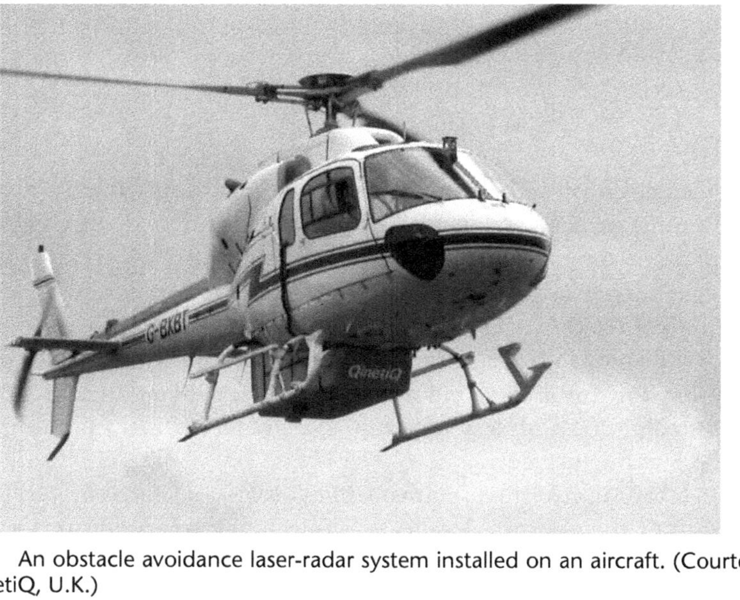

Figure 13.12 An obstacle avoidance laser-radar system installed on an aircraft. (Courtesy of Dr. P. Kightley, QinetiQ, U.K.)

Figure 13.13 A sensor head unit SferiSense 500 (HELLAS-A). (Courtesy of CASSIDIAN, Germany.)

Figure 13.14 Cassidian system installed on a helicopter. (Courtesy of CASSIDIAN, Germany.)

that is capable of high Doppler resolution of the various components in the returned beam. In particular, analysis of the vibration imagery collected by such a system offers a potential aid for long-range target classification and identification including camouflaged or partly concealed targets. This technique also allows discrimination between real targets and decoys.

The principle of operation involves the illumination of an object with a laser beam; this type of laser-radar system is very demanding on the optical source, requiring both narrow line width and high stability of the transmitted beam; the required line width is typically of the order of a few tens of kilohertz, or better. Any moving object intercepted by this beam will modify the reflected and scattered signals according to the Doppler phenomenon, producing small changes in frequency

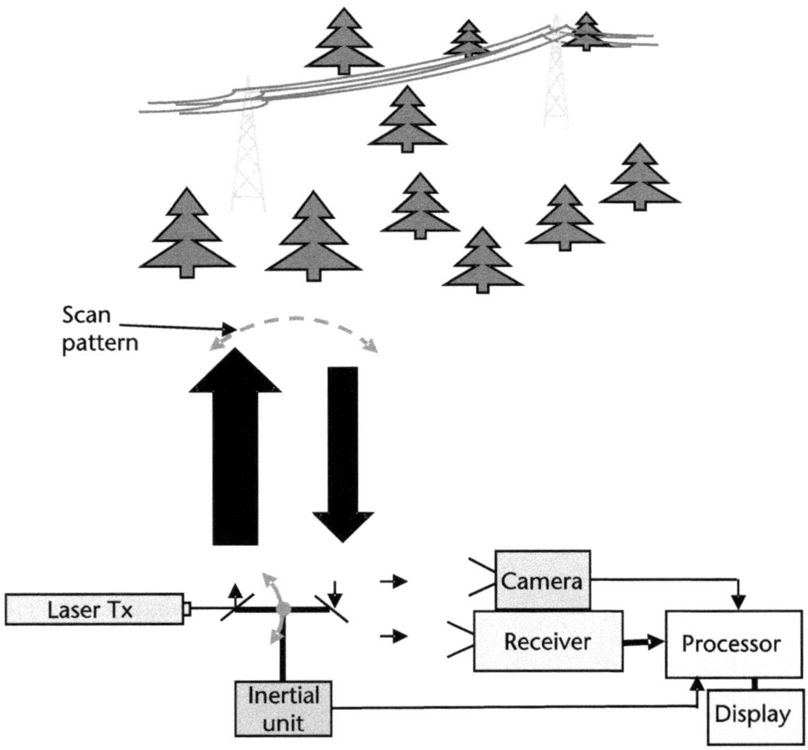

Figure 13.15 Schematic of an obstacle-avoidance system.

Table 13.3 Source Characteristics for an Obstacle-Avoidance System

Source Parameter	*Value*
Wavelength (μm)	1.55
Average power (W)	2
Divergence (mrad)	0.06
Pulse-repetition frequency (kHz)	50
Line width (nm)	Noncritical (direct detection)
Scan rate (Hz)	80
Scan angle (°)	20

(and hence wavelength) of the returned signal. Coherent processing techniques enable the Doppler components to be identified and the motion of the object calculated. A schematic diagram of a vibrometer is shown in Figure 13.16.

For airborne applications, the processing of the received data is complex, as it must be able to reject the micro-Doppler returns from the aerosols in the atmosphere being displaced by the motion of the platform. In the case of vibrometric-imaging systems, the processor requires access to an extensive library, so that the

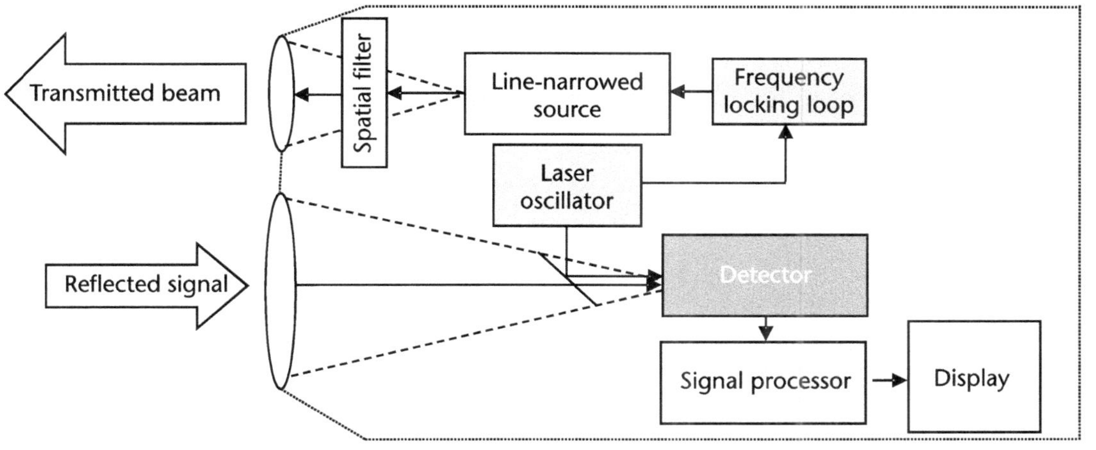

Figure 13.16 A schematic of a vibrometer.

maps of the vibration patterns deduced by the data captured by the laser-radar system from the scene may be interpreted accurately.

Vibration imagery provides additional information about targets in the field of view of the laser-radar sensor by capturing the vibration pattern over the target, potentially allowing a better classification level of each detected object. The vibration imagery may be generated by mechanically scanning the laser beam and receiver over the target zone. Reference images are made to give a better understanding of spatially unresolved vibration signatures.

Mechanical scanning techniques are time consuming, so flash-illumination techniques, as shown in Figure 13.5, have been developed to enable more rapid generation of images. An operational system will require a multi-element coherent detector array to reduce the time for capturing the vibration signatures from a scene.

Automatic data processing was first applied to the detection of buried mines by exploitation of vibrometry techniques and measurements, but then transferred successfully to the problem of classification of land vehicles. One successful automatic target-recognition algorithm used a multi-threshold approach for the segmentation of target cues collected by the receiver. This algorithm reduced the false-alarm rate by measuring the stability of possible target cues by means of object size, contrast and shape amongst the many other parameters in the frequency domain, which resulted in a three-dimensional representation (x, y, ν) of the candidate target set (i.e., position and vibrational frequency). In the case of mine detection, in addition to the amplitude of soil vibration the phase of the vibration signature was also investigated as an additional information channel to optimise the classification level and gain confidence in the output (see Table 13.4).

13.14 Acousto-Optic Mine Detection

A noncontact acousto-optical technique, linked to a laser-radar vibrometer, has been developed for the detection of buried land mines and other objects with little metal content. With this approach, acoustic waves from a transmitter, similar to a

Table 13.4 Source Parameters for a Typical Vibrometer

Source Parameter	
Wavelength (μm)	1.5 (eye-safe region)
Average power (W)	1 to 10
Divergence (μrad)	70 to 100
PRF	Continuous wave
Line width (kHz)	3

loudspeaker, are coupled into the ground around the zone of concern. If a mine or other inhomogeneities are present below the surface, the scattered field generated at the soil surface is modified by the shape and size of the buried target compared with undisturbed ground. These changes to the scattered field and other anomalous characteristics can be detected by a scanning a vibrometer over the scene.

13.15 Remote Detection of Hazardous Substances

The threat posed by the release of chemical or biological agents on a battlefield would have a dramatic effect on the ability of any military force to operate. The precautions that need to be taken in order to protect all personnel are onerous and limit the fighting capability of the forces under attack. Consequently, the wearing of such equipment as a precaution over an extended period of time is not a realistic option for mobile operations; rapid and accurate techniques are required to indicate the absolute necessity to wear appropriate protection, or undertake evasive manoeuvres.

Laser-radar systems provide noncontact sensing techniques using direct optical methods for detection of these hazards on the battlefield, without having to make close (or direct) contact with a contaminated area. These systems may offer a useful stand-off range of a few tens to hundreds of metres, depending on the limit of detection required for a given application, together with the absolute precision required.

A number of different techniques are used to determine the presence, and subsequently identify, the potential agents detected in the probed volume, particularly through vibrational spectroscopy, where a light source causes covalent chemical bonds within a sample to vibrate. The resultant feature-rich spectral signature is specific to a given chemical, enabling this technique to classify/identify chemical and explosive threat materials, and indicate the presence of biological materials. The spectral signatures of biological molecules are inherently complex owing to their larger size and the resultant increase in vibrationally active modes. These phenomena result in spectral similarities between species of the same class (i.e., proteins or carbohydrates). There exist two main forms of vibrational spectroscopy; those based upon infrared-absorption spectroscopy and those that exploit Raman scattering. Both types can be performed at stand-off ranges and will be described in detail next. Additionally, alternative techniques to vibrational spectroscopy with potential application to stand-off detection will also be discussed in the following sections of this chapter.

13.15.1 Detection Using Infrared-Absorption Spectroscopy

Chemicals within the same class contain the same functional groups, which can be bonded in different ways. When energetically excited, the bonds within this functional group vibrate at a set frequency, as determined by the mass of the atoms that form the bond and the atoms in close vicinity. Excitation is achieved when an infrared photon whose energy matches (i.e., is resonant with) the bond frequency is absorbed by the bond, resulting in promotion of an electron to a vibrationally excited level. Therefore, in infrared-spectroscopy the depletion in the amount of incident photons detected at a given energy indicates that a chemical bond exists, which can be assigned to a specific bond by matching the energy of the depleted signal (an absorption band) to information stored in an appropriate database. The identity of a material containing multiple bonds can be determined by matching its spectral signature, comprising many absorption bands, to a library of reference spectra. Clearly, the success of this process depends on:

- Whether the correct entry is present in the spectral library (database);
- The quality of the library data;
- The composition of the sample.

In mixed samples that are composed of many different chemical species, sophisticated library matching algorithms are commonly employed to provide robust and rapid identification. Infrared-absorption spectroscopy can be conducted in three main spectral regions, namely:

- Long-wave infrared (LWIR), corresponding to 6–14-μm spectral region. This region comprises many intense absorption bands owing to the fundamental frequencies of the majority of chemical bonds. The long-wave infrared is also known as the fingerprint region, as the spectral signatures in this region have many distinct spectroscopic features (i.e., the jargon is that they are feature rich) and are characteristic of particular species. Therefore, they act as spectral fingerprints for the identification of materials.
- Mid-wave infrared (MWIR), corresponding to the 2.5–6-μm spectral region, which mainly comprises the fundamental frequencies of bonds that contain hydrogen, including O-H, N-H and C-H, in addition to double and triple bonded groups, such as the cyano moiety ($C \equiv N$).
- Shortwave infrared (SWIR), corresponding to the 1–2.5-μm spectral region (as defined for this application[1]), which comprises the overtones (or higher-order harmonics) and combination bands of the fundamental frequencies found in the mid-wave and long-wave infrared regions (as defined in this context). As a result, short-wave infrared spectral signatures are generally less intense and more complex than in either of the other two spectral regions.

One key advantage of infrared-absorption spectroscopy is that it can be easily applied to species in all physical phases: solid, liquid or gas/vapour. It is worth

1. Note that other applications, such as infrared countermeasures, refer to SWIR as 1 to 2 μm or sometimes 1.4 to 2.5 μm.

noting that different experimental regimes may be needed to compensate for the low density of the gaseous samples or the high density of a solid. However, one key disadvantage of infrared-absorption spectroscopy is that a chemical bond must contain a permanent dipole moment to produce an infrared-absorption signature; this means in practise that only hetero-nuclear (e.g., X-Y) species can be detected, but homo-nuclear (e.g., X-X or Y-Y) bonds are not infrared active. Therefore, chemical species such as chlorine cannot be detected using infrared absorption-based techniques.

Figure 13.17 shows the fingerprint region for some common molecules with absorption in the infrared region between 3 and 14 μm.

13.15.2 Remote-Sensing System Operation

A two-stage optical method has been used to achieve optimum location and identification capability of the agents, involving the illumination of a volume of space with a flood-illumination beam, and viewing of this volume with an appropriate camera to locate potential positions of contaminants. These sites may then be interrogated with a suitable tuneable laser beam to enable further investigation of the suspect sites/locations. A number of techniques may be used to aid the detection and identification of potential chemical or biological agents, such as remote spectroscopy or infrared-reflectance methods.

The techniques developed for this task of locating and identifying contaminants has many similarities with the incoherent laser-radar techniques. The primary difference with the current approach is the use of the second laser for the very important identification or classification stage.

A relatively simple laser-radar technique is used to delineate potential sites where there may be toxic agents. The scene is scanned by a laser-radar system with a suitable beam to provide illumination and this scene is observed with an appropriate detector array to find any anomaly in the scene being captured by the laser-radar system, as shown in Figure 13.18. This location can then be interrogated to provide the classification of the detected agent. The use of a two-stage process is very useful for reducing the false-alarm rate, which is a critical requirement for such systems, as missing an actual event is not acceptable, whereas if there is a very high false-alarm rate the system would only provide limited benefit.

Several techniques may be used for determining the image/anomaly location based on detecting changes in the infrared reflectance of the scene being illuminated.

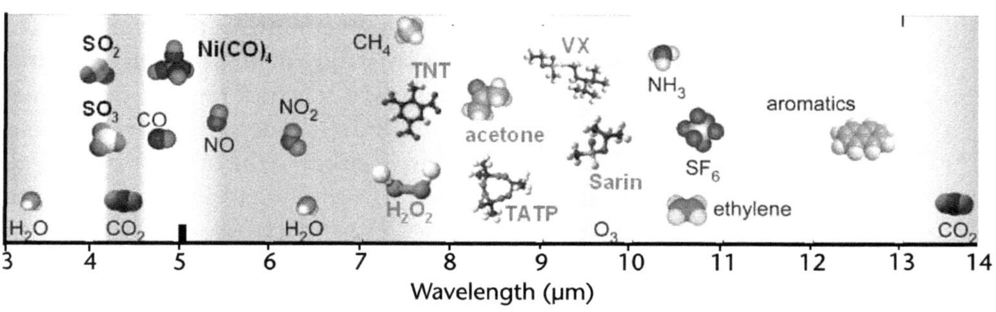

Figure 13.17 The fingerprint region for some common molecules. (Courtesy of Daylight Solutions, USA.)

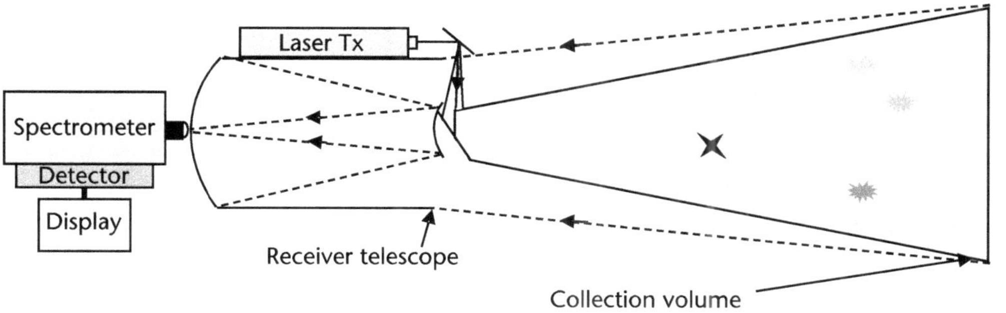

Figure 13.18 A laser-radar system with a spectrometer.

A typical generic interaction with a potentially contaminated surface is shown in Figure 13.19.

Infrared-absorption spectroscopy requires a laser source that is broadly tuneable to enable detection of a wide range of materials. Ideally, tuning should occur rapidly to negate the effects of dispersion of the target chemical over time. Pulsed sources are preferred, as this regime enables gated detection to circumvent ambient and stray light sources (as discussed in Chapter 11). The laser line width is target dependent, as it determines the maximum spectral resolution. Condensed-phase materials (solids, liquids) have broader spectral features owing to their higher density compared with gases or vapours, where lower sample density results in more degrees of freedom, hence feature-rich spectra (discussed below). Table 13.5 summarises the source-requirement characteristics for remote sensing using infrared-absorption techniques.

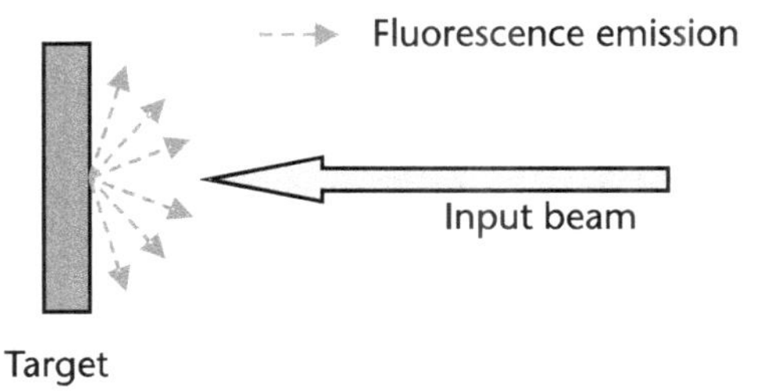

Figure 13.19 Fluorescence from an illuminated surface.

Table 13.5 Source Characteristics for Remote Detection (IR Absorption)

Source Parameter	*IR Detection*
Wavelength	SWIR/MWIR/LWIR (1–14 μm), but not simultaneously
Tuneability	Yes: rapidly or multiline; broad spectral coverage desirable
Average power	About 1W
Divergence	About 1 mrad
Pulse-repetition frequency	Pulsed (tens of hertz for spatial point to hundreds of kilohertz for imaging)
Line width	About 4 cm^{-1} (liquid/solid), 0.1 cm^{-1} (gases/vapours)

13.15.3 Detection Using Raman Spectroscopy

Raman spectroscopy exploits the inelastic scattering of photons incident upon a sample. Laser sources are critical to the application of Raman spectroscopy, as their inherent phase coherency and high photon density overcome the relatively small (low) cross section for the Raman effect (typically about 10^{-29} cm^2 molecule^{-1}, compared with approximately 10^{-17} cm^2 molecule^{-1} for infrared absorption, considered above). These incident photons impart energy to the electrons, which are promoted to a virtual excited electronic level; this intermediate level is deemed virtual as it only occurs during the laser illumination, and its energy above the ground level equals the energy imparted by the laser. The excited electrons then relax to the ground electronic level on the picosecond timescale through emission of a photon, as shown in Figure 13.20.

If this resultant photon has more vibrational energy than it had prior to excitation by the incident photons, then this photon has undergone Stokes' scattering. Conversely, if the relaxed photon has less vibrational energy after this process, then the photon has undergone anti-Stokes scattering. Light from these two limiting cases is shifted relative to the energy of the back-scattered laser radiation. As the proportion of molecules that are vibrationally excited prior to interaction with the laser is generally much less than those in the ground vibrational level (as defined by the Boltzmann distribution); the Stokes shifted light is more intense than the anti-Stokes. This effect is more pronounced in spectral features at higher Raman shifts (i.e., at greater energy differences from the laser line).

The value of these frequency shifts depends on the atoms that make up the covalent chemical bond that is excited within the sample. The Raman spectrum of each molecule within the sample is composed of several of these shifted bands, resulting in a characteristic signature that is similar to the infrared-fingerprint region, as illustrated in Figure 13.21. One advantage of Raman spectroscopy is that the entire spectral signature can be obtained from a single (or a few) laser pulses, thereby permitting rapid (subsecond to a few seconds) spectral-data acquisition. A material can be identified from comparison between the acquired data and a spectral library. For infrared spectroscopy, the success of the library matching pro-

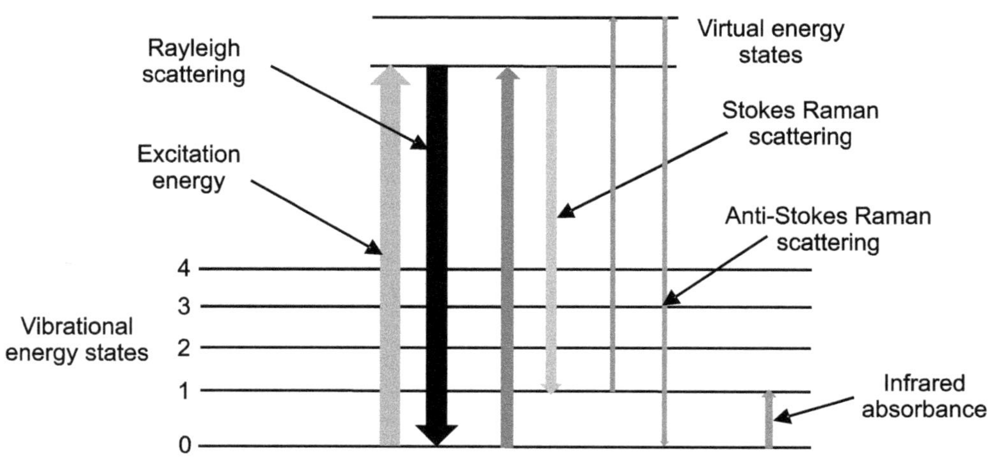

Figure 13.20 Raman process energy level scheme.

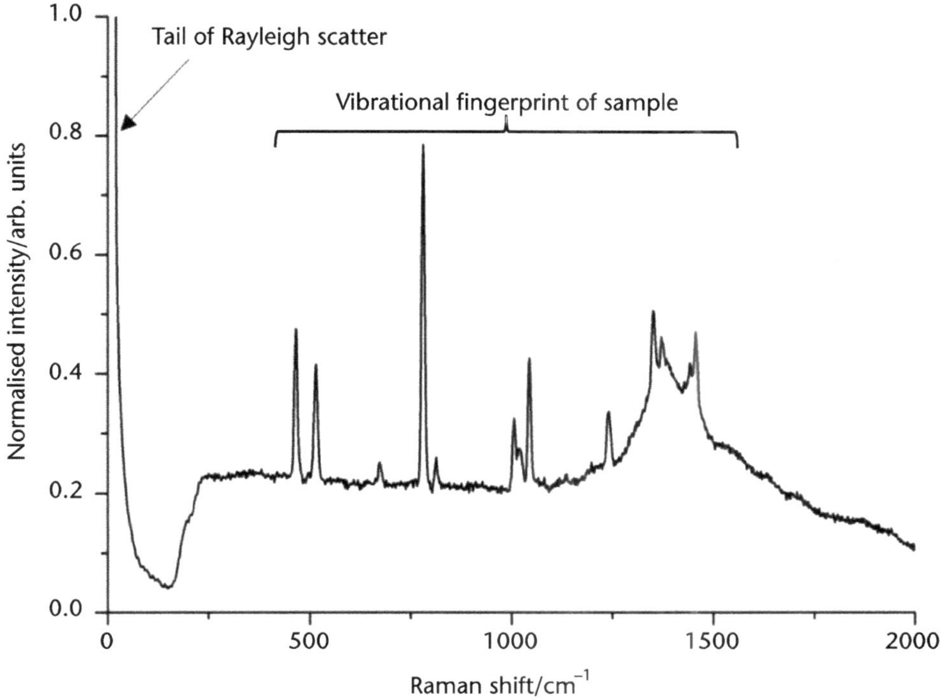

Figure 13.21 A typical Raman spectrum.

cess depends on whether the correct or appropriate entry is present in the spectral library, the quality of the library data and the composition of the sample.

Several factors adversely (negatively) affect the capability that Raman spectroscopy offers generally:

- The first is that, since the Raman effect is an inherently weak phenomenon, the sample has to be sufficiently dense so that there are enough molecular scatterers to produce a measurable signal. Raman spectra of gaseous species are difficult to obtain unless longer data acquisition times (potentially several minutes) are employed. This issue clearly precludes its use to detect gaseous species, which by their very nature are transient, especially in an outdoor environment where meteorological effects will disperse the gas cloud. However, Raman spectroscopy could have application for the detection of volatile condensed phase materials from their emanating vapour, presuming that the inherent vapour pressure of the material is sufficiently high at ambient temperature. Resonance ultraviolet Raman has been used to detect explosives, such as nitromethane and tri-nitro toluene (TNT), in the vapour phase.
- The second confounding issue is that of fluorescence, which is an alternative radiative-decay process, and consequently, competes with Raman scattering. If the incident photon energy is sufficient to excite an electron so that it populates an excited electronic level of the material, then the electron can relax to the ground electronic level by emitting a photon whose energy equals the energy difference between the excited and ground electronic levels. As the electron can populate a variety of vibrational levels in the excited state

(after photon absorption) or ground state (after relaxation), then the energy gaps between these two levels can take many values, resulting in a continuum of emitted photon energies defined by the Boltzmann equation. Therefore, fluorescence bands generally have a characteristic broad profile.

Fluorescence causes an issue for Raman spectroscopy when the excited electronic level is similar in energy to the virtual state prepared by the incident photons, determined by the laser wavelength. In this case, the broad fluorescence band will occur in the same spectral region as the Raman spectrum, obscuring the feature-rich Raman spectrum and complicating the process of comparison with stored data in the library. Even with a modest fluorescence quantum yield of 0.1 (equating to 10% of all molecules that absorb a photon will be fluorescing), the fluorescence cross section is approximately 10^{-18} cm^2 molecule^{-1}, so this process will dominate the Raman scattering. The clearest example where fluorescence competes with Raman scattering is when a coloured sample is probed using a laser wavelength in the visible spectral band. Fluorescence can occur from the sample itself, the surface the material is deposited upon or the container it is in (in the case of through-barrier measurements).

Clearly, the deleterious effects of fluorescence must be minimised in order to realise the benefits of Raman spectroscopy. The energy of excited electronic levels is usually commensurate with the ultraviolet and visible spectral regions. Fluorescence can be avoided by:

- Using a laser wavelength in the infrared spectral region (typically >1,000 nm), where the photon energy is insufficient to populate excited electronic levels. However, as the Raman scattering intensity scales as the inverse of the wavelength to the fourth power ($1/\lambda^4$), the Raman signal level drops away significantly at longer wavelengths. Therefore, it is difficult to obtain sufficient signal for detection, which results in a reduction in sensitivity for a Raman system that operates in the infrared spectral region compared with a similar system operating in the ultraviolet or visible bands. This issue is compounded by the lack of high performance detectors in this spectral region.
- Using a laser wavelength in the deep ultraviolet (DUV) spectral region (typically <260 nm), thereby shifting the Raman spectrum to a lower wavelength region to avoid overlapping with the fluorescence spectrum. Sample degradation is possible using these energetic photons through the cleavage of chemical bonds. However, fluorescence will still occur at a wavelength distinct from the Raman spectrum, as the photon energy is sufficient to populate excited electronic levels. Fluorescence spectroscopy can be useful for the detection of materials indicative of biological material, such as the amino acid tryptophan at 347 nm and the enzyme nicotinamide adenine dinucleotide (NADH) at 465 nm. The intensity of fluorescence signatures can also be enhanced via resonant effects, where the wavelength of the excitation photon coincides with an absorption feature of a given molecule.
- Using the different temporal profiles of the two signals, as fluorescence generally occurs on the nanosecond timescale, whereas Raman scatter takes place on the order of picoseconds. This time difference can be exploited

by utilising a pulsed laser source and a gated detector. Knowledge of the distance to the target is required, as the timing of the detection gate can be determined from this information. A detector that can be gated on the picosecond timescale is required for optimal rejection of fluorescence.

The laser characteristics required to induce-linear stimulated Raman scattering in materials for spectroscopic analysis are shown in Table 13.6.

Nonlinear Raman techniques that generally utilise complex excitation schemes involving multiple photons can be used to enhance the linear Raman effects described above. One example is Coherent Anti-Stokes Raman Spectroscopy (CARS), which is a four-wave parametric process in which three photons, one at a pump frequency ω_p, one at a Stokes frequency ω_s and one at a probe frequency ω_p are mixed in a sample to produce a new coherent beam at the anti-Stokes frequency $\omega_{CARS} = 2\omega_p - \omega_s$, which is shown in Figure 13.22. The pump and Stokes photons produce a coherent superposition between the vibrational ground and excited states that is converted to a Coherent Anti-Stokes Raman Spectroscopy signal by the application of the pump photon. When the difference between the pump and Stokes frequencies matches that of a particular Raman active molecular vibration (Ω), the resonant oscillators are coherently driven. As the radiation is emitted coherently, the Coherent Anti-Stokes Raman Spectroscopy signal is strongly enhanced compared with the spontaneous Raman scattering signal, for which Stokes and anti-Stokes photons are emitted incoherently. Furthermore, the Coherent Anti-Stokes Raman

Table 13.6 Source Characteristics for Linear Raman Spectroscopy

Source Parameter	*Raman Spectroscopy/Detection*
Wavelength	<260 nm, >1 μm – for fluorescence rejection
Tuneability	Not required
Average power	~300 mW – 5W
Divergence	About 0.5 mrad
Pulse-repetition frequency	cw or pulsed
Line width	1 cm^{-1}

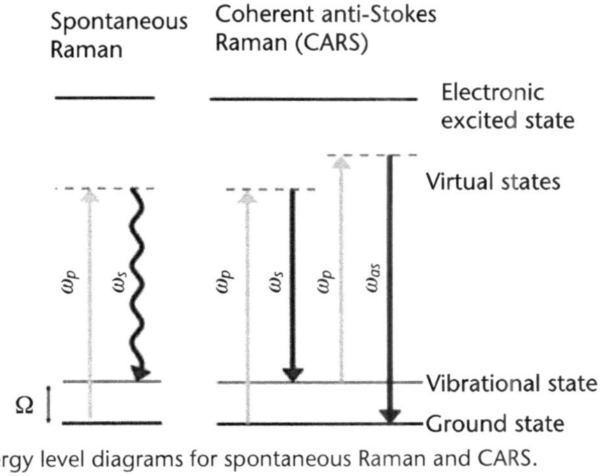

Figure 13.22 Energy level diagrams for spontaneous Raman and CARS.

Spectroscopy signal is generated at a shorter wavelength than the pump wavelength, so it is largely immune to fluorescence from the sample.

To overcome relaxation of the excited molecules, a femtosecond duration laser source is required. As a consequence of the recent development of very short-duration pulsed sources, such as Ti:Sapphire and fibre-based systems, the Coherent Anti-Stokes Raman Spectroscopy technique has gained popularity, primarily as a microscopy technique. Furthermore, for the four-wave mixing process to take place, the photons need to be overlapped in both phase and space. This situation is generally realised by using the same laser for both the pump and probe pulses, so that the photons are inherently phase locked; although Figure 13.22 illustrates the pumping scheme required to excite a single vibrational frequency (i.e., ω_p and ω_s are both narrowband). Employment of a broadband pump photon would excite a range of different molecular vibrations simultaneously, thereby accessing a broad spectral region (commensurate with the fingerprint region), which would permit the detection of biological materials from their spectral signatures. This configuration is commonly referred to as multiplex Coherent Anti-Stokes Raman Spectroscopy.

In the case of Stimulated Raman Scattering (SRS), both pump and Stokes frequencies are incident on the sample. If the frequency difference $\omega_p - \omega_s$ matches a molecular vibration of the sample Ω, stimulated excitation of vibration transitions occurs, as shown in Figure 13.23; through coherency, the output signal is enhanced relative to that produced by spontaneous Raman scattering. Pump photons are converted to Stokes photons, resulting in the intensity of the pump-field reducing (referred to as stimulated Raman loss) and the Stokes field increasing (stimulated Raman gain), as shown in Figure 13.24. The Coherent Anti-Stokes Raman Spectroscopy signal is generated at the anti-Stokes frequency (but note—it is not shown to scale in Figure 13.24).

Stimulated Raman Scattering only takes place when the difference frequency between two beams equals that of a vibrational mode. Therefore, the nonresonant background does not occur, unlike in Coherent Anti-Stokes Raman Spectroscopy. Also, the signatures acquired with Stimulated Raman Scattering are the same as those acquired using spontaneous Raman, so existing spectral libraries can be used.

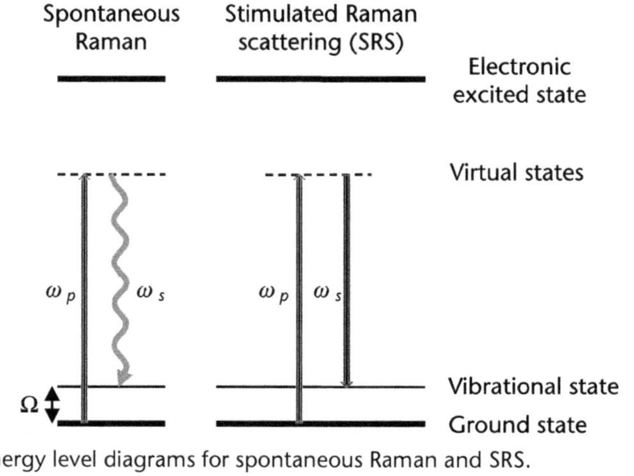

Figure 13.23 Energy level diagrams for spontaneous Raman and SRS.

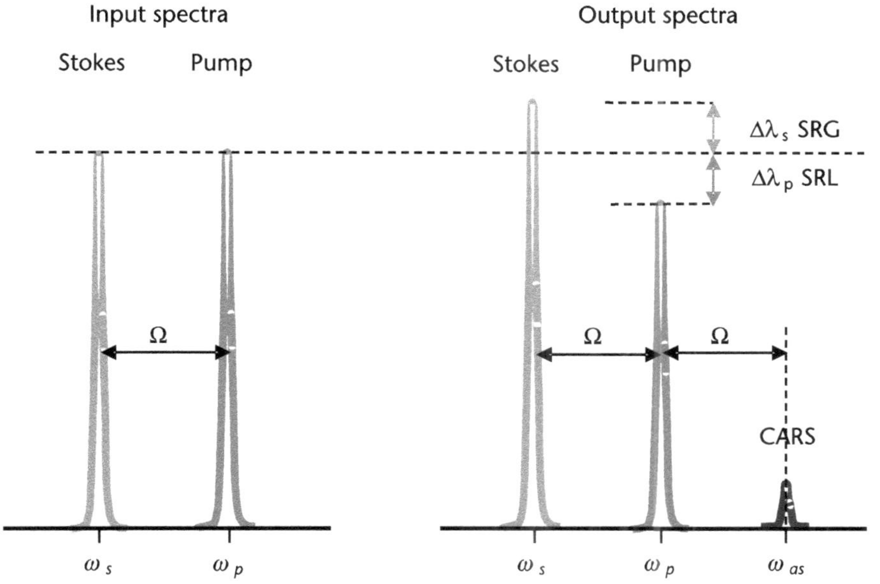

Figure 13.24 The input and output spectra for SRS spectroscopy.

Finally, the Stimulated Raman Scattering signal is linear with analyte concentration, thereby possibly enabling quantification of the amount of target material. Although Stimulated Raman Scattering uses a simplified pumping scheme compared with Coherent Anti-Stokes Raman Spectroscopy, a femtosecond duration pulsed laser is still required for the pump source. It is worth noting that the same configurations that can be used for Coherent Anti-Stokes Raman Spectroscopy can also be employed for Stimulated Raman Scattering (e.g., multiplex Stimulated Raman Scattering for the detection of a material from its spectral signature is possible).

Typical laser characteristics required to illuminate material samples to induce nonlinear Raman scattering are shown in Table 13.7.

Because Raman spectroscopy is based upon photon scattering, it offers the capability to detect substances through certain container materials, such as (semi) transparent glass and plastic. However, container fluorescence has to be minimised for coloured barrier materials, so excitation using lasers in the visible spectral band is not an option in this case. Using a different type of optical configuration, where the Raman scattered photons are detected at a spatial offset from the incident laser axis, Raman spectroscopy can be used to detect substances through completely opaque container materials; this technique is known as Spatially Offset Raman

Table 13.7 Source Characteristics for Nonlinear Raman Spectroscopy

Source Parameter	*Nonlinear Raman Detection/Spectroscopy*
Wavelength	790/800 nm
Tuneability	Desirable to generate alternative Stokes lines
Average power	100–500 mW
Divergence	About 0.5 mrad
Pulse-repetition frequency	Pulsed femtosecond emission
Line width	1 cm^{-1}

Spectroscopy (SORS). Spatially Offset Raman Spectroscopy experiments have recently been performed over stand-off distances using a 532-nm based system.

13.15.4 Super-Continuum Techniques

The extremely high peak intensities coupled with the very broad spectral emission of these sources should make them ideal sources for stand-off detection.

The generation of filaments through self-focusing of the very intense, short-duration pulses, in a super-continuum beam, permits the propagating beam to overcome the diffraction limit, so that very high intensity beams can be directed to remote locations, as discussed in Part I. Consequently, there is no need for focussing of the beam, as the self-phase modulation in the beam (along the propagation axis) does this task automatically. The main range limitation for this technique is the filamentation distance, which can exceed several kilometres with appropriate beam conditioning.

Two complementary laser-radar techniques are being developed for the remote sensing and detection of chemical and biological agents, using very short pulses from a super-continuum source. One is laser-induced fluorescence resulting from either the single-photon absorption of the third-harmonic wavelength of the fundamental wavelength (also discussed in Part I), which is present in the filaments, or the ionisation and dissociation of a molecule (agent) in the high intensity regions of the filaments. The other laser-radar technique involves the observation of the absorption of the back-scattered white-light continuum from the pollutant molecules.

13.15.5 Other Detection Techniques

A number of other techniques have been developed to provide classification of detected agents, including:

- *Laser-induced vapour emission:* a laser beam is used to vaporise the liquid agent, by increasing the rate of vapour production. The rate of absorption by the droplets needs to be rapid to increase the temperature significantly above its boiling point, so that the whole droplet vaporises rapidly. The infrared-emission spectrum from the vapour plume produced may be measured using a spectrometer.
- *Fluorescence spectroscopy:* this technique is based on the detection of the laser-induced fluorescence from the species of interest. In this case, the incident laser wavelength must correspond exactly with the peak of the fluorescence line of the molecule being investigated. These fluorescence peaks tend to occur in the ultraviolet, thus the incident laser energy needs to be tuneable in the ultraviolet.
- *DIAL:* differential-absorption Lidar (laser radar) uses two pulses at different wavelengths, one of which is strongly absorbed by the species of interest, such as a chemical agent, and the other wavelength measures the background signal. The difference between the two intensities indicates the differential concentration with respect to the background level. The range profile may be obtained by measuring the signal returns as a function of time.

- *DISC:* differential scattering is a technique that measures the difference in scattered intensity between the two wavelengths, from which the species concentration can be deduced. As in the case of the DIAL technique, the range profile may be obtained by measuring the signal returns as a function of time.
- *Photo-acoustic spectroscopy*, which is discussed in detail later.
- *Laser-induced breakdown spectroscopy*, which is also discussed in some detail later.

13.16 Evolving Remote-Detection Techniques

13.16.1 Photo-Acoustic Spectroscopy

In photo-acoustic spectroscopy, the absorption of infrared radiation causes a localised temperature rise, resulting in an expansion of air. The resultant acoustic wave can be detected using a variety of methods, including microphones and cantilevers. As this process only takes place where an infrared photon is absorbed, photo-acoustic spectra are analogous to those produced through conventional infrared-absorption techniques. This phenomenon has been tested in a variety of stand-off configurations using a suitable long-wave infrared source, such as a carbon-dioxide gas laser or quantum cascade laser.

A schematic of the operation of this evolving stand-off technique is shown in Figure 13.25.

An alternative approach for stand-off photo-acoustic spectroscopy is using a quartz crystal tuning fork for detection, where scattered light is collected and converted to sound local to the sensor. In one embodiment, tuneable infrared light is produced by two long-wave infrared quantum-cascade lasers that are incident upon a target. The light is either absorbed or reflected by target materials depending on their spectral responses. The returned light is collected by a spherical mirror and focussed on to a pair of quartz crystal tuning forks. Each of the forks is resonant with one of the quantum-cascade lasers, so that they oscillate when light from that particular laser is incident. The photo-acoustic spectrum (hence use of the infrared absorption spectrum) is recovered using a lock-in amplifier to record the signal amplitude as a function of quantum-cascade laser wavelength for each tuning fork.

13.16.2 Laser-Induced Breakdown Spectroscopy

This remote-sensing technique is evolving, owing to the availability of laser beams with very high peak powers capable of plasma formation at useful stand-off distances.

13.16.2.1 Principle of Operation

The principle of operation of laser-induced breakdown spectroscopy for remote sensing is shown in Figure 13.26; this technique is also known as laser-induced plasma spectroscopy. A highly focused laser beam from a suitable source, with a very high radiant intensity, is used to create a plasma on the surface of a target,

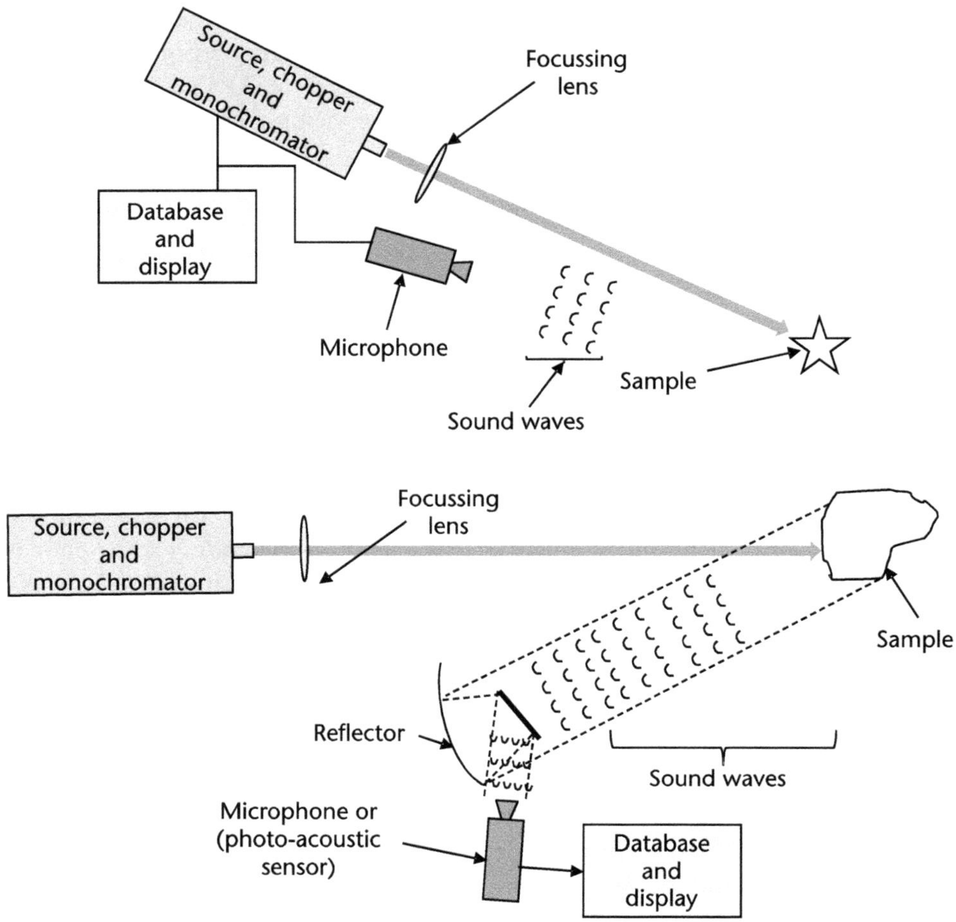

Figure 13.25 Schematic of two concepts for the photo-acoustic technique.

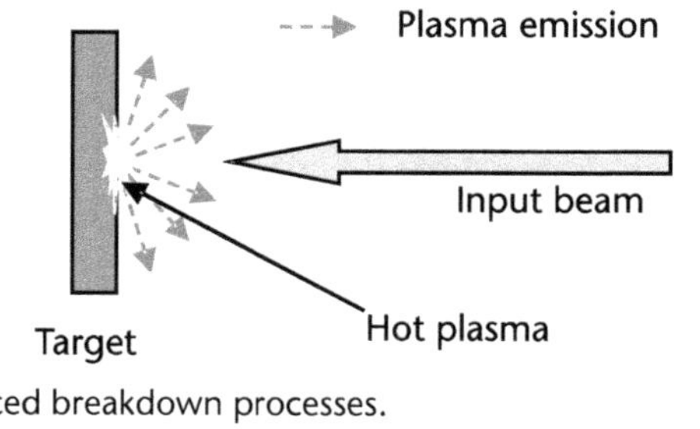

Figure 13.26 Laser-induced breakdown processes.

which could be any material. The rapid absorption of photons excites the material resulting in plasma formation and light is emitted during the relaxation processes; the wavelength of this emitted light is characteristic of the ionised material. The wavelength of the emission from the illuminated sample may be observed using a laser-radar system similar to the one shown in Figure 13.18.

Recent experimental observations showed that the frequency chirp of the incoming pulse has a significant beneficial impact on the generation of the laser-induced breakdown signals.

13.16.3 Stand-Off Terahertz Spectroscopy

The terahertz regime is beyond the long-wave infrared with a frequency between microwaves and the far infrared (i.e., 10^{12} Hz and beyond). It is particularly interesting for a number of military applications because many long-range molecular vibrations (e.g., phonons) are detectable in this part of the electromagnetic spectrum. Figure 13.27 shows the spectral response of an explosive material [Research Department Explosive (RDX)] around 1 THz.

13.16.3.1 Principle of Operation

The principle of the technique is to illuminate a sample of material with a suitable optical source coupled to an electro-optical detector to enable sampling for terahertz signals, using time-domain spectroscopic techniques. The challenge has been to develop a suitable source to achieve the appropriate emission characteristics from the material under test, so that the vibrational characteristics of the molecules can be observed and identified through comparison with a database of relevant materials.

The technique uses a very high repetition-rate source (50 to 100 MHz) with femtosecond pulse durations, emitting pairs of synchronised pulses. One of the two synchronised pulses is used to create the terahertz emission and the other is used to detect the terahertz optical field; hence, a terahertz spectrometer may be devised without any moving parts to tune the observed responses from the material under test.

13.16.3.2 Recent Development of Laser Sources for Terahertz Sensing

The spectroscopic requirement is for very short duration pulses to probe the sample under test with suitable wavelengths, so that the molecular motion of the test specimens can be investigated. The ideal source would be a compact mode-locked laser emitting very short-duration pulses in the femtosecond regime, providing an electrical field that tracks the intensity envelope of a subpicosecond pulse.

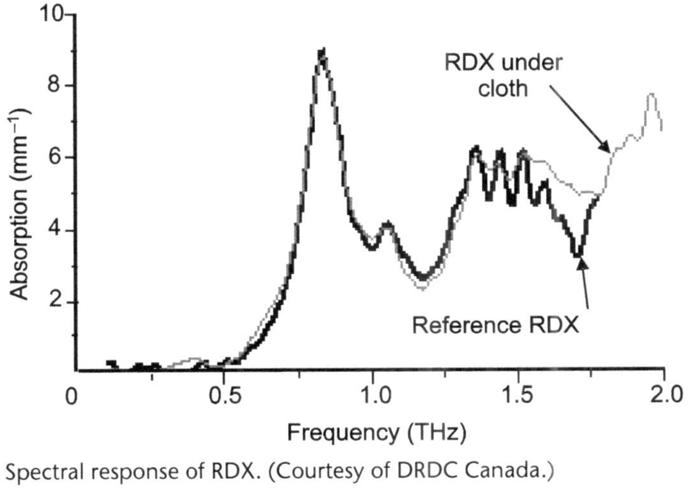

Figure 13.27 Spectral response of RDX. (Courtesy of DRDC Canada.)

Two techniques have been developed for generating terahertz radiation; the simplest is optical rectification, where a strong optical pulse propagating in a transparent medium without inversion symmetry induces a dipolar charge distribution in the medium, which moves with the pulse and emits terahertz radiation. Wide bandgap semiconductors, such as zinc telluride, are often used for this application, where a beam from a mode-locked laser is focussed into the crystal to create the emission.

This technique is relatively simple, but has some limitations on the bandwidth of the emission, owing to the absorption by the crystal; moreover, it is necessary to phase match the group velocity of the optical pulse and phase velocity of the terahertz wave. These limitations led to the development of a photo-conductive technique.

A terahertz pulse is emitted through the acceleration of photo-excited carriers generated by the ultra-short pulses from a mode-locked laser focussed into the semiconductor material such as gallium arsenide. An electric field present in the semiconductor results in a current surge that leads to a burst of terahertz emission.

A photo-conductive source requires a strong electric field to be present in the medium and the strength of this field determines the intensity of the terahertz radiation. However, simply increasing the applied field does not result in enhanced emission, owing to a number of complications with charge distributions; a solution involves an electrode design based on a metal-semiconductor-metal, interdigitated-finger structure. This electrode configuration gives a coherent super-position of the emission for the various parts of the structure in the far field, because the design masks every other gap between two electrodes. The mechanism relies on the unidirectional acceleration of the carriers in the semiconductor over the entire illuminated area; furthermore, the small gaps between the electrodes ensure that an intense electric field of tens of kilovolts per centimetre can be achieved from a modest power supply.

A new technique has been demonstrated recently based on the lateral photo-Dember effect, which avoids the need for an external electrical field to be applied to the semiconductor. In this case, the mechanism producing the time-dependent polarisation is the lateral gradient in the optically excited carrier density in the semiconductor. These intense gradients result in ultra-fast diffusion currents of electrons, compared with the holes, when the edge of an opaque material covering a semiconductor junction is excited and thus a space-charge field is created.

A very recent source development has been the demonstration of pulsed packets of terahertz radiation from a quantum-cascade laser source. This type of source offers a very practical solution to the terahertz-source requirements, although the emitted power is currently quite modest; however, the basis of the technique required to generate terahertz frequencies has been established for this type of semiconductor laser.

13.16.4 Terahertz-Induced Fluorescence

A drawback of terahertz sensing has been the limitation imposed by ambient absorption in the atmosphere, which limits the stand-off ranges that may be achieved.

A technique has been developed that exploits the omnidirectional fluorescence emission, which interacts directly with a terahertz wave. Methods have been developed to resolve temporally terahertz pulses at ranges up to 10 metres, which is adequate for a number of remote-sensing applications.

13.16.4.1 Principle of Operation

The principle of this technique involves the emission of a two-colour beam of ultra-short duration pulses from a titanium-sapphire source: the fundamental and the second harmonic, which create a plasma close to the sample under test. The use of a pair of wavelengths in the pulse enables the nitrogen in the atmosphere to be ionised asymmetrically, and thus, control the electron-drift velocity in the laser-induced plasma, enabling manipulation of the fluorescence, which interacts with the terahertz waves. The control is achieved through manipulation of the relative phase between the fundamental and the second harmonic, which is controllable to atto-second accuracy. A second emission then interacts with the terahertz wave to generate a fluorescence emission in the ultraviolet, collected via a suitable telescope and analysed with a spectrometer. The principle of this technique is shown schematically in Figure 13.28.

The time-resolved, terahertz radiation-enhanced emission of fluorescence waveforms are obtained from the observation of the fluorescence emission, as function of the time delay between optical pulses and the terahertz pulses, which is changed continuously. These waveforms are measured when the electron-drift velocities in the plasmas are parallel and orthogonal to the terahertz polarisation direction. The difference between these two waveforms is proportional to the time-dependent

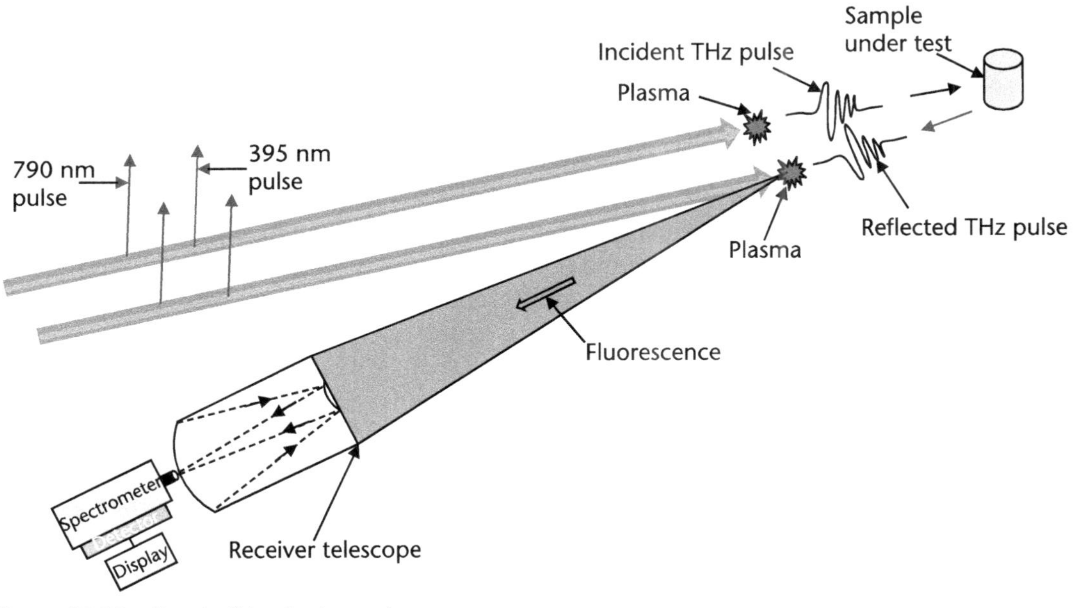

Figure 13.28 Stand-off terahertz sensing.

terahertz electrical fields. Additionally, amplitude and phase information may be deduced simultaneously from the received fluorescence signal.

13.16.4.2 Laser Requirements

The laser-device requirements centre around mode-locked short-pulse laser emissions containing the fundamental and second harmonic, with the ability to control their relative phases, so that the appropriate waveforms can be observed.

The main future requirement centres on a compact and efficient terahertz laser source with the appropriate emission wavelengths.

13.17 Noncooperative Identification or Identification Friend or Foe

There is an enduring requirement to identify assets and equipment at range, to establish whether they are friendly forces or otherwise. Various techniques have been developed using radar-based techniques, but the accuracy and/or false-alarm rates present some challenges, owing to the inherent limitations of the technologies being used in these systems. A laser-based technique has been developed.

A prototype laser system for noncooperative identification was based on the principle of vibrometry. A self-induced coded vibration frequency pattern was imposed within the vibration signature of the target using a modulated corner cube fixed on the top of a target. The vibrometer could then capture the signature from the target and use of real-time spectral analysis allowed the code to be extracted. The imposed code may be changed as frequently as required, provided it can be supplied securely to all the friendly forces in a timely fashion, or there could be a changeable code of the day distributed to a predetermined timetable. Other distribution schemes are also feasible.

13.18 Future System Requirements

An ideal solution to the majority of land-based identification and sensing tasks would be a handheld system so that all military units could have adequate remote-sensing capability and thus improved force protection or enhanced situational awareness to aid tactical planning. This system goal will be achieved through the development of more compact and efficient subsystems, not just the laser device; however, enhancement to the efficiency of the laser emission can have a very significant beneficial impact on the laser subsystem and not only on the power-supply system. Similar demands for smaller and lighter systems for airborne applications, including integration with small unmanned aircraft are known requirements.

There is a continuing requirement for highly directional inspection of distant phenomena and general effects at any time of the day or night, and under a broad range of atmospheric conditions at ever greater stand-off ranges. It is likely that commercial applications for remote laser-sensing will outpace the military applications, and will ultimately drive the technological advances in this branch of military capability.

There will be even greater demands for the measurement and analysis of objects in the atmosphere. Measurements on the surface (land or water) will only be limited by lines of sight, as new techniques using ultra-short pulses and coherent laser-radar sensors evolve to penetrate, classify and identify many classes of aerosols.

13.19 Summary

Laser-radar systems have matured significantly over the last few decades and offer a robust and effective technique for many remote-sensing applications. These developments have resulted from many technological enhancements but particularly compact, efficient laser technology and associated detector advances.

Laser-radar systems work in a similar fashion to a conventional radar system. The unique features of a laser beam leads to some significant advantages for the laser-based approach, particularly in terms of resolution and identification, but the conventional radar system has an advantage concerning beam propagation in adverse weather conditions and illumination of large volumes of space.

The uses of laser-radar technology for remote-sensing applications are many and varied, covering lots of aspects of military operations. One of the early applications concerned obstacle avoidance, but detection and recognition of potentially hazardous materials, including chemical and biological warfare agents and explosives is a fundamental requirement and of major interest. Other very important applications concern mapping, both over land as well as over water.

Another notable development during the last twenty years has been the application of infrared lasers that now have the requisite characteristics for atmospheric sensing of objects with absorption and emission characteristics in this waveband. A typical application is differential-absorption lidar, usually known as DIAL.

Modern developments with super-continuum sources and terahertz devices offer additional remote-sensing capability.

Selected Bibliography

Banakh, V. A., and V. L. Mironov, *Lidar in a Turbulent Atmosphere,* Norwood, MA: Artech House, 1987.

Cu, X., *Lidar Remote Sensing,* Boca Raton, FL: CRC Press, Taylor & Francis Group, 2005.

Demtröder, W., *Laser Spectroscopy, Vol. 1: Basic Principles*, New York: Springer-Verlag, 2008.

Fox, C. S., (ed.), *The Infrared and Electro-Optical Systems Handbook, Vol. 6*, Chapter 1 "Laser Radar," Bellingham, WA: SPIE Optical Engineering Press, 1993.

Haroche, S., et al., (eds.), *Laser Spectroscopy,* New York: Springer-Verlag, 1975.

Hollas J. M., *High Resolution Spectroscopy*, New York: Wiley-Blackwell, 1998.

Jelalian, A. V. *Laser Radar Systems,* Norwood, MA: Artech House, 1992.

Jones, J. D. C., and C. E. Webb, *Handbook of Laser Technology and Applications,* Bristol, U.K.: Institute of Physics Publishing, 2004.

Killinger, D. K., and A. Mooradian, *Optical and Laser Remote Sensing,* New York: Springer-Verlag, 1983.

Kovalev, V. A., and W. E. Eichinger, *Elastic Lidar: Theory, Practice and Analysis Methods,* New York: John Wiley & Sons, 2004.
McManamon, P. F., et al., *Laser Radar Progress and Opportunities in Active Electro-Optical Sensing,* Washington, DC: The National Academies Press, 2014.
Measures, R. M., *Laser Remote Sensing,* Malabar, FL: Krieger Publishing Company, 1992.
Osche, G. R., *Optical Detection Theory for Laser Applications,* New York: Wiley Interscience/John Wiley & Sons, 2002.
Richmond, R. D., and S. C. Cain, *Direct-Detection Ladar Systems,* SPIE Vol. TT 85, Bellingham, WA: SPIE Optical Engineering Press, 2010.
Sabins, F. F., *Remote Sensing: Principles and Interpretation,* New York: W. H. Freeman and Co, 1996.
Weitkamp, C., (ed.), *Lidar: Range-Resolved Optical Remote Sensing of the Atmosphere,* New York: Springer, 2005.

CHAPTER 14

Free-Space Optical Communication Networks

14.1 Introduction

Optical communications in one form or another is not new; in fact it probably predates the use of light as a countermeasure source, which was first attempted several millennia ago. For example, beacons have been used for many centuries, to alert surrounding areas to imminent danger and a heliograph was used to send signals from one point to another, usually from hill top to hill top, by reflecting sunlight with a suitable modulation applied to the reflected beam. Alexander Graham Bell demonstrated a photophone in 1880, which was an optical device for transmitting sound on a light beam.

The invention of Morse code enabled messages to be transmitted using an amplitude modulation-technique and this was exploited with the Aldis lamp on ships in the world wars of the last century. In this case, the optical source was a powerful electric lamp and the beam was modulated mechanically with a device that had the appearance of a Venetian blind, which could be opened and closed rapidly. Similarly, German air-defence units used a light-speaking device (Lichtsprechgerät) for optical communications. All of these techniques involve a transmitter and receiver and in these cases the receiver is a human eye.

The interest in using a laser beam to carry information arose from a desire to exploit the high bandwidth that is theoretically feasible with beams of photons in this part of the electromagnetic spectrum, owing to the very high carrier frequency of laser light; potentially high data-transfer rates of better than terabits per second are feasible. Other attributes of the laser radiation, such as the highly collimated beam, were also considered to be attractive for some applications, particularly for communication links requiring a low probability of interception by a third party. Moreover, use of the infrared wavelengths could make the transmissions quite covert.

Laser technology has been used for many types of optical-communication applications, some of which are considered below. Point-to-point techniques are most common, but other methods for short-range applications, using a broadcast-like technique, have been explored and demonstrated, whilst others are still being developed.

Recently, there have been developments with novel photonic components and techniques that offer a route to secure, high bandwidth and efficient communications in various environments. Quantum cryptography is an important application of these techniques, which offers the prospects of very rapid and secure communications over great distances, using optical methods and exploitation of the quantum-mechanical nature of light. Exploitation of these techniques provides highly secure communications, making the possibility of an undetected eaves dropping event probably impossible, as discussed further below.

14.2 Basic Principles of Communications

The use of light for a range of signalling-tasks and distribution of messages is a well-established technique, as noted above, but modern techniques for rapid and efficient operations involving radio waves generally displaced the use of the original optical techniques. The development of the laser in the 1960s renewed interest in the application of light to some special types of communications, for example, to establish a link from one point to another with the minimum use of infrastructure.

It is clear that there has been an interesting evolution in the optical techniques used for communications over the last several millennia, but it was the invention of the laser that revolutionised free-space optical communications, particularly for long-range applications. The primary reasons for this revolution are its natural emission characteristics of brightness, coherence and the ability to modulate the beam rapidly, as well as the use of invisible wavelengths, which are easily generated. The high frequency leads directly to the possibility of very high bandwidth communications, including the use to form data links. Recent space applications have seen the demonstration of the successful receipt of messages sent over tens of millions of miles.

The frequency range of optical radiation spans the range of 10^{13} to 10^{15} Hz, which may be used as the carrier frequency in any optical link, free-space or otherwise. In this band of frequencies, a modulation bandwidth of 10^{12} Hz would only occupy about 0.1% of the available spectrum, but there are still challenges to achieve this modulation bandwidth with the subsystem components, such as modulators and receivers. This large information bandwidth, along with the small divergence of laser beams in conjunction with the modern technological developments, offer significant advantages over the use of radio frequencies for free-space communication networks for providing links from one point to another.

An attraction of free-space optical communications is that there is no restriction on wavelengths or wavebands that may be used to provide the link between the transmitter and a receiver, in addition to the potentially very high data rates available and the mobility of the system. Each wavelength represents a separate communication channel, so that use of wavelength-division multiplexing techniques allows multiple channels to be transmitted simultaneously to increase capacity. Additionally, light is impervious to electromagnetic interference, is resistant to ionising radiation and there is minimal stray-light emission for an appropriately designed system.

There are many challenges for free-space optical communication systems to surmount to enable an effective system to be established. The most challenging is the impact of the atmosphere on the propagation path, as well as inclement weather; techniques have been developed to minimise the impact of turbulence on the spatial and temporal characteristics of a propagating beam. Examples of interruptions to the propagation path include:

- Atmospheric scintillation, usually with a time constant of the order of milliseconds;
- Short-duration beam blockages, such as objects flying through the beam (in this case, the time constant of the interruption could extend from milliseconds to seconds);
- Severe attenuation (absorption) of the beam resulting from inclement weather, including fog (the time scale here could extend from minutes to hours);
- Severe attenuation (scattering) of the beam resulting from dust or smoke (the time scale here could extend for up to several hours);
- Line-of-sight limitations owing to motion of the platform(s) leading to obscuration (the time scale here could be very lengthy under extreme circumstances).

Inclement weather, particularly sea fog, is often less of a limiting condition in maritime operations than is sometimes imagined. Moreover, techniques have been devised to minimise the impact of short-term interruption, but usually with a reduction in the data rate. An example is the use of Manchester coding, which tends to halve the data rate.

There has been resurgence of interest in the further development of laser-based optical communication networks, owing to the recent developments in novel photonic-component technologies. These developments offer important opportunities for many defence applications, including enhanced secure free-space communication networks and data links.

14.2.1 Fundamentals

The photons in a laser beam are used to transmit a message from one point to another location, through modulation of the beam in one form or another. The use of optical fibres has revolutionised communications worldwide through the provision of high-bandwidth transmissions and is exploited daily with high-speed Internet access, as well as more routine methods of communications. This form of optical communication is dependent on a suitable fibre-optical link to provide the conduit, or optical transmission line, to send the message as a data stream embodied within the laser beam, which is shown schematically in Figure 14.1. This figure shows a basic system that would be insecure; the lower part of the figure, below the dotted line, shows the components that would be required to achieve secure communication to a distant point.

In the case of free-space optical communications, the laser light propagates through the atmosphere, rather than along an optical fibre. The free-space

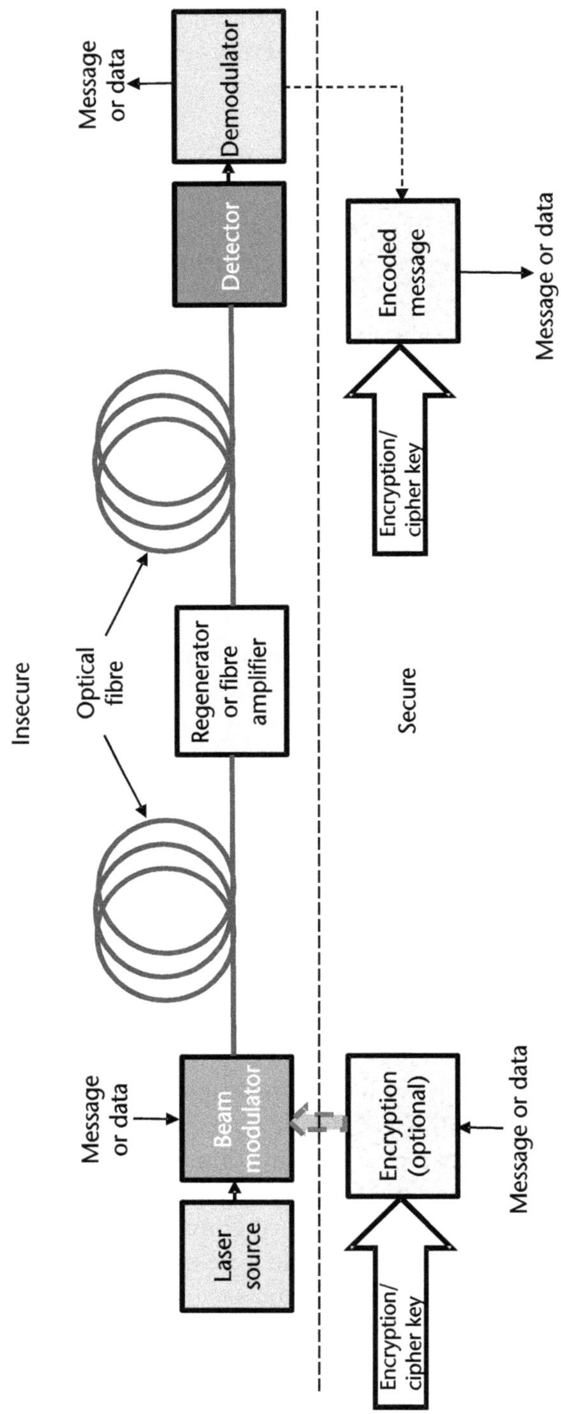

Figure 14.1 Schematic of optical-communication system using fibre cables.

transmission medium may be the atmosphere, outer space or a vacuum, and is shown schematically in Figure 14.2. As above, the insecure and the secure systems are illustrated.

The use of free-space techniques is very useful when physical connections are either not available or impractical, as is likely to occur on a battlefield or between platforms in transit. Moreover, the communications link is, in principle, only dependent on achieving and maintaining a line of sight from the transmitter to the receiver, so a physical link is not required and needs no maintenance. Consequently, links may be established rapidly as and when required, independent of any infrastructure. However, owing to the fundamental requirement of a line of sight between the transmitter and the receiver, this form of communication has limited utility in urban areas and mountainous regions. An interesting recent development concerns a nonline-of-sight technique for short-range tactical communications, which is also considered later.

14.2.2 Military Utility

Effective, secure and reliable communications are an absolutely critical component of any military operation. The availability of continuous and reliable channels of communications must be ranked equally in importance with an accurate knowledge of exactly where one's forces are positioned, as well as an accurate appreciation of the location of the enemy.

The deployment of new sensor systems on a battlefield may be used to support a range of tactical tasks, which generate significant amounts of data and information. To make the best use of the new and emerging sensor systems, a reliable communication network is required which can undertake this data transfer task accurately and rapidly. Generally, there is no substitute for real-time data; if this demanding requirement is not fully realised, then, whenever practicable, data latencies must be minimised. Information from these sensors may be used to inform command and control decisions through situational awareness, preferably in real time.

The use of laser-based communications for military applications is probably most valuable when there are long, uninterrupted sightlines. Examples include:

- Surface-to-air operations and the reciprocal path;
- Air-to-air scenarios, especially above the cloud base;
- All maritime operations.

Laser communication networks, which may be free-space optical communication or otherwise, have a number of attractions for military applications, examples include:

- An inherent very high carrier frequency to support very high bandwidth transmission;
- No licensing issues or requirements;
- Reduced cross-talk;

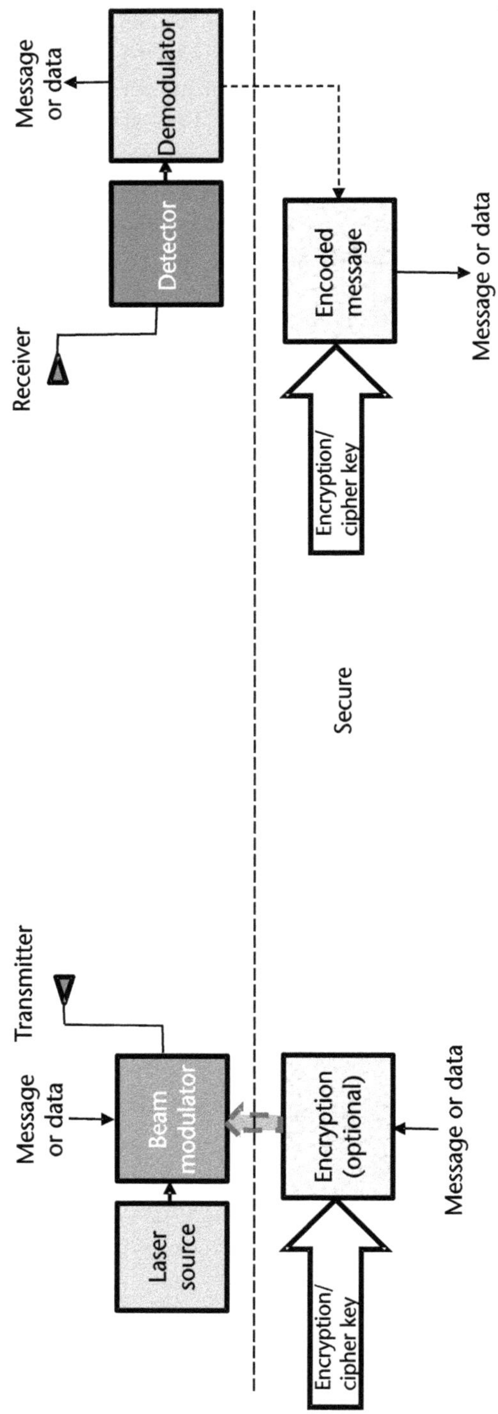

Figure 14.2 Schematic of a free-space optical-communication system.

- Smaller, lighter payloads;
- Low probability of detection or interception by an enemy;
- Established techniques that support secure communications;
- Asymmetric communications between a highly capable system and a primitive system, such as a simple transmitter;
- General immunity from interference or jamming.

However, careful system design and operational deployment are required to ensure:

- Eye safety, particularly for long-range applications, requiring emitted power of the order of tens of watts;
- A low probability of inflicting cumulative skin and eye damage, especially cataract formation, to operators and other personnel, especially when using ultraviolet wavelengths;
- Day/night operation, with particular attention to minimising cumulative exposure to system operators.

Secure communication is possible even with these simple line-of-sight techniques provided the transmitter and the receiver have access to the appropriate code and cipher. However, it is not possible to tell with these simple techniques if the message has been intercepted by an eavesdropper. An even worse scenario could involve an interception and decoding by the eavesdropper and even a retransmission with a false message.

The use of laser-based techniques offers a route to secure and efficient communications in various environments. In fact, modern techniques involving quantum cryptography seek to make the possibility of an undetected eavesdropping event impossible through the adoption of inherently very secure quantum-methods for transmission of the key, along with the message, for the cipher being used. The fundamental aspects of this technique using quantum states of a photon are discussed in more detail later.

There are a large number of scenarios where effective and efficient communication, using laser-based techniques, would provide a significant advantage over other techniques. The biggest advantages of laser communications over microwaves and radio frequencies (RF) is the highly directional nature of laser beams, which provide a much higher degree of covertness than is possible with either microwaves or RF transmissions. Other advantages are the potential for higher bit rates than are possible for these microwave or RF-based systems. The biggest disadvantage of a normal laser-based communication system is that it is truly line of sight for longer-range applications; if dense clouds or geographic features block the laser beam totally, then no message can be transmitted successfully with a conventional optical-communication system; however, a modern approach, using optical scatter, offers a short-range, nonline-of-sight, optical-communication system, as noted above and is discussed later.

The range for the secure communications in the troposphere spans from a few kilometres to hundreds of kilometres. For example, a squadron of fast jets may wish to communicate over ranges of 5 to 20 km in some circumstances, such as when flying at a low level during an interdiction mission. In contrast, unmanned air vehicles may need to communicate over much greater ranges, possibly many hundreds of kilometres, such as via satellite links. Additionally, there may be significant advantages in being able to communicate between aircraft and satellites; again, similar communication ranges may be required.

The objective of any advanced communication system is to close the loop between the vast array of sensors, available at both the tactical and strategic level, and the shooter. Various systems have different requirements for the data rate in a communications link. For example:

- A television link requires 4 to 5 Mbit/s;
- Mobile phone-type link requires 16 kbit/s;
- Raw sensor data might require 100 Mbit/s;
- Processed data can be distributed at 10 kbit/s.

The desired data rate for many applications is of the order of 1 to 10 Mbit/s; data rates of this order will provide a relatively feasible and practical system, with significant military communications utility. Examples of military operational scenarios include:

- Totally covert tactical air-to-air communications;
- Sharing of sensor data in the case of single sensors in a fighting group at television data rate for image sharing between aircrew, for example;
- Passing information back to base, for example, via unmanned aerial vehicles (UAVs) or satellite links;
- High-flying transit, which is likely to experience a low probability of cloud interference and have high reliability as well as accuracy in the communication link;
- Low-level flying, but weather can limit reliability of sustaining the link;
- Satellites links, which may communicate at 10–15 GHz;
- Other specialised military applications may include laser communications between aircraft and submarines for tactical operations. Laser-based communications systems for this application are particularly demanding, owing to the need for blue-green laser beams for the carrier wave to minimise propagation losses in the sea water.

14.2.3 System Operation

The fundamental components of a typical free-space optical communication system are:

- A transmitter subsystem, which contains the laser, a modulation device, a beam-steering element, power conditioning and control electronics;

- A receiver subsystem, which has a telescope to collect the modulated signals, an optical filter, a detector element, a demodulator, a display device and control system and some form of beacon to enable a point-to-point link to be formed rapidly. The receiver may also have a beam-steering and tracking element to ensure the integrity of the optical line of sight is maintained throughout operations.

A pointing and tracking system will be necessary to close (and maintain) the communication link; this subsystem ensures that the transmitted beam is directed on to a receiver channel (telescope) and remains on the receiver during a transmission and this is analogous to the crucial energy-on-target requirement to achieve successful countermeasure effects. A number of different beam-steering technologies are discussed in detail in Chapter 4, including non-mechanical techniques, which may be particularly relevant to this type of application.

There are essentially two modes of operation for free-space optical communications:

- The direct mode with conventional transmitter and receiver subsystems, which require relatively sophisticated pointing and tracking systems to ensure optimum performance and availability of the link with low-divergence beams and modest-power laser sources;
- Retro-reflector mode: this type of system is very attractive, as the power demands are very small and it does not require active pointing and tracking techniques. This mode of operation may be used in a network system with a transceiver (transmitter and receiver) at a node.

Hybrid (or asymmetrical) systems are also possible to enable smaller platforms, or a dismounted soldier to communicate with larger or strategic assets. In this case, the large platform would have the complete system with the transmitter and receiver integrated with a beam director, which would interrogate a smaller system operating in a retro mode.

Several types of retro system have been considered for this asymmetric-communication technique. There is often a system compromise to be made between a selection of performance parameters, such as range and data rates. For example, the returned optical power directed towards the transmitter from a retro device is determined by the size of the retro aperture. This dependency implies a large aperture reflector is required, such as a corner cube, but larger modulators achieve a slower data rate; hence, the maximum data rate that may be achieved would be limited compared with a system with a smaller aperture. One solution being investigated involves a lens to focus the incident light on to a modulator in the focal plane of the optical system. This is known as the cat's eye modulated retro reflector.

An additional influence on the configuration of the retro system is the fact that the received signal at the transmitter is also dependent on the separation of the transmitter and the retro element. Hence, as this system is operated via a reflection from the receiver, the received signal at the transmitter is inversely proportional to the fourth power of the separation between the transmitter and the receiver.

Military communications in the radio-frequency and microwave bands typically require a margin of 10 to 15 dB to compensate for atmospheric-fading effects. The desired operational range is often of the order 5 to 100 km. Laser-based communications, operating in a free-space optical environment, may require a similar margin for low-level transmission (i.e., close to the ground), owing to the various propagation effects as discussed in Chapter 5.

14.2.3.1 Transmitter

In the transmitter subsystem, the signal (message or information) to be transmitted is converted to a digital format and this is used to control or drive the laser-beam modulation device, which could be any one of a number of active emission control elements; examples include a spatial-light modulator, a Pockels' cell or Q-switch controller. The type of modulator used is dependent on the modulation scheme used to impart the signal on to the unmodulated beam, which acts as the carrier for the signal to be transmitted. Additional system considerations involve relative costs and the size, weight and power of these modulation devices, as well as the power-handling capabilities of these optical elements.

14.2.3.2 Carrier-Wave Modulation

The modulation techniques that are commonly used to impart the message on the carrier wave include:

- *Amplitude modulation:* this is a relatively common technique, which involves varying the amplitude of the carrier wave according to the information that is to be transmitted. For example, if an audio signal is to be transmitted, then the amplitude of the sinusoidal carrier wave is modulated by the audio waveform. A simple form of amplitude modulation involves on-off keying of the carrier, so that a binary signal represents the presence or absence of the carrier wave; this simple technique may be used to transmit messages in Morse code or similar codes.
- *Pulse-width modulation:* this technique requires that the amplitude of an input analogue signal is sampled periodically and the amplitude of each sample is converted into a pulse, its width being related to the amplitude of the sampled signal. The sampling process does not quantise the input signal, but varies the output-pulse duration. This technique allows multiple channels to be used in a system, through either frequency-division multiplexing or time-division multiplexing, where slow signals are interleaved to create a single fast one. The information embodied within the optical beam by this technique is carried in the baseband components; thus, the transmission of a carrier wave is not necessary, which leads to a wider useable bandwidth compared with other methods, such as the pulse-frequency modulation technique.
- *Pulse-frequency modulation:* this is a method of representing an analogue signal in a binary format (1 or 0). In this case, the amplitude of the analogue signal is encoded into the duty cycle of a square wave. A stable oscillator, which is frequency modulated, is required to create the pulse train, so that a

series of fixed-duration pulses are generated but the repetition rate is varied according to the instantaneous amplitude variation of the analogue signal. This technique is more immune to noise interference than other methods, owing to the use of pulses of a fixed amplitude and duration, but at the expense of a more complex transmitter and receiver.

- *Pulse-code modulation:* this is a technique used to represent sampled analogue signals digitally, so that a series of digits is transmitted in the carrier wave. The magnitude of the analogue signal is sampled regularly and assigned a quantised value to the nearest value in the digital range. The fidelity of this process is determined by the sampling rate and the bit depth, which determines the range of possible digital values that can be used to express the dynamic range of the input signal. Digital data-compression techniques may be applied to extend the dynamic range of the system. It is important to ensure that the sampling rate is in excess of the highest frequency in the signal, otherwise there will be aliasing of the frequencies that are greater than the sampling frequency. Additionally, as the sampling process is time dependent, an accurate clock is required in the modulation and demodulation process.

14.2.3.3 Modulation Components

The laser emission may be modulated by elements in the resonant cavity (intracavity) or beyond the output aperture (extracavity) as discussed in Chapter 2. Typical modulators are:

- *Electro-optical devices (intracavity or extracavity):* these devices may be used to rotate the plane of polarisation leading to amplitude modulation, alternatively, the device may be used to alter the phase (intracavity), which leads to frequency modulation or amplitude modulation depending on implementation. This device tends to be used for higher-power and larger aperture systems and produces a depth of modulation of the order of 99.999%.
- *Acousto-optical (intracavity or extracavity):* this type of device is generally used in lower-power systems. When used for extracavity, it generates a frequency-modulated output, whereas with intracavity operation the emitted beam is amplitude modulated. The depth of modulation is typically of the order of 90%.
- Pockels' cell (extra-cavity).
- Spatial-light modulators.
- *Mechanical modulators (extracavity)* for low-level or simple applications.
- *Direct modulation or current switching:* this is a system that lends itself to the direct modulation of laser diodes used to pump the laser cavity.

Some laser sources, such as laser diodes, may be modulated directly to form the modulated beam. This optical configuration of the source, without the need for a second laser cavity, leads to a relatively simple and compact direct-generation transmitter.

Amplitude modulation is most common, as it is compatible with direct-detection techniques. Frequency modulation requires coherent sources and coherent-detection techniques, often applied to gas-laser sources, such as the carbon-dioxide laser.

14.2.3.4 Receiver

The fundamental components of the receiver channel are:

- A light-collecting telescope, which may be reflective (Cassegrain), refractive or a combination of both (catadioptric);
- A detector, which is essentially an optical-to-electrical transducer. This sensor may be a simple photo-diode, a photo-multiplier, an avalanche photodiode, a photo-conductor, or an array of receiver elements.
- An amplifier and demodulator subsystem;
- A filter with a pass-band matched to the transmitted wavelength to optimise the signal-to-noise ratio, or the signal-to-background ratio, hence ensuring the maximum in-band signal and minimising out-of-band photons and thus providing maximum rejection of clutter.
- Tracking beacon of one form or another.
- A display unit.

As in the case of the transmitter subsystem, there are often trade-off compromises to ensure that a cost-effective subsystem is produced that does not present an undue size, weight and power burden.

14.2.3.5 Carrier-Wave Demodulation

A circuit is required to recover the original signal from the carrier wave, which is known as the detector or more commonly as the demodulator. The term demodulator is used as it is considered to be the reverse of the modulation process and ideally should provide an exact reproduction of the original signal when extracted from the modulation present in the received carrier wave. A nonlinear device is required for demodulation of the signal in the carrier wave and the detector is usually some form of semiconductor device.

The type of demodulation technique is dependent on the form of the modulation applied to the carrier wave:

- *Amplitude modulation:* the use of a diode, configured as an envelope detector, provides a simple, but effective, demodulation technique.
- *Pulse-width modulation:* the most common technique is to pass the amplified signal through a low-pass filter.
- *Pulse-frequency modulation:* the use of a low-pass filter and a suitable bias on the pulse-modulated signal provide a suitable low noise-demodulation technique. The bandwidth of the demodulator needs to be small compared with the modulation frequency, which is small with respect to the

pulse-sampling frequency. This is a challenge for implementation with a digital filter, but feasible with a state-space filter. This form of demodulation may be achieved with a phase-locked loop with its integrators reset as each new pulse is received.

- *Pulse-code modulation:* the modulation process is operated in reverse, so after each sampling period has past, the next sample is read and a signal is shifted to the new value. The new waveform is then filtered to remove the high-frequencies beyond the anticipated operational range, along with suspected aliasing frequencies.

14.2.3.6 Creating the Free-Space Link

Several techniques have been developed to enable the point-to-point link to be established rapidly and maintained even with a system on a moving platform. A typical setup involves the receiver having a beacon, which could be a retro reflector or a more traditional optical source.

The use of a retro signal from the receiver's position, enables a rapid acquisition of the receiver's telescope by the transmitter and thus completes the link. A typical rapid acquisition-system involves an initial wide-angle (divergent) beam being sent by the transmitter as an opening of the link handshake signal. On receipt of this signal, the receiver platform exposes a retro device for precision acquisition and subsequent tracking of this receiver-telescope position signal by the transmitter's beam director, to ensure the integrity of the link.

The use of this retro-return technique also allows the transmitter's beam director to centre the transmitted beam on to the receiver's retro device, or to a given offset depending on the configuration, to ensure the transmitted beam is on the axis of the receiver. This procedure permits a reduction of the divergence of this beam at the receiver to aid covertness as well as boosting efficiency, whilst maintaining the transmission beam on the receiver for the established point-to-point link.

Some free-space communication systems use separate wide-angle and low-divergence beams from the transmitter. In this case, a high bit-rate is used in the collimated (signal) beam and a low bandwidth for the tracker laser beam and its receiver; the fundamental requirement being to ensure adequate signal-to-noise ratio, which has a direct impact on the transmitted power in the beam.

One solution to the rapid and accurate establishment of the link is to use various assets on a platform, such as its inertial-navigation system, operating in conjunction with situational-awareness assets, to provide a pointing cue to the respective beam directors. The beacons may then be used to maintain the tracking functions and thus ensure the maintenance of the free-space optical link.

14.2.3.7 Communication Protocol

An example of a simple communication protocol system is a universal asynchronous receiver transmitter (UART), which operates on the premise that a multibit (usually 8-bit) data message can be sent successfully through a single channel without time synchronisation between the transmitter and receiver. This type of communication channel is known as a serial data link, because the multiple bits containing the message are transmitted in a serial format. In this case, a known bit is used to indicate the start of the message and similarly for the termination of the data stream.

A full-duplex system allows simultaneous transmission and reception, whereas a half-duplex system only allows transmission or receipt of one message at any time. Other high-capacity systems are far more complex and require time synchronisation between the transmitters and receiver, which are derived from the synchronisation bits in the data stream, defined by a pre-set algorithm.

14.2.3.8 System Configuration

Figure 14.3 shows a schematic of a single-channel, laser-based, point-to-point optical communication system. Clearly, for a complete communication channel a transmitter and receiver will be required at both ends of the link. In the case of the retro-modulation communication systems, which have a single transceiver at one location and retro-reflectors coupled to modulators at a remote site, the demand on the pointing accuracy may be relaxed, but a clear line of sight is necessary for successful communication. Other system configurations involve multiple transmitter terminals, and use of repeaters; moreover, the systems may use coherent- or incoherent-detection techniques to enhance the signal-to-noise ratio and thus reduce errors through drop out in the signal. More sophisticated configurations have a two-axis fast steering mirror in the optical train to provide the fine beam pointing, with a very similar function to the fine-track sensor and beam-pointing mechanism in a beam director, discussed in Chapter 4.

A laser-communication system may be configured as a monostatic system, with the transmitter and receiver coincident or bistatic (i.e., with a separation between the receiver and transmitter telescopes). The choice of architecture is dependent on operation of the laser-communication system, but the principles of operation are virtually identical.

For a more sophisticated tactical- or strategic-communication system, with multiple access channels, for example, there will be a number of transmitters at nodes in the complete communication system. This type of system may have multiple receivers as well as other transmitters to relay the information to other access nodes in the complete system; retro-modulation systems may also be operated in a similar fashion. Retro-communication systems are discussed in more detail later.

The type of system architecture used in point-to-point space-based systems is shown schematically in Figure 14.4 and involves multiple nodes, as well as a variety of operational techniques.

14.2.3.9 Pros and Cons of Using Optical Wavelengths

In general, the use of light in a laser beam for transmission of the signal is preferred to light from other optical sources; this is because of the fundamental characteristics of laser light, namely:

- Inherent high radiant intensity, even when using low-power emission;
- Monochromicity;

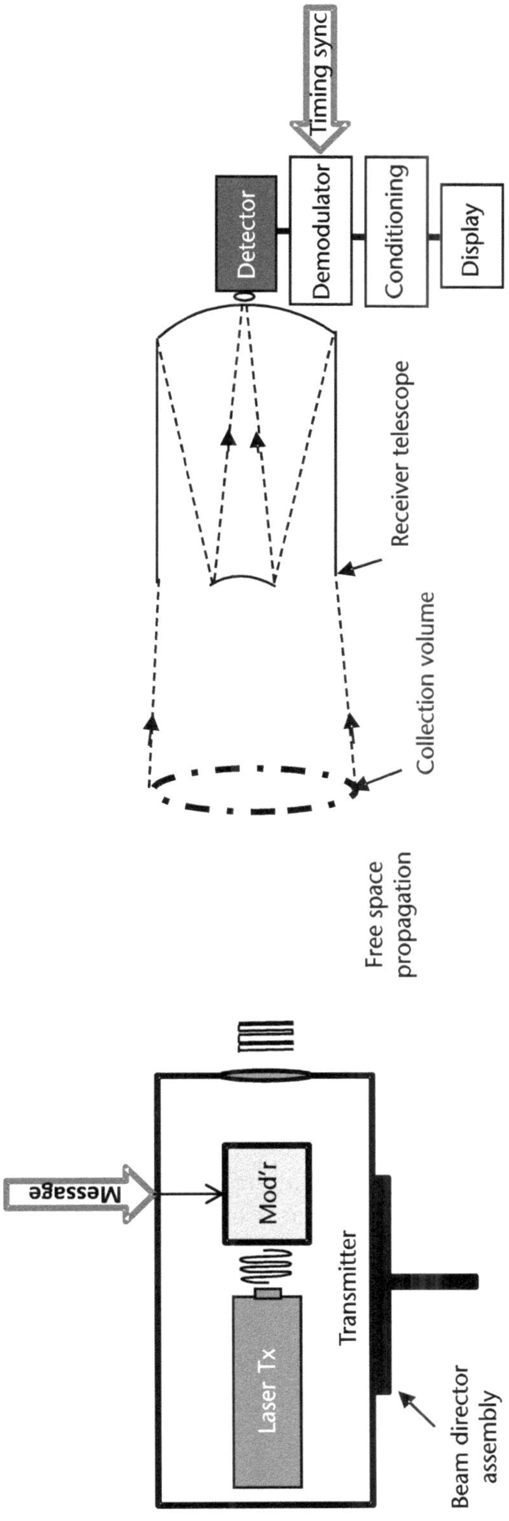

Figure 14.3 Schematic of single-channel laser-communication system.

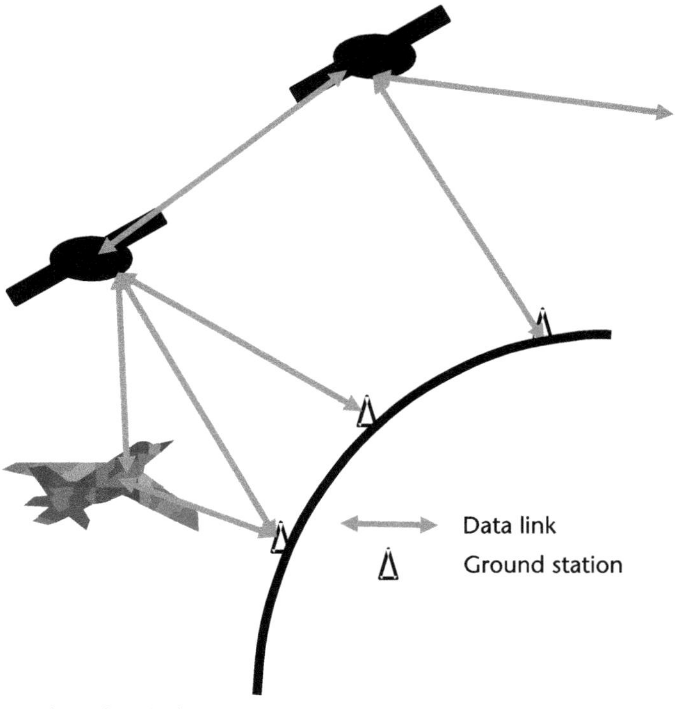

Figure 14.4 Space-based optical-communication system.

- Spatial and temporal coherence, so that its component waves are all invariant in their phase relationship;
- Opportunities for covert transmission with a low probability of interception;
- Opportunities to exploit the quantum-mechanical properties of photons;
- Application of appropriate detection techniques.

Microwaves have been used to provide a point-to-point channel for a range of communications for decades, but the use of optical wavelengths offer a number of advantages, such as:

- Bandwidth, which is directly related to the frequency of emission. Laser diodes have a potential bandwidth for transmission in excess of 10 GHz, so that the entire medium and high frequency spectrum used by amateur radio operators could be transmitted in a single light beam.
- Spot size, in which the inherent low divergence of the laser beam means that a small and compact transmitter and receiver systems are feasible and effective.
- Lower power requirements for the transmitter, owing to the small spot that can be directed at the target receiver.
- No side-lobe structure and highly collimated, so the small divergence of the beam also offers the potential of a more secure transmission channel with covert operation and low-probability of interception by an eavesdropper, particularly as a consequence of the lack of side-lobes.

- Potential for highly secure transmission, so that the presence of eavesdroppers may be detected, especially when using quantum-cryptographical techniques.

There are a number of disadvantages to using laser light for the carrier wave when compared with microwaves or radio waves, such as:

- Atmospheric attenuation and the impact of turbulence along the propagation path. The transmission wavelength(s) may be selected to minimise the extinction of the beam, its scintillation and beam-wander effects, additionally, techniques have been developed to minimise the drop-out rate.
- The higher energy per photon, in comparison with microwaves, means laser communications, with its narrow beam are really not suitable for broadcast transmissions, as a very high-power laser transmitter would be required, as well as sophisticated beam-expansion techniques.
- Eye-safety management to ensure that the operators do not get exposed to potentially hazardous optical radiation, even considering the relatively modest power transmission involved in forming the optical link.

14.2.3.10 Bit-Error Rate

Bit-error rate (or bit-error ratio) is an important parameter used to assess the potential performance of a communication system, as it transmits and receives digital signals; establishing a free-space link and achieving a very low bit-error rate is critically important so that the content in the messages is not corrupted. It is defined as the number of bit errors divided by the total number of transferred message bits during a defined period. Bit-error rate is a unitless performance parameter.

When data are transmitted from A to B over a data link, there is the possibility of errors being introduced into a data stream, which may result in the integrity of the system being compromised. Besides malfunction of the system, these errors may result from:

- Noise in the transmitter or receiver: optical and/or electronic electromagnetic interference;
- Interference within the system, such as along the propagation path or in the electronics;
- Signal fading or attenuation along the propagation path;
- Cross-talk;
- Modulation or de-modulation errors;
- Phase jitter;
- Distortion;
- Bit-synchronisation issues;
- Other imperfections within the entire system that are not compensated.

A number a techniques may be applied to limit bit-error rate to an acceptable level, such as:

- Use of an appropriate signal strength (a high signal strength is more tolerant to fading, but care is required to avoid cross-talk between channels);
- Use of a suitable modulation and demodulated scheme, which is robust and matched to the application;
- Use of channel-coding schemes, such as redundant forward-error-correction codes.

Bit-error rate test techniques have been developed to assess the potential performance of these digital systems using a structured process. A typical testing system consists of a test-pattern generator and a receiver, which can be set to the same pattern, to enable bit errors to be counted over a given period. These test devices may be used in pairs at either end of the transmission system, or in a loop mode with a single test device at the transmission end and a loop back at the remote location.

As noted earlier, various protocols may be imposed on the modulation to minimise the impact of a signal fading on the received message signal.

14.2.3.11 Laser Requirements

The critical requirement is to choose a carrier wavelength that has good propagation characteristics and a beam format so that the emission can be modulated with the appropriate technique to embody the message into the beam. Carrier wavelengths may be used in the various wavelength bands from the ultraviolet to the far infrared. The combination of high radiant intensity and narrow line width, leads to high spectral brightness, and thus, good contrast with the background luminance. The narrow line width has the potential to give enhanced signal-to-noise ratio.

One of the techniques for minimising data drop out or bit-rate errors is to transmit a flat-topped beam, with the endeavour of achieving a uniform illumination of the receiver. This may be achieved through the generation of a super Gaussian beam or avoiding transverse modes (i.e., an M^2 close to 1).

A summary of laser-emission requirements are shown in Table 14.1.

One of the challenges for many maritime operations is the communication with submerged submarines, particularly, if the submarine is to remain covert. Laser-based systems have been developed using a range of technologies; in this case the wavelength of emission has to be in the blue-green part of the spectrum, as discussed in Chapter 13. Figure 14.5 illustrates the attenuation of electromagnetic waves by sea water.

14.3 Free-Space Laser-Based Communications

A free-space, point-to-point communications link is, in principle, only dependent on establishing and maintaining a line of sight from the transmitter to the receiver, as stated above; a physical link is not required, and thus, needs neither permanent

Table 14.1 Laser-Emission Requirements for Point-to-Point, Free-Space Communications

Source Parameter	*Point-to-Point Communication Systems*
Wavelength	A wavelength that gives low attenuation; it can be from the ultraviolet to the far infrared. Designers/operators tend to favour short-wave infrared for terrestrial applications (i.e., non-visible waveband).
Average power	Range dependent; moderate (few watts) for short range, but hundreds of watts for deep space communication
Divergence	Small, but matched to the pointing accuracy
Pulse-repetition frequency	High (100 kHz or more) or continuous wave
Pulse length	<100 ns
Line width	Narrow
Beam quality	Good, usually $M^2 < 1.5$
Polarisation	Linearly preferred

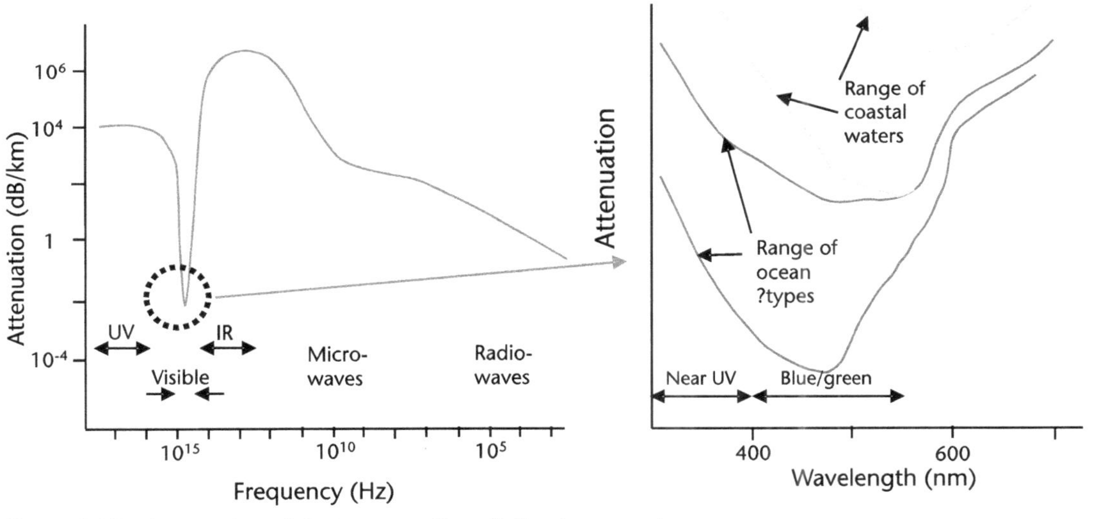

Figure 14.5 Attenuation of electromagnetic radiation by sea water.

infrastructure nor maintenance (of the physical link). Therefore, point-to-point optical communications and ad hoc data networks enable data links to be established rapidly, as and when the need arises. As these free-space optical links are independent of any infrastructure, they may be established and maintained between mobile platforms, using suitable tracking equipment at either end of the link.

A critical requirement is the maintenance of a line of sight between the transmitter and the receiver throughout the communication, which means that there is potentially limited utility in urban areas and mountainous regions, as well as with mobile units. An alternative approach has been developed for short-range applications, as discussed later. Whichever approach is chosen, it is critically important to maintain high availability and simultaneously achieve a very low bit error rate.

For point-to-point communications between moving platforms, such as ships, aircraft or ground vehicles, there are frequently stringent demands on the pointing and tracking systems required to establish and maintain the optical line of sight. As discussed earlier, a number of techniques have been demonstrated; an important trade-off consideration involves the compromise between the use of a broader transmitted beam and a less accurate pointing system, which would be cheaper. However, to maintain the equivalent range performance without incurring increased dropout effects, a more powerful beam would be required, which leads to additional hazard considerations; hence, there is a consider scope for a system trade-off to achieve an efficient and cost-effective solution.

Covert, high bandwidth, point-to-point communications are ideally suited for techniques involving laser beams, whether the laser beams travel in free space or are guided along an optical fibre. This type of optical communication contrasts sharply with optical implementation required for broadcast methods. The vast majority of laser-based communication applications will use a single beam, as this will involve point-to-point operation, whereas a broadcast would require the use of multiple laser beams, which is not really practical for most applications. Broadcasting is much easier and efficient using radio techniques.

There are significant environmental and topographical factors that limit the potential value of free-space optical communications, which have been discussed earlier and can cause some operational limitations. Various techniques have been devised to surmount the temporary loss of a sight line, many of which are specific to the application and the form of the operational limitation.

14.3.1 Summary of Point-to-Point Communication Operation

A typical point-to-point configuration is shown in Figure 14.3, as point-to-point optical communications have been used as the underlying system to describe the principles of laser-based optical communications. Moreover, the generic laser requirements are covered above, in Table 14.1.

This form of laser-based communications is capable of very long-range, free-space operation of hundreds of kilometres, as described earlier. One of the operational challenges is for the transmitter and receiver to locate each other, over very long communication paths. Directional cues from an inertial system may be used to locate the appropriate receiver, to initiate rapid acquisition and tracking of an appropriate beacon at range.

Optical links may be used to replace or complement radio-frequency (RF) links in some circumstances, especially when jam-resistant links are required. However, other operational situations may also make RF communications impossible owing to interference from other transmitters. Adoption of Ethernet protocols can ease the substitution of RF links by optical channels, but as implied above, this is only really practical for point-to-point communications; it must be emphasised that with current technology, optical links are not suitable for applications requiring a broadcast or long range, nonline-of-sight communications.

14.4 Short-Range Optical Communication System

14.4.1 Introduction

One of the major limitations of the optical communications system described above (and shown in Figures 14.3 and 14.4) is the need for a clear line of sight between the transmitter and the receiver, or even the need to have a relay station to pass on the signal. Failure to maintain a line of sight inevitably causes signal drop outs and errors or delays in signal transmission. These problems are most often encountered in mountainous terrain or urban canyons where natural topography or high buildings constrain even short range communications. These problems are exacerbated when either the transmitter or receiver or both are mobile requiring regular pointing corrections with drop outs occurring if the receiver goes out of view by perhaps one or other vehicle turning a corner. These systems often operate in the near infrared wavebands using wavelengths from 750 to 1,550 nm. Some of these system shortcomings can be overcome by transmission in the ultraviolet.

14.4.2 Short-Range Ultraviolet Nonline-of-Sight Communications

Solar ultraviolet radiation with wavelengths in the 220- to 280-nm waveband is heavily absorbed in the stratospheric ozone layer as shown in Figure 14.6. As a consequence of this phenomenon, this particular waveband has almost an absence of solar emission at ground level; this is known as the ultraviolet solar-blind region.

In addition, Rayleigh scattering of these wavelengths at ground level becomes important (as it has an inverse wavelength to the power-four dependency, λ^{-4}). This means that photons emitted from the transmitter can be scattered around objects in the scene and still be detected at the receiver (as shown in Figure 14.7), thus relaxing the line-of-sight constraint required for visible or near infrared

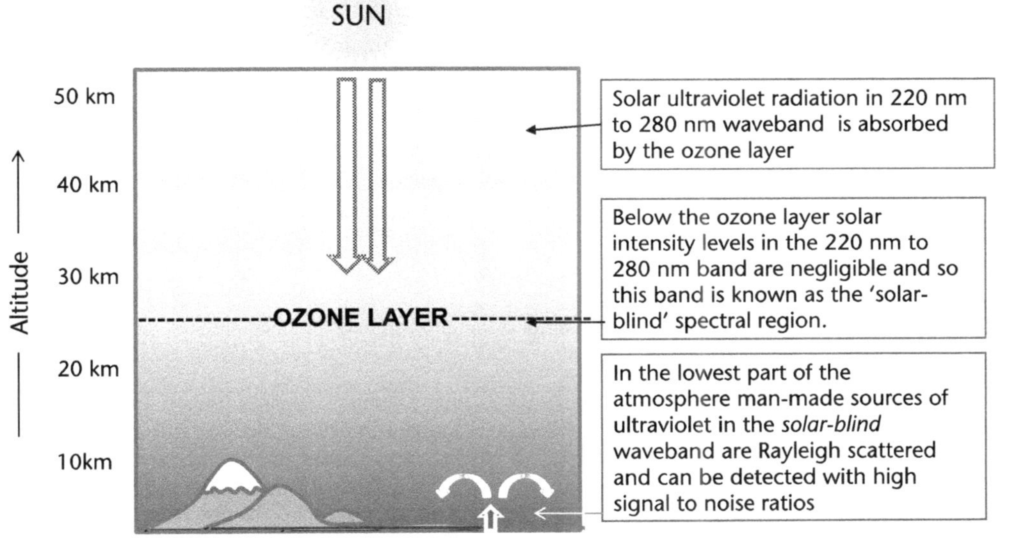

Figure 14.6 Attenuation of solar ultraviolet light by the ozone layer and scattering of low-level ultraviolet light. (Courtesy of Dr. N. J. Wood, QinetiQ, U.K.)

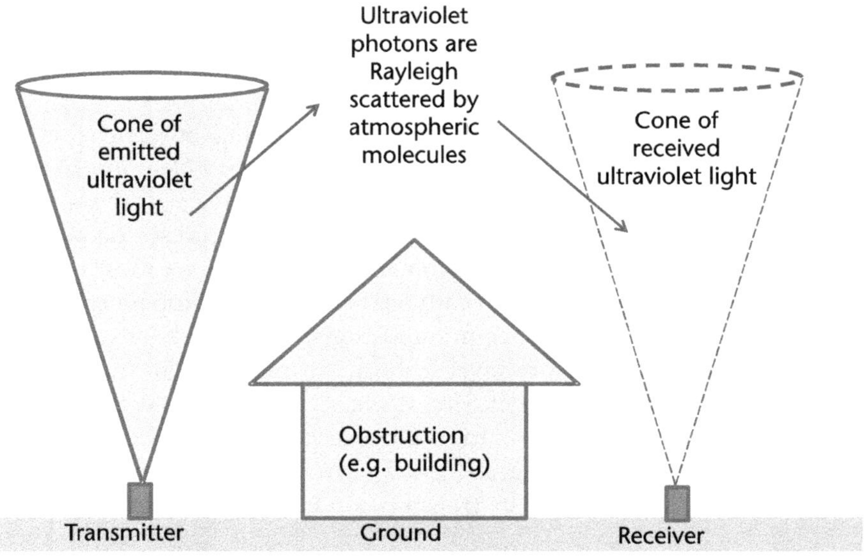

Figure 14.7 Rayleigh scattering of ultraviolet photons enables non-line-of-sight communications. (Courtesy of Dr. N. J. Wood, QinetiQ, U.K.)

optical communications links, as described in earlier sections for point-to-point communications.

Given the very low level of natural background light in the solar-blind region of the spectrum (defined above as approximately from 220 to 280 nm), along with the light scattering of photons at these wavelengths, these effects enable a low-power ultraviolet emitter to provide a nonline-of-sight communications link and the received signal to be detected with a decent signal-to-noise ratio. Such a system can be used for short-range communications perhaps building to building or vehicle to vehicle in terrain that makes it difficult for other more conventional optical communications systems to operate effectively, as illustrated in Figure 14.8.

Ultraviolet communications systems, as described above, typically use a laser source or diode-based transmitter technology coupled with a filtered photo-

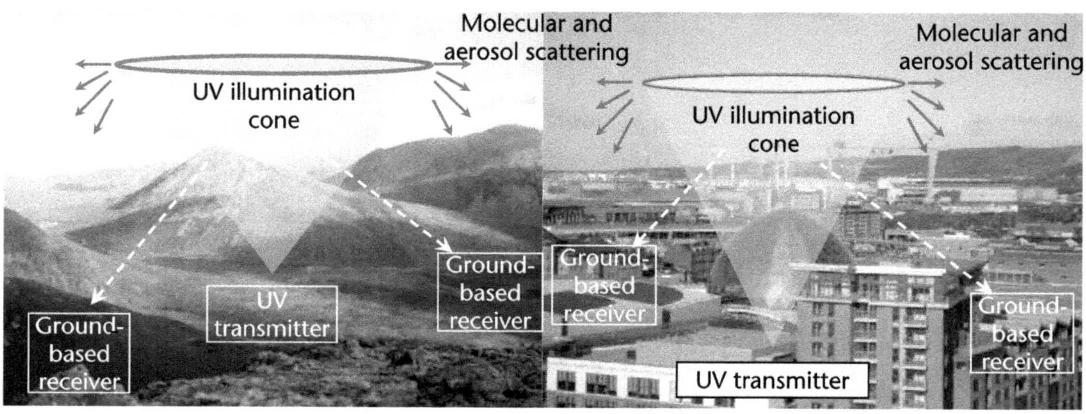

Figure 14.8 Short-range ultraviolet communications in mountainous regions and urban canyons. (Courtesy of Dr. N. J. Wood, QinetiQ, U.K.)

multiplier or avalanche photodiode as the receiver element in the communication system.

14.4.3 System Performance and Limitations

The nonline-of-sight ultraviolet communication systems described above have a typical range performance of 1 km or less when transmitting using about a 100-mW average power output source. One of the major constraints of such a system is the data-bandwidth limitations, owing to multipath effects. For a conventional line-of-sight communications system, the bandwidth can be very high and is typically limited by the rise times of the transmitter and detector. For a nonline-of-sight ultraviolet system, the performance issue is more complicated. If the transmitter emits a short pulse, then there are statistically a very large number of possible photon paths with differing path lengths between the transmitter and the detector, depending on how each solar-blind ultraviolet photon is scattered. As a result, a short pulse emitted by the transmitter is temporally elongated at the detector, which limits the effective maximum bandwidth of the system. Moreover, two-way or multiway communication can also become technically challenging, as the receivers can be optically susceptible to cross-talk, because they are sharing the same waveband.

14.4.4 Laser Requirements

Laser requirements are shown in Table 14.2.

14.5 Retro-Reflective Optical Communications

14.5.1 Introduction

The retro-communication technique is based on the physical principle that a laser beam can interrogate a retro-reflecting modulating receiver subsystem to extract information. The incident laser beam is reflected and modulated by the receiver unit and returned to the transceiver; hence, there is an inverse range to the fourth dependency (R^{-4}) on the received signal strength. The retro-reflecting modulating

Table 14.2 Laser-Emission Requirements for Short-Range, Free-Space Communications

Source Parameter	*Short Range (Broadcast)*
Wavelength (nm)	260 to 275, ultraviolet solar-blind emission (crucial)
Average power (mw)	~100
Divergence (°)	15 to 45
Pulse-repetition frequency	High kilohertz (for voice communications)
Pulse length (μs)	Few
Line width	Few nanometres
Beam quality	N/A
Polarisation	N/A

receiver consists of an array of modulators to impose the stored information or data on to the interrogating beam.

14.5.2 Military Utility

This use of asymmetrical communications is perceived to have a range of civil, as well as military applications, examples of tactical military uses include:

- Data transfer from small, lightweight airborne platforms to ground stations;
- Data transfer from temporary installed nodes in communication networks;
- Communications with buoys used by submarines, information-exchange processes;
- Interrogation of unattended sensors.

14.5.3 System Operation

A laser beam is used to interrogate a retro-reflecting modulator receiver, which may or may not have information to relay to the interrogation point. For example, this receiver or transceiver may be connected to a remote sensor-system and the stored data can be relayed to the interrogation point via a free-space optical link. The message is transferred by imposing a modulation on the retro-reflection of the interrogating beam through the modulators in the reflector subsystem.

The transmitter may be as rudimentary as a laser source and a simple beam-steering device, to maintain the beam on the retro-reflecting modulator. In many instances, the fine steering of the interrogating beam to refine its aim point on the retro-reflecting modulator may be accomplished using a solid-state, beam-steering subsystem; solid-state beam-steering techniques are discussed in some detail in Chapter 4. Alternatively, a simple two-axis, fast-steering mirror may be used to direct the interrogation beam at the retro reflector.

The retro-reflecting modulator receiver comprises a minimum of one retro-reflecting-modulator module, as shown in Figure 14.9, which can apply an on-and-

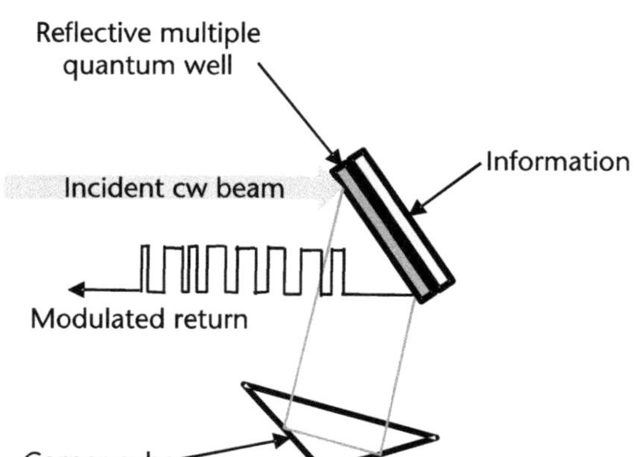

Figure 14.9 Simple retro-reflecting modulator schematic.

off modulation to the interrogation beam. Other modulation techniques may be applied to the carrier wave, as required, by any specific applications.

One of the simpler techniques for applying the modulation characteristics is a multiple quantum-well optical modulator. State-of the-art devices are capable of transferring more than 10 Mbits/s, so that real-time video links can be established over ranges well in excess of 1 km.

Multiple quantum-well optical modulators offer high-bandwidth, high-contrast modulation capability across a wide aperture, with strong electro-absorption effects, owing to the quantum confinement in the thin layers that form the quantum wells. This absorption, resulting from the quantum-confined Stark effect, may be controlled by the application of an electric field; the change from opaque to transparent in the semiconductor material takes less than a nanosecond over of wavelength band of 10 to 20 nm.

These modulators may be configured to be transmissive or reflective devices. The reflective device is manufactured as a Fabry-Perot cavity, which further enhances the absorption characteristics of the multiple quantum-well structure, and consequently, results in additional enhancement to the contrast ratio, but increases the sensitivity to incidence angle. The function of the multiple quantum-well structure in this configuration is to spoil the reflectivity of the retro-reflecting system.

In practice, it is common to have an array of retro-reflecting modulators at the interrogation point to increase the field of view of the receiver unit. The use of a wide field of view on the receiver reduces the pointing accuracy requirements of the interrogating beam director. This is shown schematically in Figure 14.10.

A cat's eye modulation device, which provides a relatively large apparent aperture, but still maintains a high modulation capability is discussed in Section 14.1.

14.5.4 Laser Requirements

The main thrust with this type of optical communication has been to use the eye-safer wavelengths, although other wavelengths in the near infrared have been used.

14.6 Quantum Cryptography

14.6.1 Introduction

This is a point-to-point (line-of-sight) application, which aims to provide a technique to permit very secure communication between two remote points, often

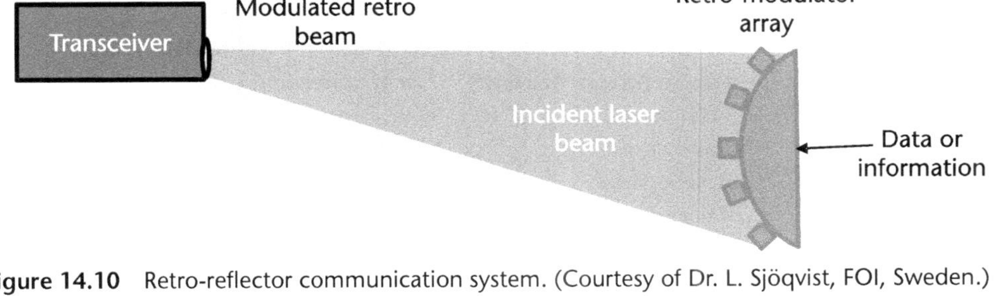

Figure 14.10 Retro-reflector communication system. (Courtesy of Dr. L. Sjöqvist, FOI, Sweden.)

termed Alice (A), the transmitter, and Bob (B), the receiver, through the frequent and simultaneous distribution of a quantum key along with the message. This communication distribution technique depends on the quantum-mechanical principles and information theory to operate effectively. It exploits the fundamental physical basis of quantum-mechanical processes to produce a practical system, where it is difficult, if not physically impossible, for any eavesdropping activity to remain undetected by the legitimate transmitter (Alice) and receiver systems (Bob). The basis for this assertion is embodied in the principle that any measurement changes the quantum state of a system, so that any measurement made of the information stream results in a perturbation and eavesdropping becomes obvious to any legitimate user (Alice or Bob).

The bits of information in the data stream comprise individual photons with different polarisation states to represent the binary conditions 1 and 0. Quantum cryptography uses the quantum bit, or qubit, which can have the normal binary states of 1 and 0 and all the quantum states in between. The data stream is detected using single-photon detectors, of a form similar to those discussed for covert rangefinding in Chapter 10.

Security of the message is achieved through encryption, using a key to generate the cipher text, in this case the quantum-key distribution (QKD) is via the optical channel. The basic optical techniques of quantum-key distribution are achieved by the ones and zeros being encoded via polarisation control of the photons and transmitted through single photons in the beam. The transmitter and receiver agree on a convention for the polarisation orientation, to enable the received data stream to be interpreted correctly.

The encryption-key convention is encoded into the data stream according to the agreed convention using non-orthogonal states of quantum systems, such as the polarisation states of photons, as indicated above. The challenge is to distribute the key with the message, so that an eavesdropper (often called Eve) is unable to capture the key and decipher the message without increasing the bit-error rate, which reveals their presence.

The recipient has to extract the same key to decipher the message embodied within the transmission. The security of any technique that uses such a key to code and decode a message is totally dependent on ensuring that the details of the key

Table 14.3 Laser-Emission Requirements for Retro, Free-Space Communications

Source Parameter	*Retro-Communications*
Wavelength (μm)	1.55 preferred but also 0.85 is used
Average power (W)	1 to 100, dependent on range
Divergence (μrad)	200 to 400 (application dependent)
Pulse-repetition frequency	Usually continuous wave
Pulse length	N/A
Line width	N/A
Beam quality	$M^2 < 1.2$
Polarisation	Random unless liquid crystal modulators used, then needs to be linearly polarised

remain secret for the duration of the message and then for however long that this key remains in use. This protection of the key is fundamental to the security of the optical channel in the link.

A server is used to manage the mechanism for the quantum-key distribution, typically distributing secure keys at the rate of one per minute to encryption appliances through an established specific protocol. Encryption engines use these keys to encrypt and decipher data transmission at rates of up to 10 Gbits/s. Each quantum-key distribution server may service a number of encryption engines.

14.6.2 Military Utility

Secure distribution of classified messages across the battle space is a key military requirement, which can range from commands for action to preflight messages for input to a system prior to operations. This form of link may also be used for a vast range of data links containing intelligence information from a range of sensors.

14.6.3 System Operation

Security in quantum-key distribution during this point-to-point communication is dependent on preparation and detection of single-photon quantum states in particular measurement bases. Most commonly, the quantum property of photon spin is exploited. Photon spin manifests itself as polarisation and possible measurement bases include rectilinear, diagonal and circular polarisation states. When the photon is prepared and measured in the same basis, the result is a deterministic measurement of the polarisation state and the two orthogonal states of any bases can be interpreted as 0 or 1 bit values. When the photon is prepared and measured in different bases, the result is a random outcome.

A secure message is generated through the use of a distributed-key code applied to form the encryption, as described earlier. The great advantage offered by this type of communication system is through the simultaneous secure broadcast of the quantum key to allow rapid deciphering of the message. Given the receipt of the key, the message can be read and suitable action taken.

When the message is deciphered and is understandable, there can be confidence that the message has not been intercepted. However, if the message is corrupted, then this may indicate the presence of eavesdroppers, as the corruption of the received message may have resulted from the interference in the quantum states by the eavesdropper. Alternatively, there many have been some problems with the integrity of the link, such as fading of the signal, leading to an unacceptable bit-error rate.

14.6.4 Laser Requirements

Source development for quantum-key distribution largely centres on methods for creating single photons. Three techniques have commonly been considered for practical systems:

- The most common is an attenuated laser pulse, which is used in the BB84 quantum-key distribution protocol, which was devised by Charles Bennett and Giles Brassard in 1984. Laser pulses can be attenuated to arbitrarily low mean photon numbers, but classical propagation analysis applies at the single-photon level. The problem with this technique is that the number of photons per pulse is given by a Poissonian statistical distribution. Hence, if the pulse contains no photons, efficiency and speed suffer, but when the pulse contains multiple photons, there is a risk an eavesdropper can capture and measure photon polarisation states without being detected.

- The second alternative is a photon-on-demand source, which in principle, would generate exactly one photon per pulse. Various research techniques have been established, typically relying on creating an energy structure in which a single electron can generate a single photon in some characteristic time.
- The third approach is entangled photon pairs. Quantum entanglement of polarisation can be used to generate photon pairs, whose polarisation states are correlated but undetermined until measurements are performed. One experimental technique involves passing blue light through a beta barium borate (BBO) crystal to generate two red photons from a single blue photon.

When using the same measurement bases, two parties can measure the polarisation states of the two photons and each know what the other measured as well. This approach is thought to have advantages but efficient sources of entangled photon pairs are still under development.

The requirements for the laser source may be summarised as follows:

- High spectral purity and fidelity;
- Maximally entangled;
- Generation of one pair at a time;
- Pairs triggered on demand;
- Miniaturised source that may be integrated into the communication system.

Table 14.4 Laser-Emission Requirements for Quantum Cryptography

Source Parameter	*Quantum Cryptography*
Wavelength (nm)	1,550 in optical fibres, but 780 to 850 for longer propagation distances in air
Average power	Single photon, but uses ~10 to 100 mW to pump the nonlinear processes used to create the entangled photon pairs
Divergence (mrad)	0.1 for minimum impact on the receiving telescope; smaller values for satellite-to-ground communications
Pulse-repetition frequency (MHz)	10 to 1,000, or CW
Pulse length (ns)	0.1 to 10
Line width (nm)	No special requirement (~1)
Beam quality	Single mode preferred
Polarisation	A high degree of polarisation and polarisation stability is critical

14.6.5 Future Requirements

The major challenge with this form of optical communication is to avoid errors being incurred owing to the various noise processes present in the received signal, which can lead to a corruption of the cipher key and subsequent loss of the message. Moreover, any corruption of the key may be assumed to be as a result of eavesdropping activity, which may not be the case with beam propagation in a noisy environment, as may occur during poor atmospheric-propagation conditions.

Suppression of side lobes is important, as an eavesdropper may extract the key through these imperfections in the beams, which may be difficult to detect.

Highly accurate direction of the transmission is required, to minimise the size of the beam and ensure covert operation over long ranges.

14.7 Summary

Laser-based, free-space optical communications have been developed for a variety of civil and military applications. This form of communication is not generally dependent on any form of infrastructure and so communication links can be established as rapidly as any changing scenario demands. Moreover, similar equipment and techniques may be used for air-to-ground, air-to-air as well as surface-to-surface communications.

The traditional approach is to use a carrier wave from an optical source, usually a laser beam, and modulate it to carry information. This beam is then pointed at a receiver, to establish a point-to-point link. The receiver may be a simple optical telescope and photo-diode, supported by the appropriate electronic systems to read the message embodied within the received carrier wave, through application of the appropriate demodulation technique.

The use of a laser beam has many attractions, which arise from the inherent low divergence, high frequency and coherence of the laser emission. These characteristics lead to some significant advantages over the more common microwave communication links for point-to-point applications. In particular, the use of a beam with a small divergence and without side lobes enables this type of communication to be suitable for covert communication with a low probability of interception by an eavesdropper. Additionally, the high frequency of the optical wavelengths leads to a potentially high-bandwidth communication system suitable for high-speed data links.

Free-space optical communication systems experience some degradation in performance in the presence of adverse weather conditions; the attenuation of optical wavelengths is normally greater than that experienced by microwaves. Free-space, point-to-point implementations require the line-of-sight link to be maintained throughout a transmission, which can lead to some limitations during operations in urban areas or mountainous regions, owing to blocking of the propagation path.

A new approach has been developed to operate in those regions or areas where establishing and maintaining point-to-point links is difficult, if not impossible, such as occurs in urban and mountainous regions. This is a nonline-of-sight technique and it is dependent for successful operation on optical scatter from molecules and aerosols in the atmosphere. When operated in the solar-blind region of the spectrum, the background luminance is suppressed, so that favourable signal-to-noise detection ratios may be achieved. This is a relatively short-range technique.

Retro-modulation communication techniques have been demonstrated for some data-link applications, which involve a transceiver with a retro-modulator device at a distant point. Interrogation of the retro device by a laser beam is used to extract the stored data, which may have been sent there, or deposited by various sensors. One constraint on the potential effectiveness of this technique is the inverse range to the fourth dependency on the receiver power when the retro device is used as a passive reflector.

Quantum cryptographical techniques have been developed to permit very secure and rapid communications between remote points. The technique has its basis in quantum-mechanical processes that are, in theory, resilient to eavesdropping; moreover, eavesdropping activities may be detected by the legitimate operators of the system. Quantum cryptography permits the transmission of the cipher key with the encoded message, so that the received message can be read rapidly and accurately.

14.7.1 Future Requirements for Optical Communication Systems

The military requirement for increased data rates and bandwidth through secure communication links are likely to demand:

- Development of techniques that will exceed 2-GHz bandwidth;
- Use of nonmechanical beam-steering techniques;
- Development of adaptive-optical techniques to provide aberration control;
- Development of a flexible means of interconnection, to permit rapid agility of movements between transmitter and receiver;
- Enhanced throughput to allow more message traffic at a faster rate;
- Enhanced encryption;
- Greater resistance to fading resulting from adverse atmospheric propagation effects;
- Greater resistance to jamming.

The development of advanced secure transmission techniques based on quantum-mechanical methods that have full resistance to eavesdropping under all operational conditions, as well as immunity to noise-induced effects is a critical emerging requirement.

Selected Bibliography

Agrawal, G. P., *Fiber Optic Communications Systems*, 3rd ed., New York: Wiley-Blackwell, 2002.

Bass, M., and E. W. Van Stryland, *Fiber Optics Handbook: Fiber, Devices and Systems for Optical Communications*, New York: McGraw-Hill Professional, 2001.

Bernstein, D. J., J. Buchmann, and E. Dahmen, *Post-Quantum Cryptography,* New York: Springer, 2009.

Bouwmeester, D., A. K. Ekert, and A. Zeilinger, (eds.), *The Physics of Quantum Information,* New York: Springer, 2000.

Lambert, S. G., and W. L. Casey, *Laser Communications in Space,* Norwood, MA: Artech House, 1995.

Kaminow, I. P., and T. L. Kock, *Optical Fiber Communications III,* San Diego, CA: Academic Press, 1997.

Majumdar, A. K. and J. C. Ricklin, (eds.), *Free-Space Laser Communications,* New York: Springer, 2008.

Pratt, W. K., *Laser Communication Systems,* New York: John Wiley & Sons, 1969.

Singh, S., *The Code Book: The Evolution of Secrecy from Mary Queen of Scots to Quantum Cryptography,* New York: Doubleday, 1999.

CHAPTER 15

Navigation Sensors

15.1 Introduction

There has long been a desire for a totally passive and self-contained technique for navigation that is accurate, reliable and cost-effective. Moreover, there has been an important requirement for rapid and accurate methods for surveying and mapping a location, so that the shoot-and-scoot technique can be executed by artillery batteries.

Inertial-navigation systems have been developed over many decades; the original systems were based on stabilised-platform technology and were capable of high-performance but were mechanically complex, large, very expensive and required careful handling. In the 1970s there was a sustained and successful effort to simplify the inertial-platform system, which resulted in the development of strapdown inertial-navigation techniques; with this system the inertial sensors are mounted in a suitable configuration to form an inertial-measurement unit and this unit is fastened directly to the body of the vehicle. The complexity of the mechanically stabilised platform is replaced in this system with complex algorithms, which resolve the various measurements of angle (or angular rate) and acceleration (specific force) made by the sensors in the inertial-measurement unit into a suitable reference frame (i.e., undertake the function of the gimbals in the mechanical stabilisation of a platform-stabilised system). A strapdown system is physically much smaller, simpler and significantly cheaper, with the complexity in the processing.

The development of strapdown inertial-navigation systems presented a number of very significant technical challenges that needed to be solved in order to make the concept feasible. The two principal challenges were:

- Development of efficient algorithmic techniques for the rapid and accurate resolution of the sensors' measurements of angular motion and specific force into the appropriate reference frame and rapid solution of the navigation equations. Modern computer technology and processing techniques have eased this burden, but in the 1970s it was a very significant challenge, particularly if the size of the computer was not going to dominate the inertial-navigation system.

- Design and development of sensors, which were compatible with the need for the inertial-measurement unit to be mounted directly on to the vehicle. In particular, this was a very significant challenge for the gyroscopes, which had traditionally been relatively fragile precision instruments that would now be subjected to the full dynamic motion of the vehicle and measure angular motion to the appropriate accuracy.

The technical challenge for the strapdown gyroscopic sensor was to combine the high-accuracy requirements of negligible drift of the reference with the dynamic range[1] of the angular motion expected to be experienced, particularly during combat flight, along with high resolution. This resulted in the requirement for a sensor with a dynamic range of the order of 10^9 for some of the more challenging aircraft applications. The initial solution to this problem was the design and development of a ring-laser gyroscope, later followed by other optical sensors such as the fibre-optical gyroscope.

The spectrum of performance of optical gyroscopes ranges from the very accurate with bias of much less than 0.001/hr, in the case of ring lasers, to tens of degrees per hour, often from the simpler fibre-optic gyroscopes. Hence, the range covered by the optical devices is very similar to that covered by the mechanical gyroscopes, but usually with more favourable operational characteristics, considered below. Generally, all these types of optical gyroscopes are suitable for various strapdown applications, depending on the demanded accuracy of the system.

It appears that the application of optics to the sensing of angular rate can offer a number of advantages over the use of the well-established mechanical sensor technology. Some of the advantages often cited are:

- Wide dynamic range;
- Instant start-up (readiness);
- Digital output;
- Output independent of some environmental conditions (acceleration, vibration or shock);
- Accurate high-rate measurement capability;
- Easy self-test;
- System-design flexibility;
- Extended running life.

15.2 Historical Development

The concept of an optical sensor is not a new idea; it was during the late nineteenth century that Michelson pioneered work with optical interferometers, although his goal was not to produce an optical gyroscope. In 1913, the Sagnac effect was reported and this is the fundamental principle on which optical gyroscopes operate.

1. Dynamic range is the maximum rotation rate divided by the minimum detectable angular rate.

The principle of the Sagnac effect is that when light travels in opposite directions (clockwise and anticlockwise) around an enclosed ring, differences arise in the apparent optical length of the two paths when the ring is rotated about an axis orthogonal to the plane containing the ring. In 1925, this concept was applied by Michelson and Gale in Chicago using a ring gyroscope with a perimeter of over 1 mile. By sending ordinary light through evacuated water pipes, they were able to detect the shift in frequency between the light in the two counterpropagating beams produced by the rotation of the Earth.

Soon after the first demonstration of a laser by Maiman in 1960, there was renewed interest in the design and development of optical sensors through exploitation of the properties of laser light. Additionally, there was the desire to capitalise on the benefits anticipated from the use of solid-state inertial sensors. In 1963, the first ring laser was demonstrated by workers at Sperry Gyroscope. This marked the beginning of the development of the ring-laser gyroscope. About a decade later the fibre-optic gyroscope was first demonstrated by a team at the Naval Research Laboratory in the United States led by Dr. Tom Giallorenzi.

The first ring-laser gyroscopes were large and somewhat delicate, owing to the use of a modular optical arrangement and bulk optical elements; however, substantial investment has led to the production of very compact monolithic devices that produce extremely low bias, of 0.001°/hour or better. Typical path lengths for the accurate sensors are about 300 mm. Very small ring-laser gyroscopes have also been produced with a path length of about 50 mm. Generally, these small sensors have a bias in the region of 1° to 5°/hour. Currently, the more accurate class of ring-laser gyroscopes are used in strap-down navigation systems in both commercial aircraft, as well as military fixed and rotary-wing aircraft.

The ring-laser gyroscope has been shown to be a very high-performance device; however, its complexity and stringent design tolerances mean that it is expensive and quite bulky for most tactical applications. These factors led to the desire for an alternative approach for realising strapdown optical gyroscopes that would be cheaper but retain the robust characteristics of solid-state devices. This effort led to the development of fibre-optical gyroscopes. However, it was recognised that there were many challenges to producing a fibre-optical sensor with true high performance inertial performance characteristics (i.e., drift values in the region of 0.01°/hour or better); consequently, the initial thrust was to produce a sensor for sub-inertial applications, particularly for use in tactical missiles. Recent advances with fibre-optical devices have seen high-performance sensors produced.

15.3 Military Utility

The accurate knowledge of one's own position is a fundamental requirement for any successful military activity. This information is normally required instantaneously and continuously throughout any engagement or preparation activity; moreover, angular and positional rate data are required for control and stabilisation functions of systems on military platforms. These applications demand a very broad range of performance from the gyroscopes particularly when used in the strapdown configuration; the approximate drift-rate errors for various applications are shown in Figure 15.1.

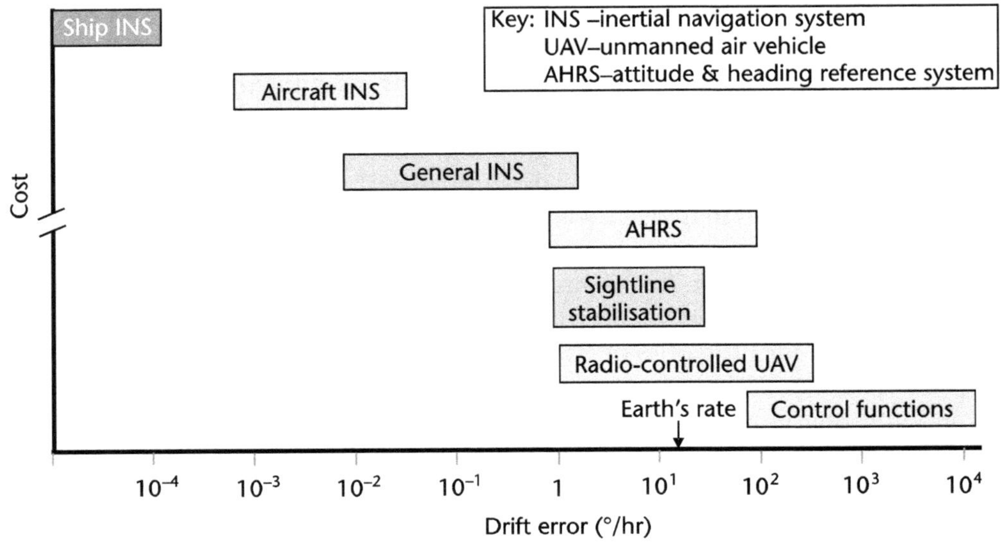

Figure 15.1 Gyroscopic performance requirement for some military applications.

There has been a critical requirement to simplify the navigation system to improve reliability and reduce cost, as well as a need to use a single inertial-sensor unit for all of the motion measurement functions. As indicated schematically in Figure 15.1, this is feasible with strapdown optical sensors, owing to their vast dynamic range. There is a dramatic increase in the cost of the high-performance sensors, but the use of a single inertial-sensor unit has important consequences for reliability, size, weight and power.

The use of inertial sensors in a strapdown configuration has led to some simplification of the navigation system, compared with the traditional stabilised-platform system. Additionally, the cost, the size and the mass of the inertial navigation system have reduced, and a single sensor (or sensor cluster) can now undertake all of the rate measurements.

Inertial navigation and sensing techniques are passive and self-contained and therefore are immune from external interference, such as jamming.

Accurate inertial-sensor systems allow rapid determination of the orientation of a platform or the position of a system on the battlefield. The inertial-navigation system may be used in conjunction with other navigational systems, such as satellite navigation systems [another name for this is Global Positioning System (GPS) and has alternative systems such as Glonass (from Russia) or Galileo (from Europe)] to provide highly accurate positional data; additionally, the use of external data may be used to calibrate systematic errors in an inertial-navigation system to give further enhancement of the navigation performance. The inertial-navigation system provides inherently high-accuracy angular rate and angular position data virtually instantaneously and continuously.

Despite the worldwide availability of satellite-navigation systems providing excellent positional accuracy with long-term stability and bounded errors, an inertial system is required in a multitude of applications to provide the high-accuracy, short-term inertial data, such as rate parameters to very high precision. Moreover, the satellite-navigation systems provide estimates of angular parameters over a short period, which are normally too slow and with significant temporal latency; hence, these estimates of angular position and rate are inadequate for many guidance and control applications, including sightline stabilisation.

Judicious use of the complementary characteristics of the data generated by the two types of navigation systems can lead to very high precision estimates of the navigation parameters, such as position and rates; this is known as integrated navigation. Moreover, for many applications it is possible to relax the performance requirements demanded from the inertial system if the satellite-navigation data are applied to enable compensation of the systematic errors inherent in an inertial-navigation system. This type of navigation system is often known as an integrated-navigation system.

15.4 Fundamental Principles

Optical gyroscopes rely upon the detection of an effective path length difference between two counterpropagating beams of light in a closed path, the Sagnac effect. It can be shown that

$$\Delta L = \frac{4A\Omega}{c} \tag{15.1}$$

where ΔL is the path length difference, A is the enclosed area of the sensor, Ω the rotation rate and c is the velocity of light. Hence, devices with a larger enclosed area are more sensitive, and the fundamental performance is essentially independent of the wavelength of the light used to detect the angular motion.

The various optical sensors described next rely on generating a path difference in an interferometer, the major differences being in how the light is generated and how the path difference is observed.

15.5 Ring-Laser Gyroscope

Operation of a ring-laser gyroscope relies on the fact that an optical frequency oscillator can be assembled as a laser, by using three or more mirrors to form a continuous light path; other configurations with four mirrors are also relatively common. The triangular-shaped light path is the simplest configuration as it minimises the number of mirrors. If a light beam is generated at any point on this path, it can travel around the closed path, being reflected in turn from each mirror, to return to its starting point. Sustained optical oscillation occurs when the returned beam is in phase with the outgoing beam (i.e., the classic requirements for laser action). Maintenance of laser action requires a constant phase at a given mirror surface after every round trip, to maintain the resonant condition. Two such travelling-wave laser beams are formed independently in the cavity, one moving in a clockwise direction and the other in an anticlockwise direction.

When the sensor is stationary in inertial space, both beams have the same optical frequency; however, when the sensor is rotated about the axis perpendicular to the plane containing the light beams, changes occur in the optical path lengths of the two beams. The frequency of each beam changes to maintain the resonant condition required for laser action, such that the frequency of the beam with the longer path length decreases, whilst the frequency of the other beam increases. This path difference is very small, no more than 1 nm; consequently, a source with high spectral purity and stability, such as a helium-neon gas laser, is required to make the laser gyroscope operate.

It can be shown that the frequency difference between the two beams ($\Delta\nu$) is inversely proportional to the wavelength (λ) of the laser light used to sense the angular motion.

$$\Delta\nu = \frac{4A\Omega\nu}{cl} = \frac{4A\Omega}{L\lambda} \tag{15.2}$$

It follows from (15.2) that the turn rate (Ω) may be determined from the frequency difference ($\Delta\nu$) generated during the rotation. The scale factor of the sensor is directly proportional to the area (A) enclosed by the optical path; therefore, the cavity path length needs to be controlled over the environmental range to within a few parts per million or less.

Substitution of typical values into (15.2) shows the beat frequency to be from a few hertz up to megahertz. This beat frequency can be detected, even for very slow rotation rates, but thermal and mechanical instabilities can cause frequency variations in the individual beams, which are far greater than the rotational beat frequency; however, the monolithic cavity design with the two beams sharing the same cavity offers a practical solution.

Detection of the rotational motion is achieved by allowing a small fraction of each beam to pass through the output mirror to form an interference pattern, which is incident on a set of photo-diodes. The frequency difference between the two beams causes the interference fringes to move across the detectors at a rate equal to the difference in frequency between the two beams, and which, is also proportional to the rotational motion.

Hence, the movement of a single fringe past the detector corresponds to an incremental rotation, $\Delta\theta$, where

$$\Delta\theta = \frac{\lambda L}{4A} \tag{15.3}$$

This equation can be used to determine the sensitivity of a ring-laser gyroscope. For a device having an equilateral triangular path of total length L, the area is given by:

$$A = \frac{1}{2}(L/3)^2 \sin 60 \tag{15.4}$$

Substituting in the preceding equation, it can be shown that the sensitivity is inversely proportional to the path length, namely,

$$\Delta\theta = \frac{3\sqrt{3}\lambda}{L} \tag{15.5}$$

For example, considering a helium-neon laser in which the optical wavelength is 0.6328 micrometre (632.8 nm), the sensitivity of a 30-cm path length device is 2.25 arc-seconds per fringe, changing to 5.6 arc-seconds for a 12-cm path length device. The latter type of ring-laser gyroscope is typical of a tactical missile-grade device.

15.5.1 Description of the Ring-Laser Gyroscope

The more commonly used ring-laser gyroscope configuration, which uses three mirrors, is illustrated in Figure 15.2. Successful configurations using four mirrors have also been produced.

The major components of the gyroscope, as shown in Figure 15.2 are the following:

- The laser block is formed in a low expansion ceramic glass such as Zerodur or Cervit (ceramic vitreous).
- The optical components are usually just the mirrors and photo-detectors. There are two types of mirror: the partially transmitting (output) mirror and the highly reflecting mirror, which both have very low scatter characteristics. One of these mirrors is attached to a piezo-electric device to provide the cav-

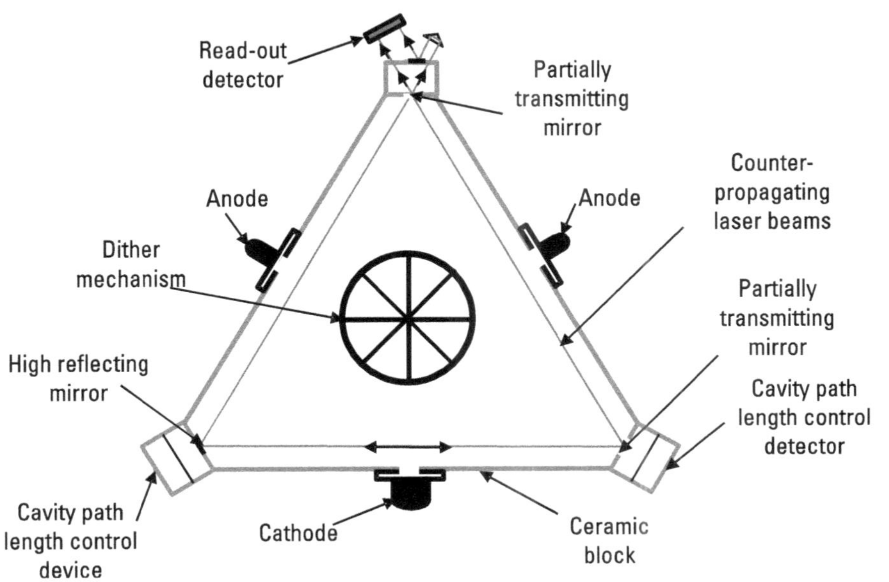

Figure 15.2 Schematic diagram of a ring-laser gyroscope.

ity path-length control. The other mirrors are firmly bonded directly to the laser block.

- The nonoptical components are used to produce a discharge to pump the gas laser.
- The biasing mechanism is required to overcome the lock-in phenomenon.

A photograph of a mechanically dithered ring-laser gyroscope is shown in Figure 15.3.

The primary challenge of this technology is the precision engineering that is needed to make and polish the faces of the laser block and the high precision technology required to produce the mirrors. This tends to make the cost of the sensor quite high, although techniques for reducing this aspect are being sought.

15.5.2 Lock-In Phenomenon

At very low rotation rates, the two laser beams in the cavity cease to oscillate at different frequencies and assume the same frequency, so there is no output signal as shown in Figure 15.4; this is the lock-in phenomenon. It is caused by the radiation of one laser beam being scattered into the other beam causing the host mode to change frequency towards that of the back-scattered energy, resulting in a common oscillation. Careful design and the use of very high quality mirrors, confine this defect to a zone close to zero rotation rate.

15.5.3 Alleviation of Lock-In

One of the most common methods used to alleviate the lock-in problem is the use of mechanical oscillation, by applying angular vibrations to the entire cavity at high frequency but at low amplitude and through small angles, thereby avoiding low-frequency outputs from the detector. Careful application of the dither motion is needed to ensure that this motion has a random component to avoid systematic errors being introduced into the measurements of angular motion.

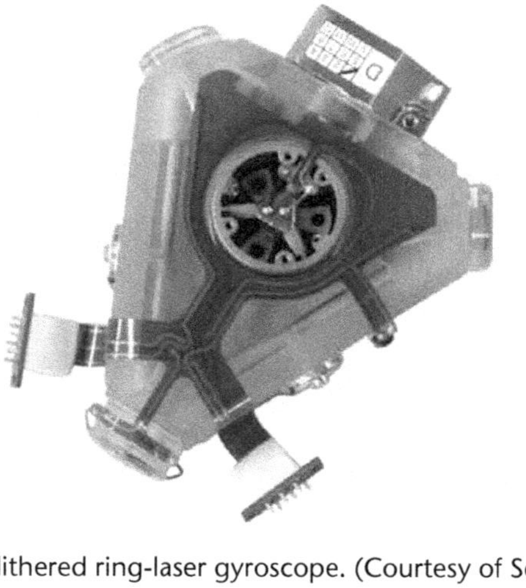

Figure 15.3 Mechanically dithered ring-laser gyroscope. (Courtesy of Selex ES, U.K.)

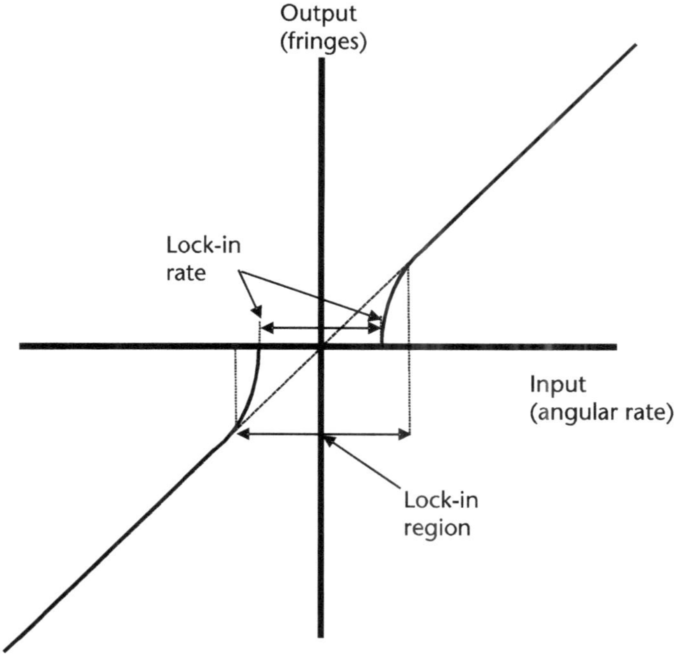

Figure 15.4 Laser-gyroscope input/output characteristic.

Other techniques for alleviating the lock-in effect have been investigated:

- Magnetic-mirror biasing exploits the magneto-optical effect to create a difference in phase delay between the two counterpropagating laser beams.
- Multi-oscillator concepts have been demonstrated, using polarised beams to allow more than one pair of beams to propagate in the same cavity, and usually four beams are used in a square configuration. Independent beams of left and right polarised modes are propagated in each direction in the cavity, giving a total of four modes. Avoiding the phenomenon of lock-in still applies, so it is necessary to bias the modes away from this zone, such as using a quartz retarder plate, or a nonplanar cavity configuration.

Other methods that have been investigated include the differential laser gyroscope (DILAG); another name for the four-frequency gyroscope with a Faraday cell in the cavity. This approach has the potential for increased precision.

15.5.4 Sources of Error

The ring-laser gyroscope is generally a very accurate sensor capable of high performance, thus fulfilling its design aim and has been used in a wide range of strapdown applications. The three dominant sources of error characteristic of a ring-laser gyroscope are:

- Nonlinearity in the scale factor, resulting from the lock-in effect;

- Null shift, an equivalent bias error resulting in a stationary sensor indicating angular motion;
- Scale-factor changes resulting from mode-pulling effects, again an error in the assumed shape of the characteristic linking the detected signals to the actual rotation rate.

15.5.5 Laser-Source Requirements

The fundamental requirement is for a laser beam with very high frequency stability, so that the very small frequency shifts induced by small or low rate angular rotations can be detected. The power of such devices is often miniscule and the divergence controlled as required, by apertures in the cavity. The source of choice has been a helium-neon laser.

15.5.6 Ring-Laser Performance

The moderate-sized ring-laser gyroscopes have a drift error of the order of 0.001°/hr, but larger devices have a performance several orders of magnitude better. The smaller ring-laser gyroscopes, aimed at tactical applications, have drift errors of the order of 0.1°/hr.

15.6 Fibre-Optic Gyroscope

This type of optical sensor also applies the principles of the Sagnac effect; however, in contrast with the ring-laser technology, the fibre-optic gyroscope senses angular motion by detecting the phase difference between the two beams passing round the fibre light path in opposite directions. This gyroscope can be constructed as a genuine solid-state sensor, even in a closed-loop mode, by the use of integrated optical components, which means that the sensor can be very compact and robust. However, extreme care and good design are required with the necessary fibre connections to avoid failure when operating in harsh environments.

These sensors have been used in many applications, particularly to provide stabilisation and sub-inertial navigation capability. Moreover, the development of modern techniques has led to high-class performance devices suitable for some inertial-quality applications.

15.6.1 Principle of Operation

In its simplest form, light from a broadband source is split into two beams that propagate in opposite directions around an optical-fibre coil. These two beams are combined at a second beam splitter to form an interference pattern, where the resultant intensity is observed using a photo-detector. This type of Sagnac interferometer gives an output when the device is rotated, in an identical fashion to that described above, for a ring-laser gyroscope. The resulting phase difference results in a change in amplitude of the interference pattern formed when the two beams are recombined.

For a rotating fibre gyroscope with a single turn of fibre, the phase difference ($\Delta\Phi$) between the counterpropagating beams of light may be expressed in terms of the path difference (ΔL) generated when the device rotates as:

$$\Delta\Phi = 2\pi \frac{\Delta L}{\lambda} \tag{15.6}$$

It can be shown that for a coil of N turns this becomes:

$$\Delta\Phi = \frac{8\pi AN}{c\lambda}\Omega \tag{15.7}$$

This may be expressed in terms of the length of the fibre ($L = 2\pi RN$) as:

$$\Delta\Phi = \frac{4\pi RL}{c\lambda}\Omega \tag{15.8}$$

The challenges for the detection of the difference in phase may be illustrated by considering a coil of radius 40 mm containing a 100-m length of fibre. If the optical wavelength is 850 nm, the phase differences that occur for rotation rates of (a) 15°/hour and (b) 500°/second are as follows: (a) $\Delta\Phi = 0.0008°$ and (b) $\Delta\Phi = 98.6°$.

Clearly, if a sensor is going to be capable of detecting Earth's rate or comparable rotations, a high level of dimensional stability will be necessary. Hence, light travelling one way around the fibre coil must travel exactly the same path as the light which travels in the opposite direction (i.e., reciprocity must be maintained).

Comparing (15.8) to the equation for the ring-laser gyroscope (15.3), the difference in sensitivity between the two sensors is obvious, owing to the occurrence of the velocity of light in the denominator of (15.8). Use of closed-loop techniques has enabled measurement of minute phase shifts needed to achieve the high performance required for accurate inertial navigation systems.

15.6.2 Description of the Fibre Gyroscope

The fundamental optical components of a simple open-loop fibre-optical gyroscope are shown schematically in Figure 15.5:

- A light source, usually a broadband source, such as a light-emitting diode, which has a coherence length chosen to minimise the scattering effects occurring within the fibre, but lasers have been used.
- Couplers that link energy into and out of the fibre. It is common to use 3-dB couplers, so they act as beam splitters.

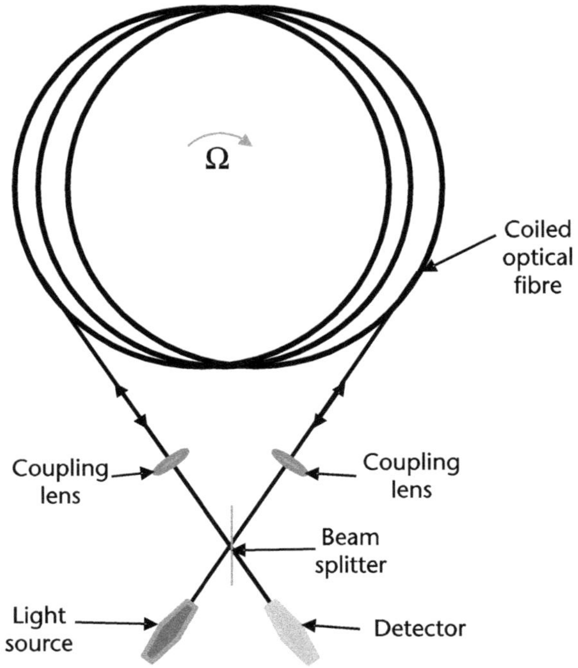

Figure 15.5 Open-loop fibre-optic gyroscope.

- The multiple turns of optical fibre in a coil, the angular-motion sensing element, are used to provide sufficient sensitivity. Special fibre may be used to provide enhanced performance.
- A photo-diode is used to detect the changes in the fringe pattern.
- The non-optical components include the former, on which the fibre coil is wound, and the electronic components.

When it is used in the simple open-loop configuration, it is particularly sensitive to any nonreciprocal effects, thereby reducing the sensitivity of the device, and consequently, the accuracy that may be achieved. This is the simplest and cheapest form of fibre-optical sensor.

15.6.3 Open-Loop Operation

A scheme has been devised that ensures that both of the returning waves have propagated along the identical path in opposite directions. The technical approach involves projecting polarised light into the interferometer through a single-mode waveguide, such as a monomode optical fibre, and observing the returning interference waves, which have been filtered by the same waveguide prior to detection. This approach is known as the reciprocal or minimum-configuration gyroscope, and is shown schematically in Figure 15.6.

The beams from the fibre coil are combined at the second beam splitter and emerge from the reciprocal port; they are exactly in phase when the fibre coil is at rest. The resultant intensity varies as a sinusoidal function with the applied rate of angular rotation sensed by the coil of optical fibre.

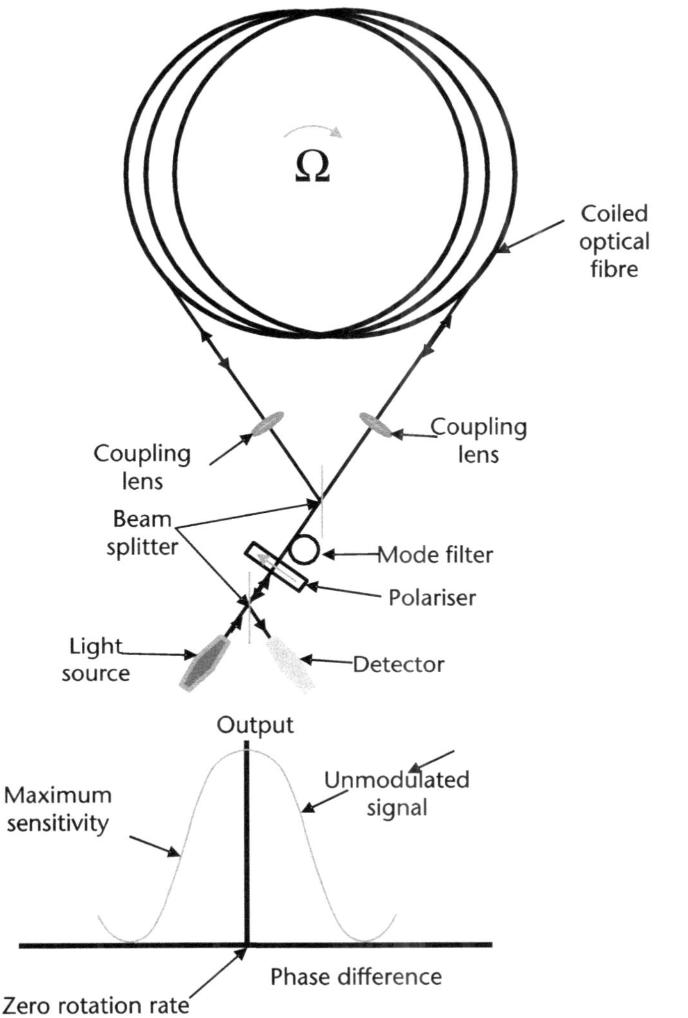

Figure 15.6 Reciprocal configuration fibre gyroscope and detector response.

The major disadvantage of this simple form of fibre gyroscope is the lack of sensitivity at small applied input rates, owing to the co-sinusoidal shape of the fringe pattern. However, the sensitivity at low rotation rates may be achieved by incorporating a phase modulator at one end of the fibre coil, which is operated asymmetrically to give a phase dither of $\pm\pi/2$ radians, which appears at the detector at twice the modulation frequency. The response is still sinusoidal, as illustrated in Figure 15.7 which also shows a schematic diagram of a phase-biased fibre gyroscope.

15.6.4 Closed-Loop Operation

There are many applications that require good accuracy over a wide angular-rate range with good response characteristics, which can be achieved using a closed-loop signal-processing approach. Two techniques based on nulling the observed phase

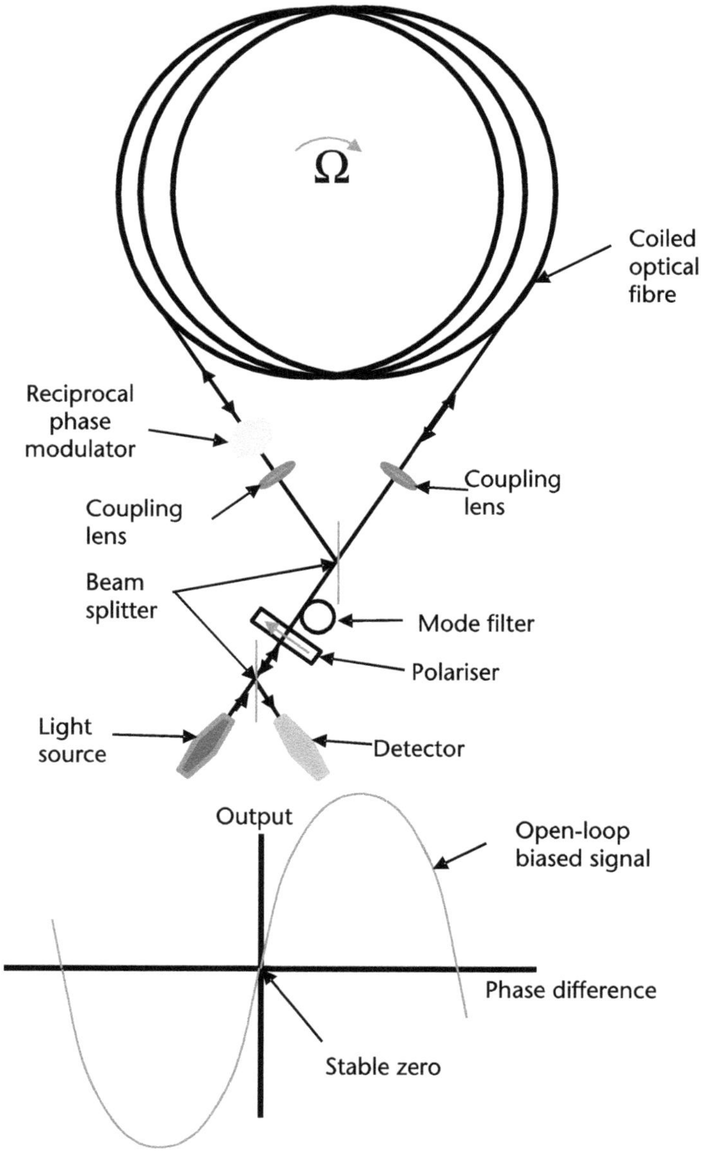

Figure 15.7 Phase-biased fibre gyroscope and detector response.

shifts have been demonstrated. The phase-nulling and frequency-nulling approach is shown in Figure 15.8.

As shown in Figure 15.8, a second phase modulator is added to the other end of the fibre coil; it is operated at twice the frequency of the dither (biasing) modulator and is used to back off, or null, the effects of the Sagnac-induced phase shifts caused by the angular motion of the fibre coil. The open-loop signal is used as an error signal to generate an additional phase difference ($\Delta\phi_n$) that has an opposite sign to the rotation-induced phase difference. Consequently, the total phase difference is arranged to be at, or very close to, the zero value, so that the system is operated about the point of greatest sensitivity. The value of the additional feedback ($\Delta\phi_n$) is used as a measurement of sensed angular rate. It has a linear response with

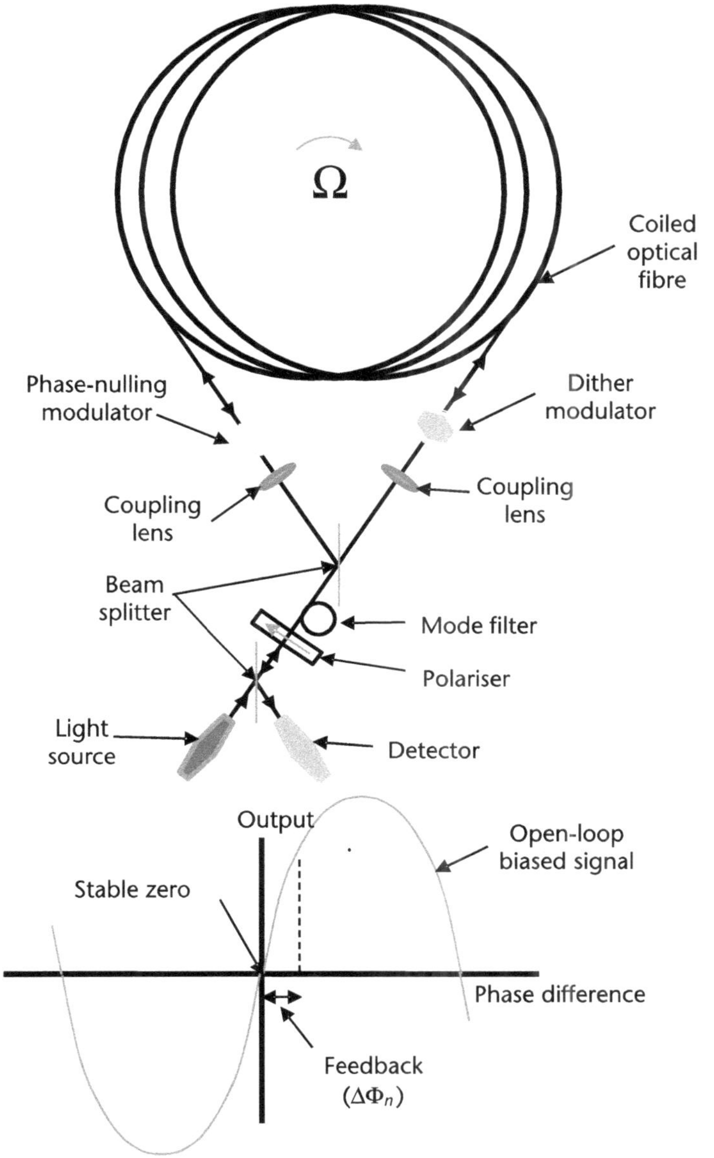

Figure 15.8 Closed-loop (phase-nulled) fibre gyroscope and detector response.

good stability, because it is independent of the power of the returning optical signal and the gain of the detection system. However, the accuracy of the scale factor does depend on the stability of the source wavelength and the geometric stability of the sensing coil of optical fibres.

An alternative technique for achieving closed-loop operation is through the use of a frequency shift generated by an acousto-optical modulator, or Bragg cell, placed at one end of the sensor coil. This frequency shift is used to produce a differential phase-shift to null that caused by the Sagnac effect during rotation of the sensor. By varying the voltage applied to the Bragg cell, the frequency shift it induces can be varied. Hence, the voltage that needs to be applied to the Bragg cell to null the detector's output is directly proportional to the applied angular-motion.

The frequency shift given to the light by the Bragg cell is chosen so that the sensor is forced to operate in the region where it is most sensitive to low rotation rates. This is dithered about the centre frequency, using a square-wave modulator, and the intensities of the light beams incident on the detector are monitored. When the sensor coil is rotated about its input axis, a small phase shift is introduced between the two beams; this phase shift gives rise to a mismatch between the intensities of the two beams on the detector. Consequently, the output signal from the detector will be modulated at the dither frequency. A phase-sensitive detection system is used to deduce the amount that the centre frequency of the Bragg cell needs to be altered to return the intensities to their original matched state, thereby nulling the output of the sensor (i.e., backing off the applied angular-rate). This change in the voltage applied to the Bragg cell is directly proportional to the applied angular-motion.

15.6.5 Summary of Laser Requirements

The principal requirement is a source with a short coherence length to mitigate interference effects induced through optical scatter within the fibre sensing coil. The power requirements are most modest and the wavelength requirement is determined by the optical fibre and the integrated optical elements.

15.6.6 Summary of Performance

Simple fibre-optic gyroscopes have drift-rate performance measured in tens of degrees per hour, but, as with all optical sensors, very stable performance parameters, such as scale factor. The more sophisticated versions are capable of inertial-grade performance, so drift rates are in the 0.01°/hr class or better.

15.7 Integrated Fibre-Optic Gyroscope

The integrated fibre-optic gyroscope uses components fused into a lithium-niobate substrate. Fibre leads are used to connect the source and detector to the integrated-optics chip. This form of gyroscope has the potential to be compact and rugged and have a long shelf life, whilst retaining the advantages of optical sensors listed above.

The closed-loop device interferometric fibre-optical gyroscope has seen very substantial increases in performance owing to:

- Developments in the super-luminescent diode technology;
- Enhanced optical fibres with very low to negligible optical loss, lower scatter and greater uniformity;
- Perfection of the functions of the integrated-circuit chip, such as the polarisation of the light, the splitting of the beam into the clockwise and counter-clockwise beams, the recombination and control of the frequency-shifting process;

- Development of microprocessor technology that can be combined within the sensor to give real-time accurate compensation of systematic errors.

This modern device technology enables each sensor to give optimum performance by use of appropriate compensation techniques applied to mitigate or cancel out systematic errors, particularly for the effects of changes in ambient temperature. Moreover, the same fundamental architecture can be used in different performance sensors, but reducing the length of optical fibre for the lower performance devices. However, the modulation frequency still has to be tracked to the actual length of fibre to give the optimum performance for the chosen fibre length.

The measurement accuracy of these interferometric fibre-optical gyroscopes is now comparable with that achieved by a standard inertial-quality ring-laser gyroscope.

15.7.1 Sources of Error

The major sources of error in the measurement of angular motion by any type of fibre-optic gyroscope are as a result of distortions to the shape of the coil of optical fibre. These distortions may be caused by environmental changes, such as temperature, as well as acceleration and vibratory forces. The changes in shape of the coil result in distortion of the scale factor of the sensor and also temperature-, acceleration- and vibration-dependent biases. Fortunately, through careful design, particularly the winding of the sensor coil, these sensitivities can be minimised.

15.8 Other Optical Sensors

A number of other laser-based sensors have been developed to implement the Sagnac effect:

- Fibre-optic ring-resonator gyroscope is a development of the fibre-optic gyroscope. This type of gyroscope uses a recirculating ring-resonant cavity to enhance the phase differences induced by the Sagnac effect when the sensor is rotated. A ring cavity is formed by joining the ends of a short optical fibre cable, typically 10m or less, to a coupler, to provide the input and output port.
- Ring-resonator rotation-rate sensors operated in a nonlinear mode. Techniques have been investigated to allow the measurement of rotation rate by the exploitation of nonlinear effects induced in the energy in the ring resonator. These effects occur when there is only a small loss, of the order of 5%, in the ring, leading to about a 20-fold increase in power. The nonlinear interactions are usually divided into two groups, intensity-dependent refractive-index effects and scattering phenomena. In the case of the fibre-optic gyroscope, these nonlinear effects are considered detrimental, as they degrade performance.
- Ring-resonator gyroscope, which is very similar to the fibre-optic ring resonator, but the fibre ring is replaced with an optical waveguide, which is

etched into a suitable substrate material. The operating principle is very similar to that described for the fibre-optic ring resonator. Small devices have been demonstrated indicating that a gyroscope on a chip can become a reality, along with all the advantages of optical sensors.

15.8.1 Laser Requirements

There are no specific requirements for these versions of fibre gyroscopes, other than the need for a short coherence length and stable emission. In fact, the development of super-luminescent diode technology has produced an ideal source for this application.

15.9 Applications

Optical sensors have found many applications in the navigation, guidance and control systems of missiles and aircraft. They have proved to be invaluable in the stabilisation of many systems. An application is described next to illustrate the value of combining sensors with complementary characteristics to produce an effective military system.

15.9.1 Artillery Pointing

A continuing and challenging requirement of military forces is to establish rapidly and precisely where they are within the battle space (i.e., having the ability to deduce a precise geographic reference for each system). This is particularly true for those units involved in the indirect battle, for example, artillery batteries using systems with unguided projectiles.

Up to the 1970s, the deployment of artillery required many hours (or possibly days) of work with theodolites to survey the position of the intended gun-battery system prior to occupation. By the late 1970s an automated survey system known as Position and Azimuth Determining Inertial Systems (PADS) was available, which allowed a mobile artillery battery to set up, aim and fire its guns in a relatively short period, about 90 minutes. This system heralded the dawn of mobile artillery operations and gave batteries the unprecedented ability to manoeuvre, set up and fire one accurate salvo, and depart before the enemy could retaliate, a philosophy known as shoot and scoot.

The development and widespread availability of inexpensive and accurate satellite-based navigation systems (GPS) has now generally superseded the use of PADS for survey; however, inertial techniques are still required for the rapid and accurate determination of azimuthal orientation to direct the gun fire.

Now, a strapdown system, aided by odometer and/or satellite-navigation data, with zero-velocity updating, can provide the required accuracy in the azimuthal plane with a cost and weight advantage of about a quarter of that of earlier-generation PADS. The modern system may also be sufficiently light and rugged to be mounted directly on the trunnion, giving accurate measurement in elevation as well. A typical example of a triple-axis system used in a gun system is shown in Figure 15.9.

A few complications associated or encountered with the typical implementation of these systems require compensation. First, there is the problem of track slip during movement of tracked vehicles, where an odometer is used as an aiding source during the navigation phase. Second, the problem of the shock (up to several hundred Gs[2]) transmitted at the instant of firing, well beyond the linear range of inertial-grade accelerometers. Algorithms to overcome these problems of track slip are now mature and robust. Compensation may be made for the error induced in the specific-force measurement by the shock from launching the projectile; a number of techniques have been developed to counter this systematic error.

In the event of the loss of the satellite-navigation information into the integrated-navigation system, the inertial navigation will have to revert to the basic mode of operation. This will be a lengthier process involving alignment of the system (i.e., determining north, east and down and then establishing bearings for triangulation, for example).

15.10 Summary

Optical sensors, particularly the inherently accurate ring-laser gyroscope, have established the credibility of strapdown inertial-navigation systems. Strapdown techniques offer very compact inertial systems, but are dependent on the availability of inertial sensors with a suitable large dynamic range.

Modern developments with other simpler forms of optical technology have proved invaluable in the development of many military systems by providing rugged, high-performance, solid-state devices for a range of navigation, guidance and control systems. Fibre-optical sensors have many attractions for these applica-

Figure 15.9 Ring-laser gyroscopic system for artillery use. (Courtesy of BAE Systems, now Selex ES, U.K.)

2. G is the acceleration of gravity; 100g is 100×9.81 ms^{-2}.

tions, as they are generally cost-effective devices and at least as rugged as ring-laser devices in compact packages.

The combination of the information provided by the global-positioning system and an inertial-navigation system has provided artillery systems with the technology required to have a cost-effective system to undertake shoot-and-scoot type of operations.

15.10.1 Future Requirements

Optical sensors provide high-performance angular-rate information, but at a price. One of the greatest challenges is to reduce the cost and size of the sensor package.

Selected Bibliography

Kayton, M., and W. R. Fried, *Avionic Navigation Systems*, New York: John Wiley & Sons, 1969.

Titterton, D. H., and J. L. Weston, *Strapdown Inertial Navigation Technology*, 2nd ed., Stevenage, U.K.: The IEE and Reston, VA: the American Institute of Aeronautics and Astronautics, 2004.

CHAPTER 16

Other Emerging Military Applications

Laser technology continues to be applied to an ever increasing number of military functions or support activities. This chapter will discuss some of these developing laser-based systems and functions.

16.1 Laser Decoys

16.1.1 Introduction

The emission from a laser may be used in a range of countermeasure applications, as described in Chapters 8 and 9. A different type of laser-based countermeasure is the creation of a decoy or false target spot to disrupt the terminal guidance of a laser-designated weapon, such as laser-guided bombs.

A laser emission may be used to decoy and defeat semi-active, laser-guided munitions, through the generation of false target spots at some safe distance from the target being attacked or asset being defended. A number of countermeasure techniques have been investigated to move or obscure the target-designation spot from an assigned targeting point, defined by the aggressor; one of the more promising methods is based on an optical emission that mimics precisely the designator emission, which is arranged to be at a safe distance from the intended victim.

The fundamental requirements of the decoy target spot are:

- Compatibility of the wavelengths;
- Rapid switch on and suitable frequency-code diversity to mimic the code used by the laser-designation system;
- Compatibility of the frequency modulation code used to define a specific target for a seeker to track;
- Similarity between the radiant intensity of the decoy and actual designated spots;
- The angular separation rate must be realistic and compatible with the likely motion of the victim.

16.1.2 System Operation

A laser-warning sensor suite, which is the normal use of the term threat warner, is used to provide an alert that a platform is under attack from a laser-designated weapon system; this is a specific type of threat-warning sensor system with similar functionality to the ones already described above, for laser-based directed-infrared countermeasures, but this sensor system aims to detect, and where possible, identify the captured laser beams. Moreover, its functionality is similar to that of a directed-countermeasure system, as a successful countermeasure event is dependent on knowledge of the sector of attack by the semi-active homing weapon.

On receipt of a laser-beam warning alert, the detected signals are analysed to decode the received beam characteristics and establish the precise timing of the illuminator-pulse pattern used by the designator system. This information is used to drive a laser-repeater unit, whose function is to mimic the illuminator beam but at a point some distance away from the platform or asset being defended, but within the field of view of the semi-active seeker on the approaching weapon.

The laser repeater is a transmitter unit that contains the laser source, a suitable modulator and a means of either delivering the laser spot to a suitable harmless position or sacrificial point for the seeker to track, or through an optical train or optical fibre, to a suitable transmitter. In the latter case, it is also necessary to have a mechanism to separate the transmitter or the sacrificial object from the platform being protected to a safe position beyond the platform. With this type of active decoy, it is a critical requirement that the decoy transmitter starts transmitting whilst the true designation point and the decoy are both within the seeker's field of view; moreover, the apparent sightline-angle rate must not exceed any values anticipated by the seeker's target-tracking logic for a normal interception, or exceed the tracking-rate limits of the seeker. Any excessive step change in tracking rates would normally negate the effect of the decoy.

Figure 16.1 shows a schematic block diagram of such a laser-based decoy system.

Some of the potential active-decoy techniques for creating a false target spot to deflect a semi-active homing weapon away from its intended victim are shown in Figure 16.2. Specifically, they are:

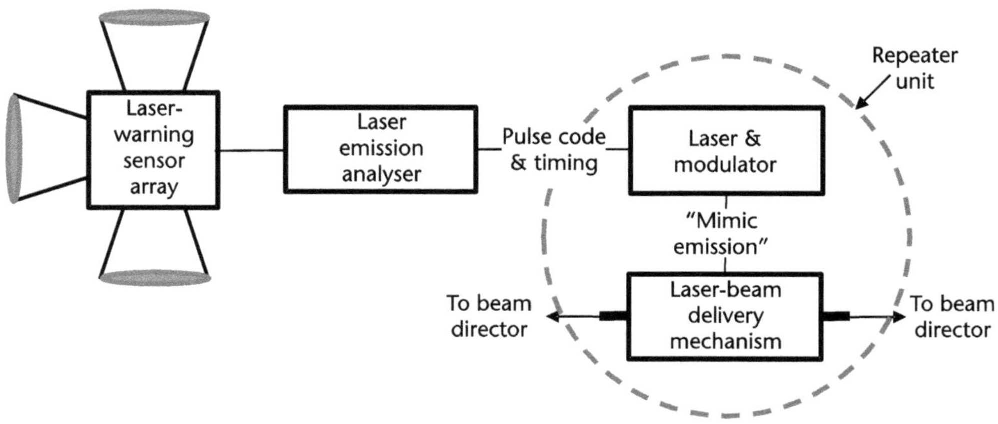

Figure 16.1 Laser-based decoy system.

Figure 16.2 Active laser-decoy techniques.

- Sea-surface reflection, in which the transmitter directs the decoy spot on to the surface of the ocean at a suitably safe distance from the platform;
- Off-board decoy, in which case there may be a sacrificial decoy target at a suitable distance from the ship or land vehicle, which is illuminated by the laser-decoy repeater unit;
- Off-board beacon, in which several techniques may be implemented, but the use of a transmitter assembly with a broad transmission beam fed by an optical fibre is a feasible and practical system;
- Beacon on a mast, which involves a transmitter assembly with a broad beam that will provide a seeker with the appropriate target characteristics for it to home on to and intercept instead of its intended victim.

This diagram of a possible naval scenario (Figure 16.2) does not show the case of the sacrificial target decoys that are illuminated from the protected platform, which may be a ship, a land vehicle or a building. An important consideration concerning this method of countering semi-active weapons is that the illuminated decoys should have good reflectivity at the wavelength of the illuminator; furthermore, in the example of a static target being protected by this method, several decoys may be required on different bearings, to ensure adequate probability of escaping a hit for threats approaching from different directions.

An active off-board beacon may be mounted on a buoy for protection of a ship. A form of transmitter in a beacon is shown in Figure 16.3. In this case, an optical fibre is positioned at a suitable point in front of a mirror assembly with the appropriate profile to generate a beam over a large solid angle. The beam may extend over 40° to 50° in elevation and over a greater angular extent in azimuth. Several of these transmitter assemblies may be required to provide adequate angular coverage in the azimuthal plane. It is worth noting that with off-board decoys, the system must be deployed and correctly functioning suitably early within the time scale of the engagement to give the countermeasure time to be effective.

Other optical configurations are feasible such as fibre-bundle arrays.

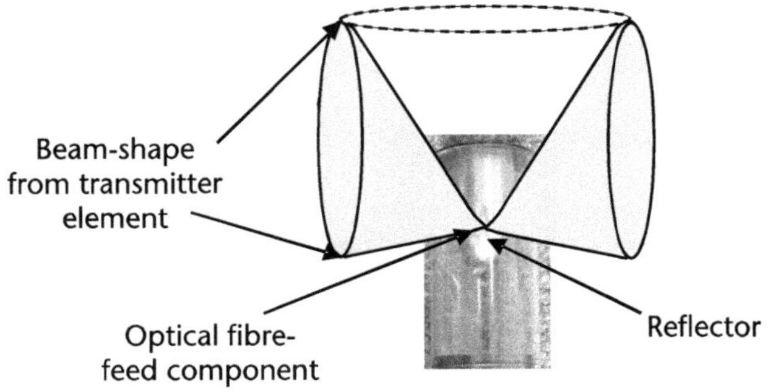

Figure 16.3 Transmitter beacon assembly.

16.1.3 Laser Parameters

There is a NATO standard (STANAG 3733) that defines the emission characteristics for designator systems. The laser in the laser-decoy transmitter is required to mimic that range of laser emission characteristics. Typically, this is a 1.064-μm pulsed emission and a few watts of average power (i.e., the emission from a Nd^{3+}:YAG laser). The transmitted beam must be compatible with any other standards that may be used by threat systems. Additionally, the beam-quality requirements are not usually demanding, other than being compatible with the beam-delivery system, which could well be based on optical-fibre technology.

16.2 Direct-Fire Simulation

16.2.1 Introduction

This is a laser-based technique used in training and as an aid in the development of tactics and procedures; for example, it may be used to indicate the accuracy of a sniper or a gunner. This technique may be extended and developed to simulate and evaluate a laser-based, directed-energy weapon performance.

This system for direct-fire simulation typically uses a semiconductor laser, such as a gallium-arsenide device mounted co-axially with the weapon, where the laser beam simulates the firing of the ordnance from the weapon. Hence, when a gun is fired there is an emission of a short burst of pulses from the semiconductor laser; detectors on the target indicate to the gunner and the controller when there is a hit, along with the location of the hit.

Various sophisticated indications of direct hits on a target can be devised, such as puffs of smoke, flashing lights or more. If a vehicle is the target, then it may be disabled for a given period or until reset by an umpire, typically used during fire-fight training.

16.2.2 Laser Requirements

The laser requirement in this case is for a covert emission that is eye-safe, although semiconductor devices operating in the near infrared are quite common, other high-efficiency sources are feasible sources. The beam-quality requirements are modest, as the main need is for a collimated beam that will provide an accurate indication of the point of impact of the projectile being simulated.

16.2.3 MILES

The multiple integrated laser engagement system (MILES) is a specific product. This system is used by many armed forces as a training aid; this system is used in conjunction with guns firing blank cartridges to simulate battlefield conditions and is similar to the direct-fire simulation technique, described above.

16.2.4 System Operation

Each soldier carries an array of small laser-beam detectors that will indicate when he or she has been "shot" by the laser-beam emitted by a weapon, such as a gun. This detector array may be in the form of a harness with suitable sensors and a halo of detectors around their helmet. A military vehicle may have an array of sensors arranged around its body and positioned about any hatches, especially where the commander is located in an armoured vehicle. When a soldier has been hit by ordnance the system emits a loud noise, which requires direct intervention by the soldier to silence the alarm.

The laser transmitters on the weapons being fired in an engagement are configured to replicate the effective lethal ranges of the various weapons in use at that time during a simulated battle. The laser beam is emitted on detection of the sound from a blank cartridge being fired. Blank-firing attachments have been developed to simulate realism in an engagement to allow soldiers to emulate the firing of their weapons as actually used on the battlefield. There is a further discussion about the injection of realism into training in Chapter 18.

The MILES systems are configured to transmit information about the weapon being fired, so that the potential performance of an engagement may be determined. For example, if an armoured vehicle is hit successfully by a light machine gun, there will be little success in terms of a kill unless the commander is engaged and hit. These data may then be used to determine the outcome of an engagement and then a battle to help determine the success of particular tactics.

16.3 Search and Rescue

16.3.1 Introduction

There are a number of approaches to fulfil this military function. Common techniques involve illumination of a search zone to detect returns from various reflectors. These devices range from the use of highly reflective tape containing an array of minute corner reflectors arranged randomly, to a sophisticated method involving interrogation of a sector of space using a coded beam that triggers a response from a transponder carried by the victim or casualty.

A laser is used to illuminate a search zone with an eye-safe beam, in a similar fashion to the laser-based targeting, but without the range gating of the camera, as described in Chapter 11. The various search zones of interest are scanned in a systematic fashion and then viewed by a suitable detector ganged to the beam director undertaking the scan. In some cases, the person needing to be rescued will carry a beacon to cue the rescuers to the co-ordinates of the search zone.

16.3.2 Laser Requirements

This function needs modest power laser emissions in the eye-safe regime. A widely divergent beam enables a large search zone to be flash illuminated, as described in Chapter 11 (laser targeting/active imaging). This application may be achieved through a suitably configured active-imaging system or a laser-radar system, although simpler dedicated systems are feasible. A wavelength that allows very good transmission through the atmosphere in adverse weather is an important requirement.

16.4 Illumination Aid

16.4.1 Introduction

There is a continuing need to enhance the night-vision capability of a range of sensors. Gallium-arsenide semiconductor lasers have been used successfully to improve the performance of image intensifiers. The wavelength of the laser is chosen to be covert to human vision, but just within the optical pass-band of the image intensifier.

In its simplest form, the laser illuminator is used as a torch or flashlight to illuminate a target zone when the natural lighting conditions are not favourable for operation of the image intensifier in a passive mode. The principle of this technique has been developed in far more sophisticated systems such as gated imaging and active-spectral imaging, discussed in detail in Chapter 11.

16.4.2 Laser Requirements

The laser requirements are broadly determined by the observation system, as the reflected light must be in-band, but not so intense that it would dazzle or damage the highly sensitive viewing aid.

16.5 Small-Arms Laser Sight

In this system, a gallium-arsenide semiconductor laser is boresighted with the optical-targeting sight used on a gun. This is an identical application of the illumination used to aid an image intensifier and is a type of active-illumination system.

16.6 Power Beaming

16.6.1 Introduction

The most common or normal technique of transferring power is through copper wires; however, there are many occasions during military operations when it would not be practical to run wires to a remote location or simply impossible owing to distance or use of celestial assets. During the 1960s some consideration was given to the use of microwaves for wire-less transfer of power, particularly to space-based assets.

The attraction of using laser beams arose because of:

- Availability of high-power emission in the near and short-wave infrared;
- Natural small angular divergence of the generated beam (i.e., availability of beams with high irradiance);
- A range of wavelengths that could be matched to the application;
- High efficiency;
- Identification of wavelengths that have high transmission through the atmosphere;
- Lack of side lobes in the beam.

It is recognised that laser beams face some challenges with efficient propagation in some regions of the electromagnetic spectrum and for some adverse weather conditions. However, there are very significant periods when propagation conditions are good and there are cloud-free lines of sight to distant objects and there have been significant advances in the wall-plug efficiency of laser devices.

The initial concept was to use a solar-powered laser for space-to-space power beaming, and so the idea for Earth-to-space power beaming developed. A number of applications were proposed, such as providing power to a lunar base during the 354-hour lunar night, which would be captured by the same solar cells that would be used during the day. A further proposal was to use power beaming from an Earth-based laser to extend the service life of communication satellites.

The use of power beaming involving lasers has only really become a viable technique since the recent enhancements in the efficiencies of the photon source and the detector/solar-cell technology. Advances with the fabrication of photo-voltaic devices have seen a significant reduction in cost in the large arrays as well as in the efficiency of conversion of captured photons to electrons.

16.6.2 Military Utility

The implementation of power beaming can lead to enhanced on-station time, with reduced installed battery load. For some operations, this technique could lead to unlimited patrol operations, such as continuous intelligence, surveillance and reconnaissance missions.

There are probably two modes of operation for laser-based power beaming:

- Near-continuous powering of the air or space-based asset, typically, for use during short-range operations with unmanned aircraft requiring the platform to be within view of its base, perhaps providing a view over the horizon or perimeter surveillance from its sensor package. These air assets could probably be irradiated/illuminated frequently, if not continuously, as required, during its various loitering activities. A number of the very small military platforms, such as unmanned air vehicles, have very limited endurance, especially if they are totally reliant on installed battery power and designed with a very limited payload. This situation arises despite modern advances with battery technology, there are still size and weight constraints, particularly for the very small, as well as with the miniature, platforms. The endurance, and hence military capability, of these platforms can be increased significantly by fitting an array of solar cells or similar detectors to augment the installed battery power.

- Intermittent recharging of the installed power system, such as batteries or fuel cells on drones or spacecraft, for longer-range activities, the aerial platform may need to return to the equivalent of a refuelling point for a recharge, before resuming operations. In this case, there may be an opportunity for establishing a data link and transferring information, depending on the mission, during the recharge. This intermittent technique is relevant to not only extended-range operations but also multimission tasks, which may involve a series of refuelling points.

This technique is dependent on the availability of a line of sight between the power source and the platform, which is a similar critical requirement for free-space optical communications. Reducing the power payload could enable a greater sensor payload to be carried or greater endurance for a similar configuration; however, it cannot be reduced to zero, as there is a finite chance of a loss of the link for a range of adverse atmospheric conditions, as described in Chapter 5. Hence, some storage capacity is important to ensure the integrity and continuous operation of a platform.

The power-beaming techniques are relevant to air assets other than unmanned aerial vehicles and spacecraft, such as balloons (lighter-than-air vehicles). A single power-beaming ground/surface station could service a number of aerial platforms, unless there is a requirement for continuous support of a single asset, in this case there has to be a dedicated beam director. Of course, the surface station could be on a ship.

The elevation for the beam could range from a few degrees above horizontal (or the local horizon) to an angle close to the zenith if required, but it is anticipated that the lower elevations would be more appropriate for many of the applications. Even with modest power laser beams it is essential to have a control system that will de-conflict the air space to avoid accidental illumination of other objects, as discussed in Chapter 7.

16.6.3 Principal Components

A ground station, which will consist of:

- A power source to activate the laser device, which could be a generator or a link to a power line;
- A laser source with beam-shaping optical components and interface to a telescope with focus control;

- A beam director with a precision target-tracking subsystem;
- A safety system to ensure that the laser beam engages the assigned targets and does not illuminate anything other than the intended aerial asset. Additionally, this subsystem will stop the transmission if an object enters the beam unexpectedly, a common problem shared with a laser-damage weapon, especially if it is to fire above the horizon.

The receiver will have an array of photo-voltaic detectors matched to the emission wavelength of the power beam, to convert the received photons back to electrical power. Moreover, as there is a likelihood of interruption of the power beam, there is a need for an electrical-storage system, which could provide auxiliary power during any short interruption in the supply of beamed power.

Specialised detector arrays need to be used in the receiver, which will support higher incident irradiances than the more conventional solar cells; the margin of the difference is about an order of magnitude. The higher irradiance, of the order of 10 W/cm^2, leads to efficiencies greater than 50% optical-to-electrical conversion.

16.6.4 Concept of Operation

A closed-loop tracking and pointing system is required that moves with the remote platform as it manoeuvres around the sky; hence the need for a precision target-tracking subsystem within the beam director. Additionally, if the source of optical power, or ground station, is moving such as on a naval ship, the sightline to the remote receiver will need to be stabilised to remove the sightline motion induced by the movement of the ground station.

An external cue may be required to enable the beam-director system to find, acquire and track the platform needing the transfer of power; moreover, an aim-point selection routine will be necessary to enable the beam director to illuminate the detector array. Accurate pointing and tracking maintenance is necessary to form an efficient power link and avoid the situation where only a small fraction of the optical power is centred on the detector array(s) on the remote platform. This is shown schematically in Figure 16.4.

Most aerial assets using this technique will be co-operative targets, so they are likely to have a beacon and possibly other tracking aids, such as satellite-navigation data (GPS) or autopilot information from the vehicle, linked directly to the tracking system. Hence, the pointing and tracking task should be straightforward, and highly accurate tracking should be feasible from a relatively simple system.

The optical requirements of the entire power-beaming system are also quite modest, owing to the range of operations involved. However, the entire optical system must have a suitable design margin to be able to handle the anticipated maximum irradiance at the transmitter as well as at the receiver, particularly taking account of the variability in range and the scintillation effects induced by the atmosphere.

It may be possible to use the power beam as part of a data link, as described in Chapter 14 and, in particular, for retro-communications. Other secondary uses of the laser beam may be possible.

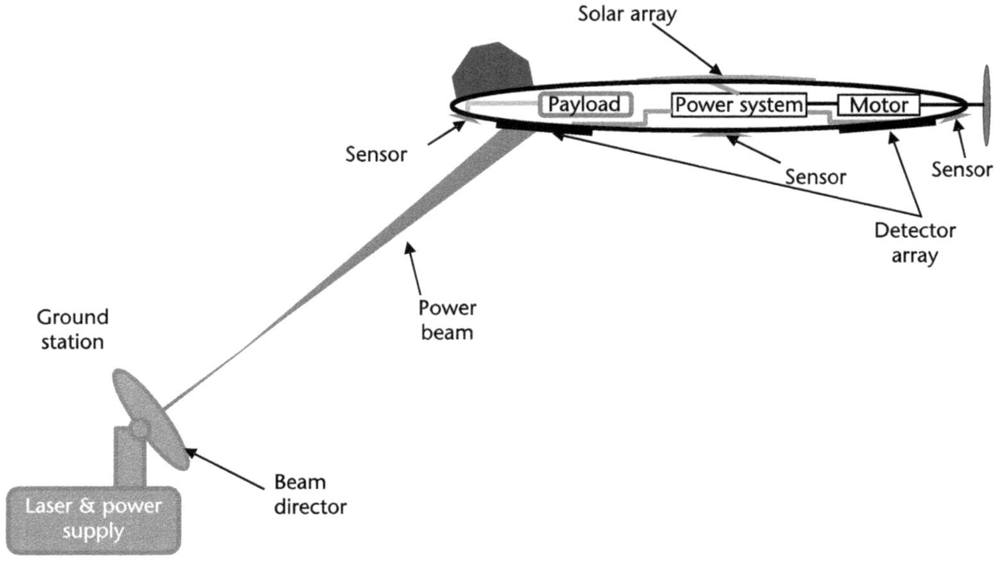

Figure 16.4 A schematic of a wireless power-beaming system.

16.6.5 Laser Requirements

The fundamental requirement is for a very efficient laser source emitting a wavelength that has very good propagation characteristics and well matched to the peak absorption band of the solar panel/detector array. Emission in the short-wave infrared could be an advantage for minimising the size as well as optimising the performance of the optical system, and in particular for minimising the potential optical hazards (i.e., the eye-safety considerations). Additionally, the ability to focus the beam on to the detector array would also enhance the efficiency of the energy transfer of the whole system, but could be more demanding on the performance requirements of the pointing and tracking system, especially for long-range, power-beaming activities.

Two laser technologies are currently considered appropriate for this type of energy transfer:

- Stacks of laser-diode arrays: multiple kilowatt-class devices are available with moderate to high efficiency (> 50%), but at the expense of beam shaping optical elements needed to achieve a well-collimated beam;
- Fibre lasers: essentially brightness convertors, so whilst the wall-plug efficiency may be less than that of a diode stack, the overall transfer of power may be greater. Additionally, the optical configuration may be simpler.

Many of the anticipated applications of this technique have an installed power requirement of the order of a kilowatt; therefore, as the efficiency of the capture-and-conversion process in the remote target could be as lower than 50%, the launched laser power requirement will be in excess of 3 kW, taking into account propagation losses. Currently, laser-diode arrays are capable of this level of performance, as are single-mode fibre lasers operated in a master-oscillator power-amplifier configuration. Currently, the power limit for this type of fibre laser is of the order of 10 kW.

One of the challenges is to match the beam profile to the shape of the detector array; generally laser beams have a circular profile with a Gaussian-intensity distribution about the propagation axis, whereas the detector arrays tend to be square and designed to have a uniform response to the incident light. Moreover, the detector arrays tend to have a filling fraction that is somewhat less than 100%; enhancing the filling fraction will improve the optical to electric conversion efficiency, as well as shaping the intensity profile of the power laser beam.

16.6.6 Summary

Infrared laser beams have many characteristics that make them ideal for providing power to remote locations without the need for electrical conductors. The original concept involving laser beams was to provide power to spacecraft when they were not receiving sunlight; more recent applications have involved powering unmanned aerial vehicles. Laser-based, power-beaming techniques have the potential to expand the capability of unmanned aerial vehicles dramatically, especially the very lightweight systems.

Several concepts of operation have been investigated for tactical applications to increase the endurance of small and miniature unmanned aircraft, as well as lighter than air platforms (balloons), by directing a laser beam on to a specifically designed detector array. One of the concepts of operation is to enable the air platform to maintain its task indefinitely (i.e., endless operation). The other principal application is the recharging of unmanned platforms undertaking long-range missions or multimission tasks.

This technique of power beaming has become feasible, as a result of improvements in efficiency of compact laser sources and detector technology. Special detector arrays have been developed to withstand higher incident irradiances compared with standard solar cells. It is possible that power beaming with these photo-voltaic arrays could exceed 50% in terms of optical-to-electrical conversion, with an overall optical-to-electrical efficiency approaching 30% to 35% for the complete system.

The candidate laser technologies are laser-diode arrays and fibre lasers. Both types of laser technology are capable of generating the kilowatt-class beams needed for this type of application with a tightly collimated beam.

The small and miniature unmanned aerial vehicles can undertake a number of situational-awareness functions, such as intelligence, surveillance and reconnaissance tasks, either close to the operating base or at distant places of interest. This type of optical power-beaming configuration could also enable other functions to be undertaken, such as the formation of data links during the power-beaming activities.

16.7 Laser Propulsion

An alternative approach to power beaming is to use a laser beam to drive a laser-based turbo jet at a remote location in a small aircraft. In this case, the photon energy in the beam is used to heat the air, which creates thrust. However, this novel use of the power in a laser beam is at a very early stage of development.

16.8 Optical Air-Data Systems

This is a system being developed to replace the pitot tubes used on aircraft and other platforms to determine air speed and potentially other parameters, such as altitude. The system exploits the vibrometer technique described in Chapter 13.

The use of pitot tubes is very well established for measuring air-flow characteristics but has a number of maintenance issues. The need for regular inspection and repair is a result of their protrusion ahead of the fuselage, which can result in frequent mechanical damage; this mechanical configuration of the pitot tube is needed to permit sampling of the undisturbed air. Another technical challenge concerns their radar cross-section, particularly for low-observable applications for both aircraft and unmanned air vehicles.

A number of techniques have been investigated to provide the requisite air-speed data; all are based on the measurement of the Doppler shift induced into the frequency of the backscattered transmission by the particles, aerosols or gasses in the air mass. The fundamental requirement for the laser is a single-frequency emission to enable the Doppler shift to be detected through a suitable processing technique. Infrared wavelengths tend to have greater utility at lower altitudes, whereas ultraviolet emissions are effective at higher altitudes, or indeed, over a greater range of altitudes. The measurement range is typically of the order of a few tens of metres. A schematic of a system is shown in Figure 16.5.

Infrared systems operate on the backscatter from dust particles and aerosols in the atmosphere, but the concentration of the photon scattering particles tends to decrease with altitude, which leads to decreasing signal-to-noise characteristics as the altitude of the platform increases. The technique based on ultraviolet wavelengths relies on the phenomenon of Rayleigh scattering, which tends to provide more consistent measurements at all altitudes, providing scattering molecules exist in a suitable concentration. The Doppler shift for a 355-nm emission is about 6 MHz for each metre-per-second of motion. Moreover, air pressure may be esti-

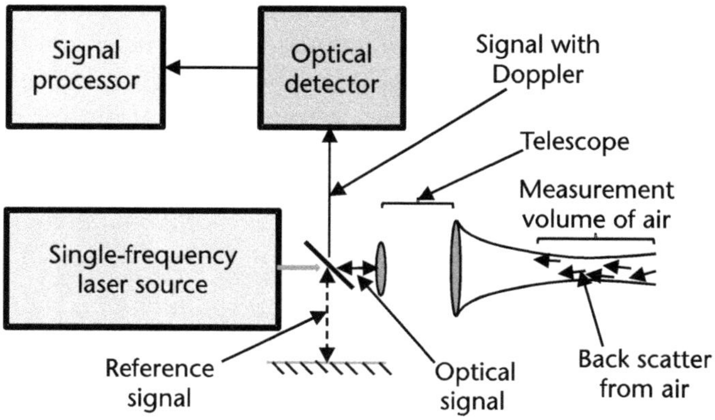

Figure 16.5 Schematic of an optical air-data system.

mated from received signal strength of the back-scattered signal. An example of a shifted spectrum is shown in Figure 16.6.

Alternative techniques are being investigated using:

- *Multimode laser emission and a Doppler filter:* multimode lasers tend to be smaller and less complex than the single-mode devices. The challenge here is to match the cavity length to the Doppler filter using an etalon.
- *Rotational-Raman scattering:* this method exploits the frequency shift in the back-scatter radiation resulting from interaction with the rotational and vibrational states of the air molecules in the atmosphere. Two interference filters are used to sample the induced frequency shifts in the received light.
- *Brillouin scattering:* the induced frequency shift in the back-scattered radiation is a function of the speed of sound, which in turn is dependent on air temperature. This is a technique that is at the research stage of investigation.

16.8.1 Laser Requirements

This is a low-power application, so the emitted optical power is of the order of a watt. Clearly, a line-narrowed output is required for some applications that attempt direct measurement of the Doppler shift.

16.9 Guided Bullets

This application is a variation of the laser-guided bomb, described in Chapter 12. A bullet has many similarities with a laser-guided bomb, so that it can home on to radiation scattered by a laser-illuminated target given a suitable sensing system.

In this case, a somewhat sophisticated bullet is launched from a smooth-bore barrel; the guided bullet has a sensor in the nose to detect the laser energy scattered from the designated target. The control fins are protected during the launch by a sabot, which is discarded as the smart bullet leaves the gun barrel.

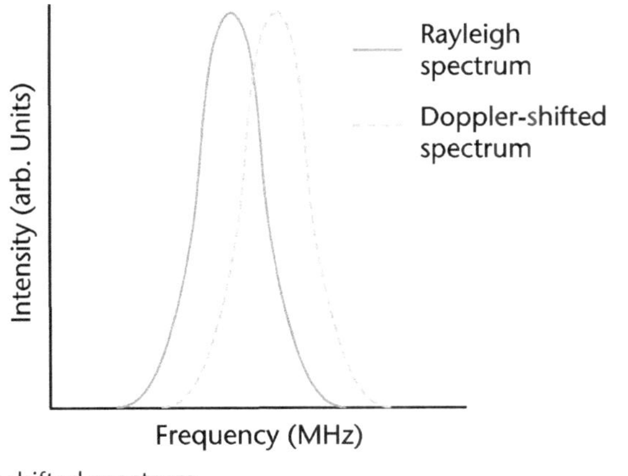

Figure 16.6 Doppler-shifted spectrum.

The moment of inertia of the bullet tends to be far less than that of guided bomb, so a guided bullet can be very agile (i.e., have a far smaller time constant in the control system). This type of responsiveness enables the trajectory to be corrected many times per second (typically tens of hertz), so the incremental corrections do not need to be very accurate to achieve high hit-point accuracy.

16.9.1 Military Utility

The military value of this application is very similar to that for the guided bomb (i.e., to improve kill-rate effectiveness and thus make better use of ordnance). In this case, there is potentially the prospect of reducing the load that that a dismounted soldier has to carry during combat, or increase the effectiveness of that soldier with the same load through enhanced fire power.

16.9.2 Laser Requirements

The anticipated range is a kilometre or so; hence a modest-power laser designator will be required, possibly in the watt class, possibily operating in the eye-safe region. It may be possible to use a well-collimated light-emitting diode to illuminate a target. Critical performance requirements include efficiency and robustness as well as size and weight.

16.10 Mine Detection

The detection and location of submerged antiship mines continues to be a high-priority requirement for navies throughout the world. Laser-radar systems have been developed using emission in the blue-green part of the spectrum to image swaths of the ocean to enable anomalies, induced by submerged objects, such as mines, to be detected.

A pulsed-laser system is used to probe the ocean with a fan-shaped beam, and in conjunction with the forward motion of the aircraft (usually a helicopter), thus avoiding the need for complex scanning equipment. The pulse rate is usually of the order of 100 Hz, but is determined by factors such as the forward speed of the aircraft and the desired resolution needed to locate a submerged mine.

16.10.1 Military Utility

Rapid and accurate all-weather methods for detecting hazards to shipping continues to be a very high-priority requirement. The use of laser-radar techniques avoids the need to trail equipment in the water behind a platform or to recover previously deployed equipment.

The use of an active system enables it to be used during the day or at night and the use of blue-green wavelengths permit all-weather operation.

16.10.2 Laser Requirements

These are very similar to the bathymetry requirements described in Chapter 13. The efficient generation of a laser line in the region of 540 nm is crucial for achieving the desired penetration of the ocean to the requisite depth. Pulse-repetition rates of the Q-switched pulses are required of the order of 100 to 500 Hz containing pulse energy in the region of 125 mJ, giving an average power of the order of 50W for a typical system.

The laser return from the ocean is detected using an array of streak tubes. The data collected by the receiver system are processed using automatic target-recognition algorithms to provide a rapid indication of anomalous returns from the searched area. A schematic of the operation of such a system is shown in Figure 16.7.

16.11 Laser Hydrophone

Laser-based devices for underwater sensing have already been discussed in Chapter 13, in the section on bathymetry. Investigations have also been undertaken to demonstrate passive-underwater sensing, using the hydrophone principle. It has been envisaged such devices could be integrated to form a towed sonar-array.

A number of techniques have been considered using optical fibres and fibre lasers. Optical fibres are known to be sensitive to changes in the bend radius from ambient pressure changes; ironically, a phenomenon that dogged the early development of fibre-optical gyroscopes. The sensitivity of a passive system depends on:

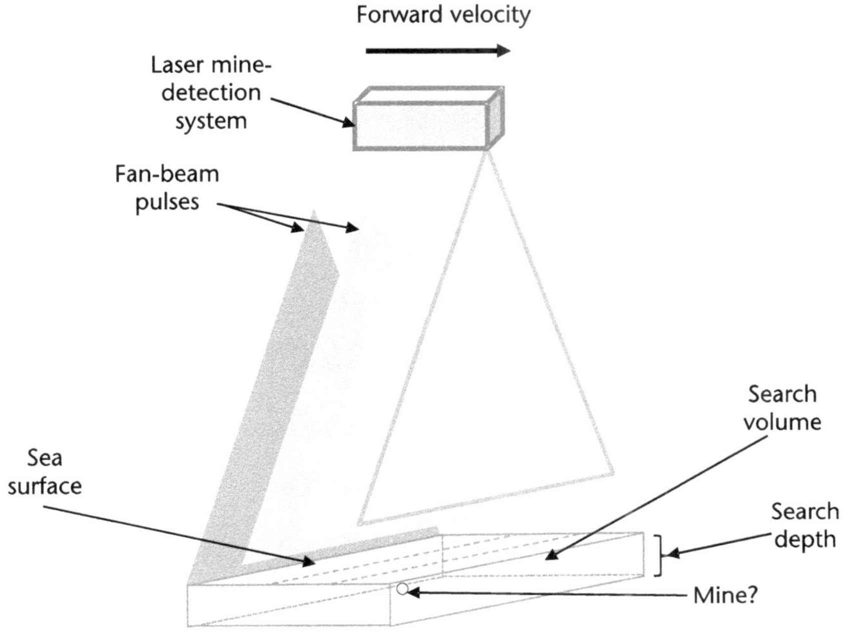

Figure 16.7 Schematic operation of a mine-detection system.

- Young's modulus and Poisson's ratio;
- Length of the sensing fibre;
- Size and shape of the encapsulating structure and material;
- Fibre-stress optical coefficient.

To achieve adequate sensitivity in a compact sensor, it is necessary to coil the fibres, in a fashion similar to the technique used in the inertial sensors (Chapter 15), and moreover, these devices are operated in a similar fashion. A major constraint imposed on this approach is the minimum bend radius to which the coil may be wound, usually of the order of 40 or 50 mm; hence, a streamlined array may not be feasible.

A linear type of array of sensors may be realised using an active fibre-laser arrangement. In this case, a number of fibre-laser cavities are arranged along an optical fibre and pumped by a narrow line width distributed-feedback laser. A schematic of this form of configuration is shown in Figure 16.8.

Some of the potential advantages of laser-fibre hydrophones are:

- Compact sensor configuration;
- Ease of fabrication and low cost;
- Low noise characteristics;
- High signal-to-noise ratio;
- Dense multiplexing capability;
- Good frequency response;
- Wide dynamic range;
- Immunity to electromagnetic interference.

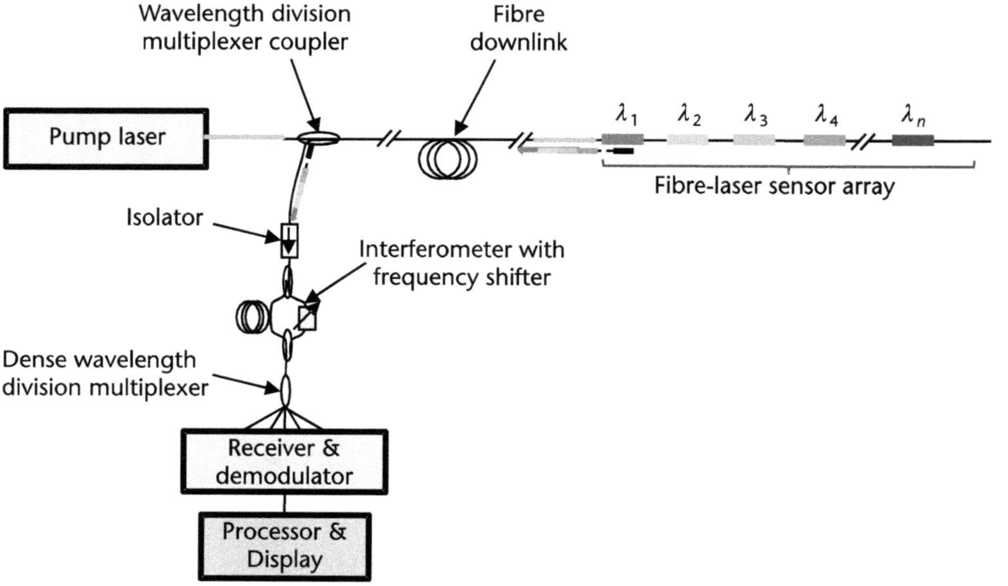

Figure 16.8 Fibre-laser hydrophone.

16.11.1 Military Utility

Passive sensing of acoustic fields, through the detection of pressure waves in the water, is an important aspect of the underwater surveillance role of many naval platforms.

16.11.2 Principle of Operation

The acoustic pressure waves induce stresses and strains in the optical fibre in the sensor array, both along and orthogonal to the fibre longitudinal-axis, which leads to a shift in the lasing wavelength (or frequency). The use of a narrow line width source enables very small changes in the lasing frequency, of the order of 1 Hz to be detected when the laser emission is analysed with an appropriate interferometer.

These fibre-laser devices may be designed to operate as frequency-modulated sensors or as intensity-modulated devices. Frequency-modulation techniques offer very high sensitivity suitable for high-performance applications, but demand careful demodulation techniques and are somewhat delicate, whereas sensors based on intensity-modulation techniques are simpler, more robust and reliable but offer lower sensitivity than the frequency-modulation devices.

An alternative device based on a laser-fibre accelerometer has been investigated. This sensor has a seismic mass fixed between two diaphragms with a fibre laser attached to the shell of the sensor and the seismic mass. The sensor is perforated to allow a flow of water into and out of the sensor shell to compensate for static pressure changes in the water.

The principle of construction of potential sensors based on distributed-feedback fibre lasers, for sensing changes in the pressure field around a sensor, is shown in Figure 16.9.

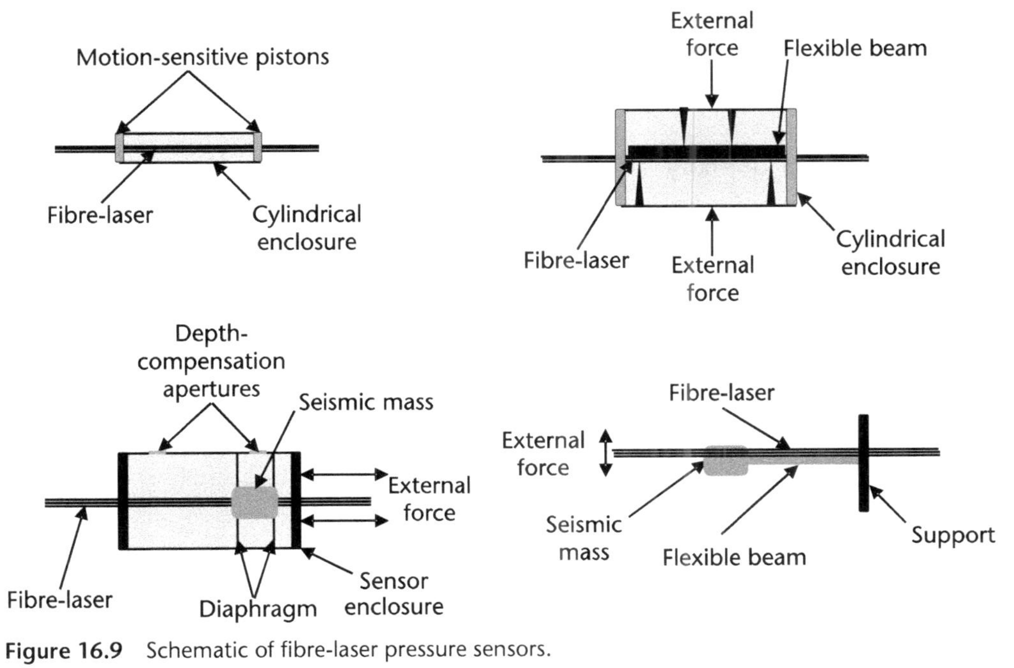

Figure 16.9 Schematic of fibre-laser pressure sensors.

When this type of sensor encounters pressure waves, the seismic mass responds and the fibre laser is either stretched or compressed resulting in a change in the operational frequency of the laser light. This type of hydrophone sensor can measure the particle velocity in the flow, its acceleration as well as the pressure of acoustic waves. A sensor based on a cantilever device has also been investigated for sensing pressure waves in the ocean.

16.11.3 Laser Requirements

Narrow-line width, distributed-feedback (DFB) fibre lasers are usually at the heart of sensors used for measuring acoustic fields, especially for underwater-surveillance and sensing applications. Strain in the fibre laser leads to a shift in the lasing frequency, as discussed above. Even very small changes in the optical frequency, down to a 1-Hz scale, can be detected with an interferometer, owing to the narrow line width of the light in this emission. This potential scale of resolution makes it possible to demodulate weak sound fields from the received signals. The laser-performance characteristics needed to support hydrophone applications are shown in Table 16.1.

A typical source is an erbium-doped fibre because of its relatively broad gain bandwidth; hence, individual lasers within the array can oscillate at an appropriate wavelength determined by the stress or strain on the gain medium within the sensor array. The distributed-feedback configuration enables longitudinal single-mode and single-frequency operation in the fibre to be realised.

The critical laser characteristics may be summarised as:

- Emission line width;
- Coherence length;
- Polarisation of the beam (single polarisation);
- Emitted power and sensitivity of the detection system.

16.11.4 Summary

The use of distributed-feedback laser technology in a hydrophone array enables a compact, rugged and efficient technique to be applied to the detection and characterisation of pressure waves within the ocean.

Table 16.1 Summary of the Laser Requirements for Hydrophone Applications

Source Parameter	*Value*
Wavelength (nm)	1,535–1,585
Average power (mW)	0.1
Divergence	Not applicable
PRF	CW
Line width (kHz)	<0.1

16.12 Summary of Novel Applications

Laser technology continues to evolve and the range of application for laser devices also continues to expand. In this chapter, some of the less high-profile applications have been considered, from countering a laser-guidance system, illumination in various forms to aid training or sustain operation of a platform, which again are illumination tasks and systems for enhancing situational awareness.

Many of these potential applications are aimed at requirements identified to fulfil gaps in current military capability or to make an existing capability more universally available.

As with the vast majority of, if not all, military applications of lasers, size weight and power are crucial considerations. Hence, the common requirements for higher efficiency as well as greater ruggedness continue to dominate the future requirements for these sources.

Part III
Safe Use, Education and Training

CHAPTER 17

Safe Use of Lasers and Legislation

17.1 Introduction

Laser safety has been a critically important, but controversial subject for decades, ever since the realisation that laser beams can contain a huge amount of power or energy, well collimated and focused to a small spot with huge irradiances.[1] Laser safety, in the opinion of the author, should not be viewed or implemented through the slavish implementation of unrealistic rules designed to prevent the effective operation of a laser device or laser-based system. The safe use and operation of lasers are really about the systematic and rigorous application of suitable and appropriate safety measures and procedures to understand then contain and mitigate potential hazards, which may arise during the use of these devices and systems. Laser safety should be seen as an integral part of any safe operating procedure. As laser technology has developed, so has the understanding of the potential hazards along with a similar evolution of the appropriate precautions.

The major emission properties and characteristics of a laser source that give rise to the need for special care when operating some of these optical sources may be summarised as follows:

- Highly collimated beam, which is highly localised in a pencil beam but with the potential for high irradiance at significant distances from the laser-beam-transmission aperture;
- The ability to generate a beam of pulses of very short duration, potentially resulting in damage over an equally short time scale;
- Monochromatic emission and, consequently, selective interaction with human tissue, which is highly dependent on emission wavelength;
- Military systems may have an irregular beam profile (non-Gaussian), which means that the beam may contain hot spots where there is an increased localised hazard.

1. In the visible band these beams would appear to be very bright, but in other wavebands they will be invisible.

The first and third characteristics mean that there is the possibility that these beams can be focused to a very small spot; if the emission has a wavelength in the visible band (or specifically between 400 nm and 1.4 μm), then the light beam can be focussed to a very small spot on the retina if the beam is incident on a human eye. In this case, even with modest incident optical powers in the laser beam, the irradiance (i.e., power density) at the retina can be huge, owing to the minute spot size. There are additional considerations if the beam is pulsed or if the wavelengths are outside the visible band, as the human aversion response will not provide any protection in these circumstances. Consequently, there have been extensive investigations of the physical issues associated with optical hazards and how to protect the human eye, so that these devices can be operated and used safely.

Any exposure of humans to high-power optical beams can present a number of other health and safety issues, whether the light source is coherent or otherwise. Some of these hazards (such as electric shocks) are associated with malfunction or irregular operation of some classes of optical devices, particularly electrical hazards, and especially, the potential to cause damage to the human body or threaten life itself. Less attention is generally given to these issues, especially the dangers of electrocution; it is likely that all of the reported fatalities associated with laser systems and other optical devices, have arisen from accidents linked with the power supply units and particularly electrocution. Fortunately, technology has progressed dramatically, so many of the modern laser devices use laser-diode technology and the need for very high currents and equally large voltages are far less common. Modern laser operations means that safety has to focus less on preventing fatalities, through electrocution, for example, and more with preservation of good vision and avoidance of injuries.

It is important to remember that the laser eye-protection equipment [goggles or personal eye protection also commonly called person-protection equipment (PPE)] is not the primary safety barrier against eye injury from a laser beam; laser-safety schemes should be designed so that eye-protection equipment is the last line of defence. The operational or experimental test and evaluation process should be designed to contain the potentially hazardous beams and make use of suitable beam stops, as well as shutters, where appropriate. In the case of evaluation testing, remote control should be considered and implemented where practical, particularly if high-power beams are involved.

Many of the techniques devised for safe operation of the laser devices have been designed for the use of lasers in a confined or contained space, such as a laboratory, or more recently, in workshops where cutting, welding or marking are the principal applications. Very different considerations are required when the laser system is to be operated in the open air on an evaluation range or battlefield. There are many critical considerations to be evaluated in these cases, such as:

- Effects of atmospheric propagation, particularly scintillation;
- Stray-light effects, including specular reflection from various surfaces, as well as wet targets;
- Pointing at, or targeting, only the tracked or designated object;

- Loss of beam control through the malfunction of a system/subsystem;
- Trespassers gaining unauthorised access to the range;
- Unauthorised viewing of a laser test by anyone on the range;
- Irradiation of unintended personnel on a battlefield.

A fundamental approach of safety management is to eliminate all hazards, a philosophy followed by some nations, but not all; however, sometimes this initial approach is neither feasible nor possible. In these cases, processes have to be devised to reduce the probability of occurrence of any hazard to an acceptably low level, so that the risk of injury to personnel is negligible. Success relies on careful consideration and accurate assessment of the potential hazards including the risk presented, particularly in terms of the likelihood of occurrence of each hazard and the severity of the impact, should it occur.

Appropriate test processes have been designed and matched by careful implementation of risk-mitigation measures and methods, to control the identified hazards even involving high-power (i.e., Class 4) devices. Safe operation of even very high power devices is possible and has been achieved multiple times.

Typically, safe operation of all types of laser-based systems in the field has been achieved through a combination of either ensuring that the accessible laser emission does not exceed the defined safety limit known as the maximum permissible exposure (MPE) or when there is a possibility that the MPE is exceeded, implementing specific and stringent safety controls to minimise the risk. Some examples of these controls for mitigation of an unavoidable risk are:

- A carefully designed set of processes for undertaking the testing, which includes the identification of appropriate safety equipment, support equipment, check lists and a prescribed count-down process that allows interruption for aborting a firing for whatever reason;
- Identification and observance of the relevant procedures for operational firing of the system;
- Training of operators to understand the relevant health and safety legislation, the limitations of the equipment, the limitations of the measures used to mitigate the identified hazards, but probably most critically, knowing at all times during the operation of a laser exactly where the beam is in its entirety, including specular reflections and scattered light;
- Adopting appropriate rules of engagement for safe operation in the chosen test arena;
- Use of appropriate safety techniques and equipment, particularly understanding their limitations;
- Refinement of engineering controls, to contain the primary beam and stray reflections;
- Provision of suitable medical support if such hazards are identified that require instant treatment.

There is a trend to encourage the use of those wavelengths that are principally absorbed by the cornea,[2] or other pre-retinal media, so that they cannot be focussed on to the retina. The common term for these wavelengths, of approximately 1.5 μm and longer is eye-safe wavelengths; however, it is important to realise that only Class 1 lasers are truly eye safe. The modern terminology is to refer to lasers emitting photons with a wavelength greater than 1.5 μm as generating eye-safer wavelengths, as the threshold for eye damage in this part of the spectrum is greater than the threshold for the visible band and near infrared, owing to the lack of transmission through the cornea. These eye-safer wavelengths are still capable of causing eye damage, or skin damage, if the incident fluence is above the damage thresholds for that part of the human anatomy.

The aim of this chapter is to describe the fundamental aspects of laser safety, including terminology, and indicate the practical considerations that are important for the safe operation of lasers and laser-based systems. The use of typical procedures that may reduce hazards during the development of the technology are listed, in addition to those associated with testing the performance of more mature devices to ensure they are achieving their operational performance, is considered in this chapter. The procedures that are outlined next are only suggestions and are offered as a starting point for developing local operating procedures, which will be suitable for the safe conduct of any specific test programme to evaluate a device or laser system.

An outline of, and reference to, the laser-safety standards in the United States and Europe is provided, along with a description of the common hazards, such as eye and skin damage. The legislation that governs the use of laser weapons, Protocol IV of the Vienna convention of the International Committee of the Red Cross, is also outlined briefly, as it is a common subject for heated discussion and misinterpretation; however, it must be stressed that this text does not attempt to provide any legal interpretation of Protocol IV.

17.2 Effective Laser Safety

The safe use of all types of laser devices is dependent on consideration of the following issues:

- Understanding and appreciating the potential hazards. As will be discussed, these hazards extend beyond the threat to human vision. It is important that all personnel involved with the use and operation of any device have a full comprehension of the range of hazards, the techniques and methods used to control these and how to mitigate potential effects, if not eliminate these entirely.
- Effective application of the identified control techniques and processes needed to eliminate these hazards as far as is practicable. In some circumstances, total elimination of all hazards will not be feasible but vigilance is

2. The 1.5-μm radiation is not completely absorbed by the cornea, just enough to mean retinal damage is minimised, by leading to a higher damage threshold.

important to ensure that the hazards are minimised to an acceptable level of probability.

- Appropriate and relevant laser-safety training.
- Vigilance by the operators and users of the devices to identify and recognise potential malfunctions of the laser technology and the safety equipment. Most accidental exposures to hazardous beams occur when there is a problem with either the normal operation of a device or the safety system.

17.2.1 The Laser-Safety Standards

The laser-safety standards cover two aspects of laser operations, embracing the safety requirements to be followed by manufacturers of laser products and by those people who use laser devices. These standards provide guidance on the hazard classification scheme for laser products and the maximum permissible exposure limits for the safe use of laser devices. In general, there are normative standards (i.e., standards that set mandatory constraints) for manufacturers of laser products, and advisory guidelines for those using those lasers. In terms of regulation, the emphasis is on the laser manufacturers making safe products and providing sufficient information to the user to ensure the product can be used safely.

An accurate and thorough understanding of the laser-safety standards that are valid in the country undertaking the testing, or use, of laser technology or systems has to be the starting point for any form of legal, as well as safe, operation of laser devices, including laser pointers. Even the common laser pointer can present a hazard if used inappropriately.

There are a number of laser-safety standards in force throughout the world; in the United States, the standards are defined by the American National Standards Institute (ANSI) and in the rest of the world outside the United States the authority is the International Electro-technical Commission (IEC). Laser users and operators should be aware that, in addition to these standards, there are mandatory regulations in force in some countries, whereas in other nations these are mere guidelines. Awareness of the current legislation is vital for legal operation of devices and systems. For example, in Europe, the Artificial Optical Radiation Directive (AORD) places obligations on employers not to expose their employees to laser radiation above the maximum permissible exposure values, which are tabulated in this document.

These laser-safety standards have many similarities and provide definitions and define processes to promote safe operation. There are defined standards for the safe use of lasers generally and then additional standards for the use of devices in particular environments. For example, during a field trial there may be:

- A standard for safe use of lasers providing the overarching information, such as general use of laser technology;
- A further applicable standard for the safe use of lasers outdoors;
- Local standards for use of lasers on a range or in a laboratory.

It is important to understand the hierarchy of these standards from a legal and legislative viewpoint, as well as the need to be conversant with all of the relevant information that will pertain to the proposed testing.

Despite the many similarities that these standards have, it is essential that operators abide by all of the safety standards that are in force in a particular country where the laser activity is undertaken. This enables the appropriate risk assessments and risk mitigation strategies to be applied, ensuring full compliance with the relevant and current health and safety legislation. When the safe-operating requirements of one standard, and its supporting documentation, are met, then those of another standard are usually achieved; however, a compatibility check is absolutely vital, and is discussed below in the section concerning host and visiting nations' responsibilities.

There has been extensive discussion and international collaboration between nations to understand the hazards and the methods for assessing the maximum permissible exposure.

Figure 17.1 provides a schematic of the interrelationship between international, continental (European) and national safety standards.

17.2.2 Other International Safety Standards

Many countries have national standards, which are based on the International Electro-technical Commission international standard or the American standard, so there is usually much similarity between definitions and requirements for safe operations. Examples of such countries following this route include many European nations plus Australia, Israel, Japan and South Africa. Other nations may have derivatives of these standards, so some care is required, and the exact basis of their standards should be investigated before any collaborative or contractual activities are undertaken.

The one nation that does appear to have an independent approach is the Netherlands. The Dutch Gezonheidsraad published a set of maximum permissible

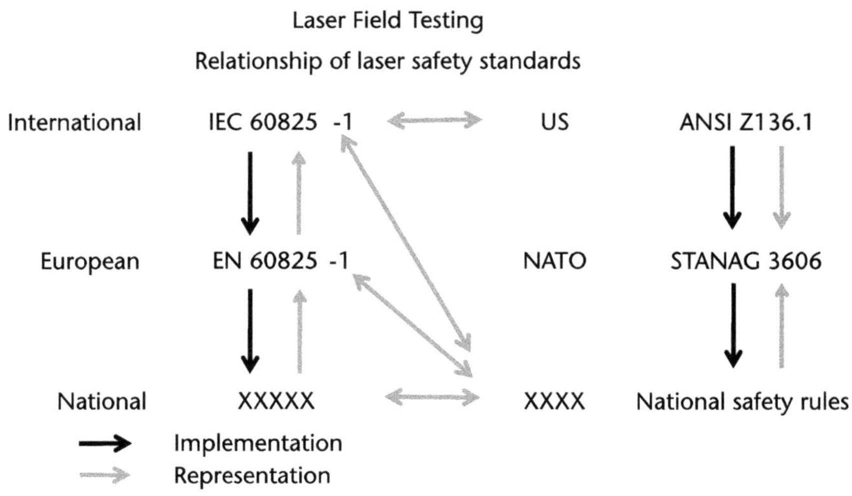

Figure 17.1 Relationship between laser-safety standards. (Courtesy of Kevin Burrett, Dstl, UK.)

exposures, which were derived separately from the International Electro-technical Commission or American National Standards Institute.

17.3 Hazard Classification of Lasers

Laser devices are classified to provide an instantaneous indication of the potential laser radiation hazard from the beam, and the control measures that are required to be implemented for safe operational use. There are four fundamental classifications of the sources based on the emission characteristics; the classes range from those with no known hazard to the most powerful that may present multiple hazards unless the emission is carefully controlled.

Lasers are classified on the basis of their accessible emission limit (AEL). For a laser device to be given a particular classification its emission must not exceed the AEL for that class, as defined in the appropriate laser standard. Class 4 does not have an upper limit; hence, some countermeasure lasers and laser-damage weapons, such as the airborne laser may all be Class 4, despite many orders of magnitude difference in power output. The difference centres on the distance the beam has to travel before its irradiance equates to the maximum permissible exposure limit; it may be measured in metres for one and in hundreds of miles for the other.

The usual classifications in ascending order of laser hazard (including the subdivisions of the four main classes) defined under IEC 60825-1 (of 2007) are:

- Class 1: these lasers are safe under all reasonably foreseeable conditions, because the AEL is believed to always be less than that which can cause eye damage, or to contain a laser that could be hazardous but is prevented from being so, through use of engineering controls and interlocks.
- Class 1M: these lasers are safe under all reasonably foreseeable conditions, except when any kind of focussing or collimating optics are used, so this class of laser can present a potential hazard if viewed directly with optical aids, such as binoculars.
- Class 2: these lasers are limited in AEL to 1 mW of output power, must emit laser radiation only in the waveband from 400 to 700 nm and will be safe for accidental exposure, owing to the natural aversion response of the eye and so generally present a very low hazard.
- Class 2M: these lasers are also limited to an AEL of 1 mW and a wavelength of 400 to 700 nm, these lasers are safe for accidental viewing as per Class 2 lasers, but are not safe if magnifying or collimating optics are used to observe the beam directly.
- Class 3R (was Class 3A): these lasers have a continuous-wave AEL limit of 5 mW (five times that of Class 2) and may be hazardous if viewed directly by the human eye. This class of laser presents a low to medium hazard requiring user awareness of the hazards presented by even incidental exposure when using magnifying optical systems; again, the natural aversion response provides protection in the visible band.

- Class 3B: these lasers have a continuous-wave AEL of 500 mW (medium-power laser devices) and can cause eye damage from both direct and reflected beams, in addition, the beam from a Class 3B device has the potential to cause skin burns. It is generally considered that these devices are likely to be safe only when viewing diffuse reflections. Operation of these devices is restricted to trained and qualified personnel.
- Class 4: this is the class of higher-power lasers and covers any laser with an output in excess of 0.5W; there is no upper limit for AEL in this classification. Class 4 lasers can cause eye and skin injuries from reflected beams even if the reflection is not particularly specular (mirror like) and may start fires or cause explosives to deflagrate or detonate.

The beams from Class 4 lasers may present multiple hazards if not properly controlled. As a general rule, the emission from these devices is not safe until the beam has propagated some distance from the aperture (i.e., the radiant intensity in any part of the beam does not exceed the maximum permissible exposure for that type of device). These devices are capable of presenting particular dangers to humans within the hazard zone/distance, for example:

- Eye hazard from photons entering the eye either directly or indirectly, even if a diffusely reflected beam is viewed (until the irradiance at the eye is below the MPE);
- Burning of exposed skin from direct or indirect exposure;
- Fire hazard from rapid absorption of photons and potential generation of toxic fumes.

Again, the use of this class of device is restricted to trained and qualified personnel, to minimise hazardous events occurring.

Special signs are usually required (mandatory) to be displayed at entry and exit points to laboratories and ranges when Class 3B and Class 4 lasers are being operated, which list the potential hazards. These signs should be brief, as shown in Figure 17.2, but indicate clearly that beyond that point, access is restricted to

Figure 17.2 Laser-warning sign.

authorised personnel following agreed procedures or using appropriate personal protection equipment. This restriction is important, because there is a hazard from exposure to the laser light of anyone not qualified/trained to work with lasers. It is normal to control access physically and/or monitor all of the access points with closed-circuit television. Many proprietary laser-safety systems shut off the beam as soon as an attempt is made to enter the restricted area in an uncontrolled manner or by unauthorised personnel.

It is the operation of Class 3B and Class 4 lasers that demand the most rigorous attention to safe operation Good safety-management practice should be applied to all laser operations, whatever the classification.

17.4 Maximum Permissible Exposure

This term expresses the irradiance (W/cm^2 for continuous wave beams or J/cm^2 for pulsed beams); the IEC definition uses W/m^2 and J/cm^2, in the optical beam that a person can be exposed to without known hazardous effects or adverse biological changes to their eyes or skin. The maximum permissible exposure (MPE) is based on the exposure duration in seconds for a specific wavelength in the range 180 to 1 million nm and other modifying factors, such as pulse duration and pulse-repetition frequency for pulsed lasers. The MPE is defined within the laser-standard documents.

Examples of accepted maximum permissible exposure values for inter-ocular viewing are:

- 1 J/cm^2 for an eye-safe pulsed device (1,540 nm);
- 2.54558 mW/cm^2 (25.4558 W/m^2) for a continuous-wave beam (0.25-second exposure duration, owing to the human-aversion response), for emission in the wavelength range from 400 to 700 nm, which equates 1 mW into a 7-mm pupil.

Values of MPE limits have been determined by experts in a range of specialist disciplines and changes to accepted values involve a lengthy and detailed process to ensure the appropriate scrutiny is applied. There has been some recent discussion concerning a revision of the currently accepted values, so these should be checked with the authorities controlling the use of laser devices and systems before use.

17.4.1 Hazard Distances

The calculation of the extent in space of the hazard zones around the aperture of a laser is the starting point for all laser-safety and personnel-protection considerations and, in particular, what safety measures will be required to ensure safe operation anywhere. The basis of these calculations is the evaluation of the irradiance of the beam from Class 3B and Class 4 lasers and establishing where this value of irradiance does not exceed the maximum permissible exposure for that wavelength and beam-emission characteristics.

There are a number of hazard zones that may be defined around a laser-emission aperture and the beam. These zones are related to the perceived hazard and the safety measures that are required to achieve an acceptable low risk to personnel.

17.4.2 Hazard Zones

The purpose of defining hazard zones is to provide an indication of the potential hazard to a human at particular distances from the laser aperture and beam-propagation direction. This assessment takes into account various physical phenomena such as reflection and scatter. The definition of hazard zones is associated with laser beams that are not enclosed and, in particular, on a range with free-space propagation.

The calculation of hazard distances is usually deterministic at first but it is necessary to consider the probability of a hazardous event occurring otherwise some of the hazard zones can cover enormous areas, making it impossible to undertake testing on any range other than the largest ones in the world. The probabilistic method does not underplay or underestimate the potential hazards, but takes into account directions where there is a very low probability of a hazard occurring, typically less than 10^{-8}.

There are two fundamental approaches that may be applied during the calculations to determine the physical extent of hazard zones, which are:

- *Deterministic assessment:* this is usually applicable to static engagements and involves exclusion zones calculated from the maximum permissible exposure and the nominal ocular hazard distance, assuming the maximum emission characteristics of the laser, given any failure of the device. This technique may also be applied to the calculation of the extended ocular hazard distance, which in some instances may lead to very large areas required to contain all possible events, however unlikely.
- *Probabilistic assessment:* this approach is usually applied to assessing the extended ocular hazard range for laser engagements of mobile targets, where the strict application of the MPE criterion is likely to lead to unrealistically large areas and consequential volumes for the extended exclusion zone to contain the laser-hazard footprint. The basis of this technique, designed not to compromise safety, is the determination of the distance beyond the range boundary at which a casual observer has an acceptably low probability of irradiation resulting in a minimal level of eye damage. This level of eye damage is defined by a minimally opthalmoscopically visible lesion, which results in a degree of damage that will not incur significant visual impairment through the creation of a retinal lesion.

Examples of some hazard zones are shown schematically in Figure 17.3 and the definition of some typical deterministic hazard zones and distances includes:

- *Nominal hazard zone (NHZ):* this term usually defines an exclusion zone around the laser aperture and possibly the target area including buffer zones,

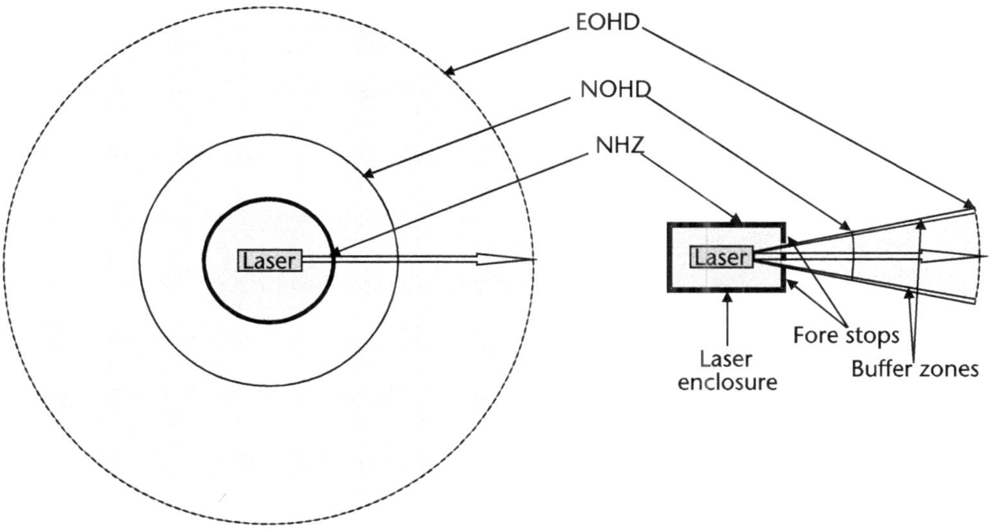

Figure 17.3 Control of hazard zones.

where direct exposure to the beam, as well as reflected or scattered radiation, would mean that the maximum permissible exposure will be exceeded. It is usually a distance expressed in metres and measures are taken, using physical barriers, to avoid accidental incursion into the zone. At distances beyond the defined nominal hazard zone, the exposure will not exceed the maximum permissible exposure.

- *Nominal ocular hazard distance (NOHD):* the distance from the laser where exposure to the beam (direct, reflected or scattered radiation) is not expected to exceed the maximum permissible exposure. It is usually defined as a contour around the test arena and laser, which is delineated by barriers. It should not be interpreted as a line where on one side it is safe to stand and watch, whereas a very short distance closer to the beam from that contour is hazardous action to view the beam. Safety factors are built into the calculations, but great care is required in the interpretation of the contour and how personnel can operate safely close to this boundary when necessary. Within the NOHD there is an increasing probability of ocular injury occurring with decreasing distance to the laser source, even when a deterministic laser-safety approach is being used, but that increase in probability of experiencing a ocular hazard is not a step function at the NOHD boundary.
- *Extended ocular hazard distance (EOHD):* this zone is defined to be beyond the NOHD and estimates a range (contour) where it would be safe to view a laser beam with an optical system with a given magnification without the observer exceeding the MPE limits. It is normal to assume a magnification of the optical system of times 10 (i.e., ×10). This zone is usually used to define the position within a controlled area, such as a field range, where the testing can take place and the extended ocular hazard distance contour can be contained within the range boundaries (i.e., so that accidental viewing by a casual observer using binoculars does not present potential hazard).

- *Laser controlled area:* an area around the laser device where personnel are subject to control to protect them from laser-radiation hazards. This zone could coincide with the nominal hazard zone, but more commonly provides a buffer zone around the nominal hazard zone.

17.4.2.1 Probabilistic Hazard Zone

This zone defines the space within which the level of exposure to the direct, reflected or scattered radiation may exceed the defined acceptable level of risk during the normal operation of the laser. The risk of encountering a hazard beyond this zone boundary is considered acceptable, despite being within a hazard zone. The probabilistic risk assessment is a method of laser-hazard evaluation based on the likelihood (i.e., probability of occurrence) of exposure and risk of injury. This technique is dependent on the establishment of an acceptable level of risk and also requires the evaluator to assume probabilities of occurrence of certain events. This form of risk assessment is used mainly when other methods cannot contain the entire laser hazard within the available range boundaries. A number of factors may affect the probabilistic hazard-zone assessment dependent on operational conditions, such as pointing accuracy, fault analysis, population density, reflections, presence of a backstop, and the physical extent of the backstop and operator reaction time.

Probabilistic risk analysis is not a get-out clause when the calculated deterministic hazard distances are problematical or impractical. This technique quantifies the risk of a hazardous exposure, by considering a range of complex factors that would make the implementation of deterministic calculations prohibitively difficult for most laser users, without any compromise in the level of safety.

17.4.3 Calculation of NOHD and EOHD

The nominal ocular hazard distance is defined as: "that distance along the axis of propagation from the laser aperture beyond which the intensity of the laser does not exceed the maximum permissible exposure (MPE) of the unaided human eye." The deterministic calculation assumes a Gaussian profile of the beam with the observer in the far field; moreover, no account is taken of atmospheric scintillation and other effects. The nominal ocular distance (NOHD) is calculated from:

$$NOHD = (1/\theta)\left[\{1.27Q/MPE\}^{1/2} - d\right] \tag{17.1}$$

where Q is the laser pulse energy or power, θ is the half-angle beam divergence and d is the emergent beam diameter at 1/e of the peak intensity.

When a beam is viewed with magnifying optical devices, the collecting power of the eye is increased, which leads to the idea of an extended ocular hazard distance (EOHD). There are three conditions to consider:

1. For a magnifying system, the increase in the hazard distance is modified typically by a factor of:

$$K = \tau \cdot M^2 \tag{17.2}$$

where τ is the transmittance of the optical aid and M is its magnifying power and not to be confused with the propagation factor M^2.

2. In the case where the size of the optical elements is known, this can be expressed in term of the size of the optical elements and then K is defined by:

$$K = \tau \cdot (\text{objective diameter}/\text{adapted pupil diameter})^2 \tag{17.3}$$

or

$$K = \tau \cdot (D_o/D_a)^2 \tag{17.4}$$

which ever K is lower.

The effect of the additional collecting power is to increase the nominal ocular hazard distance by a factor of $K^{1/2}$, hence the extended ocular hazard distance (EOHD) is given by:

$$EOHD = K^{1/2} NOHD$$

As discussed, the calculation is deterministic but it may be necessary to apply probabilistic techniques to contain the zone within available range boundaries given the appropriate risk assessment.

17.4.4 Laser Hazard Area Trace

Laser hazard area trace (LHAT) defines an area, where the risk of exposure would lead to an unacceptable chance of an eye injury occurring. This calculation may be undertaken as either a deterministic assessment or as a probabilistic evaluation. This assessment is often used to define a zone around a laser-based system operating in a military environment.

The United Kingdom often applies this technique for assessing the extent of the potential hazard zone around an installation on a platform by taking into account the probability of a given exposure event happening. Typically, the level of risk chosen to define the boundary of a laser hazard area trace is a probability of occurrence of 10^{-8} (i.e., 1 in 100 million) to give acceptably low probability of exposure per engagement, so the risk of ocular damage is acceptable in comparison with other comparable industrial injuries.

17.4.5 Engineering Controls

The linear extent of the hazard zones may be limited through suitable use of engineering controls, such as shielding. As an example, consider a nominal ocular hazard distance that would require a spherical exclusion zone around the laser aperture if the laser was free to move through 360°. This may cause some problems in finding a suitable range if the radial distance (nominal ocular hazard distance) is particularly large, perhaps hundreds of kilometres. As shown in Figure 17.3, use of a barrier or a shield can reduce this hazard zone to a relatively small arc, without increasing the probability of a hazardous event occurring. Similar approaches may be used to minimise the extended ocular hazard distance.

Buffer zones may be added to the arc defined by the fore stops to ensure a further margin of safety; the magnitude of the additional arcs is determined by the degree of control that is available over the pointing vector.

17.5 Laser-Safety Officer

The general duties of the laser-safety officer include:

- Giving advice on all aspects of laser safety to all personnel;
- Establishment of policies and procedures for the control of laser operations;
- Verification of the hazard assessment of all lasers and laser systems to be used;
- Ensuring that an accurate hazard evaluation of all laser devices has been undertaken for the environment of the testing;
- Ensuring that the appropriate control measures have been established and maintained (effective) throughout the defined test period;
- Approval of all applicable accident, incident, service, maintenance and operating procedures, prior to any testing;
- Advice on the appropriate personal-protection equipment for the proposed testing;
- Approval of the requisite signs and labels to be displayed;
- Maintenance of an up-to-date inventory of all laser devices at a given location, including those to be decommissioned;
- Ensuring that appropriate and relevant laser-safety training is provided to all personnel involved with laser testing and evaluation;
- Ensuring that appropriate and up-to-date documentation is maintained for all processes and procedures;
- Responsibility for investigation, documenting and reporting of all accidents and incidents of exposure to potentially harmful laser radiation through the appropriate management chain;
- Ensuring that all of the laser-safety features are appropriate and fully functional.

17.6 Visiting/Host Nation Interactions

An international protocol has been established concerning applicable legislation for operations by one nation (known as the visiting nation) electing to undertake testing on a range in a foreign country (host nation). It is the responsibility of the laser-safety officer of the visiting nation to ensure;

- Verification that the laser-hazards' control policies and procedures comply with the host nation's requirements;
- Provision to the host nation of the appropriate information concerning the safety and operational characteristics of all the laser devices and systems to be used during the visit;
- Provision of appropriate laser-safety training of personnel involved in the operation of the laser devices on the foreign territory and also ensuring that it is commensurate with all recognised hazards that may be encountered during normal operation of all the lasers involved during the proposed tests;
- Co-ordination with the host nation to determine the appropriate procedures for reporting accidents and incidents involving exposure to potentially harmful laser radiation, as well as agreeing what constitutes a reportable event and the required response.

The following are normally considered to be the responsibilities of the visiting nation:

- Liaise with the host nation to determine the information requirements and time constraints for the approval process to be achieved in time for the planned testing. It is usually appropriate to appoint a liaison officer to establish a single point of contact for the host nation.
- Ensure that the labels on the laser equipment are appropriate and accurate and conform to the host nation's regulations.
- Provide the appropriate protection equipment, as required.
- Provide the host nation with a list of all of the laser equipment to be used on their territory.
- Provide the host nation with the laser-safety information required to support their approval process to enable use of all of the nominated laser systems.
- Ensure that the visiting nation's regulations are adhered to, in addition to the host nation's regulations. When one nation's regulations are more stringent than those of the other nation, then the more stringent regulation normally takes priority.

In summary, the visiting users are responsible for adherence to all applicable standards, requirements and procedures relevant to laser use on the host nation's territory. If the host nation's requirements or processes are more lax than the visiting nation's rules, then the more stringent rules would expect to be applied by the visiting personnel, as noted above.

The host nation also has a number of responsibilities, such as:

- Acquiring the laser-hazard data of all laser devices and systems proposed to be used during the testing by a visiting nation;
- Provision of the range regulations and appropriate procedures for the safe use of lasers on the territory to the visiting nation;

- Provision of the appropriate safety briefing to the visiting nation's personnel, covering safety procedures, safe operating procedures, range operations and prohibited activities, including notification of out-bounds zones;
- Accurate identification of firing points, target zones, buffer angles and hazard zones;
- Provision of suitable warning signs.

17.7 Potential Hazards to Personnel

The emphasis on potential hazards is usually with respect to loss of visual acuity, or blindness. Threats to the human eye and other hazards, including skin effects and electrocution, which those personnel working with lasers may encounter are also covered below. Some of the well-established methods for controlling these hazards are discussed in the following sections, so that safe operation of a device is highly probable, but it cannot be emphasised too much that total elimination of all potential risks is often not feasible. The likelihood of occurrence of the residual risks has to be minimised to an acceptably low level of probability (typically with a probability of occurrence $< 10^{-8}$). This acceptably low-level of occurrence requires a degree of common sense to be applied because quantifying the exact risk is often extremely difficult, unless a formal probability risk assessment is undertaken for every eventuality.

The process of identifying hazards and evaluating the probability of occurrence, along with the impact (risk), has to be undertaken in a rigorous and systematic fashion. This process must consider the hazards and risks under all reasonably foreseeable conditions of operation, including all likely and potential failure mechanisms. Experience has shown that accidents usually occur when inadequate account has been taken of possible failure mechanisms or use of a laser device beyond its design limits.

The main biological damage mechanisms between a laser beam and a human tissue are:

- *Photo-thermal:* This effect is the most likely interaction involving acute injury, where some of the photon energy incident on the illuminated surface is absorbed by the underlying tissue. As a result, the temperature of the tissue rises and damage such as a burn may occur. This damage mechanism is power-dependent, so it is the rate of energy absorption that is important as well as being wavelength dependent, owing to varying amounts of water and pigments in the tissue of the human body. Damage also depends upon the beam diameter, because the central area of large spots on the skin or retina cannot cool as quickly as smaller spots. Irreversible damage to cells occurs when their temperature approaches 60°C; beyond this, there are increasingly severe effects until the tissue is vaporised.

- *Photo-acoustic:* These effects may occur when tissue is exposed to short duration (a few microseconds) high-energy pulses, resulting in significant amounts of energy being absorbed in a time scale that is short compared with the thermal-diffusion process. This process results in a rapid expansion of the tissue that forms a shock wave, which generates a mechanical reaction in the cellular structure. In this case, the damage mechanism is dependent on the incident energy.
- *Photo-chemical:* These effects occur from exposure to a laser beam where the incident flux characteristic may be insufficient to cause either of the thermally dependent effects discussed above, or where the exposure is predominately from shorter-wavelength photons (i.e., in the blue or ultraviolet regions of the spectrum). In this case, the effect is dependent on the absorbed cumulative dose, rather than the incident power or energy of the beam. The absorption of optical radiation by tissue can result in the modification of certain components in its structure, such as the tanning of skin that occurs after exposure to sunlight. This process is reversible over time if the exposure ceases, but repeated exposure can result in permanent damage. This process is photon-energy dependent, so the short wavelengths such as the ultraviolet, are potentially more damaging, whereas infrared photons are unlikely to cause photo-chemical changes to tissue. When the energy or power in the beam is sufficient to cause an effect, it will be one of the thermally-induced effects, as discussed above.

17.7.1 Eye Anatomy and Hazards

Most laser-safety training courses provide a brief overview of the anatomy of a human eye and the potential hazards presented by laser radiation. To reinforce the potential problems that may occur from such a safety lapse, a haemorrhagic lesion (vitreous haemorrhage) is shown in Figure 17.4. This image also shows other laser-bio effects, including the temporary effects of laser dazzle/laser glare.

The principal parts of the eye that are at risk from exposure to laser energy are the following:

- The cornea is the transparent front outer layer of the human eye, which is the main refracting element in the optical train of the eye and covers the iris and the crystalline lens. The cornea is the entry point for light into the eye. Its wavelength pass band is often considered to span the spectrum from 400 nm to about 1.4 μm, but it is more accurate to take the corneal transmission to start increasing from about 295 nm to the peak value and then ramps off again in the mid-wave infrared (2.5 to 3 μm).
- The retina is the main sensor of the eye that receives light from the cornea and sends its responses to the collected light to the brain, so that it can create an image of the captured scene. The retinal sensory tissue consists of rods and cones.

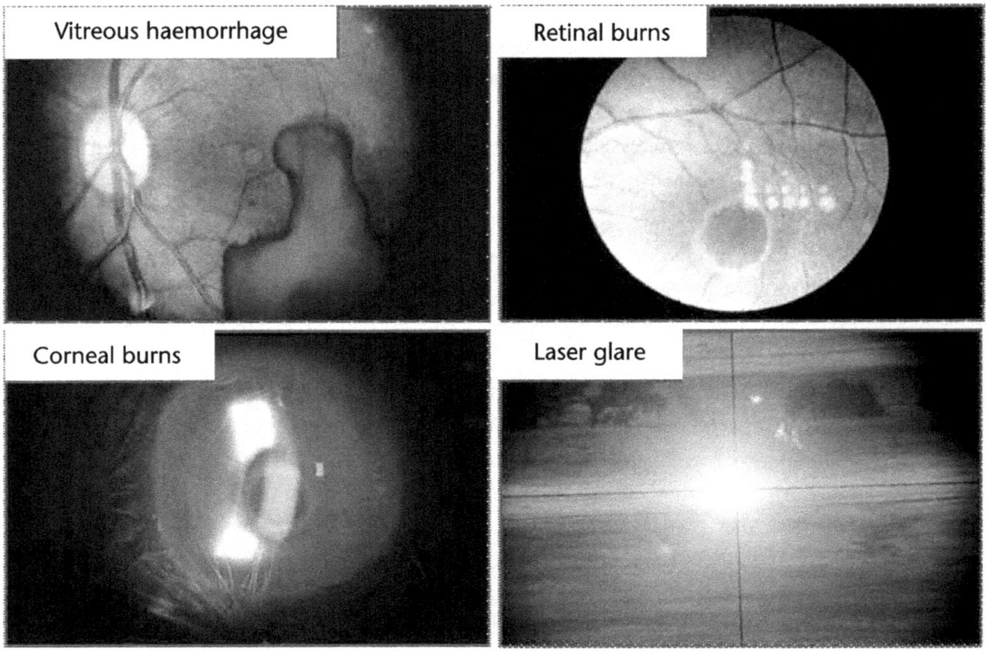

Figure 17.4 A laser-induced bio effects to an eye. (Courtesy of Mr. Bruce Stuck, Brookes AFB, USA.)

Figure 17.5 shows a simplified cross-section of a human eyeball and indicates the major components of the eye.

17.7.1.1 Hazards to the Human Eye

Hazards to the human eye come in several forms:

- Direct exposure results in corneal or retinal burns, depending in the wavelength of the incident light, as shown in Figure 17.4.
- Indirect exposure has similar results but with a greater likelihood of cumulative damage phenomena taking place including cataract formation, but indirect/diffusely reflected light is not the only cause of cataract formation.

Figure 17.6 shows the two fundamental forms of reflection that may give rise to accidental exposure to a photon beam. The specular reflection from a mirror is potentially most dangerous, owing to the point focus formed on the retina when the radiation is transparent to the optical train of the eye. The diffuse reflection appears as an extended secondary source at the reflecting surface, and so, is potentially less harmful, for a similar incident irradiance on the cornea in the incident beam.

These hazards may be mitigated, if not reduced to an infinitesimally small probability of occurrence through avoiding direct exposure to the beam and use of appropriate laser-safety eyewear [spectacles or goggles, now often called personal-protection equipment (PPE)], which blocks the emitted wavelength(s) from the device under investigation. This means knowing where the beam and its reflected

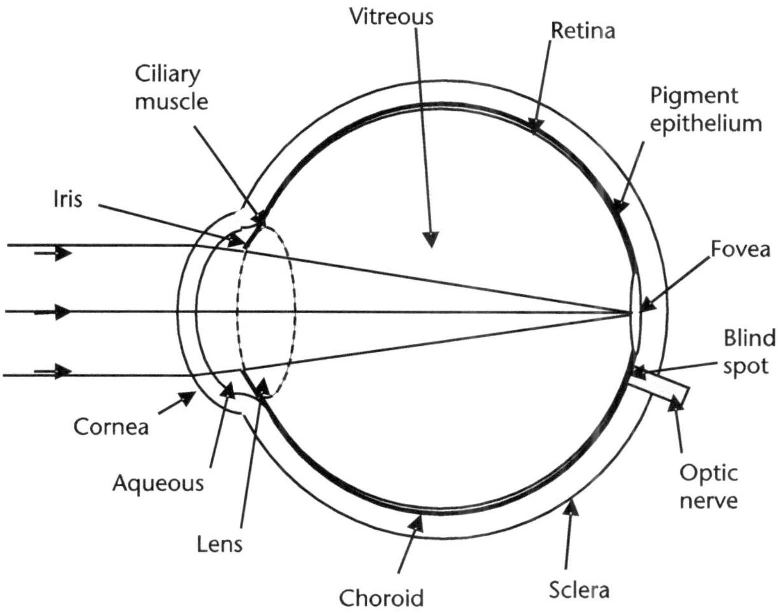

Figure 17.5 Schematic cross-section of a human eye.

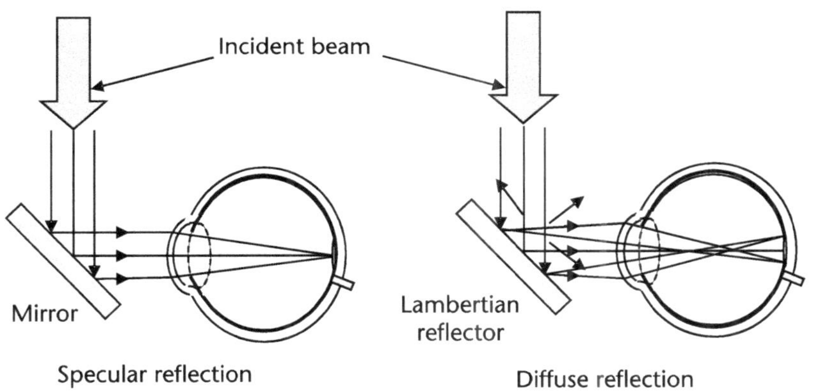

Figure 17.6 Reflections.

rays are, keeping one's head (or skin) out of the beam and wearing the correct eye protection if undertaking any operation within the nominal hazard zone.

The potential hazards presented to the human eye are dependent on the wavelength of the radiation, owing to the different optical transparencies of the various components of the eye, for example:

- Ultraviolet and infrared (beyond about 1.4 μm) radiation can damage the cornea when the maximum permissible exposure is exceeded for the incident light, because the light is absorbed by the material of the cornea in these wavebands.
- Visible-band light (from about 400 nm to about 1.4 μm, known as the VNIR) includes part of the near infrared and will damage the retina when the MPE

is exceeded, because the light is focused upon, and absorbed by, this part of the eye.

The transparency of the human eye is shown schematically as an approximate function of wavelength in Figure 17.7.

Many of the incidents of accidental exposure of eyes to laser light tend to occur during the process of alignment of a laser, the common causes include:

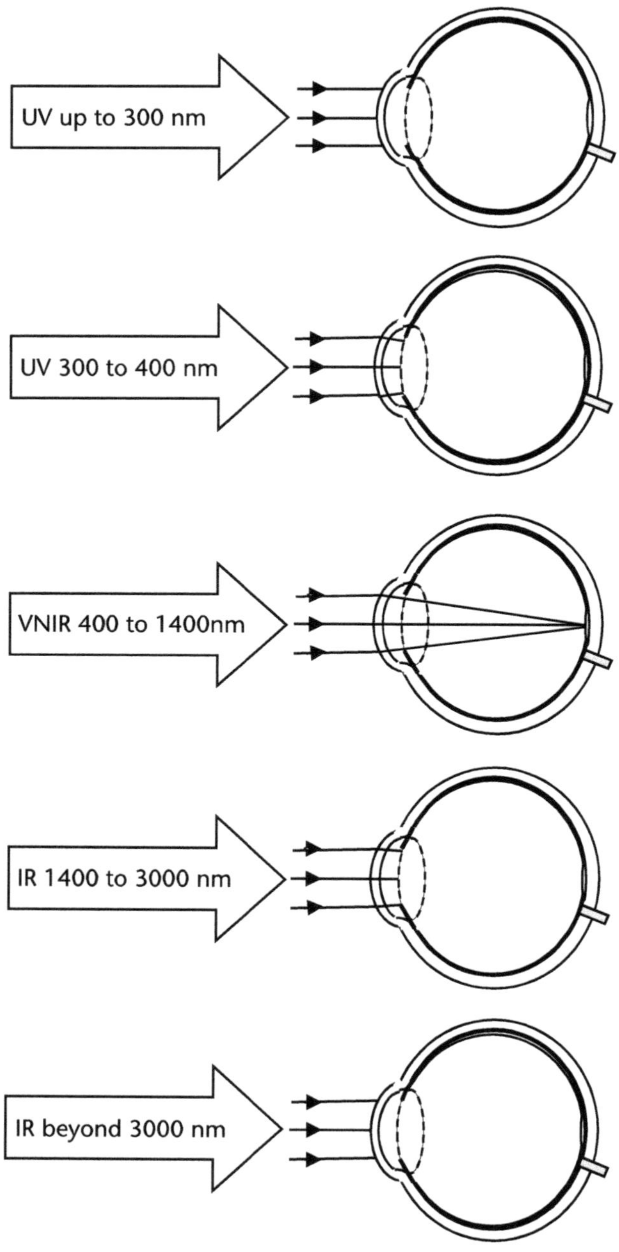

Figure 17.7 Schematic illustration of the transparency of the eye.

- Unanticipated eye exposure to the beam during the movement of a laser beam;
- Eye protection not being used, or being used inappropriately [e.g., looking over/under goggles (PPE)];
- The wrong eye protection being used; it is always worth checking the labelling on the eyewear to ensure it is compatible with the wavelength(s) being generated;
- Damaged eye-protection equipment being used;
- Misaligned optics and skew-directed beams;
- Presence of stray reflections;
- Equipment malfunction;
- Operators unfamiliar with the actual laser under test and the properties of the beam;
- The misconception that a laser has to be bright to be hazardous; for example, a 405-nm beam from a modest power Class 4 laser does not appear to very bright, but it does represent a serious hazard;
- Change in emission characteristics that may occur following repair.

The bottom line is to ensure that the operator is fully qualified to use the particular equipment under test and that he or she knows exactly where the beam is at all times and is fully conversant with the potential hazards presented by accidental exposure and the risk-mitigation strategies in their entirety. It is recommended that additional care is taken during any beam-alignment procedures.

17.7.1.2 Eye Protection

It cannot be stressed enough that laser eye-protection equipment is the last line of defence when working with lasers. Warning notices, or more usually, illuminated signs and lamps should indicate the need to wear the appropriate eye-protection equipment when operating high-power lasers The direct (intrabeam) viewing of a Class 3B and Class 4 beam must be avoided at all times (within the hazard zone) along with viewing of such beams indirectly. The laser-protection equipment must:

- Block the wavelengths emitted by the laser;
- Have adequate optical density at the hazard wavelengths;
- Fit the operator's face;
- Not be cracked, melted or otherwise damaged in any way;
- Not be sun bleached or otherwise degraded, so that the transmission characteristics are artificially changed.

Periodic inspection of the laser eye-protection equipment is vital to ensure that it is fit for purpose. The actual frequency will be dependent on the test schedule and how often the laser is operated.

It is good practice to use laser eye protection that has curved side shields to prevent stray light entering the eyes. It is recommended that the eye-protection equipment is approved (through factory testing) for providing protection against accidental exposure to the wavelengths being emitted by the laser and that it meets appropriate product standards in the country of use.

An additional consideration is that the eyewear should allow sufficient light into the eyes of the operator, so that he or she can move about safely and also operate the equipment safely; hence, it is recommended that eyewear with very low transmission across the visible waveband is avoided wherever possible. There is a potentially severe hazard to the wearer of eye protection, if they cannot get a clear view around them.

17.7.1.3 Selection of Eye Protection

It is recommended that the following steps are considered in selecting appropriate eye-protection equipment, to ensure that it is effective in the event of accidental exposure to any laser light:

- Step 1: Determine the emission wavelength, or wavelengths of the laser source and the maximum anticipated exposure time, so that the MPE can be determined for that exposure.
- Step 2: Establish the maximum irradiance to which an eye may be exposed under all possible viewing conditions, so that the requisite optical density can be determined (i.e., reduce the exposure to below the MPE limit).
- Step 3: For pulsed lasers, ascertain that the protection afforded by the eye protection remains robust during laser pulses, when there may be extremely high peak power in the incident beam.

It is also worth considering the style of eye-protection equipment as well as its comfort during use. For some of the higher-power studies it will be critically important that stray light cannot enter the eye under any circumstances. Eye-protection equipment that is uncomfortable and fogs up during extended use is a hazard; moreover, it is important that the eyewear has good transmission across the nonhazardous wavelengths of the visible band. The selected eye protection should have a high integrated visual-photopic transmission (IVPT), often referred to as luminous transmittance, and a high optical density (OD) at the wavelength(s) presenting a hazard.

17.8 Hazards to Skin

This aspect of the potential hazard is often overlooked or downplayed. Examples include:

- Burns from direct or indirect exposure to laser beams;
- Exposure to ultraviolet light (200 to 400 nm) can lead to effects similar to sunburn, resulting in increased pigmentation in the skin, accelerated ageing of the skin and possibly the onset of skin cancer;

- Photo-sensitive reactions caused by certain medication (these effects may be triggered by any waveband);

The solution is to avoid any exposure to the laser beam, either directly or indirectly. If tissue cannot be kept out of the beam then suitable clothing and gloves should be worn to screen the body. It is vital that these garments are fire resistant and block the emission wavelengths involved in the test. For some of the lower-power exposures, it may be possible to achieve adequate protection by using sun-block or sunscreen lotions. Care is required, as these wipe off and get diluted as the operator perspires, and therefore, this form of protection is not recommended for ultraviolet wavelengths, or when the exposure is likely to be over an extended period.

If skin protection is required, then eye protection will also be mandatory. In general, normal clothing has very low/negligible absorption at the wavelength(s) anticipated being used in many high-power laser tests. Therefore, special overalls may be required if there is the risk of accidental exposure of any human tissue.

Figure 17.8 shows schematically a cross-section of human skin and the penetration depth of various photon wavelengths in skin, from the hard or deep ultraviolet to the far infrared. The upper layers of human skin are a few micrometres thick, whereas the thickness of the lower layers is measured in millimetres.

17.9 Other Potential Hazards Presented by Lasers (Nonbeam Hazards)

Some laser technologies, as discussed in detail in Part I, rely on various chemicals for their successful operation. These are usually liquids or gases, but dyes are in powder form. A risk assessment will usually be required to evaluate potential hazards and how to contain them, including their safe disposal. Chemical data sheets are available for the estimation of toxicity of any materials used in a laser. Additionally, the hazards presented from any fumes generated during an interaction of a beam with target materials (e.g., burning) such as metals or dielectrics, must be assessed. Where particular hazards are identified, then suitable control measures

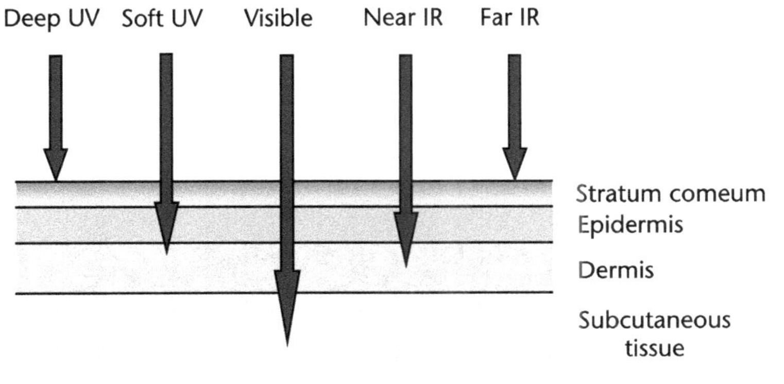

Figure 17.8 Schematic illustration of the wavelength dependence of human skin.

must be applied, such as operation in the presence of a fume extraction system, and in some circumstances, implementation of a fume capture system.

Electrical shock was a very common hazard associated with high-power devices excited by flashlamps. A typical flashlamp is activated with high voltages and rapidly rising currents from a capacitor bank. Many of these power sources involved multiple tens of kilovolts and similar scale currents.

Careful handling of the stored energy in these capacitor banks was crucial to ensure safe operation and involved separation of the power supply from the laser head. The capacitors were usually submerged in oil and supplied with a suitable earthing connection (or more accurately a bleed-down mechanism), especially for high-power devices. Another critical consideration was the efficacy of the earthing mechanism, which was the primary discharge route for removing the charge from the entire capacitor bank.

An effective earthing route is essential for the safe operation of this form of power supply; it may require a very thick copper strip connecting it to an earth pole or rod that goes directly into the ground, which is buried to a depth of several metres. Another critical consideration is the resistance of the earth pole. It is recommended that its resistance is measured frequently to ensure that it offers a short circuit to ground. Periodic watering of the earth pole is recommended and not only in very dry conditions

17.10 Summary of Potential Hazards Presented by High-Power Laser Beams

As indicated above, the potential hazard presented by a laser is indicated by its classification. It is well worth characterising a device that is to be tested to ensure its classification is correct; there have been some instances of misclassification in the recent past. A summary of the potential laser-radiation hazards resulting from exposure to a high-power beam are presented in Table 17.1.

The definition of the various wavebands referred to in the laser-safety legislation does not equate with emission-wavelength bands coinciding with the atmospheric windows. These windows are zones in the electromagnetic spectrum that define wavelengths where the propagation of optical radiation in the atmosphere is good, and consequently, exploited by all directed-energy applications. Care is required when using the various terms used to define wavebands in the electromagnetic spectrum in different contexts.

A summary of the various effects as a function of wavelengths is shown in Figure 17.9.

17.11 Common Causes of Laser Incidents

There are many reasons for a breakdown in any laser-safety regime that can lead to a laser incident, which is thankfully rare; however, the following list provides a brief synopsis of common causes of historical laser incidents:

Table 17.1 Summary of Laser-Radiation Hazards as a Function of Wavelength Band

Wavelength Band (nm)	*Radiation Range*	*Affected Organ*	*Hazard*
180 to 400	Ultraviolet	Eye	Photo-chemical change and thermal damage
180 to 400	Ultraviolet	Skin	Erythema
400 to 700	Visible	Eye	Retinal damage
400 to 600	Visible	Eye	Photo-chemical damage
400 to 700	Visible	Skin	Thermal damage
700 to 1,400	Infrared band A	Eye	Thermal damage
700 to 1,400	Infrared band A	Skin	Thermal damage
1,400 to 2,600	Infrared band B	Eye	Thermal damage
2,600 to 1 million	Infrared band C	Eye	Thermal damage
1,400 to 1 million	Infrared bands B and C	Eye	Thermal damage
1,400 to 1 million	Infrared bands B and C	Skin	Thermal damage

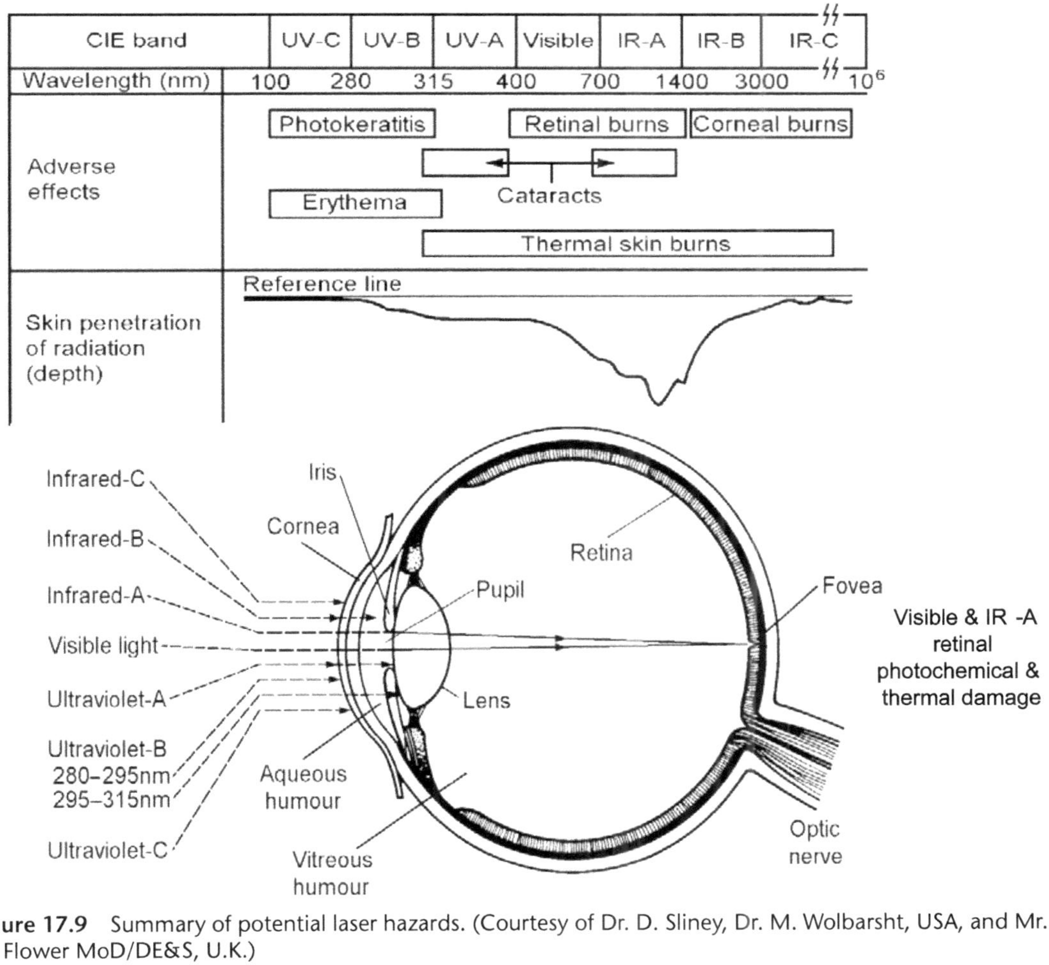

Figure 17.9 Summary of potential laser hazards. (Courtesy of Dr. D. Sliney, Dr. M. Wolbarsht, USA, and Mr. M. Flower MoD/DE&S, U.K.)

- Failure by personnel to follow defined safe-operating procedures;
- Loss of beam control during insertion of reflective optical elements into the beam path;
- Altering the beam path directly or indirectly;
- Accidentally firing the laser;
- Accidentally turning on a power supply;
- Bypassing the interlocks or negating other engineering controls;
- Use of devices by unqualified personnel;
- Reliance on incorrect eye protection or unsuitable eye protection;
- Unauthorised or inappropriate modification to the laser device(s).

Other common causes of laser-related incidents that have been documented over the last five decades, which did not result in accidental viewing of a beam, but resulted in some form of injury, include:

- Lack of protection from ancillary hazards, such as inhalation of laser-generated air contaminants and viewing of laser-generated plasmas;
- Laser-induced ignition of materials, without adequate fume-extraction systems or fire-suppression equipment;
- Incorrect or inappropriate handling of high-voltage power supplies.

A laser incident may be defined as "any unexpected exposure scenario, actual or suspected, which may have resulted in personal injury." The accident may result in some or all of the following effects:

- Damage to eyes in one form or another;
- Redness to the skin;
- Skin burns.

An immediate investigation should be initiated by an independent body to determine the fundamental cause, or causes, of the exposure incident so that immediate corrective action can be implemented. It is common for there to be several causes of accidents, so it is important to understand the principal issues, contributing effects and how they may cascade to induce other reactions and produce an overall effect.

It should be remembered that if a laser device is modified for whatever reason, it is the responsibility of the modifier to undertake a further safety assessment. Unsuitable and inappropriate modifications to laser devices are another source of accidents.

17.12 Laser-Safety Training

The principal purpose of laser-safety training is to make all personnel involved with the use of laser devices, including their operator(s) and staff supporting these studies, aware of the potential hazards and their own personal responsibility. For some personnel, a simple online course will be appropriate, whereas for others, such as laser-safety officers, a more rigorous course will be necessary. It is very important that the course content is compatible with the training requirements; perfunctory delivery of general laser-safety issues and processes is usually of very limited value. Many professional institutions offer very high-quality courses, which are well worth the investment.

Training of all personnel involved with the development, use and support of laser-related operations and research studies is a crucial duty-of-care requirement, to ensure the rigorous use of the very best laser-safety practice. Good training should enable all personnel to identify and mitigate potential hazards before they even arise. The purpose of this training is to ensure that all personnel involved with a laser test activity are familiar with the following basic safety principles:

- The safe operation of the laser device(s):
- The potential hazards;
- The risk mitigation techniques, processes and procedures;
- Safe and effective test procedures and methods;
- A thorough understanding of local requirements, national regulations and current legislation, so that they are applied during the use of the laser equipment;
- Requirements and effective use of appropriate warning notices and flashing lights;
- Medical screening requirements;
- Their own responsibilities to the other members of the test team.

It is recommended, and in many nations it is mandatory, that all personnel involved with a test undertake the appropriate laser-safety training and periodic refresher courses, which may include the following personnel:

- Operators of the laser device, whether test engineers, researchers or development technologists;
- The personnel running support equipment including instrumentation;
- Maintenance staff supporting the equipment at the test site;
- Guards controlling access to a range;
- Observers if they are within the hazard zone.

As noted previously, the training programme is usually more extensive for laser-safety officers, laser researchers, device developers and operators. The refresher courses may be a short course to remind personnel of the critical hazards, their control measures and any recent changes to legislation.

A typical more advanced laser-safety training course will cover:

- Definition of laser action and the specific potential hazards resulting from highly collimated optical beams;
- Definition of terms associated with laser action and beam propagation, such as:
  - Optical gain;
  - Temporal (pulsed and continuous wave) and spatial characteristics of emission;
  - Types of reflection (specular and diffuse) and the relevant hazards;
  - Units of measurement for expressing the various characteristics of a beam and understanding magnification of collecting optical systems.
- The purpose of classification of laser devices and terms used in the safety assessment, such as maximum permissible exposure and accessible emission limit;
- Appropriate laser-safety standards and legislation;
- Calculation of hazard distances;
- Radiometry and characterisation of optical radiation;
- Role of the laser-safety officer.

For some military personnel involved in depots, this training will require some additional instruction, particularly related to servicing of line-replaceable items and calibration for military equipment in service workshops. This aspect is covered in a little more detail in Chapter 18.

Figure 17.10 summarises the facts that influences the scale of required laser-safety training.

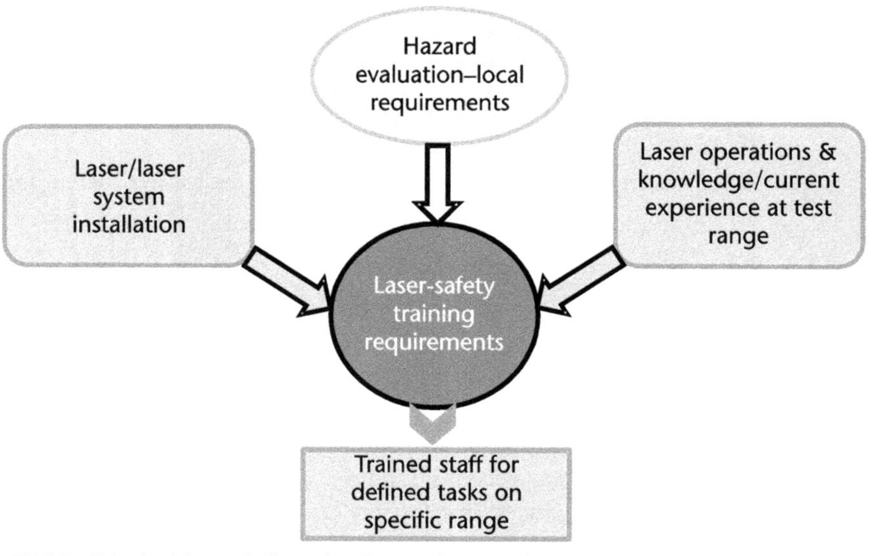

Figure 17.10 Principal factor influencing laser-safety training requirements.

17.13 Laser Safety in Practice

One of the critical aspects of laser-safety training is the translation of the laser-safety principles into practice in both the laboratory and on an open-air range. It is critical that the techniques and methods proposed for containing and mitigating the identified hazards are appropriate for both those hazards and the scale of the laser operations.

It is critical that expert advice is sought, so that the various controls are compatible with the anticipated hazards and the location for the testing, along with the associated risks. At a trivial level, recommendation of extensive use of interlock controls on a range will probably be nugatory, except for the access panels to the high-power devices or the enclosure containing the laser system, whereas in the laser laboratory, they are usually an invaluable laser-safety tool.

This section considers how laser-safety principles may be embodied in the design of a laser laboratory and the integration of engineering controls. Approaches to the development of best practice are considered, primarily through the formulation of safe-operating procedures and design of tests in a trial plan for field testing.

17.13.1 Safety-Design Considerations for Laser Research and Development Laboratory

One of the critical considerations is starting with a suitable laboratory design that is compatible with sustaining appropriate laser operations and supporting the laser-safety measures. For example, the materials used in the construction of a laboratory structure intended for high-power operations should be at least fire resistant and preferably made exclusively from noncombustible materials, such as concrete or bricks. Figure 17.11 shows an image of a modern, purpose-built laser laboratory designed for high-power operations.

Figure 17.11 High-power laser laboratory. (Courtesy of Dr. N. J. Wood, QinetiQ, U.K.)

There are probably as many thoughts on the safe design of any sort of laboratory as there are laboratories. It is essential that the laboratory provides a safe-operating environment that is fit for purpose. Some other critical considerations regarding the structure and infrastructure are:

- Stray light control (windows): there should not be any windows where stray laser light can escape from the laboratory. If the allocated building already has windows, they should be blocked with suitable screening material. Curtains and blinds are not adequate, unless the laboratory use is confined to low-power operations.
- Stray light control (doors): very careful design is required to prevent stray light escaping around the door jamb, and if double doors are used for access, the joint between them should also be staggered to ensure it is "light tight."
- Use of screening within the laboratory close to the doors to form a chicane: This will also prevent stray light leaving the laser laboratory via the doors. In some circumstances, it is possible to use an override switch on the interlock system to permit personnel to enter or leave the laboratory.
- Fitting of power-control interlocks on the doors to stop laser action if there is an unexpected entry to, or exit from, the laboratory during lasing: Interlocks may also be linked to beam shutters, but see the earlier comment concerning an override function.
- Good climatic control.
- Good fume extraction, particularly if high-power materials interaction effects are being investigated. If the high-power effects are expected to generate smoke and other combustion products, the laser beam and target interactions should be undertaken in a separate dedicated effects laboratory with a separate fume-extraction system. It is critically important to avoid smoke and/or dust contamination of optical elements of the laser and beam-delivery system.
- Lighting should be very bright, so that the light-adapted eye has a small aperture even when appropriate eye-protection equipment is being worn.
- All walls should be a matt finish (diffuse scattering) to reduce the albedo and thus avoid unwanted stray-light reflections, diffuse or otherwise.
- Avoid the use of reflective materials anywhere within the laboratory, including door and window furniture (i.e., handles and catches).
- Floor materials should be matt finish with a nonslip finish.
- Ceiling materials should be matt. Having the services exposed in the ceiling space should be avoided to prevent dust and other debris building up and eventually falling on to the optical elements or the test specimens.

The other critical hazard-mitigation strategy concerns layout of the laboratory with its optical tables and benches, to ensure adequate space to move between them without any accidental collisions and disturbing equipment. Care has to be taken regarding emission of simultaneous multiple wavebands, particularly with the need to ensure appropriate eye protection is being used. In general, it is not good practice to have simultaneous laser tests being operated in the same laboratory in different wavebands and under the control of different operators. It is recommended that if this situation arises a single operator should have executive control.

The positioning of the beam paths within the laboratory also requires thought. This consideration applies particularly to the relationship between beam heights and the provision of seating areas; careful design is recommended to avoid the risk of stray light being incident on the eyes (in particular) of personnel when seated.

Careful consideration should be given to the location of the laser laboratory and whether the laboratory is a single- or multiple-storey building. Single-storey buildings offer easier access for bulky equipment and potentially greater stability. Another useful facility is access to a longer propagation path than offered by confinement within a conventional laboratory; therefore, positioning a laboratory close to an accredited laser range with an aperture that enables direct access to that laser range and free-space propagation path is worthy of consideration.

An alternative approach to achieving an extended beam-propagation path concerns use of mirrors for multiple bounces along the length of a laboratory, or even the use of mirrors to fold a path around corridors in a building provided access can be strictly controlled. These approaches can be of great advantage, as the propagation environment can be very carefully controlled and/or remain stable over a significant distance, given multiple bounces and a reasonable length between mirrors, but it should be noted that significant levels of turbulence can be present in such a closed environment.

During the design of a new laboratory, the shape of the laboratory space should be a critical consideration. For example, a long thin laboratory, which can be subdivided into two (or more) smaller laboratories with a suitable interlocked screen may offer greater functional utility, as well as flexibility of use, compared with a square area. An example of the enhanced utility is the scope for relatively long folded beam paths discussed earlier.

Some, or all, of the optical tables may have barriers, shutters or screens surrounding them to contain stray light and to mitigate hazards further, as shown in Figure 17.12. It may be noted that the structure shown in this figure has the appropriate warning signs posted; similar signs will be on the entrance doors to the laboratory.

17.14 Laboratory Testing

Engineering controls have an important role in achieving a safe environment for testing in a laboratory and are considered next. Definition of safe-operating procedures and a coherent trial plan are critical requirements for a safe and successful laboratory testing programme.

17.14.1 Engineering Controls

There are several approaches to controlling access to a laser beam and use of techniques to avoid accidental exposure to a beam of optical radiation from a Class 3B or a Class 4 laser product, as well as mitigation of potential hazards. Examples include:

Figure 17.12 Optical table with safety screens. (Courtesy of Dr. N. J. Wood, QinetiQ, U.K.)

- Protective housings and shutters to enclose the entire beam, wherever possible, but that approach is normally only possible in the laser laboratory;
- Interlocks on the access ports to the protective system or access to the laboratory, which automatically stop the emission in a safe, controlled and rapid manner;
- Light-tight rooms with interlocked doors (linked to the laser power supply or beam shutter);
- Controlled access to all hazardous areas through key cards only issued to qualified and fully trained personnel;
- Remote firing and key-switch control;
- Separate key control of master interlocks;
- Remote surveillance of the device and beam interaction;
- Remote beam-control hardware;
- Beam shutters and tubes to control the beam-propagation direction;
- Optically filtered viewing portals, with appropriate optical density;
- Optically filtered display screens;
- Photon-collecting optical elements, with appropriate built-in filters;
- Appropriate barriers to contain or control direct impact of the laser beam or reflected and diffusely scattered beams;
- Removal of all reflective materials from the nominal hazard zone;
- A beam stop to absorb the entire main beam for a given period; it may need to be water cooled. The beam-stop material should not be combustible:
    - It may be necessary to position the beam stop at a suitable distance from the aperture for the beam's irradiance not to exceed the damage threshold of the beam-stop material.

- For out-door operations, it could be a natural feature such as a ridge for a surface-to-air firing or the ground for an air-to-ground shot.

Multiple safety systems that achieve the same purpose rarely improve the control of any hazards and do not provide any additional protection for personnel. They usually only provide additional points of failure. Reliability and confidence in the safety system are critically important.

17.15 Safe-Operating Guidelines/Procedures

There are many potentially effective approaches to the safe operation of lasers in the laboratory, but one of the crucial aspects is a clear understanding of the limitations of the laser source under test, along with the appropriate safety procedures that are both appropriate and feasible. Whether the test involves a laser-device-development activity or an evaluation phase of a laser-based project, it is recommended that a test plan is devised that defines the potential hazards and the potential mitigation plan for all identified risks and hazards. This may then be translated into a safe-operating procedure or guideline.

This safe-operating procedure should not only define the fundamental scope of the experiment and the normal operational procedure of the device under test, but list:

- All of the potential hazards and safety equipment to be used, including the appropriate laser eye-protection equipment (spectacles or goggles);
- Training requirements and identify a laser-safety officer and deputies, as required, (by name);
- Qualified operators, again named individuals;
- The procedure to be followed if any equipment malfunctions;
- The emergency shutdown procedure;
- The checks to be undertaken to ensure the entire system is safe at the end of a test event;
- Correct operation of the laboratory access controls;
- Emergency services contact details;
- The accident reporting process.

This should not be a lengthy document, possibly just a few pages for tests that do not require complex operations. Its main purpose is to provide documentation for peer review locally, to provide a mechanism to ensure all aspects of effective best practice are understood and are being applied appropriately. There may well be local rules concerning the formulation of this type of documentation that could be mandatory, so should be adhered to, covering the local safety requirements or specific operational requirements related to the laser under test.

17.16 Field Testing and Trial Planning

Field testing is the most demanding in terms of safety because of the uncontained nature of the beam-propagation path. Each nation will have its own specific procedures and processes, but there will be many similarities; hence, it will not be possible to be definitive about precise requirements, but this section provides an indication of probable requirements, procedures and guidelines for consideration.

It is normally much easier to gain approval for operation of high-power (Class 4) lasers on a range that is already certified for such activities (i.e., a laser approved, or licensed, range). If this is not possible, then it may be worth considering the use of an alternative venue either within the same country or in another country; generally, it takes a considerable amount of time to undertake all of the hazard assessments, site survey and environmental-impact investigations needed to ensure safe operation of a high-power laser on a new range. Time scales are frequently quoted in multiples of years; in some cases approval may be withheld.

When seeking a range for high-power laser operations, it is worth considering the possible need for the following characteristics at the site:

- Steep-sided valley with large (natural) backstop;
- A trial's site in a remote location, but with reasonable vehicular access, with a very low population density in surrounding areas to minimise accidental incursion by untrained people;
- No reflecting surfaces within the potential hazard area along the beam-propagation path;
- Ability to build additional backstop(s) behind the firing point(s) to contain any reflected energy;
- Possibility to erect safety screens at firing point(s) and/or use (down-range) apertures in order to limit firing arcs. These controls need to be made from a suitably resistant material that is a diffuse reflector (e.g., concrete);
- Easy access to suitable support facilities, such as electrical power, water, a crew room for shelter and briefings, a controlled area for securing equipment and good communications.

There are additional considerations when testing laser devices and systems in a maritime environment, as many of the ideal characteristics referred to above for land-based ranges will not be possible. In this case, additional considerations are necessary, particularly if the beam is directed towards the horizon, as shown in Figure 17.13; similar considerations apply for a beam pointed above the horizon.

Procedures have been established and demonstrated that enable this type of interaction or target engagement to be achieved successfully and safely, and without any accidental engagement of other assets. This is often called air-target deconfliction or accidental target engagement.

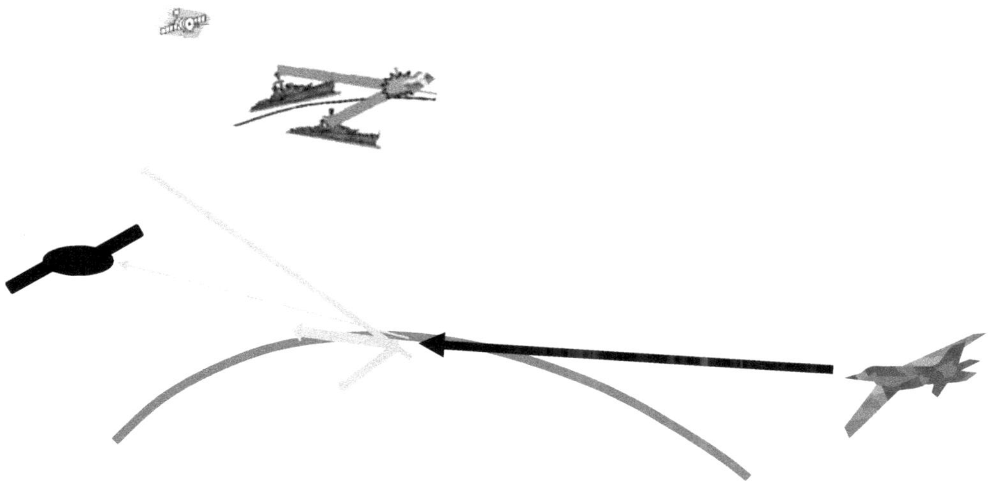

Figure 17.13 Schematic of the maritime laser-safety challenges.

17.17 Planning of Trials

Before any range/field trial can begin, it is usually necessary to write a definitive trial's plan. A trial's plan has several important functions; it not only provides a concise description of the planned testing, but equally importantly, a compilation of the processes being followed to assess the potential impact of all of the perceived hazards. This analysis aims to ensure that the trial is safe through following particular established and identified hazard-control and risk-mitigation techniques, which will form an integral part of the trial's planning activities. Chapter 6 considers trial's planning mainly from the test and evaluation viewpoint, along with the appropriate testing procedures and techniques. This chapter considered trial's planning with a concentration and emphasis on the various aspects of laser safety. Both aspects have to be an integral part of the final trial's plan, as noted earlier.

A trial's plan should provide the following information to ensure the test plan has been formulated correctly, the appropriate hazard analysis has been undertaken and risk-mitigation plans established:

- Objectives and scope of the trial plan, a definition of the trial's programme, include:
  - Parameters to be investigated;
  - Equipment requirements;
  - Ancillary/support equipment;
  - Programme timescales;
  - Data collection and storage;
  - Target control;
  - Logistics.

- Nominated trial's officer/director is where the buck stops, as far as the scientific and technological decisions are made, regarding the test programme.
- Authorisation for the trial is the audit-trail log of the approval/permission and certification process, which includes any external review and licensing for the testing.
- Description of the venue, which will include photographs of the terrain containing the proposed propagation paths.
- Dates of the proposed trial to ensure compliance with the range availability and any restrictions with the certification for the trial.
- Nominated staff involved in the trial, along with responsibilities and their competencies.
- Listing of any specific occupational health requirements to ensure that all protection measures are applied to mitigate known and perceived hazards.
- A firing procedure, including any requirements to ensure that all personnel are under cover and wearing appropriate personal-protection equipment.
- References to related documents and procedures indicating the scope of the supporting information.
- Risk assessment and analysis explain the scope of the evaluation and indicate the mitigation processes.
- Emergency and contingency planning, which is linked to the health and safety and risk assessments and the risk-mitigation strategy.
- Trial's preparation activity is a definition of the activities that must be completed before the trial's readiness review to ensure a high probability of success of all of the planned activities.
- Control and chain of command, which is a definition of the interaction of the trial's team with the external body of ultimate control, such as range control and the range-safety officer.
- Communications, for example, is the definition of the radio bands to be used and the maintenance of an open line to and from range control.
- Environmental-impact analysis, which is an in-depth investigation of the potential impact on the natural environment and procedures required to minimise the perceived impact. This part of the trial plan will also include any post-trial cleanup phase.
- Additional issues may pertain to this specific trial.
- Completion of trial requirements is a list of the tasks required to be completed to clear the trial's range prior to departure.
- A list of references is given for the supporting information;
- The management of change to a trial authorisation and processes is required, as rarely will any trial, despite very detailed planning, be executed exactly according to the original plan, hence, a record of changes is vital for the post-trial analysis.

Most nations require an independent assessment of the safety calculations, assessment of hazards and mitigation strategies, which should be referenced as a minimum requirement. Successful submissions usually result in the granting of a licence or certificate to operate on an identified range according to the defined testing regime.

A critically important consideration for all field testing is a complete understanding by the trial's team of the chain of command and a clear definition of where the ultimate control and responsibilities reside, for example:

- Range control is where the ultimate executive authority and control of the range resides for giving permission to operate a devices or instructing cessation of operations. This body will usually scrutinise the proposed activities and ensure full compliance with both the range regulations and the terms and conditions stated on the licence or certificate to operate laser-based devices and systems on that range.
- A range-safety officer's role is normally local control of operations on the range and is the principal link to range control, for example, requesting permission to fire and advising range control of any immediate safety issues, such as people trespassing in a forbidden area or a request for emergency services.
- A range laser-safety officer's role is to ensure that all of the equipment is used safety and is compliant with the operational and control details defined in the trial's plan. The range laser-safety officer is usually a specialist in laser operations and has authority to stop operations because of a laser-safety concern for whatever reason. This role and the officer's function are considered further later; it is worth noting that these roles (range-safety officer and range laser-safety officer) can be combined in certain circumstances.
- The trial's officer/director function is the control of the testing procedure and ensuring that the testing is undertaken in accordance with known and agreed practices. The technical leaders of the various supporting activities, such as laser operations, target control and instrumentation, report to the trial's officer/director, who will provide the information about readiness to fire to the range control.

It is essential that this team is aware of any need to depart from the agreed test plan, so that alternative arrangements can be considered, and where necessary, appropriate permission for the change of test plan sought.

17.18 Range Laser-Safety Officer

The role of the range laser-safety officer covers many functions to ensure that the potential hazards are fully contained or mitigated. The range laser-safety officer (or a nominated deputy) needs to be present during the setup phase to provide approval for the whole configuration and to give permission to operate the laser (i.e., fire the laser). Provided that there is a rigid reporting process, the trial's officer/director can oversee all laser operations once it has been approved by the range laser-safety officer.

Some examples of a typical check list for a range laser-safety officer include the following:

- Provide an appropriate and effective laser-safety briefing to all personnel involved in the trial who do not have current certification for working with lasers. This briefing could be in the form of a presentation or an instruction to read a particular manual.
- Provide an appropriate laser-safety briefing for visitors and observers; this briefing is likely to be more superficial than the one indicated above, covering the critical aspects from the observer viewpoint, with the important instructions not to touch equipment or stray from viewing areas, for example.
- Maintain a record of all staff that have undertaken the safety training.
- Ensure that a procedure is established to maintain an up-to-date log of all personnel on the range.
- Ensure that all supplied laser-eye protection is appropriate and fully functional.
- Agree precise positions for, and directions of, the propagation path and position of beam stops.
- Ensure that all of the engineering controls and other measures deemed appropriate are being used and are fully functional.
- Ensure that the laser beam is not penetrating restricted zones, such as areas open to the general public.
- Clear or approve the certification for each laser involved in the proposed testing.
- Check, or get independent assessment of, the calculation of hazard distances supplied by the laser suppliers or operators.
- Define the laser on/off procedures that are appropriate for the proposed tests and ensure that the countdown allows opportunities for abort commands to be given for any reason.
- Establish communication channels that are effective and ensure all of the controlling bodies agree that they are ready to fire.
- Check that appropriate optical filters, and any other necessary precautions (such as optical limiters), which are being used in any optical equipment, are in place.
- Ensure the appropriate and correct warning notices are displayed and any special warning instructions are given,
- Maintain the laser firing logs and ensure that they are completed after each test, including recording all malfunctions for whatever reason, including finger trouble.
- Maintain an up-to-date list of emergency contact numbers.

Many organisations will have additional requirements for the range laser-safety officer to follow; this list attempts to capture some of the more common, and crucially important, tasks associated with the role. If the range laser-safety officer has any doubt about the safe operation of either a device or the design of an experiment, then that officer has the executive authority to stop operations.

17.19 Range-Safety Officer

The role of the range-safety officer is a more general oversight of operations and interaction with the range-control body. This role is the critical interface with the executive control and local operations involving the test director and the range laser-safety officer. Some of the more crucial tasks undertaken by the range-safety officer are:

- Maintain an up-to-date record of personnel on site.
- Ensure that all communication channels are fully operational, audible, clear and agreed terms are being used.
- Ensure that the master emergency shutdown switch is operational.
- Ensure that a qualified person has control of the emergency shutdown switch throughout the lasing event. This should be the sole function of that operator; it may, for example, be the responsibility of the range laser-safety officer, who has a safety-observation role during a firing.
- Report to range control that all the warning notices, flags and other instructions are in place.
- Following completion of all safety preparation activities, include use of appropriate personal protection equipment and, checking that personnel are under cover, seek permission to start laser operations.
- Provide frequent radio checks with range control.
- Disseminate all instructions from range control, such as evacuation of the range for whatever reason.
- Ensure that the range-observation equipment for critical safety functions is fully functional.
- Check that the range is secure for laser operations, which includes ensuring that the appropriate physical barriers are in place and monitoring systems are functional.
- Ensure that the beam-control equipment is in place and functioning, particularly the back stops.
- Ensure that the laser operations are compliant with range-control instructions, which includes authorised targets in approved firing zones.
- Report any accidents to range control and make a report in the accident log.
- Request assistance from the emergency services when required.
- Ensure that the range hazard zone is clear of all personnel and, where possible, wildlife prior to firing.
- Stop operations if there is a subsystem or system failure that could result in loss of control of the beam-propagation direction, for example, if fore stops,

or other engineering measures are not in place to restrict the sector that may be illuminated by the laser beam.

- Cease operations if unauthorised personnel enter the range for whatever reason.
- Maintain an up-to-date log of all operations during the trial and related events on the range.
- Maintain an up-to-date list of emergency contact numbers.
- Report to range control when operations have ceased for the day.

It is common, and usually mandatory, for the range-safety officer to maintain a dedicated and open radio link with range control at all times.

It will be apparent that there are a number of functions that are common to different trial's appointments, particularly to ensure redundancy in safety-critical events.

17.20 Flight Trials

These are the most demanding form of testing, whether the laser is on the ground or on an aircraft. Additional special controls are likely to be declared to be necessary, which are briefly outlined below, as specific measures are likely to be both range and nation specific.

17.20.1 Flight Trials Involving Ground-Based Laser Trials

In the case of the laser being positioned on the ground with the intention of engaging aerial targets there are many specific requirements that have to be fulfilled. Some of the safety considerations are as follows:

- Can the laser beam position/bearing be controlled throughout the tests through the use of engineering control measures (as indicated above)? These measures include the application of fore stops or a fixed aperture either immediately in front of the laser-exit aperture or a short distance down range.
- Is a suitable back stop available, such as a mountain or a ridge that will prevent firing of the beam above the horizon? Careful positioning of the laser system close to a natural feature, such as a ridge, and application of suitable control of the elevation of the beam, again through either an aperture or a physical stop will prevent accidental firing above the horizon.
- Does range control have full control of the local flight paths that are within the calculated hazard distances?

In general, all of these issues will be required to be in place and additional measures are likely to be required as well, to comply with national standards for a constrained beam test with a back stop. If it is necessary to fire above the horizon (i.e., without a backstop), then very special safety measures will have to be taken to ensure that satellites and aircraft are not engaged accidentally and the range has full control of the local air space. This process can necessarily be very restrictive and demanding to ensure that the beam only propagates into an approved sector of space. If the pointing vector goes above the horizon, then it may be necessary to consult agencies that are available to advise on the availability of suitable firing arcs during a given period. It is recommended that there is an automatic shutdown process (or shutter closure) to terminate the emission from the laser system immediately, if the beam leaves the permitted engagement zone, where a satellite could be inadvertently engaged, for examples. Again, it is difficult to provide specific guidelines other than that it is vital to identify all of the requisite authorities at an early stage during the initial planning stage of a test that requires firing above the horizon.

17.20.2 Aircraft-Based Laser Trials

The added complication with this type of testing is gaining approval to fly the laser or laser-based system. This usually means that the laser (and its system) has to undertake some specific qualification testing to ensure it is fit for flight, and particularly, it will not endanger the safety and the integrity of the aircraft. This qualification testing may be confined to a restricted set of the full military-qualification testing, often called a special-to-test qualification and may have a number of restrictions, such as a limited number of flight hours and/or a restricted flight envelope of the aircraft. Additionally, it is well worth estimating the potential impact of the installation and flight profile on the installed performance of the laser system.

Air-to-air engagements will require extra-careful planning in order to meet the additional constraints that will inevitably be imposed on the testing regime, particularly, for firing above the horizon for whatever reason, owing to the additional degrees of freedom involving two or more aircraft during the testing. Clearly, there are multiple opportunities for the beam to miss the target through loss of beam control or the target being in the wrong place at the wrong time, for example. The latter situation has been the root cause of many nonlaser live-fire accidents.

Air-to-ground engagements may appear somewhat easier; however, there are still many hurdles to be cleared before the testing can be authorised. Many of the considerations described earlier will apply. The requirements for this type of testing may be considered to be a subset of the air-to-air requirements, as there is only one aircraft, but multiple opportunities for loss of beam control will still exist.

One of the options for air-to-ground testing is to operate the aircraft at a sufficiently great altitude above ground level, which is in excess of the hazard distance (the extended ocular hazard distance). Control measures can be implemented on air platforms that limit the field of regard of the beam director or restrict the beam-propagation direction(s). Additionally, in certain situations, probabilistic risk assessments may be required, but if adopted, this approach takes extra time and can be complex.

17.21 Some Additional Safety Thoughts

There are many additional considerations that should be reviewed as part of the general safety analysis. The following are offered as a baseline:

- Electrical supplies and earthing;
- Cryogenics and the potential impact on operations;
- Compressed gases and handling of gas cylinders;
- Presence of toxic materials or their creation during laser testing;
- Discharge arcs created in lamps and the spectrum of the emission, particularly the presence of ultraviolet radiation (awareness of colour temperature);
- Shattering of arc lamps or flashlamps;
- Potential generation of ionising radiation;
- Presence of charged capacitors;
- Explosion of capacitors;
- Generation of microwaves or radio-frequency radiation;
- Noise level and frequency content of any audio generated during the testing;
- Potential proximity of coolant and electrical supplies, the need for good maintenance of both services;
- Possibility of toxic and/or carcinogenic dyes in solution in the gain media of liquid dye lasers.

These thoughts and considerations are offered as a baseline, to help guide the laser-safety planning, which has to be an integral part of any test plan. The laser-safety aspects must be defined at the start of any evaluation programme and not added on as an afterthought, as that will rarely be effective, and in many cases, inhibit an effective test process and seriously limit the effectiveness of the evaluation project. It is prudent to get independent confirmation of the accuracy of all of the laser-safety calculations undertaken to support any laser test.

A useful sanity check is to ask the question: Would you put yourself in the position that you are asking other people to occupy during the trial and/or testing?

17.22 Summary

There are well-established techniques and methods that enable Class 4 laser technology to be used successfully and safely in a laboratory, as well as on an open air range. Safety is a critical concern, and again there are well-defined procedures and processes to enable even very high power lasers to be operated safely without inhibiting the integrity of the testing, provided that laser safety has been considered from the outset.

17.23 Legislation

The military use of laser systems that have the potential to blind personnel on the battlefield is governed by an international agreement, known as the Vienna Protocol or Protocol 4. Mistakenly, this international agreement is sometimes referred to as the Geneva Protocol, where the headquarters of the International Committee of the Red Cross is based. Moreover, European nations have adopted the artificial optical radiation (AOR) directive, which requires employers to undertake formal risk assessments to ensure that all occupational laser exposures to be below a set of agreed Maximum Permissible Exposure Limits that are published as an Annex to the European Union Directive. At the time of this writing, these limits were identical to those set by the International Committee on Non-Ionizing Radiation (ICNIRP), which are also incorporated in the American National Standard Institute and the International Electro-technical Commission laser-safety standards.

17.23.1 Protocol IV

It must be stressed that this text is not an attempt in any shape or form to provide a legal interpretation of the Protocol governing the use of laser weapons on a battlefield. Moreover, it does not represent a national view by any nation on the legality of the Protocol; the sole purpose of its inclusion in this text is to provide a summary of the Protocol, which is probably the one that is most often misquoted, and consequently misunderstood, of the international protocols related to laser operations.

Protocol IV covers the use of anti-personnel battlefield lasers and laser beams used for anti-personnel purposes. It is sometimes referred to as the Vienna Protocol on Laser Weapons, as the final negotiations took place in that city in October 1995, resulting in the publication of Protocol IV. The intention was to provide an international legal agreement to ban laser-based systems from the battlefield whose sole purpose through their design was to blind anyone that they engaged.

Protocol IV has its genesis in a proposal from Sweden in the early 1980s, which had been developed in conjunction with efforts of the International Committee of the Red Cross (ICRC). The International Committee of the Red Cross hosted four meetings of experts to explore the problem of lasers intended to blind personnel on the battlefield. Discussions involved military, technical and medical experts.

The original initiative, from Sweden, embodied a proposal concerning the control and banning of "the development of anti-personnel battlefield laser weapons and the possibility of an anti-personnel use of laser beams." A clear distinction was made concerning the use of laser beams for range-finding, designation, navigation systems (in laser gyroscopes, for example) and in [laser] radar, which were considered legitimate uses of laser systems.

The International Committee of the Red Cross proposal was similarly short and read:

1. Blinding as a method of warfare is prohibited.
2. Laser weapons may not be used against the eyesight of persons.

The rationale underlying both proposals was that the use of such types of lasers contravened the Geneva Convention concerning prevention of the use of weapons of a nature to cause unnecessary suffering or superfluous injury. This principle aims to ensure that any injury or suffering inflicted by a weapon is balanced by the military advantage gained through using it.

Protocol IV is structured in such a way that it first prohibits the use of "blinding laser weapons." It also obliges nations to take precautions to avoid causing blindness in opponents when a nation's military is using non-prohibited lasers on the battlefield. The protocol then goes on to envisage the legal blinding of persons on the battlefield whilst using nonprohibited lasers, and finally, establishes a scientific definition of blindness for the purposes of the Protocol.

Article 1 prohibits the employment of "lasers specifically designed, as their sole combat function or as one of their combat functions, to cause permanent blindness to unenhanced vision, that is to the naked eye or to the eye with corrective eyesight devices."

The main issue is to establish exactly what the combat function of the laser device actually is, as designed. In equipment that employs lasers to defeat materiel such as an aircraft, it is usually accepted that Article 1 is not relevant, as the device's combat function would be anti-materiel.

Article 2 requires that states [nations] take "all feasible precautions to avoid the incidence of blindness to unenhanced vision" in the employment of laser systems.

This is clearly aimed at those systems that are not prohibited under Article 1. However, there is no direct definition of feasible precautions within this article, although there is a definition provided elsewhere in Protocol III, which deals with incendiary weapons. In this case, feasible precautions are defined there as: "those precautions which are practicable or practically possible taking into account all circumstances ruling at the time, including humanitarian and military considerations."

The supporting documentation relates to precautions taken in the preparation and launching of attacks vis-à-vis civilians, including a recommendation specifically requiring training on the use of such laser systems.

Article 3 provides some elaboration and clarification on the issue of incidental blinding. The prohibition within Protocol IV envisages that there will be some battlefield laser systems that are not prohibited. Indeed, the International Committee of the Red Cross has acknowledged that "the main purpose of [Article 3] is to allow for the continued use of battlefield lasers, which has been common for some time." Moreover, it would appear that the purpose of Article 3 is to exclude such blinding incidents caused by the legitimate use of lasers from the scope of Protocol IV.

In fact, the clarification within Article 3 extends to the insertion of an example of the legitimate military employment of laser systems. It includes "laser systems used against optical equipment."

Article 4 provides scientific clarity on what is meant by the term permanent blindness, with the intention of avoiding any legal misunderstanding of the term.

17.24 Summary

Laser safety is a critical consideration for all forms of operation and testing of laser-based devices. It is a complex subject that requires specialist knowledge to make effective and accurate judgements about potential hazards and their effective mitigation. Laser-safety considerations must be designed into laser devices and systems from the outset; similarly, laser-safety procedures and risk-mitigation techniques must be integrated into test and evaluation techniques at the initial planning stage. Successful and incident-free testing is unlikely to occur if laser-safety procedures and protocols are added to the test and evaluation plan as an afterthought. Not to mention the rapid escalation in cost of undertaking the trial.

Over the decades since the first demonstration of the laser, there has been extensive investigation of the interaction of optical radiation from lasers with human tissue, as well as with all kinds of materials. This information has been used to formulate laser-safety standards, which forms the basis for classification of lasers according to their potential for harming a human, especially causing permanent blindness. There are a number of such laser-safety standards in force throughout the world, but their advice and requirements are, in general terms, very similar. They are based on the concept of the maximum permissible exposure, which is the amount of energy or power to which a person may be exposed with no risk of harming human tissue, particularly, the human eye.

High-power laser beams have the potential to cause severe damage to human tissue, unless suitable precautions are taken and applied, as designed. Accidental exposure to laser beams can take many forms and usually result from a number of lapses in observance of laser-safety procedures, but it would appear that the majority of incidents occur during the alignment of laser beams. Fortunately, accidents with lasers are few in number and the impact is relatively minor.

The unintentional interaction of a laser beam with the human eye is always of critical concern, as a beam from a Class 3B or Class 4 laser even with a relatively modest power in the wavelength range 400 to 1,400 nm is capable of creating a lesion on a retina and loss of visual acuity. There is some safety margin built into these standards, so that for irradiances above the maximum permissible exposure there is a gradual increase in the probability of a lesion being formed on the retina. At other wavelengths, beams from Class 3B and Class 4 devices are capable of creating eye or skin damage unless suitable protection measures are applied. Many different types of controls have been devised and used successfully to prevent eye damage from accidental exposure to a high-power laser beam. It is worth remembering that laser eye-protection equipment (personal-protection equipment) is the last line of defence against an accidental exposure to a beam; hence, other measures form the primary safety barrier.

The term eye-safe has been applied to lasers that emit a beam with wavelengths longer than about 1,400 nm; however, this should be treated as a misnomer when high power devices are being used. A high-power beam in this wavelength regime will merely attack the cornea before reaching the retina. A better term is eye-safer wavelengths, as the amount of energy or power required to cause damage to the cornea is a great deal higher than the amount of energy/power needed between 400 nm and 1,400 nm to cause damage to the retina. This term is more applicable to the reduction in the extent of the hazard zone around a target, resulting from reflected and scattered light, during high-power interaction compared with shorter wavelengths, which are transparent to the eye and would be focussed on to the retina.

Damage to skin is also an important consideration during any development test or evaluation process. Like the human eye, skin tissue is susceptible to damage and the penetration is wavelength dependent. Consideration should be given to the use of appropriate protective clothing and gloves, if skin damage is considered to be a risk. Other protective measures, such as creams and lotions, are possible but the user should be aware of their limitations, particularly through the dilution of the barrier, owing to perspiration of the wearer.

Deaths attributed directly to the use or operation of laser devices are thankfully very rare. These unfortunate events have tended to be as a result of electrocution; however, the systems that required high voltages and rapid-rise currents are increasingly rare. Special care is vital in handling power supplies of some laser-based systems.

There are now well-established techniques and methods that can be applied to the test and evaluation of laser-based devices and systems, resulting in safe operation in a range of locations and environments. These techniques and methods rely on an accurate and effective analysis of the potential hazards and then the identification and application of appropriate risk-mitigation strategies and controls. It is important to realise that the total elimination of risk in laser-based testing is often impossible. In these cases, it is crucial to identify and use the processes and protocols, which will allow the probability of a laser incident occurring to be reduced to an acceptably low probability and with minimal impact should this event occur.

The success of effective laser-safety procedures has been demonstrated on countless occasions. Some of these events have even involved the successful test and evaluation of very high-power laser-based systems, including laser-damage weapons. However, the majority of laser incidents have resulted from the failure by personnel to observe and apply appropriate laser-safety precautions, such as failure to wear laser eye-protection equipment, or the appropriate personal-protection equipment. Other accidents have resulted from inappropriate use of the laser equipment and poor communications.

Some considerations and guidelines have been suggested to help plan successful test and evaluation programmes from the laser-safety viewpoint. Probably the most important advice is to start the laser-safety planning at the outset of the project and not try to add the laser safety and hazard control at a later date.

A protocol has been written by the International Committee of the Red Cross controlling the use of laser weapons whose principal function is to blind an opponent. This is often referred to as Protocol IV of the Vienna Convention. This legislation is observed by many countries worldwide. This international legislation does not preclude or outlaw the use of high-power laser systems on a battlefield, provided that the laser system does not target the unaided human eye with the primary purpose of creating permanent blindness.

Selected Bibliography

Barat, K., *Laser Safety in the Lab,* Bellingham, WA: SPIE Optical Engineering Press, 2012.

British Standards Institute, *Safety of Laser Products, Part 1: Equipment Classification and Requirement,* 2nd Edition, with Amendment 1, BS EN 60825-1, BSI Chiswick, U.K., 2007.

British Standards Institute, *Safety of Laser Products, Part 14: A User's Guide,* PD IEC TR 60825-14, BSI Chiswick, U.K., 2004.

Henderson, A. R., *A Guide to Laser Safety*, London, U.K.: Chapman & Hall, 1997.

Henderson, A. R., and K. Schulmeister, *A Guide to Laser Safety*, 2nd ed., London, U.K.: Taylor & Francis Group/Institute of Physics Press, 2004.

Marshall, W., and D. H. Sliney, (eds.), *Laser Safety Guide,* 11th ed., Orlando, FL: Laser Institute of America, 2007.

McKinlay, A. F., et al., *Hazards of Optical Radiation: A Guide to Sources, Uses and Safety,* Bristol, U.K.: Adam Hilger, 1988.

Sliney, D. H., and M. L. Wolbarsht, *Safety with Lasers and Other Optical Sources,* New York: Plenum Press, 1980.

Sliney, D. H., *LIA Guide for the Selection of Laser Eye Protection*, 6th ed., Orlando, FL: Laser Institute of America, 2007.

Range Safety Group, Laser Safety Committee & Range Commanders Council, *Laser Range Safety,* STANAG 3606, Document 316-98, Secretariat, Range Commanders Council, White Sands Missile Range, NM, 1998.

Winburn, D. C., *Practical Laser Safety,* New York: Marcel Dekker, 1990.

CHAPTER 18

Education and Training

18.1 Introduction

The purpose of this chapter is to provide an overview of the education and training aspects associated with the development and effective use of military equipment, with a particular emphasis on the potential impact of the introduction of novel systems with advanced technology. The delivery of effective training probably represents one of the greatest challenges when new equipment approaches operational readiness and especially when centred on advanced technology, such as high-power lasers, which may require new training methods.

A common statement is that education and training are the greatest single value-added item that the military can give to its personnel, not only for progression during their military service, but for their subsequent career. It is the effective training and education of all personnel that determines whether the military achieves its objectives against an opponent; it is not the technology used to support the tasks. Inappropriate deployment or operation of the most advanced and sophisticated technology in an inappropriate way is likely to be ineffective and possibly disastrous.

With modern warfare it is absolutely vital that education and training are an integrated activity forming a continuum, rather than being delivered as separate uncoordinated and disconnected entities. Training is traditionally used to prepare personnel for specific activities and confer skills on the basis of past experience, whereas education inculcates principles to enable people to tackle unforeseen challenges and problems by creating new solutions as and when the need arises.

Training and education, when planned correctly, have always been mutually interdependent and mutually re-enforcing. Modern approaches ensure that there is effectively a continuum between education and training, so that training has an educational function and vice versa. In times of international stability, the military can emphasise the training roles and related activities, whereas when there is regional or global instability, then the emphasis needs to be on a different part of this continuum; the balance being dependent of the level of instability and the strategic importance to the nation under threat.

The current challenges centre around the fact that world order and conflict mirror modern life (i.e., they are becoming more complex and less predictable), so there is constantly much to learn and very little time in which to undertake the appropriate learning activity. Hence, the time constraints tend to force the balance in favour of education to save time. Consequently, it is critically important to ensure there is an appropriate balance that can be achieved between the two, but have the ability to change rapidly the weighting, according to changes in circumstances.

A further critical consideration relates to the balance in the education and training given by each of the armed forces, particularly with regard to joint tri-service operations, and especially, when involving emerging technology such as novel laser-based systems. This may be a critical consideration for prosecution of both modern campaigns and successful exploitation of emerging technology and associated systems. Moreover, owing to the likelihood of joint operations with coalition forces training must give due consideration to the role of novel weapons from the interoperability viewpoint.

The objective of this part of the book is to indicate some ideas of how to provide the training and education of military personnel to develop the necessary knowledge, skills, judgement and inventiveness to cope with both the normal military challenges and exploit modern laser-based systems to their full potential. It is recognised by the author that all of the ideas need to be developed in the military context by those people with an in-depth of understanding of military skills, operations and defence planning.

18.2 Doctrine and Concepts

One of the great challenges of any new technology is the definition of an appropriate and effective doctrine, so that a training programme can be defined and concepts of operation identified. Therefore, during the development of any new equipment, whether using novel technology, or otherwise, it is crucially important that:

- System thinking/value and operational deployment strategies are considered in the context of the novel technology and are sufficiently advanced in the appropriate military context, so concepts of use/employment can be defined adequately for the new equipment to enable realistic development of the training strategy and requirements, initially at the top level and then beyond.
- An understanding is developed of the training implications of future concepts.
- An assessment is undertaken of the potential impact on the logistics chain/footprint and associated maintenance requirements and schedules, along with the skills required to maintain the equipment, including appropriate servicing of line-replaceable items.
- Specific educational requirements are identified, so that the appropriate action can be undertaken at an early stage to modify or refine these aspects of the training programme.
- An assessment of the potential impact on interoperability with allies.
- Identification of the impact on joint operations, including operations with allies.

Clearly, an early definition of the doctrine and concepts of operation and use for all systems exploiting novel technology are crucial, so that the appropriate training strategy can be identified, designed and plans established to deliver it. A number of the processes, which are followed to establish training and education delivery plans from the strategies, are considered next.

18.3 Training

Training is a broad term with very many meanings as far as the military is concerned, ranging from the basic training of new recruits (civilian to soldier) to transferrable skills and accreditation. However, it may be stated simply as acquiring essential knowledge and skills to undertake a function effectively and efficiently. For example, training will cover political as well as military functions, so that peace-keeping operations may be undertaken appropriately and successfully; at the other end of this spectrum is hot war, which demands different approaches. As far as this book is concerned, training will be restricted to discussion of techniques, which may be deployed during training activities to gain in-depth experience with laser-based equipment to influence:

- Development of doctrine;
- Development of tactics;
- Development of procedures for successful and safe operation of the laser-based systems;
- Development of techniques and methods for maintaining the equipment.

All of these aspects of military activity and influence need to be considered in the context of:

- The ever-changing operational environment, including the use of sophisticated and complex equipment in the various aspects of conflict, such as the fog of war and associated human stress;
- The changing physical environment, which includes climate change, as well as the impact of the continuing movement of people to an urban environment and the impact of modern communications and the omnipresence of the media in all of its forms;
- Joint operations, both national and international, which will be across the various critical instruments of power, namely, diplomatic, economic and military force;
- Force transformation, particularly with operations in the Information Age, which brings with it a new spectrum of challenges, as well as opportunities;
- All of the support operations associated with military equipment, such as the infrastructure of logistics, maintenance and support, as well as the ever evolving training aspects.

Some of the critical principles in the development of knowledge and skills through training with any equipment are:

- Train as the forces operate, so that the processes and procedures are realistic and embrace how the equipment will be used and thus reflect operational doctrine. It is vital to avoid training to train and then training for operations, life is too short for such nugatory effort.
- Undertake training in a continuous fashion and, where appropriate, include reserve forces.
- Have flexible and adaptable training to meet the ever-changing operational environment and, in particular, rapid dissemination of all lessons learned, so that they can be embodied into the training continuum and refine the best practice.
- Have training for operations in a comprehensive multinational, joint-agency environment.
- Implement the usual moral obligations.
- Share best practice and ensuring rapid evolution when required.

To achieve the critical goals of an effective and efficient military force, there are very many forms of training undertaken by a war fighter, which will influence the scope of training with laser technology. For example:

- Hybrid foundation training, which aims to produce force elements capable of executing successful operations from conventional to irregular warfare. Its primary function must be to maintain skills and includes all levels of collective training, but does not include basic training, education or professional development.
- Mission-specific training, which is undertaken using theatre-specific equipment and operates within the force organisational structure.
- Individual training, which is a vital step in the preparation of individuals for their appointments and, consequently, does include education and professional development.

These fundamental considerations, such as the need to train as the military fights and also undertaking realistic mission-specific training, is likely to present some challenges for novel military equipment. It is highly likely that the use of simulation tools has an important role in the training of all military personnel with novel or game-changing equipment, from which fundamental tactics and procedures may be examined.

Realistic simulation offers the opportunity to train and practice for activities that cannot be undertaken solely in live training. These simulation tools face many challenges, not only in terms of their verification, validation and subsequent accreditation, but for providing the realistic military experience, for example, the introduction of fear or stress. However, modern developments in games technology may provide an opportunity to go some way to simulate some of the stressing effects of warfare.

The development of suitable and appropriate simulation methods, in combination with live fire, may enable highly effective hybrid approaches to be devised that are cost-effective, in addition to allowing some of the objectives and principles indicated above, to be achieved. One of the critical considerations of simulation is: how realistic and appropriate is the experience? Another concern is the potential introduction of simulation artefacts into the training experience; clearly, great care is required in the use of simulation techniques to ensure realistic fidelity is achieved as far as possible. Moreover, a common complaint is that the games console at home provides a more realistic scene, in which case, consideration should be given to the exploitation of such technology.

18.4 Synthetic Environments

The creation of a high-fidelity synthetic environment is becoming an increasingly important requirement to aid effective training of all classes of war fighter. Advanced and high-fidelity synthetic environment capability, may offer potential benefits in the development of systems for specific, as well as routine, military applications. These activities could include evaluation of potential deployment options and operability of new equipment.

A synthetic environment in this context aims to combine state-of-the art communications, information and visualisation technologies, to generate in real time, a realistic representation of:

- The natural environment, such as climate, weather (including fog, clouds, rain), terrain, diurnal variations in natural light, shadows, ocean surfaces;
- Man-made environment, including buildings, roads, bridges;
- Psychological factors, including noise, flashes and other disturbances.

The infrastructure of a simulator is an array of hardware and software that can be integrated in a number of ways to provide a realistic representation of military equipment, and crucially, allow realistic interaction with human participants. In this context, the participants may be operators or commanders involved with the simulated military equipment.

The hardware may well involve a significant part of an actual platform, such as a helicopter cockpit, complete with all the controls, interfaced with an advanced mathematical model to determine its motion in response to the actions of the pilots and the atmosphere. This model and other ancillary equipment then control the motion of the helicopter in relation to the projected ambient environment, which will include the natural features along with the man-made structures.

This simulation may be operated in a closed-loop mode, so that an aggressor and defender may engage each other in a fire fight, as they react to the images presented to them from their position on or over the battlefield, including the activity around them. This type of simulation enables a realistic interaction between an aggressor and a defender to be undertaken; for example, a helicopter may be over-flying some territory when engaged by hostile fire, to which the pilot and platform may react with in one way or another, which may include a manoeuvre or deployment of countermeasures, or use of other weapons. This scenario may then continue to develop with further use of ordnance until this battle has been completed. This encounter could involve operators in adjacent rooms in the simulator or in separate facilities separated by a significant distance. Locations are irrelevant; the critical requirement is the realism of the engagement and the timely and appropriate interactions in real time.

Other features may be incorporated into the simulations that influence the behaviour of the platform and its operators. Alternatively, other features may be entirely passive, enabling commanders to monitor and assess the progression of a battle from a range of viewpoints, from which various strategies may be derived, for example; sometimes this is known as operation of the stealth vehicle.

In summary, each synthetic environment assembles information and interfaces with appropriate hardware and software to represent reality in a form that is best suited to a defined activity. The critical requirements are high fidelity, so that the experience is as close to reality as is possible, without introducing simulation artefacts that would constitute negative training; moreover, the experience should be as close to real time as is feasible.

The use of simulators in training is discussed in more detail in the section on simulated training. That section emphasises the role played by simulation, in various forms, for delivering cost-effective training with real equipment, but without firing expensive weapons.

18.5 Role of Education

The critical objective of education is the introduction of the novel-technology base to the operators and commanders to understand the attributes, as well as the limitations, of the new technology and also to appreciate that a doctorate degree is not required to operate these systems. In fact, if a posse of white coats is required to ensure effective operation of the novel system in the field, then the system developers have failed miserably. Consequently, the education of the commanders of the first deployment of laser-weapon systems, for example, should be formulated to be a multifaceted project, embracing multidisciplinary skills. These skills would range from an understanding of the technology base to the appreciation of potential effects, as well as safe operation of such systems; for effective deployment, it is probably vital that these commanders have had a pivotal role in the development of the doctrine and concept of use.

Effective and flexible education of potential operators, as well as commanders, is a crucial step in the initial training of these personnel to use the new equipment, particularly with emerging technologies that are likely to be game changers. In the initial stages, there is likely to be a significant number of iterations before the appropriate syllabus can be identified, and moreover, as the experience develops, then new requirements for the education could well be identified and introduced into best practice.

Defence academies, or similar educational institutions, have an import role in the education and development of the commanders and operators of advanced equipment, such as laser systems. This educational requirement is in addition to the development of the understanding of all of the associated military and political aspects of deployment of such weapons. Such institutions can provide a unique and effective interface between the defence world and the rest of government, including political advisors, especially on issues that concern wider national and allied security.

An effective educational programme in a defence academy will serve not only to broaden the minds of the students, but to spread an understanding of the defence and national security issues into the broader community. This understanding and appreciation of the potential military value of novel equipment can extend beyond national considerations to an involvement with international partners, provided that the academy has the appropriate links.

18.6 Training Design

It is recognised that advances in technology are likely to influence the development of military equipment at an ever increasing rate. Small incremental technological advances are likely to lead to disproportionate increases in war-fighting capability and a consequent enhancement in the effectiveness of the equipment, provided its potential is exploited fully. These advances in equipment capability, fuelled by the technological advances, are likely to reduce the concept-to-capability timeframe and increase the scope for the use of more off-the-shelf devices to support operations. Clearly, future training methods and training processes need to be sufficiently flexible to exploit these opportunities fully and in a timely and cost-effective fashion.

Doctrine will probably have an even greater influence on the design of training, because of the anticipated changing characteristics of future operations. Additionally, there will be a need for a comprehensive approach to these operations, co-ordinating all inter-agency activities, to embody all the associated capabilities. This comprehensive approach is required, in order to achieve the flexibility and requisite responsiveness, so that aspect will need to be an integral part of the future training plans. Furthermore, it must be recognised that technological advances will be exploited by potential enemies, used by regular as well as by irregular forces.

This critical training function is normally undertaken by a warfare centre, whose role is to determine doctrine, and the appropriate and relevant training, for campaigns. Its most immediate mission is normally the preparation, through training, of the war fighters up to brigade level for any current and anticipated future (near-term) operations.

To fulfil these roles, it is crucial that this training body develops and directs the tactical thinking for these anticipated (or immediate) operations through the development of the appropriate doctrine. From these activities, the training requirements for the efficient and effective operation of the resources may be defined. A critical consideration is the fusing of any developing doctrine with evolving operational experience, to ensure that the latest thinking and knowledge is passed on to the next generation of operators and commanders.

A systematic process will usually be identified that will determine the long-term strategy, direction and organisation of future training requirements across the military, to equip the war fighters with the skills required to support future operations using novel, in addition to, conventional equipment. In all cases, the military objectives are identical (i.e., equipping the war fighters with the skills to ensure the equipment realises its full potential and the military objective are achieved as efficiently as possible). There are normally a number of stages in this corporate process:

- *Define:* identification of the training capabilities and techniques needed to support future operations and new equipment.
- *Direct:* identification of the training requirements, along with the training resources, needed to support the identified activities.
- *Construct:* linking the development of a training schedule and processes, involving discrete training events. To maintain its relevance, particularly when involving new equipment, this stage in training development may well require refreshing frequently.
- *Refine:* this is the stage where the personnel to be trained have an opportunity to engage with the trainers. The critical objective here, is to establish detailed training plans within the greater training plan for development of the crucial skills needed by the military personnel;
- *Execute:* this is the delivery stage for the agreed training plan, identified and defined in the previous steps;
- *Review:* this stage provides an opportunity to consider the effectiveness of the new training plan and processes, where areas for improvement can be identified and fed into the new review cycle.

It is also important to identify and define the roles of the various organisations and appointees involved in the entire range of training activities. This framework and operational structure provide another framework through which the warfare centres can deliver the capability to meet operational requirements.

A typical warfare centre will have many component parts in the training schedule. For example, with the army this is likely to involve use of ranges that span environments from tropical to arctic conditions and from treeless plains to dense jungle, as well as representative urban landscapes. These centres usually have a variety of sophisticated simulators, to enable the military personnel to develop their skills and procedures for effective operation of the weapons. It will be a challenge for users of directed-energy weapons particularly to simulate realistic effects, so that the range of defeat mechanism (kills) can be understood and fully exploited.

Training design must take account of the fundamental principle that the military should train as they operate, or in the case of novel technology, perhaps how they intend to operate the new equipment. In this case, the training design may have several iterations. Consequently, training must reflect operational doctrine and be as realistic as possible, so that the conditions of operation experienced during training are realistic with a very high degree of fidelity. It is crucially important, that war fighters are not expected to undertake tasks during operations for which they have not been trained, nor have the appropriate experience, to avoid additional risks once deployed on operations in theatre.

An additional consideration in training design is how to embody the friction and operational complexity of operating in extreme environments, particularly with nuances associated with novel technology. Appropriate training design with modern technology, as well as with all equipment using advanced technology, should minimise these issues once the concept of operation is defined and understood.

The design of any training activity must take into account the educational base and capabilities of the war fighters as well as the demands of the equipment, which ideally will be minimal. An additional consideration for the designers of training schemes, as well as the designers of advanced equipment, is how to ensure reserve forces can use the advanced equipment effectively during operations across many environments.

An additional but very important consideration in any training activity is the time spent preparing to train, as this reduces the time available for an actual training event. It is widely recognised that training should take place wherever possible on operational equipment or at least, realistic surrogate kit, which accurately mimics the actual operation of the real equipment.

To achieve the maximum benefit from any form of training, consideration must be given to:

- Effective exercise management systems that can generate appropriate scenarios rapidly and within the necessary context, which can allow exercises to be tailored to meet changing requirements; this is potentially particularly important for novel weapons to permit modifications to the training as lessons are learned.
- Opportunities for comprehensive after-action reviews to provide considered opinions on the strengths and weakness of the training experience. This process, if undertaken in a timely fashion, is likely to enhance the training experience, particularly in the context of use of equipment embodying novel technology.

The important reasons for training may be summarised as follows:

- Ensure that all relevant personnel are familiar with all of the operational processes for successful operation of any specific weapon system under any environmental and stressing conditions.
- If the military cannot practice with the actual equipment, then full system operation cannot be checked.
- Practice is required to develop techniques for effective operation;
- Timely provision of data is critically important to support development of new concepts of operation or employment.
- Investigation of the value of novel weapons, such as conducting warfare without the need to monitor the depletion of a magazine (i.e., getting attuned to the bottomless magazine), and understanding how to exploit this factor fully.

Laser weapons offer some invaluable advantages to training, such as:

- Use of an eye-safe (or eye-safer) device for training;

- No need to expend real ammunition; bullets or missiles are not required, so they are potentially very cost-effective training options;
- Ease of use through being a line-of-sight weapon; no need to calculate lead angles even in a highly dynamic engagement.

In the case of novel technologies, such as laser weapons, it is recommended that the military staff work with appropriate scientific advisors to identify and develop the necessary critical technological issues in the military context. From this baseline, the doctrine and evolution of the tactics, techniques and procedures for this modern approach to warfare may be derived. The enduring and generic tactical doctrine field manuals may then be written, for use at the divisional level and below.

These activities require various levels of experimentation, which could span activities from live trials on an appropriate range to testing in a synthetic environment of the anticipated battle space. The design and development of appropriate synthetic-environmental effects is a crucial step in training, so that commanders are provided with the tools they need to control the battle space effectively, in ever evolving and more complex operational scenarios.

It is important to realise that it is impossible, let alone affordable, for the war fighters to be trained to experience every conceivable operational eventuality. Consequently, an educated and calculated risk has to be taken against the full range of tasks, defined in any mission task list, required to support operations. This may involve:

- *Capability risk:* this risk is normally taken deliberately and as a matter of identified priorities, to concentrate on establishing capability to meet immediate operational needs and is undertaken at the expense of contingency capability.
- *Training risk:* this is usually a consequence of resource limitations, which may limit some, or constrain particular aspects, of the training plans. Once identified and the consequences are understood, then the requirements can be fed into the construct segment of the training strategy and planning cycle, considered above.

It is crucial that the military lawyers are involved in these activities, so that the correct application of international law is applied and understood by the commanders. The lawyers can then be an integral part of the definition and evolution of the rules of engagement, to ensure full compliance with international law. As noted previously, there has been much misunderstanding of the Vienna Protocol on the use of laser weapons, even going so far as naming it incorrectly.

It may be summarised that all of the facets of training must be designed to ensure that the maximum benefit can be achieved from each event. Training methods and content must be sufficiently flexible, to meet current and future operational requirements in an ever changing environment. Lessons learnt during operations must be assimilated rapidly into all aspects of training to maintain best practice throughout the training experience. Clearly, flexibility and adaptability are key aspects of training design along with the requite responsiveness.

18.7 Equipment Training

It is recognised by the military that equipment training is a critical enabling aspect of all operations, as well as a potential constraint for training activities. A major challenge with all equipment training is to ensure that all of the operators know and understand how to use the equipment efficiently and effectively; prior to operations, this activity will form part of the mission-specific training. This training will be undertaken through a mixture of simulation and live-fire activities, as discussed above. Obviously, there is limited scope for such activities with guided weapons that cost millions of dollars or pounds; however, it is less of an issue with bullets, and perhaps similarly, when training with advanced systems using novel technology, such as lasers.

New equipment entering service will require robust through-life training solutions, which may require revolutionary thinking in the case of some equipment, such as directed-energy weapons, to ensure its potential capabilities are exploited fully. Equipment training needs to be designed to enable the rapid introduction of new equipment into theatre, together with the effective use of this novel technology, so there is an opportunity to counter rapid changes in an enemy's tactics or deployment of new equipment.

Advances in weapon technology, as well as developments in doctrine, are likely to lead to:

- Increased complexity in the equipment, as well as having an impact on its deployment during operations;
- Enhanced lethality from the systems with reduced collateral damage;
- Application of tuneable and scalable effects;
- Increased responsiveness;
- Increased use of autonomous systems in defensive as well as offensive roles.

Some of these technological advances will offer opportunities for more effective training systems, some of which have been discussed in Chapter 16. Similarly, advances in computer technology, particularly with graphics, continue to enhance the simulation tools, as well as for synthetic environmental training. It would appear that in the longer term, equipment training and design will lead to the balance between simulation and live-fire training moving towards the various aspects of simulated training.

Equipment training with novel weapons must recognise that it is not a silver bullet, nor is it likely to be an exact higher-technology replacement for a standard bit of kit, so it can be trained with, and consequently, used in exactly the same way as the more conventional equipment. It is highly probable that new or revised approaches will be required, if the full potential of any novel technology is to be realised during operations. For example, an air-defence system based on novel technology may be used in a similar fashion to a missile-based system but it may be positioned in a battery in a different location compared with a guided-weapon system; however, in this case the training methods are likely to be an evolution of existing techniques and will certainly be far cheaper in live-fire training. Other deployments may require more radical thinking.

18.8 Simulated Training

Simulation techniques offer the opportunity to train and practice for those activities that cannot be undertaken solely in live-fire training. The military has an ambition for a blend of training activities involving instrumented live-fire as well as dry (use of blanks or without the full lethal system, such as a warhead or countermeasure system) events that eliminate the need for re-equipping and familiarisation training, prior to the start of training programmes. Clearly, laser technology has much to offer, both as a simulation tool on the range and as a novel-weapon technology.

The primary goal of any simulator is to render fidelity in real time of the scenes generated to match what the war fighter will experience on the battlefield, to avoid negative training. The rapid development of hardware and software in the commercial sector offers an opportunity to produce high fidelity simulators that are affordable, leading to more accessible training for all war fighters.

Modern simulators, embodying state-of-the art simulation techniques, enable advanced training to be undertaken, as well as refinement of tactics and procedures. An example is a simulator that generates and displays terrain and other features to emulate artefacts or objects observed by an advanced targeting pod, which can integrate live, virtual and construct elements to simulate increasingly complex training environments. As the overall visual fidelity in the simulators increases, so must the databases that underpin the simulation, such as the terrain database, and in the case of novel technologies, the effects generated by that equipment. It is now common for these simulators to include stored data to generate various atmospheric effects, such as fog and clouds, as well as smoke.

The advances in the hardware and software within the graphical processing units enable the simulated tasks to be reconfigured rapidly and avoid the need for pre-recorded data, owing to the techniques available for real-time scene generation, as discussed earlier. Hence, it is now possible to define a specific application using the common simulator equipment, and merely reconfigure the software and customise the platform interfaces to fit the operating environment, such as a rotary-wing or fixed-wing aircraft using particular weapons.

The initial thrust for simulated training was driven by the need to train war fighters more effectively and efficiently; however, financial and environmental pressures have limited the opportunities for the more routine live-fire training. Additionally, with the increased fidelity of simulators, exploration of the envelopes of performance of weapons, including aircraft, may be explored safely. These aspects may also be explored for advanced equipment using novel technology, including associated effects, during operations. Simulation techniques can additionally act as a tool, to identify the impact of time-critical aspects of decision made by commanders.

A sensible balance is required between live-fire events and the supporting simulated events, to ensure that the war fighters are adequately prepared for operations with the allocated equipment. One of the major challenges of any simulated training is to embody the realistic frictional aspects of operations requiring robustness, stamina, uncertainty and the ability to cope with danger, into the training events, as noted above. Currently, the only realistic option for achieving that level of preparedness is through live-fire training in the appropriate environment and this is likely to be the situation for the foreseeable future.

18.9 Laser-Safety Training

This is an example of equipment-specific training required by many organisations to ensure personnel involved with any aspects of laser operations are competent, and in particular, fully appreciate the risks associated with the operation of laser-based devices and systems, especially high-power devices. Whilst this form of training is specific to laser technology, all other technologies that present potential hazards have similar specialised mandated training requirements, such as instruction and demonstration of competency in the safe operation of electrical, explosive, toxic and radio-frequency equipment, to name but a few examples.

Laser-safety training can take many forms ranging from an in-depth understanding of the laser technology, to a general awareness of the hazards and a definition of the critical terms, such as the maximum permissible exposure, as discussed in Chapter 17. In many countries, it is a fundamental requirement that an individual, or set of individuals, undertake the appropriate detailed laser-safety course, so that they can undertake the role of the laser-safety officer (laser-safety supervisor).

It is usually a mandatory requirement that a fully trained laser-safety officer is stationed at all locations when using high-power laser technology, such as at a military maintenance depot; this requirement normally refers specifically to circumstances when operating Class 3B or Class 4 lasers (as defined in Chapter 17) with potential exposure to the beam. In the cases when the beam is fully contained and inaccessible, such as in a sealed test system, then this requirement may be relaxed. The laser-safety officer is ultimately responsible for the safe operation of laser devices and systems under his or her control, even when other suitably qualified personnel are operating the device or system; thus, it is crucially important that they have executive authority to understand and control operations.

An in-depth course is also likely to be most appropriate for personnel working on the servicing and maintenance of laser-based systems in the ordnance depots. Similarly, operators and engineers with skills in front-line maintenance are also likely to require this level of competency, which will require a periodic refresh to ensure all such personnel are fully conversant with the current regulations. Additionally, it is normal for ranges where there is any live firing of such equipment, to have either a resident laser-safety officer or have access to such designated individual, during the live-fire events.

Laser-safety awareness training is normally appropriate for the users of the laser-based equipment, so that they are cognisant of the potential hazards and risks involved with a malfunction or inappropriate use of such equipment. Just like any other weapon, inappropriate use can have potentially disastrous consequences and not only to those undertaking the inappropriate activity. It is perhaps telling that in the British Ministry of Defence, laser safety is managed within its Ordnance Safety Group; in other words, military laser-equipment, whether it uses a laser source or is a laser system, needs to be thought of as a loaded gun from the safety point of view.

Many institutions are qualified to provide this general level of training, covering the fundamental legislation and mandatory requirements stipulated by the nation concerned. These institutions provide this training on a regular basis; however, a crucial consideration is the identification of courses that will train the appropriate staff when there is a specific focus, such as the military use and maintenance of laser systems and devices. For many years the Defence Academy in the United Kingdom has provided such training for military personnel, including for her allies.

18.10 Equipment for Training

It is essential that sufficient financial provision is made during any military equipment procurement programme, for the supply of an adequate number of systems solely for training operators and commanders. These examples of the equipment, along with any training-specific analogues, must be fully representative of the theatre-ready kit, along with the service and support items. This is crucial for all types of new equipment, and moreover, this requirement is not unique or specific to novel weapon technology, or other systems using advanced technology.

The early involvement of military personnel with the development programme for new equipment can have many potential advantages, for example:

- Provision of an early indication of specific human factors in the successful operation of the equipment. There is a famous story, probably apocryphal, that an Air Force pilot made a recommendation following his maiden flight in a prototype aircraft that entry to the cockpit was very difficult and suggested that owing to the poor handling performance of the aircraft during that flight, entry to the cockpit should be made impossible. A state of affairs to be avoided through appropriate early engagement of military operators.
- Development of appropriate and effective human/machine interfaces.
- Defining the operational requirements for appropriate training projects, which is especially important for novel systems.
- Rapid development of training techniques and procedures for equipment entering service in response to a rapidly changing or evolving threat.
- Identification of the appropriate maintenance requirements.
- Understanding the needs for future training requirements and the impact on existing training equipment, in addition to an assessment of the impact on future training aids and support systems.
- Ensuring that the training burden is manageable and realistic.

It is crucial that there is an appropriate provision of real equipment-training, whether the particular military equipment involves novel technology or otherwise. The challenge presented with novel technology, such as laser weapons or directed-energy systems, is to provide the commanders and operators with the right equipment at a sufficiently early stage in the development of such a system, to enable the right training experience to be defined and delivered in a timely fashion. This type of procedure should enable the potential game-changing attributes to be exploited during operations immediately after introduction into service, if required.

18.11 Summary

Education and training are crucial components in the development and effective use of novel technology used in military equipment, if the optimum value is to be realised from the investment made in the development of the new equipment and provision of potential new military capability. A critical consideration is the early engagement of the military planners and strategists, during the development of novel weapons or applications of advanced technology; this early engagement enables concepts of use and employment to be defined and the requisite doctrine evolved to ensure appropriate realisation of the new military capability.

The deployment of advanced technology often appears to lead to complexity in a whole range of military activities: training, maintenance and support and operations in theatre. Hence, the critical role played by education and training of the war fighters, who will be responsible for ensuring the new equipment is operated as designed and is deployed correctly, so that its full operational potential can be achieved and new capability realised.

The great challenge with any new equipment is to establish an appropriate training process and procedure that is relevant to the new system, along with the provision of the timely modification to the associated military doctrine. Flexibility is required at this stage, so that the feedback from training and initial operational deployments can be injected in a timely fashion, in order for best practice to be identified and processes refined and then applied as rapidly as possible.

Military forces invest heavily in a whole range of training courses and training exercises to ensure the war fighter is as ready as possible for operations in theatre. The crux of all training is to provide the war fighters and their commanders with the essential skills, experience and knowledge to achieve their military objectives as efficiently and effectively as possible, with minimal risk to those personnel involved in those operations.

Training is traditionally to prepare personnel for specific activities and confer skills on the basis of past experience, whereas education inculcates principles to enable people to tackle unforeseen challenges and problems by creating new solutions as and when the need arises. The introduction of novel technology will require a careful analysis of the balance between these activities.

Education has a critical role to play in the training continuum, so that the commanders and other key personnel understand most if not all of the critical aspects of the technology base. This education ranges from the awareness of the performance potential of the novel technology, to the possible hazards and how to mitigate these risks. It is recommended that all commanders that are likely to apply this technology on the battle field undertake appropriate educational courses, including laser-safety training and then design the concept of operations and employments, as well as defining the doctrine for use and operation of novel technology.

Selected Bibliography

British Standards Institute, *Safety of Laser Products, Part 14: A User's Guide,* PD IEC TR 60825-14, BSI Chiswick, U.K., 2004.

Henderson, A. R., *A Guide to Laser Safety*, London, U.K.: Chapman & Hall, 1997.

Henderson, A. R., and K. Schulmeister, *A Guide to Laser Safety*, 2nd ed., New York: Taylor & Francis Group/Institute of Physics Press, 2004.

Marshall, W., and D. H. Sliney, (eds.), *Laser Safety Guide,* 11th ed., Orlando, FL: Laser Institute of America, 2007.

Sliney, D. H., and M. L. Wolbarsht, *Safety with Lasers and Other Optical Sources,* New York: Plenum, 1980.

Sliney, D. H., *LIA Guide for the Selection of Laser Eye Protection*, 6th ed., Orlando, FL: Laser Institute of America, 2007.

Winburn, D. C., *Practical Laser Safety,* New York: Marcel Dekker, 1990.

APPENDIX A

The Development of Infrared Threat-Seeker Technology

A.1 Introduction

The development of successful and effective infrared-countermeasure (IRCM) systems that will provide high-performance platform protection is dependent on a thorough understanding of the operation of the various classes of threat seeker and how to confuse their target-tracking systems. This is because the technical approach used by jammers is quite subtle, but needs to have a rapid effect on the seeker's target-tracking system, as the engagement time lines are short and this is a soft-kill technique.

The 1960s saw the emergence of heat-seeking missiles into operational service. This was the signal for the start of a technical push for much-needed robust techniques to defeat the optical seeker systems in these weapons, which are still difficult to intercept with hard-kill weapons, owing to their small size and short engagement timeline.

Many of the active-countermeasure projects were initiated in the 1960s in response to the introduction of the heat-seeking missile into operational service. On-board systems (i.e., those that are integrated with the fitted platform, and are known as jammers) were developed to confuse the target-tracking system in the missile's seeker. In this case, the defeat mechanism was through deception. This defeat mechanism is in contrast to flares or expendables, which use a seduction technique to move the missile's sensor away from the protected platform. A very brief overview of the off-board countermeasure techniques is given in Chapter 9.

Other pragmatic techniques were also devised when this threat originally appeared, including the extension of the jet pipes on fighter aircraft to mitigate the effects of an interception by a heat-seeking missile homing on to hot metal. This process of threat development and devising a countermeasure has been a classic cat-and-mouse activity between the measures and the countermeasures, which continues to this day, as shown in Figure A.1 with some examples of potential threat seeker technologies. An important operational consideration is the flexibility and adaptability of the new technology, so that the older weapon systems can still be defeated by the countermeasure system, as well as the new ones. Figure A.1 also

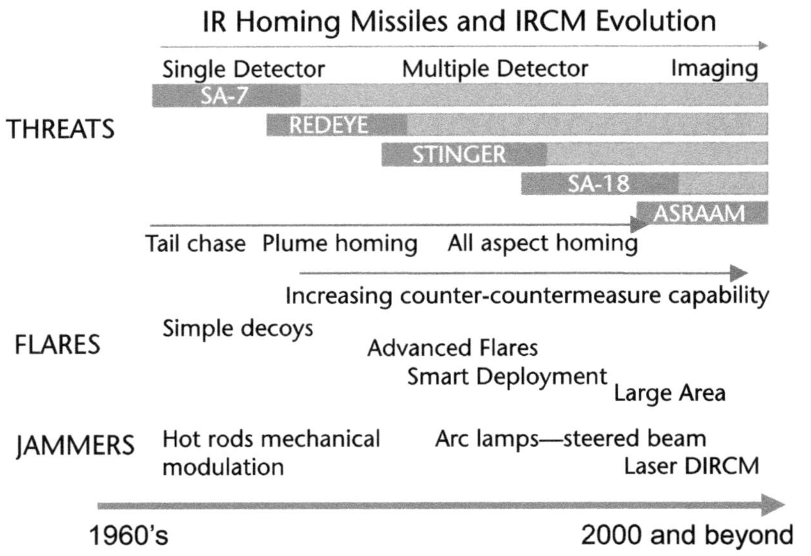

Figure A.1 Summary of threat and countermeasure development.

indicates how the versatility of the man-portable anti-air weapons has evolved from tail chase to all-aspect engagement, along with enhanced robustness to jammers and flares [i.e., increased count-countermeasure (CCM) capability], although the agreed terminology of today is somewhat different.

A.2 Generic Optical Seeker

The generic optical arrangement of an infrared seeker is shown in Figure A.2; an optical seeker contains a telescope, usually a fully reflective Cassegrain design, as shown in this figure. The whole seeker-head assembly is rotated about its optical axis to provide gyroscopic stability of the sightline to the target [i.e., to provide an inertial reference (of its pointing vector) once it is uncaged]. The rotating assembly is often known as the seeker gyroscope (or simply the gyro) or the optical gyroscope.

The infrared photons collected from the thermal scene viewed by the telescope are focussed on to the reticle, a small disc with alternative opaque and transparent zones, sometimes known as spokes, which modulate the energy incident on the detector positioned immediately behind the reticle. These modulated electrical signals generated by the detector are processed by an envelope detector and filtered to generate the guidance commands for the missile's control system to ensure interception of a target.

A.3 Target-Tracking Techniques

The fundamental approach to sensing a target's position in the early generations of infrared seekers used a single detector, which had a rotating reticle or modulator in front of it, which provided a time-referenced modulation to the detector.

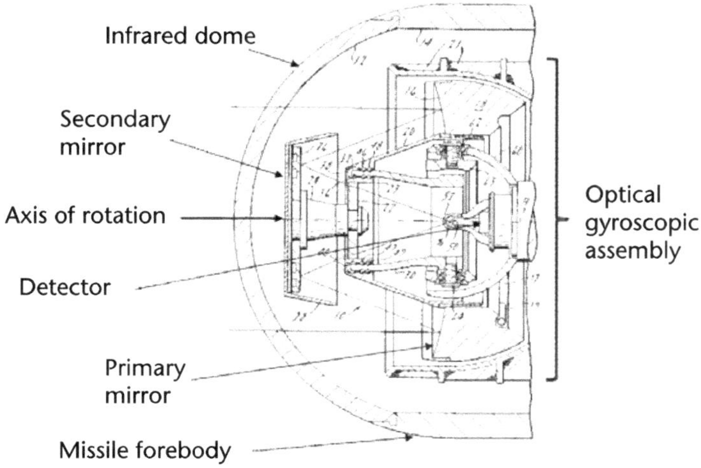

Figure A.2 Generic optical seeker. (Courtesy of Mr. B. Waggoner, NSWC Crane, United States.)

This detector, in turn, provided a modulated electrical signal to the target-tracking system. There have been a couple of variants of this fundamental approach, used to sense the displacement of the target from the optical axis of the seeker. These techniques are:

- Spin-scan methods have the rotating reticle; half of the reticle transmits 50% of the photons incident on it that are in-band to the system, whilst the remaining part of the disc of the reticle has transparent and opaque spokes. The 50% transmitting segment acts as a reference to determine the position of the target in the field of view of the seeker at this particular look-angle of the seeker. When the target is displaced from the optical axis of the seeker, there is an amplitude modulation of the signal, its phase being dependent on its angular position with respect to the datum. However, when the target approaches the optical axis, the error signal diminishes, leading to noise dominating the control signal and the guidance accuracy is degraded, especially in the terminal phase.
- Conical-scan techniques, which were developed to overcome the shortcomings of the spin-scan technique, particularly when the target was on the seeker's boresight. In this configuration, the optical train is modified slightly so that the secondary mirror is slightly offset in angle with respect to the angle of rotation of the seeker. Additionally, the reticle is fixed and only consists of transparent and opaque spokes. As a result of the motion of the optical gyroscope, the instantaneous field of view of the seeker is scanned over the reticle. When the target is on the optical axis of the seeker, the motion of the seeker's rotation causes the instantaneous field of view to trace out a circle on the reticle and there is a constant modulation frequency determined by the spin frequency of the seeker and the number of spokes in the reticle. However, as the target is displaced from the sightline, the locus of the path of the scanned scene is displaced, and as a result, the light beam is modulated with a frequency dependent on the offset of the target from the optical axis. This

technique enables guidance accuracy to be maintained, even when the target is centred on the optical axis of the seeker.

The upper part of Figure A.3 shows the fundamentals of the spin-scan technique on the left with the reticle spinning in the focal plane of the telescope. The more recent conical-scan method, with its fixed reticle and the instantaneous field of view scanned over it, is on the right; also shown on this image is the change in modulation produced by the reticle, as the target moves further away from the optical axis. The lower part of this figure is devoted to two scanned-detector techniques, which represents a more modern development prior to the development of imaging seekers. In general, the field of view of these optical seekers is of the order of a degree or two, depending on design.

More modern seekers have continued the development of the target-tracking technique using various optical arrangements such as rotating slots or mirrors to scan detector(s). These approaches were developed to enhance the robustness of a seeker's target-tracking system to optical-countermeasure techniques. These latter techniques avoid the need for a reticle, but there is a more complex arrangement to

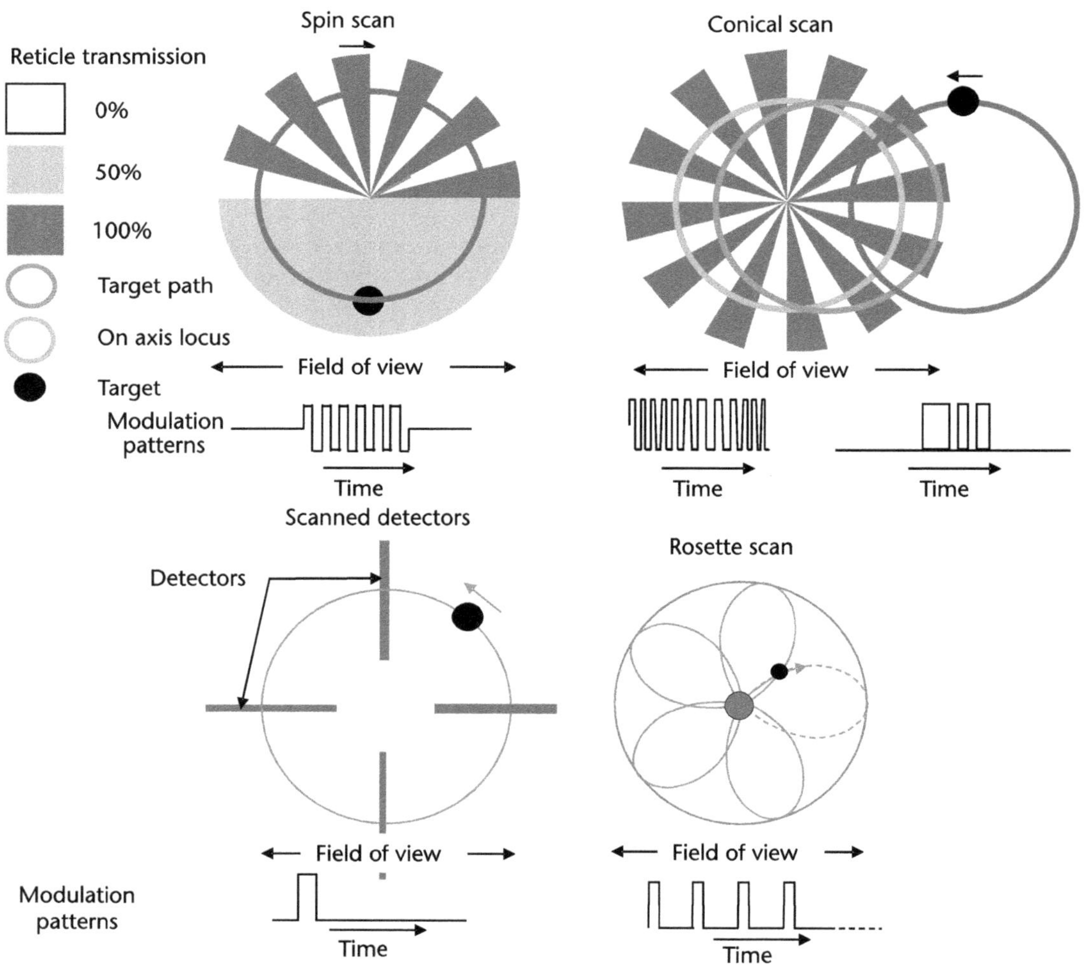

Figure A.3 Examples of some seeker-modulation techniques. (Courtesy of Dstl, United Kingdom.)

scan the focussed energy over the detectors, particularly in the case of the rosette-scan seeker.

A.4 Impact of Seeker-Technology Developments

The other important development, with regard to the seeker technology, has been the transition from uncooled detector devices such as lead salts, which are sensitive to photons with a wavelength around 2 μm (i.e., at the shorter wavelength end of the mid-wave infrared atmospheric window) to cooled detectors, such as indium antimonide. The latter detector material is sensitive to photons with a wavelength at the longer wavelength end of the mid-wave infrared window, extending from approximately 4 to 5 μm.

The change in the detector technology was made to give greater flexibility of engagement, as the first-generation missile seekers were retribution weapons, homing on to the hot-metal emissions from the jet engines, whereas the longer wavelength system allows engagement from any direction, given adequate signal-to-noise ratio. In this case, the source of infrared energy for the homing process is the exhaust plume from the engines and other sources with a temperature around 800K. The use of cooled-detector technology also enabled longer-range engagements to be achieved, thus further enhancing the utility of this type of weapon system.

The latter development had quite an impact on the potential installed performance of the various classes of jammer systems. In the case of the countermeasure systems that use omni-directional emissions from incoherent sources, the jam-to-signal ratio was reduced, and thus, the protection system became less effective and in some case the jammer only acted as a beacon for the seeker to track. This degradation in performance is a consequence of the black-body emission from this type of source, whether it is a heated rod or an arc lamp. In fact, this reduction in performance led to the development of directed-infrared countermeasure systems, where the emission is focussed into a narrow beam to increase the radiant intensity in a particular direction, but at the expense of the emission in other directions.

This change in operating band also has had an impact on the complexity of laser-based, infrared-countermeasure systems. These threat developments have meant that wavelength conversion techniques are required in a laser-based countermeasure system if a single conventional solid-state source is to be used, owing to the need for several wavelengths to be emitted simultaneously; however, modern semiconductor devices avoid this complexity, as discussed in Chapter 3.

A.5 Imaging-Seeker Technology

The most recent development in seeker technology uses imaging techniques and is generally known as the fourth-generation seeker technology. There are several approaches to forming an image of the scene, one using a scanned linear-array and another exploiting a staring method involving a focal-plane array.

This type of seeker technology is much more complex than the simple single-detector seeker but has far superior target-tracking and engagement capabilities. It operates in a similar fashion to the fine-track sensor used in a beam director, already described in Chapter 4; moreover, it will have similar subsystems to support the target detection, acquisition and tracking.

This class of seeker technology presents a significant challenge to all countermeasure techniques, owing to the availability of spatial information to the seeker's target-tracking system. Chapter 9 discusses some of the laser-based countermeasure techniques that may be applied to optical seekers to defeat the target-tracking system in this class of threat.

A.6 Summary

Infrared optical-seeker technology has evolved steadily over the last five decades, forcing a similar evolution of the platform-protection techniques. Modern, single-detector infrared-seeker technology is very capable and sophisticated countermeasure techniques are required to ensure robust platform protection. Generally, the developed countermeasure techniques have responded well to the challenges presented by the advances in seeker technology.

Each development in the seeker technology and, in particular, the techniques used has made the challenge of achieving robust platform-protection capability, or even maintaining existing protection, demanding. In general, the development and deployment of platform-protection techniques has been reactive, in all senses of the word.

The in-service deployment of imaging-seeker technology presents the next challenge for the countermeasure technology, to ensure that the new approaches will also be successful against the older systems. This is the subject of advanced research topics.

APPENDIX B

Jammer Development

B.1 Introduction

This appendix will concentrate on discussing the jammer-technology evolution and, in particular, the journey from the black bodies initially developed to defeat the first-generation seeker technology to the laser-based techniques installed on aircraft to defeat the first, second and third generations of optical seeker. The imaging-seeker developments, indicated in the Appendix A, present a definite challenge to any countermeasure technique, particularly if the jammer is to produce a robust defeat mechanism, such as optical break lock; however, the ability of modern laser-based systems to generate very flexible waveforms does form the basis for defeating this latest development in seeker technology.

The basis of the seeker-defeat mechanism is to emit a series of modulated pulses that are more intense than the modulated signals generated from the target being tracked by the seeker and close to or marginally out of phase with that signal, so that the target-tracking system in the seeker becomes confused. As a result of the confusion, false target displacements are perceived by the seeker's target-tracking system, so that erroneous commands are given to the missile's control system. It must be stressed that this in-band approach to defeating a seeker is not about dazzling that seeker but about providing spurious information to the seeker, so that its target-tracking system is defeated during this confusion process and the intended target is lost.

A summary of seeker development and projected developments is shown in Figure B.1, which also indicates a synopsis of the tracking techniques.

B.2 Development of Jammer Techniques

This countermeasure system is attached to the platform it is protecting and it uses an in-band technique to interact with the threat seeker and, ultimately, its target-tracking system. The defeat mechanism involves the jammer emitting a series of infrared pulses, which when received and processed by the seeker cause confusion in the target-tracking algorithm in the seeker's electronic system. This interaction

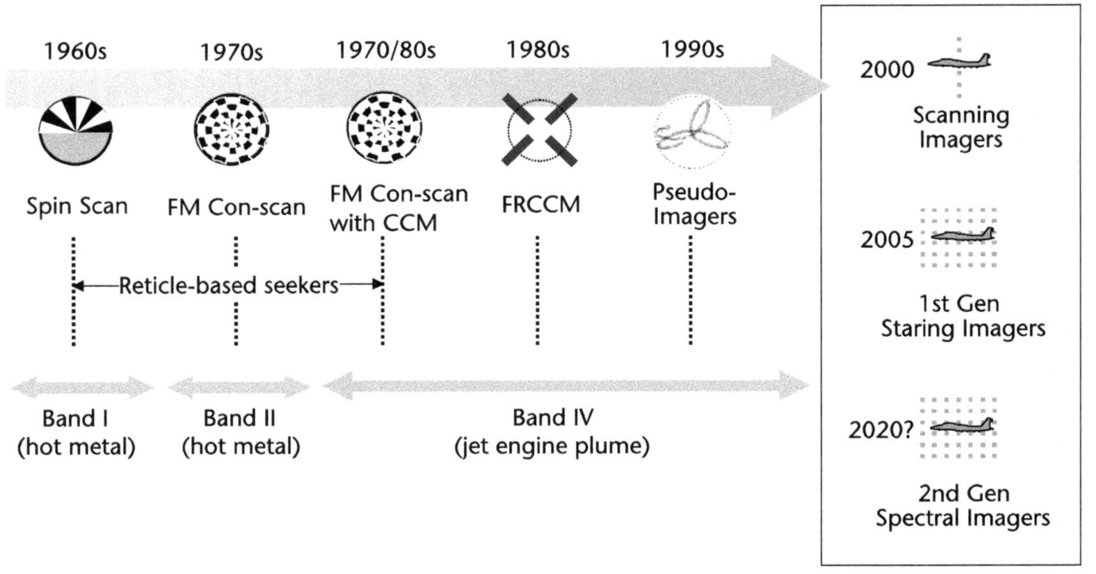

Figure B.1 Seeker operating bands and technology development.

can result in instantaneous and dramatic manoeuvres by the missile causing it to miss its intended victim, as shown in Figure 9.13.

The original infrared-countermeasure (IRCM) systems were staring devices, which contained an incoherent black-body source, whose infrared emission was modulated, or switched off and on, by a mechanical chopper system to form the modulated jam waveform. More modern approaches have used arc-lamp technology, leading to the emission of more sophisticated modulations in the jam waveforms, which, in turn, has led to enhanced platform protection.

The recent developments with jammers have seen the move to directed-infrared countermeasure (DIRCM) systems that use a narrow beam of modulated infrared energy with high radiant-intensity characteristics to defeat modern sophisticated infrared seekers, such as the second- and third-generation devices, discussed in Appendix A. The emission of a narrow beam demands the use of a beam director (or beam-pointing system), and other support systems to help locate the threats, and then to direct the countermeasure energy (i.e., the modulated infrared waveform) into the acquired threat seeker.

DIRCM systems were developed in the 1990s, using more conventional infrared-source technology, in response to the continuing enhancement of the potential lethality of infrared-homing missiles. The initial developments involved the use of arc-lamp technology and focussing optics to concentrate the photons into a narrow beam; an initial acronym was FIRCM (focussed-infrared countermeasure), but this was soon dropped in favour of directed-infrared countermeasure. The most recent development with directed infrared-countermeasure systems is the use of infrared-emitting lasers for the source of infrared photons. A very noteworthy accomplishment was the design and development of the Nemesis DIRCM system, which became operational in this decade (circa the mid-1990s), with enhanced capability when the laser-based system entered service around 2000.

This period saw large advances with compact solid-state infrared lasers for use in directed-infrared countermeasure systems. The development of periodically-poled lithium niobate, and its adaptation for use in compact mid-wave infrared systems, was a significant technological advance, together with semiconductor lasers for pumping the gain medium; moreover, more recently semiconductor devices have been designed and demonstrated for direct generation in the infrared, for a range of applications but particularly infrared-countermeasure applications.

B.3 Incoherent Sources for Infrared Jamming

B.3.1 Black Bodies

The original jammer sources were simple black-body devices that were heated to a suitable temperature to give an infrared emission of adequate radiant intensity around 2 μm to defeat the first-generation seekers, such as the early spin-scan systems. These sources were either electrically or fuel-heated carbon or tungsten radiators with an effective radiating temperature of 1,700K to 3,000K. A more recent development with incoherent jam-source technology has been the adaptation of arc lamps to generate the infrared energy in a continuum across the appropriate wavebands, to counter the later generation of optical seekers in these anti-air weapons, as noted above.

B.3.2 Arc-Lamp Technology

These electrically driven discharge devices behave like a heated black body, characterised by a colour temperature, with an emission across many wavelengths from the ultraviolet through the spectrum into the infrared, as discussed more fully in Part I. A number of gases have been used in these lamps at various pressures to provide the desired emission characteristics; common examples are alkali-metal vapours and inert gases, such as krypton or xenon. The modern directed-infrared countermeasure systems, using arc lamps, tend to use xenon or krypton gas-filled lamps, which have a typical colour temperature of 5,000K to 6,000K, whereas the metal-vapour lamps have a colour temperature around 2,000K to 3,000K, and are quite common in the staring-infrared countermeasure systems.

The development of metal-vapour arc lamps for countermeasure applications enabled significantly larger jam powers to be emitted more efficiently compared with heated rods, in the shorter wavelength region of the mid-wave infrared (Band I). Moreover, arc lamps can be modulated electronically to give better modulation depths, owing to the ability to be able to switch the lamp on and off very rapidly. This development led to a reduction of unmodulated jam energy (known as self S), which augmented the platform signature and can assist the seeker during the target-acquisition process. These developments with arc-lamp technology and, particularly, the more rapid switching ability have enabled more sophisticated and more effective jam waveforms to be generated. A photograph of a metal-vapour arc lamp is shown in Figure B.2.

Increasing the colour temperature increases the emission at all wavelengths, including longer wavelengths, as required to defeat the more modern threats, but

Figure B.2 A metal-vapour arc lamp. (Courtesy BAE Systems North America, U.S.)

the percentage of the emission in the usable region of the infrared decreases. Hence, the infrared source appears to become less efficient, as even at high colour temperatures, only a small fraction of the total emission is in the mid-wave infrared. This is illustrated in Figure B.3 showing the fraction of the infrared energy in the two principal seeker operating bands as a percentage of the total emission, for two colour temperatures of an arc lamp. Table B.1 provides additional data on radiance in the two selected bands, as well as the total emission, considered generically representative of possible operational bands for the purpose of this illustration.

The other disadvantages of arc-lamp technology are considered to be:

- Requirement for large power supplies;
- Need to shield switching circuits and also the carriage of the arc lamp(s) on the gimbal structure, to minimise interference with the beam director's control system;
- Movement of the arc between electrodes, which shifts as the lamp ages and modifies the focussing of the beam, as well as leading to a potential shortening of its operational life;
- Relatively large angular divergence, giving less than optimum radiant intensity;
- Relatively low total efficiency.

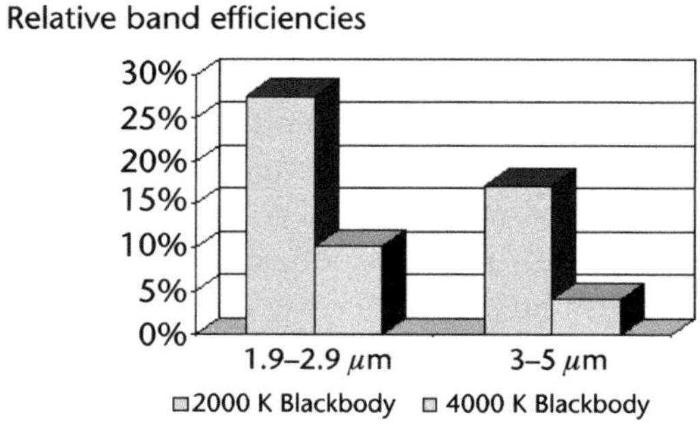

Figure B.3 Relative IR emission from arc lamps.

Table B.1 Relative Radiances

	Radiance ($Wcm^{-2}sr^{-1}$)		
BB Temperature (K)	*1.9 to 2.9 μm*	*3 to 5 μm*	*Total ($\sigma T^4/\pi$)*
2,,000	7.99	5.02	28.9
4000	46.8	18.7	462

σ is Stefan's constant and T is the absolute temperature of the source.

Table B.1 shows that the radiance from an incoherent source is directly related to the effective temperature of the infrared source and its radiating surface area. There is the potential for a trade-off by shaping the beam to provide adequate J:S over as large a volume as possible and minimising the power needed to operate the system; achieved by optically shaping the beam. An example is shown in Table B.2 for a 100-W infrared lamp, using the relationship:

$$\text{Radiant intensity} = \text{Power} \div (\theta^2/4)$$

where the emission beamwidth (θ) is in radians.

The effect is shown schematically in Figure B.4, with approximate polar plots for the aircraft and jammer signatures in a given waveband.

The superior infrared radiant intensity of arc lamps, compared with the heated rods and other black-body sources, enabled platforms with a larger thermal signature to be protected from the simpler (early generation) threat seekers. Additionally, this jam-source technology provides some capability for defeating more modern threats, which operate in the longer wavelength band of the mid-wave infrared, with emission wavelengths around 4 μm.

B.4 Coherent Sources for Infrared Jamming

Laser technology and the use of solid-state devices, in particular, represent the latest development of source technology for infrared-jammer applications. The use of laser-based techniques has many advantages over the use of incoherent sources, for example:

Table B.2 Effect of Beam Shaping

Conical Beam Size (°)	*Radiant Intensity (W/sr)*
90	160
30	1,444
9	16,000
3	144,400

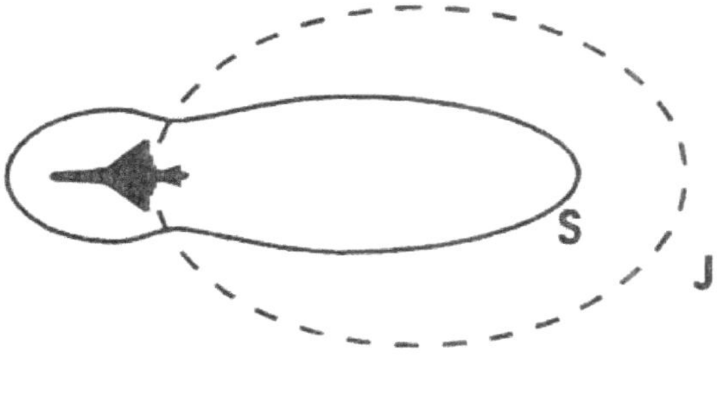

Figure B.4 Effect of shaping the IR emission from a jammer.

- Efficient photon generation;
- Emissions in the appropriate part of the seeker-operating bands;
- Inherently small divergence beam;
- High radiant intensity leading to more robust defeat mechanisms;
- Option for sophisticated jam waveforms with excellent modulation depth;
- No self S;
- Potential for long life, active or passive;
- Smaller system and smaller beam director;
- Greater system flexibility and installation;
- Option for remote positioning of the infrared source;
- Multifunctional emission potential to support addition system functionality.

The major challenges that require rapid resolution concern:

- Reduction in cost;
- Reduction in complexity;
- Enhanced reliability;
- More flexible wavelength generation.

The options for the laser-source technology are discussed in detail in Part I and the systems integration issues for laser-based countermeasures are considered in Chapter 9.

B.5 Summary

The source of the modulated waveform needed to defeat heat-seeking missiles has undergone many changes over the last four decades. The latest developments involve the use of solid-state laser technology. Implementation of modern technology has enabled very effective jam waveforms to be developed, leading to robust platform protection to be achieved.

The laser-based systems have many attributes, but require further development of the subsystem technology to produce more cost-effective devices. One of the very promising approaches concerns the development of mid-wave infrared semiconductor lasers. A number of technical approaches are being investigated, ranging from diodes to optically-pumped semiconductor devices. One of the common features is the extensive use of quantum wells in the structure to provide the desired confinement of the carriers needed for efficient photon generation. All of these developments are considered in some detail in Chapter 3.

One of the attractions of laser technology is that very effective and rapidly adaptable modulated waveforms can be generated easily from a laser source. Given infinitesimally small quantum noise in the laser cavity, virtually perfect modulation can be achieved.

APPENDIX C

Missile-Warning System Technology

C.1 Introduction

The function of the missile-warning system is to act as a bell ringer or an alerting device, which is necessary to trigger the timely deployment of the countermeasure such as a flare or operation of a jammer. These specific systems have the same function as the threat-warning or cueing systems discussed briefly in Chapters 7, 8 and 9 for the laser-based directed-energy weapons and countermeasure systems; in many cases, the missile-warning system uses the same technology base (ultraviolet solar blind or mid-wave infrared). In fact, these electro-optical sensors are capable of detecting more than just missiles, given the appropriate algorithms to process the collected scene data.

The missile-warning subsystem has an array of sensors positioned around a platform to search for potential threat weapons, which may be approaching the fitted platform (i.e., the platform being protected). These sensors may be staring or scanning devices. The function of each sensor in the subsystem is to detect and track any target within the wide field of view of its observation space around the platform and provide a pointing cue for a countermeasure system.

Each sensor, along with the supporting electronics control unit for this subsystem, is designed to sense weapon emissions in various bands of the electro-optical part of the electromagnetic spectrum; although it is worth noting that other techniques use other parts of the spectrum. Typical wavebands for this function are the ultraviolet or infrared; the missile-warning system undertakes a process starting with detection, through acquisition to tracking and, if possible, classification of the threat. It should be noted that the electro-optical sensors are capable of detecting more than just emissions from missiles, which presents a number of critical technical challenges to the defensive-aids suite, or other control techniques for moderating high-level protection systems.

These apparent spurious target features can be problematical for the directed-infrared countermeasure system, as the collected scene data may contain information that relates to potential false-alarm sources, which leads to a false alarm of the countermeasure system. Smart algorithms and processing techniques are required to process the scene data to minimise the false-alarm rate, but have to be sufficiently rigorous to avoid missing real targets (i.e., minimise the false-alarm rate but maximise the probability of detecting a target, leading to accurate and timely threat declaration). Many additional sensing characteristics and potential target-behavioural characteristics are used in the target-processing algorithms to minimise the false-alarm rate. Clearly, this is a tricky system-optimisation process, although the laser-based countermeasure (directed-infrared countermeasure) systems may be able to tolerate a slightly higher false rate, compared with other dispensable countermeasure techniques.

Modern missile-warning sensors may also be used to sense the presence of other threat-weapon systems, as discussed in Chapter 8, so sophisticated algorithms are required to extract this threat information, whilst discarding the potential false-alarm targets. Hence, the development of the term, threat-warning sensor system, as the system functionality of this sensor array increases.

Many alerting systems have a target or threat evaluation and weapon assignment routine within this processing system to ensure that the protection system engages threats in the most efficient fashion. In order to achieve this critical requirement of a high level of protection, the following processes are usually applied:

- When a platform is being attacked by a number of weapons, the threat posing the greatest or most immediate danger is engaged first.
- The appropriate defensive function is assigned to each identified threat.

The need for this additional complication, compared with a staring (omnidirectional) infrared countermeasure system, is because any directed infrared countermeasure system needs to know where the threat is in a specific reference co-ordinate system. Given adequate information about the bearing of the approaching threat, the narrow beam of jam energy can then be pointed precisely at the approaching threat, along the defined bearing once the handover to the beam director has been completed.

The number of sensors[1] in the array with a wide-angle field of view (typically greater than 90°, but a very modern system has more than 180°) required to provide the desired level of situational awareness depends on a range of factors, such as:

- The anticipated angles of arrival of the threat weapons;
- The sites available for mounting the sensors;
- The field of view of the sensors;
- The obscuration of the platform structure;
- The manoeuvre characteristics of the platform;
- The financial budget available for the platform-protection system.

Many of these features interact, leading to the inevitable pragmatic compromise; it is quite common for an aircraft to have four or five sensors, each with a

1. Other techniques may involve Doppler-radar systems; however, these are quite large and may not have the angle-of-arrival accuracy required to cue a beam director used in a DIRCM system. High-value platforms may include electro-optical and radar system to improve performance (increase probability of detection and reduce the false-alarm rate).

field of view of the order of 100°; however, if full spherical (4π sr) coverage is required, then six staring sensors would normally be fitted to the platform. This may arise from the need to protect a high-agility platform or a high-value aircraft.

A fundamental requirement defining a critical aspect of the performance of a missile-warning system is to give a high probability of a timely declaration of an approaching missile in the presence of potential false-alarm sources. The challenge of the algorithm and system designer is to ensure that:

- The declaration of a target occurs rapidly, often of the order of 1 second, and usually the faster, the better for some engagements. This requirement is designed to ensure that the countermeasure technique, be it from a flare or a jammer, can applied in a timely fashion to defeat short-range and long-range engagements with heat-seeking missiles and achieve the desired effect from the platform-protection standpoint.
- The angle-of-arrival accuracy is consistent with the error budget[2] allocated to this threat-detection process for a high probability of successful handover to the fine-track sensor in the beam director. When the handover takes place, the design aim is for the threat missile to be at, or very close to, the centre of the field of view of the fine-track (target-tracking) sensor.
- The bearing of the approaching threat is in the appropriate reference frame with minimal data latency. This requirement is particularly important when the engagement dynamics are high, for example, with a short-range shot and particularly if the aircraft and/or the missile are manoeuvring.
- The probability of declaration is as close as possible to 100% for all valid targets, given the required time lines and the relevant stochastic processes. The survivability of a platform will decrease rapidly if valid targets are not found (false negatives) and consequently missed.
- The probability of a false alarm is close to zero, but the inevitable compromise may allow a rate of a given number (usually small) false alarm per hour, which is dependent on the protection system. This is somewhat less of an issue for the operation of a laser-based, directed-infrared countermeasure system compared with a flare-based protection system, where the dispenser has a finite number of stores.

A number of techniques may be used by the declaration algorithm to meet these high-performance goals, such as:

- Use of simple inertial sensors to detect and eliminate own-platform motion from the scene on a frame-to-frame basis, to reduce any confusion in the detection of multiple threats;

2. The error budget has to take account of random effects, such as the misalignment of the warning sensors in the reference frame being used by the directed-infrared countermeasure system, the misalignment of the beam director within the same reference frame, aero-flexure of the platform and errors in computation and data-latency effects.

- Comparison of the spectral and temporal characteristics of the detected target with those of likely real targets, as well as those similar emission characteristics of clutter and other sources;
- Comparison of the spatial characteristics of the detected target with those of known valid targets, as well as those of false sources and clutter;
- An assessment of the dynamic motion of the target may be made and considered in terms of the anticipated behaviour of invalid targets.

The details of the developed and implemented algorithms and sequences are closely guarded industrial secrets (proprietary information).

Another crucial processing requirement is to have a map of the fields of view of each of the warning sensors, so that the missile-warning system can make an unambiguous single declaration of a target that may appear in the overlap region (i.e., where two or more sensors' fields of view overlap). Moreover, the system requires information regarding the bearings (and angular extent) of obstructions within the field of view of each sensor, so that these blind spots can be factored into the analysis of approaching threats and handover commands to the beam director(s).

In addition to the passive electro-optical technologies devised for the alerting function, which are consistent with emission control and low signature requirements, active systems have been used, such as pulsed-Doppler devices. A recently deployed system on a Head-of-State aircraft uses a Wide body Integrated Platform Protection Systems (WIPPS), developed by Raytheon (L3Com), which has an ultraviolet missile warning system operating in conjunction with a pulsed-Doppler warning system to increase confidence in the probability of declaration of the presence of an approaching threat.

Some electro-optical-based sensors only provide the alert of a threat within a sector of $\pm 60°$ (field of view of up to 120°), so this would not provide the accuracy required to cue any form of directed-infrared countermeasure system in the direction of the threat to ensure rapid engagement. A scanning infrared system is also in service; however, this is quite a large system, but provides significantly greater angle-of-arrival accuracy. However, most of the more successful recent threat-sensing systems are based on focal-plane arrays, designed to be compatible with all aspects of a directed-infrared countermeasure system and give timely and high accuracy angle of arrival data to the protection system. High accuracy is required on the calculated angle of arrival to give a very high probability of handover (or hand off) to the fine-track sensor in the beam director, as noted above, and to ensure a suitable small field of view can be used in the fine-track sensor to provide high resolution and other operational-performance benefits.

C.2 Ultraviolet Missile-Warning System

There are many ultraviolet-based sensors in service; common examples range from the AAR 47 to the AAR 54, AAR 57 and AAR 60 systems, but there are many others including S300 and PAWS; these devices have optical filters that restrict the sensing band of the detector elements to a part of the ultraviolet region of the spectrum. An image of a few of these sensors, which operate in the ultraviolet is shown in Figure C.1, the actual devices are about 100 mm in diameter. These and similar sensors are fabricated with very precise mounting faces, so that the sensors can be exchanged readily without having to undergo a precise measurement of sensor axes with respect to the reference frame, a process known as surveying. The potential variation in sensor mounting positions is part of the error budget allocation, discussed above.

The solar-blind region, which is usually defined as that part of the electromagnetic spectrum with a wavelength shorter than about 280 nm, is the primary operational band for these ultraviolet devices. In theory, in this part of the spectrum there should be little natural ultraviolet emission in the lower atmosphere especially close to the surface of the Earth, owing to absorption by ozone; hence, natural clutter effects should be negligible. Photon propagation in this part of the spectrum is discussed in some detail in Chapter 5.

The advantages and characteristics of ultraviolet-based missile-warning sensors may be summarised as follows:

- Successful in-service deployment of a large number of systems;
- Combat-proven performance for protection of transport and rotary-wing aircraft;
- Detects on the emission from a burning rocket motor;
- Potential for very low clutter, as the background scene has little if any solar content;
- Low processing demands for application of the detection algorithms with modern technology;
- Have been operated successfully with laser-warning receivers;
- Potential for multifunctionality.

Some of the challenges for the effective use of these sensors, without adequate compensation and appropriate processing, may be summarised as follows:

- Possible reduced performance at high altitude, owing to reduced ultraviolet-absorption effects in the solar-blind region;

Figure C.1 Electro-optical sensors for missile-warning systems (AAR 57, ARR 54 & AAR 60). (Courtesy of BAE Systems (NA), U.S., Northrop Grumman, U.S., and Cassidian, Germany.)

- Advanced processing techniques required to detect missiles that are coasting, that is, after the rocket motors have burnt out, known as the post burn-out (PBO) condition;
- Potential clutter-limited performance at extended or long ranges;
- Basic false-alarm rate sensitive to atmospheric propagation conditions, including ozone concentration, in the vicinity of the sensor.

C.3 Infrared Missile-Warning System

There is a strong interest in the development of infrared sensors for use in missile-warning systems to enhance the performance of the alerting system, especially for the protection of fast jets. The potential performance advantages concern enhanced acquisition ranges, particularly at high altitude and detection of coasting missiles (i.e., those in the PBO condition). This feature may be especially important for the detection of air-launched missiles, which can use a boost-sustain-coast thrust profile, to achieve long-range performance and, are therefore, difficult to detect when they are coasting unless the warning sensor can detect the kinetic heating of the airframe. The boost-coast profile may also be problematical for some other sensor technologies.

The big challenge for an infrared warning sensor is to find the missile signature and isolate it from amongst all of the other thermal sources on a battlefield or other environment, to permit tracking and declaration, so that an angle of arrival of the threat can be calculated rapidly and accurately. As with all warning sensors, the fundamental requirement is a very rapid declaration of a valid target, whilst maintaining a very low false-alarm rate and not missing any real targets.

The fundamental approach to declaring a missile threat using its infrared emission involves:

- Size and extent of a missile's thermal image;
- Motion of the infrared image within the thermal scene;
- Movement of the infrared image with respect to the platform, as most missiles follow a proportional-navigation guidance law;
- Temporal characteristics of the infrared signature;
- The colour ratio between the two detected wavelengths (for a two-colour system).

Devices operating at a single wavelength, as well as two-colour sensors have been investigated. Initially, the general view was that the processing burden needed to ensure a low false-alarm rate with a single-colour sensor would be prohibitive; however, modern algorithmic techniques and high-performance processors have made this approach possible as such sensors are in service. Provided techniques can be perfected and refined, then a cost-effective sensor with significant multifunctionality may be deployed. Examples of the additional functions that may be exploited include;

- Weapon aiming;
- Warning/sensing/indication of attack by other weapons;
- Pilot aid/vision;
- Navigation;
- Situational awareness.

The perceived advantages of an infrared missile-warning sensor system may be summarised as:

- Longer-range detection performance (significantly better than ultraviolet systems are claimed in some circumstances);
- Detection range increases with altitude;
- Potential to detect and track coasting missiles (i.e., in the PBO conditions);
- Use of a similar infrared waveband (Band IV) as the modern threat seekers (and the target-tracking sensor in the beam director);
- Performance insensitive to changes in ozone concentration;
- Potential for multifunctionality;
- Other potential operational benefits during multitarget engagements.

The potential technical challenges involving the use of a missile-warning system using infrared sensors may be summarised as:

- Complex techniques required to reduce or eliminate hot spots from the thermal scene;
- More complex sensor, compared with alternative techniques, that requires cooling;
- More expensive than an equivalent ultraviolet sensor.

It should be noted that as processing power developed and novel techniques evolved, many of the cited problems of any of the sensing techniques have been solved, which has led to robust sensor performance.

C.4 Two-Colour Missile-Warning Systems

A number of techniques and methods have been investigated to demonstrate a successful two-colour sensor with a low false-alarm rate. The major challenge concerns the spatial and temporal correlation of the images created, to form an accurate image of the scene, without introducing artefacts resulting from the image processing. The techniques investigated include the following:

- The use of a filter wheel, which creates filtered images in different parts of the infrared wave-band of the spectrum, usually around 4 μm and one beyond the carbon-dioxide notch in the atmosphere where the ratio of the intensities of rocket-motor signatures are somewhat different from clutter.

The main drawback of this approach, besides the additional complexity, is to get an accurate registration of the two filtered images, which is particularly challenging for rapidly manoeuvring platforms.

- Scanned linear arrays, where one line of detectors is sensitive to a specific part of the spectrum and the adjacent array of detectors is sensitive to the other wavelength of interest. Again, there is a registration issue.
- Stacked detectors, which are used to overcome the registration problem; in this case, the detector elements are superimposed, so that the registration issue should not arise, however, there is a sensitivity and integration-time problem to resolve, as well as the inherent cost.
- A similar approach uses quantum-well infrared photo-detectors, where the detectors are immediately adjacent in the array. This technique has had problems with yield and sensitivity of the array, but recent technical advances have been encouraging.
- An approach that forms two images on two adjacent arrays, where one image is in one colour and the other in another colour. The two imaging arrays are coupled and at a given fixed displacement. This is known as the multi-image multispectral registration.

Technological advances with detector technology have made the realisation of a two-colour infrared missile-warning system a viable proposition for current and future platform-protection systems. An example of an infrared threat-waning sensor is shown in Figure C.2.

C.5 Summary

The successful operation of the missile-warning system is critical to the effective operation of any platform-protection system but especially directed-countermeasure systems. This type of sub-system is normally associated with detection of

Figure C.2 An example of an IR threat-warning sensor. (Courtesy of Northrop Grumman, USA.)

approaching missiles, but the sensor technology used to provide this aspect of situational awareness is capable of detecting other objects. One of the challenges of this subsystem is to detect the real threat objects and reject the competing sources as rapidly as possible; hence, a critical system requirement is maintaining a very high probability of accurate declaration of a threat in the presence of clutter and other potential false-alarm objects. The design goal is obvious, but the sophisticated techniques required to achieve these critical objectives are very demanding.

Imaging sensors, based on sensing objects with an ultraviolet emission, have been in operational service for some time. These sensors operate in, or close to, the solar-blind part of the electromagnetic spectrum to reduce the scene complexity from which the threat has to be extracted and, hence, minimise the demands on the processor to declare the presence of a target in a timely fashion.

Despite the successful operation of these devices in service, these sensors have some well-known potential performance limitations, which have encouraged the development of sensor technology operating in the infrared. However, the operational performance of ultraviolet sensor systems has continued to develop, offering multifunctionality.

Two approaches have been investigated to provide situational awareness of an approaching missile or missile-like threats using the infrared emissions from potential targets. The clutter in the infrared part of the spectrum that is likely to be encountered in any battlefield is significant, which initially suggested that the processing burden for rejecting false targets, using a single-colour technique would be prohibitive, so a two-colour technique was developed. However, modern techniques have enabled single-colour systems to be fielded successfully.

In the case of two-colour systems, a colour ratio between an object's emission characteristics in different parts of the spectrum is used as a primary discriminator. However, efficient clutter-rejection algorithms have been developed, so that the simpler single-colour approach appears feasible and offers all of the advantages of a more complex multicolour infrared system. In the meantime, miniature sensor-detector technology has continued to develop, proving a number of realistic options for the threat-warning system.

One of the future challenges for all infrared-sensor systems is to extend the endurance and efficiency of their cooling systems. Progress continues to be made so that the mean time between failure is close to those of other avionic systems. Other developments that would enhance system performance are under way, particularly techniques that will increase efficiency of operation so that the power demand is minimised.

APPENDIX D

Rules of Thumb and Guidelines

Rules of thumb can be an absolutely invaluable tool when used appropriately. They can be applied successfully provided some care is used, particularly with respect to some appreciation or knowledge of the scale of the validity of the rule.

There are numerous instances of the use of rules of thumb in many phases of a project, spanning initial design considerations to confirmation of a successful operational test and evaluation activity on a field range. However, rules of thumb are just that, and at best, merely provide an estimate of an answer; hence, they should only be used with the utmost care, as noted above, and simply a guide to assessing the feasibility of a design or the meaning of an actual measurement. If in doubt, or when there is a safety-critical requirement, then alternative and more rigorous techniques should be used, and only then, perhaps consideration of an appropriate rule of thumb may be appropriate as a sanity check.

In many cases, the rule can be derived from first principles along with the known limitations of the assumptions or approximations. These aspects are dealt with most effectively in a number of valuable texts, two of which are referenced in the bibliography. Other statements given next are what the author considers to be common sense, but in his experience are often overlooked.

D.1 Laser-Directed Energy-Applications

The following are rules of thumb for laser-directed energy applications:

- 10 kJ of energy is required to evaporate 1 cm^3 of metal.
- The beam divergence needs to exceed the angular jitter by a factor of 5 to ensure that the energy is on target for more than 90% of the illumination period.
- Delivering 500W/cm^2 at a target zone is a potentially effective irradiance for a damage-class beam.
- Ensure that the lateral thermal conduction does not limit the temperature rise caused by the energy deposited on a target by a beam (i.e., beam size at the target beats the thermal conductivity of the material); in general, the beam needs to be at least an order of magnitude greater than the thickness for achieving rapid effects at range.

- Brightness (flux density on a target) is more important than power, but infrared beams do not appear to be bright to the human eye. However, see comment directly above, so there is a trade space.
- There are no military photons!
- Choose wavelengths that have minimal absorption in the propagating atmosphere to avoid thermal blooming.
- Jam early and jam hard (for the most robust countermeasure effects).
- Most absorbing materials, such as detectors, have a damage threshold close to $2 J/cm^2$ when illuminated with nanosecond pulses.
- Out-of-band is much harder, but more versatile and rewarding.
- Laser damage is only remote cutting and in some cases welding.
- Does nonlethality (of DE weapons) make them the weapons of first choice?
- Directed energy is not electronic warfare on steroids!

D.2 Lasers, Optics and Testing

The following are rules of thumb for lasers, optics and testing:

- A photon costs about 10^{-22} to 10^{-23} £ or \$.
- Photon lifetime in a cavity of length ℓ, round-trip loss L and output coupling reflectivity R is approximated from $\tau = 2\,\ell/\{c(L - \ell n_R)\}$, where $L = 2\,\alpha\,\ell + L_m$ where α is the absorption coefficient per unit length and L_m are other miscellaneous losses, including scattering and diffraction losses.
- It is only when the wall-plug efficiency of a laser exceeds 50% that the source generates more light than heat.
- Most laser beams rarely approximate to a genuine Gaussian profile, but there is a tendency to a Gaussian profile for a beam emitted from a low numerical aperture.
- The brightness of a single-mode laser can be estimated from the product of the emitted power and area of the radiating aperture and divided by the square of the wavelength.
- Brightness = (emitted power)/(solid angle of the beam).
- Solid angle = area/(range^2).
- On-axis intensity = Brightness/(range^2).
- Do not pump an optical parametric oscillator (OPO) at more than five times the conversion threshold if you wish to avoid back conversion.
- A singly resonant OPO generally has a higher operational threshold, but demands less stringent maintenance of alignment.
- A doubly resonant OPO has the opposite operational characteristics of a singly resonant device and has a tendency to a noisy/spiky output
- Energy conservation in an OPO gives $\lambda_p^{-1} = \lambda_s^{-1} + \lambda_i^{-1}$.

- Second-harmonic generation: $P_{SHG} \propto \frac{P_{fundamental}^2}{\omega_0^2}.L^2$, where P_{SHG} is the second-harmonic optical power, $P_{fundamental}$ is the power in the fundamental, ω_0 is the beam waist and L is the length of the nonlinear crystal.
- Second-harmonic conversion efficiency is inversely proportional to the pump line width ($\Delta\lambda$).
- A laser beam's full-angle divergence in radians approximates to the wavelength divided by the exit aperture.
- A Gaussian spherical wave spreads considerably less than a plane wave diffracted by a circular aperture.
- The M^2 of a beam is approximately equal to the square root of the number of transverse modes.
- Transverse modes determine beam quality.
- Longitudinal modes determine wavelength and line width.
- The figure of merit for a saturable absorber is the ratio of the ground- and excited-state cross-sections.
- The figure of merit of a titanium-sapphire crystal is the ratio of the absorption coefficient at the peak pumping wavelength (514 nm) to that at the peak emission wavelength.
- Doubling the beam width of an emission reduces its radiant intensity (W/sr) by a factor of 4.
- Size of the optic system required to achieve diffraction-limited performance is directly proportional to wavelength of operation.
- The far-field diffraction of light passing through uniformly illuminated opaque apertures is determined by the perimeters of those apertures.
- About 84% of the energy is contained within the central disc of a diffraction pattern.
- Field of view = (size of image plane)/(focal length).
- The scratch and dig requirements for high-power lasers are usually quite stringent at 10/5, being relaxed to 40/20 for lower-power devices.
- Use short fat fibres to avoid stimulated-Brillouin scattering; stimulated-Brillouin-scattering threshold is directly proportional to the area of the propagating medium and inversely proportional to its length.
- The laser cross-section of a target is about 10% of the target area projected towards the laser beam.
- No surface is perfectly Lambertian or specular.
- Speckle size is inversely proportional to the spot size.
- Experimental accuracy of measurement is determined by the calibration of the instrument and its dynamic range.
- The Cauchy equation enables the variation of the index of refraction with wavelength (dispersion) to be estimated.

- The diameter of the Airy disc (micrometres) for visible wavelengths approximates to the F number of the lens.
- Do not allow a glass mirror's diameter to be more than six to eight times its thickness, unless using a system with adaptive-optical compensation.
- The mass of an optical element may be estimated from the comparison with a similar element of known mass through the product of the known mass and the ratio of the diameters squared.
- The energy of a photon in electron volts is given by 1,240/wavelength in nanometres from [ΔE (eV) = (h.c $\times 10^9$)/e.λ_{nm}], so 400 nm = 3.1 eV and 750 nm = 1.65 eV.
- Wavenumbers (cm^{-1}) × wavelength (in micrometers) = 10,000 – hence, 2.5 μm = 4,000 cm^{-1}.
- Make fluorescence measurements close to the edge of a material.
- Use a series of pin holes, with different sizes, to illuminate a sample during fluorescence measurements, extrapolate to a zero diameter to estimate the intensity in the absence of trapped photons.
- Given a coefficient, the cross section for the transition can be estimated given the concentration of the species in the host.
- Light travels at approximately 1 foot per nanosecond.
- When using a plano-concave or plano-concave lens in an optical system, arrange for the curved surface to face the incident beam, which minimises aberrations, as both surfaces play a role in bending the rays in the beam; if the flat surface faced the beam, then the bending of the rays would reside with the curved surface alone.
- Thermal aberrations scale with power.
- The effect of aberration can be scaled by the field of view of the optical system and the inverse square of its F number.
- The practical minimum F number for an optical telescope (or element) is 0.7 for an imaging system.
- The total integrated scatter depends on the surface roughness and the operational wavelength.
- The deflection at the centre of a circular plate mirror is proportional to the fourth power of its diameter for a constant substrate thickness; a more useful estimate is based on r^4/t^2, where r is the radius and t is the thickness.
- In a lossless optical system, the etendue is constant at all planes crossed by the light rays.
- Glass can withstand a compressive stress of about 345 MPa before fracture and a tensile stress of about 7 MPa before damage occurs.
- Wherever possible, design optical elements to standard test-plate profiles; cost savings can be very significant.
- The induced stress birefringence in an optical element increases by about 1 nm/cm for every 5 psi of applied stress.
- Photons per second = wavelength (μm) × watts × 5 × 10^{18}.

- Reflection = $(n-1)^2/(n+1)^2$ where $n = n_1/n_2$ and transmission = $4n/(n+1)^2$.
- The peak wavelength (μm) of a blackbody (of its emitted power) is approximately 3,000 divided by the temperature (K).
- About 25% of the emitted power of a blackbody occurs at wavelengths shorter than the peak wavelength. Hence, 75% of the energy is contained in the wavelengths beyond the peak.
- Assume an emissivity of about 0.9 for any radiating surface claimed to be black.
- Most metal objects have an emissivity of about 0.2, whereas it is about 0.8 for nonmetals.
- An optical element with an output aberration/distortion of one quarter wavelength is adequate for many applications. A value of $\lambda/15$ may be required for some critical imaging systems.
- Most optical glasses that are transparent to visible light have a refractive index in the range 1.4 to 2.
- The diameter of a collimating mirror should be at least 10% to 20% larger than the optical system under test and the focal length should be 5 to 10 times that of the object under test.
- When inspecting an optical element for flatness by placing it on an optically flat surface and gently pressing the edge of the test piece, if the fringes move towards the point of pressure, then the upper surface is convex and vice versa.
- The lateral deviation of the image caused by a transparent parallel plate of thickness τ at 45° to the incident beam $\approx \tau/3$.

D.3 Scattering and Propagation

The rules of thumb for scattering and propagation are as follows:

- The total integrated scatter of a surface is determined by its surface roughness, the incident wavelength and the incidence angle.
- Raman effect is weak, about 10^8 lower than Rayleigh scattering.
- Raman intensity is proportional to $1/\lambda^4$.
- Rayleigh scattering is inversely proportional to wavelength to the power 4.
- Rayleigh scattering is virtually totally dominant when the atmospheric visibility exceeds 10 km; conversely, when the visibility is much less than 10 km when Mie scattering is dominant.
- Visual range is determined by the point where the contrast between a target and an object is 2%.
- Visibility through the atmosphere in the absence of aerosols is about 240 km, a rare condition at low altitude close to sea level at low or mid-latitudes.

- The atmospheric attenuation may be estimated from $e^{-\alpha r}$, where α is a sum of the absorption per unit length and the scattering per unit length and r is the propagation distance (Beer's Law).
- Propagation of laser beams from solid-state lasers tends to be dominated by relative humidity from the absorption of moisture by aerosols.
- Use of the U.S. standard atmosphere does not provide a realistic representation of the propagation conditions experienced by most laser beams.
- The U.S. standard atmosphere does not allow for cloud formation.
- Desert aerosols are mineral based, so changes in relative humidity do not have a direct impact on propagation performance.
- Blue bends best.
- Blue-green wavelengths propagate better than other optical wavelengths in sea water.
- The time constant of the atmosphere is about 1 ms.
- The aperture-averaged variance of the log-amplitude of a beam in the presence of turbulence seems to be proportional to $D^{-7/3}$, where D is the aperture diameter.
- Impact of turbulence for a given atmospheric condition is inversely proportional to wavelength of the beam.
- The mid-wave infrared wavelengths tend to be attenuated less than other wavebands in poor visibility conditions.
- The scattering coefficient for rain may be estimated from approximately $4/V_r$, where V_r is the visual range.
- The absorption coefficient of sea water (α) approximates to 75% to the total diffuse attenuation coefficient (K); for clear waters K/α approximates to 0.3.
- The depth resolution in a bathymeter cannot exceed one-half of the product of the pulse duration and the speed of light in water.
- The refractive index of sea water depends on its temperature and salinity.

D.4 Systems

The rules of thumb for systems are as follows:

- If you cannot specify a system's performance, then you will be unable to test it.
- Just because the laws of physics indicate that something cannot be done does not mean it can be done.
- To a first approximation, all electro-optical systems have a density of water (i.e., their weight in kilograms equals the volume in litres).
- More complex systems have a mass of about 1½ kg per litre and so will sink in water.

- Given sufficient randomness in a system, the Central Limit theorem will be valid [i.e., the effects of errors can be estimated by forming the square root of the sum of the squares of the contributors (RSS)].
- If one process dominates in a random system, then the root of the sum of the squares approximates closely to the largest value. If all of the contributors are similar, then the RSS value approximates to the largest component multiplied by the square root of the number of contributors.
- The instantaneous field of view of a system approximates to 1,000 times the pixel size (in micrometres) divided by the effective focal length of the device (in millimetres).
- It takes about five times the thermal time constant of a passive optical system for it to reach its thermal equilibrium in changing ambient conditions.
- Typically, any system will reach its stable thermal equilibrium after at least five thermal time constants, unless there are internal heat sources, whatever the conditions it appears to take a few to tens of minutes for most systems to reach their design operating temperature.
- To a first approximation, the mass of an optical system is directly proportional to the mass of the largest optical component, only useful when scaling up an optical system.
- Holmium needs thulium and thulium needs holmium for efficient pumping.
- A typical safety factor error (allowed stress/applied stress) is about 2 or 4.
- Ensure vibration isolators have a resonant frequency at least an order of magnitude less than that of the system they are protecting.
- If modelling a system with a man-in-the-loop (e.g., guiding a missile/tracking a target), do remember that the supporting simulation tools must operate in real time.
- Some form of feedback is required to control any system.
- The technology-readiness level of a system cannot exceed the technology-readiness level of the least mature subsystem or component.
- The ammunition should be appropriate to the target; if it costs $1 million to destroy a $10 target (e.g., a UAV), then you are going to lose.
- Quantity has a quality all its own.

D.5 Pointing and Tracking

The rules of thumb for pointing and tracking are as follows:

- The mass of a gimbal assembly is directly proportional to the slewed mass.
- Multiple-target resolution is determined by the Rayleigh criterion, the minimum angular separation at which two point sources can be resolved occurs when the peak of one Airy disc falls on the position of the first minimum of the other.

- A signal-to-noise ratio of about 6 is adequate to perform most sensing and tracking functions.
- The location of the blur spot on a focal plane can be estimated to an accuracy determined by resolution divided by the signal-to-noise ratio, which approximates to the tracking accuracy.
- Targets can be sensed with some certainty to a range defined by the signal-to-noise ratio and allowable false-alarm rate.
- Targets can be detected relatively easily if at least half of the photons in the pixel are from the target, the remainder being from the background.
- Good viewing/seeing from the ground is rarely much better than 5 μrad (or 1 arcsec), in ideal conditions possibly a factor of 10 better. Seeing is the combination of the aberrating effects in the atmosphere, which limit observations of distant objects.
- The number of actuators required by an adaptive-optical system to correct for aberrations is proportional to the square of the quotient of the aperture divided by the Fried parameter.

D.5.1 Analysis

The following are rules of thumb for analysis:

- To avoid aliasing in the frequency analysis of a data set, sample at least at twice the highest anticipated frequency.
- The median of a distribution is less sensitive to outlying points in a distribution than a simple average (mean) value.
- Multiply a low-order root-mean-square value by four to get the peak-to-trough (valley) error.
- For small angles $\theta \approx \sin\theta$ for angles (θ in radians) up to 14° and $\theta \approx \tan\theta$ for angles (θ in radians) up to 10°, similarly, $1 \approx \cos\theta$ for angles (θ in radians) up to 8°.

D.5.2 Human Eye

The following are rules of thumb for the human eye:

- Always undertake an ocular hazard assessment whenever an optical system/component includes a laser with a hazard classification higher than Class 1.
- It is worth remembering that incoherent laser sources can also present optical hazards.
- The optical gain of a human eye is about five orders of magnitude (10^5).
- The human eye relies on contrast for detection of objects.
- Pupil size ranges from about 1 to 8 mm; the standard dilated pupil is usually taken to be about 7 mm.

- The eye/brain can resolve short flashes (pulses of light) down to a separation of about 20 ms; the upper limit for temporal contrast sensitivity is 60 cycles/s (Hz).
- The binocular field of view of the human eye is elliptical, approximately 125° to 135° vertical and 150° to 200° horizontal, but varies with eye placement and head morphology.
- Only a very small fraction of this field of view has acute vision.
- The damage threshold of the human eye to visible light and near infrared is about 10 times the MPE, for both pulsed and continuous-wave emissions.
- The quantum efficiency of cones drops by about a factor of 10 between the centre and the edge of the retina, which is related to retinal pigment distribution.
- The eye can resolve objects to better than 1 arc-min (300 μrad).
- The eye stabilises an object by reflex movements.
- The eye is not able to perceive latencies during pursuit tasks of less than about 33 μs.
- The eye can detect a lack of flatness in a surface with a curvature of the order of 10,000 times the length being observed; under optimal conditions, foveocentric disparities of about 10 arcsec can be detected for high contrast targets centred in depth near the horopter, which equates to detecting a change in depth of 2 mm at a viewing distance of 500 mm.
- The decline in visual acuity of the human eye increases with age.
- Humans search a field of regard using a step-stare method. Typically, an approximately 5° circular zone is searched in about 0.3 second; in a minute, the human eye can fixate about 120 observations; however, it is worth noting that microsacades in target search and fixation functions can vary considerably depending on many factors such as target contours and ambient light conditions.
- Microsacades preserves the elements of contrast on the retina to enable object observation.
- Typical figures for microsacade amplitude of the eye is 1 to 23 minutes of arc and a time interval of 300 ms to 5 seconds.

D.5.3 Detectors

The following are rules of thumb for detectors:

- A cold shield should be black on the inside and shiny on the outside.
- It takes about 30W of input to achieve 1W of cooling at 80K with a Stirling cooler.
- The signal that one wishes to monitor is usually small compared with the electronic DC background; therefore, it is usually necessary to attempt to detect small changes in a large signal.

- A semiconductor detector's peak sensitivity is usually at a maximum at a wavelength of about 10% less than the cutoff wavelength (defined as 50% of the peak sensitivity).
- An optical detector's responsivity (amperes/watt of incident power) may be estimated from 80% of the quantum efficiency multiplied by the wavelength in micrometres.
- The shot noise equals the Johnson noise in a photo-diode when the photocurrent is sufficient to drop about 50 mV across the load resistor at room temperature.
- The well capacity of a readout in a focal-plane array may be assumed to have a maximum value of approximately 25,000 electrons multiplied by the area of each pixel in square micrometres.
- Lead salts increase sensitivity by about 3% for each 1K cooling below room temperature (293K).
- Mercury cadmium telluride (MCT/CMT) detectors operating in the midwave increases its sensitivity by about 7% for each 1K of cooling below 220K, whereas for long-wave devices the figure is about 15% for each 1K of cooling below 100K.
- The noise bandwidth of a photo-detector may be estimated from the reciprocal of twice the integration time.
- There is a little potential enhancement to be gained in system performance by designing a sensor with a noise value slightly lower than the noise imposed by the background in the scene to be observed. Moreover, the detector output is usually digitised to ensure the analogue-to-digital conversion unit (value) is several times smaller than the background noise.
- The diffraction limit (in milliradians) of a sensor/telescope operating in the long-wave infrared is the reciprocal of its diameter expressed in inches.
- The dark current in a typical silicon detector array doubles for every 8K or 9K increase in operating temperature.

D.5.4 Miscellany

The miscellaneous rules of thumb are as follows:

- Seventeen mrad per degree (approximately) and 1 arcsec equate to 5 μrad.
- One nautical mile is about 1.8 km.
- One knot is approximately 0.5 m/s.
- One minute of arc (latitude) on the surface of the Earth is 1 nautical mile; 1 minute of arc longitude is only equal to a nautical mile at the equator (ignoring the Earth's oblateness).
- A 3-dB reduction is to halve the power.
- No solid is perfect.
- Errors add up.
- Calibrate a spectrometer on use from a known laser standard wavelength.

- Calibrate any instrument to be used for a specific test under conditions that approximate as closely as possible to the anticipated test environment
- Laboratory black bodies are usually radiometrically accurate to a few percent.
- First-order effects dominate.
- About three-quarters (3/4) of a load is taken by the first three threads of a screw.
- The available bandwidth of a communication system is about one-tenth of the carrier frequency.
- Nyquist criterion is vital, so ensure that the sampling frequency is fast enough to capture the highest frequency of importance in a time series: $\nu_{sample} > 2\ \nu_{time}$ series.
- The lowest temperature in the atmosphere (sky temperature) is about 217K, but that does not mean the observation of a night sky will yield infrared radiation equating to a black body with an equivalent temperature of 217K.
- The surface temperature of the Sun is about 6,000K.
- The sun has a brightness of about 10^5 lux at noon at the equator.
- A full moon has a brightness of about 0.2 lux (half a million times dimmer).
- Luminance of starlight is about 10^{-3} cd/m^2, moonlight is about two orders greater depending on the phase of the moon.
- The Sun or Moon present an angle (visual angle) of about 1/2 at the surface of the Earth.
- The surface of the human body emits about 100 W/m^2.
- The surface area of a human body is about 0.2 m^2.
- Human skin has high emissivity in the short-wave infrared and is a good Lambertian surface.
- To avoid mechanical instability in a component, a material should not be stressed to more than half of its endurance limit.
- Distance to the horizon in miles is about the square root of the observer's altitude in feet.
- Most healthy plants have strong absorption features close to 0.7, 1.4 and 2 μm.
- Moore's Law is that the processing power of a micro-processor double about every 18 months.
- Murphy's (Edward Aloysius Murphy) Law is that if anything can go wrong it will, but square that probability for electrical experiments.
- The angle subtended by a thumb at arm's length is about 1½°; a fist at arm's length is about 10°.
- The ratio of specific heats (γ) of air is usually about 1.4.
- The more complex a system is, the further it has to travel to be repaired.

- Full optimisation of any system, requiring low cost, high reliability, high performance and rapid delivery, is elusive, if not impossible; the two-from-four rule often applies (i.e., any two of the four may be achieved simultaneously).
- Ensure that all system designs have an adequate design margin; the performance in the laboratory and the installed performance on a platform is usually very different. Hence, always specify system performance based on the installed requirements, not on the performance of the isolated system positioned on a laboratory bench.
- Non-recurring costs mean just that, or should, by definition.
- Most progress is made on a contract in the last few days immediately before a review, such as a quarterly progress review or a design review.
- One needs to have an audit to get something done; this follows on from the statement immediately above.
- Never go to a range to test any system without knowing the answer (to a first approximation).
- Never ever do a demonstration to a general officer without undertaking a complete test first; for this sort of test have at least one spare of everything, because if it is possible for something to go wrong it will and the probability is proportional to a function with a power law to the number of stars on the general's epaulettes (i.e., a severe form of Murphy's Law).
- No battle plan has been known to survive its first contact with the enemy, often the same for flight testing.
- Measure twice and cut once, or measure with a micrometer, mark with chalk and cut with an axe.
- If possible, undertake all countermeasure testing against a clear blue sky.
- Think twice before cleaning dirty optics; dirty optics often function surprisingly well.
- Do not use superglue type of adhesives near coated lenses (or other coated optical elements), if at all possible.
- A job is not complete until the device/system is integrated with its box and the label is on.
- Any task is not complete until the paperwork, including a report is completed.
- If at all possible during field trials, use an uninterruptable power supply; if that is not available, then use mains and as a last resort battery power; if using battery supplies, ensure that a fully charged replacement is available.
- Sit, crawl, walk and then run; the same is true for effective testing of anything from a component to a complex system.
- The cost of testing at least doubles at every stage in a test programme (i.e., going from laboratory testing to the ultimate acceptance tests); similarly, the number of people who need to be involved appears to increase exponentially. Similar rules apply if the testing involves overseas travel.
- The complexity of the approval also increases exponentially as the evaluation phase of a project progresses and a steadily increasing number of people have to be convinced.

- Test ranges are either very cold or very hot when the testing moves to outdoor facilities.
- You can only kill a target once; overkill is something else.
- There are known unknowns, also known as first-order ignorance.
- There are unknown unknowns, also called second-order ignorance.
- Things get dropped when shipped, even when labelled fragile; the typical drop distance is about 0.5 to 1m.
- Learn the trade, not the tricks of the trade.
- All humans are fallible; the trick is to avoid repeating previous errors, but difficult to simulate.
- The wheel usually does not need to be reinvented, but modifications to the size may be necessary.
- Just because something is feasible, it does not mean that it is practical.
- Without a goal, every strike is a hit.

Selected Bibliography

Friedman, E., and J. L. Miller, *Photonics Rules of Thumb, Optics, Electro-Optics, Fiber Optics and Lasers*, 2nd ed., New York: McGraw-Hill, 2004.

Schwertz, K., *Useful Estimations and Rules of Thumb for Optomechanics,* Master's thesis, Department of Optical Sciences, University of Arizona, 2010.

Vukobratovich, D., *Introduction to Optomechanical Design,* Bellingham, WA: SPIE Press, 2006.

Yoder, P. R., Jr., and D. Vukobratovich, *Field Guide to Binoculars and Scopes*, Bellingham, WA: SPIE Press, 2011.

Acronyms

ABC	Active beam combining
ACPA	All-cryo cooled power amplifier
AOA	Analysis of alternatives/angle of arrival
AOFS	Acousto-optical frequency shifter
ASE	Aperture shared element/amplified spontaneous emission
BBU	Beam-break-up unit
BDA	Battle-damage assessment
BDI/A	Battle-damage indication and assessment
BLIP	Background limited in performance (of sensors).
BPM	Beam-position monitor
BRW	Bragg reflection waveguides
C-ISR	Counterintelligence, surveillance and reconnaissance
C^3I	Command, control, communication and intelligence
C^4I	Counter C^3I
CLS	Combined logistics support
CPA	Chirped power amplifier
COEIA	Combined operational and effectiveness investment analysis
CTE	Coefficient of thermal expansion
EDFA	Erbium-doped fibre amplifier
EPDL	Edge-pumped disk laser
EP VECSEL	Electrically-pumped VECSEL
ESP	Electrostatic protection
ETU	Energy transfer upconversion
FHSS	Frequency-hopping seed source
FIR	Far-infrared waveband
FMEA	Failure mode and effects analysis
FOPA	Fibre optical parametric amplifier
FPDL	Flashlamp-pumped dye laser
FSP	Flame-spray pyrolysis
GDD	Group dispersion delay
GVD	Group-velocity dispersion
HBS	Holographic beam sampler
HNLF	Highly nonlinear (optical) fibre
HPFL	High-power fibre laser
HIS	Hyperspectral imaging

IFPC	Integrated-force protection capability
ILS	Integrated logistic support
ISR	Intelligence, surveillance and reconnaissance
IT	Integrated testing
KE	Kinetic energy
LBC	Laser-beam control
LDA	Laser-diode array
LRIP	Low-rate initial production
MOE	Measure of effectiveness
NACM	Nonlinear amplifying loop mirror
NRE	Nonrecurring engineering (costs)
OFG	Oxyfluoride glass
OPCPA	Optically pumped chirped power amplifier
OPML	Optically pumped molecular laser
OT	Operational testing
PBC	Passive beam combining
PL	Photoluminescence
PLT	Lead lanthanum titanate
PPM	Pulse-position modulation; parts per million
RIN	Relative intensity noise
S2E	Start to end
SBR	Saturable Bragg reflector
SMF	Single-mode (optical) fibre
SPAD	Single-photon avalanche detector
SRS	Stimulated-Raman scattering
SSTF	Simultaneous space and time focusing
STANAG	Standardisation agreement
STRS	Stimulated thermal Rayleigh scattering
TCO	Transmitting conducting oxides
TDF	Thulium-doped fibre
TFP	Thin-film polariser
TILL	Track-illumination laser
TIP	Test instrumentation package
TIR	Total internal reflection
TIS	Total integrated scatter
TMA	Thermomechanical analysis
UUT	Unit under test
WFE	Wavefront error
WLC	Whole-life cost
ZDW	Zero-dispersion wavelength

Glossary

AAR Designation for a warning sensor (e.g., AAR 54, AAR 57 and AAR 60).
Abbe number This parameter provides a measure of the dispersion of a material with respect to its refractive index. Also known as the constringence or the V-number of a transparent material defined as $V = (n_d - 1)/(n_f - n_c)$, where n_d, n_c and n_f are the refractive indices of the material at the Fraunhofer d-, c- and f-lines. High values of V indicate low dispersion (low chromatic aberration) and low V is the reverse; flint glass has low V, whereas crown glass has high V (typically >50).
Ac coupled A means of energy transfer in an electronic circuit; an alternating current coupling only permits transfer of ac signals; the direct current signals are blocked.
Acceptance angle Related to how much a laser beam can be focused without destroying the phase matching condition. Usually necessary to focus a laser beam in to a crystal get nonlinear effects, but this means that different parts of the beam will propagate through the crystal at different angles, so it becomes impossible to satisfy the phase matching condition for the whole of the beam simultaneously. Hence, defines a degree of focusing that can be used before the deviation away from the phase-matching condition negates the effect of focusing. It defines the degree of focusing which is beneficial (mismatch should not exceed π/l, where l is the effective length of the crystal).
Accessible-emission limit (AEL) The maximum output power (or energy) permissible for a given class of laser (hazard classification).
Accommodation The ability of the eye to change the focal length of the lens to focus on objects at different distances.
Acousto-optical modulator (AOM) An active Q-switch, which uses the formation of a diffraction grating to control the loss in a cavity through deflection of the beam within the cavity.
Adaptive-optical (AO) techniques A method that compensates for the aberration induced in a laser beam, for example, by turbulence. Distortions sensed in the received beam from a target, or a beacon, are used to create a conjugate beam by use of a segmented or deformable mirror.
Air-filling fraction Of a holey or micro-structured optical fibre. Defined as the ratio of the cross-sectional area of air holes to the bulk material cross-sectional area.
Albido Reflecting power of a surface.
ALE Designation for a missile launcher, for example, ALE 47.

Alexandrite Crystal used in tuneable solid-state lasers operating around 800 nm, invisible but capable of causing eye damage. Eye protection straightforward using narrowband filter, which causes little attenuation in the visible band. Gain medium is Cr^{3+} doped $BeAl_2O_4$); the typical emission wavelength range is 0.695 to 0.81 μm.
ALQ Designation for a jammer, for example, ALQ 144 or ALQ 157.
Aluminium oxynitride (ALON) Similar transmission characteristics to sapphire, known as poor man's sapphire.
American National Standards Institute (ANSI) For example, ANSI Z136 standard for eye safety.
Amplified spontaneous emission (ASE) Noise on the output of a laser, to be avoided, spoils the beam quality, appears as a pedestal to the laser line.
Amplifier A laser cavity with a gain medium to enhance the power of an input beam; may be single pass or several, basically an open ended cavity. Not confused with a multipass oscillator.
Anderson Tallman A theory for predicting line shape.
Angle-of-arrival Of a threat as detected by a threat-warning sensor.
Angle tuning Of an optical parametric oscillator, rotation of the nonlinear crystal to select appropriate wavelengths, according to the phase-matching condition. See Critical phase matching.
Angular subtense It is the angle formed at the detector by the physical size of a source and the distance of that source from the detector, the apparent visual angle of the source; it is not determined by the divergence of the source and not to be confused with beam divergence.
Anisoplanatic angle It is the angle between a guide star and the object being viewed by the adaptive-optical system attempting to correct any optical aberrations caused by the atmosphere. This angle needs to be small for accurate aberration correction and usually occurs when the two objects appear to have similar point-spread functions.
Anti-reflection surface structures (ARSS) See Moth eye.
Anti-Stokes lines A wavelength generated in a Raman process that is shorter than the pump beam, has a lower probability of occurrence than a Stokes line, because of the transition starting from an excited state.
Aperture An opening or hole through which electromagnetic radiation can pass.
Aperture averaging A technique used to mitigate the effects of optical turbulence, by choosing an aperture diameter that is greater than the scale of turbulence; applies to transmission or reception.
Apparent visual angle The angular subtense of the source being viewed, determined by the source size and the distance from the sensor, such as a human eye.
Arc lamp A discharge source that generates a broadband emission.
Argon-ion Electrically pumped, continuous-wave gas laser, emits lines in a narrow band around 0.5 μm, 488.89 nm (blue) 514 nm (green), inefficient laser system.
Arrhenius equation An equation that predicts the temperature dependency of reaction rates.
As low as reasonably practicable (ALARP) For example, as used in risk mitigation considerations.
Aspheric An optical element with a distribution of radii of curvature to produce a compact optical configuration; an aspheric lens or asphere is a lens whose surface profiles are not portions of a sphere or cylinder.

Atmospheric absorption Measure of power or energy loss as a laser beam passes through the atmosphere. Caused by a number of physical-interaction processes. Varies widely with weather, meteorological visibility, smoke and other obscurants for any given wavelength.

Atmospheric-attenuation coefficient The coefficient represents the total absorption per unit length and is multiplied by the range in the exponent term in Beer's law.

Atmospheric block Range of wavelengths within which light does not propagate well through the atmosphere. Has its uses, such as countermeasure-resistant short-range proximity fuzing.

Atmospheric window Range of wavelengths within which light propagates well through the atmosphere. Three important ones are visible near infrared (400 to 1,400 nm), mid-wave infrared (2 to 5 μm) and far infrared (8 to 14 m). Also known as bands.

Atomic layer deposition (ALD) A growth technique.

Attenuation The decrease in the radiant flux of any beam of electromagnetic radiation as it propagates, owing to the extinction mechanisms.

Auger losses See Auger recombination.

Auger recombination A three-body nonradiative loss in a semiconductor laser, involving electron, holes and photons.

Avalanche photo-diode (APD) A detector, particularly used in single-photon detection, which exploits photo-emission and self-amplification.

Aversion response An involuntary response to the stimulus of bright light (e.g., the blink response of the human eye when laser light is incident on the eye).

Backstop Terrain or topographical feature, such as a ridge or hill, used to terminate a beam; in the laboratory this is usually a beam dump.

Band I IR seeker operating band typically in the atmospheric window around 2 μm up to about 2.5 μm, typically associated with lead-salt detectors, such as PbS or PbSe.

Band II IR seeker operating band typically around 3 μm, up to about 3.5 μm, typically associated with cooled lead-salt detectors.

Band IV IR seeker operating band in the atmospheric windows around 4 μm and beyond the atmospheric CO_2 notch up to nearly 5 μm, typically associated with cooled semiconductor detectors, such as InSb or CMT.

Bandgap Difference in energy between states: $\Delta E(\text{eV}) = 1{,}239.851/$ wavelength (λ in nanometres).

Barium vapour A metal-vapour lasing medium. Emits on a number of lines in the mid-wave infrared.

BB84 Communications protocol devised by Charles Bennett and Giles Brassard in 1984.

Beam angle or divergence Angle subtended by a laser beam, so the beam size increases with propagation distance. Beware of the difference between half-angle and whole-angle definitions. Laser beams are not sharp-edged; beware of differences between definitions of the edge of the beam (e.g., 1/e or 1/e^2).

Beam broadening Increase in the profile (spread) of a laser beam caused by propagation through a turbulent atmosphere, dependent on wavelength and scale of turbulence.

Beam clean up Techniques for removing higher-order modes; modern techniques involve the use of optical fibres.
Beam-director assembly (BDA) The mechanical structure of an acquisition, tracking and pointing system.
Beam expander/beam-expansion telescope Telescopic optical system used to expand a beam (diameter) and consequently reduce the divergence. Uses include reducing intensities within an optical system to avoid exceeding components' damage thresholds, and to reduce the beam divergence emerging from a system. Theoretically, the aperture*divergence product of the laser is maintained at the exit to the beam expander.
Beam parameter product (BPP) Expressing the beam quality and potential brightness in units of mm.mrad.
Beam quality Degree to which the output of a laser can be focused to within theoretical precision; some lasers are better than others. See Modes. Expressed as the product of the diameter of the exit (circular) aperture and the beam divergence, units of mm.mrad, also expressed as M^2 the beam-propagation parameter. Beam quality needs to be well matched to the pointing stability, to ensure optimum time that the beam spends on target.
Beam splitter An optical element used to create two beams from a single input beam.
Beam uniformity The difference between the intensity of the peak value in a beam (I_p) and the intensity in a trough (I_t) divided by the average intensity (I_{av}): Beam uniformity = $(I_p - I_t)/I_{av}$.
Beam waist The physical position in a laser beam, where the extent of the beam profile (spot size) is a minimum, the smallest radial dimension of a beam.
Beam wander Movement of the centroid of a beam resulting from passage through a turbulent atmosphere; this effect is independent of wavelength.
Beer's law The exponential relationship used to estimate beam propagation through the atmosphere, taking account of extinction of the beam. The extinction coefficient is a summation of the molecular absorption, the aerosol absorption, the molecular scattering and the aerosol-scattering coefficients. In an atmospheric window, the absorption coefficients tend to zero.
BeL A laser crystal operating at 1.08 μm: Nd:BeL = $Nd:La_2Be_2O_5$.
Bend losses Related to optical fibre loss mechanisms, the losses incurred through light escaping from the core when a fibre is bent through a radius; it is generally more severe for the smaller radii. May be used for mode selection in an active fibre.
Beta barium borate (BaB_2O_4) (BBO) An effective nonlinear crystal for harmonic generation, slightly hygroscopic, negative uniaxial material, with large birefringence.
Binary optics Diffractive optical elements, whose diffractive properties result from microscopic patterns (the wavelength of light) on a surface rather than from a hologram, called binary from original method of manufacture.
Birefringence Double refraction for ordinary and extra ordinary rays, essential for generating nonlinear effects in a crystal.
Birefringent filter (BRF) An optical element that uses the phenomenon of birefringence (or double refraction) to produce a narrow passband for transmission of wavelengths, for example, quartz, also known as a Lyot filter.
Bistatic A separation between the transmitter and receiver systems, opposite of monostatic.

Black body An object that has the properties of complete absorbance and has an emissivity of 1.
Blink reflex Normally limits eye exposure to high power visible continuous-wave laser beams to about 0.25 second.
Blue shift Move to shorter wavelength.
Blur circle The patch of light formed by an out-of-focus point source; its radius is determined by the size of the aperture and the distance from the focus.
Boltzmann statistics/distribution Defines the normal population of energy levels; the distribution is proportional to $1/k_BT$ where k_B is the Boltzmann constant and T is the absolute temperature.
Boresight error DC component (bias) of a tracking or pointing error.
Born-Oppenheimer approximation The assumption that the electronic motion and the nuclear motion in molecules can be separated.
Bottle necking A built-up population of the lower laser level, owing to the long lifetime of this level; leads to loss of population inversion and termination of laser action. Becomes self-terminating in a continuous-wave device and requires a mechanism to deplete the lower laser level.
Bottom-emitter OPSL Type of OPSL; after growth, the structure is flipped over and bonded to a heat spreader, followed by removal of the substrate (by etching).
Boule The crystal created during a crystal-growth technique, such as Bridgman or Czochralski.
Bow-tie cavity A description of the path followed by the laser beam through the cavity; shape of a ray traced on a trip around the cavity. Often used to minimise the potential reflection path to the pump source.
Bragg grating A modulated distribution in the refractive index of a material with a particular period to the grating, which leads to the reflection of a specific wavelength; more correctly known as a distributed Bragg grating or reflector.
Bragg mirror A distributed Bragg reflector (in a fibre laser); can be designed to select a preferred wavelength on reflection form the surface.
Bragg reflector See Bragg mirror.
Branching ratio The ratio of products formed in a chemical reaction (e.g., HF/DF in the reaction between F and HD).
Brewster's angle $\theta_B = \tan^{-1}(n_1/n_0)$, where n_1 is the refractive index of the window and n_0 is the refractive index of the air. For glass it is 57°, with n =1.54.
Brewster plate (BP) A polarising element.
Brewster windows A surface (window) at an angle that will transmit one polarisation without loss, the other plane of polarisation is reflected; occurs when beam is incident at Brewster's angle.
Bridgman crystal growth A crystal-growth technique involving a seed crystal and a molten flux. A thermal gradient in the furnace is moved through the reservoir of molten flux so that the temperature on one side of the gradient is below the melting point and above the melting point on the other side. There are two types, horizontal and vertical, which define the motion of the thermal gradient. It takes several weeks to grow a boule which used to grow a boule of ZGP.
Brightness Energy (or Power)/(beam divergence)^2.
Brillouin Name given to some nonlinear optical processes for amplification and phase conjugation of laser beams in systems.

Brillouin amplifier Arises from light scattering from thermal acoustic waves in a medium and, under normal conditions, is about one part in 10^{10}. The acoustic mirror acts as a moving mirror, so get a Doppler shift of 2 v/c, known as the Brillouin shift. Can use two beams, which have a frequency difference equal to the Brillouin shift and make them interfere in a medium (e.g., CS_2 or $TiCl_4$). This forms an interference pattern, travelling at the speed of sound, creating an acoustic wave. With a suitably intense pump beam, a strong acoustic wave is created, which acts as a strong grating, scattering light from the higher frequency pump beam to the lower frequency signal beam. The lower frequency (Stokes shift) beam is effectively amplified, gains of up to 10^9 possible.
Brillouin-enhanced four-wave mixing (BEFWM) A nonlinear process.
Broadband protection Type of sensor protection, in particular of eyes in the visible, against laser threats whose wavelengths are unknown and can be anywhere in the spectrum. See Wavelength diversity. Robust, convenient broadband protection is feasible.
Broadband source An emission with a range of wavelengths such as an arc lamp or flashlamp.
Broadening (of a laser transition) Many mechanisms results in an increase in the line width of a transition; common examples are collision, Doppler, pressure (in a gas laser).
Buffer angle The angular dimension added to the beam divergence to account for uncertainties in the beam-pointing angle; for a fixed steering mirror, the buffer angle will be small. Used in defining safety zones.
Buffer zone The volume of space around a propagating beam defined by the buffer angle.
Bulk damage Laser-induced damage in the body of the material, contrast with surface damage; bulk damage tends to have a higher damage threshold than surface damage.
Burn in A process used to establish and stabilise the performance of a component, often a laser diode.
Burst-illumination laser (BIL) A system that uses gated-detector techniques to reject clutter.
Cadmium mercury telluride (CMT) Equivalent of MCT, a detector material for the infrared.
Cadmium silicon phosphide ($CdSiP_2$) (CSP) A modern chalcopyrite nonlinear crystal.
Cascade lasing Laser action resulting from a previous lasing event, which creates a population inversion for this event. Can have a chain reaction.
Catastrophic optical damage (COD) Usually of an optical element in a device or system.
Cavity Abbreviation for laser cavity or resonant cavity.
Cavity dumping Technique used to produce short pulses (e.g., tens of nanoseconds) from pulsed lasers. Particularly useful when a specific laser does not have energy storage capability needed for Q-switching, such as a dye laser.
Cell on wheels (COW) A mobile communication system.
Chalcogenide glass A type of glass transmitting in the mid-wave infrared; the composition contains at least one chalcogen (sulphur, selenium or tellurium) element.

Chalcopyrite Type of crystal used in nonlinear conversion; of the form $M1M2X_2$, where M1 and M2 are metals (e.g., $ZnGeP_2$). A common example is iron pyrites.
Charge-coupled device (CCD) Imaging sensor in a digital camera.
Chemical laser A laser device using chemical reaction to achieve population inversion in the stimulation emission process. Examples include DF and COIL.
Chemical oxygen iodine laser (COIL) The source for the airborne laser (ABL) of the USAF. Emission $\lambda = 1.315\ \mu m$. Excitation of iodine by singlet oxygen delta.
Chemical-vapour deposition (CVD) A technique for growing semiconductors; a chemical process in which one or more two-chemical species (precursors) react together to form the desired material on the surface of a substrate. It is used to produce high-purity, high-performance solid materials.
Chirally-coupled core (CCC) A type of fibre laser that maintains single-mode operation.
Chirp Variation of emission wavelength with time, often as a consequence of a change in temperature of the emitter.
Chirped-pulse amplifier (CPA) Used in ultrashort pulse generation.
Chromatic aberration Spread in the focus of a lens or mirror when illuminated with different wavelengths, or with a broadband source.
Chromophore An atom or group of atoms that serve as a unit in light absorption; in the rhodamine class of organic dyes it is the N-N backbone.
Circular polarisation The polarisation vector traces out a circular path, either to the left direction or to the right as it propagates. Caused by the light being formed by two linearly polarised components of equal amplitude, with one lagging the other by 90°. See also Elliptical polarisation.
Cladding mode A mode in an optical fibre that is confined to the cladding region because of the higher refractive index of this region (i.e., total internal reflection).
Cladding-mode suppressor A high refractive index outer coating of the fibre, which removes the higher-order modes.
Cladding pumping Injection of the pump light into the cladding region of a fibre laser.
Cleavage planes The planes in a structure that separate easily, owing to the weakness of the bonds along those particular directions.
Close-coupled cavity A cavity design with the flashlamp(s) in close proximity to the gain medium, usually using diffuse reflectors to couple the excitation from the indirect path, from the lamp to the gain medium, into the gain medium. Provides a compact design.
Clutter Target-like returns detected by a receiver; unwanted echoes.
C_n^2 Refractive-index structure parameter. Indicates magnitude of refractive turbulence in the atmosphere, has units of $m^{-2/3}$.
Coatings Whole selection: antireflection, colour filter (dyes that absorb at some wavelengths and emit at others), neutral density (absorb/attenuate uniformly over a range of wavelengths), metal film (to form mirrors), protective. Coating can be a weak link in a military laser system.
Coelostat An optical device that uses two mirrors to direct a beam, as used in a beam director.
Coherence Characteristic of light output from laser; types are temporal and spatial coherence.

Coherence length Defined as $\ell = c/\Delta\nu$, $\nu = \Delta\lambda/\lambda^2$, where $\Delta\nu$ is the frequency spread, $\Delta\lambda$ is the laser line width and λ is the wavelength, and is the distance travelled in Δt. (see below).

Coherence time (Δt) $\Delta t = 1/\Delta\nu$. Coherence length.

Coherent anti-Stokes Raman spectroscopy (CARS) Used for chemical analysis.

Coherent beam combination (CBC) Of lasers, a technique for making a source with high optical power from components with lower optical power (e.g., fibre lasers).

Coherent change detection (CCD) A method for finding difficult targets.

Cold shield A physical aperture, usually in a Dewar, which restricts/minimises the amount of stray light entering the detector.

Collateral damage Incidental or superfluous damage occurring whilst the military objective is pursued.

Collecting optics Optical system used for magnified viewing. The systems usually produce an increased optical hazard, as more photons can enter an eye or sensor.

Collimation The process of aligning the boresights (axes) of optical systems to one another (e.g., tracking and pointing axes). Carried out automatically in sophisticated systems.

Collimator Optical device for generating a beam of parallel light.

Colloids A substance consisting of extremely small particles, which are intermediate in size between a true solute and a suspension.

Collquirities Crystal hosts such as LiCAF ($LiCaAlF_6$), LiSAF ($LiSrAlF_6$) and LiSGAF ($LiSrGaAlF_6$), named after a city in Bolivia where the minerals were first found.

Colour centre A defect where an electron occupies a vacancy in a ionic crystal and interacts with the lattice vibrations to give energy states not usually present in the crystal: ΔE_{abs} = few electron volts (i.e., in the visible and near infrared).

Colour-centre laser A solid-state laser using the energy levels created by a colour-centre (defect) in a lattice; emission is in the visible and near infrared.

Colour temperature This relates to a broadband emitter, such as an arc lamp and is the temperature of a black body that would generate the equivalent energy distribution.

Coma Aberration of a mirror or lens: image of an off-axis point has a comet-like shape.

Commercial of the shelf (COTS) An item readily available.

Complementary metal oxide semiconductor (CMOS) Used to form integrated circuits.

Conduction band Part of the structure of a semiconductor above the valence band, where there are free electrons.

Confocal parameter Defined as twice the Rayleigh length. This parameter indicates the potential for the overlap between the pump and the signal (and idler) beams in a nonlinear crystal, and hence, the prospects for efficient nonlinear conversion with a tightly focused pump beam.

Confocal resonator A stable design, which has two concave mirrors with identical radii of curvature and arranged so that their focal points coincide.

Congruent PPLN Normal type of periodically poled lithium niobate (PPLN), at little deficient in Li, owing to its volatility. A useful nonlinear crystal for wavelength conversion using quasi-phase matching techniques.

Conjugate (beam) A conjugate beam has the exact opposite phase variations to those of the aberrated beam it is attempting to correct.
CO_2 laser Gas laser, which emits a number of lines in the far infrared. Relatively efficient. Can be made in a number of forms up to large powers and pulse energies. Eye-safe in low power applications (e.g., range finding). Primary candidate for countering forward-looking infrared (FLIR) systems and damaging dielectric materials.
Concepts of operations (CONOPS) A description of how a system may be operated effectively.
Continuous wave A laser with a continuous emission for a period of at least 0.25 second. Not a pulsed beam.
Control measure(s) A means to mitigate potential hazards in three categories: engineering, process/procedural and personal.
Controlled zone A volume of space with controls on operations to protect personnel from laser hazards.
Copper vapour A metal-vapour lasing medium. Emits lines in the green and yellow (511 and 578 nm).
Coude path An elbow-shaped optical path, often used to direct a beam on to a beam steering element.
Cost function A technique applied to the optimisation of a system with a number of variables. Often used to select the most appropriate element from a range of alternatives.
Couplets Crowns and flints for chromatic aberration correction; material characteristics are mutually compensating.
Covalent bond Sharing of the electrons in a bond between two atoms.
Critical phase matching The $\Delta k = 0$; conservation of momentum during the annihilation of the pump photon. As used in an optical parametric oscillator, it is usually tuned by rotating the nonlinear crystal.
Cross relaxation Virtually the opposite of energy-transfer upconversion. In this case, one ion has a ground-state absorption to an upper level and there is a relaxation with an emission that has a longer wavelength (smaller ΔE), this photon is then absorbed by a second ion with an appropriate energy-level structure.
Cross-section This term has many meanings, such as the equivalent area of an optical system that would give rise to an equivalent isotropic emission, or the observed population of an event. In the case of a laser transition, it is the probability of an active ion absorbing or emitting a photon (i.e., proportional to its probability cross-section).
Cryogenic lasers Lasers operating at a depressed temperature, typically around 80K; at this temperature many gain media have good thermal conductivity, enhanced cross-sections, reduced ground-state absorption and a refractive index less sensitive to thermal fluctuations.
Crystal field Interacts with degenerate energy levels in crystals to give distinct broad energy levels, which leads to energy-level splitting. A consequence of the formation of electric and magnetic interactions when a host is doped with rare-earth or transition metal ions.
Cut-back method Technique used to estimate absorption in an optical fibre; a transmission measurement is made through a given length of fibre and repeated

through a shorter length. The difference in transmission equates to the additional absorption occurring in the length cut back from the original length of fibre.

Current guiding Use of an applied current in a diode to form a waveguide; used to confine the zone for laser action in an effort to increase efficiency.

Chemical vapour deposition (CVD) A technique for growing semiconductors.

Cyclo octo tetraene (COT) A triplet quencher.

Czochralski crystal growth Crystal-growth technique involving a seed crystal pulled from the surface of a molten flux; the seed crystal is rotated as it is pulled up from the molten flux. Crystal growth is slow taking several weeks or more to produce a crystal.

Damage Permanent effects of lasers on electro-optical sensors. Usually damage thresholds are easy to establish. The higher the repetition rate, the better the hit chance with a pulsed beam.

Damage threshold of infrared materials Normally related to mean-power dissipation, rather than electric-field intensities, so commonly quoted as a beam fluence. Thresholds for optical parametric oscillator action rise in mid infrared, owing to the increase in refractive index and the product of the output wavelengths. Can use very short synchronous pulses, to avoid this problem. In this case, the intensity in each picosecond pulse is very high, but the energy is low, thus avoiding damage. Careful control of the position of the focus of the pump beam will also avoid the onset of damage.

Dazzle Temporary (reversible) effect of lasers on electro-optical sensors, including the eye. Includes sensor saturation and distortion of scene contrast. Mechanisms involved include internal reflections and forward scatter within sensor and forward scatter from atmosphere, windscreens, visors and other optical devices. The eye takes time to recover (seconds/minutes) from severe dazzle.

Dazzler Laser designed to induce temporary effects in sensors, such as glare or flash blindness.

de Broglie wavelength (of an electron) The de Broglie wavelength is inversely proportion to the momentum (p) of a particle ($\lambda = h/p$) or $\lambda = hc/pc$ where pc is expressed in electron volts (where hc ≈ 1,240). For an electron with kinetic energy of 1 eV and rest mass of 0.511 MeV, $\lambda = 1.23$ nm.

Decay A transition from a higher to a lower energy state; the rate of this process is important for laser action.

Deep ultraviolet (DUV) For example, $\lambda < 300$ nm.

d_{eff} Nonlinear coupling coefficient; an indication of the conversion potential of a nonlinear interaction in a particular crystal direction.

Defensive aids suite (DAS) A suite of detectors (e.g., RWR, LWR, ESM and effectors for self-protection).

Deflagration Subsonic combustion, where propagation of the heat is through thermal conductivity.

Deformable mirror Used in an adaptive-optical system to create the conjugate beam shape; the surface of the mirror is continuous, contrasts with a segmented mirror.

Depleted beam The remnants of a pump beam after it passes through another medium, such as a nonlinear crystal; the unused part of the pump beam, which could be recycled.

Depletion layer The zone between the p and n sides of a pn junction. Zone in semiconductor without mobile carriers, resulting from forward bias, where the laser action takes place in an injection laser.
Depth of field The extent of the zone of focus in front and behind the point that is in focus to an optical system; objects beyond this zone will be out of focus (i.e., blurred); the depth of field for a fast optical system is small.
Deterministic risk assessment (DRA) Method of laser-hazard evaluation based on the physical properties of the beam and its propagation compared with the exposure limits defined by the maximum permissible exposure.
Detonation A supersonic effect, which drives a shock front.
Deuterium fluoride (DF) Chemical, high-power gas laser, producing lines around 3.8 to 4.2 μm; currently, about the only state-of-the-art laser for weapons in the mid-infrared. These lasers are continuous wave and similar to rocket motors in operation, producing toxic, corrosive efflux. Electrically-pumped, sealed, pulsed DF lasers exist, but are much less powerful.
Diamond field-effect array (DFEA) cathode For free-electron lasers, a source of electrons.
Dielectric edge mirrors (DEM) Used in wavelength-beam-combining systems.
Difference frequency-generation (DFG) A nonlinear wavelength conversion process ($\lambda = (\lambda_s^{-1} - \lambda_p^{-1})^{-1}$, where λ is the new wavelength and λ_p and λ_s are the pump and signal wavelengths, respectively.
Diamond field-effect array (DFEA) cathode An array of diamond pyramids, an alternative to photo-cathodes. A high-brightness electron source.
Dielectric An insulator capable of sustaining an electrical stress (charge).
Diffraction A consequence of the wave nature of light, resulting in shadows generated by an obstruction in a beam not having a sharp edge. Two forms: Fresnel and Fraunhofer.
Diffraction grating A device for generating spectra through diffraction of the incident beam and also for measuring wavelength. Often consists of a structure with a series of very closely spaced, equidistant parallel lines.
Diffraction limit The smallest spot size that can be achieved from a given aperture size (d), determined by $\theta = 1.22\ (\lambda/d)$, resolving power of a telescope. Often used as a measure of beam quality, where λ is the wavelength.
Diffractive optical element (DOE) Can be used as a beam splitter, or if operated backwards, it acts as a beam combiner.
Diffuse reflection Opposite of a specular reflection, so different part of a beam incident on a surface are reflected/scattered at different angles, governed by Lambert's Law (or the Cosine Law).
Diffusion bonding A technique for joining crystals to make larger structures. Dependent on high-precision polishing to form flat surfaces and held together by van der Waal forces. In some circumstances superseded by the use of ceramic hosts.
Dimer A molecule consisting of two identical atoms, but this molecule only exists in the excited state. Originally called an excited dimmer.
Diode-pumped alkali laser (DPAL) Attempts to combine the best attributes of gas and solid-state lasers.
Diode-pumped solid-state laser (DPSSL) Usually very efficient form of a solid-state laser.

Diode pumping A technique for exciting laser ions in a solid-state host or some gases, using the emission form a laser diode, usually very efficient.

Diode smile Distortion in a diode source owing to misalignment between diode bars, named from the appearance of the emission from the diode array.

Direct generation A laser emission that occurs without using nonlinear conversion techniques.

Direct-vision optics (DVO) A type of sighting system without magnification.

Directed-infrared countermeasure (DIRCM) A jammer designed to defeat heat-seeking missiles.

Dispersion Change of refractive index with wavelength (or frequency) in a material and consequently a change in velocity of each wavelength.

Distributed Bragg reflector (DBR) See Bragg reflector or Bragg grating. This is the correct term for these devices.

Distributed feedback (DFB) A form of seed laser providing a particular characteristic.

Diurnal variation Variation of a property or parameter throughout the day (e.g., diurnal variation of C_n^2).

Divergence The spread in the diameter of a beam as it propagates from the exit aperture; a consequence of the fact that even a well-collimated beam has an angular extent. Beware of multiple definitions.

Dopant A species added to a semiconductor to achieve the desired conductivity of the n or p structure.

Doppler shift Change in frequency resulting from the motion of a source and/or receiver; blue shift in light is from a closing geometry and red shift is a receding observation.

Double-clad core An optical fibre used in a fibre laser with two regions of cladding around the central doped core; enables multimode pumping of the inner cladding layer that passes into the doped core of the optical fibre.

Double-clad fibre Optical fibre with two cladding layers around the core, often used to enhance pumping efficiency.

Doubling Meaning frequency doubling through a nonlinear process, which halves the wavelength (doubles the frequency) of a laser (e.g., generating 532 nm from the fundamental line from a neodymium YAG laser). Can be an inefficient way of generating green light. In this nonlinear process, both generated waves have the same wavelength. See also Harmonic generation.

Doubly resonant (OPO) The optical parametric oscillator (OPO) cavity is arranged so that both the signal and idler wavelengths are in a resonance; tends to have a lower threshold than a singly resonant device, but at the expense of stability of output.

Doubly resonant oscillator (DRO) Both the signal and idler wavelengths are in resonance in the OPO.

Down conversion or down shift Reduce the frequency of a transition via transitions in the gain medium; i.e., longer wavelength emission.

Dual ion-beam sputtering (DIBS) A coating technique, can produce coatings with high LIDT (see below).

Duty cycle The fraction of the time that a system is active compared with the time it is quiescent.

Dye Organic compound with an extended π-electron system, whose application is dependent on one or more properties of this π system.
Dye bleaching Absorption mechanism that is a result of a single transition in a dye molecule. As the incident intensity increases, more and more dye molecules absorb light, raising them to an excited state; eventually, an incident intensity is reached where they cannot absorb any more photons. Although the excited states eventually decay at some point, the incident light intensity reaches a level at which as many species are in a lower state as are in an excited state. Under these pumping conditions, emission is occurring at the same rate as absorption, so no net absorption takes place; the dye has saturated or is bleached. Useful for passive Q-switching and some electro-optical protection measures (i.e., a reverse saturable absorber).
Dye laser Family of lasers giving wavelength diversity and relatively broad individual lines. Flashlamp-pumped dye lasers are energetic, but high-power liquid dye lasers require a dye solution to be circulated. Devices can provide more than one colour simultaneously. Safe solvents are now well established for the liquid devices, and the solid-state version is now available.
Eaton lens This lens bends light through 90° from all incident directions.
ED(50) A level of laser interaction with the eye, which will give a 50% probability that a minimum opthalmoscopically detectable lesion of the retina will be caused to a white Caucasian. The level depends on lighting conditions and the degree of adaptation of the eye (pupil size) and position on the retina.
Edge-emitting laser (EEL) Semiconductor device.
Efficiency Care is required interpreting statements of laser efficiency. Many common lasers are very inefficient at converting electrical energy to light, around 1 to 10%. Wall-plug efficiency includes power supply and all support equipment, an accurate measure of the demand on the power supplies in a system. Other common descriptors include optical-to-optical and slope efficiency (of the laser action) may be as high as 70%.
EH(90) Level of laser interaction with the eye, which will give a 90% probability that a haemorrhagic (bleeding) lesion of the retina will be caused. The threshold for this event depends on ambient lighting, degree of adaptation of the eye and laser parameters.
Electromagnetic compatibility (EMC) A test to ensure that the laser-based device can work in the environment of a platform and also not disrupt the operation of other systems. Some lasers use high voltages and fast current rise time, potentially leading to electromagnetic-pulse problems.
Electromagnetic (EM) radiation Radiation that has varying orthogonal electric and magnetic fields, which sustain its propagation at the speed of light.
Electron-hole recombination The mechanism that gives rise to photon generation in a semiconductor diode.
Electron-paramagnetic resonance (EPR) A technique that identifies charge defects in a material under test.
Electron volt (eV) Energy needed to move an electron through a potential of 1V.
Electronic energy levels Those energy levels in a laser species that involve an electron changing its energy status. Usually involve photon emissions in the visible and near infrared.
Electro-optical (EO) Usually refers to sensor operating in the visible or near infrared.

Electro-optical countermeasures (EOCM) Includes low-power laser directed-energy weapons, also jammers, and flares.
Electro-optical modulator (EOM) A Q-switch using electro-optical properties of a crystal to control the loss in a cavity, such as modification of the refractive index of one plane of polarisation with respect to another in response to an external field.
Elliptical cavity A simple reflective pump cavity with the flashlamp at one focus and the gain medium at the other.
Elliptical polarisation A state of polarisation where stationary observer would see the polarisation vector tracing out an ellipse as it propagates. Caused by the light being formed by two linearly polarised components of unequal amplitude, with one lagging the other by 90°. With circularly polarised light, the two components are equal.
Emerald Laser vibronic gain medium, $Cr^{:3+}Be_3\ Al_2(SiO_3)_6$ emission from 0.69 to 0.83 μm; Cr^{3+} is the lasing ion.
Encircled energy Refers to the concentration of energy in an optical image or a projected laser spot at a given range.
End cap An undoped region attached to a gain medium, such as a fibre or a laser rod. Its purpose is to mitigate some of the potential damage mechanisms, such as those resulting in bulk damage resulting from imperfections close to the surface, resulting from crystal fabrication or polishing.
End-to-end testing (E2E) Of a complete system.
Energy-distribution curves (EDC) Emitted electron intensity as a function of electron energy, describes cathode performance.
Energy level Quantum state of an atom or molecule (species).
Energy-recovery linac (ERL) Linear accelerator in an FEL (see below).
Energy-transfer upconversion An energy-transfer mechanism between two ions (inter-ionic process), which have similar energy states, so that the relaxation emission from an upper state in one ion can be absorbed by an excited state in another ion, leading to a higher energy level being populated. Longer wavelength emission results from the relaxation from this higher state to the lowest level by skipping the intermediate level.
Epi growth Epitaxial growth (layers of a semiconductor), where the pn junctions are created.
Epitaxy A method of growing thin layers of semiconductor material on a substrate, so that the crystal structure is identical to that of the crystalline substrate.
Etalon A structure of layers of transparent materials with frequency-selective characteristics (e.g., a filter).
Etendue Brightness = Power ÷ (Area × Divergence Angle).
Evanescent wave An evanescent (tending to vanish) wave is produced when a propagating wave is incident on a classically forbidden region. For example, light reflected from an interface at an angle greater than the critical angle, so that total internal reflection occurs, forms an evanescent wave on the opposite side of the interface. The intensity of an evanescent wave decays exponentially with distance from the interface at which it was formed.
Excimer laser Type of laser that typically emits in the ultraviolet, laser action as the result of formation of dimers. From excited dimer: gases mix but do not react (e.g., XeCl).

Exciplex technique A technique used to broaden the very narrow absorption feature in atomic species such as the alkali metals; a mechanism involves collision broadening by adding gases such as the noble gases, often involving three-body interactions.

Exciplex-pumped alkali laser (XPAL) Attempts to combine the best properties of gas and solid-state lasers. Excitation through collisions of dissociated alkali atoms.

Excitation Raising of a species to a higher energy state; process can be absorption of photons, recombination of current carriers in a semiconductor, collision of electrons and species, chemical action, collision between atoms and molecules, recombination of free electrons and ionised atoms or acceleration of atoms.

Excited-state absorption (ESA) An intra-ion process, which involves further absorption of the pump light by states already in an excited state, so that upper energy levels become populated. Usually a loss mechanism.

Explosion Detonation of a chemical substance (can contain a single molecular species), get a shock wave and a chemical reaction; a technique used to excite some gain media.

Explosively-pumped iodine laser (EPIL) An amplifier.

Extended cavity See Long laser.

Extended ocular hazard distance (EOHD) Takes account of the increase in the nominal ocular hazard distance (NOHD) where a laser beam may be viewed safely by magnifying optics.

Extended source This is a source that can be resolved into a geometrical image by a human eye; the opposite of a point source.

Extinction Summation of absorption and scattering effects on the intensity of a propagating beam.

Eye-safe (r) wavelengths Those wavelengths that are generated in the eye-safe regime, where in theory the laser light is absorbed before it reaches the retina. Typically wavelengths beyond about 1.4 μm. Care is required, as powerful laser light will only be slightly delayed as it boils off the liquids on the cornea.

F number (F#) Aperture size (D) of an optical system in relation to its focal length (f), giving F# = f/D; the larger the aperture, the more photons are collected, but the depth of field decreases. A large F# equates to a fast optical system.

Fabry-Perot cavity Usually a semiconductor laser with plane cleaved faces.

Fabry-Perot etalon Pair of parallel plates close together, so that the resonant modes are far. Spacing between the plates can be changed by tilting the etalon or moving the plates apart in order to select wavelengths. A pair of parallel flat optical surfaces used to select wavelengths.

Facet Face or surface of a crystal; particularly related to semiconductor devices.

Factory-acceptance test (FAT) The testing undertake prior to delivery of a system.

Faraday rotation The change in the plane of polarisation of an electromagnetic beam in a material in response to the application of a magnetic field.

Faraday rotator An optical component that rotates the plane of polarisation by a given angle in response to an applied potential; is nonreciprocal and hence is used as an isolator.

Far field Beyond the near field of a radiation pattern where the angular-field distance is independent of range from the transmitter (source). May be estimated from far field = 2d2/λ for a 45° phase taper across the wavefront. See Fresnel zone.

Far infrared (FIR) See Atmospheric windows.

Fast axis The emission of a semiconductor that diverges most rapidly; occurs from the narrowest part of the emitter, orthogonal to the slow axis.
Fast axis (crystal) Light travels with highest phase velocity along the optical axis with the smallest refractive index; opposite of slow axis.
Fast-Fourier transform (FFT) A spectral analysis technique based on the Fourier theorem.
Fast optics Those with a large F number (F#).
Fermi level A hypothetical level of potential energy for an electron in a crystalline solid; it is used to describe the top of the electron-energy levels at 0K.
Fibre Bragg grating (FBG) A mirror induced in a fibre to help form the resonant cavity, have high reflectivity (HR) and lower reflectivity (LR) or output coupler (OC). A distributed Bragg reflector in an optical fibre, typically used to form the resonant cavity in a fibre laser. It may be configured to have very high reflectivity and lower reflectivity for the output coupler.
Fibre laser Long thin solid-state laser. Laser action occurs in the fibre core, often considered to be a brightness enhancer and wavelength convertor.
Field of regard (FoR) The solid angle over which, the point of regard moves; the angle through which a sensor can move and provide data.
Field of view (FoV) The angles within which a sensor sees or can detect light; less simple in definition for eyes than sensors. Not to be confused with point of regard.
Figure of merit (FOM) A function derived to quantify the potential benefits of a technique or method, not to be confused with a cost function.
Filament/filamentation The form of the emission in free space during super-continuum generation and is related to plasma formation as the beam self focuses, as result of the intense peak power of these very short duration pulses. The emission from the filament is broad band and does not diverge. Multiple filaments may be formed as the beam propagates and extend for many metres.
Fill factor (FF) Usually the percentage of an aperture that is utilised. The fraction of an aperture that is being used.
Filter A device used to attenuate laser radiation; various types are fixed line and rugate.
Findlay-Clay analysis A technique to determine the losses in a laser system, so that the lasing threshold may be determined.
Finite-element analysis (FEA) Used in the structural analysis of a system or body, particularly useful for calculating temperature rises in a gain medium.
Fixed-line filter Used in personal-protection equipment. Many common lasers' outputs are pure, monochromatic light at a single precisely-defined wavelength. Such light is easily filtered out, using fixed-line filters, which will be transparent to most of the rest of the spectrum. Multiline filters have been produced, some including narrow-band elements as well as for the visible and near-infrared.
Flare A pyrotechnic store ejected from a platform to seduce a heat-seeking missile.
Flashlamp Excites the lasing medium by illuminating it with light from a high-power xenon flashlamp. Used, for example, in early solid-state lasers such as neodymium and ruby, also used to excite dye lasers. Used in pulsed lasers; requires tens of kilovolts input with fast current rise times. See EMC and power supply. Colour temperature rises with increasing current density. λ_{max} of the black-body emission moves towards the ultraviolet.

Flash blindness Temporary loss of visual acuity owing to a bright beam of light incident on the eye.

Fluorescence Emission of light from a material illuminated at a different wavelength, usually a longer wavelength than the excitation wavelength. The illuminated material appears to glow; the emission ceases when the excitation source is removed.

Fluorescence conversion Typically the use of a dye in solution between the flashlamp and the gain medium to convert some of the pump light into a band that leads to more efficient pumping of the gain medium. Not necessary when using laser diodes as a pump source.

Fluorescence lifetime This term refers to the average time a molecule remains in an excited state before emitting a photon. Long lifetimes are required for energy storage.

Focal length (f) Distance from the principal plane to the focal plane, $1/f = [(n - 1)\{1/r_1 + 1/r2\}]$, where r_1 and r_2 are the radii of curvature of the two surfaces and n is the refractive index.

Focal-plane array An array of detectors used in a thermal camera. Array sizes of 256×256 and 512×512 common, also $1{,}024 \times 1{,}024$ available.

Focus The process of obtaining a sharp image.

Forbidden transition A transition of an atom or molecule that is not permitted by the quantum-mechanical selection rules, owing to the need for a change in the quantum number. These transitions can occur, but the probability is low.

Forsterite A vibronic gain medium. Host crystal Mg_2SiO_4, lasing ion Cr^{4+}, emission from $\lambda = 1.15$ to $1.3\ \mu m$.

Four-wave mixing This is used in phase conjugation and has some similarities with real-time holography. The technique requires two auxiliary pump beams, in addition to the signal beam being conjugated. The two pump beams are counter-propagating and mix with the signal beam in a medium such as CS_2, which has an intensity-dependent refractive index. The signal interferes with the pump beam and forms interference fringes in the liquid (nonlinear medium). The intensity variations in the interference pattern produce a variation in the refractive index, thus forming refractive-index gratings. These gratings act like holographic mirrors. The grating formed by the signal and one pump beam scatters the other pump beam to form a fourth beam, which is the phase conjugate of the signal. The power in the phase conjugate comes from the pump beams, so it may appear that the mirror has more than 100% reflectivity.

Fourier cavity A cavity that performs a Fourier transform on the initial field distribution during a single round trip of the cavity.

Fourier plane The focal plane of a converging lens.

Fourier series Any single-valued periodic function can be represented as the sum of sinusoidal components, with frequencies that are multiples of the frequency of that function.

Fourier transform The technique for moving between the frequency and time domains, based on the fact that any single-valued periodic function can be represented as the sum of sinusoidal components, with frequencies that are multiples of the frequency of that function.

Fourier-transform infrared (FTIR) A spectrometer for characterising materials by measuring transmission through a sample and reflection from its surfaces. Sophis-

ticated devices make measurements from the ultraviolet to the far infrared or the terahertz region.
Four-level laser Those lasers that use four energy levels in the laser species during the stimulated emission—the most common example is the neodymium laser.
Four-wave mixing (FWM) A nonlinear conversion process.
Fovea Area of the retina providing the central region of vision, about one degree wide, about the eye's point of regard.
Forward-looking infrared (FLIR) A thermal camera.
Free-electron laser Excitation from the passage of a relativistic electron beam through an oscillating magnetic field converting kinetic energy into light. Not classical laser action, but involves a resonant cavity.
Free-spectral range (FSR) Of a resonant cavity.
Fresnel equation Expresses the dependence of the refractive index on the propagation direction in birefringent materials.
Fresnel number (F_N or N_F) $F_N = A^2/\lambda L$ where a high No. equates to short fat laser, typical for a dye laser or a thin disk, solid-state device. Indicates the potential that a laser cavity has for generating a beam with high beam quality, the lower the number the better, as in the case of a fibre laser. May be interpreted as a ratio between the reflected power of a cavity mirror to the diffraction loss at that mirror.
Fresnel reflection The reflection caused by the mismatch between the refractive indices of two media as light passes from one to the other. Effects of these reflections may be mitigated through the use of (refractive) index-matching techniques.
Fresnel zone Distance = $2d^2/\lambda$ for a 45° phase taper across the wavefront; defines the distance to the far field, where d is the diameter of an aperture.
Fried parameter (r_0) Provides an indication or measure of the optical quality of the atmosphere, expressed as a distance (usually in centimetres) that corresponds to an area over which the root mean square (rms) error is less than 1 rad.
Full wave half maximum (FWHM) A width of a profile at half of its maximum intensity.
g-tensor Used to identify the type of defect in a material—part of the electron-paramagnetic-resonance technique.
Gadolinium scandium gallium garnet (GSGG) Laser crystal host crystal gadolinium scandium gallium garnet ($Gd_3Sc_2Ga_3O_{12}$) with Cr^{3+} emission λ = 0.74 to 0.9 μm.
Gain curve This is a function that indicates the gain as a function of frequency (or wavelength), for a photon, of a given frequency, making a round trip of a resonant cavity in a laser.
Gain medium The part of the laser cavity containing the species that will be excited to provide the laser action. Can be solid, liquid or gas. The phase of the gain medium is often used to classify lasers (e.g., a gas laser has a gaseous gain medium).
Gain switched A method of pulse generation by modulating the laser gain via the pump source.
Gain-switched oscillations Pulse generation by the modulation of the gain in the cavity, through changes to the pump power applied to the gain medium.
Gaussian beam Description of the intensity profile of an emission; this type of beam has its intensity concentrated along the resonator axis. Many beam shapes approximate loosely to a Gaussian.
$Gd_3Sc_2Al_3O_{12}$ (GSAG) Laser crystal host.

General laser analysis and design (GLAD) A software tool for predicting optical performance.
Geo-spatial and temporal referencing (G&TR) An inertial reference.
Giant pulse Another term for a Q-switched pulse.
Gillette As of razor blade, an old unit used to measure the damage potential of a pulsed laser. For example, 5 gillettes meant a laser pulse would cut a hole through a stack of 5 razor blades.
Glare Veiling glare, caused by stray light in an optical system, resulting from multiple reflections between surfaces. The consequence is a reduction of contrast or loss of an image.
Glint A highly directional specular reflection (e.g., a solar glint from a reflective surface). Also, the movement of the radar-echo centre in a radar/radio-frequency context.
Graded-index fibre An optical fibre where the core has a refractive index that decreases with increasing distance from the fibre axis; gives decreased modal dispersion.
Graded refractive-index profile (GRIN) In a lens, for example.
Green problem Noise on the output caused by multimode interference. Avoid by using many modes or a long cavity.
Greenwood frequency Of an adaptive-optical system, it is the bandwidth required for optimal correction by an adaptive-optical system.
Grey body Similar to a black body but with lower intensity.
Grey-body radiator A radiator whose emissivity is independent of wavelength.
Grey tracking A damage effect in nonlinear crystals; typical problem with potassium titanyl phosphate (KTP). Caused by trapped carriers of Ti^{3+}.
Ground-state absorption (GSA) A loss mechanism.
Group velocity Some forms of wave motion experience the phase velocity varying with wavelength. In this case a nonsinusoidal wave appears to travel at a velocity that is distinctly different from the phase velocity. A particular problem in nonlinear conversion.
Guided-mode resonance filter (GMRF) A material is patterned to drive a resonance at a particular wavelength. Often used as a line-narrowing element or a filter. A substrate can be patterned to give an output coupler and a grating in a single optical element.
Harmonic generation A nonlinear process involving mixing of waves in a nonlinear crystal to generate a harmonic of the beam incident on the nonlinear crystal. Common harmonic is the second, used for frequency doubling to give half the wavelength; other common harmonics are the third and fourth. Frequency doubling often used with a solid-state laser to generate visible-band light (e.g., doubling the neodymium output to generate green light).
Half-wave plate (HWP) A polarising element.
Hazard classification Provides an indication of the potential hazard presented by a laser during normal operation; assigning a laser to a specific class determined by the potential hazard presented by the flux in the beam and protection possibly needed for safe operation.
Haze The constituent particles of the atmosphere, such as dust, smoke and water vapour, which reduce the optical transmission of the atmosphere.

Heat-capacity laser A laser without additional cooling; it relies on the ability of the host to store the waste heat generated during laser action. In theory, the host should be free from cooling stresses, so the gain medium can expand uniformly without distortion.
Heavy, expanded mobility tactical truck (HEMETT) A test bed for U.S. Army L-DEW system.
Helium neon (He-Ne) A gas laser, emits at 0.6329 nm in the red, and less powerfully in green and orange/yellow in the visible but also in the near infrared (1.15 μm) and mid infrared (3.39 μm). Highly stable but inefficient.
Hermite polynomials These are a classical orthogonal polynomial sequence used to describe the rectangular modes emitted from gain media with a strong polarisation characteristic.
Hetero-junction An interface at a pn junction where the compositions are different, used in interband semiconductors, which means little confinement; keeps the holes and electrons together long enough for recombination and photon generation. With a double hetero-junction the energy of the lasing photon is only determined by the band edges (valence band and conduction band) in the bulk semiconductor.
Heterodyne detection A method of detecting radiation by nonlinear mixing of the incident radiation with a reference frequency.
High operating temperature (HOT) Usually related to operation of detectors.
High reflector (HR) A mirror, sometimes called the back reflector (i.e., the back of a laser where the energy does not emerge); opposite end of the source to the laser aperture. The mirror with close to 100% reflectivity.
Higher-order modes (HOM) Of a laser beam.
Highly accelerated life testing (HALT) Designed to stress the system during testing to discover potential weakness and interdependencies.
Highly accelerated stress screening (HASS) Related to HALT for system testing.
Holographic optical elements (HOE) Use of a hologram to scatter light in the same way as a conventional optical element (reflection or refraction). Can be quite compact, may replace several optical elements in a bulky system. Need high diffraction efficiency (~80% achieved) and monochromatic light (single line) if energy to be focused to a single point.
Holey fibre A class of optical fibre that has a periodic array of holes to provide the appropriate refractive index difference in the cladding structure around the central core. Beware of varying definitions.
Homogeneous broadening This is a type of emission spectrum broadening in which all of the atoms radiating from a specific laser level do so with an equal opportunity, so the line width is the natural line width with a Lorentzian shape.
Host Laser crystal that is doped to form the gain medium.
Hot isostatically pressed (HIP) A technique to remove pores during sintering process to make highly transparent, low scatter ceramic laser host materials.
Hufnagel-Valley A turbulence model, gives a good representation of higher-altitude turbulence.
Hund's rule Every orbital in a sub-shell is by occupied a single electron before any orbital is occupied by a second electron. All of these single-occupancy electrons have the same spin.
Hydride vapour-phase epitaxy (HVPE) For rapid growth of QPM crystals (e.g., OPGaAs).

Hydro-thermal growth A way of growing bulk crystals.
Hypergolic reaction An event when a fuel ignites spontaneously when in contact with an oxidiser; the basis of liquid rockets and chemical lasers.
Hyper focal distance The range and beyond where objects are in focus (e.g., targets).
Hyper-spectral More than 20 channels of spectral information, as in hyper-spectral imaging.
I-cavity Shape of resonant cavity—a linear cavity, often used with disk lasers.
Idler A component of the nonlinear parametric process with the longer wavelength.
Image dancing The movement of an image formed in the focal plane of a sensor as a result of turbulence in the atmosphere.
Image plane The plane in an optical system containing the image.
In-band Coincidence of the pass band of an optical system or optical sensor and the incident wavelength, essential for laser-based DIRCM systems. With an in-band damage system the optical train will focus the incident radiation. This intensifies the energy at the detector(s) and requires relatively low powers to achieve a defeat mechanism.
Index guiding Use of changes in refractive index to form a waveguide, used to confine the zone for laser action in an effort to increase efficiency.
Index matching Refractive-index matching, by use of an intermediary (fluid), is used to suppress the effect of Fresnel reflection.
Indirect generation A conversion stage is used to generate the desired wavelengths, using a nonlinear process.
Infrared (IR) A general term for those wavelengths not visible to the unaided eye.
Inhomogeneous broadening Opposite of homogeneous broadening. Hence, there is an increase in the natural line width, caused by effects that act differently on different radiating or absorbing atoms.
Injection laser Original name for a semiconductor-diode laser; the name arises from the injection of carriers across the pn junction to create the population inversion.
Inner scale Of turbulence, relates to the length of the dissipation of the kinetic energy (KE) in the atmospheric eddies into the atmosphere.
Indium antimonide (InSb) A semiconductor used for detection of infrared photons.
Integrated surveillance, target acquisition and reconnaissance (ISTAR) A system to convey information from a sensor to a shooter.
Integrated test, evaluation and acceptance (ITEA) A detailed test plan to support a system development and entry into operational service.
Integrated-visual photopic transmission (IVPT) A transmission corrected for the response of the eye (i.e., green counts for more than red), in daylight, the peak of the eye's sensitivity is in the green at 555 nm). Under these lighting conditions the cones of the retina predominate.
Integrated-visual scotopic transmission (IVST) A dark-adapted eye, the peak response is at 507 nm. Under these lighting conditions the rods of the retina predominate; under these conditions, all cats appear to be black.
Inter-band transition A transition between the conduction band and the valence band in a semiconductor. Use of this approach limits the useful wavelengths that may be achieved.

Intra-beam viewing Not recommended, as it is the condition where an eye is exposed to all or a part of a (laser) beam.
Interaction length Of a nonlinear crystal in an optical parametric oscillator; the effective length in the crystal where the annihilation of the pump photon occurs. With walk-off, this is usually less than the length of the crystal.
International Electro-technical Commission (IEC) For example, IEC 60825 for eye safety.
Inter-range instrumentation group (IRIG) Definition of interfaces, function and protocol. A definition of standards.
Intrinsic layer Used in a pin diode; layer of undoped semiconductor material, where the electron and hole densities are equal. Between the "p-" and "n-" doped regions, hence the name.
Ion-beam milling (IBM) A type of dry-etching technique, useful for large substrates.
Ion-beam sputtering (IBS) A coating technique.
Ionic bond Transfer of electrons from one atom to another in the formation of a bond between two atoms.
IR-A The part of the spectrum extending from 700 to 1,400 nm.
IR-B The part of the spectrum extending from 1,400 to 2,600 nm.
IR-C The part of the spectrum extending from 2,600 to 1,000,000 nm.
IR seekers operating bands See Band I, II and IV. Primarily defined by atmospheric windows.
Iron sight A simple sighting device to acquire (and track) a target (e.g., a ring and bead arrangement).
Irradiance Radiant flux per unit area or power density on a surface (W cm^{-2}).
Isolator An optical component to prevent back reflections into the laser (e.g., between an amplifier and an oscillator).
Isoplanatic angle The range of angles over which the wavefront of a system is approximately constant.
Jahn-Teller effect It is a description of the geometric distortion of a nonlinear molecular system that reduces its symmetry and energy.
Jammer A system used to defeat the target-tracking algorithm in an optical seeker by emitting a series of pulses of energy. The jamming process causes confusion in the seeker's target-tracking system resulting in false steering commands being sent to the missile's guidance and control system. Typically designated as ALQ XX by NATO.
Jam-to-signal ratio (JSR or J:S) An indication of jamming potential of a system.
Jamming The process of defeating an optical seeker through the emission of a pulsed waveform.
Jitter AC (time varying) component of tracking or pointing error.
K Thermal conductivity, $W.cm^{-1}.K^{-1}$.
k_B Boltzmann's constant (1.380622×10^{-23} Jk^{-1})
Kerr cell Application of the Kerr effect; use of two parallel plates immersed in a liquid that permits changes in the polarisation state when an electrical potential is applied; hence when polarised light is incident on such a cell the beam can be switched on or off; a shutter.
Kerr effect The electro-optical effect is where certain fluids become birefringent (double refracting) when placed in an electrical field orthogonal to the incident light

beam. The increase in the refractive index of a material with increasing incident intensity, which gives rise to self-phase modulation.

KGW Crystal host potassium gadolinium tungstate—$KGd(WO_4)^2$.

Kolmogorov theory Theory of turbulence, based on fluid-velocity turbulence. The key idea is the hypothesis that the kinetic energy associated with the larger eddies is redistributed without loss to successively smaller and smaller eddies until it is finally dissipated by viscosity. In the atmosphere, the solar energy generates atmospheric kinetic energy over scale sizes of a few metres to global scales. Kinetic energy leaves the atmosphere by frictional generation over scales of less than a centimetre. The intervening range (from a few millimetres to a few metres) is called the subinertial range.

Knife-edge test Use of a sharp-edged component, such as a razor blade, that is scanned across a beam in conjunction with a detector to determine the beam profile.

Krypton ion Electrically pumped continuous-wave gas laser, many laser lines, but most powerful is in the red (647 nm).

KTP Potassium titanyl phosphate ($KTiOPO_4$), nonlinear material for optical parametric conversion in the infrared, also commonly used for frequency doubling, but now largely superseded for frequency doubling by BBO and LBO.

KTP growth Hydrothermal (high pressure) or flux (top-seed solution). Hydrothermal appears to be more grey-track resistant; grey tracks are a laser-induced defect that reduces the conversion efficiency.

KYW Crystal host potassium yttrium tungstate, $KY(WO_4)_2$.

Laguerre polynomial A solution to the Laguerre equation, a second-order linear differential equation. Laguerre polynomials are used to describe the modes from a fibre laser.

Laguerre-Gaussian (LG) modes Spiral modes, often used in sources for quantum cryptography.

Lambertian surface Ideal diffuse reflector, where reflected radiation is independent of viewing angle.

Lambert's law The radiant intensity observed from a Lambertian surface or a Lambertian radiator is directly proportional to the cosine of the angle θ between the observer's line of sight and the surface normal.

Laporte's rule Transitions between states with the same parity are forbidden.

Large mode-area (LMA) fibre An optical fibre with a large diameter core that will sustain a number of transverse modes.

Laser Source of light, which can be focused with unique precision to cause overload and ultimately damaging effects in sensors, at all ranges. The coherent nature of the emission enables the beam to have negligible divergence from a small aperture. Lasers are often named by their lasing media, for example, Nd, dye or CO_2.

Laser cavity The optical structure defined by the laser mirrors and contains the gain medium, also called the resonant cavity. The shape of the mirrors defines the type of cavity such as a plane-plane or confocal cavity.

Laser cross-section Calculated from $(\rho \cdot A_t)/\Omega$, where ρ is the reflection coefficient, A_t is the size and Ω is the solid angle.

Laser dye An organic molecule used in a dye laser.

Laser-induced damage threshold (LIDT) Average power at which damage occurs at 50% of the test sites. Expresses minimum power/energy density necessary to cause specified type of damage. Tends to be less than the intrinsic material value owing to imperfections in the structure or film.

Laser-induced fluorescence (LIF) A sensing technique. Use of a laser to excite species, typically in the atmosphere, to cause them to fluoresce, developed as a remote sensing technique.

Laser-induced media perturbation (LIMP) In a gas laser typically, which leads to beam distortion.

Laser light A product of stimulated emission.

Laser line Description of a laser emission, related to its monochromicity.

Laser photo-acoustic sensing (L-PAS) A sensing technique.

Laser slope efficiency Governed by mode overlap and ratio of losses to gain.

Laser-warning receiver (LWR) Passive receiver with fibre optic links to a detector array to give arrival direction. Endeavours to determine the emission characteristics of the illumination source.

Lasing threshold A measure of the input power required to overcome the losses in the cavity prior to the onset of laser action; beyond the lasing threshold the laser action proceeds as the pump intensity is increased. The lowest level of excitation where stimulated emission exceeds spontaneous emission (i.e., laser action).

Layers in a semiconductor Cap layer, light absorbing/active layer, buffer and substrate.

Leakage-channel fibre (LCF) Larger-core fibre but maintains M^2 and increased SBS threshold.

Lidar Laser radar; like radar, often a coherent system, uses a narrow linewidth source to reduce noise and allow use of lower-power emissions, at the expense of a more complex receiver; used to sense the presence of soft targets.

Light emitting diode (LED) A semiconductor being operated in a forward-biased regime, but below threshold for laser action.

Lifetime The radiative lifetime of an electronic state, where only radiative processes depopulate the state (level).

Ligand An ion or molecule (functional group) that binds to a central metal atom to form a coordination complex.

Limiting angular subtense The apparent visual angle that divides intra-beam viewing from extended-source viewing.

Limiting aperture The maximum circular area over which irradiance and radiant exposure can be averaged for comparison with the maximum permissible exposure.

Linear optics An optical effect related to the real part of χ (and is a linear function), its imaginary part (related to the absorption properties of the material) is close to zero in this regime, except when close to an atomic resonance.

Line by line (LBL) Used in propagation modelling. Use of laser-wavelength data.

Lines Spectral emission from a laser, referred to as lines, because of the monochromatic nature of the emission; colours or wavelengths. Some lasers operate on one line; others will emit on more than one line, but designed to operate on just one.

Line width Of a laser emission; a result of the transition plus the effects of broadening mechanisms; hence the use of the symbol $\Delta\lambda$.

Lithium tri-borate (LBO (LiB_3O_5)) Nonlinear crystal for conversion via frequency doubling from the ultraviolet to visible and near infrared.

Littrow cavity A resonant cavity with a diffraction grating, which generates a line-narrowed output.

Littrow configuration An optical configuration in which light is diffracted back towards the source.

LIV (curve) Laser emission, current and voltage plot, usually used to illustrate lasing performance from a diode laser. A multi-axis plot of performance.

LOCSET A technique for phase locking multiple independent beams at a remote target; a self-referenced piston-clamping technique used to align phases of the various components.

Longitudinal modes Leads to different wavelengths in the output beam (i.e., as a result of different resonances along the cavity); hence, single longitudinal mode is a single-frequency emission.

Long laser Use of the reflection from a target to seed the laser cavity; application of the phase conjugation technique. May be considered to be the equivalent of an extended cavity.

Long-period grating (LPG) A form of cavity mirror written into a fibre laser to form the cavity.

Lorentzian Description of the beam shape resulting from different homogeneous line broadening processes occurring simultaneously.

LP_{xy} Description of fibre modes; circular modes owing to the lack of a strong polarising influence in the gain medium. From Laguerre polynomial.

LuAG $Lu_3AL_5O_{12}$; a laser host.

Luminescence An emission of electromagnetic energy from a material as a result of any non-thermal excitation process; if the emission ceases as soon as the excitation ceases, then this is fluorescence, if the emission persists, then it is phosphorescence.

Lumophore An atom, or group of atoms, that gives light emission.

Luneburg lens This type of lens, with its graded refractive index profile, focuses light from all directions equally well.

Lyot filter A type of optical filter that uses birefringence to produce a narrow passband of transmitted wavelengths; used to tune an emission.

M^2 Known as the propagation factor; a parameter for expressing beam quality. Its value indicates the number of times the beam size exceeds the diffraction limit. $M^2 = 1$ is a perfect beam (diffraction limited).

Mach-Zehnder interferometer A device used to determine the relative phase shifts between two collimated optical beams from a coherent light source.

Macula Area of the retina 9 × 25 degrees outside the fovea (peripheral vision). Significantly less susceptible to damage by laser irradiation than the fovea.

Mangin mirror A divergent meniscus lens, silvered on the outer convex surface; used to give a parallel beam; a combined reflection and refraction in a system corrected for spherical aberration and coma.

Maragoni effect Describes the mass transfer along an interface between two fluids owing to a surface tension gradient.

Master-oscillator, power-amplifier (MOPA) Potentially gives high brightness. Pulse duration, temporal profile, spatial profile, beam divergence and spectral width are determined by laser (master) oscillator, pulse energy determined by the amplifier. Idea is to amplify a beam and maintain high beam quality (e.g., M^2 close to 1).

Maximum permissible exposure (MPE) Exposure of personnel to laser output (0.2 mJ intra-ocular energy). A safety limit. Multivalued, depending on characteristics of laser. Believed to be suitably conservative; at least a factor of 10 less than the least exposure shown to produce any damage to an eye. The maximum level of laser radiation to which a person may be exposed, during normal operation, with incurring any physiological effects. Fundamental to deterministic risk assessment (DRA).
MCT See CMT; mercury cadmium telluride the same alloy as CMT.
Melanin Pigment in retina, determines absorption and thus the susceptibility of the retina to damage by light from intense sources.
MESA Shape of active region in a stripe semiconductor, named after the physical feature.
Meta optics Meta optics is an extension of micro-optics where a number of optical functions, such as spectral filtering, are combined on the same optical element. Meta-optical techniques rely on subwavelength features to engineer the electromagnetic fields within the structure, which enable specific nonclassical effects such as negative refractive indices to be realised. One of the issues is that many of the meta-optical elements are created in thin metal films, so the losses in the visible band are large.
Metal organic chemical vapour phase deposition (MOCVD) A semiconductor-growth technique.
Meta-stable state Meta-stable levels are where a species becomes delayed, as there is no readily allowable transition (i.e., becomes trapped). Can be the cause of a bottle neck in a transition process.
Metal vapour Laser gain medium. Examples include copper, barium or gold. Gas lasers, but the need to vaporise metal presents engineering problems, warm-up times and need to keep the laser tube close to the horizontal.
Michelson interferometer Similar to a Mach Zehnder interferometer, but with the beams folded back on themselves, so that the beam-splitting element is also used to recombine the beams.
Micro-channel plate A form of amplifier, which may be used to provide a two-dimensional intensified image. This is an array of curved, hollow tube-like channels coated with a material to provide electron amplification, used in conjunction with a photo-emissive surface.
Micron One micron is one millionth of a metre. Used to state the longer laser wavelengths, abbreviated to μm or (mu)m. One micron is one thousand nanometres, nm.
Micro-structured fibre A type of active or passive optical fibre with an array of holes in the core part of the structure to reduce the refractive index in comparison with the core or waveguide. Also known as holey fibres.
Milliradian Measure of angle, one thousandth of a radian, 0.0573 degree (or approximately 17 mrad per degree).
Mid IR (MIR) Mid-infrared part of the spectrum. See Atmospheric windows.
Mie scattering Scattering by particles with size comparable with the wavelength; intensity proportional to the presented area (and the number); elastic scattering process.
Military off the shelf (MOTS) Military equivalent of COTS.
Mission defeat Use this term with care. Frustration of a military mission, entirely context dependent.

Mission-defeat criterion Weapon deliverable (e.g., pulse of energy, bullet) and associated delivery definition required to achieve a mission defeat. See Mission defeat.
MIXSEL A mode-locked VECSEL. See Mode locking and VECSEL.
Modal instability (MI) In a beam. The absorbed heat changes the refractive index of the gain medium and the change in the refractive index causes a change in absorption.
Mode competition With a relatively broad gain curve, it is possible for a number of modes to resonate in a cavity. The strongest modes tend to draw energy from the weaker ones, and as there is a finite flux in the cavity, the stronger ones grow at the expense of the weaker ones, leaving the most dominant modes oscillating.
Mode-field adaptors A fibre device to transfer photons efficiently from a single-mode fibre to a large-mode-area fibre or a low-order mode multimode fibre.
Mode hopping Change of wavelength, owing to change in the resonator condition (e.g., the cavity warming up); spacing between modes depends on cavity length. Can be controlled by insertion of a Fabry-Perot etalon into cavity.
Mode locking Group of photons clumped together and aligned in phase as they oscillate through the cavity. Each time they arrive at the output coupler part of the light escapes as an ultrashort pulse. The clump of photons then makes another round trip (2L where L is the cavity length) and is emitted. The pulses are separated by the cavity-round-trip time (2L/c). The basic idea is that the modulation of the phase of the photons in a cavity by an optical element causes the different longitudinal modes to oscillate together in phase. Need the modulation frequency to correspond to the cavity round-trip time, and requires a laser that oscillates in many longitudinal modes. Pulse length is inversely proportional to the laser's oscillating bandwidth; hence, need a source with broad gain bandwidth, such as a dye laser.
Mode stripper A device for reducing the number of transverse modes emitted by a laser.
Modes Light transiting a laser cavity can do so in any number of modes. Few modes (or single) are required for more precise applications (i.e., high beam quality), multimodes are acceptable for the more crude systems requiring raw high power. Its oscillation characteristic (and hence frequency) are dependent on its resonance characteristics in the laser cavity. Single longitudinal mode, a single-frequency laser.
Modified chemical vapour deposition (MCVD) A type of semiconductor-growth technique.
Modulation transfer function (MTF) Gives the response of a system as a function of the spatial resolution in lines per spectral width, a method to describe image quality.
Modulator An optical element that changes the fraction of the incident light it transmits in response to an external signal.
Monostatic A system in which all the optical functions work through one objective aperture. Design aim for compact, self-aligning system configuration.
Monochromatic Single wavelength; hence, the use of the term laser line.
Moth-eye coating A form of an anti-reflection optical coating consisting of a surface pattern impressed on the surface of an optical element, named after the surface pattern on a moth's eye.
MPR Multiphonon relaxation (rate), as in bridging an energy gap.
MTV Decoy flare, composition magnesium teflon viton.

Multi-layer dielectric mirrors Laser mirrors with potentially very high reflection characteristics at particular wavelengths, but with very low optical scatter. Made by depositing alternative layers of dielectrics with high and low refractive index, each layer $\sim\lambda/4$ thick. Using constructive interference to build up reflection. Need to be deposited on a very smooth substrate.

Multi-mode combiner (MMC) This is the combiner that puts the pump light into the cladding and the signal light into the core of the double-clad gain fibre.

Multi-phonon absorption A loss mechanism, which occurs when two or more phonons simultaneously interact to produce electric dipole moments that couples with the incident radiation to cause absorption of the incident light. These dipoles can absorb energy from the incident radiation, reaching a maximum coupling with the radiation when the frequency is equal to the vibrational mode of the dipole in the infrared.

Multi-phonon edge A result of intrinsic vibrations in a glass, which limits the long-wavelength pass-band, resulting from polar modes of the lattice vibrations.

Multi-static A system where each of the various optical systems works through its own aperture.

Nadir A pointing singularity that occurs with a two-axis beam-director system as the gimbals become aligned. A consequence of the inner gimbal being carried by the outer gimbal, only the rotational component of the outer gimbal normal to the sightline can contribute to the sightline rotation (Cosine of the inner gimbal deflection); at ±90° the cosine of the angle → 0; therefore can only rotate the inner gimbal to align itself with the desired vector rotation. Hence, small line-of-sight deflections require large outer gimbal movements.

Nanometres (nm) See Microns. Used to state the shorter laser wavelengths in the visible band of the electromagnetic spectrum.

Narrow-band filter Used to block a small part of an otherwise usable spectrum (e.g., to exclude a closely spaced group of lines). These can block an appreciable amount of the visible spectrum (e.g., a filter stopping blue passes mainly red light). Fixed filter types are readily available and often used in personal-protection equipment (PPE).

Nd The rare-earth ion neodymium.

N_d Thermal-blooming distribution number.

Near field Between the source transmitter and the far field. In this region there is typically a complex relationship between the electrical and magnetic fields of an electromagnetic wave and the angular-field distribution is definitely a function of distance from the transmitter.

Near infrared (NIR) Beware of definitions, often from 0.8 to 1.4 μm.

Negative electron affinity (NEA) In a free-electron laser beam.

Negative uniaxial crystal A uniaxial crystal is negative if $n_e < n_o$, so the different polarisations travel at different speeds through the crystal.

Neodymium YAG (Nd: YAG) Neodymium laser ion in an yttrium aluminium garnet host; for continuous wave/high repetition-rate emission (has higher thermal conductivity than glass); for high energy emission and better energy storage, owing to broadening of fluorescence line. Common type of solid-state laser, producing a fixed line at 1.064 μm wavelength in the near infrared, but other lines are possible. High pulse energies are possible. Has been the basic workhorse for many military applications.

Nephalometer An instrument for measuring the suspended particulates in a liquid or gas colloid.

Network-enabled capability (NEC) A system of systems.

Neutral density Attenuates transmitted light equally at all wavelengths in the spectrum of interest (e.g., good sunglasses). Expressed as log (proportion of transmitted power); ND2 transmits 1/100th of incident light. See Optical density.

N_F Fresnel number of a cavity.

Noise-equivalent irradiance (NEI) The limit of performance of a sensor.

Nominal ocular hazard distance (NOHD) Equates to a safety trace radius for safe laser operation, often exceeds damage ranges substantially. Application makes selection of trials range for directed-energy weapons very difficult for the very powerful systems. See Safety standards.

Non-critical phase matching Of an optical parametric oscillator; use of temperature to change the indices of the crystal and thus the wavelengths of the signal and idler beams.

Non-lethal weapons (NLW) Less-than-lethal weapons, not intended to kill or cause irreversible effects.

Non-linear coefficient Relates the ability of a non-linear crystal to convert light from one wavelength to another. Want it to be as large as possible, so get conversion with as low an incident intensity as possible. Means the crystal has to have a large value for the polarisability, so some of the bonds may be weak. The optical-gain coefficient is proportional to λ^2, but polarisation independent on λ; therefore, nonlinear optical effects in the mid infrared is more of a challenge. Units of pm/V.

Non-linear crystal As used in an optical parametric oscillator for wavelength conversion; has polarisability, so the $\chi^{(2)}$ effects can be exploited. See Nonlinear optics.

Non-linear optics Reliant on the very intense optical field generated by a laser beam, where χ is not linear and the electric field in the material is comparable with the inter-atomic field, $P = \varepsilon(\chi^{(1)}E + \chi^{(2)}E2 + \chi^{(3)}E^3 ...)$, used for wavelength conversion. Can have mixing of optical waves, for example, three-wave mixing in a nonlinear crystal or opto-mechanical (vibration) interaction as in stimulated Raman inelastic scattering.

Non-radiative loss A loss mechanism in laser action that reduces efficiency and usually creates heat within the device.

Normal-conducting radio-frequency (NCRF) injector Used in a free-electron laser.

Normalised frequency (of optical fibres) See V parameter. Determines the number of linearly polarised modes in a fibre.

Numerical aperture (NA) A parameter used to indicate the range of angles over which an optical system can accept light. A dimensionless term that indicates the range of angles (θ_{max}) over which an optical device, such as an optical fibre, can accept light. $NA = n \sin \theta_{max} = (n_{core}^2 - n_{clad}^2)^{1/2}$, where n is the refractive index of the medium.

n-type material Semiconductor material doped to provide an excess of electrons (negative side of the pn junction).

Operational analysis (OA) Used to establish the potential military value of a concept.

OPO crystal A nonlinear, anisotropic (noncentro-symmetric) crystal in a resonant cavity, transforms energy from the pump beam into two other wavelengths. The two other wavelengths are called signal (shorter λ) and idler (longer λ) and build up from amplified noise.

Optic axis Direction in a crystal where the two polarisations experience the same refractive index.

Optical augmentation (OA) See Retro-reflection.

Optical parametric amplifier (OPA) Similar to OPO, but without the cavity to provide the oscillation.

Optical path difference (OPD) May be expressed as a phase difference (ϕ = OPD $\times 2\pi/\lambda$).

Optical parametric oscillator (OPO) For wavelength conversion, equates to energy splitting of incident pump photons; generates two new wavelengths, although they may be degenerate (i.e., of the same wavelength) and then used for wavelength doubling.

Optical absorption To be avoided in a laser system, as it has many deleterious effects: for example, heats up the optical components leading to changes in operating characteristics owing to thermo-optical effects, leads to less than optimum conversion in a nonlinear crystal; temperature detuning.

Optical cross-section A measure of the optical energy returned to a transmitter, analogous to radar cross-section. Measured in m^2/sr. In effect, it is the size of the sphere placed in a beam that will create the same optical return at the receiver. It is not related to the physical size of an object.

Optical density (OD) The attenuation produced by a filter OD = $\log_{10}(P_I/P_O)$, or $-\log_{10}$ (transmission), where P_I is the incident power and P_O is the output. OD^2 means 0.01 of incident light is transmitted.

Optical gain of the eye The magnification of the human eye is about 10^5.

Optical indicatrix Refractive index ellipsoid for birefringent materials.

Optical isolator An isolator placed between optical components to prevent back reflections, an optical valve.

Optical quality Measured in terms of transmitted-wave distortion per unit length.

Optical radiation Often defined as EM radiation extending from 100 nm to 1 mm, divided into wave bands, such as ultraviolet, visible, near infrared, short-wave infrared, mid-wave infrared, far (or long-wave) infrared. Beware of different definitions of the bands.

Optical switch Electro-optical protection device, the transmission of which, is reduced in response to incident intensity when illuminated by laser light (e.g., under attack or accidental exposure). Examples include: photo-chromic (modified liquid) or as part of an active system, including a sensor (liquid crystal).

Optical-to-optical efficiency An expression of the efficiency of the conversion of the input optical power to the output power. See Slope efficiency.

Optically patterned gallium arsenide (OPGaAs) A type of nonlinear crystal where the crystal domains in the structure GaAs are reversed from domain-to-domain during the epitaxial growth process; used in a quasi-phased matched condition.

Optically pumped semiconductor laser (OPSL) Uses photons to inject the holes into the n-region.

Optically pumped VECSEL (OP VECSEL) See VECSEL.

Organic metal vapour phase epitaxy (OMVPE) Technique for growing semiconductor materials.

Ormosil Organically modified silicates. A polymer with SiO_2 linking the other components together. Used as the host in plastic dye laser gain medium.
Oscillator Laser device where the photon beam makes multiple passes to generate the photon beam, often modest power, but high beam quality. See MOPA. Also known as a multi-pass oscillator
Outer scale Of turbulence, relates to the solar heating scale creating the eddies in the atmosphere. Close to the ground, it approximates to h/2, where h is height above ground.
Out of band Attack of optical elements or structure with radiation, to which they are not transparent (e.g., windscreens or TV camera lenses with long-wavelength radiation). Transparencies are rendered opaque or break when illuminated with such a powerful beam.
Output coupler Partially reflecting mirror or other arrangement, through which the laser beam emerges from the laser cavity.
Overtone Occurs in a chemical laser when the transition in over two vibrational states; halves the emission wavelength of the fundamental.
P^3I Pre-planned programme improvement.
P^5 Prior planning prevents poor performance – there are higher-order variants involving additional adjectives starting with *p*.
Paraxial approximation Use of optical rays close to the optical axis.
Passive Q-switch Use of a passive optical element to control the loss in a cavity and thus form the Q-switched pulses.
P branch The long-wavelength branch of emissions.
Peltier effect Part of the thermo-electric effect. The liberation of heat at one junction and absorption of heat at the other when a current flows around a circuit consisting of two different metals; the basis of a thermo-electric cooler. Converse of the Seebeck effect and is reversible.
Perovskite structure Material with the same type of crystal structure as calcium titanium oxide ($CaTiO_3$).
Persistent wide-area surveillance (PWAS) Looking for changes over time.
Personal-protection equipment (PPE) Worn by operators of laser equipment and bystanders to protect from potentially hazardous effects of laser beams (e.g., laser goggles with appropriate fixed-line filters). It is important to ensure that the PPE provides the appropriate protection for all of the laser emissions that may be encountered during a test.
Periodically poled lithium niobate (PPLN) A nonlinear material with a grating where the crystal planes are reversed in each domain throughout the grating, used in quasi-phase matching.
Phase conjugation Reflection process with special characteristics. Implemented inside laser systems by a number of physical processes, promised to enable substantial improvement in the general precision of laser weapons by automatically correcting for aberrations, so may be used to improve beam quality in high-power systems and also for laser-pointing over a small angle of regard. Long-laser concept based on this phenomenon. Principle of phase conjugation is based on the reversibility of light, through the generation of a conjugate beam with the exact opposite phase variations to those of the aberrated beam. Hence, when the conjugate beam passes through the aberrating medium, a near-perfect beam (in terms of the wave-

front distortion) is recreated. A corner cube is a crude form of phase conjugation mirror, but it is imperfect, owing to the small displacement of the beams.

Phase matching A requirement resulting from the dispersion in the refractive index of nonlinear crystals to achieve conservation of energy and momentum in a nonlinear process. Efficient optical nonlinear conversion requires an initially small amplitude wave (usually noise), which should be amplified with a parallel depletion of the intensity of the pump beam. The greatest optical gain occurs when the three optical waves travel in the gain medium with a particular (and well-defined) phase relationship. Hence, the velocities of the waves must be matched in some way, despite the fact that the refractive index varies with wavelength. The effects of dispersion may be overcome by the use of birefringence (double refraction) in anisotropic crystals (i.e., Type-II phase matching where the two generated beams have orthogonal polarisations with respect to each other; other combinations are feasible). Hence, can find pairs of wavelengths where the group velocities are matched for particular directions of propagation through the nonlinear crystal (i.e., angle tuning). Temperature tuning is also possible.

Phase velocity The velocity that the crests and troughs of a wave travel through a medium.

Phonon A lattice vibration.

Phonon broadening The presence of the crystal-field modulations (from atom-atom interactions) and ion-phonon interactions that modulate the atom-atom separation, resulting in modulation of the atomic transition frequencies.

Phonon quenched The rate of multiphonon transitions decrease exponentially with the increasing number of phonons in a transition.

Photo-conductivity Enhanced conductivity in some semiconductors when exposed to electromagnetic radiation (such as photons). Occurs when an electron is raised from the valence band to the conduction band. However, if the energy of the absorbed photon is greater than the work function of the material, results in the photo-electric effect (liberation of electrons).

Photo-conductor A form of detector that changes its resistance (hence its conductivity) when exposed to light.

Photo-darkening A phenomenon that occurs in fibre lasers, which reduces the optical emission over time. The mechanism is a result of the formation of colour centres, by higher energy photons.

Photo-detector A detector for measuring the photon flux; a broad class of detectors.

Photo-diode A pn junction operated in reverse bias, below breakdown voltage. Its conductivity is modulated by the absorption of light in or near the depletion layer. If absorbed elsewhere get recombined without generating any current in the external circuit, hence use of intrinsic layer in pin diodes. See Pin diode.

Photo-electric effect Liberation of electrons from absorbed light when the energy of the photon exceeds the work function of the material; Albert Einstein received the Nobel Prize for this discovery.

Photo-emission The process that leads to the emission of an electron from a surface when exposed to light.

Photo-multiplier A very sensitive form of detector that exploits the photo-electric effect and the amplification of the subsequent electron emission from the photo-cathode.

Photo-voltaic (PV) detector A detector that generates a potential difference between two electrodes in response to the incident photon flux.
Photonic bandgap An energy range (and corresponding wavelength range) for which a photonic crystal neither absorbs light nor allows light propagation. It is analogous to the bandgap in a semiconductor.
Photonic-crystal fibre An optical fibre using holey periodic structure (an air array in the bulk material) to form the cladding layer around the central core, which is normally solid.
Photo-conductive (PC) detector Modification of conductivity owing to the absorption of incident light. For detection, energy of photon absorbed must exceed energy of bandgap, hence leads to small bandgaps and significant dark current so cool to 80K to reduce noise.
Photo-electromagnetic (PEM) A type of detector.
Photo-multiplier tube (PMT) A detector.
Photopic transmission Percentage transmission (e.g., of a filter) integrated over the spectral sensitivity of the daylight-adapted human eye.
Photo-thermal common-path interferometry (PCI) A technique for measuring absorption in materials.
Photo-thermal refractive (PTR) (glass) Photosensitive, used to make volume Bragg gratings (VBG).
Plane-parallel resonator One of the simplest laser resonators, but very difficult to maintain alignment. Uses two plane mirrors to form the resonator. A stable resonator, as the rays are contained in the cavity.
Plank's law Gives spectral radiant emittance of a target (fluence {W/cm^2} per μm), leads to black-body curves.
Plankian radiator A black-body radiator.
Plasma A region of ionised gas, made up of charged particles such as ions and electrons.
Plasma flash A damaging strike by a pulsed CO_2 laser, for example, on a transparency is accompanied by a bright visible flash.
Plasmon Plasma oscillations can occur, for example, when the electron gas in a metal oscillates against the background of ionised atoms. A plasmon is a quantum of such an oscillation and is usually often used to describe the collective oscillations of electron gases in conductors.
Plasmon laser These devices couple electromagnetic waves with electrons that oscillate in the surface of a metal to confine light into nanoscale dimensions that exceed the accepted diffraction limit. Typically used to control the beam divergence of semiconductor lasers, especially the fast-axis divergence.
PLZT Lead lanthanum zirconium titanate, a fast-acting optical switch.
pn junction The interface between p-type and n-type material in a semiconductor.
Pockels' cell An optical component used to modulate a beam; implementation of the Kerr effect in a piezo-electrical material.
Pointing accuracy Ability to point at a target, dependent on a vast array of variables (e.g., harmonisation, sightline stabilisation, jitter, target-tracking accuracy, single point failures and others).
Point of regard Direction of the optical axis of a sensor (e.g., where the eye is looking).

Point source A source with infinitesimally small dimensions; it is an isotropic radiator.
Point-spread function (PSF) Of an optical system (i.e., its blur circle). The point-spread function describes a two-dimensional distribution of light in a focal plane of an optical system. A perfect optical system would form an Airy pattern in the focal plane.
Poisson distribution Describes the occurrence of independent random events.
Poisson's ratio When a material is stretched in one dimension, it tends to shrink in the other two orthogonal dimensions. Hence, it is the ratio of the contraction (or transverse strain perpendicular to the applied load) to the axial strain (in the direction of the applied load). Used to predict mechanical structural performance.
Polarisation $P = \chi E$, where χ is the dielectric susceptibility and becomes nonlinear at high optical intensities, and where E is the electric field.
Polarisation contrast Calculated from the difference in the intensities in the two orthogonal polarisation planes divided by the sum of the intensities $[(I_{\parallel} - I_{\perp})/[(I_{\parallel} + I_{\perp})]$.
Polarisation-extinction ratio (PER) A measure of the preservation of the plane of polarisation, especially with reference to polarisation-preserving fibres.
Polarisation-maintaining (PM) fibre As on a special type of optical fibre to preserve the plane of polarisation of the emitted light.
Population inversion Population of a higher energy level is greater than that of a lower energy level, which is needed for stimulated emission. But when a species is in thermodynamic equilibrium the reverse is true, so species normally have a larger population in their lower states than the higher energy states; hence, spontaneous emission dominates.
Positive-branch unstable resonator An unstable resonator with a focus outside of the cavity.
Positive uniaxial crystal $n_e < n_o$ in a uniaxial crystal.
Post burn out (PBO) When a rocket motor stops thrusting – important for threat detection.
Potential false alarm sources (PFAS) Determines potential false alarm rate of a sensor system.
Power Of a continuous wave laser, usually stated in watts (1W = 1J/s) or fractions or multiples thereof. The rate of emission, transfer or receipt of energy.
Power-conversion efficiency (PCE) Electrons to photons.
Power-in-the-bucket (PIB) A type of power measurement. The amount of optical power collected by a detector through a given aperture.
Power-in-the-bucket efficiency (PIBE) An expression of the laser-emission efficiency when capturing the beam in a bucket.
Power radiated $W = \varepsilon\sigma T^4$ W.m^{-2}, where σ is Stefan's constant, ε is the emissivity and T is the absolute temperature.
Power spectral density (PWD) A function indicating the variations in a measured quantity as a function of frequency.
Power supply Discrete subsystem to provide the particular requirements of lasers for voltage and current. These can be extreme; packaging, reliability and safety can present problems.
Poynting vector Direction of energy flow, normal to a wavefront.

Poynting-vector walk-off Critical phase matching demands the creation of at least an e-ray which propagates with an elliptical wavefront, rather than a spherical one. Hence, the direction of energy flow (Poynting vector) differs from the normal to the wavefront. Thus, although the ordinary (*o*) and extra-ordinary rays (*e*) may start off in the same direction, the e-ray will follow the Poynting vector and deviate from the o-ray, so giving less than optimum energy transfer, owing to the de-coupling of the waves. Larger focus spots reduce the effect, but create a smaller optical field with reduced conversion efficiency.

Probabilistic hazard zone (PHZ) A description of the space within which the risk of hazardous exposure of the direct, reflected or scattered beam may exceed the accepted level of risk during normal operation. Beyond the PHZ the hazard risk is considered acceptable.

Probabilistic risk assessment (PRA) A method of laser-hazard evaluation based on the likelihood of exposure and risk of injury. The PRA requires the establishment of an acceptable level of risk and the assessor to assume probabilities for the occurrence of certain events; this method is used when other methods cannot contain the extended hazard boundaries within defined range boundaries; assumes a probability of certain events occurring.

Propellant Uses a controlled deflagration or chemical reaction: it can contain a single molecular specie.

Pseudo-random bit sequence (PRBS) A type of random number generator.

PST Lead scandium tantalite, detector material for a low-cost thermal imager.

p-type material Semiconductor material doped to give an excess of holes (positive-type material).

Pulse compressor Reverse of a pulse stretcher.

Pulse duration Also known as pulse width; the duration of the coherent emission, usually measured at a particular point on the intensity curve.

Pulse energy Energy per pulse put out by a pulsed laser, usually stated in joules (J) or fraction or multiples thereof.

Pulse-forming network (PFN) The circuitry for powering flashlamps, design is critical to pulse shape and length (duration) achieved. Can have multiple flashlamps connected in parallel or series.

Pulse-repetition frequency (PRF) Measured in hertz. Describes the frequency of occurrences of the pulses in a pulse train.

Pulse stretcher Use of a dispersive optical element (such as a grating) to redistribute the optical energy in a pulse. Chirped volume Bragg grating used for this purpose.

Pulse width See Pulse duration.

Pulsed laser A device that delivers its photon energy in pulses, either single or a train of pulses, usually designed to give high peak power.

Pumping A means of exciting the gain medium within a laser to create the population inversion.

Pyro-electric A slow-response detector with a wide spectral response (sensitivity).

Pyromethene A family of laser dyes, which come in various colours (e.g., pyromethene 567 or pyromethene 580).

Pyrophoric A pyrotechnic store; a solid or liquid which has the properties of instantaneous combustion.

Pyrotechnic An off-board decoy (expendable store), which burns to give local thermal source. It is always a mixture of chemicals (e.g., Mg and $NaNO_3$).
PZT Lead zirconate titanate; a piezo-electric material.
Q (quality factor) Q = (E stored per pass)/(energy dissipated per pass).
Q-switch Technique used to produce short pulses (e.g., tens of nanoseconds) from pulsed lasers. A loss mechanism which spoils the resonant (Q) condition of a laser cavity. Four basic types. Rotating mirror (e.g., 3×10^4 rpm synchronised to the pulse-forming network), electro-optical (Kerr cell or Pockels' cell), which uses a current to change the polarisation of a crystal (aligns the polarisation vector to that of a polarising filter), acousto-optical (use of a diffraction grating to deflect the beam), or photo-chemical with saturable dyes, which bleach and then return to an absorbing state. Q = (E stored per pass)/(energy dissipated per pass). It will only work in gain media capable of storing energy for a time longer than the Q-switch pulse duration, a result of suppressing the stimulated emission, so needs the lasing ion to have an upper state lifetime of at least a few microseconds and usually a few milliseconds.
Quantum-cascade laser (QCL) Also known as intersubband laser. Has a series of quantum wells in the conduction band. Get gain from the cascading effect of an electron passing through the various subbands and liberating photons at each transition in the series of quantum wells to form the cascade. Recycling electrons increases efficiency. These devices emit in the mid-infrared and longer wavelengths to the terahertz region.
Quantum defect The difference in energy between the pump energy and the generated photon, also known as the heating ratio, as that is the consequence.
Quantum dots A combination of typical inorganic semiconductor elements, such as In, Ga, As and P, to form nanometre-scale structures on a substrate, via self-organisation. Dispersion in the dot size leads to a laser structure with many longitudinal modes. Owing to the scale of the dots, the structure can be considered to be dimensionless as far as constraint is concerned, so the electrons are concentrated in the ground state; as a result, the devices have a low-noise emission, relatively insensitive and can be modulated rapidly.
Quantum efficiency (QE) Efficiency of the conversion of photons to electrons in a detector. A figure of merit indicating the efficiency of the generation of electrons from the incident light.
Quantum number An integer value that indicates the value of a particular state of an atom, such as its angular momentum.
Quantum well A semiconductor layer with active layer ~30 nm thick (see the de Broglie wavelength of an electron) that gives quantisation of allowed energy states in valence and conduction bands, determined by the layer thickness. Allows control of bandgap energy (alternative to using different dopants). Resultant equalisation of energy levels enhances carrier recombination; lower threshold currents compared with bulk semiconductor.
Quantum well infrared photo-detector (QWIP) A photo-conductor.
Quarter-wave plate (QWP) A polarising element.
Quasi-phase matching (QPM) Uses the natural birefringence of some crystals. Input and output beams polarised along the same axis (extraordinary ray) and so can use the d_{33} nonlinear coefficient. Many materials have the appropriate nonlinear optical coefficient, but lack the ability to phase match, because they are

centro-symmetric crystals (i.e., are not birefringent). These crystals can give some small amount of conversion over a short distance, known as the coherence length, and then return to zero over the next coherence length. If the crystallographic orientation (crystal plane) of the crystal is reversed after one (or an odd number of) coherence length(s), then the nonlinear conversion process can build up, because the phase error, which builds up in one domain, is reversed in the next. The crystallographic domain reversal can be achieved mechanically or by periodic polling of a ferroelectric crystal, such as lithium niobate (see PPLN) or use of optically patterned techniques. See OPGaAs.

Quenching A process that reduces the excited-state population; many mechanisms, such as collisions or interactions with phonons.

Radiance The radiant flux output or power density per unit solid angle (W cm^{-2} sr^{-1}).

Radiant emittance Radiant power per unit area at a given temperature.

Radiant energy Energy emitted (J) by the device.

Radiant exposure The total energy per unit area incident on a given surface in a given period, quoted as a surface density of radiant energy received (J cm^{-2}).

Radiant flux The rate of flow of radiant energy (W).

Radiant intensity The radiant flux emitted by a surface in a given direction of propagation per unit solid angle, quoted in (Wsr^{-1}).

Radiation balance heat-capacity laser Similar to a heat capacity solid-state laser, but the absorption and emission characteristics are selected to result in negligible heating of the host, by pumping the red side of the fluorescence peak, so the average spontaneous emission removes the heat.

Radiometry The study of the absorption, emission and transmission of electromagnetic energy.

Radius of curvature (ROC) Usually of a mirror.

Raman effect The Raman effect is a two-photon process involving an incident photon and a scattered photon at a different wavelength. This effect can be induced in various wavelength bands given suitable energy levels in the nonlinear materials; however, the Raman effect is inversely proportional to the fourth power of the incident wavelength (λ^{-4}); hence, the use of ultraviolet wavelengths is attractive. Therefore, halving the incident wavelength leads to a factor of 16 times the Raman signal but at all wavelengths Rayleigh scattering predominates, usually with 30 to 40 dB more signal; Raman is a weak inelastic effect.

Raman lasers Class of nonlinear device in which the wave mixing occurs between an electromagnetic wave and a mechanical vibration (phonon), a $\chi^{(3)}$ process. It is inelastic scattering and occurs in a suitable medium (solid, liquid or gas) where the phonon energy is added to the incident beam (anti-Stokes) or subtracted from the incident pump (Stokes). Offers inherent high-efficiency conversion, particularly when using very high pressure gases such as H_2 or D_2. In the mid-infrared, they are less attractive. The Raman coefficient varies inversely with the incident wavelength; so at 4 μm is about 1/8 of the value for visible light, hence more sensitive to losses and thereby needs greater optical intensities, which can lead to breakdown.

Raman shift Also known as Stokes shift; the change in wavelength of the output compared with the incident wavelength.

Raman shifting Non-linear optical technique for converting laser output at a particular wavelength to others to provide frequency diversity. Generally considered to be inefficient. Active component is a Raman cell if using a gaseous medium for the inelastic process. Solid-state material available such as barium nitrate.

Raman spectroscopy A sensing technique exploiting the Raman effect. The wavelengths of the Stokes (and anti-Stokes) lines emitted by a substance when illuminated by a laser are determined and compared with a database to identify the unknown substance.

Rare-earth ions Ionised member of the lanthanide series elements in the periodic table. Used for laser ions to dope a host.

Rate equations An expression of the dynamics of energy-level populations in a laser gain medium by using a number of differential equations; the rate equations, which describe the temporal evolution of the population in these levels under the influence of pumping and nonradiative decay.

Ray tracing A geometrical technique for generating an image formed by an optical system by tracing the paths followed by rays of light.

Rayleigh length The distance from the beam waist where the mode radius increases by a factor of $\sqrt{2}$, so for a circular beam the mode area doubles by this point.

Rayleigh scattering Scattering of light by particles much smaller than the wavelength of the light, intensity proportional to λ^{-4}. Gives rise to blue sky.

Rare-earth element (RE) The lanthanide series in the periodic table. Used for laser ions.

Read-out integrated circuit (ROIC) In a focal-plane array.

Red green blue (RGB) As used to create a projected colour image.

Reflectance The ratio of the reflected radiant flux to the incident radiant flux.

Reflectivity See Reflectance.

Refractive index Expresses the velocity of a light beam through a medium, with respect to the value in a vacuum, so indicates the change of light velocity in a dense medium. The frequency does not change, only the wavelength. The refractive index of air is a function of temperature, pressure, humidity and optical wavelength.

Regenerative amplifier An optical amplifier often used with ultrashort pulse lasers, such as femtosecond devices. The resonator is configured so that the light pulses can make several round trips before leaving the amplifier.

Resonance Occurs when cavity length L ($= n\lambda/2$), and oscillation only allowed within the gain bandwidth of the species; therefore, an infinite number of emission wavelengths are not possible.

Resonance-enhanced multiphoton ionisation time-of-flight mass spectroscopy (REMPI ToF MS) A sensing technique.

Resonant cavity The optical structure defined by the laser mirrors and contains the gain medium, also called the laser cavity, or more simply the cavity. The shape of the mirrors defines the type of cavity, such as a plane-plane or confocal cavity.

Resonant-energy transfer A technique for transferring an excitation from one species to another, similar to fluorescence resonant energy transfer. In this case, there is also a donor specie, which is easy to excite and an acceptor that undertakes the laser action.

Reticle A chopper, which modulates the energy incident on an optical seeker's detector, acts as a spatial filter, so rejects extended objects. Spin-scan system gives amplitude modulation; conical scan gives frequency modulation signal. Basically, depth of modulation gives angle off the seeker's optical axis (point of regard), that is, off-axis position and phase, with respect to a datum, indicates orientation (in polar co-ordinates).

Retinal pigment epithelium (RPE) A sensitive part of the retina.

Retro reflection When light enters a focusing optical system (e.g., optical sight or binoculars) from within the field of view, it will be reflected from any optical element at the focus (e.g., graticule, detector) and return from whence it came. Thus, a laser system can detect such systems that are looking at it from considerable ranges. This technique is used in defensive systems to detect electro-optical systems and point laser countermeasures. A powerful technique, but relies on being within the victim sensor's field of view.

Retro-reflection cross-section Units of square metres. The measurement of the size of a retro return given by an optical system. Values bear no relation to the physical size of equipments concerned. Defined in a similar way to radar cross-section (i.e., that area of a sphere which needs to be interposed in a beam to give an equivalent received signal).

Reverse saturable absorber (RSA) A technique for optical limiting. A broadband filter, using a dye, which transmits well at low intensities, but becomes opaque at high intensities, owing to excited state absorption. Have broadband absorption across desired protection region, but with very high excited-state absorption. This filter uses dyes with high intersystem crossing rates and uses the triplet manifold to assist the absorption, opposite of laser dyes.

Rhodamine A family of laser dyes.

Rigrod analysis A technique for optimising the emission from a laser cavity, through determining the optimum reflectivity of the output coupler.

r_0 Fried parameter. An indication of the optical quality of the atmosphere.

ROC curve Relative operating characteristic curve.

Rotational energy levels The energy levels in gaseous molecules associated with rotational motion of the components of the molecule. Tend to have small energy differences and provide the fine structure to vibrational energy levels; there presence leads to the multiple lines available from gas lasers.

RS 232 Recommended standard 232, used in telecommunications for a series of standards for serial binary, single ended data and control signals.

RS 422 Recommended standard 422, used in telecommunications for data transmission using balanced or differential signals. Associated with twisted pairs.

RS 423 Recommended standard 423, used in telecommunications for serial communications.

Ruby The original solid-state laser; produces a fixed line at 0.6943 μm, in the red. Used in older range finders, potentially dangerous to eyes even at low powers, but stopped by a fixed-line filter.

Rugate Blocking filter, capable of blocking (stopping) one or more fixed lines. Consists ideally of transparent material, the refractive index of which varies continuously across its thickness in a waveform to give the required behaviour. Rugate filters with at least three lines have been demonstrated.

Rytov number A fundamental scaling parameter used to indicate the impact of turbulence on a propagating laser beam in that turbulent medium. A Rytov number > 0.2 indicates a strong impact of turbulence causing scintillation in the intensity profile of the beam.

Safety standards Currently, IEC (BS EN) 60825-1 and ANSIZ136.

Safire Small arms fire—e.g., bullets from a light machine gun.

Schottky contact A Schottky barrier contact refers to a metal to semiconductor contact having a large barrier height and low doping concentration that is less than the density of states in the conduction and valence bands. Can be used to form a rectifying junction. Used to minimise injection of holes.

Scintillation Temporal fluctuation (flickering, in continuous-wave emission) of laser beam intensity caused by coherence of the laser beam and the "focusing and defocusing" of the beam by the atmospheric turbulence leading to interference. Causes instantaneous levels higher than average, as well as lower values. The intensity of the effect varies as a function of beam parameters, range and atmospheric conditions.

Scintillation index The normalised variance of the fluctuating intensity.

Scintillation parameter An expression of the level of scintillation experienced by a propagating laser beam.

Scotoma Part of the retina, which is inactive after damage - either temporarily (after image) or permanently.

Scotopic transmission Percentage transmission (e.g., of a filter) integrated over the spectral sensitivity of the dark-adapted human eye.

Scraper mirror A type of output coupler often used in chemical lasers—usually positioned around the edge of the resonant cavity.

Scratch and dig Description of the surface finish of an optical component—expressed as x/y. A scratch is a mark along the surface and is expressed in units of 1/10,000 of a millimetre, whereas a dig is a rough spot (e.g., pit or hole) in the polished surface—a bubble is considered to be a dig defect. Its units are different; the units are 1/100 of a millimetre. Hence, a 10/5 definition is a scratch of width 1 μm and a pit of diameter 0.05 mm (50 μm).

Second-harmonic generation (SHG) A nonlinear process for halving the wavelength of an incident beam (pump laser beam) when it passes through a suitable nonlinear crystal. See Doubling or harmonic generation.

Seebeck effect Part of the thermo-electric effect. The direct conversion of a temperature difference between two dissimilar metals into a voltage, and vice versa. Exploited in a thermocouple.

Seed laser A laser used to generate laser light that is injected into another laser such as a master oscillator or an amplifier. The injected light from the seed laser usually has special properties, such as single-frequency characteristics.

Seeker operating bands For the IR see Band I, II and IV.

Segmented mirror Used in an adaptive-optical system to create the conjugate beam to deliver an aberration-corrected beam to a target. A mirror that can be distorted by the various segments moving with respect to each other to provide the appropriate shape to the wavefront of the conjugate beam.

Selection rules The rules that follow from quantum mechanics that specify which transitions may occur between energy states in a system.

Self-phase modulation (SPM) A nonlinear process that leads to filament formation during propagation of very intense laser beams. See Filamentation.

Self-terminating (of laser action) A mechanism, such as bottle necking that leads to a termination of laser action through the dynamics of the population inversion, and hence, population of states. Occurs when the lifetime of the lower state is longer than the upper state.

Sellmeier Refers to an empirical equation for calculating refractive indices of materials as a function of wavelength.
Semiconductor A composition that is neither a conductor nor an insulator.
Semiconductor-disk laser (SDL) The active component in a VECSEL.
Semiconductor laser A forward-biased pn junction, so holes are injected into the n region and recombine with electrons to give light. The forward bias creates a shift in the energy levels (Fermi), so creating the population inversion. With very high current densities (5×10^3 A/cm^2) get negative absorption. Resonant cavity formed from cleaved facets.
Semiconductor saturable absorber mirror (SESAM) A form of passive Q-switch.
Sesquioxides Metal oxides such as alumina (Al_2O_3) or yttria (Y_2O_3).
Sum frequency generation (SFG) A nonlinear technique.
Self-phase modulation (SFM) See Filamentation.
Shack-Hartmann wavefront sensor An optical device for sensing the distortion in a wavefront. It uses a lenslet array to form a spot field, which is focussed on to a detector array to assess the relative angular displacement of each spot in the array, with respect to the anticipated regular periodic spatial distribution.
Short-wave infrared (SWIR) Beware of definition, but usually from about 1.4 to 2 μm.
Shot noise A type of electronic noise that may dominate a signal when the finite number of energy-carrying particles (e.g., electrons) is sufficiently small that uncertainties, owing to the Poisson distribution, become significant and may dominate.
Side-mode suppression (SMS) A technique for improving beam quality.
Single-frequency emission The laser line width ($\Delta\lambda$) is less than the Brillouin gain bandwidth.
Single-mode filter (SMF) An optical element used to control lateral mode—typically an aperture.
Single-photon counting Physical process based on quantum technology and exploits the ability of specialised laser sources to emit packets of energy with a small number of photons.
Single-side polished (SSP) Of an optical element.
Singlet levels The energy levels in a dye laser that give rise to laser transitions.
Singly resonant oscillator (SRO) Only one of the wavelengths in an optical parametric oscillator is in resonance.
Signal wavelength A component of the nonlinear parametric process with the shorter wavelength.
Sintering Densification process used in the manufacture of ceramics.
Size, weight and power (SWAP) An increasingly important parameter in system development.
Slab-coupled optical waveguide laser (SCOWL) A super-large optical cavity.
Slope efficiency A measure of the efficiency of a laser. Calculated from the gradient of the line linking the input optical power to the generated output power, but does not take account of the threshold for laser action, hence the desire for a very low threshold condition.
Slow axis The direction of the emission from a semiconductor laser that diverges more slowly.
Slow axis (crystal) Light travels with slowest phase velocity along the crystal axis with the larger refractive index.

Smile The distortion in a laser-diode array; the vertical bowing of the emitters in an array in a transverse direction during packaging.

Solar blind That part of the ultraviolet where the ozone absorbs the solar energy, from about 280 nm and down through shorter wavelengths.

Solarisation Formation of colour centres by ultraviolet photons, leads to bulk or surface damage. In titanium-sapphire (Ti:S) laser rods, it is caused by formation of Ti^{2+} ions.

Sol gel Glass-like structure with holes. Exploits the room temperature hydrolysis of metal alkoxides into metal oxides, such as silica, titania and zirconia. Manufactured by dissolving a metal alkoxide in alcohol. A controlled amount of water in the solution hydrolyses the alkoxides to produce a sol, an even distribution of small clusters of molecules or colloids in the alcohol. As the colloids grow (encouraged by moisture), they form a jelly-like mass called a gel, a three-dimensional network of metal oxides with solvent in its pores. Heating to 200°C to 500°C (composition dependent) removes the solvent and sinters the network, so the gel hardens, shrinks and densifies into a glass or ceramic. Shrinkage causes stresses and hence cracking; the addition of organic materials helps to relieve the stresses.

Solid angle The ratio of the area on the surface of a sphere to the radius squared of that sphere, units of steradians (sr) and 4π sr per sphere.

Solid state Laser class. In these devices the gain medium is a solid transparent material doped with the active element (e.g., neodymium in glass or YAG crystal). Also applied to doubling and Raman shifting elements (if a crystal). Convenient, robust construction, but waste heat causes optical distortion, and cooling is a significant problem.

Soliton A wave packet or pulse that maintains its shape as it propagates through a medium at constant speed, owing to the cancelling out of dispersive (frequency-dependent) and nonlinear (intensity-dependent) effects. Solitons in optical fibres have been the subject of much investigation, because of their potential application in long-distance communication.

Spatial coherence Ability to predict phase (shape) of a wavefront as it propagates. Perfect spatial coherence means the shape is unaltered with time, as relative phase difference between two points, orthogonal to the direction of propagation, remains constant.

Spatial-hole burning A distortion in the gain curve of a laser gain medium, caused by saturation effects of standing waves in the cavity.

Spatial-light modulator (SLM) An optical element that imposes some form of spatially varying modulation on a light beam; devices that vary their transmission according to an external electrical current. For example, a liquid crystal device, used to modulate the spatial profile of a beam.

Speckle Produced when laser light is scattered from a diffusing surface, is like noise. Caused by phase variations in the reflected light resulting from the microscopic unevenness of the reflector; consequence of the coherence of the incident and reflected light interfering, leading to the speckled interference pattern on the surface.

Spectral beam combining (SBC) Combination of multiple laser beams with slightly different wavelengths; comparable with using a prism in reverse, or the reverse of a rainbow.

Spectral luminous efficiency function A function defining the optical spectral response of the human eye, designated by $V(\lambda)$.
Spectralon A dielectric material with high albido, used in a close-coupled cavity.
Specular reflection A mirror-like reflection and maintains the directional properties of a beam.
Spherical aberration The spread (blur) at the focus of a lens or mirror resulting from imperfections in its shape, or changes in its refractive index (lens), when illuminated with monochromatic light.
Spin selection rule (for transitions) Transitions between different states with differing multiplicities are forbidden, so those transition that violate this rule are generally weak and may be ignored.
Spinel Magnesium aluminate ($MgAl_2O_4$); a low absorbing ceramic, highly transparent in the visible to mid-wave infrared. A window material for high-power laser systems.
Spontaneous emission Release of photon by an excited atom in the absence of an external stimulus. Exact instant or amount cannot be predicted. $dN/dt = N/T$ where N is the population of excited (isolated) atoms and T is the mean life (~10^{-8} seconds), about the time it takes to emit a photon, so about 3-m long photon chain, so $\Delta f \sim 10^8$. Hence, green light (5×10^{14} Hz) has a photon with ~5×10^6 cycles. Only have coherence within a single pulse and for many atoms, the $T \sim 10^{-11}$ so L ~1 mm (owing to pressure broadening).
Stable resonator A ray of light that is initially parallel to the cavity axis will be reflected forever, compared with an unstable resonator. A stable resonator concentrates light along the resonator's axis and allows efficient energy extraction from this zone, but not from the outer edges, thus leads to Gaussian-shaped beams.
Standard operating procedure (SOP) A description of the safety and administrative procedures to be followed when undertaking a task.
Stark effect The splitting of atomic spectral lines in response to the application of an electric field.
State-of-the-art (SOA) Constantly changing.
Step-index fibre (SIF) A description of the refractive-index gradient in an optical fibre. The core has the highest refractive index and the cladding has a lower refractive index to create the conditions required for total internal reflection of light travelling along the core.
Strehl ratio (or strehl) Ratio of the peak of the intensity compared with that of the theoretical diffraction limit value of the peak intensity.
Stimulated-Brillouin scattering (SBS) Focusing a single pump laser beam into a suitable material (often a liquid) causes the light to be scattered by the background thermal acoustic fluctuations. With intense pump beams, the scattered light will be amplified by the Brillouin process until the amplified noise extracts all of the energy from the pump beam. The new beam is the phase conjugate of the incident beam. The phonon lifetime is proportional to the square of the wavelength (λ^2), but the gain is independent of wavelength; hence, at longer wavelengths it is necessary to use longer pulses or very short pulses with high intensities. However, competing nonlinear mechanisms tend to limit the maximum energy density that can be achieved before damage occurs; hence, most suitable for visible and near infrared, more difficult in the mid-infrared. The Brillouin medium needs to be highly transparent and dense, and the acoustic wave needs to build up quickly, comparable

with the laser pulse length. At 1.06 μm can use fused silica, but the Brillouin and damage thresholds are close and need to avoid liquids with C-H bands, owing to weak absorption.

Stimulated emission Occurs when an atom is in an excited state and it absorbs another photon with the same energy as the atom would radiate in a spontaneous emissive event. The two emitted photons have the same wavelength and importantly, have the same phase, direction of travel and polarisation. Like negative absorption. Probability of stimulated emission occurring depends on the relative populations in the two energy levels involved in the event, hence the need for population inversion.

Stochastic parallel gradient descent (SPGD) A technique for optimising parameters such as the on-axis optical intensity of a laser.

Stoichiometric PPLN Periodically poled lithium niobate (PPLN) produced to give the exactly balanced lithium niobate crystal, i.e., no deficiency of lithium.

Stokes lines The laser lines resulting from transitions invoking the Raman effect (longer wavelength), the opposite of the anti-Stokes lines.

Stokes shift Also known as Raman shift, frequency reduction is fixed and is independent of the frequency (or λ) of the incident radiation. Occurs with the phonons of ionic groups, such as CO_3, SO_4 NO_3. Shifts in solids are usually a few hundred wave numbers. Difference in wavelength between the absorption and emission peaks.

Strength Rule of thumb, approximately half Young's Modulus.

Substrate A substrate is a planar slice taken from a large single crystal of a particular material. Substrates are normally used as the basis for the subsequent growth of high quality semiconductor materials.

Sum and difference generation Can occur in solids or gases: the principle is to mix the frequencies of two input frequency waves (ω_1 and ω_2) to give additional waves with frequencies $\omega_1 + \omega_2$ and $\omega_1 - \omega_2$.

Superconducting interference device (SQUID) A sensor.

Superconducting radio-frequency (SCRF) injector/gun Used as an electron emitter in a free-electron laser.

Super continuum (SC) For example, as in a filament.

Super Gaussian A profile with positive bulging (kurtosis).

Super lattices A super lattice is a periodic structure comprising layers of two (or more) different materials. In a semiconductor super lattice, where the thickness of one layer is normally of nanometre dimensions, the periodic potential created by the super lattice (in one dimension) creates new energy bands and bandgap known as a mini band.

Super radiant Laser operation where the gain is very high, so mirrors not needed (e.g., excimer laser).

Surface damage Laser-induced damage on, or very close to, the surface of a crystal. Careful control of the beam size on the surface is required, as the surface damage threshold is often lower than the bulk damage threshold.

Surface modes Electromagnetic surface modes are present at all surfaces and interfaces between materials with different dielectric properties and have an important influence of a range of physical effects, such as surface tension.

Survivability Fundamental to protection principles, routes to damage and being hit; the aims are: don't be seen, don't be engaged, don't be hit and don't be killed.

Susceptibility A part of a target that may be damaged. See Vulnerability.
Synchronous pumping A technique for avoiding damage to nonlinear crystals by pumping with very short duration pulses, with low energy but at a high pulse-repetition frequency, such as those from a mode-locked laser. Thresholds for optical parametric oscillator action rise in mid-infrared, owing to the increase in refractive index and the product of the output wavelengths. Can use very short synchronous pulses, to avoid this problem. In this case, the intensity in each picosecond pulse is very high, but the energy is low, thus avoiding damage.
T_0 and T_1 Characteristic of a semiconductor laser emission, where T_0 is a measure of the laser-threshold temperature sensitivity. The laser threshold, I, is given by $I = I_0 \exp(T/T_0)$, where T is usually in kelvin and T_1 refers to the slope-efficiency temperature sensitivity, $\exp(-T/T_1)$ and I_0 is a fitting parameter.
Tactical high-energy laser (THEL) An experimental L-DEW system.
Tactics, training and procedures (TTP) Describing the use of equipment to ensure optimum performance.
Talbot cavity An external cavity with a mirror often used with a diode array to achieve coherent combination of the emissions from the various lasers in the array; the feedback into the facet forces the emission from each laser in the array into a mode-locked condition. The mirror is positioned at half the Talbot distance from the output facet.
Tapered laser Description of the shape of the cavity; gives good spatial and temporal beam shape.
Technology-readiness level (TRL) A method for estimating maturity of a device.
Temperature tuning See Noncritical phase matching. Tuning of an OPO output using heating or cooling of the nonlinear crystal.
Temporal coherence Concerns the relative phase difference between two points along the direction of travel, with perfect temporal coherence the phase difference remains constant with time.
Terahertz (THz) Part of the electromagnetic spectrum used for sensing, beyond the far infrared.
Terbium gallium garnet (TGG) A laser crystal.
Tetramethylorthosilicate (TMOS) A precursor used in the manufacture of ormosils.
Thallium arsenic selenide (Tl_3AsSe_3) (TAS) A nonlinear crystal with the thermal properties comparable with those of balsa wood. Little used.
Thermal conductivity (TC) See K.
Thermal impedance Of a heat-transfer system, units of (K/W).
Thermal lens A lens created in a gain medium as a result of changes in the ambient temperature, particularly when thermal gradients appear; a consequence of the change in refractive index, owing to the change in ambient conditions.
Thermal resistance Of a cooling system ($R_{th} = \Delta T/Q$, where ΔT is the temperature rise and Q is the waste-heat generated).
Thermo-electric cooler (TEC) Exploitation of the Peltier effect.
Thermo-optical properties The change in the optical characteristics (such as refractive index) of optical elements as a result of changes in temperature (e.g., thermal lens in a gain medium).
Thin-disk laser (TDL) A type of solid-state rod laser. Description of the gain medium, also called an active mirror.

Third-harmonic generation (THG) A nonlinear process; a process generating a wavelength one-third of the incident wavelength.

Threat evaluation and weapon assignment (TEWA) Part of a weapon control system.

Three-level laser Those lasers that use three energy levels in the stimulated emission process; careful design is required if sustained operation required, as the ground-state population increases rapidly as the crystal heats up, and therefore, population inversion is no longer possible. Examples are ruby and ytterbium.

Time-correlated, single-photon counting (TCSPC) See Single-photon counting.

TMSPM Another precursor used in the manufacture of ormosils, 3(Trimethoxysilyl) propyl methyl methacrylate.

Training filter An attenuating filter used to reduce laser-hazard levels for training purposes with a system.

Training mode Mode of operation, usually with reduced optical power to lower the laser hazard, used during training exercises.

Transcendental equation An equation containing a transcendental function (i.e., not algebraic); such an equation cannot be solved for one factor in terms of another.

Transform-limited line width Short pulses cannot give very narrow spectral lines, owing to the short coherence time of such a pulse. Related to a Fourier transform from time to frequency domain.

Transparency Wavelength bands where the interaction of the optical energy with the energy states in a crystal is negligible; needs to be as low as possible for all of the wavelengths involved in the nonlinear process, for example. Crystals containing oxygen have a large absorption band starting at about 4 μm, caused by the effect of the frequency of the internal vibrations (phonons) in the crystal. For higher transparency in the longer wavelength bands, need crystals with smaller frequency phonons (i.e., with heavier component atoms and consequently weaker bonding). Unfortunately, these crystals are often weaker and have lower thermal conductivity than the shorter-wavelength equivalents.

Transition A shift of a species from one energy level to another. Inner shell electronic transitions involve higher energies and so lead to hard ultraviolet or X-ray emission. Outer shell transitions are of the order of 1 eV and correspond to visible or near infrared light. Vibration transitions are a fraction of an electrovolt and rotational transitions are <0.1 eV. Forbidden transitions are unlikely, but not impossible. Metastable levels are where a species becomes delayed as there is no readily allowable transition (i.e., becomes trapped).

Transmissivity See Transmittance.

Transmittance The fraction of the total transmitted radiant power to the total incident radiant power.

Transverse modes Of a laser beam, manifest themselves in the cross-sectional profile of a beam (i.e., its intensity profile). Hence affects beam quality. Characterised by the number of nulls that appear across a beam in two directions. Hence, TEM_{00} has its intensity peak on the central axis.

Transversely-excited atmospheric (TEA) laser An atmospheric-pressure laser.

Triplet quencher A species added to a dye-laser solution to depopulate the triplet levels; cyclo octo tetraene (COT) is a common additive to quench the triplet states.

Triplet states Of a dye molecule, for example. A parallel set of energy levels to the singlet states, but tend to have slightly lower energy, so can soak up the pump photons, but have an extended lifetime so quench laser action.
Tuneable Of a laser. See also wavelength agile. Uses a dispersive element to alter the output wavelength of a laser under its gain profile, to be effective, require a broad gain bandwidth.
Turbulence Local variations of refractive index within the atmosphere, owing to temperature variations with wide variation in scale. Causes scintillation, beam spreading and beam wander. Effect varies widely with conditions. See C_n^2.
Two-for-one pumping A method of pumping that increases the efficiency, owing to the effective use of a cross-relaxation process.
Two-photon absorption The simultaneous absorption of two photons of identical or different wavelengths to excite a molecule from one state to another. The energy difference between the upper and lower energy levels of the absorbing molecule is equal to the sum of the energies of the two photons. Two-photon absorption is a third-order process several orders of magnitude weaker than linear absorption; however, it differs from linear absorption because the strength of absorption depends on the square of the light intensity; hence, it is a nonlinear optical process.
Type-I phase matching Critical phase matching, but with the polarisation of the output beams in the same plane.
Type-II phase matching Critical phase matching, but with the polarisation of the output beams in orthogonal planes.
Type-I semiconductor A form of bandgap engineering in semiconductor materials where there is no displacement between the valence and conduction bands.
Type-II semiconductor In this form of bandgap engineering in a semiconductor materials there is a displacement between the top of a valence band and the bottom of the conduction band.
Type-III semiconductors In this form of bandgap engineering in the semiconductor material there is some overlap between the bottom of the conduction band and the top of the valence band.
Ultra-low expansion (**ULE**) As in zerodur glass.
Ultrashort-pulse laser (**USPL**) See Filamentation.
Ultraviolet (**UV**) A part of the electromagnetic spectrum, from about 10 nm to approximately 400 nm.
Uni-axial crystal A crystal that has one crystallographic axis that is different from the other two, often called the optic axis (or extraordinary axis), which can be a fast axis or a slow axis.
Unmask The instant a target becomes visible to a weapon system (e.g., by crossing its horizon).
Unstable resonator Spreads light throughout the gain volume, but gives annular profile in near field, two types: positive branch with focus outside of the cavity and negative branch (focus inside).
Up-conversion Changes colour and increases frequency (reduces wavelength) from the energy-transfer upconversion between ions in the gain medium.
Up-shift Increases the frequency and lowers the wavelength.
Urbach tail Electronic absorption edge; absorption increases sharply with increasing wavelength.
UV-A The part of the spectrum extending from 315 to 400 nm.

UV-B The part of the spectrum extending from 280 to 315 nm.
UV-C The part of the spectrum extending from 100 to 280 nm.
$V(\lambda)$ Spectral luminous efficiency function for photopic vision; the spectral response of the human eye.
V cavity The shape of a cavity; the path taken by the laser light inside the cavity.
V number See Abbe number. Indicates a material's dispersive characteristics.
V parameter Determines the number of modes in a step-index fibre. $V = k_o a\sqrt{(n_1^2 - n_2^2)}$, where k is the wave number and a is the radius. Or $V = 2\pi a\mathrm{NA}/\lambda$. When V < 2.405, there is single-mode operation.
Valence band The valence band of a semiconductor with the bound carriers.
Validation This process generates objective evidence that the capability enabled by the equipment or system satisfies the needs defined by the prospective user. Successful validation is dependent on identifying the correct validation criteria and effectiveness envelope at the requirements capture stage. Validation generates the evidence that informs and enables authorisation of an in-service date (ISD). It is vital that each individual candidate user requirement must be testable/measurable. Hence: "Did we build the right system?"
Variable birefringence filter (VBF) Used for tuning a laser with a broad emission line.
Vertical-cavity surface-emitting laser (VCSEL) A semiconductor device.
Vertical external-cavity surface-emitting laser (VECSEL) Semiconductor laser with external cavity, usually using a disk laser configuration that is optically pumped.
Veiling glare Caused by stray light in an optical system, resulting from multiple reflections between surfaces. The consequence is a reduction of contrast or loss of an image.
Verdet constant A measure of the rotation of the electric vector in materials that exhibit Faraday rotation of the polarisation of an electromagnetic beam, exploited in optical isolators.
Verification This process confirms that the characteristics and behaviour of an equipment or system comply with the performance requirements specified during the planning of a project. Also provides the information required to guide remedial action to correct shortcomings in the equipment or system. The compliance arguments of each individual candidate system requirement must be verifiable. Hence, "Did we build the system correctly?"
Vibrational energy levels The energy levels associated with the various vibrational modes of a molecule, as a change in the vibrational motion within a molecule results in a change in its energy. A series of energy levels with relatively small energy differences, smaller than electronic transitions but larger than rotational transitions.
Vibrometer A laser-radar system using coherent-detection techniques to sense motion of a target(e.g., vibrations).
Vibronic laser Strong interaction between electronic and vibrational states (phonons) leading to a manifold of energy states, similar to those on a laser dye, gives tuneable output beam.
Vidicon (tube) An imaging detector that uses a scanning electron beam operating with a photo-conductor to create an image of a viewed scene.
Visible spectrum Often defined to be from 380 to 750 nm; many can see longer wavelengths.

Visibility Generally dependent on the concentration of water vapour and aerosols in the atmosphere; a quantitative measure of the range that a human eye can observe standard objects through various meteorological conditions.
Visible and near infrared (VNIR) See Atmospheric window.
Visual interference levels Levels of irradiance that can cause temporary effects, such as distraction, glare and after images. Exposures at these fluence levels can impair the performance of critical tasks such as tracking a target or flying an aircraft.
Visual range An estimate of the distance at which objects can be observed by the human eye using passive techniques. In the case of meteorological range, there is a formal definition of the reduction in transmission to 2%.
Volume Bragg grating (VBG) Used to select wavelengths, or line narrow.
Volume Bragg reflector A reflecting structure with a periodic modulation of its refractive index. Used to control the line width ($\Delta\lambda$) of an emission.
Vulnerability The response of a target to a susceptibility (i.e., when a target gets hit).
Walk-off Poynting-vector walk-off. Caused by the different phase velocities in the nonlinear medium.
Walk-off compensation A technique for compensating for the loss of efficiency in the nonlinear conversion, as a result of Poynting-vector walk-off. Essentially, involves using several shorter crystals that are positioned at suitable angles to alleviate the effect of the separation of the two beams (signal and idler).
Wall-plug efficiency (WPE) Expresses the efficiency of an entire system (e.g., from electrical power input to the optical power generated, taking into account all supporting subsystems such as cooling units). Used to describe the efficiency of a system taking account of the applied power and the emitted power (e.g., the optical power emitted to the applied electrical power).
Wavefront Used to characterise the quality of an optical beam, measured in wavelengths of distortion. A line joining points with equal phase.
Wavefront sensor (WFS) An optical system, such as a Shack-Hartmann device, for determining the distortion in a wavefront. An important component of an adaptive-optical system.
Waveguide A structure that guides waves, such as electromagnetic waves. The rule of thumb is that the width of a waveguide needs to be of the same order of magnitude as the wavelength being propagated.
Wavelength agile The ability to change the wavelength transmitted, usually by changing a component such as the liquid dye-solution in a dye laser, or the solid-state dye gain medium. Can also be achieved through rapid tuning of the output of a laser with a broad spectral band, by using a dispersive element, but the tuning range is usually more limited than a fully wavelength-diverse system.
Wavelength beam combining (WBC) Of independent lasers (e.g., fibre lasers), to make a high-power beam.
Wavelength diverse Changing the wavelength of the output beam, which can be altered by changing some aspect of the gain medium, such as the laser dye.
Wavelength doubler Conversion of input (pump) wavelength to twice its input wavelength, by operating an optical parametric oscillator at degeneracy. One snag is the very broad spectral spread, caused by the very sharp change in wavelength with angle, which the beam makes to the (optic) axis of the crystal, can be mitigated

by using opposite polarisations, but then has problems of fluctuations in intensity, owing to variations in phase matching.

Wave numbers Measurement of frequency (cm^{-1}), for example, 2 μm is 5,000 cm^{-1}, 4 μm = 2,500 cm^{-1}; λ in wave numbers ($cm^{-1} = 10^4/\lambda(\mu m)$). Commonly used by chemists.

Wave optics A technique for simulating optical systems, which takes in account diffraction effects.

Wave plates Delay of the phase of the part of the input light with one linear polarisation by a specified phase with respect to light in an orthogonal polarisation. For quarter-wave plates it is 90° delay.

Weibull distribution This is one of the most widely used lifetime distributions in reliability engineering.

Whispering-gallery modes Modes that travel around the edge of a resonant cavity, usually as a result of total internal reflection. Usually occur at a particular resonant wavelength, or wavelengths, in a circular or elliptical cavity. From the whispering gallery below a dome or vault.

White noise Noise that is not frequency-dependent, so it has a flat power spectral density (PSD).

Wurtzite A crystal structure named after the mineral wurtzite (Zn, FeS) and is a crystal structure of various binary compounds. An example of a hexagonal crystal system.

Xanthophyll Yellow pigment in fovea; absorbs blue (i.e., colours complementary to yellow).

XLMA Extra-large mode area – description of the mode area in the gain medium.

X-ray diffraction (XRD) A technique for determining atomic structure.

YAG Yttrium aluminium garnet ($Y_3Al_5O_{12}$); a solid-state crystalline host.

YALO Yttrium ortho-aluminate ($YAlO_3$); a solid-state crystalline host, has little depolarisation of the beam in the isotropic materials.

YGG Yttrium gallium garnet ($Y_3Ga_5O_{12}$); solid-state crystalline host.

YLF Yttrium lithium fluoride ($LiYF_4$); a solid-state crystalline host, naturally birefringent.

Young's Modulus Also known as the tensile modulus, is a measure of the stiffness of elastic materials.

YVO Yttrium vanadate (YVO_4), a solid-state crystalline host.

Zeolite Scrubber for spent DF in a chemical laser.

Zernike coefficients Used in Zernike polynomials to describe the performance of an optical system. Often used to describe wavefronts, but not particularly useful in the presence of optical turbulence.

Zero-point dispersion In a single-mode fibre the zero-dispersion wavelength is that wavelength (or wavelengths) where the material and the waveguide dispersion cancel.

Zig-zag slab Optical designs of the laser gain medium involving total internal reflection within the slab, so that the effects of thermally induced inhomogeneities are mitigated, or at least averaged out. A description of the optical path through the gain medium.

Zirconium barium lithium aluminium sodium fluoride composition (ZBLAN) A fluoride-glass composition with transparency from the ultraviolet to the mid-infrared (beyond 4 μm).

χ Dielectric susceptibility; a dimensionless proportionality constant that indicates the degree of polarisation of a dielectric material in response to an applied electric field.

ε A measure of the degree to which a material can resist the flow of charge.

λ Wavelength of a beam.

μ Permeability; the ratio of the magnetic flux density in a body or medium to the external magnetic field strength inducing it; μ = B/H.

χ Polarisation field $(P) = \varepsilon(\chi^{(1)}E + \chi^{(2)}E^2 + ...)$, where χ is the dielectric susceptibility, ε is the permittivity and E is the electric field.

$\chi^{(1)}$ The linear component of the dielectric susceptibility.

$\chi^{(2)}$ Nonlinear dielectric susceptibility, used to determine the magnitude of the polarisation field. $P = \varepsilon(\chi^{(1)}E + \chi^{(2)}E^2 + ...)$.

About the Author

Professor David H Titterton BSc., Ph.D., FIET, CEng, graduated from the University of Sheffield with a first-class honours degree in physics in 1969. He was awarded a doctorate three years later by the same university for his research into the atomic structure of copper salts, using neutron diffraction techniques. He was a Dstl Fellow and technical leader of laser-based systems within the British MoD. He is also a visiting professor at Cranfield University in the UK, a Fellow of the Institute of Engineering and Technology, a Fellow of the SPIE and a Chartered Engineer.

He joined the then RAE, now Dstl, Farnborough, in 1972 and investigated techniques for predicting and assessing the performance of guided weapons. During the mid-1970s he started his involvement with high-power lasers, particularly to assess laser-induced damage and target-interaction effects. The laser theme continued with the development of a miniature ring laser gyroscope, which started his interest in strapdown inertial navigation, which led to the publication of his book (with John Weston) on inertial navigation. Towards the end of the 1980s he became leader of the dye-laser group at DERA (now Dstl). Currently, he has a broad interesting in the development of laser-based systems for a range of military applications and the chairman of the SPIE symposium on Security and Defence. He is now a consultant to the Defence Academy of the UK.

He has published more than 150 papers and reports on a range of subjects, as well as the book on strapdown inertial navigation technology; the second edition of "*Strapdown Inertial Navigation Technology*" was published in 2004. He wrote the chapters on liquid lasers and solid-state dye lasers for "*The Handbook on Lasers and Applications*" published by the UK Institute of Physics. He wrote the chapter on infrared countermeasures in the book on mid infrared optronics, published in 2005. Dave was awarded the Society of Dyers and Colourists Centenary medal for his article on "*Dye Lasers- a non-colour use of dyes*" and has been recognised nationally and internationally for his collaboration with Australia and NATO.

Index

A

AAR 54, 560, 561, 583
AAR 57, 560, 561, 583
AAR 60, 560, 561, 583
Absorption, 4, 6, 20, 21, 22, 31, 36, 53, 55, 65-67, 80, 90, 91, 101, 108, 109, 120, 125-127, 142, 156, 165-167, 169, 170-172, 174, 175, 177, 185-187, 193, 194, 221, 247, 257, 331, 359, 361, 362, 370, 380, 388-392, 394, 398-400, 402, 409, 431, 468, 488, 496, 497, 503, 561, 568, 569, 572, 577, 585
Absorption cross-section, 21, 120
Accelerometers, 457
Accessible emission limit, 487, 508, 583
Acousto-optical modulation/modulator, 30, 33, 417, 453, 583, 618
Adaptive optics, 152, 153, 163, 181-184, 193, 194, 243, 246, 253, 254, 328, 436, 570, 573, 574, 583, 584, 592, 601, 602, 631
Aero flexure, 144, 559
Aero-optical effects, 192-194
Aerosol absorption, 165, 171, 175, 586
Aerosol scattering, 165, 170-172, 586
Aerosols, 165, 167, 169, 172, 173, 183, 185, 188, 193, 225, 309, 362, 365, 367, 369, 386, 405, 435, 470, 571, 572, 631,
Aerosol scattering, 165, 170-172, 586
AHRS, see attitude and heading reference system
Aim-point refinement/selection, 9, 136, 141, 142, 146, 147, 149, 150, 154, 299, 319, 331, 338, 339, 341-344, 357, 366, 374, 430, 467
Airborne laser, 99, 487, 589
Airborne laser laboratory, 97, 235
Airy diffraction factor/pattern, 256, 570, 573, 616
Aldis lamp, 407
Alexander M Prokhorov, 6
Alice, 432
Amplifiers, 17, 26, 38, 68, 79, 83, 129
Amplitude modulation, 407, 416-418, 547, 621
Angle of arrival, 356, 562, 584
Angle-of-arrival accuracy, 143, 558-560
Angle tuning, 43, 584, 614
Anomaly detection, 334, 335, 390
ANSI, see American National Standards Institute
Anti-Stokes lines, 46, 392, 395-397, 584, 590, 619, 620, 626
AORD, see Artificial Optical Radiation Directive
Aperture averaging, 180, 181, 184, 572, 584
Apollo 12, 8, 242, 363
Arc lamps, 4, 88, 171, 191, 287, 290, 293, 522, 549, 552-555, 584, 588, 590
Archimedes, 233, 241, 266
Artificial Optical Radiation Directive, 485
ASRAAM, 546
ATI, see automatic target identification
Atmospheric windows, 165, 167, 170-173, 176, 246, 255, 288, 504, 549, 585, 586, 597, 604, 608, 631
Atomic-fluorescence scattering, see resonance scattering
ATR, see automatic target recognition
Attenuation, 4, 35, 166, 167, 169, 173, 175, 357, 370, 409, 423-425, 427, 435, 572, 584, 585, 612

Attitude and heading reference system, 149
Auger (effects), 102, 107, 108, 112, 114, 115, 585
Automatic target identification, 9, 365
Automatic target recognition, 9, 365, 368, 371, 373-377, 387
Axis alignment, see harmony
Axis transformation, 377

B

Back scatter, 143, 145, 147, 167, 279, 308, 321, 325, 331, 359, 370, 379, 380, 298, 446, 471
Balloons, 383, 466, 469
Band I, 170, 553, 585, 604, 622
Band II, 170, 585
Band IV, 170, 563, 585
Bandwidth, 23, 26, 29, 32, 34, 36, 37, 54, 55, 60, 88, 91, 106, 176, 188, 253, 297, 371, 402, 407-409, 411, 416, 418, 419, 422, 426, 429, 431, 435, 436, 476, 576, 577, 601, 609, 620, 623
Barium vapour, 277, 585
Bartoli damage criterion, 246, 247
Bathymetry, 9, 331, 363, 368, 379-382, 474, 572
Battle-damage assessment, 153, 258, 259, 262, 262, 374
Battle-damage indication, 153, 163, 242, 258, 260
Battle-space de-confliction, 251, 258
BB84 quantum-key protocol, 433, 585
BBO, see beta barium borate
Beacons, 174, 338-341, 344, 347, 407, 415, 418, 418, 426, 461, 462, 464, 467, 549, 583
Beam blockages, 409
Beam break up, 180, 191, 327
Beam broadening/spread, 176, 178, 180, 181, 184, 191, 585, 629
Beam combination, 17, 80, 82, 84, 86, 255, 590, 593, 625, 632
Beam control, 142, 150, 152, 163, 184, 233, 253, 258, 262, 349, 483, 506
Beam quality, 17, 25, 25, 26, 29, 48, 49, 53, 61, 66, 68-75, 79, 80, 82-84, 86, 87, 92, 94-96, 99, 100, 102, 105, 111, 113, 115, 116, 118, 120, 123, 127-129, 224, 249, 425, 429, 432, 434, 462, 569, 584, 586, 593, 600, 607, 609, 613, 623, 629
Beam-riding techniques, 338, 339, 347-352
Beam stability, 55, 66, 84, 140, 180
Beer's law, 173, 174, 572, 585, 586
Beta barium borate, 434, 586
Bias, 21, 103, 104, 381, 418, 440, 441, 446-448, 451, 452, 455, 587, 592, 606, 614, 623
Bi-directional pumping, 80
Binary format, 416
Biological agents, 56, 125, 368, 388, 390, 394, 396, 398, 405, 489, 496
Birefringence, 17, 31, 43, 45, 570, 586, 607, 614, 619, 630
Bistatic system, 309, 366, 420, 586
Bit-error rate, 423, 424, 432
Black-body emission, 68, 549, 598
Blazed-grating beam steering, 160
Blind arcs, 142, 150, 191
Blobs, 178
Blooming, 271
Bob, 432
Boltzmann distribution/equation/statistics, 18, 53, 392, 394, 587, 604
Boresighting, 150-152, 237, 252, 273, 312
Bragg cell, 453, 454
Broadband, 5. 65, 123, 124, 126, 128, 170, 171, 176, 178, 188, 221, 267, 274, 329, 333, 370, 396, 448, 449, 584, 588-590 598, 621
Built-in test, 151, 252, 279, 295
Burst illumination, 183, 308, 315, 320-327, 333, 588

C

Calibration, 151, 213, 220, 295, 306, 442, 508, 569, 576, 577
Canards, 346
Candela, 199, 201-203, 206
Cannon lasers, 121

Carbon-dioxide laser, 55, 95, 96, 121, 171, 2325, 254, 255, 277, 307, 364, 418
Carrier wave, 288, 414, 416-418, 423, 424, 431, 435
CARS, see coherent anti-Stokes Raman spectroscopy
Catadioptric, 418
Cat's eye effect, 145, 274, 279, 306, 415, 431
Cavity dumping, 33-35, 91, 588
Cavity stability (criteria), 28, 364
Ceilometer, 359
Ceramic lasers, 6, 75, 602
Cervit, 445
Chalcogenide glass, 588
Charles H Townes, 6
Chemical agents, 398
Chemical lasers, 8, 96-99, 101, 121, 235, 241, 277, 589, 603, 613, 622, 632
Chemical oxygen iodine laser, 7, 98-100, 129, 235, 255, 58
Chirp, 67, 246, 248, 260, 272, 274, 295, 364, 372, 400, 589
CIE V(λ) function, 200
Cipher, 413, 432, 433, 435, 436
Class 1, 213, 312, 484, 487, 574
Class 1M, 487
Class 2, 487
Class 2M, 487
Class 3A, see Class 3R
Class 3B, 488, 489, 501, 511, 525, 541
Class 3R, 487
Class 4, 225, 312, 483, 487-489, 501, 511, 514, 522, 525, 541
CLIRCM, see closed-loop infrared countermeasures
CLOS, see command-to-line-of-sight guidance
Closed-loop infrared countermeasures, 295, 297, 299-301, 303
Clutter, 9, 145, 155, 158, 171, 191, 225, 306, 309, 315, 320, 322, 323, 325, 329, 357, 365, 367, 372, 375, 376, 418, 560-565, 588, 589
C_n^2, 179, 180, 190, 225, 589, 594, 629
Coatings, 35, 36, 52, 55, 67, 103, 109, 123, 127, 142, 156, 173, 221, 589, 594, 604, 609
Coherence, 5, 15, 34, 45, 54, 184, 326, 327, 368, 408, 422, 435, 449, 454, 456, 476, 589, 622, 624, 625, 627, 628
Coherency, 54, 392, 396
Coherent anti-Stokes Raman spectroscopy, 395-397, 590
Coherent beam combination, 82-84, 590
Coherent detection, 361, 369, 373, 418, 420, 631
COIL, see chemical oxygen iodine laser
Colour temperature, 205, 206, 522, 553, 554, 590, 598
Command-to-line-of-sight (guidance technique), 338, 348, 350
Commercial-off-the shelf, 131
Common laser incidents, 504, 506, 526
Communication protocol, 419
Concept of operation, 63, 226, 259, 290, 467, 537, 543, 591
Concepts, 62, 121, 400, 447, 469, 530, 531, 537
Conduction band, 103, 107, 108, 110-114, 590, 602, 603, 614, 618, 629
Cones (of human eye), 201, 202, 575, 603
Conical scan, 547, 548, 621
Contact fuze, 350
Co-pumping, 80
Cornea, 267, 484, 497-499, 525, 597
COTS, see commercial-off-the shelf
Counter pumping, 80
Covert transmission, 422
Cross talk, 411, 423, 424, 429
Cryogenic lasers, 120. 591

D

Damage routes, 233, 239, 242-246, 250, 596
DARPA, 256
DAS, see defensive-aids suite
Data latency, 143, 144, 411, 559
Data link(s), 408, 409, 423, 425, 433, 435, 436, 466, 467, 469, 606
Data rates, 408, 409, 414, 415, 436
Data transfer, 407, 411, 430, 433, 621
De-centred lens beam steering, 160
Decoy(s), 145, 284, 285, 320, 376, 385, 459-462, 609, 615

Defensive-aids suite, 291, 557, 592
Demodulation, 367, 415, 417-419, 424, 435, 475, 476
Depth bias, 381
Depth resolution, 314, 383, 572
Designation, 56, 96, 310, 311, 319, 320, 324, 337-350, 459, 460, 523
Deterministic risk assessment, 593, 608
Deuterium-fluoride laser, 8, 37, 97, 98, 121, 122, 169, 171, 254, 255, 277, 593
Development test and evaluation, 198, 208, 209, 226
DIAL, see differential-absorption lidar
Differential laser gyroscope, 447
Differential phase shift, 453
Differential scattering, 331, 399
Differential-absorption lidar, 331
Diffraction limit, 49, 82, 83, 94, 104, 193, 249, 398, 569, 576, 593, 607, 615, 625
Diffusion bonding, 70, 593
DILAG, see differential laser gyroscope
Diode lasers, 22, 88, 102, 105, 116, 129, 313, 341, 371, 603, 607
Diode pumping, 6, 7, 22, 64, 65, 68-72, 74, 84, 85, 88, 94, 95, 98, 117, 307, 311, 340, 341, 347, 363, 364, 595
Diode-pumped alkali laser, 94, 101, 503, 597
DIRCM, see directed-infrared countermeasure
Direct detection, see incoherent detection
Directed-infrared countermeasure, 137, 138, 191, 273, 281, 283-303, 460, 549, 552, 553, 558-560, 594
Disability glare, 266
DISC, see differential scattering
Disk lasers, 7, 69, 84-86, 116-118, 276, 277, 603, 623
Dispersive elements, 55, 124, 221, 332, 617, 629, 630, 631
Distributed key code, 433
Distributed-Bragg reflectors, 106, 127, 587, 594, 598
Distributed-feedback, 106, 372, 474-476, 594
Distributed-feedback devices, 475, 476
Dither, 446, 451, 452, 454
Diurnal variation, 166, 179, 180, 533, 594
Doctrine, 61, 226, 530-536, 538, 539
Doppler (shift), 4, 37, 360, 363, 363, 366, 368, 371, 372, 384-386, 470, 471, 558, 560, 588, 59Double refraction, see birefringence
DPAL, see diode-pumped alkali laser
Drift rate, 441, 454
Drop out, 420, 423, 424, 427
Dual-axis scanning micro-mirrors beam steering, 160
Duplex system, 420
Dwelling, 139
Dwell time, 244, 261
Dye lasers, 18, 34, 55, 73-75, 88-92, 121, 254, 261, 276, 522, 588, 595, 598, 600, 605, 609, 613, 623, 629, 631
Eaves dropper(ing), 408
Eddies, 178, 179, 603, 605, 613
Edge-emitting arrays, 104, 106, 595
Efficiency, 7,10, 16, 17, 23, 27, 44, 50, 51, 53, 61, 65-71, 73, 75-77, 80, 82, 86, 87, 90-92, 94-96, 99, 100, 102, 104, 111, 113, 115, 116, 120, 123, 286, 288, 310, 342, 462, 468, 568, 594, 595, 606, 612, 616, 618, 624, 631
Electronic levels, 17, 18, 91, 392-394
Electro-optical prisms beam steering, 160, 162
Encryption key, 432
Engineering controls (laser safety), 483, 487, 493, 506, 509, 511, 518, 520
Entanglement, 434
Envelope detector, 418, 546
Environmental cover, 137, 140, 155-157, 245, 367, 440, 457
EOHD, see extended ocular hazard distance
EPIL, see explosively pumped iodine lasers
Equipment training, 539, 542
Erbium, 67, 276 277, 307, 311, 324, 364
Error budget, 62, 63, 144, 216, 559, 561, 587
Eve, 432
Explosively pumped iodine laser, 21, 101, 121-123, 235, 597
Extended ocular hazard distance, 490-494, 521, 597
Extinction, 68, 165, 169, 173, 174, 193, 259, 330, 380, 423, 586, 597
Eye anatomy, 497

Eye hazards, 316, 498
Eye protection, 482, 499, 501-503, 506, 510, 513, 518, 525,
Eye response, 201, 202, 497, 500, 505, 575, 603
Eye-safe (wavelengths), 71, 79, 275, 311-314, 316, 324, 329, 333, 364, 379, 382, 388, 462-464, 468, 472, 484, 489, 525, 537, 597
Eye-safer wavelengths, 311, 312, 324, 329, 341, 346, 347, 364, 379, 484, 525, 537, 597

F

Fabry-Perot cavity, 26, 44, 55, 106, 431, 597
Fabry-Perot etalon, 32, 597, 609
Facetted dome, 156, 157
False-alarm rate, 302, 365, 387, 390, 404, 557-559, 562-565, 574
FAR, see false-alarm rate
Far field, 24, 25, 49, 217, 224, 225, 256, 257, 402, 492, 569, 597, 600
Faraday cell, 447,
Fascode, 175, 176
Fibre Bragg grating, 31, 81, 598
Fibre-laser, 7, 17, 29, 35, 48, 69, 76-85, 126-129, 163, 235, 257, 277, 307, 364, 371, 468, 469, 473-476, 589, 594, 598, 605, 614, 632
Fibre gyroscope, 441, 448-455, 473
Fibre-optic ring resonator, 455, 456
Field of regard, 137, 139, 141, 143-147, 151, 156, 160, 193, 252, 273, 366, 372, 377, 378, 383, 521, 575, 598
Filaments, 5,123,188-190, 194, 398, 598
Filter wheel, 563
Fine-track sensor, 145, 149, 151, 152, 154, 193, 292, 295, 299, 420, 549, 560
Finger-print region, 389, 390, 392, 393, 396
Fire-control solution, 278, 281, 342, 343
Firing post, 350-353
Flares, 171, 285, 545, 546, 595
Flash blindness, 266, 270, 280, 592, 598
Flash illumination, 363, 371, 373, 387, 464
Flash laser-radar, 372
Flashlamp, 7, 8, 21, 22, 64, 66-74, 87, 88, 91, 92, 254, 261, 276, 310, 312, 504, 522, 588, 589, 596, 598, 617
Fluorescence conversion, 68, 91, 599
Fluorescence quantum yield, 394
Fluorescence spectroscopy, 394, 398
Focal-plane array, 144, 222, 322, 325, 329, 330, 362, 363, 549, 560, 576, 599
FOG, see fibre gyroscope
Forward scatter, 167, 592
Four-level laser, 29, 307, 600
Free-spectral range, 600
FPA, see focal-plane array
Free-electron laser, 118-120, 128, 186, 593, 600, 610, 611, 626
Frequency nulling, 452
Frequency stability, 6, 34, 444, 448
Fuze, 170, 337, 338, 346, 350, 353-358
Fuze-warhead matching, 353, 354, 356

G

GaAs, see gallium arsenide
Gain guiding, 105
Gain medium, 8, 16-22, 25-30, 33-37, 48-53, 55, 59, 68-72, 75-78, 84-95, 101-104, 108, 109, 120, 121, 126-128, 157, 235, 246, 254, 261, 277, 310, 311, 340, 341, 364, 476, 522, 553, 584, 589, 591, 594, 596-602, 605, 607-609, 613, 614, 617-620, 624, 628, 630-643
Gain switching, 34, 600
Gallium arsenide, 38, 104, 109, 112, 219, 274, 307, 320, 327, 402, 462, 464, 612
Gas laser, 6, 8, 16, 17, 37, 52, 56, 92-97, 99, 100, 129, 235, 255, 276, 277, 361, 363, 371, 399, 418, 444, 446, 584, 588, 591, 593, 600, 602, 605, 606, 608, 621
Gas-dynamic laser, 6, 96-100, 235, 255
Gated detector, 322, 391, 395, 588
Gated imaging, 3, 9, 171, 320, 323, 325, 326, 329, 333, 334, 464
Gathering, 226, 350

Gaussian (beam shape), 25, 27, 29, 51, 185, 219, 224, 424, 469, 481, 492, 568, 569, 600, 605, 625, 626
Geiger mode, 309, 313
Geneva Convention, 319, 523
Geometric beam combination, 81, 82, 255
Geospatial and temporal referencing (G&TR), 251
Gimbal, 136, 137, 140-142, 147-150, 155, 156, 160, 162, 342, 378, 439, 554, 573, 610
Gimbal functions, 141
Global-positioning system, 308, 442, 456, 458, 467
GLOW, 269
Gordon Gould, 18
GPS, see global-positioning system
Guidance beam, 348-351
Guidance corridor, 349-351
Guidance-gain degradation/reduction, 289
Guidance law, 346, 351, 562

H

HALT, see highly accelerated life testing
Handover, 143-145, 251, 274, 294, 558-560
Harmonic generation, 7, 38, 45-47, 72, 117, 125, 569, 586, 594, 601, 622, 628
Harmony, 141, 144, 147, 150, 151, 157, 252, 257, 273, 296, 312
Hazard distances, 489-494, 508, 518, 520, 521, 597, 611
Heliograph, 407
Helium-neon laser, 6, 445, 448
Heating ratio, see quantum defect
HELSTF, 235, 236
Hetero junction, 105, 602
Hetero structures, 105, 108
Highly accelerated life testing, 223, 602
Hitran (data base), 175, 176
Hittile, 338
Holey fibre, see photonic-crystal fibre or photonic band-gap fibre
Holmium, 67, 254, 277, 573
Homo junction, 104
Horopter, 575
Humidity, 171, 172, 174, 572, 620
Hydrophone, 473-476
Hyper-hemispherical dome, 156
Hyper-spectral imaging, 334, 603

I

I-cavity, 603
Idler beam, 41-45, 590, 594, 603, 611, 612, 631
IEC, see International Electro-technical Commission
IEDs, see improvised explosive devices
Illuminance, 199, 203-206
Image dancing, 176, 325, 326, 603
Image intensifier, 174, 320, 321, 331, 334, 350, 464
Image processing, 149, 155, 159, 324, 326, 328, 334, 365, 375, 384, 563
Image sharpening, 326
Image stitching, see lucky imaging
Imaging seeker, 288, 548-550
Improvised explosive devices, 240, 249, 331
IMU, see inertial-measurement unit
In-band (effects), 8, 237-246, 251, 254, 257, 259, 261, 267, 270, 272, 275, 277, 279, 281, 285-288, 296, 302, 322, 331, 547, 551, 603
Incoherent beam combination, 81, 82, 255
Incoherent detection, 420
Inertial-measurement unit, 377, 378, 439, 440
Inertial-navigation system, 29, 147, 149, 280, 377, 419, 439, 442, 443, 448, 449, 457, 458
Infrared-absorption spectroscopy, 388-391, 399
Infrared countermeasures, 110, 114, 283-285, 288, 290, 389, 545, 552, 553, 558
Infrared missile-warning sensor, 562-564
Inner scale (of turbulence), 178, 603
INS, see inertial-navigation system
Installed performance, 62, 63, 190, 191, 197, 198, 208-210, 217, 223, 229, 521, 549, 578
Integrated fibre gyroscope, 454
Inter-band cascade laser, 114,115
Interferometer, 361, 440, 443, 448, 450, 475, 476, 607, 608

International electro-technical commission, 485-487, 523, 604
Ionisation, 124, 188, 189, 398, 620
IRCM, see infrared countermeasures
Irradiance, 124, 158, 166, 171, 175, 184, 186-188, 194, 199, 203-306, 219-221, 244, 266, 267, 341, 352, 382, 465, 467, 469, 481, 482, 487-489, 498, 502, 512, 525, 567, 604, 606, 611, 631
Isolator, 29, 105, 573, 597, 604, 612, 630

J

J:S, 170, 172, 288-290, 295, 298-301, 555, 604
Jam code, 302
Jam waveform, 150, 285, 287, 289, 292, 295-300, 303, 552, 553, 556
Jamage, 302
Jamming, 265, 283, 285, 286, 288, 289, 292, 295, 298-302, 357, 413, 436, 442, 553, 555, 604
Jam-to-signal ratio, see J:S
J-MUSIC, 156, 293

K

Keep-out range, 250
Keratitis, 267
Kerr cell, 34, 604, 618
Kolmogorov theory, 177, 178, 605
Krypton lamp, 553
KTP, see potassium titanyl phosphate
Kurtosis, 626

L

Ladar, 8, 9, 360
Lambertian, 220, 569, 577, 605
Lapse rate, 166, 183, 281
Large mode-area fibre, 76, 609
Laser cavities, see resonators
Laser dazzle, 245, 265, 273, 497
Laser dyes, 55, 68, 73, 75, 66-92, 605, 618, 621, 632
Laser hazard area trace, 493
Laser information field, 348
Laser modes, 23, 24, 105, 609
Laser splash, 353
Laser transitions, 22, 112, 588, 591, 623
Laser warning, 265, 269, 353, 460, 561, 606
Laser Weapon System (ANSEQ-3(XN-1)), 238
Laser-beam warning, 460
Laser-guided bomb, 5, 8, 340, 342-347, 459, 471
Laser incident, 504, 506, 526
Laser-induced breakdown spectroscopy, 154, 399, 400
Laser-induced fluorescence, 398, 606
Laser-induced plasma spectroscopy, 399
Laser-induced vapour emission, 398
Laser-radar equation, 360
Laser-radar seeker, 377-379
Laser-safety officer, 214, 215, 494, 507, 508, 513, 517-519, 541
Laser-safety standards, 12, 484-486, 508, 523, 525
Laser-safety training, 485, 494, 495, 497, 506-508, 541
Lasing threshold, 51, 66, 598, 606
LaWS, see Laser Weapon System (ANSEQ-3(XN-1))
Lead lanthanum zirconium tantalate, 160, 615
Lead selenide, 170
Lead sulphide, 170
LED emission, 103, 105
Less than lethal effects, see nonlethal effects
LGB, see laser-guided bomb
LHAT, see laser hazard area trace
LIBS, see laser-induced breakdown spectroscopy
LiCAF, 87, 590
Lichtsprecgerät, 407
Lidar, 8, 9, 136, 158, 331, 339, 360, 363, 398, 606
Line width, 29, 55, 80, 89, 106, 274, 364, 379, 382, 385, 386, 388, 391, 395, 397, 424, 425, 429, 432, 434, 474-476, 569, 588, 590, 602, 603, 608, 623, 628, 631
LiSAF, 87, 275, 590
LMA fibre, see large mode-area fibre
Lock in, 446, 447

Longitudinal modes, 23, 29, 34, 55, 364, 569, 607, 609, 618
Lorentzian (beam shape), 602, 607
Low-light television, 143, 331, 342
Low observable, 322, 371, 470
Lowtran, 175, 176
LSO, see laser-safety officer
Lucky imaging, 182-184, 328
Lumen, 199, 202, 203, 205, 206
Luminance, 199, 201, 203-206, 424, 436, 577
Luminous exposure, 204-206
Luminous flux, 199, 202-206
Luminous intensity, 199, 202-206
Lunar night, 465
Lux, 199, 203-206, 577

M

M^2, see propagation factor
Macro-optical lens beam steering, 160, 161
Magnetic mirror, 447
MANPADS, 283
Maritime boundary layer, 175
Master oscillator power amplifier, 26, 71, 79, 80, 83, 607
Maximum permissible exposure, 265, 266, 270, 272, 273, 279, 382, 483, 485-493, 499, 508, 523, 525, 541, 593, 606, 608
Metal-vapour arc lamp, 553, 554
Metal-vapour laser, 276, 585, 591, 608
Meteorological visibility, see visibility
Microsacades, 575
Mid-infrared chemical laser, 235,236
Mie scattering, 167, 169, 171, 173, 369, 571, 608
MILES, see multiple integrated laser engagement system
Mine detection, 331, 366, 387, 472, 473
MIRACL, see mid infrared chemical laser
Missile-approach warning (system/sensor), 291
Mode locking, 33, 34, 69, 73, 87, 91, 401, 402, 404, 609, 627
Modtran, 175
Modulated jam waveform, 285, 287, 552
Modulation depth, 295, 296, 417, 553, 556
Molecular absorption, 125, 165, 171, 172, 586
Molecular scattering, 169, 175, 393, 586
Monochromatic (emission), 5, 25, 37, 55, 202, 205, 309, 481, 598, 602, 606, 609, 625
Monostatic system158, 309, 367, 420, 586, 609
MOEMS, 161
MOPA, see master oscillator power amplifier
Morse code, 407, 416
MPE, see maximum permissible exposure
Multi oscillator (gyroscope), 447
Multi-dimensional measurements (laser radar), 136, 330, 361-364 373
Multiple integrated laser engagement system, 463
Multiple quantum-well optical modulators, 431
Multi-spectral imaging, 329, 330, 332, 334, 335, 564
MUSIC, 156, 293

N

NACL, see Navy ARPA chemical laser
NADH, see nicotinamide adenine dinucleotide
Nadir, 136, 137, 139, 141, 610
Nadir avoidance, 136, 141
Navstar, see global-positioning system
Navy ARPA chemical laser, 235, 236
NCID, see non-cooperative identification
N-dimensional imaging, 373
Near field, 25-27, 217, 224, 597, 610, 630
Neodymium, 6, 23, 45, 46, 66-68, 71, 82, 275, 276, 307, 311, 340, 341, 344, 347, 594, 598, 600, 601, 610, 624
Neodymium 'YAG', 344, 594, 610
Nicolay G Basov, 6
Nicotinamide adenine dinucleotide, 394
Night-vision goggles, 331, 334
NOHD, see nominal ocular hazard distance
Noise reduction, 326
Nominal ocular hazard distance, 490-493, 597, 611
Non-contact sensing, 388
Non-cooperative identification, 366, 368, 404
Non-cooperative identification friend or foe, 404
Non-lethal effects, 265, 266

Non-linear crystal, 7, 38, 40-47, 71, 277, 569, 584, 586, 588, 590-592, 601, 604, 606, 611, 612, 614, 622, 627
Non-linear Raman scattering, 395, 397
Non-linear Raman spectroscopy, 397
Non-line-of-sight communication, 411, 413, 426-429, 435
Non-mechanical beam steering, 137, 159-163, 372, 436
Non-radiative effects, 22, 102, 107, 112, 114, 115, 585, 620
Non-reciprocal effects, 450, 597
Numerical aperture, 76, 126, 127, 568, 611

O

OBL, see optical break lock
Obstacle avoidance, 5, 0, 158, 366, 368, 372, 383-386
Off-board (systems), 285, 461, 545, 618
Operational test and evaluation, 198, 208-211, 215, 229, 567
Operator-in-the-loop (guidance), 348
OPO, see optical parametric oscillator
OPSL, see optically pumped semiconductor laser
Optical air-data system, 470
Optical break lock, 290, 292, 298, 301, 303, 551
Optical cross-section, 279, 612
Optical fibre, 47, 76, 77, 81, 124, 127, 157, 273, 278, 297, 371, 409, 426, 434, 448, 450, 453-455, 460-462, 473-475, 583, 586, 589, 591, 594, 598, 601, 602, 605, 608, 611, 615, 616, 624, 625
Optical gyroscope, 9, 440-443, 449, 454, 455, 473, 546, 547
Optical parametric oscillator, 7, 27, 38, 40-44, 55, 72, 277, 287, 332, 333, 341, 568, 584, 591, 592, 594, 604, 611, 612, 623, 627, 632
Optically pumped semiconductor disk laser, 86, 116
Optically pumped semiconductor laser, 85, 107, 109, 110, 288, 612
Organically modified silicates, 613, 627, 628
Ormosil, see organically modified silicates
Oscillators, 25-27, 29, 44, 71, 79, 80, 104, 119, 364, 368, 395, 416, 443, 447, 584, 594, 604, 613, 623
Outer scale (of turbulence), 178, 179, 613
Out-of-band (effects), 150, 151, 239, 240, 242-245, 247, 251, 254, 257, 259, 261-263, 568
Own-ship motion, 136, 148, 222
Ozone, 125, 126, 167, 168, 170, 427, 5661-563, 624

P

PADS, see position and azimuth determining inertial system
Panda fibre, 77
PAS, see photo-acoustic spectroscopy
PAVE Spike/Way, 340
PAWS, 560
PBG fibre, see photonic band-gap fibre
PbS, see lead sulphide
Pbse, see lead selenide
PCF, see photonic-crystal fibre
P_{Dec}, see probability of detection
Periodically poled lithium niobate, 38, 41, 553, 590, 613, 626
Personal-protection equipment, 489, 494, 498, 516, 519, 525, 526, 598, 610, 613
Phase conjugation, 587, 599, 607, 613, 614, 626
Phase locking, 125, 255, 314, 396, 419, 607
Phase matching, 42-45, 402, 583, 584, 591, 611, 614, 617, 627, 629, 632
Phase nulling, 452, 453
Phased-array beam steering, 161
Photo-detector, 200, 309, 445, 448, 564, 576, 614, 618
Photo-acoustic effects, 497, 606
Photo-acoustic spectroscopy, 399, 400
Photo-chemical effects, 497, 505, 618
Photometry, 199-204, 229
Photo-multiplier tube, 309, 418, 615
Photon spin, 433
Photonic-crystal fibre, 124, 615
Photopic vision, 200, 201, 630
Photo-thermal effects, 496, 615

PIB, see Power-in-the-bucket (measurement)
Picture element (pixel), 366
Piezo-electric devices, 445, 616, 618
Pitot tube, 4, 470
Pixel, 144, 154, 155, 316, 322, 324-330, 363, 372, 373, 376, 573, 574, 576
Planck, 16, 205, 206
Plastic lasers, 73, 74
Platform survey, 142, 144
PLZT, see lead lanthanum zirconium tantalate
Pn junction, 103, 104, 592, 596, 602, 603, 611, 614, 615, 623
Pockels' cell, 34, 56, 416, 417, 616, 618
Point cloud, 159, 326, 329, 338, 364, 367
Pointer lag, 146
Pointing stability, 49, 122, 137, 158, 191, 246, 248, 275, 278, 281, 295, 296, 344, 586
Pointing system, 55, 137, 146, 147, 151, 246, 252, 278, 281, 299, 309, 343, 426, 467, 552, 586
Pointing vector, 44, 137, 147-152, 155, 157, 193, 257, 273, 295, 299, 494, 521, 546
Poisson distribution, 434, 616, 623
Poisson's ratio, 474, 616
Polarisation, 5, 7, 31, 34, 35, 40, 44, 56, 76, 77, 330-334, 362, 364, 373, 402, 403, 417, 425, 429, 432-434, 454, 476, 587, 589, 595-597, 602, 604, 610-612, 614, 616, 618, 626, 629, 630, 632, 633
Population inversion, 7, 16, 19-23, 30, 34, 36, 50, 70, 77, 84, 88, 93, 94, 97, 99-103, 109, 112, 121, 122, 587-589, 603, 616, 617, 623, 626, 628
Position and azimuth determining inertial system, 456
Post burn out (condition), 562, 616
Potassium titanyl phosphate, 39-41, 42, 277, 601, 605
Power beaming, 464-469
Power-in-the-bucket (measurement), 616
Power stability, 66
Power-to-mass factor, 255
Poynting-vector walk off, see Walk off
PPE, see personal-protection equipment
PPLN, see periodically poled lithium niobate
Precision Avionics Vectoring Equipment (see PAVE Way/Spike),
Probabilistic hazard zone, 492, 617
Probabilistic risk assessment (laser safety), 492, 521, 617
Probability of detection, 413, 558-560
Probability of escaping hit ($P_E(H)$), 250, 461
Probability of intercept, 306, 307, 407, 413, 422, 435
Propagation factor (M^2), 24, 49 82, 86, 104, 128, 224, 249, 324, 333, 424, 492, 493, 569, 586, 606, 607
Proportional navigation, 292, 562
Protocol IV, 12, 262, 268, 270, 279, 281, 484, 522, 523, 524
Proximity fuze, 338, 350, 353-358, 585
Pulse coding (range-finding), 310, 347
Pulse duration/length, 32-35, 52, 125, 219, 275, 311, 312, 316, 324, 333, 383, 401, 416, 489, 572, 607, 617, 618
Pulse-code modulation, 417, 419
Pulsed Doppler, 560
Pulse-frequency modulation, 416, 418
Pulse-repetition frequency, 32, 52, 69, 71, 73, 243, 246, 248, 275, 295, 296, 309-312, 324, 333, 353, 367, 378, 379, 382, 386, 388, 391, 395, 397, 476, 617
Pulse-width modulation, 416, 418
Push-broom scanning, 158, 377, 378

Q

QCLs, see quantum-cascade lasers
QKD, see quantum-key distribution
Q-switch(ing), 8, 18, 30-34, 41, 46, 68, 1202, 270, 307, 310, 363, 416, 473, 585, 588, 595, 601, 613, 618, 623
Quantum-cascade lasers, 7, 110, 112, 113, 129, 277, 288, 399, 402, 618
Quantum cryptography, 408, 413, 423, 431-434, 436, 605
Quantum defect, 20, 53, 65, 101, 121, 618
Quantum efficiency, 575, 576, 618
Quantum electronics, 16
Quantum key, 432, 433

Quantum-key distribution, 432, 433
Quantum well infrared detector, 564, 618
Quantum wells, 106-109, 111-113, 115, 116, 276, 277, 296, 372, 431, 556, 618
Quasi phase matching, 42, 44, 45, 590, 613, 619
QWIP, see quantum well infrared detector

R

Radiance, 175, 198, 199, 203-206, 220, 221, 326, 330, 554, 555, 619
Radiance temperature, 205, 206
Radiant exitance, 204, 206
Radiant exposure, 204-206, 606, 619
Radiant flux, 199, 202-206, 221, 585, 604, 619, 620
Radiant intensity, 3, 5, 10, 15, 25, 48-51, 82, 169, 184, 194, 199, 202, 203, 206, 221, 248, 267, 270, 272, 278, 286-290, 295, 303, 347, 399, 420, 424, 459, 488, 549, 552-556, 569, 605, 619
Raman conversion, 46
Raman laser, 332, 619
Raman spectroscopy, 331, 392-398, 590, 620
Range ambiguity, 306, 313
Range laser-safety officer, 517-519
Range profiling, 308, 367, 398, 399, 585
Rare earth ions, 79, 120, 610, 620
Rayleigh range, 24, 25
Rayleigh scattering, 125, 126, 166, 167, 169, 172, 173, 369, 427, 428, 470, 571, 619, 620
RDX, see Research Department Explosive
Redeye (missile), 546
Refraction, 160, 165, 175, 177, 193, 380, 569, 586, 602, 607, 614
Refractive index, 31, 35, 42, 43, 46, 47, 51, 74, 76, 105, 109, 126, 161, 165, 176-178, 185, 186, 192, 383, 455, 571, 572, 583, 587, 589, 591, 592, 594, 595, 598-603, 605, 607-612, 614, 620, 621, 624, 625, 627, 629, 631
Refractive-index structure parameter (C_n^2), 178, 179, 190, 225, 589
Release to service, 227
Research Department Explosive, 401
Resonance fluorescence scattering, see resonance scattering
Resonance scattering, 370
Resonant absorption, 166, 167
Resonators, 16, 25-29, 33, 35, 41, 48, 51, 55, 66, 70, 85, 96, 98, 100, 103-107, 119, 123, 127, 246, 590, 615, 616, 620, 625, 630
Reticle, 286, 288, 289, 298, 299, 302, 303, 546-548, 621
Retina, 201, 266, 267, 482, 484, 490, 496-499, 525, 575, 595, 597, 600, 603, 607, 608, 621, 622
Retinal damage, 484, 505
Retro-communication techniques, 420, 429, 432, 467
Retro-reflecting modulator, 429-431
Retro-reflection, 145, 295, 299, 415, 420, 429-431, 612, 621
Reversible defeat mechanisms/effects, 266, 268, 281, 592
Ring resonator, 55, 455, 456
Ring-laser gyroscope, 6, 9, 440-449, 455, 457
Risley prisms beam steering, 159, 160, 372
RLG, see ring-laser gyroscope
Rod lasers, 68-72, 76, 85, 628
Rods (of human eye), 201, 202, 497, 603
RoE, see rules of engagement
Roll off, 20
Rosette scan, 548
Rotational levels, 18
Ruby laser, 6-8, 22, 275, 307, 316, 340, 363, 598, 621, 628
Rules of engagement, 226, 244, 319, 320, 483, 538

S

S300 (missile warner), 560
SA 18 (missile), 546
SA 7 (missile), 546
Sacrificial decoy, 461
Safe-operating guidelines/procedures, 214, 481, 496, 506, 509, 511, 513
Safety management, 423, 483, 489
Sagnac effect, 440, 441, 443, 448, 453, 455

Sagnac interferometer, 448
SBS, see stimulated Brillouin scattering
Scale factor, 444, 447, 448, 453-455
Scale of testing, 216
Scanned detectors, 548
Scanned linear arrays, 549, 564
Scanning, 135, 136, 139, 158-160, 273, 305, 313-316, 324, 332, 351, 354, 362, 366, 372, 373, 377-379, 387, 388, 472, 557, 564, 631
Scattering, 29, 38, 46, 47, 83, 112, 125-127, 158, 165-175, 193, 247, 267, 331, 332, 361, 366, 369, 370, 379, 380, 388,392,-399, 409, 427, 428, 449, 455, 470, 471, 510, 568, 569, 571, 572, 586, 588, 597, 608, 611, 619, 620, 625
Scintillation, 165, 176, 178, 180, 184, 194, 267, 325-327, 409, 423, 467, 482, 492, 622, 629
Scintillometer, 179
Scotopic vision, 201
Search and rescue, 383, 463
Secure communications, 408, 409, 413, 431, 436
Seeker operating wavebands, 287, 290, 552, 554, 556, 585, 622
Seeker telescope, 547, 548, 551
Self S, 287, 290, 295, 553, 556
Semi-active guidance, 339, 340
Semiconductor laser, 16, 37, 102-117, 126, 277, 288, 296, 357, 371, 402, 462, 464, 553, 556, 585, 597, 612, 615, 623, 624, 627, 630
Semiconductor saturable absorber mirror, 31, 623
SESAM, see semiconductor saturable absorber mirror
Shack-Hartmann (wave-front sensor), 182, 623, 631
Shoot and scoot, 308, 439, 456, 458
Side lobe(s), 161, 422, 435, 465
Sidewinder (AIM 9)
Sightline isolation, 122, 148, 342
Signal beam, 41, 44, 368, 419, 588, 599, 612, 623, 631
Signal-to-noise (ratio), 145, 158, 299, 309, 357, 367, 418-420, 424, 428, 470, 474, 549, 574
Simulated training, 534, 540
Single-photon avalanche detector, 314
Single-photon range finding, 312, 313
Single-photon techniques, 312, 315
Singlet oxygen, 91, 99-101, 589
Skin damage/hazards, 12, 413, 484, 488, 489, 496,-499, 502-506, 525
Slab laser, 69-72
Slip rings, 137, 141
Slod laser, 72, 73
Solar blind, 429, 557, 565, 624
Solar-blind region, 427-429, 435, 561, 565
Solid-state laser, 6, 7, 16, 17, 22, 37, 38, 45, 51, 53, 64-76, 85, 86, 88, 101, 104, 117, 129, 159, 169, 235, 356, 276, 287, 288, 296, 307, 311, 333, 363, 364, 373, 556, 572, 584, 590, 593, 598, 601, 610, 619, 621
SOP, see safe-operating procedures/guidelines
SPAD, see single-photon avalanche detector
Spatial coherence, 5, 351, 589, 624
Specific emission factor, 255
Specific force, 439, 457
Speckle, 54, 325-328, 371, 569, 625
Spectral beam combination, 82, 255, 593, 625, 632
Spectral luminous efficiency function, 200, 201, 625, 630
Spectral reflectance, 330
Spectralon, 625
Spectrometer, 391, 398, 401, 403, 576, 599
Spin by carrier (waveform), 288, 298
Spin scan, 547, 548, 553, 621
Spinel, 36, 157, 625
Spontaneous emission, 21, 103, 104, 118, 126, 606, 616, 619, 625
SRS, see stimulated-Raman scattering
Stability, 44, 90, 148, 219, 220, 295, 295, 385, 387, 434, 442, 444, 449, 453, 511, 546, 594, 609
Stabilised platform, 279, 309, 439, 442
Stacked detectors, 564
STANAG, 341, 344, 462

Stand-off detection, 331, 388, 398
Stand-off tera-hertz sensing, 401, 403
Sarrk effect, 431, 625
Startle (dazzle effect), 266
Stimulated-Brillouin scattering, 29, 83, 569, 625
Stimulated emission, 22, 25, 26, 35, 37, 40, 41, 56, 64, 67, 77, 88, 95, 96, 103, 104, 119, 123, 600, 606, 616, 618, 626, 628
Stimulated-Raman scattering, 46, 83, 395-397, 611
Stinger (missile), 546
Stokes lines, 47, 397, 584, 620, 626
Strained-engineered diodes, 106, 115, 372
Strapdown (navigation), 9, 378, 439-442, 447, 456, 457
Stratosphere, 166, 175, 427
Stray light, 142, 266, 270, 391, 408, 482, 501, 502, 510, 511, 590, 601, 630
Streak tube(s), 473
Strehl, 193, 625
Sum and difference frequency mixing, 38, 47, 626
Super continuum, 125, 188, 398
Super-continuum generation, 5, 7, 38, 47, 48, 88, 123-126, 188, 189, 267, 272, 332-335, 398, 405, 598, 626
Super-continuum laser, 123, 125, 267, 272, 333, 335, 398, 405
Super lattices, 106, 626
Surface emitting (devices), 7, 104, 106, 107, 116, 276, 277, 313, 630
Synthetic environment, 533, 534, 538, 539
Syracuse, 233, 234, 243

T

Target discrimination, 145, 330, 375, 376, 385
Target tracking, 4, 135-155, 181, 184, 237, 241, 243, 253, 262, 269, 271, 273, 274, 278-280, 283, 285-289, 294, 299, 302, 303, 338, 340, 344-352, 372, 374, 460, 467, 545, 550, 551, 559, 563, 604, 616
Target-defeat techniques, 244
TEA lasers, see transversely excited atmospheric (CO_2) lasers
Technology insertion, 217
Technology-readiness levels, 208-212, 224-226, 229, 573, 627
TEM_{00}, see laser modes
Temperature tuning, 43, 614, 627
Temporal coherence, 5, 15, 54, 368, 422, 627
Tera-hertz spectroscopy, 401
Tera-hertz-induced fluorescence, 402
Test and evaluation, 178, 206, 208-212, 215, 217, 226, 228, 229, 482, 515, 524, 526, 567
Test planning, 188, 212-215, 229, 515, 517, 522, 603
Theodore H Maiman, 3, 6, 235, 441
Thermal blooming, 165, 171, 184-186, 194, 239, 246, 259, 271, 568, 610
Thermal crossover, 166, 179, 180, 321, 533
Thermal Imaging & Air-borne Laser Designator, 345
Thermal lens, 17, 27, 51, 53, 55, 68, 74, 75, 79, 120, 127, 627, 628
Threat evaluation and weapon assignment (TEWA), 250, 251, 293, 558, 628
Threat-alerting system, 143, 150
Threat-approach warning (sensor), 273, 274
Three-level laser, 22, 23, 84, 120, 628
Thulium, 67, 82, 86, 276, 277, 364, 573
TIALD, see Thermal Imaging & Air-borne Laser Designator
Time to full brightness, 73
Time-of-flight (techniques and methods), 52, 305, 313, 315, 316, 322, 367, 620
Tip-tilt correction, 153, 154, 181, 253
TNT, see tri-nitro toluene
Topographical mapping, 4, 331, 363, 368, 379-382
Topography, 331, 363, 368, 382, 427
Toxic agent, 365, 390
Training, 61, 62, 64, 198, 208, 316, 348, 353, 462, 463, 483, 529-543, 628
Transverse modes, 29, 49, 105, 424, 569, 605, 609, 629
Transversely-excited atmospheric (CO_2) lasers, 95, 364, 629

Tri-nitro toluene, 393
Triplet quenching, 91, 92, 592, 629
Triplet states (energy levels), 92, 621, 629
TRL, see technology-readiness levels
Turbulence, 145, 165, 169, 176-184, 189-194, 225, 240, 243, 246, 267, 275, 321, 326-328, 361, 363, 365, 371, 409, 423, 511, 572, 583-586, 589. 602-605, 613, 622, 629, 632
Turbulence mitigation, 180-182
Two-colour sensor, 562, 563
Type-I phase matching, 44, 629
Type-I semiconductor, 108, 629
Type-II phase matching, 44, 629
Type-II semiconductor, 108, 629
Type-III semiconductor, 108, 113, 629

U

UART, see universal asynchronous receiver transmitter
UAV, see unmanned air vehicles
Ultraviolet missile-warning sensor, 560, 561
Universal asynchronous receiver transmitter, 419
Unmanned air vehicles, 237, 268, 404, 414, 465, 466, 469, 470
Unstable cavities/resonators, 25-27, 616, 625, 630

V

V cavity, 630
VCSEL, see vertical-cavity surface-emitting laser
VECSEL, see vertical external-cavity surface-emitting laser
Vee diagram, 207, 208, 224
Veiling glare, 266, 270, 280, 601, 630
Vertical external-cavity surface-emitting laser, 7, 276, 277, 630
Vertical-cavity surface-emitting laser, 7, 104, 107, 116, 630
Vibrational levels/states, 17, 47, 86, 94, 97, 98, 392, 393, 471, 613, 631
Vibrometer, see vibrometry
Vibrometry, 29, 363, 371, 384-387, 404
Vibronic laser, 18, 67, 69, 71, 73, 86-88, 91, 631
Video processor, 144, 145, 147, 154, 342
Visibility, 118, 171, 174, 306, 350, 383, 571, 572, 585, 631
Volume Bragg grating, 615, 617, 631
Volume Bragg reflector, 631

W

W-cavity, 85
W-structures (semiconductor laser), 107, 108, 109, 114
Walk off, 44, 804, 617, 631
Wall-plug efficiency, 7, 50, 61, 108, 248, 256, 311, 465, 468, 568, 595, 631
Waveguide lasers, 104, 126, 127, 364, 623
Wavelength diversity, 55, 74, 75, 88, 92, 123, 272, 276, 280, 357, 408, 569, 588, 595, 631, 632
Wavelength tuneability, 10, 18, 37, 38, 41, 55, 69, 71, 73, 75, 82, 86-88, 91, 92, 96, 99, 100, 106, 111, 113-116, 120, 123, 128, 172, 239, 259, 272, 276, 280, 332-334, 390, 391, 305, 397-399, 539, 584, 629, 631
Weather effects, 166, 171, 174, 194, 225, 262, 361, 376, 383, 405, 409, 414, 435, 464, 465, 472, 533, 585
Wedge pairs, see Risley prisms
Wide-body integrated platform-protection system, 560
Wind effects, 183, 184
Wind shear, 365
WIPPS, see wide-body integrated platform-protection system

X

Xenon lamp, 293, 553, 598

Y

Young's modulus, 474, 626, 632
Ytterbium, 67, 79, 80, 120, 254, 628

Z

ZBLAN, 633
Zerodur, 445, 629
ZGP, see zinc germanium diphosphate
Zig-zag slab, 8, 25, 51, 70, 71, 85, 633
Zinc germanium diphosphate, 39, 40, 46, 587
Zinc telluride, 402

Milton Keynes UK
Ingram Content Group UK Ltd.
UKHW052103170824
447067UK00001B/10

9 781608 077786